Insulation Coordination for Power Systems

POWER ENGINEERING

Series Editor

H. Lee Willis

*ABB Electric Systems Technology Institute
Raleigh, North Carolina*

1. Power Distribution Planning Reference Book, *H. Lee Willis*
2. Transmission Network Protection: Theory and Practice, *Y. G. Paithankar*
3. Electrical Insulation in Power Systems, *N. H. Malik, A. A. Al-Arainy, and M. I. Qureshi*
4. Electrical Power Equipment Maintenance and Testing, *Paul Gill*
5. Protective Relaying: Principles and Applications, Second Edition, *J. Lewis Blackburn*
6. Understanding Electric Utilities and De-Regulation, *Lorrin Philipson and H. Lee Willis*
7. Electrical Power Cable Engineering, *William A. Thue*
8. Electric Power System Dynamics and Stability, *James A. Momoh and Mohamed E. El-Hawary*
9. Insulation Coordination for Power Systems, *Andrew R. Hileman*

ADDITIONAL VOLUMES IN PREPARATION

Insulation Coordination for Power Systems

Andrew R. Hileman

Taylor & Francis
Taylor & Francis Group
Boca Raton   London   New York

A CRC title, part of the Taylor & Francis imprint, a member of the
Taylor & Francis Group, the academic division of T&F Informa plc.

Published in 1999 by
CRC Press
Taylor & Francis Group
6000 Broken Sound Parkway NW, Suite 300
Boca Raton, FL 33487-2742

© 1999 by Taylor & Francis Group, LLC
CRC Press is an imprint of Taylor & Francis Group

No claim to original U.S. Government works

10 9 8 7 6 5 4 3

The disks mentioned in this book are now available for download on our Web site at:
http://www.crcpress.com/e_products/downloads/default.asp

International Standard Book Number-10: 0-8247-9957-7 (Hardcover)
International Standard Book Number-13: 978-0-8247-9957-1 (Hardcover)

This book contains information obtained from authentic and highly regarded sources. Reprinted material is quoted with permission, and sources are indicated. A wide variety of references are listed. Reasonable efforts have been made to publish reliable data and information, but the author and the publisher cannot assume responsibility for the validity of all materials or for the consequences of their use.

No part of this book may be reprinted, reproduced, transmitted, or utilized in any form by any electronic, mechanical, or other means, now known or hereafter invented, including photocopying, microfilming, and recording, or in any information storage or retrieval system, without written permission from the publishers.

Trademark Notice: Product or corporate names may be trademarks or registered trademarks, and are used only for identification and explanation without intent to infringe.

Library of Congress Cataloging-in-Publication Data

Catalog record is available from the Library of Congress

T&F informa

Taylor & Francis Group
is the Academic Division of T&F Informa plc.

Visit the Taylor & Francis Web site at
http://www.taylorandfrancis.com

and the CRC Press Web site at
http://www.crcpress.com

The disks mentioned in this book are now available for download on our Web site at: http://www/crcpress.com/e_products/downloads/default.asp

Series Introduction

Power engineering is the oldest and most traditional of the various areas within electrical engineering, yet no other facet of modern technology is currently undergoing a more dramatic revolution in technology and industry structure. Deregulation, along with the wholesale and retail competition it fostered, has turned much of the power industry upside down, creating demands for new engineering methods and technology at both the system and customer levels.

Insulation coordination, the topic of this latest addition to the Marcel Dekker, Inc., Power Engineering series, has always been a cornerstone of sound power engineering, since the first interconnected power systems were developed in the early 20th century. The changes being wrought by deregulation only increase the importance of insulation coordination to power engineers. Properly coordinated insulation strength throughout the power system is an absolute requirement for achieving the high levels of service customers demand in a competitive energy market, while simultaneously providing the long-term durability and low cost required by electric utilities to meet their operating and financial goals.

Certainly no one is more the master of this topic than Andrew R. Hileman, who has long been recognized as the industry's leader in the application of insulation coordination engineering methods. His *Insulation Coordination for Power Systems* is an exceedingly comprehensive and practical reference to the topic's intricacies and an excellent guide on the best engineering procedures to apply. At both introductory and advanced levels, this book provides insight into the philosophies and limitations of insulation coordination methods and shows both a rich understanding of the structure often hidden by nomenclature and formula and a keen sense of how to deal with these problems in the real world.

Having had the pleasure of working with Mr. Hileman at Westinghouse for a number of years in the 1980s, and continuously since then in The Pennsylvania State

University's power engineering program, it gives me particular pleasure to see his expertise and knowledge included in this important series of books on power engineering. Like all the books planned for the Power Engineering series, *Insulation Coordination for Power Systems* presents modern power technology in a context of proven, practical application. It is useful as a reference book as well as for self-study and advanced classroom use. The Power Engineering series will eventually include books covering the entire field of power engineering, in all its specialties and sub-genres, all aimed at providing practicing power engineers with the knowledge and techniques they need to meet the electric industry's challenges in the 21st century.

H. Lee Willis

Preface

This book is set up as a teaching text for a course on methods of insulation coordination, although it may also be used as a reference book. The chapter topics are primarily divided into line and station insulation coordination plus basic chapters such as those concerning lightning phenomena, insulation strength, and traveling waves.

The book has been used as a basis for a 3 credit hour, 48 contact hour course. Each chapter requires a lecture time of from 2 to 6 hours. To supplement the lecture, problems assigned should be reviewed within the class. On an average, this requires about 1 to 2 hours per chapter. The problems are teaching problems in that they supplement the lecture with new material—that is, in most cases they are not considered specifically in the chapters.

The book is based on a course that was originally taught at the Westinghouse Advanced School for Electric Utility Engineers and at Carnagie-Mellon University (Pittsburgh, PA). After retirement from Westinghouse in 1989, I extensively revised and added new materials and new chapters to the notes that I used at Westinghouse. Thus, this is essentially a new edition. There is no doubt that the training and experience that I had at Westinghouse are largely responsible for the contents. The volume is currently being used for a 48 contact hour course in the Advanced School in Power Engineering at Pennsylvania State University in Monroeville. In addition, it has been used for courses taught at several U.S. and international utilities. For a one-semester course, some of the chapters must be skipped or omitted. Preferably, the course should be a two-semester one.

As may be apparent from the preceding discussion, probabilistic and statistical theory is used extensively in the book. In many cases, engineers either are not familiar with this subject or have not used it since graduation. Therefore, some introduction to or review of probability and statistics may be beneficial. At

Pennsylvania State University, this Insulation Coordination course is preceded by a 48 contact hour course in probability and statistics for power system engineers, which introduces the student to the stress–strength principle.

The IEEE 1313.2 Standard, Guide for the Application of Insulation Coordination, is based on the material contained in this book.

I would like to acknowledge the encouragement and support of the Westinghouse Electric Corporation, Asea Brown Boveri, the Electric Power Research Institute (Ben Damsky), Duke Energy (Dan Melchior, John Dalton), and Pennsylvania State University (Ralph Powell, James Bedont). The help from members of these organizations was essential in production of this book. The education that I received from engineers within the CIGRE and IEEE committees and working groups has been extremely helpful. My participation in the working groups of the IEEE Surge Protective Devices Committee, in the Lightning working group of the IEEE Transmission and Distribution Committee, and in CIGRE working groups 33.01 (Lightning) and 33.06 (Insulation Coordination) has been educational and has led to close friendships. To all engineers, I heartily recommend membership in these organizations and encourage participation in the working groups. Also to be acknowledged is the influence of some of the younger engineers with whom I have worked, namely, Rainer Vogt, H. W. (Bud) Askins, Kent Jaffa, N. C. (Nick) AbiSamara, and T. E. (Tom) McDermott. Tom McDermott has been especially helpful in keeping me somewhat computer-literate.

I have also been tremendously influenced by and have learned from other associates, to whom I owe much. Karl Weck, Gianguido Carrara, and Andy Ericksson form a group of the most knowledgeable engineers with whom I have been associated. And finally, to my wife, Becky, and my children, Judy, Linda, and Nancy, my thanks for "putting up" with me all these years.

Andrew R. Hileman

Contents

Series Introduction	H. Lee Willis	*v*
Preface		*vii*
Introduction		*xi*
1.	Specifying the Insulation Strength	1
2.	Insulation Strength Characteristics	31
3.	Phase–Ground Switching Overvoltages, Transmission Lines	89
4.	Phase–Phase Switching Overvoltages, Transmission Lines	135
5.	Switching Overvoltages, Substations	163
6.	The Lightning Flash	195
7.	Shielding of Transmission Lines	241
8.	Shielding of Substations	275
9.	A Review of Traveling Waves	313
10.	The Backflash	373
	Appendix 1 Effect of Strokes within the Span	425
	Appendix 2 Impulse Resistance of Grand Electrodes	433
	Appendix 3 Estimating the Measured Forming Resistance	441

	Appendix 4	Effect of Power Frequency Voltage and Number of Phases	447
11.	The Incoming Surge and Open Breaker Protection		**461**
12.	Metal Oxide Surge Arresters		**497**
	Appendix 1	Protective Characteristics of Arresters	549
13.	Station Lightning Insulation Coordination		**557**
	Appendix 1	Surge Capacitance	625
	Appendix 2	Evaluation of Lightning Surge Voltages Having Nonstandard Waveshapes: For Self-Restoring Insulations	627
14.	Line Arresters		**641**
15.	Induced Overvoltages		**677**
16.	Contamination		**701**
17.	National Electric Safety Code		**733**
18.	Overview: Line Insulation Design		**753**

Appendix   Computer Programs for This Book   *762*
Index   *765*

Introduction

1 GOALS

Consider first the definition of insulation coordination in its most fundamental and simple form:

1. Insulation coordination is the selection of the insulation strength.

If desired, a reliability criterion and something about the stress placed on the insulation could be added to the definition. In this case the definition would become

2. Insulation coordination is the "selection of the insulation strength consistent with the expected overvoltages to obtain an acceptable risk of failure" [1].

In some cases, engineers prefer to add something concerning surge arresters, and therefore the definition is expanded to

3. Insulation coordination is the "process of bringing the insulation strengths of electrical equipment into the proper relationship with expected overvoltages and with the characteristics of surge protective devices" [2].

The definition could be expanded further to

4. Insulation coordination is the "selection of the dielectric strength of equipment in relation to the voltages which can appear on the system for which equipment is intended and taking into account the service environment and the characteristics of the available protective devices" [3].

or

5. "Insulation coordination comprises the selection of the electric strength of equipment and its application, in relation to the voltages which can appear

on the system for which the equipment is intended and taking into account the characteristics of available protective devices, so as to reduce to an economically and operationally acceptable level the probability that the resulting voltage stresses imposed on the equipment will cause damage to equipment insulation or affect continuity of service" [4].

By this time, the definition has become so complex that it cannot be understood by anyone except engineers who have conducted studies and served on committees attempting to define the subject and provide application guides. Therefore, it is preferable to return to the fundamental and simple definition: the selection of insulation strength. It goes without saying that the strength is selected on the basis of some quantitative or perceived degree of reliability. And in a like manner, the strength cannot be selected unless the stress placed on the insulation is known. Also, of course, the engineer should examine methods of reducing the stress, be it through surge arresters or other means. Therefore, the fundamental definition stands: it is the selection of insulation strength.

The goal is not only to select the insulation strength but also to select the *minimum* insulation strength, or minimum clearance, since minimum strength can be equated to minimum cost. In its fundamental form, the process should begin with a selection of the reliability criteria, followed by some type of study to determine the electrical stress placed on the equipment or on the air clearance. This stress is then compared to the insulation strength characteristics, from which a strength is selected. If the insulation strength or the clearance is considered to be excessive, then the stress can be reduced by use of ameliorating measures such as surge arresters, protective gaps, shield wires, and closing resistors in the circuit breakers.

As noted, after selection of the reliability criteria, the process is simply a comparison of the stress versus the strength.

Usually, insulation coordination is separated into two major parts:

1. Line insulation coordination, which can be further separated into transmission and distribution lines
2. Station insulation coordination, which includes generation, transmission, and distribution substations.

To these two major categories must be added a myriad of other areas such as insulation coordination of rotating machines, and shunt and series capacitor banks. Let us examine the two major categories.

2 LINE INSULATION COORDINATION

For line insulation coordination, the task is to specify all dimensions or characteristics of the transmission or distribution line tower that affect the reliability of the line:

1. The tower strike distances or clearances between the phase conductor and the grounded tower sides and upper truss
2. The insulator string length
3. The number and type of insulators
4. The need for and type of supplemental tower grounding
5. The location and number of overhead ground or shield wires

Introduction

6. The phase-to-ground midspan clearance
7. The phase–phase strike distance or clearance
8. The need for, rating, and location of line surge arresters

To illustrate the various strike distances of a tower, a typical 500-kV tower is shown in Fig. 1. Considering the center phase, the sag of the phase conductor from the tower center to the edge of the tower is appreciable. Also the vibration damper is usually connected to the conductor at the tower's edge. These two factors result in the minimum strike distance from the damper to the edge of the tower. The strike distance from the conductor yoke to the upper truss is usually larger. In this design, the strike distance for the outside phase exceeds that for the center phase. The insulator string length is about 11.5 feet, about 3% greater than the minimum center phase strike distance.

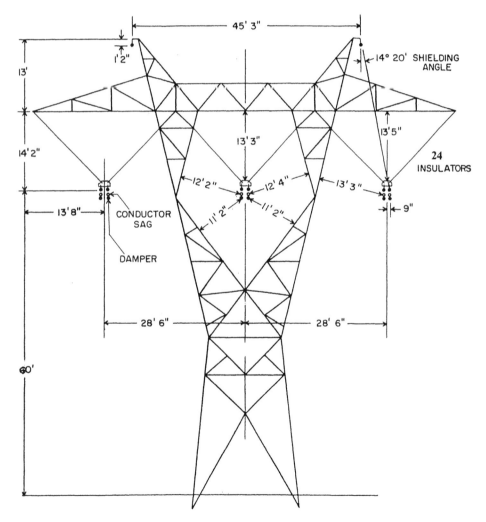

Figure 1 Allegheny Power System's 500-kV tower.

3 STATION INSULATION COORDINATION

For station insulation coordination, the task is similar in nature. It is to specify

1. The equipment insulation strength, that is, the BIL and BSL of all equipment.
2. The phase–ground and phase–phase clearances or strike distances. Figure 2 illustrates the various strike distances or clearances that should be considered in a substation.
3. The need for, the location, the rating, and the number of surge arresters.
4. The need for, the location, the configuration, and the spacing of protective gaps.
5. The need for, the location, and the type (masts or shield wires) of substation shielding.
6. The need for, the amount, and the method of achieving an improvement in lightning performance of the line immediately adjacent to the station.

In these lists, the method of obtaining the specifications has not been stated. To the person receiving this information, how the engineer decides on these specifications is not of primary importance, only that these specifications result in the desired degree of reliability.

It is true that the engineer must consider all sources of stress that may be placed on the equipment or on the tower. That is, he must consider

1. Lightning overvoltages (LOV), as produced by lightning flashes
2. Switching overvoltages (SOV), as produced by switching breakers or disconnecting switches

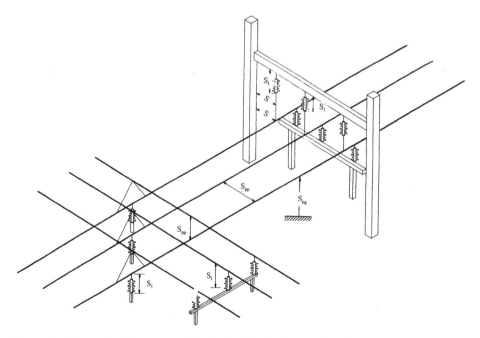

Figure 2 The strike distances and insulation lengths in a substation.

3. Temporary overvoltages (TOV), as produced by faults, generator overspeed, ferroresonance, etc.
4. Normal power frequency voltage in the presence of contamination

For some of the specifications required, only one of these stresses is of importance. For example, considering the transmission line, lightning will dictate the location and number of shield wires and the need for and specification of supplemental tower grounding. Considering the station, lightning will dictate the location of shield wires or masts. However, subjective judgment must be used to specify whether shield wires or masts should be used. The arrester rating is dictated by temporary overvoltages. In addition, the number and location of surge arresters will primarily be dictated by lightning. Also, for the line and station, the number and type of insulators will be dictated by the contamination.

However, in many of the specifications, two or more of the overvoltages must be considered. For transmission lines, for example, switching overvoltages, lightning, or contamination may dictate the strike distances and insulator string length. In the substation, however, lightning, switching surges, or contamination may dicatate the BIL, BSL, and clearances.

Since the primary objective is to specify the minimum insulation strength, no one of the overvoltages should dominate the design. That is, if a consideration of switching overvoltages results in a specification of tower strike distances, methods should be sought to decrease the switching overvoltages. In this area, the objective is not to permit one source of overvoltage stress to dictate design. Carrying this philosophy to the ultimate results in the objective that the insulation strength be dictated only by the power frequency voltage. Although this may seem ridiculous, it has essentially been achieved with regard to transformers, for which the 1-hour power frequency test is considered by many to be the most severe test on the insulation.

In addition, in most cases, switching surges are important only for voltages of 345 kV and above. That is, for these lower voltages, lightning dictates larger clearances and insulator lengths than do switching overvoltages. As a caution, this may be untrue for "compact" designs.

4 MODIFICATION OF STRESSES

As previously mentioned, if the insulation strength specification results in a higher-than-desired clearance or insulation strength, stresses produced by lightning and switching may be decreased. Some obvious methods are the application of surge arresters and the use of preinsertion resistors in the circuit breakers. In addition, methods such as the use of overhead or additional shield wires also reduce stress. In this same vein, other methods are the use of additional tower grounding and additional shielding in the station.

5 TWO METHODS OF INSULATION COORDINATION

Two methods of inuslation coordination are presently in use, the conventional or deterministic method and the probabilistic method. The conventional method consists of specifying the minimum strength by setting it equal to the maximum stress. Thus the rule is *minimum strength = maximum stress*. The probabilistic method consists of selecting the insulation strength or clearances based on a specific reliability criterion. An engineer may select the insulation strength for a line based on a lightning flashover rate of 1 flashover/100 km-years or for a station, based on a mean time between failure (MTBF), of 100 or 500 years.

The choice of the method is based not only on the engineer's desire but also on the characteristics of the insulation. For example, the insulation strength of air is usually described statistically by a Gaussian cumulative distribution, and therefore this strength distribution may be convolved with the stress distribution to determine the probability of flashover. However, the insulation strength of a transformer internal insulation is specified by a single value for lightning and a single value for switching, called the BIL and the BSL. To prove this BIL or BSL, usually only one application of the test voltage is applied. Thus no statistical distribution of the strength is available and the conventional method must be used.

It is emphasized that even when the conventional method is used, a probability of failure or flashover exists. That is, there is a probability attached to the conventional method although it is not evaluated.

The selected reliability criterion is primarily a function of the consequence of the failure and the life of the equipment. For example, the reliability criterion for a station may be more stringent than that for a line because a flashover in a station is of greater consequence. Even within a station, the reliability criterion may change according to the type of apparatus. For example, because of the consequences of failure of a transformer, the transformer may be provided with a higher order of protection. As another example, the design flashover rate for extra high voltage (EHV) lines is usually lower than that for lower-voltage lines. And the MTBF criterion for low-voltage stations is lower than for high-voltage stations.

6 REFERENCES

1. IEEE Standard 1313.1-1996, IEEE standard for insulation coordination—definitions, principles, and rules.
2. ANSI C92.1-1982, American national standard for power systems—insulation coordination.
3. IEC 71-1-1993-12, Insulation coordination Part 1: Definitions, principles and rules.
4. IEC Publication 71-1-1976, Insulation coordination, Part 1: Terms, definitions and rules.

1
Specifying the Insulation Strength

As discussed in the introduction, insulation coordination is the selection of the strength of the insulation. Therefore to specify the insulation strength, the usual, normal, and standard conditions that are used must be known. There also exist several methods of describing the strength, such as the BIL, BSL, and CFO, which must be defined. It is the purpose of this chapter to describe the alternate methods of describing the strength and to present the alternate test methods used to determine the strength. In addition, a brief section concerning generation of impulses in a laboratory is included.

1 STANDARD ATMOSPHERIC CONDITIONS

All specifications of strength are based on the following atmospheric conditions:

1. Ambient temperature: 20°C
2. Air pressure: 101.3 kPa or 760 mm Hg
3. Absolute humidity: 11 grams of water/m^3 of air
4. For wet tests: 1 to 1.5 mm of water/minute

If actual atmospheric conditions differ from these values, the strength in terms of voltage is corrected to these standard values. Methods employed to correct these voltages will be discussed later.

2 TYPES OF INSULATION

Insulation may be classified as internal or external and also as self-restoring and non-self-restoring. Per ANSI C92.1 (IEEE 1313.1) [1,2].

2.1 External Insulation

External insulation is the distances in open air or across the surfaces of solid insulation in contact with open air that are subjected to dielectric stress and to the effects of the atmosphere. Examples of external insulation are the porcelain shell of a bushing, bus support insulators, and disconnecting switches.

2.2 Internal Insulation

Internal insulation is the internal solid, liquid, or gaseous parts of the insulation of equipment that are protected by the equipment enclosures from the effects of the atmosphere. Examples are transformer insulation and the internal insulation of bushings. Equipment may be a combination of internal and external insulation. Examples are a bushing and a circuit breaker.

2.3 Self-Restoring (SR) Insulation

Insulation that completely recovers insulating properties after a disruptive discharge (flashover) caused by the application of a voltage is called self-restoring insulation. This type of insulation is generally external insulation.

2.4 Non-Self-Restoring (NSR) Insulation

This is the opposite of self-restoring insulators, insulation that loses insulating properties or does not recover completely after a disruptive discharge caused by the application of a voltage. This type of insulation is generally internal insulation.

3 DEFINITIONS OF APPARATUS STRENGTH, THE BIL AND THE BSL

3.1 BIL—Basic Lightning Impulse Insulation Level

The BIL or basic lightning impulse insulation level is the electrical strength of insulation expressed in terms of the crest value of the "standard lightning impulse." That is, the BIL is tied to a specific waveshape in addition being tied to standard atmospheric conditions. The BIL may be either a statistical BIL or a conventional BIL. The statistical BIL is applicable only to self-restoring insulations, whereas the conventional BIL is applicable to non-self-restoring insulations. *BILs are universally for dry conditions.*

The *statistical BIL* is the crest value of standard lightning impulse for which the insulation exhibits a 90% probability of withstand, a 10% probability of failure.

The *conventional BIL* is the crest value of a standard lightning impulse for which the insulation does not exhibit disruptive discharge when subjected to a specific number of applications of this impulse.

In IEC Publication 71 [3], the BIL is known as the lightning impulse withstand voltage. That is, it is defined the same but known by a different name. However, in IEC, it is not divided into conventional and statistical definitions.

3.2 BSL—Basic Switching Impulse Insulation Level

The BSL is the electrical strength of insulation expressed in terms of the crest value of a standard switching impulse. The BSL may be either a statistical BSL or a conventional BSL. As with the BIL, the statistical BSL is applicable only to self-restoring insulations while the conventional BSL is applicable to non-self-restoring insulations *BSLs are universally for wet conditions*.

The *statistical BSL* is the crest value of a standard switching impulse for which the insulation exhibits a 90% probability of withstand, a 10% probability of failure.

The *conventional BSL* is the crest value of a standard switching impulse for which the insulation does not exhibit disruptive discharge when subjected to a specific number of applications of this impulse.

In IEC Publication 71 [3], the BSL is called the switching impulse withstand voltage and the definition is the same. However, as with the lightning impulse withstand voltage, it is not segregated into conventional and statistical.

3.3 Standard Waveshapes

As noted, the BIL and BSL are specified for the standard lightning impulse and the standard switching impulse, respectively. This is better stated as the standard lightning or switching impulse *waveshapes*. The general lightning and switching impulse waveshapes are illustrated in Figs. 1 and 2 and are described by their time to crest and their time to half value of the tail. Unfortunately, the definition of the time to crest differs between these two standard waveshapes. For the lightning impulse waveshape the time to crest is determined by first constructing a line between two points: the points at which the voltage is equal to 30% and 90% of its crest value. The point at which this line intersects the origin or zero voltage is called the virtual origin and all times are measured from this point. Next, a horizontal line is drawn at the crest value so as to intersect the other line drawn through the 30% and 90% points. The time from the virtual origin to this intersection point is denoted as the time to crest or as the virtual time to crest t_f. The time to half value is simply the time between the virtual origin and the point at which the voltage decreases to 50% of the crest value, t_T. In general, the waveshape is denoted as a t_f/t_T impulse. For example

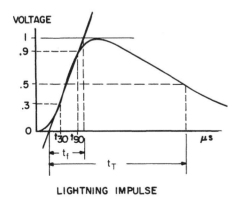

Figure 1 Lightning impulse wave shape.

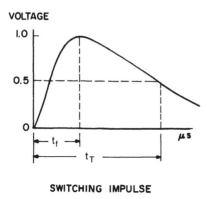

Figure 2 Switching impulse wave shape.

with a 1000-kV, 2.0/100-μs impulse, where the crest voltage is 1000 kV, the virtual time to crest or simply the time to crest is 2 μs and the time to half value is 100 μs. In the jargon of the industry, t_f is more simply called the front, and t_T is called the tail. The front can better be defined by the equation

$$t_f = 1.67(t_{90} - t_{30}) \tag{1}$$

where t_{90} is the actual time to 90% of the crest voltage and t_{30} is the actual time to 30% of crest voltage.

The standard lightning impulse waveshape is 1.2/50 μs. There exists little doubt that in the actual system, this waveshape never has appeared across a piece of insulation. For example, the actual voltage at a transformer has an oscillatory waveshape. Therefore it is proper to ask why the 1.2/50 μs shape was selected. It is true that, in general, lightning surges do have short fronts and relative short tails, so that the times of the standard waveshape reflect this observation. But of importance in the standardization process is that all laboratories can with ease produce this waveshape.

Although the tail of the switching impulse waveshape is defined as the time to half value, the time is measured from the actual time zero and not the virtual time zero. The time to crest or front is measured from the actual time zero to the actual crest of the impulse. The waveshape is denoted in the same manner as for the lightning impulse. For example, a 1000 kV, 200/3000μs switching impulse has a crest voltage of 1000 kV, a front of 200 μs, and a tail of 3000 μs. The standard switching impulse waveshape is 250/2500 μs. For convenience, the standard lightning and switching impulse waveshapes and their tolerances are listed in Table 1.

3.4 Statistical vs. Conventional BIL/BSL

As noted, the statistical BIL or BSL is defined statistically or probabilistically. For every application of an impulse having the standard waveshape and whose crest is equal to the BIL or BSL, the probability of a flashover or failure is 10%. In general, the insulation strength characteristic may be represented by a cumulative Gaussian distribution as portrayed in Fig. 3. The mean of this distribution or characteristic is

Specifying the Insulation Strength

Table 1 Standard Impulse Wave Shapes and Tolerances

Impulse Type →	Lightning	Switching
Nominal Wave Shape	1.2/50 µS	250/2500 µs
Tolerances		
front	±30%	±20%
tail	±20%	±60%

Source: Ref. 4.

defined as the critical flashover voltage or CFO. Applying the CFO to the insulation results in a 50% probability of flashover, i.e., half the impulses flashover. Locating the BIL or BSL at the 10% point results in the definition that the BIL or BSL is 1.28 standard deviations, σ_f, below the CFO. In equation form

$$\text{BIL} = \text{CFO}\left(1 - 1.28\frac{\sigma_f}{\text{CFO}}\right)$$
$$\text{BSL} = \text{CFO}\left(1 - 1.28\frac{\sigma_f}{\text{CFO}}\right) \qquad (2)$$

Sigma in per unit of the CFO is properly called the coefficient of variation. However, in jargon, it is simply referred to as sigma. Thus a sigma of 5% is interpreted as a standard deviation of 5% of the CFO. The standard deviations for lightning and switching impulses differ. For lightning, the standard deviation or sigma is 2 to 3%, whereas for switching impulse, sigma ranges from about 5% for tower insulation to about 7% for station type insulations, more later.

The conventional BIL or BSL is more simply defined but has less meaning as regards insulation strength. One or more impulses having the standard waveshape and having a crest value equal to the BIL or BSL are applied to the insulations. If no flashovers occur, the insulation is stated to possess a BIL or BSL. Thus the insulation strength characteristic as portrayed in Fig. 4 must be assumed to rise from zero probability of flashover or failure at a voltage equal to the BIL or BSL to 100% probability of flashover at this same BIL or BSL.

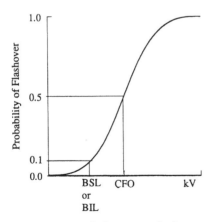

Figure 3 Insulation strength characteristic for self-restoring insulation.

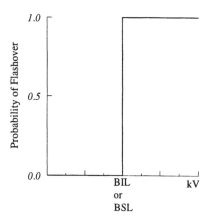

Figure 4 Insulation strength characteristic for non-self-restoring insulation.

3.5 Tests to "Prove" the BIL and BSL

Tests to establish the BIL or BSL must be divided between the conventional and the statistical. Since the conventional BIL or BSL is tied to non-self-restoring insulation, it is more than highly desirable that the test be nondestructive. Therefore the test is simply to apply one or more impulses having a standard impulse waveshape whose crest is equal to the BIL or BSL. If no failure occurs, the test is passed. While it is true that some failures on the test floor do occur, the failure rate is extremely low. That is, a manufacturer cannot afford to have failure rates, for example on power transformers, that exceed about 1%. If this occurs, production is stopped and all designs are reviewed.

Considering the establishment of a statistical BIL or BSL, theoretically no test can conclusively prove that the insulation has a 10% probability of failure. Also since the insulation is self-restoring, flashovers of the insulation are permissible. Several types of tests are possible to establish an estimate of the BIL and BSL. Theoretically the entire strength characteristic could be determined as illustrated in Fig. 3, from which the BIL or BSL could be obtained. However, these tests are not made except perhaps in the equipment design stage. Rather, for standardization, two types of tests exist, which are

1. The n/m test: m impulses are applied. The test is passed if no more than n result in flashover. The preferred test presently in IEC standards is the 2/15 test. That is, 15 impulses having the standard shapes and whose crest voltage is equal to the BIL or BSL are applied to the equipment. If two or fewer impulses result in flashover, the test is passed, and the equipment is said to have the designated BIL or BSL.

2. The $n + m$ test: n impulses are applied. If none result in flashover, the test is passed. If there are two or more flashovers, the test is failed. If only one flashover occurs, m additional impulses are applied and the test is passed if none of these results in a flashover. The present test on circuit-breakers is the 3 + 3 test [5]. In IEC standards, an alternate but less preferred test to the 2/15 test is the 3 + 9 test [6].

These alternate tests can be analyzed statistically to determine their characteristic. That is, a plot is constructed of the probability of passing the test as a function of the actual but unknown probability of flashover per application of a single impulse. The characteristics for the above three tests are shown in Fig. 5. These should be compared to the ideal characteristic as shown by the dotted line. Ideally, if the actual probability of flashover is less than 0.10, the test is passed, and ideally if the probability is greater than 0.10 the test is failed. The equations for these curves, where P is the probability of passing, p is the probability of flashover on application of a single impulse, and q is $(1 - p)$, are

$$\text{For the 2/15 test} \quad P = q^{15} + 15pq^{14} + 105p^2q^{13}$$
$$\text{For the 3+3 test} \quad P = q^3 + 3pq^5 \qquad (3)$$
$$\text{For the 3+9 test} \quad P = q^3 + 9pq^{11}$$

Per Fig. 5, if the actual (but unknown) probability of flashover for a single impulse is 0.20, then even though this probability of flashover is twice that defined for the BIL or BSL, the probabilities of passing the tests are 0.71 for the 3 + 3, 0.56 for the 3 + 9, and 0.40 for a 2/15. That is, even for an unacceptable piece of equipment, there exists a probability of passing the test. In a similar manner there exists a probability of failing the test even though the equipment is "good." For example, if the probability of flashover on a single impulse of 0.05, the probability of failing the test is 0.027 for the 3 + 3 test, 0.057 for the 3 + 9 test, and 0.036 for the 2/15 test. In general then, as illustrated in Fig. 6, there is a manufacturer's risk of having acceptable equipment and not passing the test and a user's risk of having unacceptable equipment and passing the test. A desired characteristic is that of discrimination, discriminating between "good" and "bad." The best test would have a steep slope around the 0.10 probability of flashover. As is visually apparent, the 2/15 is the best of the three and the 3 + 3 is the worst. Therefore it is little wonder that the IEC preferred test is the 2/15. The 3 + 9 test is a compromise between the 3 + 3 and the 2/15 tests included in the IEC Standard at the request of the ANSI circuit breaker group. The unstated agreement is that ANSI will change to the 3 + 9 test.

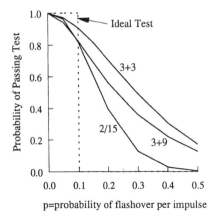

Figure 5 Characteristics for alternate test series.

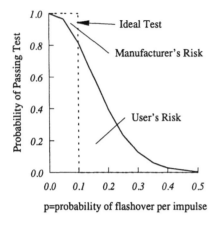

Figure 6 Manufacturer's and user's risk.

3.6 Standard BILs and BSLs

There exists a standard number series for both BILs and BSLs that equipment standards are encouraged to use. In the USA, ANSI C92 and IEEE 1313.1 lists the values shown in Table 2, while IEC values are shown in Table 3.

These values are "suggested" values for use by other equipment standards. In other words, equipment standards may use these values or any others that they deem necessary. However, in general, these values are used. There are exceptions. For any specific type of equipment or type of insulation, there does exist a connection between the BIL and the BSL. For example, for transformers, the BSL is approximately 83% of the BIL. Thus given a standard value of the BIL, the BSL may not be a value given in the tables. In addition, in IEC, phase–phase tests are specified to verify the phase–phase BSL. The phase–phase BSL is standardized as from 1.5 to 1.7 times the phase–ground BSL. Thus, in this case, the BSL values are not the values listed.

Table 2 Standard Values of BIL and BSL per ANSI C92, IEEE 1313.1

30	300	825	1925
45	350	900	2050
60	400	975	2175
75	450	1050	2300
95	500	1175	2425
110	550	1300	2550
125	600	1425	2675
150	650	1550	2800
200	700	1675	2925
250	750	1800	3050

Source: Ref. 7.

Specifying the Insulation Strength

Table 3 Standard Values of BIL and BSL per IEC 71.1

20	325	1300	2550
40	450	1425	2700
60	550	1550	2900
75	650	1675	
95	750	1800	
125	850	1950	
145	950	2100	
170	1050	2250	
250	1175	2400	

Source: Ref. 5.

In the ANSI Insulation Coordination Standard, C92, no required values are given for alternate system voltages. That is, the user is free to select the BIL and BSL desired. However, practically, there are only a limited number of BILs and BSLs used at each system voltage. For the USA, these values are presented in Tables 4 and 5 for transformers, circuit breakers, and disconnecting switches. For Class I power transformers, the available BILs are 45, 60, 75, 95, 110, 150, 200, 250 and 350 kV. For distribution transformers, the available BILs are 30, 45, 60, 75, 95, 125, 150, 200, 250 and 350 kV.

BSLs are not given in ANSI standards for disconnecting switches. The values given in the last column of Table 5 are estimates of the BSL. Note also that BSLs for circuit breakers are only given for system voltages of 345 kV or greater. This is based on the general thought that switching overvoltages are only important for these system voltages. Also, for breakers, for each system voltage two BSL ratings are given, one for the breaker in the closed position and one when the breaker is opened. For example, for a 550-kV system, the BSL of the circuit breaker in the closed position is 1175 kV, while in the opened position the BSL increases to 1300 kV.

BILs/BSLs of gas insulated stations are presented in Table 6, and BILs of cables are shown in Table 7. In IEC, BILs and BSLs are specified for each system voltage. These values are presented in Tables 8 and 9, where BSL_g is the phase–ground BSL, and BSL_p is the phase–phase BSL. Note as in ANSI, BSLs are only specified for maximum system voltages at and above 300 kV. Phase–phase BSLs are not standardized in the USA.

3.7 CFO and σ_f/CFO—"Probability Run Tests"

An alternate method of specifying the insulation strength is by providing the parameters of the insulation strength characteristic, the CFO and σ_f/CFO. This method is only used for self-restoring insulations since flashovers are permitted: they do occur. This method of describing the insulation strength characteristic is primarily used for switching impulses. However, the method is equally valid for lightning impulses although only limited data exist. For example, the switching impulse insulation strength of towers, bus support insulators, and gaps are generally specified in this manner.

Table 4 Transformer and Bushings BILs and BSLs

System nominal/ max system voltage, kV	Transformers BIL, kV	Transformers BSL, kV	Transformer bushings BIL, kV	Transformer bushings BSL, kV
1.2/-	30, 45		45	
2.5/-	45, 60		60	
5.0/-	60, 75		75	
8.7/-	75, 95		95	
15.0/-	95, 110		110	
25.0/-	150		150	
34.5/-	200		200	
46/48.3	200, 250		250	
69/72.5	250, 350		350	
115/121	350	280	450	
	*450	375	50	
	550	460		
138/145	450	375	450	
	*550	460	550	
	650	540	650	
161/169	550	460	550	
	*650	540	650	
	750	620	750	
230/242	650	540	650	
	*750	620	750	
	825	685	825	
	900	745	900, 1050	
345/362	900	745	900	700
	*1050	870	1050	825
	1175	975	1175, 1300	825
500/550	1300	1080	1300	1050
	*1425	1180	1425	1110
	1550	1290	1550	1175
	1675	1390	1675	1175
765/800	1800	1500	1800	1360
	1925	1600	1925	—
	2050	1700	2050	—

* Commonly used.
Source: Ref. 7, 8.

Specifying the Insulation Strength

Table 5 Insulation Levels for Outdoor Substations and Equipment

Rated max voltage, kV	NEMA Std, 6, outdoor substations		Circuit breakers		Disconnect switches	
	BIL, kV	10s power frequency voltage, kV	BIL, kV	BSL, kV	BIL, kV	BSL, kV estimate
8.25	95	30	95		95	
15.5	110	45	110		110	
25.8	150	60	150		125	
					150	
38.0	200	80	200		150	
					200	
48.3	250	100	250		250	
72.5	350	145	350		350	
121	550	230	550		550	
145	650	275	650		650	
169	750	315	750		750	
242	900	385	900		900	
	1050	455			1050	
362	1050	455	1300	825	1050	820
	1300	525		900	1300	960
550	1550	620	1800	1175	1550	1090
	1800	710		1300	1800	1210
800	2050	830	2050	1425	2050	1320
				1500		

Source: Ref. 5, 9.

Table 6 BILs/BSLs of Gas Insulated Stations

Max system voltage, kV		IEC [10]		ANSI [11]	
IEC	ANSI	BIL, kV	BSL, kV	BIL, kV	BSL, kV
72.5	72.5	325	—	300, 350	—
100		450	—		
123	121	550	—	450, 550	—
145	145	650	—	550, 650	—
170	169	750	—	650, 750	—
245	242	950	—	750, 900	—, 720
300		1050	850		
362	362	1175	950	900, 1050	720, 825
420		1300	1050		
525	550	1425	1175	1300, 1550	1050, 1175
765	800	1800	1425	1800	1425

Table 7 BILs of Cables (No BSLs provided), AEIC C54-79

Rated voltage, kV	BIL, kV
115, 120, & 130	550
138	650
161	750
230	1050
345	1300
500	1800

Source: Ref. 12.

Table 8 IEC 71.1: BILs are Tied to Max. System Voltages for Max. System Voltage from 1 to 245 kV

Max system voltage, kV	BILs, kV	Max system voltage, kV	BILs, kV
3.6	20 or 40	52	250
7.2	40 or 60	72.5	325
12	60, 75 or 95	123	450 or 550
17.7	75 or 95	145	450, 550, or 650
24	95, 125 or 145	170	550, 650, or 750
36	145 or 170	245	650, 750, 850, 950, or 1050

Source: Ref. 3.

Table 9 IEC BIL/BSLs, from IEC Publication 71.1

Max. system voltage, kV	Phase–ground BSL, BSL_g, kV	Ratio BSL_p/BSL_g	BIL, kV
300	750	1.50	850 or 950
	850	1.50	950 or 1050
362	850	1.50	950 or 1050
	950	1.50	1050 or 1175
420	850	1.60	1050 or 1175
	950	1.50	1175 or 1300
	1050	1.50	1300 or 1425
550	950	1.70	1175 or 1300
	1050	1.60	1300 or 1425
	1175	1.50	1425 or 1550
800	1300	1.70	1675 or 1800
	1425	1.70	1800 or 1950
	1550	1.60	1950 or 2100

Source: Ref. 3.

Specifying the Insulation Strength

The procedure for these tests can be provided by an example. Assume that in a laboratory, switching impulses are applied to a post insulator. First a 900-kV, 250/2500-μs impulse is applied 100 times and two of these impulses cause a flashover, or the estimated probability of flashover when a 900-kV impulse is applied is 0.02. Increasing the crest voltage to 1000 kV and applying 40 impulses results in 20 flashovers, or a 50% probability of flashover exists. The voltage is then increased and decreased to obtain other test points resulting in the data in the table. These test

Applied crest voltage, kV	No. of "shots"	No. of flashovers	Percent of flashovers
900	100	2	2
1000	40	20	50
1050	40	33	82.5
1075	100	93	93
960	40	7	17.5
980	40	16	40
960	40	10	25

results are then plotted on normal or Gaussian probability paper and the best straight line is constructed through the data points as in Fig. 7. The mean value at 50% probability is obtained from this plot and is the CFO. The standard deviation is the voltage difference between the 16% and 50% points or between the 50% and 84% points. In Fig. 7, the CFO is 1000 kV and the standard deviation σ_f is 50 kV. Thus σ_f/CFO is 5.0%. If the BSL is desired, which it is not in this case, the value could be read at the 10% probability or 936 kV. These two parameters, the CFO and the standard deviation, completely describe the insulation characteristic using the assumption that the Gaussian cumulative distribution adequately approximates the insulation characteristic. For comparison, see the insulation characteristic of Fig. 8.

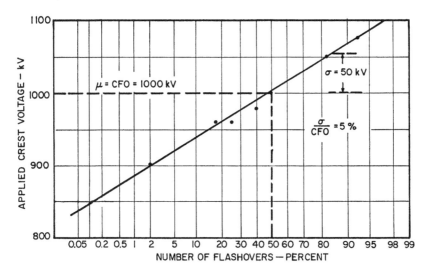

Figure 7 Insulation strength characteristic plotted on Gaussian probability paper.

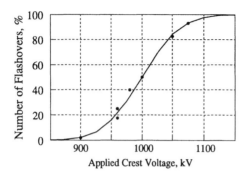

Figure 8 Data plotted on linear paper.

To be noted and questioned is that use of the Gaussian cumulative distribution assumes that the insulation characteristic is unbounded to the left. Of course this is untrue, since there does exist a voltage at which the probability of flashover is zero. However, the insulation characteristic appears valid down to about 4 standard deviations below the CFO, which is adequate for all applications. Recently, the Weibull distribution has been suggested as a replacement for the Gaussian distribution since it may be bounded to the left. However, all available data have been obtained using the assumption of the Gaussian distribution and there exists little reason to change at this time.

In concept, these types of tests may also be performed for non-self-restoring insulations. However, every flashover or failure results in destruction of the test sample. Thus the test sample must be replaced and the assumption made that all test samples are identical. Thus using this technique for non-self-restoring insulation is limited to purely research type testing.

The number of shots or voltage applications per data point is a function of the resultant percent flashovers or the probability of flashover. For example, using the same number of shots per point, the confidence of the 2% point is much less than that of the 50% point. Therefore the number of shots used for low or high probabilities is normally much greater than in the 35 to 65% range. 20 to 40 shots per point in the 35 to 65% range and 100 to 200 shots per point outside this range are frequently used.

As mentioned previously, this type of testing is normally performed only for switching impulses. Limited test for lightning impulse indicates that σ_f/CFO is much less than that for switching impulses, that is, in the range of 2 to 3%.

3.8 CFO

In many cases, an investigator only desires to obtain the CFO. This is especially true when testing with lightning impulses. The procedure employed is called the up and down method:

1. Estimate the CFO. Apply one shot. If flashover occurs, lower the voltage by about 3%. If no flashover occurs, increase the voltage by about 3%. If upon application of this voltage, flashover occurs, decrease the voltage by 3% or if no flashover occurs, increase the voltage by 3%.

Specifying the Insulation Strength

2. Continue for about 50 shots. Discard the shots until one flashover occurs. The CFO is the average applied voltage used in the remaining shots.

This up and down method in a modified form may also be used to determine a lower probability point. For example, consider the following test:

1. Apply 4 shots. Denote F as a flashover and N as no flashover.
2. If NNNN occurs, increase the voltage by 3%.
3. If F occurs on the first shot or on any other shot, and as soon as it occurs, lower the voltage by 3%. That is, if F, NF, NNF, or NNNF occurs, lower the voltage.
4. Continue for from 50 to 100 tests.

The probability of increasing the voltage is $(1-p)^4$, where p is the probability of flashover at a specific voltage. Therefore for a large number of 4-shot series,

$$(1-p)^4 = 0.5 \quad or \quad p = 0.16 \tag{4}$$

That is, the average applied voltage is the 16% probability of flashover point. This method has been found to have a low confidence and is not normally used; the probability run tests are better.

3.9 Chopped Wave Tests or Time-Lag Curves

In general, in addition to the tests to establish the BIL, apparatus are also given chopped wave lightning impulse tests. The test procedures is to apply a standard lightning impulse waveshape whose crest value exceeds the BIL. A gap in front of the apparatus is set to flashover at either 2 or 3 µs, depending on the applied crest voltage. The apparatus must "withstand" this test, i.e., no flashover or failure may occur. The test on the power transformers consists of an applied lightning impulse having a crest voltage of 1.10 times the BIL, which is chopped at 3 µs. For distribution transformers, the crest voltage is a minimum of 1.15 times the BIL, and the time to chop varies from 1 to 3 µs. For a circuit breaker, two chopped wave tests are used: (1) 1.29 times the BIL chopped at 2 µs and (2) 1.15 times the BIL at 3 µs. Bushings must withstand a chopped wave equal to 1.15 times the BIL chopped at 3 µs.

These tests are only specified in ANSI standards, not in IEC standards. Originally, the basis for the tests was that a chopped wave could impinge on the apparatus caused by a flashover of some other insulation in the station, e.g., a post insulator. Today, this scenario does not appear valid. However, the test is a severe test on the turn-to-turn insulation of a transformer, since the rapid chop to voltage zero tests this type of insulation, which is considered to be an excellent test for transformers used in GIS, since very fast front surges may be generated by disconnecting switches. In addition, these chopped wave tests provide an indication that the insulation strength to short duration impulses is higher than the BIL. The tests are also used in the evaluation of the CFO for impulses that do not have the lightning impulse standard waveshape. In addition, the chopped wave strength at 2 µs is used to evaluate the need for protection of the "opened breaker."

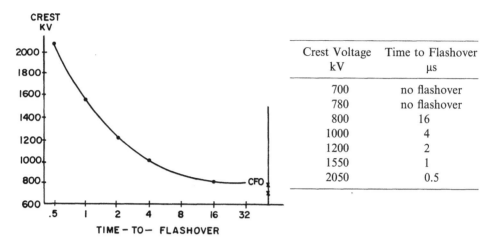

Figure 9 A sample time-lag curve.

To establish more fully the short-duration strength of insulation, a time-lag or volt–time curve can be obtained. These are universally obtained using the standard lightning impulse wave shape, and only self-restoring insulations are tested in this manner. The procedure is simply to apply higher and higher magnitudes of voltage and record the time to flashover. For example, test results may be as listed in Fig. 9. These are normally plotted on semilog paper as illustrated in Fig. 9. Note that the time-lag curve tends to flatten out at about 16 μs. The asymptotic value is equal to the CFO. That is, for air insulations, the CFO occurs at about a time to flashover of 16 μs. Times to flashover can exceed this time, but the crest voltage is approximately equal to that for the 16 μs point that is the CFO. (The data of Fig. 9 are not typical, in that more data scatter is normally present. Actual time-lag curves will be presented in Chapter 2.)

4 NONSTANDARD ATMOSPHERIC CONDITIONS

BILs and BSLs are specified for standard atmospheric conditions. However, laboratory atmospheric conditions are rarely standard. Thus correlation factors are needed to determine the crest impulse voltage that should be applied so that the BIL and BSL will be valid for standard conditions. To amplify, consider that in a laboratory nonstandard atmospheric conditions exist. Then to establish the BIL, the applied crest voltage, which would be equal to the BIL at standard conditions, must be increased or decreased so that at standard conditions, the crest voltage would be equal to the BIL. In an opposite manner, for insulation coordination, the BIL, BSL, or CFO for the nonstandard conditions where the line or station is to be constructed is known and a method is needed to obtain the required BIL, BSL, and CFO for standard conditions. In a recent paper [13], new and improved correction factors were suggested based on tests at sea level (Italy) as compared to tests at 1540 meters in South Africa and to tests at 1800 meters in Mexico. Denoting the voltage as measured under nonstandard conditions as V_A and the voltage for standard conditions as V_S, the suggested equation, which was subsequently adopted in IEC 42, is

Specifying the Insulation Strength

$$V_A = \delta^m H_c^w V_S \tag{5}$$

where δ is the relative air density, H_c is the humidity correction factor, and m and w are constants dependent on the factor G_0 which is defined as

$$G_0 = \frac{\text{CFO}_S}{500S} \tag{6}$$

where S is the strike distance or clearance in meters and CFO_S is the CFO under standard conditions.

By definition, Eq. 5 could also be written in terms of the CFO or BIL or BSL. That is,

$$\text{CFO}_A = \delta^m H_c^w \text{CFO}_S$$
$$\text{BIL}_A = \delta^m H_c^w \text{BIL}_S \tag{7}$$
$$\text{BSL}_A = \delta^m H_c^w \text{BSL}_S$$

The humidity correction factor, per Fig. 10, for impulses is given by the equation

$$H_c = 1 + 0.0096\left[\frac{H}{\delta} - 11\right] \tag{8}$$

where H is the absolute humidity in grams per m^3. For wet or simulated rain conditions, $H_c = 1.0$. The values of m and w may be obtained from Fig. 11 or from Table 10.

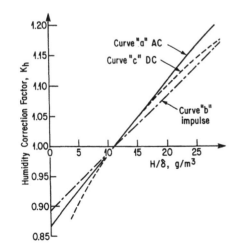

Figure 10 Humidity correction factors. (Copyright IEEE 1989 [13].)

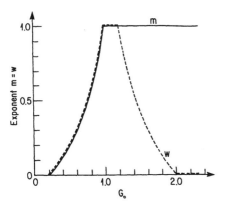

Figure 11 Values of m and w. (Copyright 1989 IEEE [13].)

Lightning Impulse

For lightning impulses, G_0 is between 1.0 and 1.2. Therefore

$$V_A = \delta H_c V_S$$
$$CFO_A = \delta H_c CFO_S \tag{9}$$
$$BIL_A = \delta H_c BIL_S$$

In design or selection of the insulation level, wet or rain conditions are assumed, and therefore $H_c = 1.0$. So for design

$$V_A = \delta V_S$$
$$CFO_A = \delta(CFO_S) \tag{10}$$
$$BIL_A = \delta(BIL_S)$$

Switching Impulse

For switching impulses, G_0 is between 0.2 and 1, and therefore

$$m = w = 1.25 G_0 (G_0 - 0.2) \tag{11}$$

Table 10 Values of m and w

G_0	m	w
$G_0 < 0.2$	0	0
$0.2 < G_0 < 1.0$		$m = w = 1.25 G_0 (G_0 - 0.2)$
$1.0 < G_0 < 1.2$	1	1
$1.2 < G_0 < 2.0$	1	$w = 1.25(2.2 - G_0)(2 - G_0)$
$G_0 > 2.0$	1	0

For dry conditions

$$V_A = (\delta H_c)^m V_S \qquad (12)$$

However, in testing equipment, the BSL is always defined for wet or simulated rain conditions. Also in design for switching overvoltages, wet or rain conditions are assumed. Therefore, $H_c = 1$ and so

$$V_A = \delta^m V_S$$
$$\text{CFO}_A = \delta^m \text{CFO}_S \qquad (13)$$
$$\text{BSL}_A = \delta^m \text{BSL}_S$$

The only remaining factor in the above correction equations is the relative air density. This is defined as

$$\delta = \frac{P T_0}{P_0 T} \qquad (14)$$

where P_0 and T_0 are the standard pressure and temperature with the temperature in degrees Kelvin, i.e., degrees Celsius plus 273, and P and T are the ambient pressure and temperature. The absolute humidity is obtained from the readings of the wet and dry bulb temperature; see IEEE Standard 4.

From Eq. 14, since the relative air density is a function of pressure and temperature, it is also a function of altitude. At any specific altitude, the air pressure and the temperature and thus the relative air density are not constant but vary with time. A recent study [14] used the hourly variations at 10 USA weather stations for a 12- to 16-year period to examine the distributions of weather statistics. Maximum altitude was at the Denver airport, 1610 meters (5282 feet). The statistics were segregated into three classes; thunderstorms, nonthunderstorms, and fair weather. The results of the study showed that the variation of the temperature, the absolute humidity, the humidity correction, and the relative air density could be approximated by a Gaussian distribution. Further, the variation of the multiplication of the humidity correction factor and the relative air density δH_c can also be approximated by a Gaussian distribution.

The author of Ref. 14 regressed the mean value of the relative air density δ and the mean value of δH_c against the altitude. He selected a linear equation as an appropriate model and found the equations per Table 11. However, in retrospect, the linear equation is somewhat unsatisfactory, since it portrays that the relative air density could be negative—or more practically, the linear equation must be limited to a maximum altitude of about 2 km. A more satisfactory regression equation is of the exponential form, which approaches zero asymptotically. Reanalyzing the data, the exponential forms of the equations are also listed in Table 11.

These equations may be compared to the equation suggestion in IEC Standard 71.2, which is

$$\delta = e^{-A/8.15} \qquad (15)$$

Table 11 Regression Equations, A in km

Statistic	Linear equation for mean value	Exponential equation for mean value	Average standard deviation
Relative air density, δ			
Thunderstorms	$0.997-0.106A$	$1.000\,e^{-A/8.59}$	0.019
Nonthunderstorms	$1.025-0.090A$	$1.025\,e^{-A/9.82}$	0.028
Fair	$1.023-0.103A$	$1.030\,e^{-A/8.65}$	0.037
δH_c			
Thunderstorms	$1.035-0.147A$	$1.034\,e^{-A/6.32}$	0.025
Nonthunderstorms	$1.023-0.122A$	$1.017\,e^{-A/8.00}$	0.031
Fair	$1.025-0.132A$	$1.013\,e^{-A/7.06}$	0.034

Either form of the equation of Table 11 can be used, although the linear form should be restricted to altitudes less than about 2 km. The exponential form is more satisfactory, since it appears to be a superior model.

Not only are the CFO, BIL, and BSL altered by altitude but the standard deviation σ_f is also modified. Letting x equal δH_c, the altered coefficient of variation $(\sigma_f/\text{CFO})'$ is

$$\left(\frac{\sigma_f}{\text{CFO}}\right)' = \sqrt{\left(\frac{\sigma_f}{\text{CFO}}\right)^2 + \left(\frac{m\sigma_x}{\mu_x^m}\right)^2} \tag{16}$$

Considering that for switching overvoltages, the normal design is for wet conditions, Eq. 13 is applicable with the mean given by the first equation in Table 11, where the average standard deviation is 0.019. For a strike distance S of 2 to 6 meters, at an altitude of 0 to 4 km, the new modified coefficient of variation increases to 5.1 to 5.3% assuming an original σ_f/CFO of 5%. For fair weather, Eq. 12 applies, and the last equation of Table 11 is used along with the standard deviation of 0.034. For the same conditions as used above, the new coefficient of variation ranges from 5.4 to 5.8%. Considering the above results, the accuracy of the measurement of the standard deviation, and that 5% is a conservative value for tower insulation, the continued use of 5% appears justified. That is, the coefficient of variation is essentially unchanged with altitude.

In summary, for insulation coordination purposes, the design is made for *wet* conditions. The following equations are suggested:

(1) For Lightning

$$\text{BIL}_A = \delta(\text{BIL}_S)$$
$$\text{CFO}_A = \delta(\text{CFO}_S) \tag{17}$$

(2) For Switching Overvoltages

$$\text{BSL}_A = \delta^m(\text{BSL}_S)$$
$$\text{CFO}_A = \delta^m(\text{CFO}_S)$$

either $\quad \delta = 0.997 - 0.106A \quad$ or $\quad \delta = e^{-(A/8.6)} \quad\quad (18)$

$$m = 1.25 G_0(G_0 - 0.2)$$
$$G_0 = \frac{\text{CFO}_S}{500 S}$$

where the subscript S refers to standard atmospheric conditions and the subscript A refers to the insulation strength at an altitude A in km. Some examples may clarify the procedure.

Example 1. A disconnecting switch is to be tested for its BIL of 1300 kV and its BSL of 1050 kV. In the laboratory, the relative air density is 0.90 and the absolute humidity is 14 g/m^3. Thus the humidity correction factor is 1.0437. As per standards, the test for the BIL is for dry conditions and the test for the BSL is for wet conditions. The σ_f/CFO is 0.07. The test voltages applied for the BIL is

$$\text{BIL}_A = (\delta H_c)\text{BIL}_S = 1221 \text{ kV} \quad\quad (19)$$

Thus to test for a BIL of 1300 kV, the crest of the impulse should be 1221 kV. For testing the BSL, let the strike distance, S, equal 3.5 m. Then

$$\text{BSL}_A = \delta^m \text{BSL}_S$$
$$\text{CFO}_s = \frac{\text{BSL}_s}{1 - 1.28\, \sigma_f / \text{CFO}_s} = 1153 \text{ kV}$$
$$G_0 = \frac{\text{CFO}_S}{500 S} = 0.6591 \quad\quad (20)$$
$$m = 1.25 G_0(G_0 - 0.2) = 0.3782$$
$$\text{BSL}_A = 0.90^{0.3782}(1050) = 1009 \text{ kV}$$

Thus to test for a BSL of 1050 kV, the crest of the impulse should be 1009 kV.

An interesting problem occurs if in this example a bushing is considered with a BIL of the porcelain and the internal insulation both equal to 1300 kV BIL and 1050 kV BSL. While the above test voltages would adequately test the external porcelain, they would not test the internal insulation. There exists no solution to this problem except to increase the BIL and BSL of the external porcelain insulation so that both insulations could be tested or perform the test in another laboratory that is close to sea level.

An opposite problem occurs if the bushing shell has a higher BIL/BSL than the internal insulation and the laboratory is at sea level. In this case the bushing shell cannot be tested at its BIL/BSL, since the internal insulation strength is lower. The solution in this case would be to test only the bushing shell, after which the internal insulation could be tested at its BIL/BSL.

Example 2. The positive polarity switching impulse CFO at standard conditions is 1400 kV for a strike distance of 4.0 meters. Determine the CFO at an altitude of 2000 meters where $\delta = 0.7925$. Assume wet conditions, i.e., $H_c = 1$.

$$G_0 = \frac{1400}{4.0(500)} = 0.700$$

$$m = 1.25 G_0 (G_0 - 0.2) = 0.4375 \tag{21}$$

$$\text{CFO}_A = (1400) 0.7925^{0.4375} = 1265$$

Example 3. Let the CFO for lightning impulse, positive polarity at standard atmospheric conditions, be equal 2240 kV for a strike distance of 4 meters. Assume wet conditions, i.e., $H_c = 1$. For a relative air density of 0.7925, the CFO is

$$\text{CFO}_A = 0.7925(2240) = 1775 \, \text{kV} \tag{22}$$

Example 4. At an altitude of 2000 meters, $\delta = 0.7925$ and the switching impulse, positive polarity CFO for wet conditions is 1265 kV for a gap spacing of 4 meters. Find the CFO_S. This problem cannot be solved directly, since m is a function of G_0 and G_0 is a function of the standard CFO. Therefore the CFO for standard conditions must be obtained by iteration as in the table. Note that this is the exact opposite problem as Example 2 and therefore the answer of 1400 kV coordinates with it. This example represents the typical design problem. The required CFO is known for the line or station where it is to be built, i.e., at 2000 meters. The problem is to determine the CFO at standard conditions. Alternately, the required BIL/BSL is known at the altitude of the station, and the BIL/BSL to be ordered for the station must be determined at standard conditions.

Assumed CFO_S, kV	G_0	m	δ^m	$\text{CFO}_S = 1265/\delta^m$
1300	0.650	0.3656	0.9185	1377
1377	0.689	0.4204	0.9069	1395
1395	0.698	0.4338	0.9040	1399
1399	0.700	0.4368	0.9034	1400
1400	0.700	0.4375	0.9033	1400

5 GENERATION OF VOLTAGES IN THE LABORATORY

Lightning impulses are generated by use of a Marx generator as shown schematically in Fig. 12. The same generator is used, except in the former USSR, to generate switching impulses. In the former USSR, the switching impulse is generated by discharge of a capacitor on the low-voltage side of a transformer.

Specifying the Insulation Strength

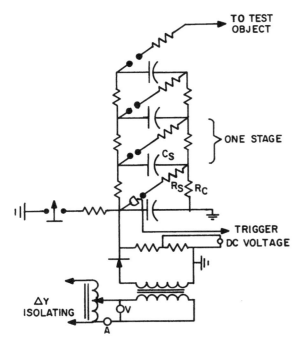

Figure 12 The Marx impulse generator.

The Marx generator consists of several stages, each stage consisting of two charging resistors R_c, a capacitor C_S and a series resistor R_S. A DC voltage controllable on the AC side of a transformer is applied to the impulse set. The charging circuit of Fig. 13 shows that the role of the charging resistors is to limit the inrush current to the capacitors. The polarity of the resultant surge is changed by reversing the leads to the capacitors.

After each of the capacitors has been charged to essentially the same voltage, the set is fired by a trigatron gap. A small impulse is applied to the trigatron gap that fires or sparks over the first or lower gap. This discharge circuit, neglecting for the moment the high-ohm charging resistors, is shown in Fig. 14. To illustrate the procedure, assume that the capacitors are charged to 100 kV. If gap 1 sparks over, the voltage across gap 2 is approximately 200 kV, i.e., double the normal voltage across the gap. Assuming that this doubled voltage is sufficient to cause sparkover, 300 kV appears across gap 3—which sparkover places 400 kV across gap 4, etc. Thus gap sparkover cascades throughout the set placing all capacitors in series and

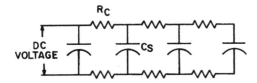

Figure 13 The charging circuit.

Figure 14 The discharging circuit.

producing a voltage that approaches the product of the number of stages and the charging voltage.

The simplified equivalent circuit of the discharge circuit is shown in Fig. 15, where n is the number of stages and L is the inherent inductance of the set. The capacitor C_b represents the capacitance of the test object, and the voltage divider is illustrated as either a pure resistance divider R_D, which can be used to measure lightning impulses, or a capacitor divider C_D, which can be used to measure switching impulses.

First examine the equivalent circuit using the resistor divider and assume that the inductance is zero. The values of $nR_c/2$ and R_D are much greater than nR_S. Therefore the circuit to describe the initial discharge is simply an RC circuit as illustrated in Fig. 16. The voltage across the test object E_0 is given by the equation

$$E_0 = \frac{C_S}{C_b + (C_S/n)} E\left(1 - e^{-t/\alpha}\right) \qquad (23)$$

where

$$\alpha = \frac{R_S C_b C_S}{C_b + (C_S/n)} \qquad (24)$$

which illustrates that the shape of the front is exponential in form and is primarily controlled by the series resistance of the set.

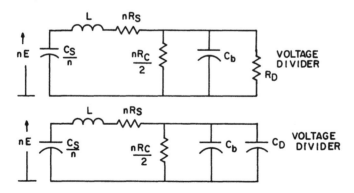

Figure 15 Equivalent discharge circuit.

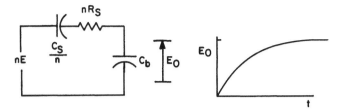

Figure 16 Simplified discharge circuit and initial voltage.

The tail of the impulse occurs by the action of the discharge of the capacitance through the voltage divider resistance and the charging resistance $nR_c/2$ and again is exponential in form. Neglecting the inductance, the voltage across the test object has the so-called double exponential shape, i.e.,

$$E_0 = A\left(e^{-at} - e^{-bt}\right) \tag{25}$$

The analysis of the lower circuit of Fig. 15 with the capacitor divider is similar in nature to the above except that the tail of the impulse will be longer.

The inductance of the set and any inductance of the leads connected to the test object may lead to oscillations on the wave front if nR_S is small. Therefore when attempting to produce short wave fronts, nR_S is adjusted to minimize oscillations.

The series resistance may be supplemented by series resistors external to the generator to produce longer wave fronts, i.e., for switching impulses.

At the former Westinghouse laboratory in Trafford, PA, the outdoor impulse generator had the following constants: 31 stages, 200 kV per stage, $C_S = 0.25\,\mu F$, $L = 200\,\mu H$, and $R_c = 40\,k\Omega$, which produced a maximum open circuit voltage of 6200 kV and an energy of 165 kJ.

To obtain a wavefront of 1.2 µs and a time to half value of 50 µs, nR_S is set at about 400 ohms. Table 12 illustrates the required total resistance nR_S for other wave fronts. the generator efficiency is the crest output voltage divided by the open circuit voltage nE. As noted in Table 12, the efficiency sharply decreases for longer wave fronts.

Impulse voltages are measured with a voltage divider that reduces the voltage to a measurable level. For lightning impulses, a resistor divider is normally used. The resistance of this divider in combination with the charging resistors produces a 50 µs

Table 12 Series Resistance Required and Generator Efficiency

Front, µs	Internal resistance, ohms	External resistance ohms	Generator efficiency, %
10	930	0	80
150	16,275	0	77
630	116,250	0	66
1200	116,250	150,000	59

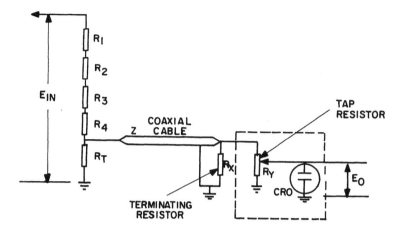

Figure 17 Divider and measuring circuit for lightning impulse.

tail. The voltage divider and the measuring circuit are shown in Fig. 17. The resistor R_x is set equal to the surge impedance of the cable to eliminate reflections. The voltage across the terminating resistor is

$$\frac{R_T Z}{R_T Z + R_S R_T + R_S Z} E_{IN} \tag{26}$$

This resistor ratio is called the divider ratio, where R_S is the sum of the resistors R_1, R_2, R_3, etc. or the resistance of the upper arm of the divider. Voltage to the cathode ray oscilloscope (CRO) is varied by the tap resistor.

To measure switching impulses, a capacitor divider is used so as to decrease the loading on the generator. In this case, the coaxial cable is not terminated. The capacitance of the cable is added to the capacitance of the lower arm of the divider to determine the voltage divider ratio.

Another type of divider having the ability to measure both lightning and switching impulses and also power frequency voltages is often used. This RC divider consists of resistance and capacitance in series: the resistance for high-frequency measurements, the capacitors for low-frequency measurements.

6 OTHER MISCELLANEOUS ITEMS

6.1 Standard Current Impulse

Impulse currents are used to test surge arresters to determine their discharge voltage and their durability. The waveshapes are 8/20 μs and 4/10 μs. The fronts are determined in a similar manner as for the lightning impulse waveshape except that the 10 and 90% points are used [15].

6.2 Apparatus Standards and Effects of Altitude

All apparatus standards state that the equipment maintains its electrical strength up to altitudes of 1000 meters. However, the tests prescribed by these standards require that the BIL and BSL be given for sea level conditions. That is, no increase in the BIL or BSL is prescribed for 1000 metres. Therefore it is concluded that the statements concerning the altitude in apparatus standards are incorrect and that the BIL or BSL decreases at 1000 meters.

7 SUMMARY

7.1 BIL/BSL

1. The BIL and BSL are defined for

    (1.) Standard atmospheric conditions, i.e., sea level, relative air density $\delta = 1$.

    (2.) Standard lightning or switching impulse waveshape, i.e., 1.2/50 μs or 250/2500 μs.

2. The BIL or BSL is equal to the crest value of the standard impulse.
3. The BIL is defined for dry conditions.
4. The BSL is defined for wet conditions.
5. There are two types of BIL and BSL:

    (1.) Statistical: The probability of flashover or failure is 10% per single impulse application. Used for self-restoring insulations. The BIL or BSL is 1.28 standard deviation below the CFO, i.e.,

$$\text{BSL} = \text{CFO}\left(1 - 1.28\frac{\sigma_f}{\text{CFO}}\right)$$
$$\text{BIL} = \text{CFO}\left(1 - 1.28\frac{\sigma_f}{\text{CFO}}\right) \quad (27)$$

    (2.) Conventional: Insulation must withstand one to three applications of an impulse whose crest is equal to the BIL or BSL. Used primarily for non-self-restoring insulations. The probabilistic insulation characteristic is unknown.

6. Tests to establish the statistical BIL or BSL are (1) 3 + 3, (2) 3 + 9, and (3) 2/15 test series. The 2/15 is an IEC test and is best. The 3 + 3 is the IEEE circuit breaker test and is poorest, the 3 + 9 is a compromise test.

7.2 CFO and σ_f/CFO

1. The CFO is universally defined at standard atmospheric conditions.
2. The insulation strength characteristic for self-restoring insulations may be approximated by a cumulative Gaussian distribution having a mean defined as the CFO and a coefficient of variation σ_f/CFO.
3. Tests may be performed to obtain the entire characteristic or just the CFO.
4. These tests are used mainly to establish the switching impulse CFO of air gaps or air–porcelain insulations as a function of the strike distance and other variables.

5. The coefficient of variation differs for lightning and switching impulses. For switching impulses it is about 5% for towers, 6 to 7% for station insulations. For lightning impulses it is about 2 to 3%.

7.3 Chopped Waves

1. A 1.2/50 µs impulse chopped at a specific time is applied to transformers and circuit breakers. These apparatus must withstand these impulses: For breakers, 1.15 times the BIL chopped at 3 µs, and 1.29 times the BIL chopped at 2 µs. For power transformers, 1.10 times the BIL chopped at 3 µs. For distribution transformers, about 1.15 times the BIL chopped at between 1 and 3 µs.

2. Time-lag or volt–time curves are used to show the insulation strength for short duration impulses.

7.4 Atmospheric Correction Factors

1. With the subscript A signifying the strength at an altitude A in km, or the insulation strength at nonstandard atmospheric conditions, and the subscript S indicating the strength at standard conditions,

$$V_A = \delta^m H_c^w V_S \tag{28}$$

where δ is the relative air density and H_c is the humidity correction factor, which is

$$H_c = 1 + 0.0096 \left[\frac{H}{\delta} - 11 \right] \tag{29}$$

where H is the absolute humidity in grams of water per m³ of air.

2. m and w are constants that depend on G_0 defined as

$$G_0 = \frac{\text{CFO}_S}{500S} \tag{30}$$

where S is the strike distance in meters.

3. For wet or rain conditions $H_c = 1$. For design of lines and stations, assume wet conditions.

4. Using the proper m and w, for lightning design

$$V_A = \delta V_S \tag{31}$$

5. Using the proper m and w, for switching overvoltage design

$$V_A = \delta^m V_S$$
$$m = 1.25 G_0 (G_0 - 0.2) \tag{32}$$

6. The mean value of the relative air density δ is related to the altitude A in km by the equation

Specifying the Insulation Strength

$$\delta = e^{-A/8.6} \tag{33}$$

or by linear equation

$$\delta = 0.997 - 0.106A \tag{34}$$

The latter equation should be limited to altitudes of 2 km and therefore Eq. 33 is preferred. Both equations refer to thunderstorm conditions.

7. σ_f/CFO is slightly affected by altitude. However it may be neglected. Only consider the mean value per Eqs. 33 and 34.

8. V_S may be the standard BIL, BSL, or CFO. V_A are these same quantities at an altitude A.

8 REFERENCES

1. ANSI C92.1–1982, "Insulation Coordination," under revision.
2. IEEE 1313.1, "IEEE Standard for Insulation Coordination, Principles and Rules," 1996.
3. IEC Publication 71.1, "Insulation Coordination Part I, Definitions, Principles and Rules." 1993-12.
4. IEC Publication 60, "High-Voltage Test Techniques" and IEEE 4–1978, "IEEE Standard Techniques for High-Voltage Testing."
5. ANSI/IEEE C37.04–1979, "IEEE Standard Rating Structure for AC High-Voltage Circuit Breakers Rated on a Symmetrical Basis."
6. IEC Publication 71.2, "Insulation Coordination Part II, Application Guide, 1996-12.
7. ANSI/IEEE C57.12.00–1987, "IEEE Standard General Requirements for Liquid-Immersed Distribution, Power, and Regulating Transformers."
8. IEEE C37.12.14, "Trial Use Standard for Dielectric Test Requirements for Power Transformer for Operation at System Voltages from 115 kV through 230 kV."
9. ANSI/IEEE C37.32–1972, "Schedules of Preferred Ratings, Manufacturing Specifications, and Application Guide for Air Switches, Bus Supports, and Switch Accessories."
10. IEC Publication 517, "Gas Insulated Stations."
11. ANSI/IEEE C37.122–1983, "IEEE Standard for Gas-Insulated Stations."
12. AEIC C54-79, "Cables."
13. C. Menemenlis, G. Carrara, and P. J. Lambeth, "Application of Insulators to Withstand Switching Surges I: Switching Impulse Insulation Strength," *IEEE Trans. on Power Delivery*, Jan. 1989, pp. 545–60.
14. A. R. Hileman, "Weather and Its Effect on Air Insulation Specifications," *IEEE Trans. on PA&S*, Oct. 1984, pp. 3104–3116.
15. IEEE C62.11, "IEEE Standard for Metal-Oxide Surge Arresters for AC Power Circuits."

9 PROBLEMS

1. At a high-voltage laboratory, the ambient atmospheric conditions at the instant of test of a 3-meter gap are $H = 14\,\text{g/m}^3$, temperature = 15°C, pressure = 600 mm Hg.

(A) The CFO for a 1.2/50 μs impulse for dry conditions is found to be 1433 kV. Find the CFO for standard atmospheric condtions.

(B) Same as (A) except the CFO for a 250/2500 μs impulse for dry conditions is 1000 kV.

(C) A post insulator is rated as having a BSL of 1175 kV for wet conditions and has a strike distance of 4.23 meters. What voltage magnitude should be applied to the post insulator to "prove" the BSL rating? Assume $\sigma_f/\text{CFO} = 0.06$.

(D) Same as (C) except that the BSL is for the internal insulation of a transformer.

2. The rated BIL/BSL of a transformer bushing, both of the external porcelain and the internal part of the bushing, is 1300/1050 kV. Assume dry conditions for the BIL, wet conditions for the BSL. Determine the BIL/BSL of the bushing at an altitude of 1500 meters. Assume the bushing strike distance is 2.3 meters and that σ_f/CFO is 0.06 for switching impulses and 0.03 for lightning impulses. The relative air density is 0.838 and $\delta H_c = 0.814$.

3. Assume the Trafford impulse generator is used to generate a switching impulse. All 31 stages are used and charged to 200 kV. Assume the inductance is zero and the series resistance is 400 ohms per stage. Assume that the parallel combination of the capacitance voltage divider and the test object capacitance is 2000 pF.

(A) Determine the crest voltage, the actual time to crest, and the actual time to half value of the switching impulse.

(B) Find the generator efficiency.

(C) Calculate the virtual front time if this impulse were assumed to be a lightning impulse.

(D) Use approximations to calculate the front and tail time constants and the generator efficiency. Compare with the exact value per (A). Show the two circuits.

4. The suggestion has been made to use the Weibull distribution instead of the Gaussian distribution to approximate the strength distribution. Using the Weibull distribution in the form of

$$F(V) = p = 1 - e^{-\left(\frac{V-\alpha_0}{\alpha}\right)^\beta} \tag{35}$$

find the parameters assuming

(1) $p = 0.5$ for $V = \text{CFO}$.
(2) $p = 0$ for $V = \text{CFO} - 4\sigma$.
(3) $p = 0.16$ for $V = \text{CFO} - \sigma$.
(4) $Z = \dfrac{V - \text{CFO}}{\sigma}$.

2
Insulation Strength Characteristics

1 INTRODUCTION

As discussed in the previous chapter, the insulation strength is described by the electrical dielectric strength to lightning impulses, switching impulses, temporary overvoltages, and power frequency voltages. The purpose of this chapter is to present the characteristics of air–porcelain insulations subjected to lightning and switching impulses. In addition, the lightning impulse strength of wood or fiberglass in series with air–porcelain insulation is discussed. Because of the primary importance of switching surges in the design of EHV systems, and because the investigations of the switching impulse (SI) strength lead to an improved understanding of the lightning impulse (LI) strength, the SI strength of insulation is presented first.

Prior to the advent of 500-kV transmission in the early 1960s, little was known about switching surges as generated by and on the system, and therefore little was also known about the insulation strength when subjected to switching impulses. Prior to 500-kV transmission, insulation strength was defined only by its lightning impulse and power frequency voltage strengths. However, some field tests [1–4] were performed in the late 1950s that produced the first quantitative information on switching surges.

The first modern basic or fundamental investigations of the switching impulse insulation strength is credited to Stekolinikov, Brago, and Bazelyan [5] and to Alexandrov and Ivanov [6]. These authors startled the engineer world by showing that the switching impulse strength of air was less than that for lightning.

Thus there appeared to be adequate information to indicate that switching surges may be a problem for 500-kV systems. Studies were performed using an analog computer (known as a transient network analyzer or TNA) to determine the *maximum* magnitude and shape of the switching surges [7, 8]. The remaining task was simply to determine the *minimum* strength of insulation. That is, the design criterion was simply to set the maximum stress or maximum switching overvoltage equal to the minimum insulation strength. Given the maximum switching overvoltage, the next task was to find the SI strength of transmission tower insulation.

2 SWITCHING IMPULSE STRENGTH OF TOWERS

To determine the SI strength of a tower, a full-scale simulated tower is created in a high-voltage laboratory. This simulated tower shown in Fig. 1 is constructed of 1 inch angle iron covered with 1 inch hexagon wire mesh (chicken wire) to simulate the center phase of a transmission tower [17]. A two-conductor bundle is hung at the bottom of a 90-degree V-string insulator assembly. Switching impulses are then applied to the conductor with the tower frame grounded. First note the parameters of the test: (1) the strike distance, that is the clearance from the conductor to the tower side and the clearance from the yoke plate to the upper truss, (2) the insulator string length (or the number of insulators), (3) the SI waveshape (or actually the wavefront), and (4) wet or dry conditions.

Before proceeding to examine the test results, examine briefly the flashovers as shown in Figs. 2 to 5. These flashovers occurred under identical test conditions, i.e., dry, identical crest voltage, identical waveshape, and for the same strike distances and insulator length. The strike distance to the tower side and insulator length are approximately equal. First note that the flashover location is random, sometimes terminating on the right side of the tower (Fig. 2), sometimes on the left side (Fig. 3),

Figure 1 Tower test set-up.

Insulation Strength Characteristics

Figure 2 Switching impulse flashover.

Figure 3 Switching impulse flashover.

Figure 4 Switching impulse flashover.

Figure 5 Switching impulse flashover.

Insulation Strength Characteristics

Figure 6 Lightning impulse flashover.

sometimes upward to the truss (Fig. 4), and sometimes part way up the insulator string and then over to the tower side (Fig. 5). Thus the tower is not simply a single gap but a multitude of air gaps plus two insulator strings, all of which are in parallel and any of which may flash over.

To develop fully the concepts and ideas, return to those early days of the 1960s when testing with switching impulses was new. Until this time, all testing knowledge was based on lightning impulses. For the lightning impulse, the concept in vogue was that there existed a critical voltage such that a slight increase in voltage would produce a flashover and a slight decrease in voltage would result in no flashover, i.e., a withstand. This critical voltage is called a critical flashover voltage or CFO. In testing with switching impulses, we were amazed to find that this same concept could not be applied. For example, apply a 1200 kV impulse. A flashover occurs. Next decrease the voltage to 1100 kV. Another flashover occurs. Searching for that magical CFO, decrease the voltage again to 1000 kV, and at last, a withstand. But now increase the voltage back to 1100 kV—and a withstand occurs whereas before a flashover occurred! Now, apply the 1200 kV 40 times, to get 8 flashovers and 32 withstands. That is 20% flashed over. And if the voltage is decreased and another 40 impulses are applied, a lower percentage flashed over.

Not to belabor the point, it was found that at any voltage level there exists a finite probability of flashover between 0 and 100 percent. If the percent flashover is now plotted as a function of the applied voltage, an S-shaped curve results as shown in Fig. 7 [9]. (In detail, the upper and lower data points are for 100 "shots"; or voltage applications, while the data points in the center of the curve are for 40 shots.)

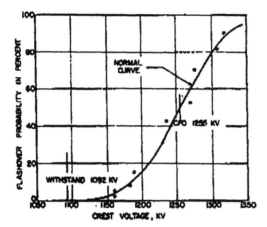

Figure 7 Best fit normal cumulative distribution curve for SI data points shown, center phase, positive, dry, 24 insulators [9].

When these data are plotted on normal or Gaussian probability paper, as shown by the upper curve of Fig. 8, the S-curve becomes a straight line, showing that the insulation strength characteristic may be approximated by a cumulative Gaussian distribution having a mean or 50% point that is called the CFO and a standard deviation or sigma σ_f [9]. Usually the standard deviation is given in per unit or percentage of the CFO, which is formally known as the coefficient of variation. In engineering jargon, an engineer might state that the sigma is 5%, which is interpreted as 5% of the CFO.

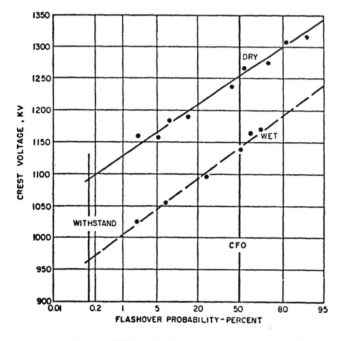

Figure 8 Data of Figure 7 plotted on normal probability paper [9].

This development was interesting and important, but it did not relieve the problem of searching for the minimum insulation strength, since, as stated in section 1, the design criterion was to equate the minimum strength to the maximum stress. So a withstand or minimum strength was still required. In a somewhat arbitrary manner but realizing that a low probability value was necessary, the withstand, or perhaps better, the "statistical withstand" voltage for line insulation V_3, was set at 3 standard deviation below the CFO, or in equation form,

$$V_3 = \text{CFO} - 3\sigma_f = \text{CFO}\left(1 - 3\frac{\sigma_f}{\text{CFO}}\right) \tag{1}$$

With the strength characteristic defined by two parameters, CFO and σ_f/CFO, investigation of the effect of other variables could proceed—testing to determine the effect of these other variables on the CFO or on the σ_f/CFO. For completeness, the equation for the cumulative Gaussian distribution is

$$p = F(V) = \frac{1}{\sqrt{2\pi}\sigma_f}\int_{-\infty}^{V} e^{-\frac{1}{2}\left(\frac{V-\text{CFO}}{\sigma_f}\right)^2} dV \tag{2}$$

where p or $F(V)$ is the probability of flashover when V is applied to the insulation. In more condensed form,

$$p = \frac{1}{\sqrt{2\pi}}\int_{-\infty}^{Z} e^{-\frac{Z^2}{2}} dZ \tag{3}$$

where

$$Z = \frac{V - \text{CFO}}{\sigma_f} \tag{4}$$

As noted in the above equations, the lower limit of integration is minus infinity—which is physically or theoretically impossible since this would mean that a probability of flashover existed for voltages less than zero. Detail tests on air–porcelain insulations have shown that the lower limit is equal to or less than about 4 standard deviations below the CFO [10].

2.1 Wave Front

The effect of the wave front or time to crest on the CFO is shown in Fig. 9 for a strike distance of about 5 meters, for wet and dry conditions and for positive and negative polarity [11]. First note the U-shaped curves showing that there exists a wave front that produces a minimum insulation strength. This is called the critical wave front or CWF. Next, wet conditions decrease the CFO, more for negative than for positive polarity. Also, positive polarity wet conditions are the most severe. In fact, for towers, the negative polarity strength is sufficiently larger than that for positive polarity that only positive polarity needs to be considered in design. Thus only positive polarity needs to be considered for further testing.

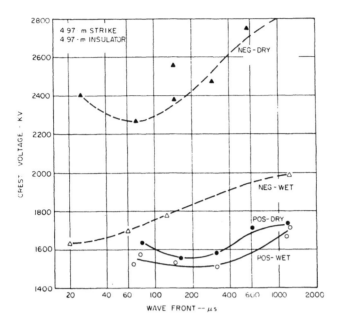

Figure 9 Effect of wave front on the CFO [11].

In immature EHV systems where switching of the EHV line is done from the low-voltage side of the transformer, the predominant wave front is not equal to the CWF but is much larger, of the order of 1000 to 2000 μs. From the test results shown in Fig. 9 and from other tests, the CFO for these longer fronts is about 13% greater than the CFO for the CWF. As is discussed later, for application, this value of 13% is reduced to 10% since the standard deviation also increases with the wave front.

Additional U-curves for other strike distances are shown in Fig. 10, where it is evident that the critical wave front increases with strike distance. Using these data, the CWF is plotted in Fig. 11 for positive polarity. Approximately, for positive polarity,

$$\text{CWF}^+ = 50(S - 1) \approx 50S \tag{5}$$

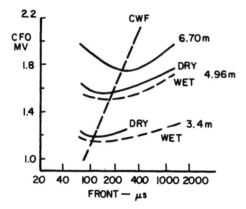

Figure 10 Critical front depends on strike distance. Data for tower window.

Insulation Strength Characteristics

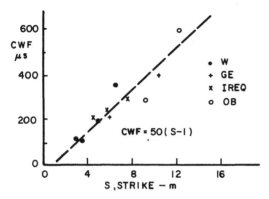

Figure 11 Critical wave front, CWF, for tower window, positive polarity.

where S is the strike distance in meters and the CWF is in microseconds. For negative polarity,

$$\text{CWF}^- = 10S \tag{6}$$

2.2 Insulator Length

Maintaining the strike distance at 4.97 meters and using the CWF, Fig. 12 presents the effect of the length of the insulator string [11]. For dry conditions, as the

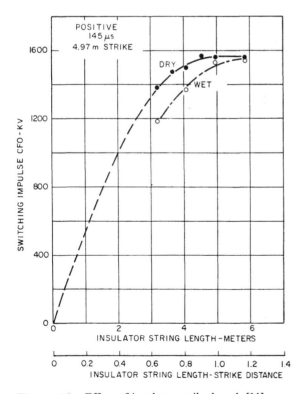

Figure 12 Effect of insulator strike length [11].

insulator length increases, the CFO increases until the insulator length is equal to the strike distance. That should be expected, since if the insulator length is less than the strike distance, flashovers will occur across the insulators, and thus the insulator string limits the tower strength. Oppositely, if the strike distance is less than the insulator length, flashovers will occur across the air strike distance, and the strike distance is limiting.

For wet conditions, this saturation point increases to a level where the insulator length is 1.05 to 1.10 times the strike distance. Thus to obtain the maximum CFO within a tower "window", or for a fixed strike distance, the insulator length should be 5 to 10% greater than the strike distance. The explanation for this wet-condition behavior appears to be that wet conditions degrade the CFO of the insulators more than that of the air.

Performing this same test for other strike distances, a family of curves results, as shown in Fig. 13.

2.3 Strike Distance

Using the results of Fig. 13, the maximum CFO for each strike distance is plotted in Fig. 14, the curve being denoted as "Tower" and compared to that for a rod–plane

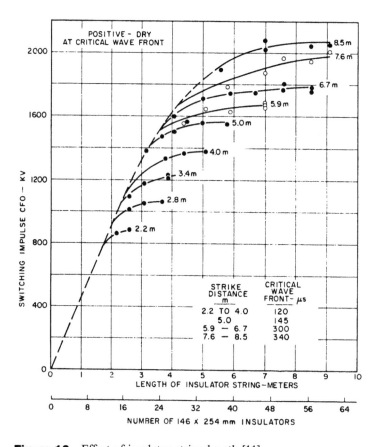

Figure 13 Effect of insulator string length [11]

Insulation Strength Characteristics

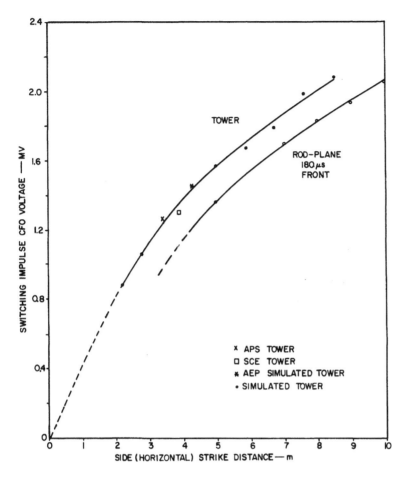

Figure 14 Maximum CFO of a tower window [11].

gap. As will be discussed later, this relationship between the CFO and the strike distance can be approximated by the following equation proposed by Gallet et al. [12].

$$\text{CFO} = k_g \frac{3400}{1 + (8/S)} \qquad (7)$$

where S is the strike distance in meters and the CFO is in kV. The variable k_g is called the gap factor, a term originally proposed by Paris and Cortina [13]. The gap factor for the center phase of a tower [14, 15] is given by the equation

$$k_g = 1.25 + 0.005\left(\frac{h}{S} - 6\right) + 0.25\left(e^{-\frac{8W}{S}} - 0.20\right) \qquad (8)$$

where (as illustrated in Fig. 15) h is the conductor height and W is the tower width.

To compare the results of the tests as shown in Fig. 14 to the values obtained from Eqs. 7 and 8, some adjustments are necessary. The upper four data points in Fig. 14 were obtained using a tower width of 12 feet (3.6 m), while the other data points are

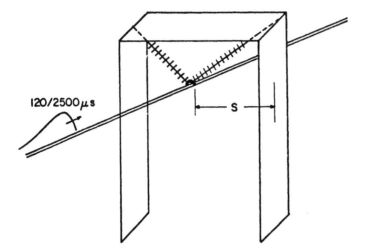

Figure 15 Definitions, a tower window.

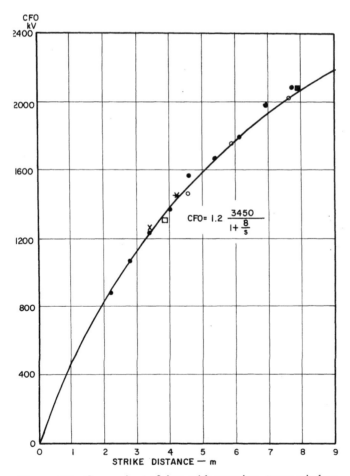

$$CFO = 1.2 \frac{3450}{1 + \frac{8}{S}}$$

Figure 16 Comparison of data with equation, tower window.

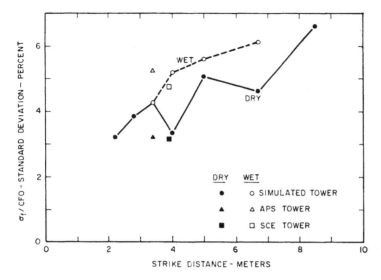

Figure 17 Effect of strike distance on σ_f/CFO [11].

for a tower width of 6 feet (1.9 m). These data points corrected to a base of W/S of 0.20 and h/S of 6 are shown in Fig. 16 along with the plot of the above equation, thus illustrating the excellent fit of the equation to the data. Note that for W/S of 0.20 and h/S of 6, $k_g = 1.25$. Usually, the gap factor is approximately 1.20 for lattice type towers and may increase to 1.25 for steel poles where the tower width is small.

2.4 Standard Deviation of Flashover

The standard deviation of the strength characteristic in per unit of the CFO, for the present series of tests, is shown as a function of strike distance in Fig. 17. The tendency of σ_f/CFO to increase with increased strike distance as shown in this figure has not been totally verified by other investigators, and therefore an average value of 5% is normally used for both wet and dry conditions. Actually, for dry conditions, the average value of σ_f/CFO is 4.3%, for wet conditions, 4.9%.

Menemenlis and Harbec [16] have shown that σ_f/CFO also varies with the wave front; their results are shown in Fig. 18. Since $V_3 = \text{CFO} - 3\sigma_f$ the curve of V_3 as a function of wave front will show a smaller variation with wave front than the CFO. As shown in Fig. 18, σ_f/CFO increases by about 10% from the CWF to a wave front

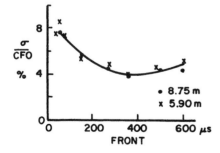

Figure 18 Effect of wave front on σ_f/CFO [16]. (Copyright IEEE, 1974.)

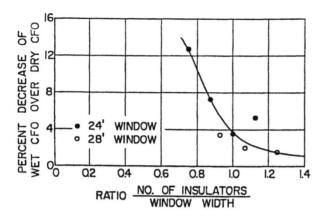

Figure 19 Effect of wet conditions [17].

of 600 µs. If, as stated previously, the CFO increases by 13% for a 2000 µs front, the value of V_3 will increase by about 10 to 11%. Therefore for general application the suggested increase in V_3 for long wave fronts is 10%, i.e., multiply Eq. 7 by 1.10 for long wave fronts.

2.5 Wet/Dry Conditions

Figure 19 [17] shows the effect of wet conditions in decreasing the CFO for dry conditions. For an insulator length/strike distance ratio of 1.05 to 1.10, which represents the design condition, wet conditions decrease the dry CFO by only 1%. However, from other tests, shown in Fig. 20 [9], larger values have been measured. For application, a value of 4% is suggested, i.e., multiply Eq. 7 by 0.96.

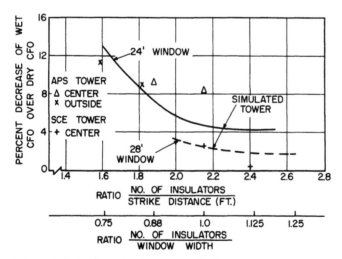

Figure 20 Effect of wet conditions [9].

Insulation Strength Characteristics

2.6 The Outside Phase

The CFO of the outside phase with V-string insulator strings should be expected to have a larger CFO than that of the center phase, since there exists only one tower side. From test data, the outside phase CFO is about 8% greater than that of the center phase, so multiply Eq. 7 by 1.08 [9, 11, 17].

2.7 V-Strings vs. Vertical or I-Strings

Only limited tests have been made on vertical or I-string insulators [17]. While dry tests showed consistent results, tests under wet conditions were extremely variable. For insulators in vertical position, water cascades down them so much that it may be said that water and not insulators is being tested. Only when the string is moved about 20° from the vertical position does water drip off each insulator so that test results become consistent. (The V-string is normally at a 45° angle.)

This should not be construed to mean that I-strings have a lower insulation strength than V-strings. Rather, the CFO of I-strings is difficult to measure for practical rain conditions. It is suggested that Eq. 7 multiplied by 1.08 be used to estimate the CFO. The strike distance S to be used is the smaller of the three distances as illustrated in Fig. 21: S_H to the upper truss, S_V to the tower side, and $S_I/1.05$, where S_I is the insulator string length. The factor 1.05 is used for the insulator string since the insulator string length should be a minimum of 1.05 times the strike distances per Section 2.2. For practical designs, usually, the insulator string length is controlling.

3 SUMMARY—INSULATION STRENGTH OF TOWERS

Before proceeding to discuss SI insulation strength of other insulation structures, a summary of the insulation strength of towers followed by a sample design problem appears appropriate. The summary:

1. The insulation strength characteristic can be approximated by the equation for a Gaussian cumulative distribution having a mean denoted by the CFO and a standard deviation σ_f. The statistical withstand voltage for line insulation V_3 is defined as

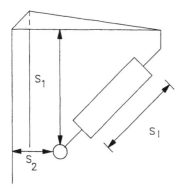

Figure 21 Outside arm strike distances.

$$V_3 = \text{CFO} - 3\sigma_f = \text{CFO}\left(1 - 3\frac{\sigma_f}{\text{CFO}}\right) \qquad (9)$$

where σ_f/CFO is 5%.

2. The CFO for the center phase, dry conditions, the critical wave front (CWF), positive polarity, and V-string insulators is

$$\text{CFO}_S = k_g \frac{3400}{1 + (8/S)} \qquad (10)$$

where S is in meters, CFO_S is the CFO in kV under standard atmospheric conditions, and

$$k_g = 1.25 + 0.05\left(\frac{h}{S} - 6\right) + 0.25(e^{\frac{8W}{S}} - 0.2) \qquad (11)$$

where h is the conductor height and W is the tower width.

3. For other conditions

Wet conditions decrease the CFO_S by 4%, i.e., multiply Eq. 10 by 0.96.
Outside phase has an 8% higher CFO_S, i.e., multiply Eq. 10 or Eq. 11 by 1.08.
The CFO_S and V_3 should be increased by 10% for wave fronts of 1000 µs or longer, i.e., multiply Eq. 10 by 1.10.
The insulator string length should be a minimum of 1.05 times the strike distance.
For I-string insulators, the CFO_S may be estimated by Eq. 10 multiplied by 1.08.
S is the minimum of the three distances (1) the strike distance to the tower side, (2) the strike distance to the upper truss, and (3) the insulator string length divided by 1.05.

4. The usual line design assumes thunderstorm or wet conditions. Also the line should be designed for its average altitude. Therefore the CFO under these conditions, CFO_A, may be obtained by the equation from Chapter 1,

$$\text{CFO}_A = \delta^m \text{CFO}_S \qquad (12)$$

For application, the CFO_S is changed to that for wet conditions. That is,

$$\text{CFO}_S = 0.96 k_g \frac{3400}{1 + (8/S)} \qquad (13)$$

and therefore CFO_A becomes

$$\text{CFO}_A = \delta^m \text{CFO}_S = 0.96 k_g \, \delta^m \frac{3400}{1 + (8/S)} \qquad (14)$$

or if the strike distance is desired, then

Insulation Strength Characteristics

$$S = \frac{8}{\frac{3400(0.96)k_g\delta^m}{\text{CFO}_A} - 1} \quad (15)$$

where

$$m = 1.25G_0(G_0 - 0.2)$$
$$G_0 = \frac{\text{CFO}_S}{(500)S} \quad (16)$$

and the relative air density δ is

$$\delta = e^{-A/8.6} \approx 0.997 - 0.106A \quad (17)$$

where A is the altitude in km.

4 DETERMINISTIC DESIGN OF TRANSMISSION LINES

Using the information and equations of the previous sections, a method called the deterministic method can now be developed. This method was employed to design the first 500 kV and 765 kV lines. Only during the last 10 years has the improved probabilistic method been adopted. This probabilistic method will be discussed in the next chapter.

To develop the simple deterministic design equations, assume that an EMTP or TNA study has been performed to determine the *maximum* switching surge E_m. The design rule is to equate V_3 to E_m:

$$V_3 = E_m \quad (18)$$

substituting V_3,

$$\text{CFO}_A = \frac{E_m}{1 - 3(\sigma_f/\text{CFO})} \quad (19)$$

Thus from E_m the CFO_A and the strike distance can be determined. To illustrate, consider the following example.

Example. Determine the center phase strike distance and number of standard insulators for a 500 kV (550 kV max) line to be constructed at an altitude of 1000 meters. The maximum switching surge is 2.0 per unit (1 pu = 450 kV) and $W = 1.5$ meters and $h = 15$ meters. Assume that all surges have a front equal to the critical wave front. Design for wet conditions and let $\sigma_f/\text{CFO} = 5\%$.

The CFO_A, which is the CFO required at 1000 meters, is from Eq. 19, $900/0.85 = 1059$ kV. Also the relative air density from Eq. 17 is 0.890. Because the gap factor and G_0 are both functions of the strike distance, the strike distance cannot be obtained directly. Rather an iterative process is necessary. To calculate S, Eq. 15 is used, i.e.,

$$S = \frac{8}{\frac{3400(0.96)k_g\delta^m}{\text{CFO}_A} - 1} \qquad (20)$$

As a first guess, let $k_g = 1.2$ and $m = 0.5$ and therefore $S = 3.2$. Iterating on S,

S	k_g	CFO$_S$	G_0	m	S
3.2	1.199	1118	0.699	0.436	3.18
3.18	1.199	1113	0.700	0.438	3.18

k_g is obtained from Eq. 11, CFO$_S$ from Eq. 13, G_0 and m from Eq. 16, and finally S from Eq. 15 or 20. As noted, only two calculations are necessary. Usually no more than three iterations are required.

Therefore, for the center phase, the strike distance is 3.18 meters (10.4 feet) and the minimum insulator length is 5% greater or 3.34 meters, which translates to 23 insulators ($5\frac{3}{4} \times 10$ inches).

The strike distance for the outside phase will be less than that for the center phase. Since the strength is 8% greater than that of the outside phase, the outside phase strike distance is approximately 3.18/1.08 or 2.94 meters. However this assumes a linear relationship, which is untrue. The proper procedure is to perform the above calculation with $k_g = 1.08$ times the value of k_g for the center phase. Performing this calculation results in a 2.91-meter strike distance for the outside phase, which in turn requires a minimum of 20 insulators.

5 SWITCHING IMPULSE STRENGTH OF POST INSULATORS

The CFO of station post insulators is presented in Fig. 22 for positive and negative polarity and dry conditions [18]. The parameter of the curves is the steel pedestal height, since at this time some authors suggested the use of a higher pedestal height to increase the SI strength. This suggestion prompted these tests.

As shown, as the pedestal height increases, the positive polarity strength increases but the negative polarity strength decreases. This implies the possibility that for some steel pedestal height, the positive and negative CFOs are equal. Per Fig. 22, this does not occur for practical pedestal heights. However, for a 1000 µs wave front, Fig. 23, the negative polarity CFO is only 3% above that for positive polarity for a pedestal height of 20 feet. (see also [19].)

For the CWF of about 120 µs used in these tests and a steel pedestal height of 8 feet (2.4 meters), an approximated equation for the CFO is

$$\text{CFO} = k_g \frac{3400}{1 + (8/S)} \qquad (21)$$

where $k_g = 1.4$ for positive polarity and $k_g = 1.7$ for negative polarity. The coefficient of variation σ_f/CFO is about 7%.

As for wet tests on the vertical insulator strings, wet tests on these vertical columns produced erratic results. Other investigations showed similar results in

Insulation Strength Characteristics

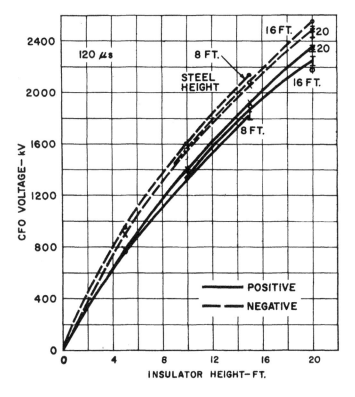

Figure 22 CFO of station posts [18].

this erratic behavior and indicated that the insulation strength is a function of the number of post units that compose the complete unit. That is, a post insulator column composed entirely of porcelain, i.e., without intervening metal caps, showed a higher CFO.

In 1988, IEC Technical Committee 36 proposed a revision of Publication 273 to provide a list of standard BIL/BSLs of post insulators along with the height of the

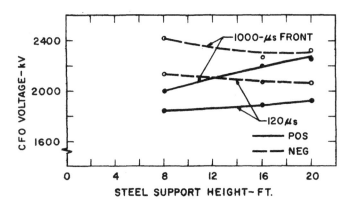

Figure 23 Effect of steel pedestal height, post insulator height = 15 ft [18].

Table 1 BIL/BSLs of Post Insulators, IEC 273-1990

			Creep distance, m	
			Class I	Class II
850	NA	1.90	3.10	4.40
950	750	2.10	3.40	4.90
1050	750	2.30	4.00	5.65
1175	850	2.65	4.60	6.50
1300	950	2.90	5.10	7.00
1425	950	3.15	5.60	7.80
1550	1050	3.35	6.20	8.50
1675	1050	3.65	6.35	9.40
1800	1175	4.00	6.90	10.25
1950	1300	4.40	7.65	11.35
2100	1300	4.70	8.25	12.25
2250	1425	5.00	8.70	13.20
2400	1425	5.30	9.20	14.10
2550	1550	5.70	9.80	15.00

column and creepage distances; see Table 1. These should be interpreted as BSLs under wet conditions.

The values in this table are ambiguous in that the same BSL is given for two different heights of insulators. Obviously, the BSL associated with the lower height would appear correct. These BSLs are plotted as a function of the height or strike distance S in Fig. 24. A regression line through the uppermost points results in the equation

$$\text{BSL}_S = 1.07 \frac{3400}{1 + (8/S)} \qquad (22)$$

or using a σ_f/CFO of 7%

$$\text{CFO}_S = 1.18 \frac{3400}{1 + (8/S)} \qquad (23)$$

where the BSL_S and CFO_S are the BSL and CFO at standard wet-weather conditions. As noted, the equation using the CFO provides a gap factor of 1.18 for wet conditions. From the test results presented previously, a gap factor of 1.40 was obtained for post insulators under dry conditions and positive polarity. Comparing these gap factors indicates that wet conditions decrease the CFO by about 16%, a not unreasonable value. Also from Table 1, as shown by Fig. 24, the standard BIL_S is approximately 450 kV/m of insulator length, i.e.,

$$\text{BIL}_S = 450S \qquad (24)$$

where S is the insulator height or strike distance in meters and the BIL_S is the BIL for standard atmospheric conditions in kV.

Insulation Strength Characteristics

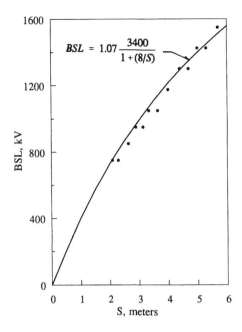

Figure 24 BSL of post insulators per Table 1.

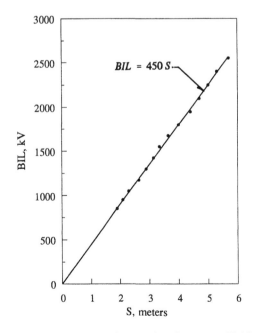

Figure 25 BIL of power insulators per Table 1.

6 A GENERAL APPROACH TO THE SWITCHING IMPULSE STRENGTH

A general approach to the estimation of the positive polarity CFO for alternate gap configurations was suggested by Paris and Cortina in 1968 [13]. They noted that all curves of the CFO as a function of gap spacing S had essentially the same shape and that the rod–plane gap had the lowest CFO. For example, note the shapes of the rod–plane and tower curves of Fig. 14. Therefore, they proposed the following general equation for positive polarity and dry conditions:

$$\text{CFO} = 500 k_g S^{0.6} \qquad (25)$$

As before, S is the gap spacing or strike distance in meters with the CFO in kV. The parameter k_g is the gap factor and is equal to 1.00 for a rod–plane gap. For other gap configurations, the gap factor increases to a maximum of about 1.9 for a conductor-to-rod gap. To be carefully noted is that in developing the above equation, the authors used a *250-µs front*—that is, it is not the critical wave front for all strike distances—so that the equation is not directly applicable to the minimum strength or minimum CFO.

Paris and Cortina suggested several gap factors, and in a 1973 ELECTRA paper, Paris et al. [20] proposed the gap factors shown in Table 2. As noted, the maximum gap factor of 1.9 is listed for a conductor-to-rod gap. From Table 2, two gap factors were selected for calculation of the phase–ground clearances in IEC Publication 71-2, 1976; (1) $k_g = 1.30$ for a conductor-to-structure gap, where, for example, the structure is a tower leg, and (2) $k_g = 1.10$ as a conservative gap factor for a rod–structure gap, where, for example, the gap configuration could be considered as the top of an apparatus bushing with a small or no grading ring to a tower leg. These clearances are given as a function of the BSL using a σ_f/CFO of 6%. (This will be more fully discussed and used in Chapter 5 concerning substation insulation coordination.)

Subsequently, Gallet et al. [12] further investigated the gap factor concept. They realized that the Paris–Cortina equation was valid only for a 250-µs wave front, and therefore they sought an alternate equation to express the CFO for the critical wave front, that is, an equation for the minimum CFO. Their proposed equation, which is now used exclusively, is

$$\text{CFO} = k_g \frac{3400}{1 + (8/S)} \qquad (26)$$

which again is valid for positive polarity and is normally applied only for dry conditions. As is recognized, this form of the equation was used for the tower insulation strength, where k_g was normally 1.2, and was also used for the post insulator.

Using the Gallet equation, the gap factors of Table 3 apply and as noted they do not differ greatly from those of Table 2. With further study, it became obvious that the gap factor was not simply a specific number but did vary with the specific parameters of the gap configuration. For example, note the equation of a rod–rod or conductor–rod gap of Table 3. Most recently, a CIGRE working group published a guide [15] in which general equations are presented for gap factors. But before

Insulation Strength Characteristics

Table 2 Gap Factors Proposed in Ref. 11 for Use with the Paris–Cortina Equation

Electrode configurations	Diagram	k_g
Rod–plane		1.00
Rod–structure (under)		1.05
Conductor–plane		1.15
Conductor–window		1.20
Conductor–structure (under)		1.30
Rod–rod ($h = 6$ m, under)		1.30
Conductor–structure (over and laterally)		1.35
Conductor–rope		1.40
Conductor–crossarm end		1.55
Conductor–rod ($h = 3$ m, under)		1.65
Conductor–rod ($h = 6$ m, under)		1.90
Conductor–rod (over)		1.90

examining these equations, consider the Gallet equation and note that as S approaches infinity, for a rod–plane gap, the CFO approaches 3400 kV, which would seem to indicate that a maximum CFO exists for any gap configuration. This is totally untrue and points out the limit of the equation. In general, the Gallet equation appears valid for a gap spacing in the range of about 15 meters. Beyond this spacing, Pigini, Rizzi, and Bramilla [21] proposed the following equation for a rod–plane gap for S in the range of 13 to 30 meters:

$$\text{CFO} = 1400 + 55S \tag{27}$$

Comparing at a gap spacing of 15, 20, and 25 meters, for a rod–plane gap, the Gallet equations gives CFOs of 2217 kV, 2429 kV, and 2579 kV, whereas Eq. 27 results in 2225 kV, 2500 kV, and 2775 kV.

Another equation for the CFO, positive polarity, appears in IEC Publication 71 [41], which is stated to be applicable for rod–plane gaps up to 25 m:

$$\text{CFO} = 1080 k_g \ln(0.46S + 1) \tag{28}$$

Table 3 Gap Factors for Gallet Equation

Configuration	Diagram	k_g
Rod–plane		1.00
Rod–rod		$1 + 0.6 \dfrac{h}{h+S}$ or $\dfrac{h}{h+S} e^{0.5}$
Conductor–plane		1.10
Conductor–rod		$1.1 + 1.4 \left(\dfrac{h}{h+S}\right)^{1.62}$ or $1.1 \dfrac{h}{h+S} e^{0.7}$
Conductor–structure		$1.1 + \dfrac{0.3}{1 + \dfrac{W}{S}}$
Conductor–large structure		1.30
Conductor–guy wire		1.45
Rod–structure		1.05

The standard deviation is stated to be about 5% to 6% of the CFO.

For negative polarity, Publication 71 provides the following equation applicable for spacing from 2 to 14 m, which is stated to have a standard deviation of about 8% of the CFO:

$$\text{CFO} = 1180 k_g S^{0.45} \qquad (29)$$

Comparing the CFO as determined by Eqs. 28 and 26, Eq. 28 results in essentially the same CFO as Eq. 26 for an S of 3 m and a CFO that is about 1.8% greater than that of Eq. 26 for an S of 6 m. Therefore there exists little reason to alter the equation for the basic rod–plane gap. That is, Eq. 26 is valid.

6.1 The Conductor–Window Gap—Center Phase

The equation Fig. 26, [15] for the gap factor, which was used in a previous section is

$$k_g = 1.25 + 0.005 \left(\dfrac{h}{S} - 6\right) + 0.25 \left(e^{-\frac{8W}{S}} - 0.2\right) \qquad (30)$$

Insulation Strength Characteristics

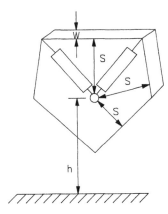

Figure 26 Tower window.

This equation is applicable in the range of $S = 2$ to 10 meters, $W/S = 0.1$ to 1.0, and $h/S = 2$ to 10. Again, W is the tower width, h is the conductor height, and S is the minimum value of S in Fig. 26. Usually this minimum distance is to the lower portion of the tower where the conductor exits the tower window. If a vibration dampener is used, the minimum distance is usually from this point.

Some observations: For the previously described tests, h/S is equal to or greater than 2. For the usual conditions, h/S is 4 to 5. For the normal lattice tower, W/S is 0.5 to 0.6. For a steel pole, W/S is about 0.2. Therefore for the lattice tower, k_g is about 1.20, and for a steel pole, k_g is about 1.25. There is not much of a variation.

6.2 Conductor–Crossarm—Outside Phase

The gap factor equation Fig. 27 [15] is

$$k_g = 1.45 + 0.015\left(\frac{h}{S_1} - 6\right) + 0.35\left(e^{-\frac{8W}{S_1}} - 0.2\right) + 0.135\left(\frac{S_2}{S_1} - 1.5\right) \quad (31)$$

This equation is applicable for $S_1 = 2$ to 10 meters, $S_2/S_1 = 1$ to 2, $W/S_1 = 0.01$ to 1.0, and $h/S_1 = 2$ to 10.

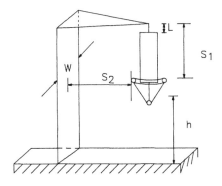

Figure 27 Conductor-crossarm.

Some observations: For these tests, S_2 is greater than or equal to S_1. This is the normal case for no wind-swing of the conductor and will normally result in all flashover occurring across the insulator string. For $h/S_1 = 4$ to 5, $W/S_1 = 0.5$, then $k_g = 1.35 + 0.135(S_2 - 1.5)$. If $S_1 = S_2$, then $k_g = 1.28$. Using the previous test results, the suggested gap factor for the outside phase was 1.08 times the gap factor for the center phase. Thus $1.08(1.20) = 1.30$ which is essentially equal to 1.28. Therefore the suggestion remains valid, i.e., multiply the gap factor for the center phase by 1.08 to obtain the gap factor for the outside phase.

6.3 Conductor–Lower Structure

The complex gap factor equation Fig. 28 [15] is

$$k_g = 1.15 + 0.81\left(\frac{h'}{h}\right)^{1.167} + 0.02\left(\frac{h'}{S}\right) - A\left[1.209\left(\frac{h'}{h}\right)^{1.167} + 0.03\left(\frac{h'}{S}\right)\right]\left(0.67 - e^{-\frac{2W}{S}}\right) \tag{32}$$

where $A = 0$, if $W/S < 0.2$, otherwise $A = 1$.

The equation is applicable in the range of $S = 2$ to 10, $W/S = 0$ to infinity, and $h'/h = 0$ to 1.

Some observations: If $h' = 0$ and $W = 0$, then the gap reverts to a conductor-to-plane gap with a gap factor of 1.15, which checks with Table 2 but not with Table 3. Next, assume that a truck is under the line with $W = 8$ meters and $h' = 3$ meters. Let $h = 10$ meters and $S = 7$ meters. Then $k_g = 1.181$, which is applied to the distance S, giving a CFO of 1875 kV. If the truck is not present, then $k_g = 1.15$ but is applied to $S = 10$ meters giving a CFO of 2172 kV. Thus the truck only decreases the CFO by 14% even though the strike distance is decreased by 30%.

6.4 Conductor–Lateral Structure

The equation Fig. 29 [15] is

$$k_g = 1.45 + 0.024\left(\frac{h}{S} - 6\right) + 0.35\left(e^{-\frac{8W}{S}} - 0.2\right) \tag{33}$$

The equation is applicable for $S = 2$ to 10 meters, $W/S = 0.1$ to 1.0, and $h/S = 2$ to 10.

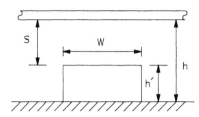

Figure 28 Conductor-lower structure.

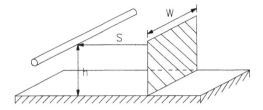

Figure 29 Conductor-lateral structure.

Some observations: Assume that the lateral structure is a tower leg with $W/S = 0.5$ to 0.6 and $h/S = 4$ to 5; then $k_g = 1.41$–1.43. Compare this to the gap factor for the crossarm, which is 1.28 to 1.30. The gap factor for the crossarm should be less than the 1.41 to 1.43 calculated here since the crossarm case adds an additional "arm" to the conductor–lateral structure case. Check is OK.

6.5 Rod–Rod Structure

This is a very complex arrangement Fig. 30 [15] having two different gap factors, k_{g1} for gap spacing S_1 and k_{g2} for S_2.

$$k_{g1} = 1.35 - 0.1\frac{h'}{h} - \left(\frac{S_1}{h} - 0.5\right)$$
$$k_{g2} = 1 + 0.6\frac{h'}{h} - 1.093A\frac{h'}{h}\left(0.549 - e^{-\frac{3W}{S_2}}\right)$$
(34)

where $A = 0$ if $W/S_2 < 0.2$ and otherwise $A = 1$.

For gap factor k_{g1}, S_2 must be greater than S_1, and the applicable range is $S_1 = 2$ to 10 meters and $S_1/h = 0.1$ to 0.8. For gap factor k_{g2}, S_1 must be greater than S_2, and the applicable range is $S_2 = 2$ to 10 meters and $W/S_2 = 0$ to infinity.

If $h' = 0$ and $W = 0$, then $k_{g2} = 1$, which checks the equation for a rod–plane gap. To obtain the gap factor for a vertical rod–rod, let W/S_2 be small. Then $A = 0$ and $k_{g2} = 1 + 0.6[h'/(h' + S_2)]$, which is the same equation as shown in Table 3. For a horizontal rod–rod gap, set $h' = 0$ and therefore $k_{g1} = 1.35 - (S_1/h - 0.5)$. If S_1/h is small, then k_{g1} becomes 1.4.

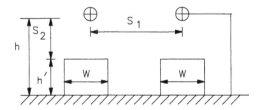

Figure 30 Rod–rod with lower structures.

6.6 Other Data

The results of other investigations of the switching impulse strength are provided in Refs. 22 to 26. Reference 26 is an excellent summary and analysis of the switching impulse strength data up to 1982.

7 LIGHTNING IMPULSE STRENGTH

As discussed previously, the lightning impulse (LI) strength is usually specified only by the CFO. However, benefiting from the results from switching impulse strength investigations, it is quickly realized that the LI strength characteristic can also be modelled as a cumulative Gaussian distribution having a mean value equal to the CFO and a standard deviation. However, in this case, the standard deviation is much smaller than that for switching impulses, usually in the range of 1 or 3% of the CFO, although values as high as 3.6% have been obtained for specific cases. However, seldom is the LI strength characteristic employed. Rather, the LI insulation strength is thought of as a single value, i.e., the CFO or the BIL. Voltages applied to the insulation that are below the BIL or CFO are assumed to have a zero probability of flashover or failure. Alternately, applied voltages that are greater than the CFO or BIL are assumed to have a 100% probability of flashover or failure. The LI insulation strength can also be given by the time-lag (or volt–time) curve.

In general, the curve of the CFO as a function of strike distance is linear, i.e., a straight line, and therefore the CFO can be given by a single value of gradient at the critical flashover voltage, or a CFO gradient in terms of kV per meter.

Considering the applied voltage waveshape, for switching impulses, the CFO is primarily a function of the wave front, while the tail is sufficiently long that it does not significantly alter the SI CFO. For lightning impulses, the CFO is primarily a function of the wave tail, and the front is only of importance when considering very short wave tails.

7.1 CFO of Insulators and Gaps—McAuley's Data

In 1938, McAuley published sets of curves giving the LI strength of suspension and apparatus insulators and rod gaps [27]. These curves have been frequently reproduced by many authors and therefore are still in use today to provide an estimate of the CFO. The following equations for the CFO are obtained from McAuley's curves.

For Rod–Rod Gaps

For positive and negative polarities, from 10 to 100 inches or from 0.25 to 2.5 meters, with S in meters,

$$CFO^+ = 60 + 581S$$
$$CFO^- = 87 + 623S$$
(35)

For Suspension Insulators

For positive and negative polarities, from 3 to 20 insulators, with S in meters,

$$\text{CFO}^+ = 130 + 561S$$
$$\text{CFO}^- = 171 + 489S \tag{36}$$

A word of caution is necessary when using these curves.

1. *Tower Representation*: The curves were obtained without representation of any nearby grounded objects. That is, the test arrangement, called a T-bar test, consists of a string of insulators hung by a crane (which is grounded). A pipe representing the conductor is placed at the bottom end of the insulator string. Usually this type of test set up results in a higher CFO than if grounded objects surrounded the insulator string. Thus the T-bar test only tests the insulator string and does not consider flashovers to grounded objects or the effect of grounded objects in altering the electric field. As an example of the problem, consider the V-string in the center phase of a tower. Depending on the strike distance to the tower sides, flashovers may occur either along the insulator string or to the tower sides. In addition, the grounded metal tower sides alter the electric field along the string and thus alter the insulation strength. As an example, for 15 insulators, these curves indicate a CFO of 1350 kV, whereas if the string were part of a V-string in the center phase of a tower, a CFO of 1225 kV would be expected.

2. *Wave Shape*: These curves were obtained with the then standard waveshape of 1.5/40. At that time period the 1.5 μs front was defined in a different manner from the now standard 1.2 μs-front. Accounting for this difference, the present 1.2 μs front is essentially equal to the old 1.5-μs front. However, the 40 vs. 50 μs time to half value does represent a difference, and since the CFO is primarily a function of the time to half value, the CFO provided by these curves should be slightly lower than for the now standard 50-μs tail. However, this difference is normally not of great significance.

3. *Positive and Negative Polarity CFO*: From the above equations, for rod–rod gaps, the negative polarity CFO is greater than that for positive polarity. However, for insulators, the positive polarity CFO is greater than that for negative polarity. As will be shown, for all practical cases, the CFO for positive polarity is less than that for negative polarity.

For these reasons, the CFO for insulators is highly suspect and should not be used except to obtain a crude value. The data for rod gaps appears better, but again if other data are available, they should be used.

7.2 CFO of Insulators and Gaps—Present Day Data

Illustrative of currently available data is presented in the aforementioned CIGRE Technical Brochure [15]. Figure 31 shows the LI CFO for a rod–plane gap for positive and negative polarity. CFOs for wet and dry conditions result in essentially the same CFO. While the positive polarity curve is linear, the negative polarity only becomes linear at gap distances above about 2 meters.

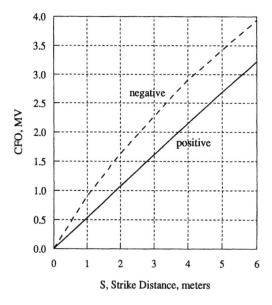

Figure 31 LI CFO for rod–plane gaps without insulators [15].

The LI CFO for the more practical case of the outside phase of a transmission tower, i.e., the crossarm case, is shown in Fig. 32 for dry conditions. While the CFO of this gap without insulators is independent of polarity, polarity affects the CFO when insulators are in the gap. To be noted is that all flashovers for this arrangement are across the insulators, i.e., none occurred to the side of the tower.

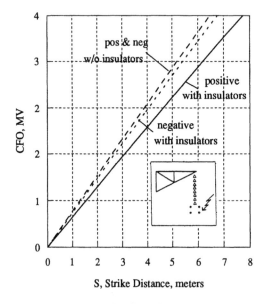

Figure 32 LI CFO of conductor–crossarm [15].

Insulation Strength Characteristics

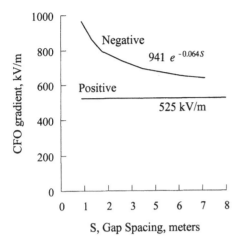

Figure 33 CFO gradient of rod–plane gaps [15].

The relationship between the gap spacing and the CFO gradient for rod–plane gaps is shown in Fig. 33. While the CFO gradient for positive polarity is constant at 525 kV/m, for negative polarity the CFO gradient varies with the gap spacing. The equation shown on the curve provides a crude estimate of the CFO gradient. Reference 15 also provides curves of Fig. 34 showing the ratio of the LI CFO gradient for a specific gap to the LI CFO gradient of a rod–plane gap as a function of the SI gap factor k_g. Equations for these curves are shown in Fig. 34.

Combining the data in Figs. 33 and 34, the negative polarity CFO gradient as a function of the gap factor can be obtained and is shown in Fig. 35 for gap spacing of 2, 3, and 5 meters. The dotted curve is that previously presented by Paris and Cortina [13]. It appears that the Paris–Cortina curve is applicable to a gap spacing of between

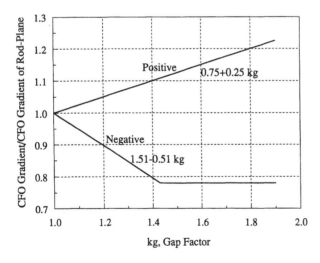

Figure 34 CFO gradient in per unit of CFO gradient for a rod–plane gap [15].

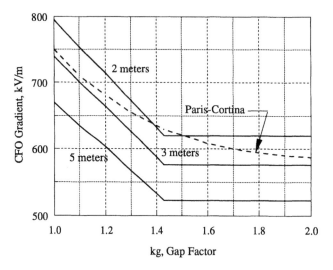

Figure 35 CFO gradient, negative polarity.

2 and 3 meters. The results for positive polarity are shown in Fig. 36 along with the curve for negative polarity for a 3-meter gap. The curve marked CIGRE TB 72 is that obtained from the present analysis using Figs. 33 and 34. Again the Paris–Cortina curve [13] is also shown and is a close match to the other curve. The equations for these curves for positive polarity are

Paris–Cortina:

$$\text{CFO}^+ = 383 + 147k_g \tag{37}$$

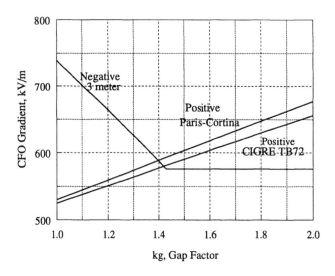

Figure 36 CFO gradient, positive polarity with a single negative polarity curve.

Insulation Strength Characteristics

CIGRE TB 72:

$$\text{CFO}^+ = 394 + 131 k_g \tag{38}$$

In addition, Ref. 15, i.e., CIGRE Technical Bulletin 72, also gives another curve having the equation

$$\text{CFO}^+ = 354 + 154 k_g \tag{39}$$

IEC Publication 71:
Other equations appear in IEC 71 [41]. For positive polarity with spacings up to 10 m,

$$\text{CFO}^+ = 530 S (0.74 + 0.26 k_g) \tag{40}$$

having a standard deviation of 3% of the CFO. For negative polarity up to 6 m, for k_g from 1 to 1.44,

$$\text{CFO}^- = 950 S^{0.8} (1.5 - 0.5 k_g) \tag{41}$$

and for k_g greater than 1.44,

$$\text{CFO}^- = 741 S^{0.8} \tag{42}$$

For negative polarity, the quoted standard deviation is 5% of the CFO. These equations are compared for positive polarity in the table.

k_g	Paris–Cortina Eq 37	TB 72 Eq 38	TB 72 Eq 39	IEC 71 Eq 40
1	530	525	508	530
1.2	560	551	538	558
2	677	656	662	668

For the important gap factor of 1.2, which approximates that for the tower, the values range from 538 to 560, which as will be shown compare favorably with full-scale tower test results of 560 kV/m.

For negative polarity, from Fig. 35, for a gap factor of 1.2, for 2-, 3-, and 5-meter gaps, the CFO gradients are 714, 664, and 603 kV/m. Using the IEC 71 equation for 2-, 3-, and 5-meter gaps, the CFO gradients are 744, 686, and 620 kV/m, whereas results from the aforementioned tower tests show a CFO gradient of about 605 kV/m for 3.4- to 3.5-meter gaps, which is about 10% lower.

As shown by Fig. 36, for large gap factors, the negative polarity CFO may be less than that for positive polarity. However, gap factors greater than about 1.4 seldom occur in practical gap configurations found in transmission lines and substations.

As a guide to the LI strength for alternate gap configurations, Table 4 presents ranges of the CFO gradient as obtained from a survey of the literature, primarily from Refs. 13 and 15 but also from 26–32. The asterisk in this table signifies that the CFO vs. distance curve is nonlinear; the value given is that for a distance of 4 meters. Before reaching any conclusion relative to these values, a brief review of data obtained for tests on practical tower insulation is necessary.

7.3 LI Strength of Towers

During the SI testing of the 500 kV towers, as presented previously, the LI characteristics were also obtained. An example of the test results is shown in Table 5 for the Allegheny Power System tower [9]. V-string insulator assemblies are on all phases, and the tower strike distances for the center and outside phase are as given in Table 5. Varying the number of insulators in the string for the center phase, dry conditions, a type of saturation point is reached at about 24 to 25 insulators (an insulator length of 3.51 to 3.65 meters), at which point the flashovers to the tower (air) are 78 to 100% of the total. Therefore the strike distance controls the CFO, and of interest is that the ratio of the insulator length (IL) to the strike distance is 1.03 to 1.07, about the same as for switching impulse. For wet conditions, for 24 insulators, the majority of flashovers revert to the insulator string, thus indicating that to achieve maximum strength in the center phase, at least one additional insulator should be used. The strength using an additional unit is estimated to be approximately equal to the strength for dry conditions. Turning to the negative polarity, dry condition, all flashovers were across the insulator string for both 24 and 25 insulators, again showing that this is the weak link for these conditions. However, for all conditions

Table 4 Lightning Impulse CFOs for Gaps with and without Insulators

Gap configuration	Diagram	Positive polarity CFO, kV/m		Negative polarity CFO, kV/m	
		w/o ins.	with ins.	w/o ins.	with ins.
Rod–plane		540	520	660	*375
			540		*500
Outside arm		600	500	600	595
		625	520	625	620
Conductor–upper structure		575	560	625	610
Conductor–upper rod		655	*500	595	585
Rod–rod		560	500	640	*425
					*475

* CFO vs. distance curve nonlinear. Value given is for 4m.
For gaps with insulators, all flashovers occur across the insulator, i.e., the insulation strength is limited by the insulators.

Insulation Strength Characteristics

Table 5 Lightning Impulse CFO for APS 500 kV Tower

Phase	Tower strike, m	Pol. + or −	Dry or wet	No. of ins.	IL/S	CFO kV	kV/m ins. length	Flashover location Air	Flashover location Ins.
Center	3.40	+	Dry	21	0.90	1840	600	0	100
		+	Dry	24	1.03	1950	556	78	22
		+	Dry	25	1.07	1960	537	100	0
		+	Wet	24	1.03	1890	539	29	71
		−	Dry	24	1.03	2025	578	0	100
		−	Dry	25	1.07	2070	567	0	100
Outside	4.04	+	Dry	21	0.76	1800	587	0	100
		+	Dry	24	0.87	2050	585	0	100
		+	Wet	24	0.87	1970	562	0	100
		−	Dry	24	0.87	2200	628	0	100

the minimum CFO appears to be 1950 kV or about 570 kV per meter of strike distance.

The above characteristic for positive polarity, dry conditions, can be portrayed as being much the same as for switching impulses [11]. Using other additional test data, Fig. 37 shows the CFO as a function of insulator length with strike distance as

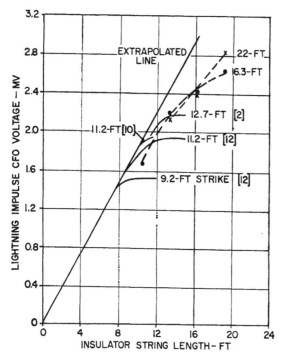

Figure 37 LI CFO versus insulator length, strike distance as parameter [11].

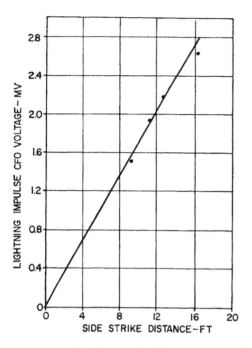

Figure 38 Maximum LI CFO in a tower window [11].

a parameter. Plotting the maximum obtainable CFO for the center phase as a function of strike distance produces the linear relationship per Fig. 35 where the CFO gradient is approximately 560 kV per meter of strike distance.

Returning to Table 5 to examine the outside phase, because of the larger strike distance, all flashovers occurred across the insulators. Therefore the CFO gradient should be analysed in terms of the CFO per meter of insulator length and is comparable to the values of Table 4. The comparison is shown in Table 6 where suggested values for the outside phase are given.

For these tower tests, for positive polarity, the standard deviation was 1.0% of the CFO, which increased to 3.6% of the CFO for negative polarity.

Table 6 Comparison of CFO for Outside Arm of Table 4 with Values in Table 5 for Outside Phase.

	Table 5 kV/m of insulator length	Table 4 kV/m of insulator length	Suggested kV/m of insulator length
CFO, pos., dry	585	500–520	560
CFO, pos., wet	562	500–520	560
CFO, neg., dry	628	595–620	605

CFO in kV per meter of insulator length.

Insulation Strength Characteristics

7.4 Suggested Values for Air Gaps and Insulators

By use of the curves and data presented, the CFO for gaps and tower insulation can be estimated. For tower insulation (wet), for either the center or the outside phase, these data show that the CFO gradients are approximately

For positive polarity: 560 kV/m (170 kV/ft)
For negative polarity: 605 kV/m (185 kV/ft)

For V-strings in the center phase, the relevant distance is the tower strike distance, and the insulator string length should be a minimum of 1.05 times the tower strike distance. For the outside phase using V-strings or other phases using vertical or I-strings, the distance should be the insulator length or the strike distance, whichever is smaller.

Gap configurations within a substation vary but may be typified by the outside phase or crossarm and by the conductor–upper structure configurations. For these gaps, the positive polarity CFO ranges from 575 to 625 kV/m and the negative polarity CFO ranges from 600 to 625 kV/m. Therefore, for substation clearances, the same values as above are suggested for use, i.e.

For positive polarity: 560 kV/m (170 kV/ft)
For negative polarity: 605 kV/m (185 kV/ft)

7.5 Time-Lag (Volt–Time) Curves

Time-lag or volt–time curves vary significantly with gap configuration. As the gap configuration approaches a uniform field gap, the upturn of the time-lag curve becomes less pronounced: the curve becomes flatter. Oppositely, as the gap configuration approaches a more nonuniform field gap, the upturn at short times becomes greater. Some typical time-lag curves obtained from tower testing [9] are presented in Figs. 39 and 40, and typical time-lag curves obtained from other sources [27, 33] are

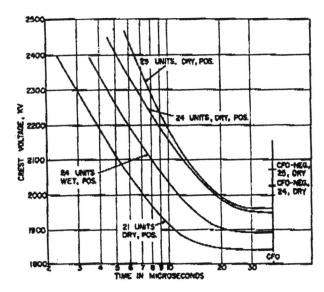

Figure 39 Time-lag curves for center phase of APS tower, 3.4-m strike distance [9].

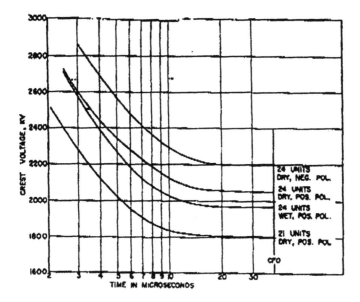

Figure 40 Time-lag curves for outside phase of APS tower, 4-m strike distance [9].

shown in Fig. 41. An equation that crudely represents the time-lag curve from about 2 to 11 μs is

$$\frac{V_B}{\text{CFO}} = 0.58 + \frac{1.39}{\sqrt{t}} \tag{43}$$

where V_B is the breakdown, flashover, or crest voltage, and t is the time to breakdown or flashover.

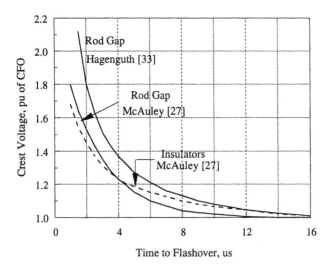

Figure 41 Typical time-lag curves.

Insulation Strength Characteristics

Table 7 Comparison of Ratios of V_B/CFO

Time to breakdown, μs	Rod gaps, Hagenguth [33]	Rod gaps, McAuley [27]	Insulators, McAuley [27]	Tower tests	Eq. 43
2	1.53	1.80	1.45	1.40–1.73	1.56
3	1.35	1.51	1.31	1.24–1.45	1.38

Table 7 presents a comparison of the data for tower insulations from Figs. 39 and 40 and from Eq. 42. Suggested values for tower insulation are

Breakdown voltage at 2 μs = 167(CFO)
Breakdown voltage at 3 μs = 1.38(CFO)

For apparatus porcelain insulations, the 3 μs breakdown voltage varies from about 1.22 to 1.31 per unit of the CFO, and the 2 μs breakdown voltage varies from about 1.32 to 1.48. However, standard chopped wave tests, if specified, normally use a 3 μs test value of 1.15 times the BIL, and for the circuit breaker at 2 μs, 1.29 times the BIL. Therefore for apparatus, the latter two values are frequently used.

8 LIGHTNING IMPULSE STRENGTH OF WOOD AND PORCELAIN

Important features of the LI characteristic of wood or wood and porcelain in series are (1) that the variability of the dielectric strength, an inherent characteristic of wood, reaches about plus and minus 15 to 20% and is primarily a function of the moisture content of wood and (2) that wood and porcelain in series produce a dielectric strength that may be greater than the strength of either of the insulations but is less than their sum. Another advantage or characteristic of wood that is important in distribution line design is the ability of wood to extinguish the power frequency arc that follows the lightning flashover, thus limiting breaker tripping. This ability of wood is discussed in Chapter 10. Considering the first feature of wood, since the dielectric strength of wood is dependent on the moisture content of the wood, the strength varies with the seasoning of the wood and with wet or dry atmospheric conditions.

The impedance equivalent circuit of Fig. 42, as present in an AIEE Committee Report [34], provides an explanation of the strength of wood and porcelain in series. Figure 43 gives estimates of the capacitance and resistance of wood [35]. The

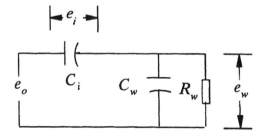

Figure 42 Equivalent impedance circuit of wood and porcelain in series.

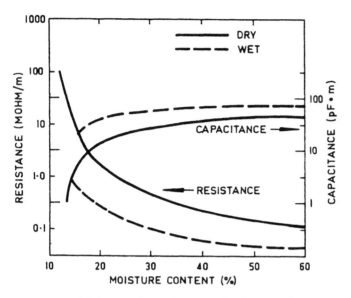

Figure 43 Estimates of capacitance and resistance of wood [35].

insulator capacitance is approximately 36 pF per insulator or better 36 pF-insulator. For an exponential impulse of the form

$$e_0 = Ee^{-t/\tau_A} \tag{44}$$

the voltage across the wood is

$$e_w = \frac{C_i}{C_i + C_w} \left[\frac{\tau_A e^{-t/\tau_B} - \tau_B e^{-t/\tau_A}}{\tau_A - \tau_B} \right] E \tag{45}$$

and the voltage across the insulators is

$$e_i = \frac{C_w}{C_i + C_w} \left[\frac{\tau_A e^{-t/\tau_B} - \tau_B e^{-t/\tau_A} + \tau_A(e^{-t/\tau_A} - e^{t/\tau_B})}{\tau_A - \tau_B} \right] E \tag{46}$$

where

$$\tau_B = R_w(C_i + C_w) \tag{47}$$

and R_w is the resistance of wood, C_i is the capacitance of the insulator, and C_w is the capacitance of the wood.

If the applied voltage e_0 is more simply a square wave or infinite rectangular wave, then the voltages become easier to understand and are

Insulation Strength Characteristics

$$e_w = \frac{C_i}{C_i + C_w} E e^{-t/\tau_B}$$

$$E_i = \frac{C_w}{C_i + C_w} E + (1 - e^{-t/\tau_B})E \qquad (48)$$

Using the ATP program, the voltages across the wood and insulators are shown in Fig. 44 for an applied voltage of 1000 kV, 1.2/50 µs. The values of the parameters are based on four insulators plus 2 meters of wood. Thus

$$C_i = 9\,\text{pF} \qquad C_w = 25\,\text{pF} \qquad R_w = 1\,\text{M}\Omega$$

Therefore the time constant τ_B is 34 µs

The voltage across the wood reaches a crest voltage of 722 kV at 1.2 µs and then decays to half value at 16.5 µs (time constant of 25 µs). The voltage across the insulators initially rises to a voltage of 280 kV at 1.2 µs and then because of the long tail of the applied voltage continues to increase reaching a crest of 526 kV at 37 µs.

The voltages in Fig. 44 assume an applied voltage having a waveshape of 1.2.50 µs, the standard lightning impulse, and thus relate to a laboratory test condition. This waveshape would also be appropriate when considering a shielding failure. For the event of a lightning stroke to the ground wire, the tail of the impulse would be much smaller. This situation is illustrated in Fig. 45 where voltages are shown for an applied voltage of 1000 kV, 1.2/14 µs. All other parameters are the same as for Fig. 44. In this case, both the voltage across the wood and the voltages across the insulators achieve crest at 1.2 µs, 722 kV for the wood and 278 for the insulators.

Therefore, as an approximation, the crest voltages are equal to the initial voltages. From either form of the equations, the initial voltage across the wood e_w and the voltage across the insulator e_i are

$$e_w = \frac{C_i}{C_i + C_w} E \quad \text{and} \quad e_i = \frac{C_w}{C_i + C_w} E \qquad (49)$$

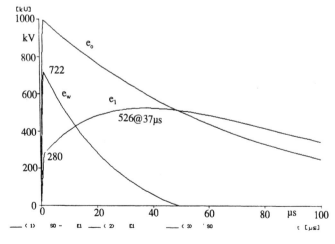

Figure 44 Voltages across wood and insulators for applied surge of 1000 kV, 1.2/50 µs.

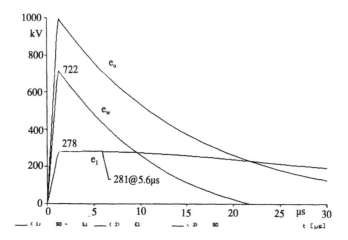

Figure 45 Voltages across wood and insulators for applied surge of 1000 kV, 1.2/14 μs

Considering again the voltages shown in Figs. 44 and 45, it is noted that the waveshapes across the wood or insulators (1) are not equal to the standard 1.2/50 μs and (2) are not similar. Because of these factors, the CFO of each insulation will vary. However, to continue in this development, assume that the CFOs of each insulation can be specified as those for a 1.2/14 μs wave.

Then, to illustrate the flashover of the combination of wood and insulator, assume that the CFO of the combinations of four insulators and 1 and 2 meters of wood is desired. Assume that the CFO of wet wood alone, CFO_w, is 300 kV/m and that the CFO of four insulators alone is 325 kV. Let the time-lag curve of each insulation be flat, i.e., only the magnitude of the voltage is important.

Case 1. Assume wood length $L_w = 1$ meter, $C_w = 50$ pF, $C_i = 9$ pF, and $R_w = 0.5$ MΩ. Then per Eq. 49, $e_w = 0.1525$ pu of the applied voltage and $e_i = 0.8475$ pu. Since the voltage is greater across the insulator, flashover will occur there first. So for critical conditions, let $e_i = 325$ kV, which translates to an applied voltage of 383 kV. Therefore the flashover sequence is (1) flashover of the insulator and (2) flashover of the wood, since after the insulator flashover, the voltage across the wood is the applied voltage of 383 kV, which exceeds the CFO of wood.

Case 2. Assume wood length $L_w = 2$ meter, $C_w = 25$ pF, $C_i = 9$ pF, and $R_w = 1.0$ MΩ. Then per Eq. 49, $e_w = 0.264$ pu and $e_i = 0.735$ pu. As in Case 1, the insulator will flash over first for which the crest voltage of the applied surge is $325/0.735 = 442$ kV. However, since the wood CFO is now 600 kV, no flashover occurs across the wood. Therefore the applied surge must be increased to 600 kV for the combination to flash over.

The conclusion to this crude example is that the CFO of the combination, CFO_c is either (1) the CFO of the insulator multiplied by a capacitance ratio, i.e.,

$$CFO_c = \left(1 + \frac{C_i}{C_w}\right) CFO_i \tag{50}$$

Insulation Strength Characteristics

but if a c_w is defined as the capacitance per length (or more properly in capacitance length, e.g., μF-m) of wood, then $C_w = c_w/L_w$ and

$$\text{CFO}_c = \left(1 + \frac{C_i L_w}{c_w}\right)\text{CFO}_i \tag{51}$$

or (2) the CFO of wood. If the CFO of wood is defined as a critical gradient, CFO_{wg}, e.g., 300 kV/m,

$$\text{CFO}_c = \text{CFO}_{wg} L_w \tag{52}$$

Below some critical length, Eqs. 50 or 51 apply, and above this critical length, Eq. 52 applies. Equating Eqs. 51 and 52 to find this critical length L_{cw}, we obtain

$$L_{cw} = \frac{\text{CFO}_i}{\text{CFO}_{wg} - \left(\frac{C_i}{c_w}\right)\text{CFO}_i} \tag{53}$$

For the example with $c_w = 50\,\text{pF/m}$, $C_i = 9\,\text{pF}$, $\text{CFO}_i = 325\,\text{kV}$, and $\text{CFO}_{wg} = 300\,\text{kV/m}$, the critical length L_{cw} is 1.35 meters.

The results of this crude example are shown in Fig. 46. The curve of the combined CFO linearly increases at a rate of 59 kV/m below the critical length, that is at

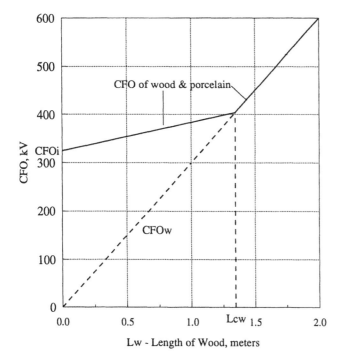

Figure 46 Results of the example.

a rate of $(C_i/c_w)CFO_i$, and at a rate of 300 kV/m above the critical length, that is at a rate of CFO_{wg}.

Although the above two cases are crude examples and not all factors are properly considered, the result is, in general and in concept, true. However, as should be expected, in the actual case, there does exist a more gradual change from the insulator CFO to the wood CFO. For example, Darveniza [35], following an extensive review of the literature and after extensive testing, suggests the use of the curves of Figs. 47 and 48 to estimate the CFO of the combination of wood and porcelain. The curves of Fig. 47 show a gradual increase to the wood CFO, whereas the curves of Fig. 48 show that the combined CFO is always greater than the wood CFO. However note the large confidence limits in Fig. 48.

If we examine Fig. 48, some very general rules appear:

1. The critical length of wood L_{cw} is approximately equal to twice the insulator length. (The critical length in feet is equal to the number of "standard" insulators, $5\frac{3}{4} \times 10$ inches.)
2. At this critical length, wood adds approximately 100 kV per meter of wood (30 kV/ft) to the insulator CFO, which agrees with the aforementioned AIEE Committee report [34].
3. Above this critical length of wood, the CFO is equal to the CFO of the wood alone, at a CFO gradient of about 300 kV/m (90 kV/ft).
4. At about half the critical length, wood adds only about 40 kV/m (10 kV/ft) to the insulator CFO.

As an overall general concept, the insulation strength of wood and porcelain in series is either the CFO of the insulator alone or the CFO of the wood alone. Per item 2 above, this is not strictly true, but the idea is good as a general concept. Because the length of wood for transmission lines is usually at or slightly above the critical length, wood only adds marginally to the CFO. However, for lower voltage lines,

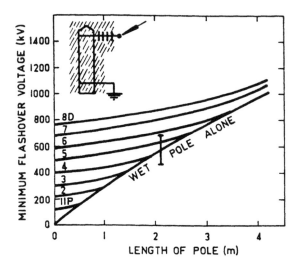

Figure 47 LI flashover of wet wood pole and porcelain in series [35].

Insulation Strength Characteristics

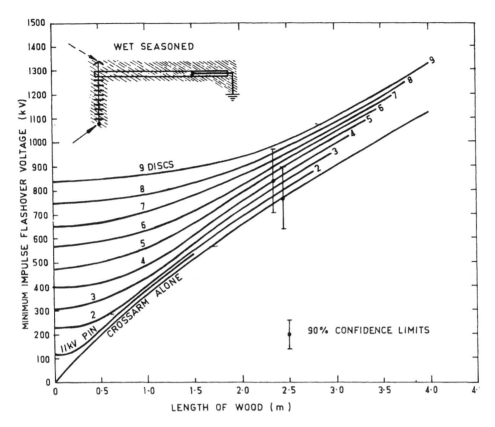

Figure 48 LI flashover of wet wood and porcelain in series [35].

e.g. 34.5- and 69-kV lines and distribution lines, the length of wood usually exceeds the critical length and therefore the CFO is usually that of wood alone.

As a final note, the book published by M. Darveniza titled *Electrical Properties of Wood* is highly recommended [35]. It contains a complete review and evaluation of previous literature, further tests performed by the author and his associates, design curves, and examples of design. Other interesting data are contained in Refs. 36–39.

9 LIGHTNING IMPULSE STRENGTH OF FIBERGLASS

Few test data are available concerning the LI strength of fiberglass rods. Of help is a paper on fiberglass crossarms, the results of which are presented in Fig. 49 for 1 to 4 feet of crossarm length in series with three suspension insulators [40]. For the fiberglass crossarm alone, for negative polarity, the CFO gradient is 605 and 700 kV/m for wet and dry conditions, respectively, which compares favorably with the CFO gradient for air suggested previously of 605 kV/m. The author shows that the CFO of the three insulators is 374 kV and 363 kV, dry, negative and positive polarity, respectively, and 347 kV and 340 kV, wet, negative and positive polarity, respectively. The averages of these values are plotted in Fig. 49.

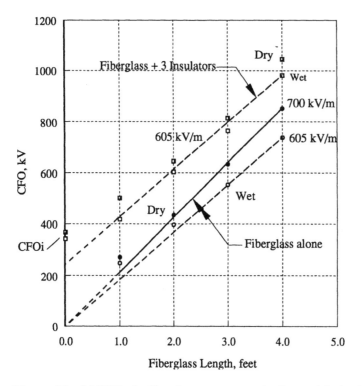

Figure 49 LI CFO of a fiberglass crossarm and a 3 porcelain insulators in series.

The dotted line for fiberglass plus wood is draw for 605 kV/m and intersects the CFO axis at about 265 kV, which amazingly is the CFO for three insulators if evaluated at 605 kV/m. Therefore, as an excellent approximation, the CFO of fiberglass and insulators in series is simply the length of fiberglass plus the length of the insulators multiplied by 605 kV/m for negative polarity. For positive polarity, use 560 kV/m. That is, in equation form, for positive polarity.

$$\text{CFO}_c = 560(L_f + L_i) \tag{54}$$

and for negative polarity,

$$\text{CFO}_c = 605(L_f + L_i) \tag{55}$$

where CFO_c is the CFO of the combination of fiberglass and porcelain, L_f is the fiberglass crossarm length, and L_i is the length of the insulator string.

10 EFFECT OF TAIL ON CFO

In Chapter 10, when considering the backflash, an equation is derived that shows the effect of the time to half value, or the tail time constant, on the CFO, i.e.,

$$\text{CFO}_{NS} = \left(0.977 + \frac{2.82}{\tau}\right)\text{CFO}_S \tag{56}$$

Insulation Strength Characteristics

Table 8 Effect of the Tail on the CFO

Time to half value, μs	Tail time constant, γ, μs	CFO_{NS}/CFO_S
100	144	0.997
50	72	1.016
40	58	1.025
20	29	1.075
10	14	1.172

where CFO_S is the CFO for the standard 1.2/50 μs wave shape and the CFO_{NS} is the nonstandard CFO for the tail time constant τ in μs.

The equation is valid for time to crests between 0.5 and 5 μs and for tail time constants between 10 and 100 μs. Thus, as noted, the front is of minor importance if below 5 μs; the tail is of greater importance.

Table 8 shows the results of Eq. 56 for times to half value from 10 to 100 μs. For the standard tail of 50 μs, the ratio is 1.016, an error of 1.6%. As the tail decreases, the CFO increases, reaching about 17% greater than the standard CFO for the short tail of 10 μs.

11 LI FLASHOVER MECHANISM

The time-exposure photograph of a lightning flashover, shown in Fig. 6 and repeated here as Fig. 50, clearly illustrates the three stages of the breakdown mechanism: (1) the corona stage, (2) the channel stage, and (3) the actual flashover. As the voltage increases along the front of the impulse, a voltage is reached such that

Figure 50 LI flashover illustrating the flashover mechanism [23].

corona streamers are emitted into the interelectrode space, producing a dense waterfall effect as seen at the lower right of the photograph. This corona formation does not constitute breakdown but is merely the precursor of breakdown. Ascribing a personality to the corona streamers, it could be said that their job is to change the nonuniform field existing in the gap into a more uniform field, thus preparing the way for the next stage, the channels.

As the voltage increases above the corona inception level, a voltage is achieved that initiates the channel from the conductor (positive polarity). This channel travels at ever-increasing velocity toward the grounded tower side, as seen in the lower left of Fig. 50. When this channel reaches about midgap, another channel is initiated from the grounded tower side. These channels continue to travel toward each other until they finally meet, and actual flashover occurs as seen across the upper left insulator string. In clarification, the initial two stages of breakdown also occurred across the insulator string, and for this path, the channels met before the channels at the lower left could meet.

As a final comment, the comparison of the CFOs for lightning and switching impulses is shown in Fig. 51 and graphically illustrates the reason that engineers were concerned with the switching impulse strength of towers and the switching surge design of transmission lines. For example, to achieve a LI CFO of 1600 kV, a strike distance of about 3 meters is required. However, to obtain this same CFO for switching impulse, a strike of about 5 meters is required.

12 POWER FREQUENCY

For transmission lines or substations, the insulation strength under power frequency voltage for clean or noncontaminated conditions is seldom a determinant for insulator design or in determining the strike distance. Rather, it is the performance of external

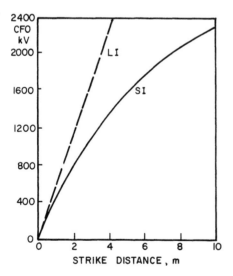

Figure 51 Comparison of SI and LI strength of towers. Lightning CFO at 560 kV/m and switching CFO for a gap factor of 1.2.

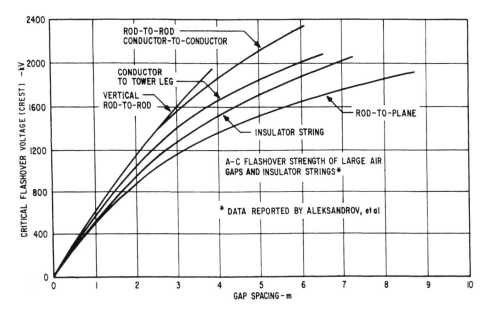

Figure 52 Power frequency CFO of large gaps and insulators [46].

insulation under contaminated conditions that may dictate the insulation design. However, for completeness and for comparison to the insulation strength under contaminated conditions, the power frequency performance is briefly presented. Reference 45 presents the data from Alexandrov et al. [46], which is reproduced here as Fig. 52. These values are for dry conditions where the standard deviation is about 2%. For insulators, rain may substantially reduce the insulation strength dependent on the rate of rainfall, the conductivity, and the insulator configuration (V-string, I-string, or horizontal). For I-strings or vertical insulator strings, this decrease may approach 30%. The effect of rain on air gaps is negligible. IEC Publication 71 [41] provides an approximate equation for the power frequency CFO:

$$\text{CFO}_{\text{PF}} = 750(1.35 k_g - 0.35 k_g^2) \ln(1 + 0.55 S^{1.2}) \tag{57}$$

which is valid for spacings greater than or equal to 2 meters. In general, this equation agrees with the plots in Fig. 52. Although not stated in the IEC publication, it appears that this CFO is in kV, rms. The following tabulation compares this power frequency CFO with impulse CFOs for a 3-meter gap and a gap factor of

CFO, kV	Per unit
$\text{CFO}_{\text{SI}}^{+} = 1113$	1.00
$\text{CFO}_{\text{SI}}^{-} = 1935$	1.74
$\text{CFO}_{\text{LI}}^{+} = 1680$	1.51
$\text{CFO}_{\text{LI}}^{-} = 1815$	1.63
$\text{CFO}_{\text{PF}} = 1322$ peak	1.19

1.2. As noted, the power frequency CFO is 19% greater than the switching impulse, positive polarity CFO.

13 COMPARISON WITH IEC

Throughout this chapter various comparisons were made to IEC Standard 71 [41]. However, one item remains to be discussed. IEC 71 recommends the use of the Weibull distribution to represent the insulation strength characteristic, i.e., to replace the Gaussian distribution. It has been noted in this chapter that the Gaussian distribution is unbounded to the right and left. That is, it is defined between plus and minus infinity. To explain further, the limit of minus infinity indicates that there does exist a probability of flashover for a voltage equal to zero. While it is known that the Gaussian distribution is valid to at least four standard deviations below the CFO, it is reasonable to believe that there exists a nonzero voltage for which the probability of flashover is zero. A distribution that possesses this attribute is the Weibull whose cumulative distribution function is

$$p = F(V) = 1 - e^{-\left(\frac{V-\alpha_0}{\alpha}\right)^\beta} \qquad \infty > V > \alpha_0 \tag{58}$$

At $V = \alpha_0$, $F(V) = p = 0$. To adapt this to the features of the Gaussian, the following is specified:

1. Let the truncation value be set at the CFO $- 4\sigma_0$.
2. Let $p = 0.5$ at $V = $ CFO.
3. Let $p = 0.16$ at $V = $ CFO $- \sigma_f$.

Considering the first requirement,

$$\alpha_a = \text{CFO} - 4\sigma_f \tag{59}$$

Considering the second requirement,

$$0.5 = e^{-\left(\frac{4\sigma_f}{\alpha}\right)^\beta}$$
$$\alpha = \frac{4\sigma_f}{(\ln 2)^{1/\beta}} \tag{60}$$

Considering the last requirement,

$$0.84 = e^{-\ln 2 (3/4)^\beta}$$
$$\beta = 4.80 \tag{61}$$

Insulation Strength Characteristics

In IEC 71, the value of β was rounded to 5.0, and therefore

$$p = F(V) = 1 - e^{-(\ln 2)\left(\frac{Z}{4}+1\right)^5}$$
$$= 1 - 0.5^{\left(\frac{Z}{4}+1\right)^5} \tag{62}$$

where

$$Z = V - \frac{\text{CFO}}{\sigma_f} \tag{63}$$

This Weibull equation closely approximates the Gaussian in the important voltage region below the CFO. The use of this equation is discussed further in Chapter 3.

14 SUMMARY

1. For self-restoring insulations such as the tower insulation, the insulation strength characteristic may be mathematically represented by a cumulative Gaussian distribution having a mean, denoted as the CFO, and a standard deviation denoted by σ_f. For switching impulses (SI), σ_f/CFO is 5% for tower insulations and between 6 and 7% for substation insulations. For lightning impulses, (LI) σ_f/CFO is small, between 1 and 3%, and is generally ignored; only the CFO is used.

2. For SI, the CFO for standard conditions and positive polarity, CFO_S, may be estimated by the equation

$$\text{CFO}_S = k_g \frac{3400}{1 + (8/S)} \tag{64}$$

where S is the strike distance in meters, CFO_S is in kV, and k_g is the gap factor. For tower insulations, the CFO_S is modified for wet conditions to

$$\text{CFO}_s = 0.96 k_g \frac{3400}{1+(8/S)} \tag{65}$$

The gap factor for tower insulations, center phase position, is

$$k_g = 1.25 + 0.005\left(\frac{h}{S} - 6\right) + 0.25\left(e^{-\frac{8W}{S}}\right) \tag{66}$$

where W is the tower width and h is the height of the conductor. This gap factor is usually 1.2 for the normal lattice tower and about 1.25 for steel poles. The gap factor for the outside phase position is 1.08 times the above gap factor. For substations, the practical gap factor is 1.3, although lower gap factors of about 1.1 have been suggested. For bus support insulators (post), the gap factor, wet conditions, is 1.18 as applied to Eq. 64.

For non-standard atmospheric weather conditions, the CFO is altered by the relative air density δ and is

$$\text{CFO}_A = \delta^m \text{CFO}_S \tag{67}$$

where m is a factor dependent on G_0, i.e.,

$$G_0 = \frac{\text{CFO}_S}{500(S)} \qquad m = 1.25 G_0 (G_0 - 0.2) \tag{68}$$

3. Negative polarity SI strength is sufficiently higher than that for positive so that the negative polarity strength is not of primary importance; it can usually be ignored.

4. The deterministic design of tower insulations consists of equating the maximum switching surge overvoltage E_m to the minimum insulation strength defined as V_3. In equation form,

$$\begin{aligned} V_3 &= E_m \\ V_3 &= \text{CFO}_A \left(1 - \frac{3\sigma_f}{\text{CFO}_A}\right) \end{aligned} \tag{69}$$

5. For LI, the CFO may be estimated by use of a CFO gradient, i.e.,

$$\begin{aligned} \text{CFO}^+ &= 560(S) \\ \text{CFO}^- &= 605(S) \end{aligned} \tag{70}$$

where the superscript + denotes positive polarity and the superscript − denotes negative polarity. Again the CFOs are in kV, and S is in meters. These equations apply for both strike distances and insulator string length, i.e., substitute the insulator string length for S.

6. For LI, for wood in series with porcelain insulators, the CFO is dependent on the critical length of wood, which is equal to twice the insulator string length. At this critical length wood adds 100 kV/m of wood to the CFO of the insulators. Above this length, wood acts as the primary insulation, and the total CFO is equal to the CFO of wood alone, 300 kV/m of wood. Below the critical length, wood is not effective, only adding about 40 kV/m of wood.

7. For LI, fiberglass insulation has the same CFO as air–porcelain, and Eq. 70 can be used, where S is the combined length of the insulator string and the fiberglass.

15 REFERENCES

1. A. J. McElroy, W. S. Price, H. M. Smith and D. F. Shankle, "Field Measurements of Switching Surges on 345 kV Unterminated Transmission Lines," *IEEE Trans. on PA&S*, Aug. 1963, pp. 465–487.
2. W. S. Price, A. J. McElroy, H. M. Smith and D. F. Shankle, "Field Measurements on 345-kV Lightning Arrester Performance," *IEEE Trans. on PA&S*, Aug. 1963, pp. 487–500.
3. A. J. McElroy, W. S. Price, H. M. Smith and D. F. Shankle, "Field Measurements on Switching Surges as Modified by Unloaded 345-kV Transformers," *IEEE Trans. on PA&S*, Aug. 1963, pp. 500–520.

4. W. H. Croft, R. H. Hartley, R. L. Linden and D. D. Wilson, "Switching Surge and Dynamic Voltage Study of Arizona Public Service Company's Proposed 345-kV Transmission System Utilizing Miniature Analyzer Techniques," *IEEE Trans. on PA&S*, Aug. 1962, pp. 302–312.
5. I. S. Stekolinikov, E. N. Brago, and E. M. Bazelyan, "The Peculiarities of Oblique Wave Front Discharges and Their Role in the Estimation of EHV Transmission Line Insulation," presented at the International Conference, Leatherhead, Surrey, England, Paper 39, Session Ib, 1962.
6. G. N. Alexandrov and V. L. Ivanov, "Electrical Strength of Air Gaps and Insulator Strings under the Action of Switching Surges," *Elec. Technology USSR*, Vol. 3, 1962, pp. 460–473.
7. J. H. Sabath, H. M. Smith, and R. C. Johnson, "Analog Computer Study of Switching Surge Transients for the 500-kV System," *IEEE Trans. on PA&S*, Jan. 1966, pp. 1–9.
8. W. C. Guyker, A. R. Hileman, H. M. Smith, and G. E. Grosser, Jr., "Line Insulation Design for APS 500-kV System," *IEEE Trans. on PA&S*, Aug. 1967, pp. 987–1014.
9. W. C. Guyker, A. R. Hileman, and J. F. Wittibschlager, "Full Scale Tower Insulation Test for APS 500 kV System," *IEEE Trans. on PA&S*, Jun. 1966, pp. 614–623.
10. K. Anjo, I. Kishijima, Y. Ohuch, and T. Sizuki, "Parallel Multi-Gap Flashover Probability," *IEEE Trans. on PA&S*, Aug. 1968, pp. 1814–1823.
11. J. K. Dillard and A. R. Hileman, "UHV Transmission Tower Insulation Tests," *IEEE Trans. on PA&S*, Apr. 1965, pp. 1772–1784.
12. G. Gallet, G. LeRoy, R. Lacey, and I. Kromel, "General Expression for Positive Switching Impulse Strength up to Extra Long Air Gaps," *IEEE Trans. on PA&S*, Nov./Dec. 1975, pp. 1989–1973.
13. L. Paris and R. Cortina, "Switching and Lightning Impulse Discharge Characteristics of Large Air Gaps and Long Insulator Strings," *IEEE Trans. on PA&S*, Apr. 1968, pp. 947–957.
14. L. Thione, "Evaluation of the Switching Impulse Strength of External Insulation," *ELECTRA*, May 1984, pp. 77–95.
15. CIGRE Working Group 07, Study Committee 33, "Guidelines for Evaluation of the Dielectric Strength of External Insulation," CIGRE Technical Brochure 72, 1992.
16. C. Menemenlis and G. Harbec, "Coefficient of Variation of the Positive-Impulse Breakdown of Long Air-Gaps," *IEEE Trans. on PA&S*, May/Jun. 1974, pp. 916–927.
17. A. W. Atwood, Jr., A. R. Hileman, J. W. Skooglund, and J. F. Wittibschlager, "Switching Surge Tests on Simulated and Full Scale EHV Tower Insulation Systems," *IEEE Trans. on PA&S*, Apr. 1965, pp. 293–303.
18. A. R. Hileman and H. W. Askins, "Switching Impulse Characteristics of UHV Station Post Insulators," *IEEE Trans. on PA&S*, Jan./Feb. 1973, pp. 139–144.
19. J. H. Moran, "Switching Surge Study of EHV Posts: II," *IEEE Trans. on PA&S*, Mar. 1969, pp. 238–244.
20. L. Paris., A. Taschini, K. H. Schneider, and K. H. Weck, "Phase-to-Ground and Phase-to-Phase Air Clearances in Substations," *ELECTRA*, Jul. 1973, pp. 29–44.
21. A. Pigini, G. Rizzi, and R. Brambilla, "Switching Impulse Strength of Very Large Air Gaps," 3rd I.S.H. Milan 1979, Article 526–15.
22. J. W. Kalb, "How the Switching Surge Family Affects Line Insulation," *IEEE Trans. on PA&S*, Aug. 1963, pp. 1024–1033.
23. J. A. Rawls, J. W. Kalb, and A. R. Hileman, "Full Scale Surge Testing of VEPCO 500-kV Line Insulation," *IEEE Trans. on PA&S*, Mar. 1964, pp. 245–250.
24. A. F. Rolfs, H. E. Fiegel, and J. G. Anderson, "The Flashover Strength of Extra-High-Voltage Line and Station Insulation," *AIEE Trans.*, Aug. 1961, pp. 463–471.
25. W. R. Johnson, J. B. Tice, E. G. Lambert, and F. J. Turner, "500-kV Design II: Electrical Strength of Towers," *IEEE Trans. on PA&S*, Aug. 1963, pp. 581–587.

26. K. J. Lloyd, and L. E. Zaffanella, "Insulation for Switching Surges," Chapter 11 of *Transmission Line Reference Book—345 kV and Above*, 2nd ed., Electric Power Research Institute, 1982.
27. P. H. McAuley, "Flashover Characteristics of Insulation," *Electric Journal*, Jul., 1938.
28. A. F. Rohlfs and H. E. Fiegel, "Impulse Flashover Characteristics of Long Strings of Suspension Insulators," *AIEE Trans.*, 1957.
29. B. E. Kingsbury, "Suspension Insulator Flashover under High Impulse Voltages," *AIEE Trans.*, 1957.
30. T. Udo, "Switching Surge and Impulse Sparkover Characteristics of Long Gap Spacings and Long Insulator Strings," *IEEE Trans. on PA&S*, Apr. 1965, pp. 304–309.
31. B. Hutzler, "Behavior of Long Insulator Strings in Dry Conditions," *IEEE Trans. on PA&S*, May/Jun. 1979, pp. 982–991.
32. T. Udo, "Sparkover Characteristics of Large Gap Spaces and Long Insulator Springs," *IEEE Trans. on PA&S*, May 1964, pp. 471–483.
33. J. H. Hagenguth, A. F. Rohlfs, and W. J. Degnan, "Sixty-Cycle and Impulse Sparkover of Large Gap Spacings," *AIEE Trans.* 71, 1952.
34. AIEE Committee Report, "Impulse Flashovers of Combinations of Line Insulators, Air Gaps, and Wood Structural Members," *AIEE Trans.*, Apr. 1956, pp. 16–21.
35. M. Darveniza, *Electrical Properties of Wood and Line Design*, St. Lucia, Queensland, Australia; University of Queensland Press, 1980.
36. J. M. Clayton and D. F. Shankle, "Insulation Characteristics of Wood and Suspension Insulators in Series," *AIEE Trans.* 73, no. 3, Dec. 1955, p. 1305.
37. J. T. Lusignan and C. J. Miller, "What Wood May Add to Primary Insulation for Withstanding Lightning," *AIEE Trans.*, 59, Sep. 1940, pp. 534–540.
38. J. J. Trainer and L. B. LeVesconte, "230 kV Wood-Pole Transmission Line Design," *AIEE Trans.*, 73, no. 3A, Jun. 1954, pp. 522–528.
39. Paul M. Rosa, "Burning of Wood Structures by Leakage Currents," *AIEE Trans.*, 66, 147, pp. 279–287.
40. S. Grzybowski and E. B. Jenkins, "AC and Lightning Performance of Fiberglass Crossarms Aged in 115 kV Transmission Line, *IEEE Trans. on Power Delivery*, Presented at the 1993 Winter Meeting, Columbus, Ohio, Jan./Feb. 1993.
41. IEC Standard 71-2 "Insulation coordination – Part 2: Application guide," 1996-12.
42. A. Hauspurg, V. Caleca, and R. H. Schloman, "765-kV Transmission Line Insulation Testing Program," *IEEE Trans. on PA&S*, Jun. 1969, pp. 1355–1365.
43. C. Menemenlis and G. Harbec, "Switching Impulse Breakdown of EHV Transmission Towers," *IEEE Trans. on PA&S*, 1964, pp. 255–263.
44. A. R. Hileman, "Transmission Line Insulation Coordination," *Transactions of the South African Institute of Electrical Engineers*, 71 (6), Jun. 1980.
45. K. J. Lloyd and H. M. Scheider, "Insulation for Power Frequency Voltage," Chapter 10 of *Transmission Line Reference Book*, 2d ed., Palo Alto, CA: Electric Power Research Institute, 1982.
46. G. N. Aleksandrov, V. I. Ivanov, and V. P. Radkov, "Electrical Strength of Air Gaps Between EHV Line Conductors and Earth During Surges," *Elektrichestro* No. 4, 1965, pp. 20–24.

16 PROBLEMS

1. Using the deterministic method, determine the tower strike distance and the insulator length for a single circuit 500-kV (550-kV max) transmission line under the following conditions. Note: 1 per unit = 450 kV.

Insulation Strength Characteristics

Towers have V-strings on all phases.
The fronts of all switching surges are equal to the critical wave front.
Line length = 200 km with three towers per km.
Line altitude = 1000 m.
Maximum switching surge = 2.052 per unit.
$\sigma_f/\text{CFO} = 0.05$.
Wet conditions decrease the dry CFO by 4%.
Tower width = 1.8 m; conductor height = 20 m.

2. For the line and conditions of problem 1, calculate the probability of at least one flashover on the line for a switching surge of 2.052 per unit and also for a switching surge of 900 kV.

3. Figure 53 shows a general wood-pole line design using polymer line post insulators. The overhead ground wire is grounded by a downlead that is offset from the pole by fiberglass rods. The dimensions per Fig. 53 are provided in Table 9. All ground wires are 7 No. 8 Alumoweld having a diameter of 0.385 inches. The downlead is a No. 4 AWG three-strand copper having a diameter of 0.254 inches. The wood pole diameter is 1 foot. The dry arcing distance is given for the

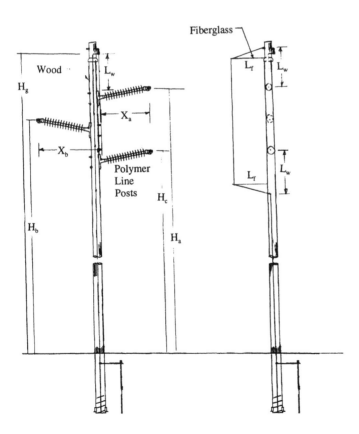

Figure 53 General wood-pole, polymer line post insulator design.

Table 9A Distances in Feet

System voltage	H_g	H_a	H_b	H_c	L_w	L_f	X_a	X_b	Post dry arcing
34.5	40	33	30	27	7	3.8	3.3	4.3	2.42
69	52	45	42	39	8	1.5	3.8	4.8	2.94
115	70	60	55	50	12	2.0	6.7	7.7	4.5

Table 9B Phase Conductors

System voltage	Name	kCM	Stranding Al/St	Diameter, in.
34.5	Merlin	336	18/1	0.684
69	Drake	795	26/7	1.108
115	Drake	795	26/7	1.108

System voltage, kV	Span length, ft.	Shield wire sag, ft.	Phase conductor sag, ft.
34.5	300	2.0	4.0
69	500	4.0	6.0
115	800	9.0	14.0

post insulator which is the distance used to determine the CFO. Determine the lightning impulse positive and negative polarity CFO for each design. As a crude approximation, the sag may be estimated as follows:

$$\text{Conductor sag} = 7 \times 10^{-5} (\text{span length})^2$$
$$\text{Shield wire sag} = \frac{2}{3} (\text{conductor sag})$$

(71)

with the sag and span length in meters.

4. Figure 54 shows a 115 kV H-frame wood-pole design with a downlead attached directly to the wood pole. The shield wire is 7 No. 8 Alumoweld, diameter 0.385 inches. The phase conductors are two-conductor bundled Linnet subconductors, 336.4 kCM, having a diameter of 0.721 inches. The subconductor spacing is 18 inches. The downlead is a No. 4 AWG, three-strand copper having a diameter of 0.254 inches. Twelve insulators, $5\frac{3}{4} \times 10$ inch units, are used. The span length is 1000 feet and the phase conductor and shield wire sags are 21 and 14 feet respectively. Determine the lightning impulse positive and negative polarity CFO.

Insulation Strength Characteristics

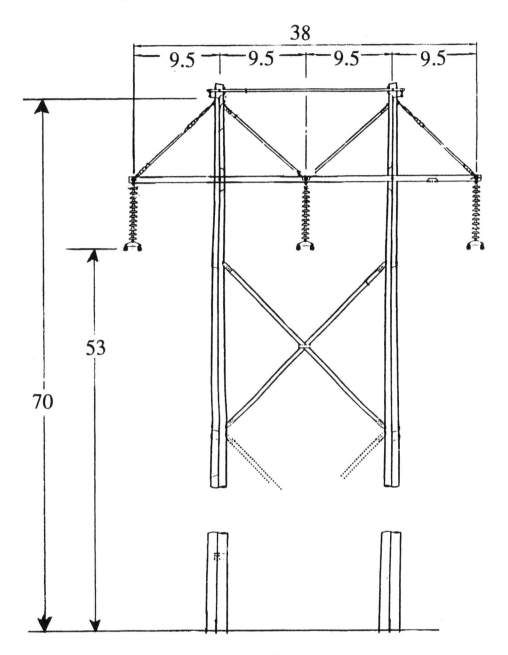

Figure 54 115 kV H-frame wood pole tower (dimensions in feet).

3
Phase–Ground Switching Overvoltages, Transmission Lines

1 INTRODUCTION

As discussed in Chapter 2, prior to the advent of 500 kV transmission in the early 1960s, little was known about switching surges or switching overvoltages (SOV), either as to their generation or as to the strength of insulation. Some field measurements had been made on 138 kV and 345 kV systems that indicated that SOVs might be important, and some laboratory tests had been made that indicated that the switching impulse insulation strength was less than that for lightning impulses. Nevertheless, until the time of the first 500 kV transmission, no lines had been designed using SOVs as a design criterion.

The design of the first 500-kV lines was performed using the deterministic method as described in Chapter 2. The primary reasons that a probabilistic method was not used were simply that (1) the random nature of SOVs had not been considered, or the SOV probability distribution was unknown, and that (2) the theory and application of a probabilistic method had not been developed. These two problems were quickly overcome. The SOV distribution was obtained by use of the transient network analyzer (TNA) by random switching of the circuit breaker [1]. Theoretical methods and their application were developed by borrowing from methods used in generation expansion and in structural engineering [1–8]. Interestingly, the probabilistic method was not immediately adopted by the industry. Until about 1979 when BPA redesigned their 500-kV lines [8], all 500-kV and 765-kV lines were designed using the deterministic method. It appeared that the industry did not 'trust' the new method, and there appeared little need to adopt it. However, in this era, when obtaining right-of-way for new lines was difficult, the idea of uprating lines for a higher voltage level became popular. That is, for example, the question was asked, "Could 138-kV lines be operated at 230 or 345 kV?" To answer this question, consideration of SOVs was necessary, and the use of the probabilistic method was mandatory to prove that uprating could be done. A need arose; and to prove that uprating was possible, the probabilistic method was accepted.

Today, virtually all EHV lines are designed using the probabilistic method. The method is used at 345 kV and sometimes at 230 kV. However, at 230 kV and below, SOVs are not considered to be a design problem, so that, as shown in Chapter 1, switching impulse insulation levels, BSLs, are not given for these voltage levels.

As with all engineering design, the switching surge design of lines uses the concepts of stress and strength. The stress here is the SOVs applied to the line for which the SOVs can be described statistically as a probability distribution. The strength is the tower electrical insulation strength, which can also be described statistically by a Gaussian cumulative distribution per Chapter 2. From these two statistical descriptions of stress and strength, the probability of a flashover can be determined. It is the purpose of this chapter to develop this concept, the resulting equations, and the practical application.

The presentation of the switching surge design of lines begins with a theoretical development. For those who desire a clear understanding of the method and are not totally comfortable with continuous probability density functions, Section 2 attempts to develop the methodology using SOV histograms. For those who are familiar with probability concepts, Section 2 may be skipped. The third section develops the concept using continuous probability density functions to describe the random nature of SOVs. Simplified methods of calculation are then developed and practical application problems of line design considered.

The development as presented in this chapter is for transmission lines. Use of these methods is equally applicable to substations and is considerably simpler. However, the application to substations will be delayed until Chapter 5 where it becomes more directly usable.

2 THE CONCEPT—USING HISTOGRAMS OF SWITCHING OVERVOLTAGES

By use of a TNA or a computer transient program, the random nature of the SOVs is developed by randomly switching the circuit breakers within their pole closing span. Assume that this distribution of SOVs for a 500 kV line is described by the bar chart of Fig. 1, where 1 per unit, 1 pu, is equal to 450 kV, i.e., the crest line-to-neutral voltage for a maximum system voltage of 550 kV. Per Fig. 1, 1% of the voltages are equal to 1.9 pu and 5% are equal to 1.8 pu. Also assume that the insulation strength characteristic can be approximated as a cumulative Gaussian distribution (Fig. 2) having a CFO of 900 kV and a standard deviation of 45 kV, or a coefficient of variation of 5%. The detail calculations of the probability of the line flashover are shown in Table 2.

First consider the highest SOV of 1.9 pu or 855 kV. The probability that this occurs is 1% or 0.01. Entering Fig. 2 with 855 kV, the probability of flashover given this voltage, $P(FO|V)$, is 0.1587. Or better, let Z be the reduced variate for the Gaussian distribution. Then

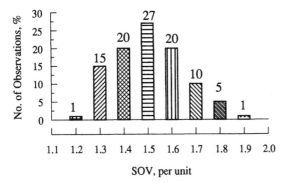

Figure 1 Bar chart of SOVs.

$$Z = \frac{V - \text{CFO}}{\sigma_f} = \frac{855 - 900}{45} = -1.0 \tag{1}$$

and entering Table 1 with $Z = 1.0$, $F(Z) = 0.8413$, and $F(-1.0) = 0.1587$, which can also be obtained using a hand calculator program. Thus the probability of flashover is 0.1587. The probability of flashover for this voltage is simply the probability of occurrence of 0.01 times the probability of flashover given this voltage, or 0.001587.

Consider the next voltage, and the next, etc., and perform the same calculation. To obtain the total flashover rate or the switching surge flashover rate SSFOR, add the numbers to obtain 0.00288, which is better stated as 0.288 flashovers per 100 breaker (or switching) operations or per 100 breaker closings.

As noted from Table 2, only four voltage levels were considered, since the probabilities for SOVs below 1.6 pu are insignificant. The message is simply that it is the upper tail of the switching overvoltage distribution and the lower tail of the strength distribution that are important. The mean or modal value of the SOVs of 1.5 per unit and the CFO are of no importance!

Although not stated, the strength distribution of Fig. 2 is actually for one tower as obtained from Chapter 2, and a line consists of several towers. In this case, the generated SOVs are applied to all these towers simultaneously. The probability of flashover given a SOV changes to

$$P(\text{FO}|V) = 1 - q^n = 1 - (1-p)^n \tag{2}$$

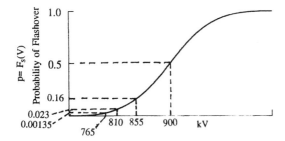

Figure 2 Strength characteristic for one tower.

Table 1 The Cumulative Normal Distribution Function, Values of $F(Z)$ Given Z
$F(Z) = 1/\sqrt{2\pi} \int_{-\infty}^{Z} e^{(-Z^2/2)} dZ$ $F(-Z) = 1 - F(Z)$ $F(2.02) = 0.97831$ $F(-2.02) = 0.02169$
$F(3.57) = .9^3 8219 = .9998215$

Z	.00	.01	.02	.03	.04	.05	.06	.07	.08	.09
.0	.5000	.5040	.5080	.5120	.5160	.5199	.5239	.5279	.5319	.5359
.1	.5398	.5438	.5478	.5517	.5557	.5526	.5636	.5675	.5714	.5753
.2	.5793	.5832	.5871	.5910	.5948	.5987	.6026	.6064	.6103	.6141
.3	.6179	.6217	.6255	.6293	.6331	.6368	.6406	.6443	.6480	.6517
.4	.6554	.6591	.6628	.6664	.6700	.6736	.6772	.6808	.6844	.6879
.5	.6915	.6950	.6985	.7019	.7054	.7088	.7123	.7157	.7190	.7224
.6	.7257	.7291	.7324	.7357	.7389	.7422	.7454	.7486	.7517	.7549
.7	.7580	.7611	.7642	.7673	.7703	.7734	.7764	.7794	.7823	.7852
.8	.7881	.7910	.7939	.7967	.7995	.8023	.8051	.8078	.8106	.8133
.9	.8159	.8186	.8212	.8238	.8264	.8289	.8315	.8340	.8365	.8389
1.0	.8413	.8438	.8461	.8485	.8508	.8531	.8554	.8577	.8599	.8621
1.1	.8643	.8665	.8686	.8708	.8729	.8749	.8770	.8790	.8810	.8830
1.2	.8849	.8869	.8888	.8907	.8925	.8944	.8962	.8980	.8997	.90147
1.3	.90320	.90490	.90658	.90824	.90988	.91149	.91309	.91466	.91621	.91774
1.4	.91924	.92073	.92220	.92364	.92507	.92647	.92785	.92922	.93056	.93189
1.5	.93319	.93448	.93574	.93699	.93822	.93943	.94062	.94179	.94295	.94408
1.6	.94520	.94630	.94738	.94845	.94950	.95053	.95154	.95254	.95352	.95449
1.7	.95543	.95637	.95728	.95818	.95907	.95994	.96080	.96164	.96246	.96327
1.8	.96407	.96485	.96562	.96638	.06712	.96784	.96856	.96926	.96995	.97062
1.9	.97128	.97193	.97257	.97320	.97381	.97441	.97500	.97558	.97615	.97670
2.0	.97725	.97778	.97831	.97882	.97932	.97982	.98030	.98077	.98124	.98169
2.1	.98214	.98257	.98300	.98341	.98382	.98422	.98461	.98500	.98537	.98574
2.2	.98610	.98645	.98679	.98713	.98745	.98778	.98809	.98840	.98870	.98899
2.3	.98928	.98956	.98983	$.9^2 0097$	$.9^2 0458$	$.9^2 0613$	$.9^2 0863$	$.9^2 1106$	$.9^2 1344$	$.9^2 1576$
2.4	$.9^2 1802$	$.9^2 2024$	$.9^2 2240$	$.9^2 2451$	$.9^2 2656$	$.9^2 2857$	$.9^2 3053$	$.9^2 3244$	$.9^2 3431$	$.9^2 3613$
2.5	$.9^2 3790$	$.9^2 3963$	$.9^2 4132$	$.9^2 4297$	$.9^2 4457$	$.9^2 4614$	$.9^2 4766$	$.9^2 4915$	$.9^2 5060$	$.9^2 5201$
2.6	$.9^2 5339$	$.9^2 5473$	$.9^2 5604$	$.9^2 5731$	$.9^2 5855$	$.9^2 5975$	$.9^2 6093$	$.9^2 6207$	$.9^2 6319$	$.9^2 6427$
2.7	$.9^2 6533$	$.9^2 6636$	$.9^2 6736$	$.9^2 6833$	$.9^2 6928$	$.9^2 7020$	$.9^2 7110$	$.9^2 7197$	$.9^2 7282$	$.9^2 7365$
2.8	$.9^2 7445$	$.9^2 7523$	$.9^2 7599$	$.9^2 7673$	$.9^2 7744$	$.9^2 7814$	$.9^2 7882$	$.9^2 7948$	$.9^2 8012$	$.9^2 8074$
2.9	$.9^2 8134$	$.9^2 8193$	$.9^2 8250$	$.9^2 8305$	$.9^2 8359$	$.9^2 8411$	$.9^2 8462$	$.9^2 8511$	$.9^2 8559$	$.9^2 8605$
3.0	$.9^2 8650$	$.9^2 8694$	$.9^2 8736$	$.9^2 8777$	$.9^2 8817$	$.9^2 8856$	$.9^2 8893$	$.9^2 8930$	$.9^2 8965$	$.9^2 8999$
3.1	$.9^3 0324$	$.9^3 0646$	$.9^3 0957$	$.9^3 1260$	$.9^3 1553$	$.9^3 1836$	$.9^3 2112$	$.9^3 2378$	$.9^3 2636$	$.9^3 2886$
3.2	$.9^3 3129$	$.9^3 3363$	$.9^3 3590$	$.9^3 3810$	$.9^3 4024$	$.9^3 4230$	$.9^3 4429$	$.9^3 4623$	$.9^3 4810$	$.9^3 4991$
3.3	$.9^3 5166$	$.9^3 5335$	$.9^3 5499$	$.9^3 5658$	$.9^3 5811$	$.9^3 5959$	$.9^3 6103$	$.9^3 6242$	$.9^3 6376$	$.9^3 6505$
3.4	$.9^3 6613$	$.9^3 6752$	$.9^3 6869$	$.9^3 6982$	$.9^3 7091$	$.9^3 7197$	$.9^3 7299$	$.9^3 7398$	$.9^3 7493$	$.9^3 7585$
3.5	$.9^3 7674$	$.9^3 7759$	$.9^3 7842$	$.9^3 7922$	$.9^3 7999$	$.9^3 8074$	$.9^3 8146$	$.9^3 8215$	$.9^3 8282$	$.9^3 8347$
3.6	$.9^3 8409$	$.9^3 8469$	$.9^3 8527$	$.9^3 8583$	$.9^3 8637$	$.9^3 8689$	$.9^3 8739$	$.9^3 8787$	$.9^3 8834$	$.9^3 8879$
3.7	$.9^3 8922$	$.9^3 8964$	$.9^4 0039$	$.9^4 0426$	$.9^4 0799$	$.9^4 1158$	$.9^4 1504$	$.9^4 1838$	$.9^4 2159$	$.9^4 2468$
3.8	$.9^4 2765$	$.9^4 3052$	$.9^4 3327$	$.9^4 3593$	$.9^4 3848$	$.9^4 4094$	$.9^4 4331$	$.9^4 4558$	$.9^4 4777$	$.9^4 4988$
3.9	$.9^4 5190$	$.9^4 5385$	$.9^4 5573$	$.9^4 5753$	$.9^4 5926$	$.9^4 6092$	$.9^4 6253$	$.9^4 6406$	$.9^4 6554$	$.9^4 6696$

Table 1 (continued)

Z	.00	.01	.02	.03	.04	.05	.06	.07	.08	.09
4.0	$.9^4 6833$	$.9^4 6964$	$.9^4 7090$	$.9^4 7211$	$.9^4 7327$	$.9^4 7439$	$.9^4 7546$	$.9^4 7649$	$.9^4 7748$	$.9^4 7843$
4.1	$.9^4 7934$	$.9^4 8022$	$.9^4 8106$	$.9^4 8186$	$.9^4 8263$	$.9^4 8338$	$.9^4 8409$	$.9^4 8477$	$.9^4 8542$	$.9^4 8605$
4.2	$.9^4 8665$	$.9^4 8723$	$.9^4 8778$	$.9^4 8832$	$.9^4 8882$	$.9^4 8931$	$.9^4 8978$	$.9^5 0226$	$.9^5 0655$	$.9^5 1066$
4.3	$.9^5 1460$	$.9^5 1837$	$.9^5 2199$	$.9^5 2545$	$.9^5 2876$	$.9^5 3193$	$.9^5 3497$	$.9^5 3788$	$.9^5 4066$	$.9^5 4332$
4.4	$.9^5 4587$	$.9^5 5065$	$.9^5 5065$	$.9^5 5288$	$.9^5 5502$	$.9^5 5706$	$.9^5 5902$	$.9^5 6089$	$.9^5 6268$	$.9^5 6349$
4.5	$.9^5 6602$	$.9^5 6759$	$.9^5 6908$	$.9^5 7051$	$.9^5 7187$	$.9^5 7318$	$.9^5 7442$	$.9^5 7561$	$.9^5 7675$	$.9^5 7784$
4.6	$.9^5 7888$	$.9^5 7987$	$.9^5 8081$	$.9^5 8172$	$.9^5 8258$	$.9^5 8340$	$.9^5 8419$	$.9^5 8494$	$.9^5 8566$	$.9^5 8634$
4.7	$.9^5 8699$	$.9^5 8761$	$.9^5 8821$	$.9^5 8877$	$.9^5 8931$	$.9^5 8983$	$.9^6 0320$	$.9^5 0789$	$.9^5 1235$	$.9^5 1661$
4.8	$.9^6 2067$	$.9^6 2453$	$.9^6 2822$	$.9^6 3173$	$.9^6 3508$	$.9^6 3827$	$.9^6 4131$	$.9^6 4420$	$.9^6 4696$	$.9^6 4958$
4.9	$.9^6 5208$	$.9^6 5446$	$.9^6 5673$	$.9^6 5889$	$.9^6 6094$	$.9^6 6289$	$.9^6 6475$	$.9^6 6652$	$.9^6 6821$	$.9^6 6981$

where p is the probability of flashover for a single tower insulation (the value in column three of Table 1), q is the probability of no flashover and equal to $(1 - p)$, and n is the number of towers.

Equation 2 shows that the probability of flashover is equal to 1 minus the probability of no flashover on each of the towers. To amplify, consider two towers. The probability of no flashover on the one tower is q and the probability of no flashover on both towers is q times q or q^2. The probability of a flashover on the first tower but no flashover on the second tower—or the probability of no flashover on the first tower and a flashover on the second tower—is pq. The probability of a flashover on both towers is p times p or p^2. All these probabilities must add to one, since one of these events must happen. That is, $p^2 + 2pq + q^2 = 1$. However, the value desired is the probability of a flashover on either the first or second tower or both, that is $p^2 + 2pq$. An easier way to obtain this is simply by noting that $1 - q^2 = p^2 + 2pq$, or for n towers, $1 - q^n$. Figure 3 illustrates this increase in probability of flashover for n towers.

Using Eq. 2, the calculation of the SSFOR proceeds as before and is shown in Table 2 for a 100-tower line. As expected, the SSFOR increases to 7.09/100, but again only the upper tail of the SOV distribution is important.

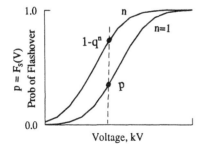

Figure 3 Increase in probability of flashover for n towers.

Table 2 Calculation of the Probability of Flashover or SSFOR (see Figs. 1, 2, and 3)

V, pu/kV	P(V)	n = 1 tower		n = 100 towers	
		P[FO|V]	P[FO]	P[FO|V]	P[FO]
1.9/855	0.01	.1587	.00159	1.00000	0.01000
1.8/810	0.05	.02275	.00114	.8999	0.04499
1.7/765	0.10	.001350	.00014	.12636	0.01264
1.6/720	0.21	.00003267	.00001	.00797	0.00326
1.5/675	0.27	.00000028	.00000	.000028	0.00001
		Total	0.00288		0.07090
		SSFOR	0.288/100		7.09/100

To complete this presentation, consider one additional factor, the voltage profile along the line. In the prior calculation, the SOV at each of the towers was considered constant. That is, the SOV at tower 1 was assumed equal to the SOV at tower 100. However, usually the SOV at each tower is different, as the SOV near the switched breaker is lower than the SOV at the end of the line. This voltage profile is illustrated in Fig. 4 by the solid line curve. Assume for purposes of simplification that V_s/V_R is equal to 0.9 and that only three locations will be considered: the voltage at the end of the line V_R, the voltage at the midpoint of the line V_{mid}, and the voltage at the switched end of the line V_s. Also, assume that for each of these voltages there are 33 towers, i.e. $n = 33$, as illustrated in Fig. 4 by the dotted lines. Now consider the calculation of the probability of flashover given a SOV of 1.9 per unit at the end of the line. The probability of no flashover given V_R (1.9 per unit) is q_R^n, the probability of no flashover given V_{mid} of $0.95(1.9) = 1.805$ per unit is q_{mid}^n, and the probability of no flashover given V_s of $0.9(1.9) = 1.71$ per unit is q_S^n. The probability of flashover given all these values is then

$$P(\text{FO}|V) = 1 - q_R^n q_{mid}^n q_s^n \qquad (3)$$

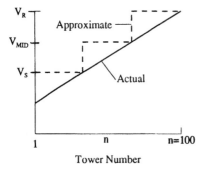

Figure 4 SOV profile along line.

Table 3 Calculation of the Probability of Flashover or SSFOR $V_s/V_R = 0.9$ (see Fig. 4)

V, pu/kV	P(V)	q_R	q_{mid}	q_s	P[FO|V] $1 - q_R^n q_{mid}^n q_S^n$	P[FO]
1.9/855	0.01	.84134	.97441	.99813	.99866	.00999
1.8/810	0.05	.97725	.99813	.99993	.56111	.02806
1.7/765	0.10	.99865	.99994	1.00000	.04549	.00455
1.6/720	0.21	.99997	1.00000	1.00000	.00108	.00023
					Total	.04283
					SSFOR	4.28/100

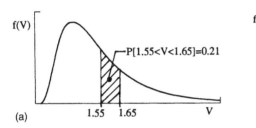
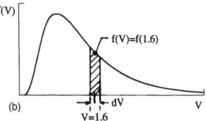

Figure 5 Changing to a continuous distribution for the SOVs.

The probability of flashover for a voltage of 1.9 per unit at the end of the line is the probability of occurrence of 0.01 times the probability of flashover given this voltage or 0.00999 per Table 3. As noted, the SSFOR decreases slightly from 7.09 to 4.28/100.

Hopefully the presentation in this section has succeeded in the providing an understanding of the calculation of the SSFOR when the SOVs are described by a histogram. The move to describing the SOV distribution as a continuous distribution can be viewed as simply the conversion of the histogram to a continuous curve as shown in Fig. 5. This curve is called the probability density function $f(V)$, and probabilities are obtained from it in terms of areas. For example, consider the histogram of Fig. 1, where the probability of occurrence of 1.6 per unit is 0.21. In terms of the probability density (Fig. 5a), the probability of V between 1.55 and 1.65 is 0.21 or $P(1.55 < V < 1.65) = 0.21$. Whereas previously probabilities were assigned in "chunks", $P(V) = 0.21$, now they are assigned in terms of areas. Therefore, per Fig. 5b, the probability of occurrence of exactly 1.6 per unit is $f(1.6)\,dV$ where dV is an incremental small value. Upon closing the breaker, some voltage must occur, and therefore the area under the $f(V)$ curve must be 1.

With this new method of obtaining probabilities, the next section considers the calculation of the SSFOR using continuous probability distributions.

3 THE CONCEPT—CONTINUOUS DISTRIBUTIONS

3.1 With Strength Distribution for One Tower

As mentioned in Section 1 and in Chapter 2, the strength distribution or strength characteristic of one tower or one insulation can be described by a cumulative Gaussian distribution $F_S(V)$ having a mean denoted as the CFO and a standard

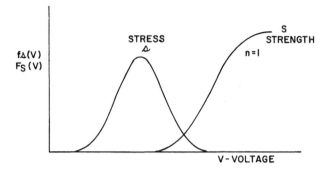

Figure 6 Stress–strength diagram.

deviation σ_f approximately equal to 5% of the CFO or $\sigma_f/\text{CFO} = 0.05$. The distribution of the stress or the SOV distribution can be represented by any probability density function $f_s(V)$; see Fig. 6. The problem is to determine the probability that the stress exceeds the strength, or the probability that the strength is less than the stress. This can only occur in the region where the two distributions overlap. Figure 7 is an expanded view of this area. The probability that a voltage V occurs is $f_s(V) dV$, and the probability of a flashover given that V occurs is $F_S(V)$ or more simply p. In equation form

$$\text{Probability that } V \text{ occurs} = P(V) = f_s(V) dV$$
$$P(\text{FO}|V) = F_S(V) = p \tag{4}$$

The incremental probability of a flashover for a voltage V is denoted as dP and is therefore the multiplication of these values or

$$dP = pf_s(V) dV = F_S(V) f_s(V) dV \tag{5}$$

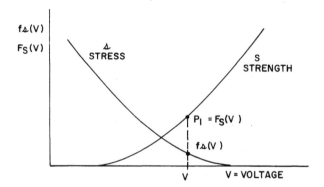

Figure 7 Expanded view of Figure 6.

The total probability of flashover considering all SOVs is the sum of Eq. 5 for all SOVs or

$$\text{SSFOR} = P(F) = \frac{1}{2} \int_{E_1}^{E_M} p f_s(V) \, dV \tag{6}$$

where SSFOR is the switching surge flashover rate. The integration is taken from E_1, the minimum voltage, which is usually 1.0 per unit of system line-to-neutral voltage, to E_m, which is the maximum SOV. To amplify, since the SOV is, by definition, an overvoltage, then it must be 1.0 per unit or greater. Also, for any system with its circuit breakers, there does exist a maximum SOV. Thus the integral is practically limited between these two values.

The integral of Eq. 6 is multiplied by 1/2. The distribution of SOVs is composed of all positive and negative. That is, half of the values are positive polarity and half are negative polarity. Per Chapter 2, the insulation strength for negative polarity is significantly larger than that for positive polarity. Therefore negative polarity SOVs can be neglected, and to obtain the SSFOR the integral must be multiplied by 1/2.

Equation 6 may be further expanded to

$$\text{SSFOR} = \int_{E_1}^{E_m} f_s(V) \left[\int_{-\infty}^{V} f_S(V) \, dV \right] dV \tag{7}$$

since

$$F_S(V) = \int_{-\infty}^{V} f_S(V) \, dV \tag{8}$$

3.2 Stress and Strength Both Normal—For One Tower

Although, as will be shown later, this is not a practical case, since only one tower is assumed, the following development is presented because it is used in other areas of engineering and because it permits an insight into the calculation of probability of failure. Assume that both distributions of stress and strength are Normal or Gaussian. Then, in short notation form with the letter N signifying a Normal distribution

$$\begin{aligned} f_s(V) &= N(\mu_0, \sigma_0) \\ f_S(V) &= N(\mu_S = \text{CFO}, \sigma_f) \end{aligned} \tag{9}$$

where S denotes the insulation strength and s denotes the stress or the SOVs.

The probability of a flashover or a failure, $P(\text{FO})$ is defined, as before, where the strength is less than the stress or

$$P(\text{FO}) = P(S < s) = P[(S - s) < 0] \tag{10}$$

Now let $Z = (S - s)$, or more formally, let the random variable Z equal the random variable S minus the random variable s. Then since both distributions are Normal, the distribution of Z will be Normal, of $f(Z)$ is Normal with the parameters

$$\mu_Z = \text{CFO} - \mu_0$$
$$\sigma_Z = \sqrt{\sigma_f^2 + \sigma_0^2} \tag{11}$$

The resultant density function $f(Z)$ is illustrated in Fig. 8 and as noted extends below zero. The area of interest is that below zero, since this is where $(S - s) < 0$. Therefore

$$P(\text{FO}) = \int_{-\infty}^{0} f(Z)\,dZ \tag{12}$$

or in terms of the standardized Normal distribution,

$$\text{SSFOR} = P(\text{FO}) = 1 - F\left[\frac{\text{CFO} - \mu_0}{\sqrt{\sigma_f^2 + \sigma_0^2}}\right] \tag{13}$$

As an example of the use of this equation assume a CFO of 900 kV, a σ_f of 45 kV, a μ_0 of 675 kV, and a σ_0 of 90 kV. Then

$$\text{SSFOR} = \frac{1}{2}P(\text{FO}) = \frac{1}{2}[1 - F(2.236)] = 0.0064 \tag{14}$$

As discussed previously, $P(\text{FO})$ is multiplied by $1/2$, and therefore the SSFOR is 0.64 flashover per 100 breaker closing operations.

3.3 More Than One Tower

Returning to the general case, the development in Section 3.1 and Eq. 6 assumes only one tower. If there are n towers, the probability of flashover of at least one tower as depicted in Fig. 9 is

$$P_n = 1 - (1 - p)^n = 1 - q^n \tag{15}$$

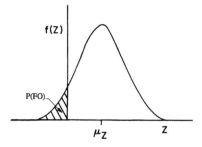

Figure 8 Probability of flashover, stress and strength Gaussian.

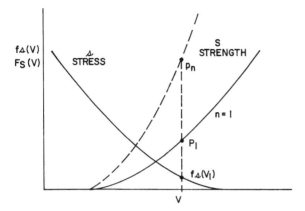

Figure 9 Probability of flashover increases for more than one tower.

where p is the probability of flashover of one tower and q is the probability of no flashover on one tower.

Therefore Eq. 6 must be modified by replacing p with $(1 - q^n)$ or

$$\text{SSFOR} = \frac{1}{2} \int_{E_1}^{E_m} [1 - q^n] f_s(V) \, dV \tag{16}$$

3.4 Voltage Profile

Again Eq. 16 needs further modification since it assumes that the SOV is constant at every tower along the line, when in reality the SOV is usually lower at the switched end of the line and is maximum at the opened end of the line. To include this effect, consider a transmission line, illustrated in Fig. 10, composed of n towers. For a single case of breaker closing, the voltages along the line are V_1 (or V_s) at tower 1, V_2 at tower 2,..., and V_n (or V_R) at the last tower. The probability that the SOV is equal to V_1 at tower 1 is $f(V_1)\,dV_1$, the probability that the SOV is equal to V_2 at tower 2 is $f(V_2)\,dV_2$, and the probability that the SOV is equal to V_n at the last tower is $f(V_n)\,dV_n$. However, $V_1, V_2, \ldots, v_n$ are dependent or exactly correlated. That is, for one switching operation, if V_n occurs at tower n, the V_1 *will* occur at tower 1, V_2 *will* occur at tower 2, etc. Therefore the probability of occurrence of V_1 is equal to the probability of occurrence of V_2, is equal to the probability of occurrence of V_n, etc. Or in equation form

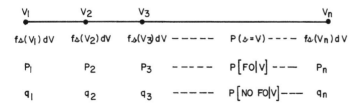

Figure 10 SOV profile along line.

$$f_s(V_1)\,dV_1 = f_s(V_2)\,dV_2 = \cdots = f_s(V_n)\,dV_n \tag{17}$$

To simplify, we will use $f_s(V)\,dV$ and note that $f_s(V)$ is the probability density function at the opened end of the line.

Returning to Fig. 10, the probability of a flashover at tower 1 for the SOV of V_1 is p_1; at tower 2 for the SOV of V_2 the probability of flashover is p_2, etc. However, since the probability of at least one flashover (on one tower) is desired, we must first calculate the probability of no flashovers (on any of the towers) and then subtract this from 1. The probability of no flashover on tower 1 is q_1, the probability of no flashover on tower 2 is $q_2, \ldots$, and the probability of no flashover on tower n is q_n. Therefore the probability of *no flashover* on the line for a single switching operation is

$$[q_1 q_2 q_3 \cdots q_n] \tag{18}$$

and the probability of at least one flashover is

$$[1 - q_1 q_2 q_3 \cdots q_n] \tag{19}$$

Considering all switching operations, the SSFOR is

$$\text{SSFOR} = \frac{1}{2}\int_{E_1}^{E_m}\left[1 - \prod_{i=1}^{n} q_i\right] f_s(V)\,dV \tag{20}$$

To be noted is that if the SOV is constant along the line, then $q_1 = q_2 \cdots q_n$ and Eq. 20 is the same as Eq. 16.

Unfortunately, there is no unique solution for Eq. 20, and therefore it must be solved numerically (easily accomplished with the aid of a digital computer). However, there exist simplified methods that can be used to obtain quickly an acceptable estimate of the SSFOR. Before presenting this simplified method, a sensitivity analysis of the SSFOR will assist in understanding the relative importance of the parameters or variables.

4 SOV (STRESS) DISTRIBUTIONS

4.1 Case Peaks or Phase Peaks

The SOVs are usually obtained by use of a transient computer program. Breakers are random switched throughout their pole closing span and the SOVs obtained. For each switching case, a SOV occurs on each phase and the probability of flashover $P(F)$ may be calculated for each case as

$$P(F) = \frac{1}{N}(1 - q_A q_B q_C) \tag{21}$$

where q_A is the probability of no flashover on phase A, q_B is the probability of no flashover on phase B, q_C is the probability of no flashover on phase C, and N is the number of cases. The addition of the $P(F)$ values for the N cases is then the SSFOR.

This method is often called the brute force method. However, it represents the most exact method since it accounts for voltages on all three phases as they occurred for each case. The advantage and disadvantage of this method are the same in that it is specific to the exact system considered. It does not permit a general evaluation of the effect of the parameters and does not provide an understanding of the phenomena.

Note that in the solution of Eq. 21, the SOVs may be positive or negative. If the SOV is negative, for practical designs, the value of q is essentially unity. Also, in the same manner, if the SOVs are small in magnitude, the value of q will be essentially unity. Usually, the value of $P(F)$ is controlled by only one of the values of q, the one obtained from a SOV that is positive and has the highest value of the three phases.

To circumvent the problem associated with the brute force method, the data may be collected and analyzed by two methods:

1. Case Peak Method. For each switching operation, the SOVs are collected. Only the SOV with the largest crest value, either positive or negative polarity, is used. This SOV is treated as positive since, if negative, the exact opposite breaker switching sequence would produce an opposite polarity SOV. This method, primarily used in the USA and Canada, as developed in Ref. 1, assumes that only one SOV predominates. In terms of Eq. 21, two of the qs, for example q_B and q_C, are essentially equal to unity, so that

$$P(F) = \frac{1}{N}(1 - q_A) = \frac{1}{N} p_A \tag{22}$$

Note that this probability should be multiplied by $1/2$ since with equal likelihood, either a positive or a negative polarity may occur and the negative polarity SOV is neglected since the negative polarity strength is significantly greater than that for positive polarity. The SSFOR calculated by this method is the SSFOR per three-phase breaker operation or the SSFOR for the line.

2. Phase Peak Method. The phase peak method consists of using all the three SOVs from each phase, and each of these are assumed as positive polarity. The $P(F)$ is calculated individually for each of the three SOVs. Thus the SSFOR calculated by this method is the SSFOR per phase. In terms of Eq. 21, the $P(F)$ is

$$P(F) = \frac{1}{3N}[(1 - q_A) + (1 - q_B) + (1 - q_C)] = \frac{1}{3N}[p_A + p_B + p_C] \tag{23}$$

where the equation is divided by $3N$ since three times as much data is collected. As for the case peak method, the probability should also be divided by two. Usually, two of the values of p are essentially zero so that, except for the $3N$, Eqs. 22 and 23 are identical. However, to obtain the SSFOR or the sum of $P(F)$ each calculation must be multiplied by 3. The problem occurs when a continuous distribution is employed to represent these three values of SOV. In this case, as an approximation, to obtain the SSFOR, the following equation is used:

$$\text{SSFOR} = 1 - [1 - (\text{SSFOR}_p)]^3 \approx 3(\text{SSFOR}_p) \tag{24}$$

where $SSFOR_p$ is the SSFOR calculated using the phase peak method. For the normal low values of SSFORP, the SSFOR is simply three times the $SSFOR_P$.

4.2 Case Peaks To Be Used

Because the case peak method appears to be a superior approximation, the developments in this chapter assume that the case peak method is employed. (However, the brute force method represents the only method that considers all variations of the SOV distribution.)

5 THE CONTINUOUS DISTRIBUTIONS

The random nature of the SOVs may be described by any distribution function. The most popular or most used distribution is the Gaussian or Normal distribution. However, the extreme value positive skew distribution has been frequently employed. Each of these is briefly described below.

5.1 Gaussian Distribution

The Gaussian probability density function has the familiar bell shape as shown in Fig. 11, the equation for which is

$$f_s(V) = \frac{1}{\sqrt{2\pi}\sigma} e^{-\frac{1}{2}\left(\frac{V-\mu_0}{\sigma_0}\right)^2} \tag{25}$$

This is usually placed in the form of the reduced variate Z, where

$$Z = \frac{V - \mu_0}{\sigma_0} \tag{26}$$

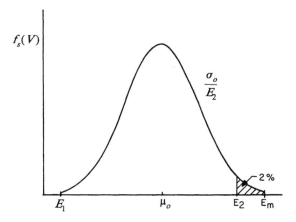

Figure 11 The Gaussian density function.

and the resultant probability density, called the standard normal density, is

$$f_s(Z) = \frac{1}{\sqrt{2\pi}} e^{-\frac{Z^2}{2}} \qquad (27)$$

The cumulative distribution function is the integral of the density function or

$$F_s(V) = \int_{-\infty}^{V} f_s(V)\,dV$$
$$F_s(Z) = \int_{-\infty}^{Z} f(Z)\,dZ \qquad (28)$$

As seen by these equations, the distribution has two parameters, the mean μ_0 and the standard deviation σ_0. For use as a SOV distribution, the distribution is defined by two related parameters, E_2 and σ_0/E_2. E_2 is the "statistical switching overvoltage." The probability that the SOV equals or exceeds E_2 is 0.02, or in other words, 2% of the SOVs equals or exceeds E_2. In equation form, where P is the probability,

$$P(\text{SOV} \geq E_2) = 0.02 \qquad (29)$$

and therefore E_2 in terms of μ_0 and σ_0 is

$$E_2 = \mu_0 + 2.054\sigma_0 \qquad (30)$$

or given E_2 and σ_0/μ_0, then

$$\mu_0 = E_2\left(1 - 2.054\frac{\sigma_0}{E_2}\right) \qquad (31)$$

One of the disadvantages of the Gaussian distribution is that Eq. 28 cannot be implicitly integrated and therefore tables, such as Table 1, or approximate equations must be used. For example, for $E_2 = 900\,\text{kV}$, $\sigma_0/E_2 = 0.10$, the probability that the SOV will equal or exceed 800 kV is 0.0173, or about 17.3% of the SOVs equal or exceed 800 kV.

Another disadvantage of the Gaussian distribution is that it is untruncated to both the left and the right, that is, it is defined from $-$ infinity to $+$ infinity, whereas the SOVs are limited between 1.0 pu to a maximum SOV of E_m. This is not a severe limitation, since E_m is usually sufficiently large that, for conservative calculations, it can be assumed as infinity.

Estimates of E_2, in pu of the maximum system crest line-neutral voltage, are presented in Table 4 as compiled by the CIGRE Working Group 33.02 [12]. In general, E_2 is about 2.8 pu for high-speed reclosing without closing resistors in the breaker and about 1.8 pu with closing resistors.

Estimates of σ_0, in pu, are

$$\sigma_0 = 0.17(E_2 - 1) \qquad (32)$$

Table 4A Estimates of E_2 for Shunt Compensation Equal to or Greater than 50%

Operation	Energizing				3-Phase reclosing			
Feeding network	Complex		Inductive		Complex		Inductive	
Closing resistor	Yes	No	Yes	No	Yes	No	Yes	No
Max	1.24	2.17	1.88	2.78	1.94	2.45	2.20	3.54
Average	1.18	1.85	1.51	2.24	1.74	2.00	1.63	2.74
Min	1.11	1.62	1.31	1.81	1.62	1.52	1.32	1.89

Table 4B Estimates of E_2 for Shunt Compensation Less than 50%

Operation	Energizing				3-Phase reclosing			
Feeding network	Complex		Inductive		Complex		Inductive	
Closing resistor	Yes	No	Yes	No	Yes	No	Yes	No
Max	1.99	2.59	2.20	2.90	1.80	3.48	2.14	3.66
Average	1.55	1.90	1.77	2.31	1.52	2.55	1.72	2.90
Min	1.27	1.41	1.35	1.66	1.20	1.46	1.37	2.14

For E_2 between 1.8 pu and 2.8 pu, σ_0/E_2 ranges from 0.08 to 0.11. The maximum SOV, E_m, is about one standard deviation above E_2 or

$$E_m = E_2 + \sigma_0 = E_2\left(1 + \frac{\sigma_0}{E_2}\right) \tag{33}$$

Therefore, for E_2 of 1.8 and 2.8 pu, E_m is 1.94 and 3.11 pu.

5.2 The Extreme Value Positive Skew Distribution

As shown in Fig. 12, the probability density function is given by the equation

$$f_s(V) = \frac{1}{\beta}e^{-\frac{V-u}{\beta}}e^{-e^{-\frac{V-u}{\beta}}} \tag{34}$$

where V is between + and − infinity. The cumulative distribution function is

$$F_s(V) = e^{-e^{-\frac{V-u}{\beta}}} \tag{35}$$

If the reduced variate y is defined as

$$y = \frac{V-u}{\beta} \tag{36}$$

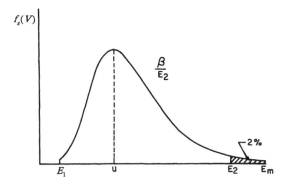

Figure 12 The extreme value, positive skew distribution.

then

$$f_s(V) = e^{-y} e^{-e^{-y}}$$
$$F_s(V) = e^{-e^{-y}} \tag{37}$$

where y is between $+$ and $-$ infinity.

The parameters of the distribution are u, the modal value, and β, the slope parameter, which is similar to the standard deviation. For use as a SOV distribution, the parameters are E_2, as defined previously, and β/E_2. If the mean μ or the standard deviation σ is desired, they may be obtained from the equations

$$\mu = u + \gamma\beta$$
$$\sigma = \frac{\pi}{\sqrt{6}}\beta \tag{38}$$
$$\gamma = 0.57722$$

where γ is Euler's constant.

The value of E_2 may be estimated as before using Table 4. The value of β/E_2 is between 0.05 and 0.09, slightly less than σ_0/E_2 for the Gaussian. The upper truncation point, i.e., E_m, is about

$$E_m = E_2 + 2\beta = E_2\left(1 + 2\frac{\beta}{E_2}\right) \tag{39}$$

For this distribution

$$E_2 = u + 3.902\beta \tag{40}$$

or

$$u = E_2\left(1 - 3.902\frac{\beta}{E_2}\right) \tag{41}$$

As for the Gaussian, this distribution suffers since it is untruncated either to the left or to the right. However, for conservative calculations, the upper truncation point may be assumed as infinite, $E_m = \infty$.

6 SENSITIVITY ANALYSIS OF SSFOR

In order of importance, the parameters that affect the SSFOR, as determined by the proceeding equations, are

1. The strength V_3 to stress E_2 ratio, where per Chapter 2,

$$V_3 = \text{CFO}\left(1 - 3\frac{\sigma_f}{\text{CFO}}\right) \tag{42}$$

2. The change of the SOVs along the line, denoted as the SOV profile.
3. σ_0/E_2 or β/E_2
4. The number of towers n

Before presenting the results of the sensitivity study, the SOV profile requires some additional definition. For this analysis, the SOV profile is assumed to be linear along the line, the voltage at the switched end is defined as E_S, and the voltage at the opened or receiving end is E_R, as illustrated in Fig. 13. With L as the total line length and x the length from the switched end, the voltage E is

$$E = E_S + (E_R - E_S)\frac{x}{L} \tag{43}$$

or

$$\frac{E}{E_R} = \gamma + (1 - \gamma)\frac{x}{L} \tag{44}$$

where

$$\gamma = \frac{E_S}{E_R} \tag{45}$$

Usually, the span length is considered constant and the number of towers is used instead of the line length. Referring to Fig. 13, the SOV profile is

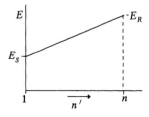

Figure 13 Assumed SOV profile.

$$\frac{E}{E_R} = \gamma + (1-\gamma)\frac{n'-1}{n-1} \tag{46}$$

Therefore, for a linear SOV profile, the SOVs along the line may be described simply by γ or E_S/E_R ratio and the SOV at the opened end of the line E_R.

To clarify further, the SOV distribution $f_s(V)$ always refers to the voltage at the opened end of the line. The ratio described by γ is used to find the voltages at other locations along the line. With this definition, let us return to the discussion of the sensitivity analysis.

6.1 Strength–Stress ratio V_3/E_2

Fig. 14 shows an example of the calculations of the SSFOR for a Gaussian SOV distribution for $n = 500$ towers, $\sigma_0/E_2 = 0.10$, $\sigma_f/\text{CFO} = 0.05$, and $E_2 = V_3 = 2.0\,\text{pu}$. The upper portion of the figure shows the stress and strength distributions; the lower portion shows the integrand of Eq. 20. The maximum SOV is assumed to be greater than 2.4 pu. The area under these curves (the integral) multiplied by 1/2 is the SSFOR as shown.

The principal observation is that the curves in the lower figure only extend from about 1.9 pu to 2.3 pu. That is, it is only within a narrow region that the SSFOR is produced, and only the upper tail of the stress distribution and the lower tail of the strength distribution are of importance. Therefore it appears reasonable to describe these distributions not by their means or CFO but by values on the tails of the distributions. Thus the stress or SOV distribution is specified by E_2—and σ_0/E_2 if

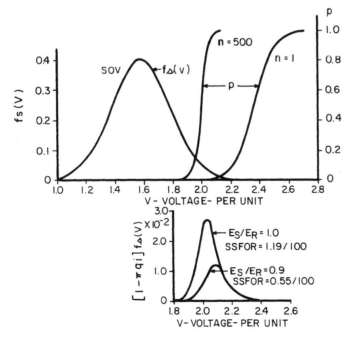

Figure 14 Tails of the distribution determine the SSFOR.

it is a Gaussian distribution or by β/E_2 if it is an extreme value distribution. The strength distribution is specified by V_3 and by σ_f/CFO.

Using E_2 and V_3, the strength–stress ratio is expressed as V_3/E_2, and the SSFOR is given by the curves of Fig. 15 for alternate values of σ_0/E_2 or β/E_2. The value of σ_f/CFO is 0.05, and n is 500 towers, which represents a line length of about 170 km or 100 miles ($E_S/E_R = 1.00$).

As noted in Fig. 15, the SSFOR decreases as V_3/E_2 increases, and all curves tend to cross at a SSFOR of about 1.0/100 and $V_3/E_2 \approx 1.0$. This fortuitous circumstance permits an immediate observation, that if a design SSFOR of 1.0/100 is desired, then independent of σ_0/E_2, or β/E_2, the simple design rule of

$$V_3 = E_2 \qquad (47)$$

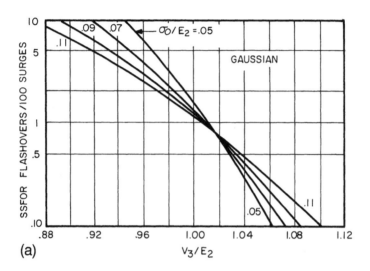

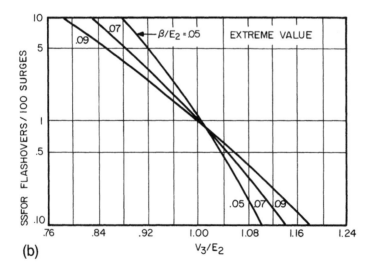

Figure 15 SSFOR for Gaussian and extreme value distributions, $n = 500$, $E_S/E_R = 1.00$.

applies. This design rule, based on a probabilistic approach, should be compared to that for the deterministic approach which per Chapter 2 is $V_3 = E_m$.

6.2 σ_0/E_2 or β/E_2

From Fig. 15, the SSFOR varies considerably with σ_0/E_2 or β/E_2 if V_3/E_2 differs from unity. For example, considering the Gaussian distribution, for a V_3/E_2 of 0.95, the SSFOR varies from 3.0/100 to 8.7/100 for σ_0/E_2 of 0.05 and 0.11, respectively. For a V_3/E_2 ratio of 1.05, the SSFOR varies from 0.19/100 to 0.38/100 for σ_0/E_2 of 0.05 and 0.11, respectively.

6.3 Number of Towers

The effect of the number of towers on the required V_3/E_2 ratio to achieve a SSFOR of 1.0/100 is presented in Fig. 16. For these curves, it is assumed that a SSFOR of 1.0/100 occurs at a V_3/E_2 ratio of 1.0. that is, the curves are per unitized on this basis. Also, $n = 500$, $\sigma_f/CFO = 0.05$, $E_S/E_R = 1.00$, and a Gaussian SOV distribution is assumed. The value of σ_0/E_2 was varied between 0.05 and 0.11. The curves illustrate that the strength–stress ratio varies only between ±1% for n between 300 and 1000 towers, or for lines between 60 to 300 km (40 to 190 miles). Thus the value of n within this range is not a sensitive parameter.

6.4 σ_f/CFO

The effect of σ_f/CFO on the required V_3/E_2 ratio for a SSFOR of 1.0/100 is shown in Fig. 17. As for Fig. 16, these curves are per unitized on the basis of a SSFOR of 1.0/100 at a V_3/E_2 ratio of 1.0. Also assumed is that $n = 500$, $E_S/E_R = 1.00$, $\sigma_0/E_2 = 0.05$, and that the SOV distribution is Gaussian. The value of σ_f/CFO of 5% is considered conservative, but even if the value changes from 3 to 7%, the required V_3/E_2 ratio only varies by ±1%. That is V_3/E_2 is insensitive to σ_f/CFO.

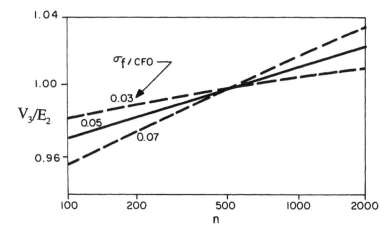

Figure 16 Effect of the number of towers.

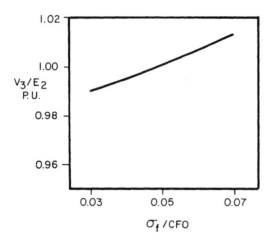

Figure 17 Effect of σ_f/CFO.

6.5 SOV Profile

Next to the strength–stress ratio, the SOV profile along the line is the most sensitive parameter, as illustrated in Fig. 18. This figure also presents the required strength–stress ratio to obtain a SSFOR of 1.0/100. As shown, for a σ_f/CFO of 5% and a σ_0/E_2 of 11%, a change in E_S/E_R from 1.0 to 0.9 decreases the required V_3/E_2 ratio by about 4% to maintain a SSFOR of 1.0/100, while a change to E_S/E_R of 0.6 only decreases the required strength–stress ratio by another 2%. Thus the curves of Fig. 15 should be shifted to the left for an E_S/E_R less than 1.00, and the approximate design of $V_3 = E_2$ becomes conservative.

The effect of decreasing E_S/E_R can also be obtained by comparing Figs. 15 and 18. For an E_S/E_R of 0.9, the SSFOR decreases from 1.19/100 to 0.55/100.

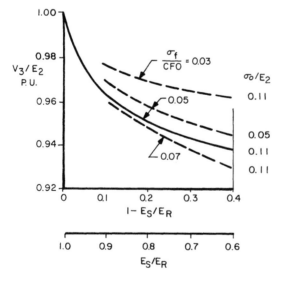

Figure 18 Effect of SOV profile.

7 ESTIMATING THE SSFOR

7.1 Brown's Method

In the previous sections, the general equations were developed and solved by numerical methods, since no closed form solution is possible. Several attempts have been made to obtain an approximate and simplified method so that computer programs using numerical methods could be circumvented. Two such methods are those of Alexandrov [2] and Brown [3]. Alexandrov's method is limited in application but was one of the first attempts; Brown's is more general and represents an excellent approximation. Further, Brown's method permits the calculation of both the SSFOR and the required V_3/E_2 ratio given the SSFOR. Therefore only Brown's method will be presented here, first the method of calculating the SSFOR, then the method of estimating the V_3/E_2 ratio. Another recent method is that presented in IEC Publication 71-2 [11]. This will be fully discussed later in Section 13.

First, assume that for every switching operation, the SOVs are identical at each of the towers, i.e. $E_S/E_R = 1.00$. As illustrated in Fig. 19, as the number of towers increases, the strength characteristic becomes steeper, or the standard deviation becomes smaller. That is, for any specific voltage, the probability of flashover increases from p to $(1-q^n)$. If the strength can be represented by a single-valued function located at a voltage equal to CFO_n, as illustrated by Fig. 20, the probability of flashover or the SSFOR is simply

$$\text{SSFOR} = \frac{1}{2} \int_{\text{CFO}_n}^{E_m} f_s(V)\, dV \tag{48}$$

which can easily be evaluated for any SOV distribution. For example, for the Gaussian distribution,

$$\text{SSFOR} = \frac{1}{2}\left[F\left(\frac{E_m - \mu_0}{\sigma_0}\right) - F\left(\frac{\text{CFO}_n - \mu_0}{\sigma_0}\right) \right] \tag{49}$$

where the first term is approximately 1.00. For the extreme value positive skew distribution,

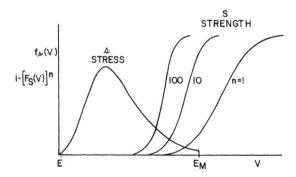

Figure 19 Steepness of strength characteristic increases with the number of towers.

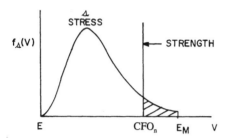

Figure 20 Simplifying calculation if strength is a single valued function.

$$\text{SSFOR} = \frac{1}{2}\left(e^{-e^{-y_e}} - e^{-e^{-y}}\right) \tag{50}$$

where

$$y_e = \frac{E_m - u}{\beta} \qquad y = \frac{\text{CFO}_n - u}{\beta} \tag{51}$$

and again, the firm term is approximately 1.00.

The CFO_n is the CFO for n towers and can be obtained from a knowledge of the strength characteristic for a single tower as illustrated in Fig. 21. As for the CFO of a single tower, the CFO_n for n towers is defined at a probability of 0.5. The objective is to find the probability p on the single-tower strength characteristic that is equivalent to the CFO for n towers. Therefore

$$0.5 = 1 - (1 - p)^n$$
$$p = 1 - \sqrt[n]{0.5} \tag{52}$$

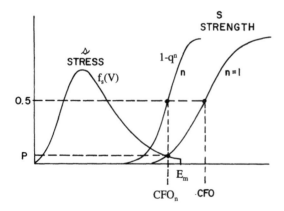

Figure 21 Effect of number of towers.

From the value of p, the body of Table 1 is entered to obtain the value of the reduced variate Z, which in this case is denoted Z_f. By the formula for the reduced variate, i.e.,

$$Z_f = \frac{\text{CFO}_n - \text{CFO}}{\sigma_f} \tag{53}$$

the CFO_n is obtained as

$$\text{CFO}_n = \text{CFO}[1 + Z_f \frac{\sigma_f}{\text{CFO}}] \tag{54}$$

To be observed is that the value of Z_f is usually negative.

To illustrate by example, for $n = 200$, $p = 0.003460$. Entering Table 1 with $F(Z)$ of $1 - p$ or 0.996540, Z is about 2.70 and therefore Z_f is approximately -2.70. Therefore CFO_n is 0.865 CFO assuming a σ_f/CFO of 5%. To continue, if the CFO is 1000 kV and the SOV distribution is Gaussian with $E_2 = 900$ kV, $E_m = 999$ kV (one standard deviation above E_2), and $\sigma_0/E_2 = 0.11$, the $\mu_0 = 696.67$ kV, the SSFOR per Eq. 49 is

$$\begin{aligned}\text{SSFOR} &= \frac{1}{2}\left[F\left(\frac{999 - 696.67}{99}\right) - F\left(\frac{865 - 696.67}{99}\right)\right] \\ &= \frac{1}{2}[F(3.05) - F(1.70)] = \frac{1}{2}[0.998856 - F(1.70)] \\ &\approx \frac{1}{2}[1 - F(1.70)] = 2.23/100\end{aligned} \tag{55}$$

where the $F(Z)$s are obtained from Table 1. As noted, the first term of the calculation can be conservatively assumed as 1.00, which illustrates that E_m seldom needs to be considered or evaluated.

If in the above example the SOV distribution is an extreme value positive skew distribution with the same E_2 and with $\beta/E_2 = 0.11$, then $u = 513.7$ kV and the reduced variate y is

$$y = \frac{865 - 513.7}{99} = 3.548 \tag{56}$$

Then neglecting E_m, per Eq. 50, the SSFOR is

$$\text{SSFOR} = \frac{1}{2}\left[1 - e^{-e^{-y}}\right] = \frac{1}{2}\left[1 - e^{-e^{-3.548}}\right] = 1.42/100 \tag{57}$$

To consider the SOV profile, and equivalent number of towers n_e is calculated and then used in Eq. 52. The value of n_e is the number of towers having an $E_S/E_R = 1.00$, which gives the same SSFOR as the actual number of towers with the specified E_S/E_R, as illustrated in Fig. 22. The equivalent number of towers may be estimated from the equation

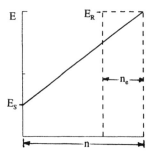

Figure 22 Equivalent number of towers is at constant SOV.

$$n_e = \frac{k_n}{1-\gamma} \frac{\sigma_f}{\text{CFO}} n \quad \text{or} \quad n_e = n \quad \text{whichever is less} \tag{58}$$

where

$$\gamma = \frac{E_S}{E_R} \tag{59}$$

and k_n is a function of n_e as shown in Fig. 23. Theoretically, k_n should be determined iteratively. However, over the practical range of 30 to 500 towers, an average value of k_n of 0.4 may be used, since the exact number of towers is insensitive to the required V_3/E_2 ratio. Therefore.

$$n_e = \frac{0.4}{1-\gamma} \frac{\sigma_f}{\text{CFO}} n \quad \text{or} \quad n_e = n \quad \text{whichever is less} \tag{60}$$

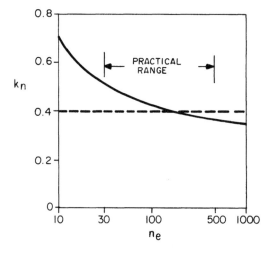

Figure 23 k_n in practical range is 0.4. Copyright IEEE, 1978 [4].

As an example, consider the previous example for the Gaussian distribution with 200 towers but now assume an E_S/E_R of 0.9. Thus $n_e = 40$. Continuing, the CFO_n is 895 kV, and therefore the SSFOR reduces to 1.15/100.

7.2 Discussion of Estimating Method

As developed in Section 2.2, if the stress, the SOV, distribution and the strength characteristic are both Gaussian, the SSFOR is

$$\text{SSFOR} = \frac{1}{2}\left[1 - F\left(\frac{\text{CFO} - \mu_0}{\sqrt{\sigma_0^2 + \sigma_f^2}}\right)\right] \tag{61}$$

Note carefully that this assumes that the strength distribution is Gaussian. It is for $n = 1$ but it is not for $n > 1$. But if the strength characteristic were Gaussian for any value of n with a CFO_n and σ_{fn}, then the SSFOR would be

$$\text{SSFOR} = \frac{1}{2}\left[1 - F\left(\frac{\text{CFO}_n - \mu_0}{\sqrt{\sigma_0^2 + \sigma_{fn}^2}}\right)\right] \tag{62}$$

The primary reason why Brown's estimating method produces good estimates of the SSFOR is that σ_f for n or n_e towers, σ_{fn}, is much less than σ_0 of the SOV distribution, and in this case

$$\text{SSFOR} = \frac{1}{2}\left[1 - F\left(\frac{\text{CFO}_n - \mu_0}{\sigma_0}\right)\right] \tag{63}$$

which is identical to Eq. 49. As a word of caution, the criterion that σ_{fn} is much less than σ_0 is not always true. For example if $n = 1$, σ_f may be important. This is considered in Chapters 4 and 5. Also see the discussion of the IEC method in Section 13.

8 ESTIMATING THE V_3/E_2 GIVEN THE SSFOR

8.1 Insulation Strength, Gaussian

An overwhelming advantage of the method proposed by Brown is that it can be used to obtain directly an estimate of the required value of V_3/E_2 given a value of SSFOR—which is the usual design problem. To develop this method, note that Eq. 54 may be placed in terms of V_3, i.e.,

$$\text{CFO}_n = \frac{V_3}{K_f} \tag{64}$$

where

$$K_f = \frac{1 - 3(\sigma_f/\text{CFO})}{1 + Z_f(\sigma_f/\text{CFO})} \tag{65}$$

8.2 Gaussian SOV Distribution

Neglecting E_m, the SSFOR is

$$\text{SSFOR} = \frac{1}{2}\int_{\text{CFO}_m}^{\infty} f_s(V)\,dV = \frac{1}{2}[1 - F(Z_e)] \tag{66}$$

or

$$2(\text{SSFOR}) = 1 - F(Z_e) \tag{67}$$

where

$$Z_e = \frac{\text{CFO}_n - \mu_0}{\sigma_0} \tag{68}$$

To obtain Z_e, as given by Eq. 67 and portrayed by Fig. 24, Table 1 must be entered with 1–2 (SSFOR). For example, if the desired SSFOR is 1/100, Table 1 is entered with 0.98 from which Z_c is approximately 2.054. Continuing with the development, the substitution of Eq. 64 for the CFO_n, using the relationship between μ_0 and E_2, produces

$$\frac{V_3}{E_2} = K_f K_G \tag{69}$$

where

$$K_G = 1 - (2.054 - Z_e)\frac{\sigma_0}{E_2} \tag{70}$$

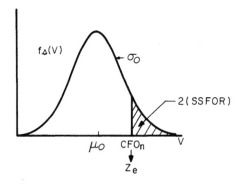

Figure 24 Use twice the SSFOR to find CFO_n

As noted, the ratio V_3/E_2 is obtained directly from K_f, which is only a function of the strength, and K_G, which is only a function of the stress.

To clarify, assume that the ratio of V_3/E_2 is desired for an SSFOR of 1.5/100. Let $n = 200$, $\sigma_0/E_2 = 0.08$, and $\sigma_f/\text{CFO} = 0.05$. Then $p = 0.003460$, $Z_f = -2.70$, and $K_f = 0.9827$. For a SSFOR of 1.5/100, the value of Z_e is obtained for $F(Z_e) = 1 - 2(0.015) = 0.97$. From Table 1, $Z_e = 1.88$ and therefore $K_G = 0.9862$. Thus V_3/E_2 must be 0.9691.

The effect of the SOV profile is considered as before. For example, if $E_S/E_R = 0.90$ and $\sigma_f/\text{CFO} = 0.05$, then $n_e = 40$. Proceeding as in the last paragraph, $K_f = 0.9507$ and $V_3/E_2 = 0.9376$.

8.3 Extreme Value Positive Skew SOV Distribution

For an extreme value positive skew SOV distribution, Eq. 67 becomes

$$2(\text{SSFOR}) = 1 - e^{-e^{-y_e}} \tag{71}$$

where

$$y_e = \frac{\text{CFO}_n - u}{\beta} \tag{72}$$

and the y_e must be obtained for twice the SSFOR as illustrated in Fig. 25. For example if the SSFOR is 2/100, then the y_e from Eq. 71 is 3.902.

Since K_f is unchanged, then

$$\frac{V_3}{E_2} = K_f K_E \tag{73}$$

where

$$K_E = 1 - (3.902 - y_e)\frac{\beta}{E_2} \tag{74}$$

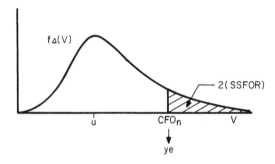

Figure 25 Use twice the SSFOR to find CFO_n.

8.4 Tables of Values of K_f, K_G, and K_E

For convenience of use, Tables 5, 6, and 7 provide values of the previous parameters. From these tables:

1. For $n = 500$, K_f is approximately 1.00 or any of the three values of σ_f/CFO.
2. For a SSFOR of 1.0/100 and for 500 towers, K_G and K_E are equal to 1.0000.

Therefore for an SSFOR of 1.0/100 and 500 towers, $V_3/E_2 = 1.00$, which agrees with Fig. 15.

9 EXAMPLES

Determine the strike distance and insulator string length for a 500 kV tower for a design SSFOR of 1.0/100. V-strings are used on all phases. Line altitude is 1500 meters. SOV (Stress): Gaussian, $E_2 = 810$ kV, $\sigma_0/E_2 = 0.07$, $E_S/E_R = \gamma = 0.90$. Strength: $\sigma_f/\text{CFO} = 0.05$, $n = 250$, tower width $W = 1.6$ m, $h = $ conductor height $= 18$ m.

Table 5 K_f

		K_f for σ_f/CFO of		
n or n_e	Z_f	0.020	0.050	0.070
1	0	0.9400	0.8500	0.7900
5	−1.129	0.9617	0.9009	0.8578
10	−1.499	0.9691	0.9189	0.8826
20	−1.825	0.9756	0.9353	0.9057
50	−2.204	0.9833	0.9553	0.9341
100	−2.463	0.9887	0.9693	0.9545
200	−2.701	0.9937	0.9827	0.9742
500	−2.992	0.9998	0.9996	0.9993
1000	−3.198	1.0042	1.0118	1.0178
2000	−3.393	1.0084	1.0236	1.0360

Table 6 K_G for Gaussian SOV distribution SSFOR in Flashovers/100 Switching Operations

		K_G for σ_0/E_2 of			
SSFOR	Z_e	0.05	0.07	0.10	0.15
10	0.8415	0.9394	0.9151	0.8787	0.8181
5	1.2817	0.9614	0.9459	0.9228	0.8841
1	2.0542	1.0000	1.0000	1.0000	1.0000
0.5	2.3268	1.0136	1.0191	1.0273	1.0409
0.2	2.6525	1.0299	1.0419	1.0598	1.0897
0.1	2.8785	1.0412	1.0577	1.0824	1.1236

Phase-Ground SOVs, Transmission Lines

Table 7 K_E for Extreme Value Positive Skew SOV Distribution SSFOR in Flashovers/100 Switching Operations

SSFOR	y_e	K_E for β/E_2 of		
		0.05	0.07	0.10
10	1.4999	0.8799	0.8319	0.7598
5	2.2504	0.9174	0.8844	0.8348
1	3.9019	1.0000	1.0000	1.0000
0.5	4.6002	1.0349	1.0489	1.0698
0.1	6.2136	1.1156	1.1618	1.2312

1. Basic

$$\delta = e^{-A/8.6} = 0.840 \quad n_e = 50 \quad K_f = 0.9553$$
$$K_G = 1.0000 \quad V_3/E_2 = 0.9553 \quad (75)$$
$$V_3 = 0.9553(810) = 774\,\text{kV} \quad \text{CFO}_A = 774/0.85 = 910\,\text{kV}$$

2. Center Phase

$$k_g = 1.25 + 0.005[(18/S) - 6] + 0.25[(e^{-(12.8/S)} - 0.2] \quad (76)$$

$$m = 1.25 G_0 (G_0 - 0.2) \quad G_0 = \frac{\text{CFO}_S}{500\,S} \quad (77)$$

$$\text{CFO}_S = 0.96 k_g \frac{3400}{1 + (8/S)} \quad \text{CFO}_A = \delta^m \text{CFO}_S \quad (78)$$

$$S = \frac{8}{\dfrac{0.96(3400) k_g\, \delta^m}{\text{CFO}_A} - 1} \quad (79)$$

As an initial guess, let $m = 0.5$, $k_g = 1.20$. Then per Eq. 79, $S = 2.72$.

Therefore, for the center phase (Table 8), $S = 2.70$ meters (8.86 feet) and the minimum insulator length is 5% greater than the strike distance or 2.84 meters, which is 19.4 standard insulators, $5\frac{3}{4} \times 10$ inches.

3. Outside Phase

$$\text{CFO}_S = 0.96(1.08) k_g \frac{3400}{1 + (8/S)} \quad \text{CFO}_A = \delta^m \text{CFO}_S \quad (80)$$

Table 8 Iteration for Center Phase

S	k_g	CFO_S	G_0	m	S
2.72	1.205	998.2	0.734	0.490	2.70
2.70	1.205	998.2	0.734	0.490	2.70

$$S = \frac{8}{\dfrac{0.96(1.08)(3400)k_g \delta^m}{CFO_A} - 1} \tag{81}$$

As an initial guess, let $k_g = 1.2$, $m = 0.5$. Then $S = 2.46$. Also the initial estimate of S could be the S for the center phase divided by 1.08 or 2.50.

For the outside phase (Table 9), $S = 2.51$ meters (8.23 feet). The minimum insulator length is 2.64 meters or 18.0 standard insulators.

10 ESTIMATING THE SSFOR FROM V_3/E_2

From the development presented in Section 7, the SSFOR may be estimated in terms of V_3/E_2 and K_f as follows:

For the Gaussian SOV distribution

$$\begin{aligned} \text{SSFOR} &= \frac{1}{2}[1 - F(Z_e)] \\ Z_e &= 2.054 - \frac{1 - [(V_3/E_2)/K_f]}{\sigma_0/E_2} \end{aligned} \tag{82}$$

For the extreme value positive skew SOV distribution

$$\begin{aligned} \text{SSFOR} &= \frac{1}{2}[1 - F(y_e)] = \frac{1}{2}[1 - e^{-e^{-y_3}}] \\ y_e &= 3.902 - \frac{1 - [(V_3/E_2)/K_f]}{\beta/E_2} \end{aligned} \tag{83}$$

To illustrate, let $K_f = 0.955$ and $\sigma_0/E_2 = 0.07$. For ratios of V_3/E_2 of 0.95, 1.00, and 1.05, the SSFOR is 1.19/100, 0.16/100, and 0.013/100, respectively, thus showing that a change of V_3/E_2 by 5% alters the SSFOR by a decade step.

Table 9 Iterating for Outside Phase

S	k_g	CFO_S	G_0	m	S
2.46	1.208	1001	0.814	0.625	2.51
2.51	1.207	1016	0.810	0.618	2.51

11 EFFECT OF WIND ON I-STRINGS

As illustrated in Fig. 26, for insulator strings not constrained from movement, the I-string, wind can move the conductor closer to the tower side, thus decreasing both the strike distance to the tower side and the strike distance to the upper truss. This strike distance decreases the CFO, thus increasing the SSFOR.

For a wind speed v impinging on the conductor, the swing angle α_S is [5]

$$\alpha_S = \tan^{-1}(k_1 v^{1.6}) \tag{84}$$

where

$$k_1 = (1.138 \times 10^{-4}) \frac{D/W}{V/H} \tag{85}$$

and D = conductor diameter, cm; W = conductor weight, kg/m; V = vertical or weight span length; H = horizontal or wind span length; and v = wind speed, km/h.

As shown in Fig. 27, the horizontal or wind span length is measured between adjacent midspans, and the vertical or weight span is measured between minimum sag locations of adjacent spans. Thus on level terrain, V/H is 1.00, while in mountainous areas V/H is less than 1.00.

The wind speed, its distribution, and its parameters were investigated by a study of hourly wind speeds at six USA weather stations [5]. The investigations showed that the variations of wind speed at a specific location could be described by a Weibull distribution

$$F(v) = 1 - e^{-4.6(v/v_{100})^\beta} \tag{86}$$

where v_{100} is the 100-hour wind speed. For thunderstorm conditions, the exponent β varied between 1.5 and 2.2, with an average of 1.9. For nonthunderstorm conditions and fair weather conditions, β varied between 1.4 and 2.1 with averages of 1.9 and 1.7, respectively.

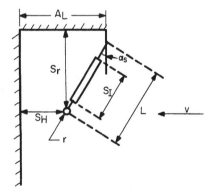

Figure 26 Wind moves the conductor toward the tower.

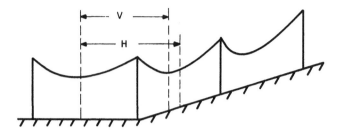

Figure 27 Illustrating vertical and horizontal span lengths.

For the six stations, the 100-hour wind speed varied between 21 and 30 knots for thunderstorm conditions (1 knot = 1.85 km/h = 1.15 mile/h). For the other two weather conditions, the 100-hour wind speed varied from 19 to 28 knots. Thus the maximum 100-hour wind speed was 56 km/h.

To gain a better understanding, consider that $v_{100} = 50$ km/h, $(D/W)/(V/H) = 1.5$, and $\beta = 1.9$. Then, for $v = v_{100}$, $F(v) = 0.99$ and $\alpha_s = 5.1$ degrees. Therefore

$$P[\alpha_s \geq 5.1 \text{ degrees}] = 0.01 \tag{87}$$

Now consider Fig. 28, where the arm length A_L is

$$A_L = S_H + r + L \sin \alpha_s \tag{88}$$

where r is the conductor support radius. For α_s of 5.1 degrees and a connection length L of 3.0 m, $(A_L - S_H - r) = 0.267$ m which is labeled Z in Fig. 28. Thus the probability that the strike distance S_H will be reduced by greater than 0.267 m is 0.01 or

$$P(A_H - S_H - r) \geq 0.267 \text{ m}) = 0.01 \tag{89}$$

If L is 3 m, the value of A_L is usually about 3 m and therefore S_H is reduced to about 2.73 m, about a 9% reduction.

The actual method used to calculate the SSFOR for an I-string is to consider each wind speed and its probability of occurrence. That is, in Fig. 29, consider a single wind speed v_1 whose probability of occurrence is $f(v_1)\,dv$. The swing angle α_s is found using v_1, and, knowing the tower arm, etc., dimensions, the strike distances S_H and S_V can be found. Using the minimum of these two strike distances and insulator length S_I divided by 1.05, the CFO is determined, from which the SSFOR can be calculated. This SSFOR is multiplied by $f(v_1)\,dv$. The process is continued for all values of v, and finally the results are added to obtain the total SSFOR.

Note that S_I the insulator length is divided by 1.05, and that the CFO of the outside phase I-string is 8% larger than that of the center phase. See Chapter 2.

The above method requires a considerable amount of calculation. To circumvent this effort, approximately the same SSFOR can be determined by using only one

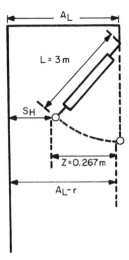

Figure 28 Example.

wind speed v_d, denoted as the design wind speed, which is equal to 60% of the 100-hour wind speed, or [5]

$$v_d = 0.60 v_{100} \qquad (90)$$

To clarify the procedure, the steps of the calculations are

1. Proceed in exactly the same manner as for the outside phase V-string design. That is, for a given SSFOR, find V_3/E_2 and the strike distance S.
2. Using a design wind speed of 60% of the 100-hour wind speed, calculate the swing angle.
3. As illustrated in Fig. 30, knowing the insulator string length S_I, the hanger length H_L, and the lower connection length to the conductor, the total connection length L is known. A circle of radius $S + r$ is drawn from the center of the conductor to describe the clearance circle. All grounded steel tower

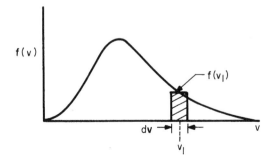

Figure 29 Using the wind speed distribution.

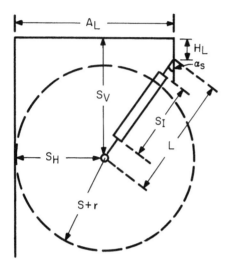

Figure 30 The clearance circle for an I-string insulator.

members must be outside this circle. In addition, S_I must be 1.05 times the minimum strike distance. In equation form,

$$A_L \geq L \sin \alpha_s + S_H + r$$
$$H_L \geq S_V + r - L \cos \alpha_s \qquad (91)$$
$$S_I \geq 1.05 S$$

As a note of caution, the radius r is not the conductor radius but the radius of the conductor support—better known as the conductor "shoe."

For some tower structures such as the vertical double-circuit tower, another phase is directly below the top phase position. In this case, the lower arm should be located so as to maintain the specified strike distance within the design swing angle. That is, the distance S must be maintained between the conductor and the lower tower arm for swing angles between 0 and the design swing angle. In concept, the strike distance or clearance required is that of a rolling ball that rolls or travels between the swing angle of 0 and the design swing angle. This means that the lower arm can be tangent to circles in Fig. 31. Two possible locations of the arm are illustrated. However, many other locations or positions are possible.

As a final note to this section, although hopefully the methodology is clear, there still exists the problem of determining the 100-hour wind speed. Usually the 100-hour wind speed can be obtained at nearby airports where weather statistics are recorded. However, the 100-hour wind speed at the "nearby" line locations is desired. Usually, the 100-hour wind speed at the airport is greater than that at the line location. To explain, consider Fig. 32. If v_1 is the recorded wind speed at the airport at height h_1, then the wind speed v_2 at the height h_2 is

$$\frac{v_2}{v_1} = \left(\frac{h_2}{h_1}\right)^\alpha \qquad (92)$$

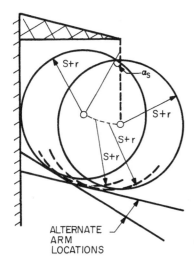

Figure 31 Clearance "rolling ball" for lower arm.

where α is the roughness factor and is considered to be dependent only on the terrain. For airports α is about 0.14, whereas for surfaces such as forest α is much larger. At some height h_2, a "gradient" wind exists such that the wind speed v_2 can be considered constant over a general area. This wind speed v_2 is then decreased to v_3 at the tower height h_3, using an appropriate value of α, i.e., an α greater than that at the airport. Thus

$$v_3 = \left(\frac{h_3}{h_2}\right)^{\alpha_F} \left(\frac{h_2}{h_1}\right)^{\alpha_A} v_1 \tag{93}$$

where α_A and α_F refer to the values for the airport and the line location, respectively. Since usually the α at the line location is greater than that at the airport, v_3 is less than v_1 and it would appear conservative to assume that the airport wind speed could be used conservatively for design. However, no firm recommendation can be given, since so-called wind tunnels do exist, which increase the wind speed. Of

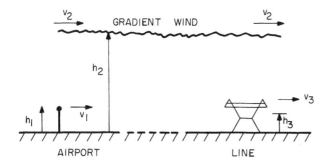

Figure 32 Wind speed at line location is less than at airport.

course, the best way is to measure the wind speeds along the line location, but normally construction times do not permit this luxury and therefore some engineering judgment must be used to establish the 100-hour wind speed.

12 FACTORS OF DESIGN

The results of studies to determine the strike distance may be portrayed as a curve as illustrated by Fig. 33, thus permitting a sensitivity analysis. Also the strike distance could be translated into cost so that the cost of various designs can be evaluated.

However, the SSFOR is normally not considered a stand-alone design criterion. For example, in areas of low lightning activity, the SSFOR may be selected as high as 1/10, since the probability of lightning faults, which cause the breaker to reclose, is low, while in areas of very high lightning activity, the SSFOR may be selected as low as 1/1000. To place this concept in mathematical terms, an unsuccessful reclosure rate could be calculated. This is sometimes called the Storm Outage Rate (SOR) and is the SSFOR multiplied by the lightning flashover rate for the line. That is, the lightning flashover rate is normally given as the number of flashovers per 100 km-years, which for this calculation must be converted to the number of flashovers per year. For example, if the SSFOR is 1.5/100 and the lightning flashover rate is 2.0 flashovers per 100 km-year for a 200-km line, then the SOR is 6.0 per 100 years. Denoting the lightning flashover rate in units of flashovers per year as LFOR, in equation form we have

$$\text{SOR} = (\text{SFOR})(\text{LFOR}) \tag{94}$$

The units of the equations are SOR: unsuccessful reclosures per year; SSFOR: switching surge flashovers per year; LFOR: lightning flashovers per year. Thus a lightning flashover causes a fault that results in the breaker opening and reclosing—which causes a switching surge, which results in a flashover. For EHV systems, the breaker is reclosed only once, after which it is locked open. Thus the SOR is the

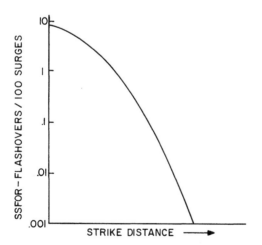

Figure 33 SSFOR vs strike distance.

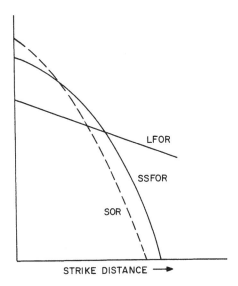

Figure 34 Using a storm outage rate as a design criterion.

outage rate. If a breaker is reclosed twice, the outage is then the SOR as calculated by Eq. 94 multiplied again by the SSFOR.

Curves of the SSFOR and the LFOR and the resultant SOR are illustrated in Fig. 34.

13 COMPARISON WITH IEC

The IEC Standard 71-2 [11] primarily considers substations and therefore deals with a selection of substation insulation levels, the BIL and the BSL, and clearances. However, in this document there exist several technical ideas and proposals that require discussion and understanding. Therefore these items are considered here and will be further considered in Chapters 4 and 5.

13.1 Strength Characteristics for *n* Parallel Insulations

As discussed in Chapter 2, the proposed revision of IEC 71-2 suggests that the insulation strength characteristics be modeled as a Weibull distribution instead of as a Gaussian distribution as is done in this chapter. From Chapter 2, the resultant equation for the Weibull distribution is

$$p = F(V) = 1 - 0.5^{\left(1+\frac{Z}{4}\right)^5} \qquad (95)$$

where

$$Z = \frac{V - \mathrm{CFO}}{\sigma_f} \qquad (96)$$

This formulation has the advantage that for n parallel insulations, the distribution remains a Weibull distribution. To explain, first consider the original Weibull distribution per Chapter 2, i.e.,

$$p = F(V) = 1 - e^{-\left[\frac{V-\alpha_0}{\alpha}\right]^\beta} \qquad \alpha_0 < V < \infty \tag{97}$$

Then the probability of flashover for n insulations is

$$p_n = 1 - q^n = 1 - e^{-n\left[\frac{V-\alpha_0}{\alpha}\right]^\beta} = 1 - e^{-\left[\frac{V-\alpha_0}{\alpha'}\right]^\beta} \tag{98}$$

where

$$\alpha' = \frac{\alpha}{\sqrt[\beta]{n}} \tag{99}$$

Thus the strength characteristic for n parallel insulations is also given by a Weibull cumulative distribution. The parameters of this distribution are CFO_n, σ_{fn}, and β. As before let $p = 0.0$ at $V = CFO_n - 4\sigma_{fn}$. Then

$$\alpha_0 = CFO_n - 4\sigma_{fn} \tag{100}$$

Then let $p_n = 0.5$ when $V = CFO_n$ and using Eq. 98,

$$\alpha' = \frac{4\sigma_{fn}}{(\ln 2)^{1/\beta}} = \frac{\alpha}{\sqrt[\beta]{n}} = \frac{4\sigma_f}{\sqrt[\beta]{n}(\ln 2)^{1/\beta}} \tag{101}$$

where the equation for α is obtained from Eq. 60 of Chapter 2. Therefore

$$\sigma_{fn} = \frac{\sigma_f}{\sqrt[\beta]{n}} = \frac{\sigma_f}{\sqrt[5]{n}} \tag{102}$$

Continuing, but skipping some steps since this same procedure was performed in Chapter 2,

$$p_n = 1 - 0.5^{\left[\frac{Z_n}{4}+1\right]^5} \tag{103}$$

where

$$Z_n = \frac{V - CFO_n}{\sigma_{fn}} \tag{104}$$

To find CFO_n, note that

$$CFO_n - 4\sigma_{fn} = CFO - 4\sigma_f \tag{105}$$

and therefore

$$\text{CFO}_n = \text{CFO}\left[1 - 4\frac{\sigma_f}{\text{CFO}}\left(1 - \frac{1}{\sqrt[5]{n}}\right)\right]$$

$$\frac{\sigma_{fn}}{\text{CFO}_n} = \frac{\sigma_f}{\text{CFO}}\left[\frac{1}{\sqrt[5]{n}\left(1 - 4\frac{\sigma_f}{\text{CFO}}(1 - \frac{1}{\sqrt[5]{n}})\right)}\right] \qquad (106)$$

13.2 SOV Distribution

IEC 71 also suggests that the Weibull distribution may be a superior distribution for SOVs. The objection to the Gaussian distribution is that it is defined between plus and minus infinity, whereas there does exist a maximum SOV, which is here denoted as E_m. The Weibull distribution can overcome this objection but, in contrast to the strength characteristic, the SOV distribution must be truncated on the high end. The form of this Weibull distribution is the reverse of that used for the strength characteristic and is therefore called the reverse Weibull, that is,

$$F(V) = e^{-\left[\frac{\alpha_0 - V}{\alpha}\right]^\beta} \qquad -\infty < x < \alpha_0 \qquad (107)$$

To make this distribution similar to the Gaussian and use the same parameters we stipulate

1. At $F(V) = 1.0$, let $V = E_m = \mu_0 + 3\sigma_0$
2. At $F(V) = 0.5$, let $V = \mu_0$
3. At $F(V) = 0.98$, let $V = E_2 = \mu_0 + 2.054\sigma_0$

Using the three simultaneous equations derived from these three definitions, then

$$F(V) = 0.5^{\left[1 - \frac{1}{3}\frac{V - \mu_0}{\sigma_0}\right]^3} \qquad (108)$$

This equation is not entirely satisfactory, since an arbitrary value is given to the maximum SOV. Because of this, the Weibull distribution is not totally recommended in the IEC document. That is, the Gaussian SOVs distribution is still used.

13.3 Stress–Strength and Estimating the SSFOR

Per the development in Section 3.2, if SOV distribution and the strength characteristic is Gaussian, then

$$\text{SSFOR} = \frac{1}{2}\left[1 - F\left(\frac{\text{CFO} - \mu_0}{\sqrt{\sigma_0^2 + \sigma_f^2}}\right)\right] \qquad (109)$$

Note carefully that this assumes that the strength distribution is Gaussian. It is for $n = 1$ but it is not for $n > 1$.

But if the strength characteristic were Gaussian for any value of n with a CFO_n and σ_{fn}, then the SSFOR would be

$$\text{SSFOR} = \frac{1}{2}\left[1 - F\left(\frac{CFO_n - \mu_0}{\sqrt{\sigma_0^2 + \sigma_{fn}^2}}\right)\right] = \frac{1}{2}[1 - F(Z_e)] \tag{110}$$

where

$$Z_e = \frac{CFO_n - \mu_0}{\sqrt{\sigma_0^2 + \sigma_{fn}^2}} \tag{111}$$

Now assume that the stress distribution is Gaussian but the strength distribution is Weibull. Note that the Weibull distribution has been characterized so as to approximate the Gaussian. Therefore it appears possible to approximate the SSFOR per Eq. 110 by the use of the equations for the CFO_n and σ_n. To demonstrate

Example. Let $n = 500$, $\sigma_f/CFO = 0.05$, $CFO = 941 \text{ kV}$, $E_2 = 808 \text{ kV}$, $E_S/E_R = 1.00$ and $\sigma_0/E_2 = 0.10$. Then

$$\sigma_{fn} = \frac{\sigma_f}{\sqrt[5]{500}} = 13.57$$

$$CFO_n = CFO\left[1 - 4\frac{\sigma_f}{CFO}\left(1 - \frac{1}{\sqrt[5]{500}}\right)\right] = 0.8577 CFO = 807 \text{ kV} \tag{112}$$

$$\frac{\sigma_{fn}}{CFO_n} = 0.0168$$

Then

$$\text{SSFOR} = \frac{1}{2}\left[1 - F\left(\frac{807 - 642}{81.93}\right)\right] = \frac{1}{2}[1 - F(2.015)] = 1.10/100 \tag{113}$$

Using a computer program, the SSFOR = 1.50/100, and Brown's method gives a SSFOR of 1.26/100. Thus it appears that the method provides a good approximation. The advantage of the method is that the CFO_n can be directly calculated without using a Gaussian table of probabilities. However to calculate the SSFOR, a Gaussian table is necessary. So, it is a little help.

As discussed in Section 7.2, Brown's method essentially neglects the σ_{fn}, that is, it assumes that $\sigma_{fn} \lll \sigma_0$ and thus can be neglected. This is normally true as demonstrated by the previous example where σ_{fn} is only 1.7% of the CFO_n or 13.6 kV, where σ_0 is about 81 kV. If σ_{fn} is neglected, then the value of Z is 2.043, which gives an SSFOR of 1.26/100, Brown's value.

Thus the method, which is called the IEC method, is viable. However, Brown's method is also viable, and it may also be used for distribution other than the Gaussian. That is, either may be used when the SOV distribution is Gaussian.

For values of E_S/E_R less than 1.00, Brown's equation for the equivalent number of towers n_e can be used with the IEC method. For the same parameters as before

Phase-Ground SOVs, Transmission Lines

except that $E_S/E_R = 0.9$, the IEC method results in a SSFOR of 0.63/100, while a computer program shows a SSFOR of 0.72/100, and Brown's method gives a SSFOR of 0.59/100. Thus in this case all three methods agree.

13.4 Estimating the Strike Distance

In this chapter, the estimate of the strike distance was performed using Brown's method, which was placed in terms of K_G and K_f. The IEC method may also be placed in terms of these quantities. That is,

$$K_f = \frac{1 - 3(\sigma_f/\text{CFO})}{1 + Z_f(\sigma_f/\text{CFO})}$$

$$Z_f = -4\left[1 - \frac{1}{\sqrt[5]{n}}\right]$$ (114)

$$K_G = 1 - \frac{\sigma_0}{E_2}\left[2.054 - Z_e\sqrt{1 + \left(\frac{\sigma_{fn}}{\sigma_0}\right)^2}\right]$$ (115)

Where Z_e is from Table 6. To use this information, the value of σ_{fn} must be known, which is given by Eq. 102. To illustrate, assume the desired SSFOR is 1/100 and from Table 6, $Z_e = 2.0542$. Also assume the data in the last example, $E_2 = 808\,\text{kV}$, $\sigma_0/E_2 = 0.10$, $\sigma_f/\text{CFO} = 0.05$, and $E_S/E_R = 1.00$. For this example let $n = 200$. To obtain a solution, the equation for K_G must be iterated, since the value of σ_{fn} depends on the CFO. First assume that $\sigma_{fn} = 0$, then iterate per the following table. For this case,

$$Z_f = -2.6137; \quad K_f = 0.9778; \quad \text{and} \quad \sigma_{fn} = 0.34657\,\sigma_f \quad (116)$$

σ_{fn}	K_G	V_3/E_2	V_3	CFO	σ_{fn}
0	1.0000	0.9778	790	930	16.1
16.1	1.0040	0.9817	793	933	16.2

Another method is to set Z_e of Eq. 112 to the value given in Table 6, i.e., 2.0542. Thus

$$2.0542 = \frac{\text{CFO}_n - \mu_0}{\sqrt{\sigma_{fn}^2 + \sigma_0^2}}$$

$$\text{CFO}_n = \mu_0 + 2.0542\sqrt{\sigma_{fn}^2 + \sigma_0^2}$$ (117)

From the last example, $\sigma_0 = 80.8$, and $\mu_0 = 642$. Thus

$$\text{CFO}_n = 0.8693\,\text{CFO} \quad (118)$$

First assume that σ_{fn} is zero, then

σ_{fn}	CFO_n	CFO	σ_{fn}
0	808	929	16.1
16.1	811	933	16.2

For either method, the CFO is 933. For $k_g = 1.2$, $S = 2.37$ m. Brown's method gives 2.38 m, and the computer program gives 2.40 m. Thus both approximation methods are viable.

14 REFERENCES

1. A. R. Hileman, P. R. LeBlanc, and G. W. Brown, "Estimating the Switching Surge Performance of Transmission Lines," *IEEE Trans. on PA&S*, Sept./Oct. 1970, pp. 1455–1456.
2. T. H. Frick, J. R. Stewart, A. R. Hileman, C. R. Chowaniec, and T. E. McDermott, "Transmission Line Design at High Altitude," *IEEE Trans. on PA&S*, Dec. 1984, pp. 3672–3680.
3. K. J. Lloyd, and L. E. Zaffanella, "Insulation for Switching Surges," Chapter 11 of Transmission Line Reference Book—345 kV and Above, 2d ed. Palo Alto, CA: Electric Power Research Institute, 1982.
4. G. W. Brown, "Designing EHV Lines to a Given Outage Rate—Simplified Techniques," *IEEE Trans. on PA&S*, Mar./Apr. 1978, pp. 379–383.
5. A. R. Hileman, "Weather and Its Effects on Air Insulation Specifications," *IEEE Trans. on PA&S*, Oct. 1984, pp. 3104–3116.
6. Charles Lipson and Narendra J. Sheth, *Statistical Design and Analysis on Engineering Experiments*, McGraw-Hill, 1973.
7. Roy Billington, *Power System Reliability Evaluation*, Gordon and Breach, 1970.
8. E. J. Yasuda and F. B. Dewey, "BPA's New Generation of 500 kV Lines," *IEEE Trans. on PA&S*, Mar./Apr. 1980, pp. 616–624.
9. G. N. Alexandrov, "Methods of Choosing Insulation for Distribution Networks with the View to Reliable Operation in the Presence of Internal Voltage Surges," *Energetika*, No. 7, 1962, pp. 16–24.
10. J. Elovaara, "Risk of Failure Determination of Overhead Line Phase-to-Earth Insulation under Switching Surges," *ELECTRA*, Jan. 1978, pp. 69–87.
11. IEC Standard 71-2, "Insulation Coordination – Part 2: Application guide," 1996-12.
12. CIGRE Working Group 13.02, "Switching Overvoltages in EHV and UHV Systems with Special Reference to Closing and Reclosing Transmission Lines," *ELECTRA*, Oct. 1973, pp. 70–122.

15 PROBLEMS

1. Using the simplified method, determine the strike distances and insulator length for the center and outside phases of a 500-kV, 550-kV maximum transmission line for the following conditions (*note*: 1 per unit = 450 kV). Use the estimating method as described in this chapter. Following this, check the answer with the appropriate computer program.

1. V-string insulators on all phases.
2. Switching surge wave fronts are equal to the critical wave front.
3. Line length is 200 km with three towers per km.
4. Line altitude is 1000 m.
5. Gaussian SOV distribution with E_2 of 1.8 per unit, σ_0/E_2 of 0.07, and $E_S/E_R = \gamma = 0.88$. the maximum switching overvoltage occurs at $E_2[1 + 2(\sigma_0/E_2)]$.
6. $\sigma_f/CFO = 0.05$, and assume wet conditions, decrease the CFO by 4%.
7. The line is to be designed for an SSFOR of one flashover per 100 switching operations.
8. Height of the phase conductor is 20 m, and the tower width is 1.8 m.

2. Assume a line has been designed for a SSFOR of 1 flashover per 100 breaker reclose operations. Also assume that the line has lightning flashover rate of 0.5 flashovers per 100 km-years and that the line length is 200 km. Find the storm outage rate SOR (1) if only a single breaker reclose operation is permitted and (2) if two reclosures are used.

3. The SOV distribution at the line entrance to a 230 kV, 242 kV maximum station may be approximated by an extreme value positive skew distribution with an E_2 of 2.6 per unit and a β/E_2 of 0.09. Between the line entrance tower and the opened breaker there exists an equivalent of 10 post insulators. Determine the SSFOR for these 10 post insulators assuming that the CFO of a single post insulator is 644 kV and σ_f/CFO is 0.07. Assume the SOVs are equal on all insulators and that 1 per unit is 198 kV (*Note*: the CFO of 644 kV is an estimate for a 900 kV BIL post insulator). Use the estimating method as described in this chapter, then check the answer with the appropriate computer program.

4. Determine the crossarm lengths A_1 and the crossarm separation distance Y for the 345 kV, 362 kV maximum tower shown in Fig. 35. Design the line for a SSFOR of 1 flashover per 100 breaker operations. The arm length for phase C must be 1.0 m longer than the arm length for phase A. The insulator string must contain a minimum of 18 insulators ($5\frac{3}{4} \times 10$ inches or 146×254 mm). Use the following data:

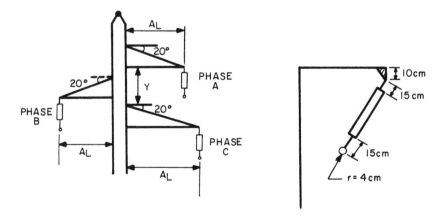

Figure 35 A 345-kV tower of Problem 4.

1. Gaussian SOV distribution with E_2 of 2.8 pu, σ_0/E_2 of 0.07, and E_S/E_R of 0.90. Assume all wave fronts are equal to the critical wave front.
2. Altitude = 2000 m, tower width = 1.3 m, I-string insulators on all phases, $\sigma_f/\text{CFO} = 0.05$, 100-hour wind speed = 60 km/h, β for wind = 1.9, $D/W = 1.3$, $V/H = 1.0$; assume all conductor heights = 15 m, number of towers = 500.

Use the estimating method as presented in this chapter, then check the answer with the appropriate computer program.

5. Determine the SSFOR of a 500 kV line, 550-kV maximum, for the following conditions (assume 1 pu = 450 kV). Use the estimating method as described in this chapter, then check this with the appropriate computer program.

1. Single-circuit, horizontal-phase configuration, having a strike distance of 2.6 m for both the center and outside phases. Altitude is 1000 m. Phase conductor height is 20 m, and the tower width is 1.5 m. Number of towers = 100. $\sigma_f/\text{CFO} = 0.05$.
2. Gaussian SOV distribution, $E_2 = 2.0$ pu, $\sigma_0/E_2 = 0.10$ at the opened end of the line. SOV profile: $E_S/E_R = 0.90$.

6. Calculate the SSFOR for a 500-kV line (1 pu = 450 kV) for (1) 1 tower and (2) 200 towers for the following conditions:

1. CFO = 900 kV, $\sigma_f/\text{CFO} = 0.05$, flat SOV profile: $E_S/E_R = 1.00$.
2. SOV distribution given by the following table.

V, pu	No. of observations	V, pu	No of observations
1.2	1	1.6	21
1.3	15	1.7	10
1.4	20	1.8	5
1.5	27	1.9	1

Use the above table directly. Do not use the data to approximate or determine the SOV continuous SOV distribution. That is, assume that the probability of occurrence of 1.2 pu is 0.01.

4
Phase–Phase Switching Overvoltages, Transmission Lines

1 INTRODUCTION

For phase–ground insulation coordination, the objective is either to determine the phase–phase SSFOR or, given the SSFOR, to determine the phase–phase strike distance or clearance. In this chapter, the application to transmission lines is considered, and in Chapter 5, the application to stations is studied. However, not all transmission line designs require a phase–phase specification. For towers where grounded tower members are located between phases, i.e., today's normal tower, flashover will occur from phase to ground before occurring phase–phase. That is, the phase–ground strike distance is controlling. However, for some towers, such as those of Fig. 1, only air separates the phase conductors, and flashovers may occur either to ground or between phases. In this case the phase–phase strike distance and SSFOR must be considered. Fig. 1A is a low-profile double-circuit tower, which is frequently referred to as a German Delta and is used throughout Europe and by some USA utilities. The tower of Fig. 1B is of recent design and is known as the Chainette, first used for the 765 kV line in Canada. The tower of Fig. 1C is only a conceptual design.

Phase–phase insulation naturally occurs in stations—between buses and from one piece of equipment to another. Therefore the phase–phase strike distance and SSFOR should always be considered in station insulation coordination studies. This special situation is considered in Chapter 5.

Because the application of phase–phase insulation coordination methodology is primarily dependent on the description of the phase–phase insulation characteristic, the insulation strength is discussed first. As noted in Chapters 1 and 2, the phase–phase strength was not discussed there, under the premise that it was best left to this chapter where it is needed and first used.

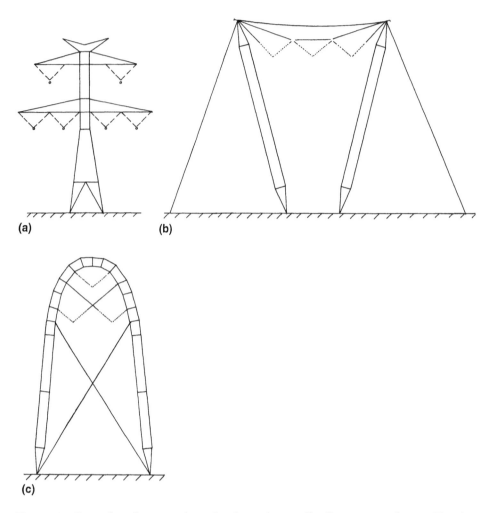

Figure 1 Examples of towers where the phase-phase strike distance must be considered.

2 PHASE–PHASE INSULATION STRENGTH

Insulation strength characteristics for self-restoring insulation can be obtained by a setup as illustrated in Fig. 2. This single-tower setup can be expanded into one or more spans to simulate a transmission line. A negative voltage V^- is applied to one of the conductors and a positive voltage V^+ to the other conductor, resulting in a phase–phase voltage of V_p. The negative voltage is considered as a positive quantity, so that

$$V_p = V^+ + V^- \tag{1}$$

To develop the methodology of test, assume first that the strike distance to ground S_g is much larger than that between the conductors, so that only flashovers between

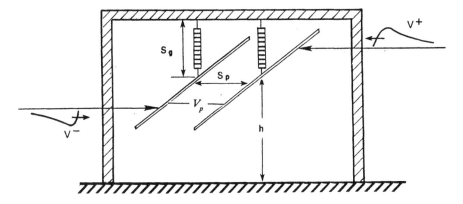

Figure 2 Opposite polarity switching impulses are applied to each conductor.

the conductors can occur. Two methods of determining the phase–phase insulation strength exist, the alpha method and the V^+–V^- method.

2.1 The Alpha Method

The applied voltages to the conductors are

$$V^+ = (1-\alpha)V_p \qquad V^- = \alpha V_p \tag{2}$$

and α is held constant throughout the test. Therefore

$$\alpha = \frac{V^-}{V^+ + V^-} = \frac{V^-}{V_p} \tag{3}$$

The voltage V_p is varied, and the Gaussian strength characteristic is obtained, as shown in Fig. 3. Thus the CFO is the phase–phase CFO or CFO_p, and the standard deviation is denoted as σ_{fp}. Although α may be varied between zero and essentially

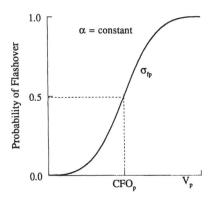

Figure 3 Obtaining the insulation characteristics for the α method.

one, usually the CFO_p is obtained for only two values of α, 0.33 and 0.5, that is, where $V^+ = 2V^-$ and $V^+ = V^-$, respectively, or

$$V^+ = \frac{1-\alpha}{\alpha} V^- \tag{4}$$

The CFO_p is given by the equation

$$CFO_p = k_{gp} \frac{3400}{1 + (8/S_p)} \tag{5}$$

where S_p is the phase–phase strike distance and k_{gp} is the gap factor. As illustrated in Fig. 4, plotting the CFO_p or k_{gp} as a function of α results in an approximate linear relationship [1].

The gap factors as obtained by Gallet et al. [1] are presented in Table 1 for five types of gaps. The value of σ_{fp}/CFO_p was 5% except for the conductor-to-conductor gap, where this value decreased to 3.5% for a 10 m length of conductor. For "long" conductor lengths, the authors suggest decreasing this coefficient of variation to 2%. However, the gap factor should also be reduced on the basis that several 10 m lengths form, for example, a 400 meter span. Using a σ_{fp}/CFO_p of 3.5%, the gap factor for a 400 m span should be about 1.50 for α = 0.5 and 1.41 for α = 0.33. The calculated value of σ_{fp}/CFO_p is about 0.019, thus verifying the authors' value. The results per Ref. 1 of Table 1 apply for gap distances of 3 to 8 m and for heights of 8.2 to 9.0 m. The authors found that the critical wave front was between 25 and 30 times the strike distance.

In testing a 360 meter conductor–conductor gap, having a height of 26 meters, gap factors of about 1.68 and 1.54 were obtained length for α = 0.5 and 0.33, respectively [2]. σ_{fp} was about 3% of the CFO_p. Although rain or wet conditions do not significantly decrease the strength of air gaps, if vertical or I-string insulators are present, the phase–phase strength decreases by about 12% [2].

Recent tests on a jumper-to-shielding-ring and a short-length-conductor-to-conductor gap resulted in gap factor of 1.68 and 1.57 for α = 0.5 and 0.33, respectively, where σ_{fp} was 4 to 5% of the CFO_p [3]. These values are also shown in Table 1.

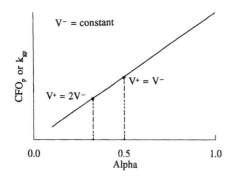

Figure 4 The CFO_p as a function of α.

Table 1 Gap Factors for Phase–Phase Switching Impulses

	Alpha method			V^+–V^- Method		
Gap configuration	α	k_{gp}	σ_{fp}/CFO_0	K_L	K_{GP}	σ_{FP}/CFO_0
Ring–ring or large smooth	0.33	1.70	0.05	0.70	1.53	0.05
electrodes [1]	0.50	1.80	0.05			
Crossed conductors [1]	0.33	1.53	0.05	0.62	1.34	0.05
	0.50	1.65	0.05			
Rod–rod [1]	0.33	1.52	0.05	0.67	1.35	0.05
	0.05	1.62	0.05			
Conductor–conductor,	0.33	1.52	0.035	0.67	1.35	0.035
10 m length [1]	0.50	1.62	0.035			
Conductor–conductor,	0.33	1.41	0.02	0.68	1.26	0.02
400 m length	0.50	1.50	0.02			
Supported busbar fittings [1]	0.33	1.40	0.05	0.65	1.23	0.05
	0.50	1.50	0.05			
Asymmetrical geometries, e.g.,	0.33	1.36	0.05	0.67	1.21	0.05
rod–conductor [1]	0.50	1.45	0.05			
Jumper-shield ring and	0.33	1.57	0.04	0.66	1.39	0.04
conductor–conductor (short	0.50	1.68	0.04			
length) [3]						
Conductor–conductor	0.33	1.56	—	0.68	1.40	—
(46 m length) [4]	0.50	1.66	—			
Conductor–conductor	0.33	1.54	0.03	0.65	1.38	0.03
(360 m length) [2]	0.50	1.68	0.03			

Grant and Paulson tested a conductor–conductor gap with conductor lengths from about 46 to 365 meters [4]. They found only small effects of height and length of the conductor (doubling the height increases the CFO_p by 1.3%, doubling the length decreases the CFO_p by 1.4%). The estimated gap factor equation was

$$k_{gp} = 1.36 + 0.6(\alpha) \tag{6}$$

which provides gap factors of 1.56 and 1.66 for $\alpha = 0.33$ and 0.5, respectively, applicable for heights of 9 meters and conductor lengths of 46 meters. These values are also shown in Table 1. Interestingly, these authors suggested that the strength characteristic could be modeled as a Weibull distribution, a distribution adopted by IEC [5].

2.2 The V^+-V^- Method

In this method, the V^- component is held constant and V^+ is varied to obtain the Gaussian strength characteristic of Fig. 5. Thus the CFO for the positive impulse CFO^+ is obtained and is a function of V^-. If the CFO^+ is plotted versus V^-, the linearly falling characteristic of Fig. 6 is obtained and therefore

$$CFO^+ = CFO_0 - K_L V^- \tag{7}$$

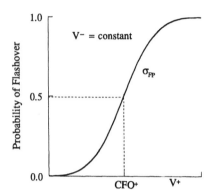

Figure 5 Obtaining the insulation characteristic for the V^+–V^- method.

where CFO_0 is the CFO when the negative component is zero and K_L is a constant dependent on the gap configuration. CFO_0 is given by the equation

$$CFO_0 = K_{GP} \frac{3400}{1 + (8/S_p)} \tag{8}$$

The standard deviation, σ_{FP}, in per unit of CFO_0 for all single or simple gaps, ranges from 4 to 9% [6.2] and averages 6%. However, for long-length conductor–conductor gaps, σ_{FP} is lower, from 2 to 3% of the CFO_0. Cortina et al. [7] tested 500-m lengths of a conductor-to-conductor gap with heights from 10 to 18 m and strike distances between 7 and 10 m, using positive polarity wave fronts of 300 µs. σ_{FP}/CFO_0 was between 2 and 3%. The authors suggested the following equations for CFO_0 and K_L:

$$CFO_0 = 640 S_p^{0.6}\left(1 - \frac{S_p}{4h}\right)$$
$$K_L = 0.86 - 0.54\left(\frac{S_p}{h}\right) \tag{9}$$

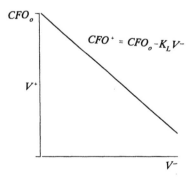

Figure 6 The CFO^+ is dependent on V^-.

Equation 9 for the CFO_0 may be changed to

$$CFO_0 = K_{GP} \frac{3400}{1 + (S_p/4h)} \tag{10}$$

where

$$K_{GP} = 1.30[1 - (S_p/4h)] \tag{11}$$

Another estimate of K_L may be obtained from Ref. 6.2. The limits of the data shown in [6.2] are shown by the two lines in Fig. 7 along with the value of K_L as obtained from Eq. 9. These data lead to other equations for K_L, i.e.,

$$K_L = 1.0 - 0.8\left(\frac{S_p}{h}\right) \tag{12}$$

and

$$K_L = 0.8 - 0.8\left(\frac{S_p}{h}\right) \tag{13}$$

for the maximum and minimum values, respectively.

2.3 Correlating the Methods

Because the alpha method was first employed, many test results are available using this method. However, these test results may be transcribed or changed so as to apply to the other method. That is

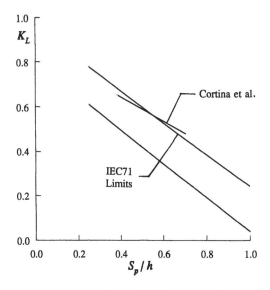

Figure 7 Data from IEC 71 leads to alternate equations.

$$\text{CFO}_p = \text{CFO}^+ + V^- = \text{CFO}_0 + (1 - K_L)V^- \qquad (14)$$
$$V^- = \alpha \text{CFO}_p$$

Then

$$\text{CFO}_0 = [1 - \alpha(1 - K_L)]\text{CFO}_p \qquad (15)$$

which upon substituting for the gap factors becomes

$$K_{GP} = k_{gp}[1 - \alpha(1 - K_L)] \qquad (16)$$

Knowing two values of α and k_{gp}, the values of K_L and K_{GP} can be determined. The resultant gap factors for the two methods are presented in Table 1. As noted, the values of σ_{FP}/CFO_0 are assumed equal to values of σ_{fp}/CFO_p; more about this later.

2.4 Phase–Phase and Phase–Ground

In the practical case, a phase–phase insulation always exists with a phase-to-ground insulation. For example, for a transmission line, there exists a strike distance to ground and a strike distance between phases. This general case may be easily portrayed by the V^+–V^- diagram as illustrated in Fig. 8. The diagram is composed of three straight line segments: (1) the horizontal line at the positive polarity phase-to-ground CFO, CFO_g^+, (2) the linearly falling line representing the phase–phase CFO, CFO^+, and (3) the vertical line representing the negative polarity CFO to ground, CFO_g^-. For low and high values of V^-, flashover to ground is more probable than

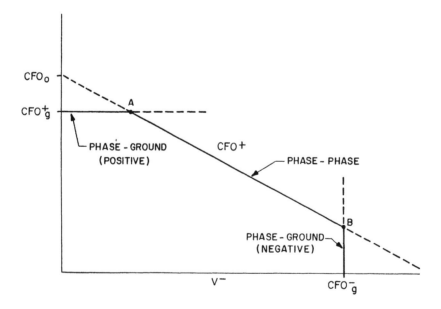

Figure 8 Switching impulse strength characteristics for phase-phase and phase-ground.

flashover between phases. However, between these high and low values of V^-, flashover between phases is most probable.

As noted from this portrayal, phase–phase flashover can be essentially eliminated if the phase–ground strike distance is sufficiently lowered—or the phase-to-ground flashover can be essentially eliminated if the phase-to-ground strike distance is increased. Thus the V^+–V^- method can handle both phase–phase and phase–ground insulations simultaneously.

The diagram of Fig. 8 is general in nature in that it can also be used to portray the insulation strength of internal gas or solid insulation such as is employed in a transformer, GIS, or cable. In this case, the strength of the insulation is only dependent on the magnitude of the phase–phase voltage and not on the division of the phase–phase voltage into positive and negative components. Therefore $K_L = 1.0$, and from Eq. 15, the phase–phase CFO, CFO_p, is equal to CFO_0.

In general, for small air gaps of less than about 2 to 3 meters, i.e., for nominal system voltages less than 500 kV, K_L is also equal to one. Thus only for large air clearance does the separation of the phase–phase voltage into separate components become important, and for this case K_L is less than one.

Effect of S_p on CFO_g

In Ref. 6.2, the authors evaluate the effect of the strike distance between phases on the positive CFO to ground, CFO_g^+. The indicate that as S_p/h decreases below 1.0, the CFO_g^+ decreases. For example, if S_p/h is 0.5, then the CFO_g^+ decreases by about 20 to 25%.

Using the data from Ref. 6.2 for the conductor–conductor gap when $V^- = 0$ and all flashovers are to ground, Table 2 indicates that the CFO_g^+ remains constant for S_p/h of 0.89 and 0.23. that is, the phase–ground gap factor remains constant at 1.17.

The authors also consider the phase–phase gap called screen to screen with a phase–ground gap to pedestals with a pedestal height of 2.5 meters. For $V^- = 0$, all flashovers are to ground. From the CFO_g^+, the phase–ground gap factor is found. Then using Eq. 32 of Chapter 2 with $A = 0$, the phase–ground gap factor is calculated. Since the calculated and actual gap factor are equal, the conclusion is that the CFO_g^+ is not affected by the values of S_p/h of from 0.81 to 1.0; see Table 3.

Thus the overall conclusion is that using the presently available data, the CFO_g^+ is unchanged for S_p/h as low as 0.23. It is to be hoped that future laboratory studies will closely study this issue.

Table 2 Effect of S_p/h on CFO_g^+ on Conductor–Conductor-to-Ground Gap

h, meters	S_p, meters	S_p/h	CFO_g^+, kV	k_g
9	4	0.89	2120	1.17
11	10	0.23	2300	1.17

Table 3 Effect of S_p/h on CFO_g^+ for Screen–Screen-to-Grounded-Pedestal Gap

h, meters	S_p, meters	S_g, meters	S_p/h	CFO_g^+, kV	k_g	Calc k_g
7	7	4.5	1.0	1810	1.48	1.41
9	9	6.5	1.00	1524	1.34	1.33
11	9	8.5	0.81	1752	1.29	1.30

Effect of the Third Phase

To simplify the presentation, only two phases were considered. The effect of the third phase is minor, since for the higher SOVs the voltage on this phase will be significantly less than that on the negative polarity second phase [6.4].

Effect of Time Delay Between Times to Crest

One other point needs some discussion. The data presented assumes that the two impulse waveshapes have the same time to crest and are synchronized so that they are applied at the same instant. That is, viewing Fig. 9, ΔT is zero. If the positive voltage precedes the negative voltage as in Fig. 9, ΔT has no effect, i.e., there is no decrease in the CFO. However, if the negative voltage precedes the positive voltage, a decrease may occur, this decrease being from 10 to 15% if ΔT is several milliseconds. For the higher SOVs which are of primary importance, ΔT is small or nonexistent, and therefore this decrease is normally not considered [6.4, 8].

The Value of K_L

For transmission lines there exist several values of S_p/h. Since h is highest at the tower, the minimum value occurs at this location, whereas since h is lowest at the midspan, the maximum values occur there. For present designs of 500- and 765-kV transmission lines at the tower, S_p/h ranges from about 0.23 to 0.43 with an average of about 0.40. At the midspan, the ratio S_p/h ranges from about 0.70 to 1.45 and averages about 0.95. Using Eqs. 9 and 12, K_L at the tower is about 0.64 to 0.68 and at the midspan is 0.24 to 0.34.

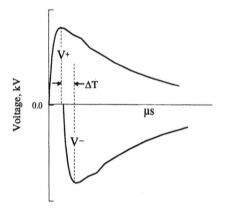

Figure 9 Non-synchronous impulses may decrease the CFO.

For present substations S_p/h is about 1.2, although future substations may decrease this to about 0.80. Also, S_p/S_g is about 2.0 for present stations but may decrease to 1.4 for future substations. If Eqs. 9 and 12 are used, K_L becomes less than zero, an unrealistic value.

In view of these inconsistencies and those regarding the effect of S_p/h on the CFO to ground, it is recommended that K_L be selected from Table 1, i.e., independently of S_p/h. It goes without saying that for any future installation, tests should be made, since the values of Table 1 are to be considered as estimates. From Table 1, for transmission lines whose span lengths are in the range of 300 to 400 meters, the average value of K_L is 0.68, the average value of σ_{FP}/CFO_0 is 0.02, and the average value of K_{GP} is about 1.26. These values are recommended for use for transmission lines.

More on the CFO_0

When V^- is equal to zero, the CFO_0 is the phase–phase CFO with the second phase grounded. That is, assume a conductor–conductor-to-ground arrangement. $V^- = 0$ means that one of the conductors is grounded. Thus the CFO_0 is the same as a phase-to-ground CFO of a conductor-to-grounded conductor gap. Thus the value of K_{GP} is equal to k_g. For example, for a rod–rod phase–phase arrangement, Table 1 shows that K_{GP} is 1.35. From Eq. 34 of Chapter 2, for $h' = 0$

$$k_g = 1.35 - \left(\frac{S_p}{h} - 0.5\right) \tag{17}$$

which is practically identical to K_{GP}. This attribute or equality is used in IEC Standard 71-2 [5], which is discussed in Chapter 5. This also indicates that σ_f/CFO_g^+ is equal to σ_{FP}/CFO_0, which is assumed in Table 1.

3 PHASE–PHASE SWITCHING OVERVOLTAGES

As for phase–ground overvoltages, there exist three phase–phase overvoltages. For simplicity of presentation, two of these are shown in Fig. 10. At each instant in time, the insulation is stressed by a different overvoltage. Therefore, ideally or theoretically, the insulation strength and the probability of flashover should be evaluated at each instant in time. From this, the total probability of flashover is 1 minus the probability of no flashover at each of the time instants.

To circumvent this laborious procedure, only two time instants are usually considered: (1) the time T^+, at which the maximum positive ground–ground voltage occurs and (2) time T_{12}, at which time the maximum phase–phase voltage occurs. The SSFOR is then calculated for each of these time instants and the larger SSFOR is used.

In general, for air clearances greater than bout 3 meters (500-kV systems and above), the time of the maximum positive SOV is more severe, whereas for non-self-restoring insulations, such as transformer insulation, cables, for GIS, and for small air clearances (below 500 kV systems), the time for the maximum phase–phase SOV is more severe.

At present, the digital transient programs, EMTP or ATP, are not equipped to obtain these data conveniently. That is, the programs easily obtain the maximum phase–ground voltage at T^+ and the maximum phase–phase voltage at T_{12}.

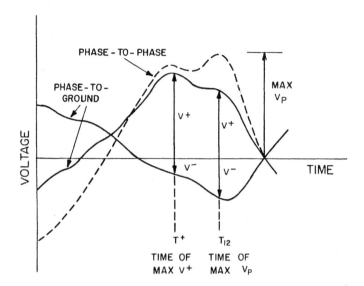

Figure 10 The voltages at two time instants, T^+ and T_{12} are considered to evaluate the SSFOR.

Therefore engineers are prone to collect this mixture of data. That is, the maximum phase–phase SOVs are collected, i.e., using time T_{12}, and the maximum phase–ground SOVs are collected using time T^+. Again, collecting the data in this manner is incorrect. However, this method is inherently conservative since it involves collection of the "worst" data at each time instant.

Further, in many cases, the SOV densities for the higher SOVs, which are of primary interest, are sufficiently close for both time instants so that, as an approximation, either time instant may be employed.

For further development of the methodology, the simple assumption is made that the SOV distribution is composed of voltages V^+ and V^- or voltages V^+ and V_p—and that these are obtained at one of the two time instants, or in any manner.

Since there are three voltages of interest, V_p, V^+, and V^-, any two of these can be collected, and from the distribution of the two voltages, the distribution of the third can be obtained. Today the trend is to collect the V^+ data, since it is necessary for evaluation of the SSFOR for phase–ground insulations and to collect also the V_p data. The other option of collecting the V^+ and V^- data is infrequently employed.

In most cases, the Gaussian or normal distribution is used to approximate the random values of SOVs. In this case, the following sections apply.

3.1 Gaussian SOV Distributions, V_p and V^+

Given

$$f(V_p) = N(\mu_p, \sigma_p)$$
$$f(V^+) = N(\mu^+, \sigma^+) \tag{18}$$

$$\rho_{p^+} = \text{correlation coefficient between } V_p \text{ and } V^+$$

then, as for the strength, assuming V^- as a positive voltage,

$$\begin{aligned} V_p &= V^+ + V^- \\ f(V^-) &= N(\mu^-, \sigma^-) \\ \mu^- &= \mu_p - \mu^+ \\ \sigma^- &= \sqrt{(\sigma^+)^2 + (\sigma_p)^2 - 2\rho_{p+}\sigma^+\sigma_p} \end{aligned} \tag{19}$$

Note that if $\rho_{p+} = 1.00$, then

$$\sigma^- = |\sigma^+ - \sigma_p| \tag{20}$$

3.2 Gaussian SOV Distributions, V^+ and V^-

Given

$$\begin{aligned} f(V^+) &= N(\mu^+, \sigma^+) \\ f(V^-) &= N(\mu^-, \sigma^-) \\ \rho_{+-} &= \text{correlation coefficient between } V^+ \text{ and } V^- \end{aligned} \tag{21}$$

Then

$$\begin{aligned} V_p &= V^+ + V^- \\ f(V_p) &= N(\mu_p, \sigma_p) \\ \mu_p &= \mu^+ + \mu^- \\ \sigma_p &= \sqrt{(\sigma^+)^2 + (\sigma^-)^2 + 2\rho_{+-}\sigma^+\sigma^-} \end{aligned} \tag{22}$$

Note that if ρ_{+-} is equal to 1.00, then

$$\sigma_p = \sigma^+ + \sigma^- \tag{23}$$

3.3 The Correlation Coefficients

The relationship between the correlation coefficients can be obtained by use of the previous equations for σ^- and σ_p. The resulting equations are

$$\rho_{+-} = \frac{\rho_{p+}\sigma_p - \sigma^+}{\sigma^-} \qquad \rho_{p+} = \frac{\sigma^+ + \rho_{+-}\sigma^-}{\sigma_p} \qquad \rho_{p-} = \frac{\sigma^- + \rho_{+-}\sigma^+}{\sigma_p} \tag{24}$$

Note that if $\rho_{+-} = 1.00$, then $\rho_{p+} = 1.00$ and $\rho_{p-} = 1.00$.

3.4 Estimating the Value of E_{2p}

To date, sufficient studies have not been performed to estimate accurately the parameters of the phase–phase switching overvoltage distribution. However, from IEC 71 [5], the value of the statistical phase–phase SOV, E_{2p}, can be estimated from the value of the statistical phase–ground SOV, E_2. The ratio E_{2p}/E_2 only varies from about 1.60 for an E_2 of 2.0 per unit to about 1.50 for an E_2 of 3.0 per unit. Thus a ratio of 1.55 appears reasonable for all values of E_2. Thus for an E_2 of 1.8 and 2.8 per unit, E_{2p} is 2.8 and 4.3 per unit, respectively. Also, in general, the values of σ_p/E_{2p} are equal to the values of σ_0/E_2 (or σ^+/E_2^+). The maximum phase–phase SOV is approximately 1 to 2 standard deviations above E_{2p}. In addition, the values of the voltage profile factor γ_p are similar to those for phase–ground.

4 CALCULATION OF THE PHASE–PHASE SSFOR

4.1 Phase–Phase SSFOR

Properly and theoretically, the SSFOR of the usual insulation system, which is composed of both phase–ground and the phase–phase, should be calculated as a single system. To develop the equations for this calculation, first consider only the phase–phase SOVs and their insulation strength. That is, neglect the phase–ground SOVs and the phase–ground insulation strength. The phase–phase SSFOR, denoted as SSFOR_p is

$$\text{SSFOR}_p = \frac{1}{2}\int_{-\infty}^{+\infty}\left[\int_{-\infty}^{+\infty} p_p[f(V^+|V^-)]\,dV^+\right]f(V^-)\,dV^- \quad (25)$$

This formidable equation may be visualized by use of Fig. 11. Figure 11 is a three-dimensional drawing since there are two random variables, V^+ and V^-. The joint distribution of these variables is contained within this space. The phase–phase insulation CFO, CFO$^+$, is given by the linear line. At any value of V^-, a straight dotted line is drawn parallel to the V^+ axis. The SOV conditional density function of $V^+|V^-$ exists along this line. As shown by the inset figure, the conditional density and the strength characteristic p_p are convolved to obtain the conditional probability of flashover, i.e., the probability of flashover or the SSFOR given the negative voltage V^-. In equation form,

$$\text{SSFOR}_p|V^- = \int_{-\infty}^{+\infty} f(V^+|V^-)p_p\,dV^+ \quad (26)$$

where p_p is the probability of a phase–phase flashover.

Next, this value of $\text{SSFOR}_p|V^-$ is multiplied by the density $f(V^-)$ and placed on the V^- axis. This process is repeated for other values of V^- until a curve is established. The area under this curve is twice the value of SSFOR_p. By equation

$$\text{SSFOR}_p = \frac{1}{2}\int_{-\infty}^{+\infty}[\text{SSFOR}_p|V^-]f(V^-)\,dV^- \quad (27)$$

which is identical to Eq. 25.

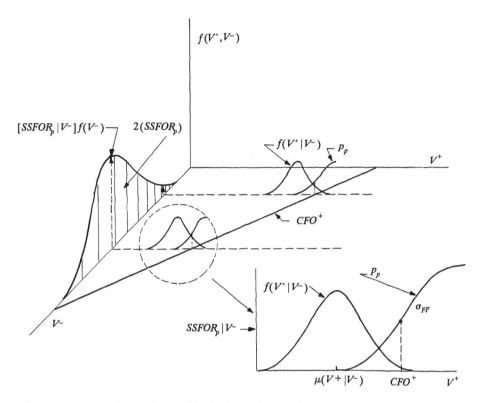

Figure 11 Calculating the combined phase-phase and phase-ground SSFOR.

4.2 Combined Phase–Phase and Phase–Ground SSFOR

The process of considering both the phase–phase and the phase–ground insulations simultaneously is illustrated in Fig. 12 and is similar to that described in Fig. 11 except that the combinations of probabilities of flashovers to ground and flashovers between phases must be considered. Referring to the inset of Fig. 12, for any voltage V^+ the probability of a phase–phase flashover, a phase-to-ground flashover, or both is $[1 - (1 - p_p)(1 - p_g)]$ or $[1 - q_p q_g]$ where p_p is the probability of a phase–phase flashover and p_g is the probability of a phase–ground flashover. As before, the qs are the probabilities of no flashovers. Thus the SSFOR becomes

$$\text{SSFOR}_p = \frac{1}{2} \int_{-\infty}^{+\infty} \left[\int_{-\infty}^{+\infty} [1 - q_p q_g] f(V^+|V^-) \, dV^+ \right] f(V^-) \, dV^- \quad (28)$$

Even this equation is not totally complete since it applies to only a single insulation, whereas lines are composed of 100 to 1000 towers or spans. Therefore, as for the phase–ground insulations in Chapter 3, assuming there are n parallel insulations and that the SOVs may differ at each tower or span, the SSFOR becomes

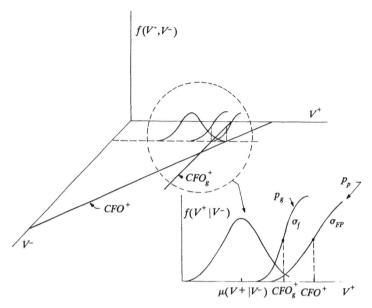

Figure 12 Calculating the combined phase-phase and phase-ground SSFOR.

$$\text{SSFOR}_p = \frac{1}{2} \int_{-\infty}^{+\infty} \left[\int_{-\infty}^{+\infty} \left[1 - \prod_{i=1}^{n} q_{pi} q_{gi} \right] [f(V^+|V^-)] \, dV^+ \right] f(V^-) \, dV^- \qquad (29)$$

Using typical values, the SSFOR was calculated for phase–phase and phase–ground separately and also combined per Eq. 27 and Eq. 29 and Eq. 20 of Chapter 3. The results are shown in Fig. 13. Per Eq. 29, the total SSFOR for both phase–phase

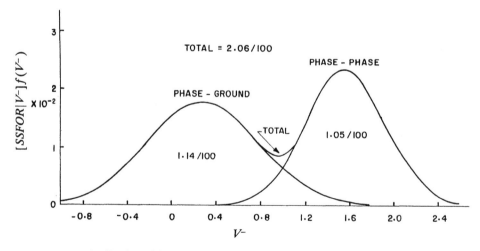

Figure 13 Distribution of flashover probabilities illustrating the overlap of the phase-phase and phase-ground SSFOR.

and phase–ground is 2.06/100, whereas the phase–phase SSFOR is 1.05/100 and the phase–ground SSFOR is 1.14/100. Thus, within a few percent, the total SSFOR is equal to the phase–phase SSFOR plus the phase–ground SSFOR. In other words, the phase–phase SSFOR and the phase–ground SSFOR can be calculated separately. It is not necessary to use Eq. 29. That is, only Eq. 27 needs to be used to find the phase–phase SSFOR. The remaining step is to simplify the procedure. But first, to complete the analysis, consider the inclusion of the negative polarity strength and the use of "reversed parameters."

4.3 Including the Phase–Ground, Negative Polarity

In Fig. 8, the strength characteristic also shows the negative polarity phase–ground CFO, CFO_g^-. To include this in the calculation, Eq. 29 must be modified to

$$\text{SSFOR}_p = \frac{1}{2} \int_{-\infty}^{+\infty} \left[\int_{-\infty}^{+\infty} \left[1 - \prod_{i=1}^{n} q_{pi} q_{gi}^+ q_{gi}^- \right] f(V^+|V^-) dV^+ \right] f(V^-) dV^- \tag{30}$$

where the probability of no flashover for negative polarity is obtained as illustrated in Fig. 14. The negative polarity strength characteristic is only a function of the negative polarity voltage, so that the probability of flashover is simply obtained for any value of V^-. As for the phase–ground SSFOR, including the negative polarity CFO is usually unnecessary, since the insulation strength for negative polarity is much greater than that for positive polarity.

4.4 Reversed Parameters

With equal likelihood, the positive SOVs can be negative and the negative SOVs can be positive. Therefore, theoretically, the SSFOR is the sum of the SSFOR with the parameters as initially given plus the SSFOR with the parameters of the SOV distributions reversed. This is seldom necessary, since the SSFOR with reversed

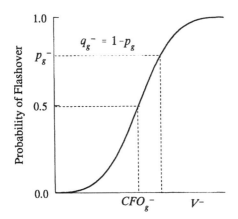

Figure 14 Including the negative polarity phase-ground strength.

5 SIMPLIFYING THE STRESS—FROM TWO VARIABLES TO ONE VARIABLE

5.1 General

Note that the equation for the strength can be rearranged as

$$\text{CFO}_0 = \text{CFO}^+ + K_L V^- \tag{31}$$

showing that the phase–phase strength as described by the CFO_0 and its standard deviation is a function of a positive voltage CFO_0 plus a negative voltage $K_L V^-$. If the phase–phase stress can be stated in terms of these same variables, i.e.,

$$V_Z = V^+ + K_L V^- \tag{32}$$

then the problem of calculating the phase–phase SSFOR can be reduced from the consideration of two random variables V^+ and V^- to the consideration of only a single random variable V_Z. That is, per Fig. 15, the density $f(V_Z)$ is convolved with the strength F_f or p_p, which is defined by the CFO_0 and σ_{FP}/CFO_0. Therefore, considering only a single insulation, the phase–phase SSFOR is

$$\text{SSFOR}_p = \frac{1}{2}\int_{-\infty}^{+\infty} p_p\, f(V_Z)\, dV_Z \tag{33}$$

For n insulations,

$$\text{SSFOR}_p = \frac{1}{2}\int_{-\infty}^{+\infty} \left(1 - \prod_{i=1}^{n} q_{ip}\right) f(V_Z)\, dV_Z \tag{34}$$

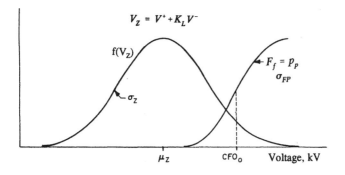

Figure 15 Using a single random variable, V_Z, to calculate the SSFOR.

This equation indicates that the stress or SOVs should be collected in terms of V_Z per Eq. 32 or since

$$V_p = V^+ + V^- \tag{35}$$

in the form

$$V_Z = (1 - K_L)V^+ + K_L V_p \tag{36}$$

To explain further, for each random switching operation, the voltages, V_p and V^+ or the voltages V^+ and V^- are measured and then combined using Eq. 32 or Eq. 36 to form V_Z. As for the phase–ground case, the data are usually approximated by a continuous distribution, and this distribution may be Gaussian (normal), extreme value, or some other distribution. Regardless of the type of distribution, Eqs. 33 and 34 apply. Further, the same simplifications and approximations used for calculating the phase–ground SSFOR, e.g., Brown's method can be used to calculate the phase–phase SSFOR.

Since the Gaussian distribution is frequently selected as the SOV distribution and also since manipulation of the Gaussian distribution is comparatively easy, the use of the Gaussian distribution is considered as a special case in the next section.

5.2 With SOVs Density as Normal or Gaussian

Assume that for each switching operation, the voltages V^+ and V^- are collected and the resulting distributions approximate Gaussian distributions. Thus the densities, $f(V^+)$ and $f(V^-)$ are normal, i.e.,

$$\begin{aligned} f(V^+) &= N(\mu^+, \sigma^+) \\ f(V^-) &= N(\mu^-, \sigma^-) \end{aligned} \tag{37}$$

Then, using Eq. 32, the density, $f(V_Z)$ is also normal with the mean and standard deviation given by the equations

$$\begin{aligned} \mu_Z &= \mu^+ + K_L \mu^- \\ \sigma_Z &= \sqrt{(\sigma^+)^2 + (K_L \sigma^-)^2 + 2\rho_{+-} K_L \sigma^+ \sigma^-} \end{aligned} \tag{38}$$

or if V_p and V^+ are collected and the densities are approximated as normal, i.e.

$$\begin{aligned} f(V_p) &= N(\mu_p, \sigma_p) \\ f(V^+) &= N(\mu^+, \sigma^+) \end{aligned} \tag{39}$$

then per Eq. 36, the density $f(V_Z)$ is normal with the mean and standard deviation given by the equations

$$\mu_Z = (1 - K_L)\mu^+ + K_L\mu_p$$

$$\sigma_Z = \sqrt{[(1 - K_L)\sigma^+]^2 + (K_L\sigma_p)^2 + 2\rho_{p+}(1 - K_L)K_L\sigma^+\sigma_p} \tag{40}$$

Note that if $K_L = 1.00$, then

$$\sigma_Z = \sigma_p \quad \mu_Z = \mu_p \quad E_{2Z} = E_{2p} \tag{41}$$

6 ADAPTING BROWN'S SIMPLIFIED METHOD

To estimate the phase–phase SSFOR or estimate the strike distance for a given SSFOR, the same method as used for calculating the phase–ground SSFOR or in calculating the phase–ground strike distance can be used, except that the voltage profile for V_Z must be used. Defining the voltage profiles as

$$\gamma^+ = \left(\frac{E_S^+}{E_R^+}\right) \quad \gamma^- = \left(\frac{E_S^-}{E_R^-}\right) \quad \gamma_p = \left(\frac{E_{pS}}{E_{pR}}\right) \tag{42}$$

then

$$\gamma_Z = \left(\frac{E_{ZS}}{E_{ZR}}\right) = \frac{E_S^+ + K_L E_S^-}{E_R^+ + K_L E_R^-} \tag{43}$$

Dividing the numerator and denominator by E_R^+,

$$\gamma_Z = \frac{\gamma^+ + \gamma^- K_L(E_R^-/E_R^+)}{1 + K_L(E_R^-/E_R^+)} \tag{44}$$

The best estimates of the Es are their E_2 values. Thus

$$\gamma_Z = \frac{\gamma^+ + \gamma^- K_L(E_2^-/E_2^+)}{1 + K_L(E_2^-/E_2^+)} \tag{45}$$

In a similar manner, γ_Z can also be given in terms of γ_p:

$$\gamma_Z = \frac{\gamma_p + \gamma^+ \dfrac{1 - K_L}{K_L} \dfrac{E_2^+}{E_{2p}}}{1 + \dfrac{1 - K_L}{K_L} \dfrac{E_2^+}{E_{2p}}} \tag{46}$$

And as for the phase–ground insulation, the equivalent number of towers or spans n_e is

$$n_e = \frac{0.4}{1 - \gamma_Z} \frac{\sigma_{FG}}{\text{CFO}_0} n \tag{47}$$

6.1 Example 1

An example may help. Calculate the SSFOR for the following conditions:
Strength

$$S = 3.4 \, \text{m} \quad K_L = 0.68 \quad K_{GP} = 1.26 \quad \frac{\sigma_{FP}}{\text{CFO}_0} = 0.02 \tag{48}$$

Stress
The distributions of V^+ and V_p were obtained and are assumed to be Gaussian. Following are the data

$$E_2^+ = 2.0 \, \text{pu} \quad \frac{\sigma^+}{E_2^+} = 0.10 \quad E_{2p} = 3.20 \, \text{pu} \quad \frac{\sigma_p}{E_{2p}} = 0.10$$
$$\gamma_p = 0.90 \quad \gamma^+ = 0.90 \quad \rho_{p+} = 0.80 \tag{49}$$
$$n = 625 \quad \text{Max. system voltage} = 550 \, \text{kV} \quad 1 \, \text{pu} = 449 \, \text{kV}$$

Then

$$\mu^+ = 1.589 \quad \sigma^+ = 0.20 \quad \mu_p = 2.543 \quad \sigma_p = 0.32 \tag{50}$$

From Eq. 40,

$$\mu_Z = 2.238 \, \text{pu} = 1004.8 \, \text{kV} \quad \sigma_Z = 0.272 \, \text{pu} = 121.9 \, \text{kV} \tag{51}$$

Continuing

$$\gamma_Z = 0.9 \quad n_e = 50 \quad Z_f = -2.204 \, \text{(Chapter 3)}$$
$$\text{CFO}_n = \text{CFO}_0 + Z_f \sigma_{FG} = 1221.4 \, \text{kV} \tag{52}$$
$$\text{SSFOR}_p = \frac{1}{2} P(V_Z > 1221.4 \, \text{kV})$$

$$Z = \frac{1221.4 - 1004.8}{121.9} = 1.7768$$
$$\text{SSFOR}_p = 1.89/100 \tag{53}$$

Computer, 1.94/100; with reversed parameters, 2.24/100

If $\rho_{p+} = 1.00$, then

$$\sigma_Z = (1 - K_L)\sigma^+ + K_L\sigma_p = 0.279\,\text{pu} = 126.5\,\text{kV}$$

$$Z = \frac{1221.4 - 1004.8}{126.5} = 1.7123 \quad \text{SSFOR} = 2.17/100 \tag{54}$$

Computer, 2.22/100; with reversed parameters, 2.41/100

6.2 Using $\rho_p+ = 1.00$

As noted, the assumption that $\rho_{p+} = 1.0$ is a conservative assumption. This is shown more clearly in Table 4 for which the same parameters are used as in the above example, except that $\gamma_p = \gamma^+ = 1.0$. Changing the correlation coefficient from 0.7 to 1.0 only changes the SSFOR by 7 to 16%.

Theoretically, the total SSFOR is the sum of the calculation as performed in the example plus the SSFOR when the input stress parameters are reversed. That is, the positive voltages become negative and the negative voltages become positive. Table 4 shows that the effect of this is to add an insignificant value to the SSFOR and therefore this effect may be neglected.

6.3 Reversed Parameters

The conservative nature of considering that $\rho_{p+} = 1.0$ is also shown in Table 5 when calculating the strike distance. These strike distances were calculated for a SSFOR of 1.0/100 and considered both the inputed and reversed stress parameters. The second row of the table shows the division of the SSFOR between these alternate SSFORs. Again, the SSFOR for the reversed parameters is insignificant. More importantly, changing the correlation coefficient from 0.7 to 1.0 only changes the strike distance

Table 4 Phase–Phase SSFOR for $\gamma_o = \gamma^+ = 1.0$

Condition	$\rho_{p+} = 1.0$	$= 0.9$	$= 0.8$	$= 0.7$
For only as inputed	3.17	3.01	2.84	2.67
Also for reversed parameters	3.51	3.42	3.33	3.25

Table 5 Strike Distance for SSFOR $= 1.0/100$ and $\gamma_p = \gamma^+ = 1.00$

	$\rho_{p+} = 1.0$	$= 0.9$	$= 0.8$	$= 0.7$
Strike distance, m	3.69	3.68	3.66	3.65
Division of SSFOR				
Parameters as inputed	0.948	0.921	0.890	0.852
Reversed parameters	0.052	0.078	0.110	0.148
With no reversed parameters	3.68	3.66	3.64	3.62

by 0.04 meters, 1%. The last row provides the strike distances when no reversed parameters are used, again illustrating the close agreement between the strike distances calculated with and without reversed parameters.

6.4 Sensitivity

The sensitivity of the SSFOR to the strength-to-stress ratio is presented in Fig. 16. The strength is V_{30} where

$$V_{30} = \text{CFO}_0 \left(1 - \frac{3\sigma_{\text{FP}}}{\text{CFO}_0}\right) = 0.925\text{CFO}_0 \quad (55)$$

and the stress E_{2Z} is

$$E_{2Z} = \mu_Z + 2.0538\sigma_Z \quad (56)$$

These curves are constructed for $n = 500$ towers or spans and $\gamma^+ = \gamma^- = \gamma_Z = 1.00$, i.e., the voltage is constant along the line. The parameter of the curves is the per unit standard deviation of V_Z or σ_Z/E_{2Z}. As noted, the three curves cross at a SSFOR of 1.0/100 for which the ratio V_{30}/E_{2Z} is 1.00. In the more practical case, where the voltage profiles are less than 1.00, the ratio V_{30}/E_{2Z} required for a SSFOR of 1.0/100 will decrease. Thus, conservatively, as for the phase–ground case, for a SSFOR of 1.0/100, set V_{30} equal to E_{2Z}.

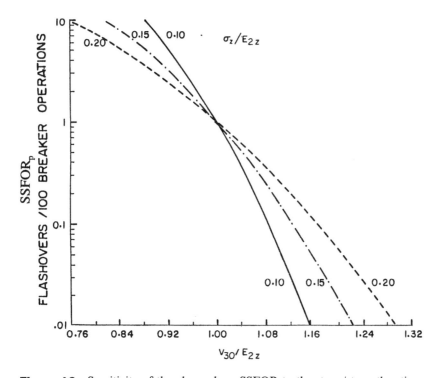

Figure 16 Sensitivity of the phase-phase SSFOR to the stress/strength ratio.

For other desired values of SSFOR, Tables 5, 6, and 7 of Chapter 3 can be used to estimate the required ratio of V_{30}/E_{2Z}. That is, from Table 5, the value of K_f is obtained using a σ_f/CFO of 0.02. The value of K_G or K_E is then obtained from Table 6 or 7. As before, V_{30}/E_{2Z} is equal to either $K_f K_G$ or $K_f K_E$.

7 CALCULATING THE STRIKE DISTANCE—SIMPLIFIED METHOD

The technique in calculating the strike distance given the desired SSFOR is exactly the same as for the phase–ground case as developed in Chapter 3. In fact, Table 5 of Chapter 3 contains the value of K_f for $\sigma_f/CFO = 0.02$, which is the value required for σ_{FP}/CFO_0. Of course, γ_Z should be used. Also the same weather or altitude correction factors as for the phase–ground case should be used. The use of the simplified method in calculating the strike distance is illustrated in a homework problem.

8 INTERNAL INSULATIONS

The presentation in this chapter has been directed primarily to self-restoring or external insulations such as exist on transmission lines and in portions of substations. For internal insulations such as in transformers, cables, and GIS, the insulation strength is only a function of the phase–phase voltages and is not a function of the division of this voltage into positive and negative polarities. Thus the insulation coordination procedure is simplified to that shown in Fig. 17. The density function is that for phase–phase overvoltages. The strength of the insulation is usually not statistically known. That is, the strength is specified only by the conventional BSL. Since the statistical strength characteristic is unknown, the only viable assumption is that at the BSL, the probability of failure increases instantaneously form 0 to 100%. The SSFOR for this situation as illustrated in Fig. 17 is

$$\text{SSFOR} = \int_{\text{BSL}}^{\infty} f(V_Z) \, dV_Z = \int_{\text{BSL}}^{\infty} f(V_p) \, dV_p \tag{57}$$

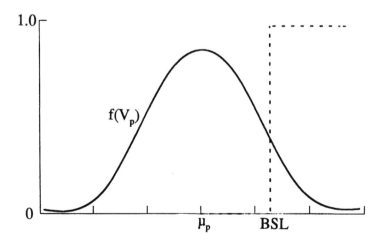

Figure 17 Calculating the phase-phase SSFOR for internal insulation.

Because the strength is identical for both the original and the reversed parameters, the integral is not multiplied by 1/2. However, usually, the phase–phase voltage is modified by a surge arrester and will be considered in Chapter 5.

9 EXTERNAL INSULATIONS WITH $K_L = 1.00$

As stated previously, for small gaps in the order of 3 meters or less, $K_L = 1.0$. Therefore $\sigma_Z = \sigma_p$, $\mu_Z = \mu_p$, $E_{2Z} = E_{2p}$. Also $CFO_0 = CFO_p$. Therefore, the SSFOR is

$$\text{SSFOR} = \int_{CFO_n}^{+\infty} f(V_Z)\,dV_Z = \int_{CFO_n}^{+\infty} f(V_p)\,dV_p \tag{58}$$

As for internal insulations, the integral is not multiplied by 1/2 since both the original and the reversed parameters result in the same SSFOR. To calculate the strike distance, the input value of the SSFOR must be equal to 1/2 of the desired value of the SSFOR. For example, if the desired SSFOR is 1.0/100, then in the calculation, a value of 0.5/100 is used.

10 IEC AND CIGRE

A short background may be helpful. In 1976, IEC published the insulation coordination standards 71-1 and 71-2. However, these standard publications did not include phase–phase insulation coordination. Therefore, in about 1977, IEC Technical Committee 28 began work on a new standard on phase–phase. At this time, there existed much confusion in the technical understanding of the process of phase–phase insulation coordination. In an attempt to mitigate the maelstrom, CIGRE Committee 33 on Insulation Coordination published four articles on the subject in *ELECTRA* [6]. This did little to relieve the confusion. Few engineers understood the concepts and methodology and few engineers had faith in the process.

Work continued within CIGRE Working Group 33.06, and in 1985 [9] a report was presented that suggested the method presented in this chapter. Recently, the IEC Technical Committee began revision of Publication 71. Both parts 71-1 and 71-2 are now complete and available. IEC 71-1, "Definitions, Principles, and Rules," was written primarily by Gianguido Carrara, while IEC 71-2, "Application Guide," was written primarily by Karl Weck.

The IEC application guide only considers insulation coordination of substations, and therefore the comparison of the techniques here presented with those presented in the IEC application guide will await Chapter 5.

11 REFERENCES

1. G. Gallet, B. Hutzler, and J. Riu, "Analysis of the Switching Impulse Strength of Phase-to-Phase Air Gaps," *IEEE Trans. on PA&S*, Mar./Apr., 1978, pp. 485–494.
2. K. J. Lloyd and L. E. Zaffanella, "Insulation for Switching Surges," Chapter 11 of *Transmission Line Reference Book*, 2d ed, Palo Alto, CA: Electric Power Research Institute, 1982.

3. M. Miyake, Y. Watanabe, and E. Ohasaki, "Effects of Parameters on the Phase-to-Phase Flashover Characteristics of UHV Transmission Lines," *IEEE Trans. on Power Delivery*, Oct. 1987, pp. 1285–1291.
4. I. S. Grant and A. S. Paulson, "Phase-to-Phase Switching Surge Design," *EPRI EL-3147*, Project 1492, Jun. 1983.
5. IEC Standard 71-2—"Insulation Coordination, Part 2: Application Guide," 1996.
6. CIGRE Study Committee 33, Insulation Coordination, *ELECTRA*, May 1979, pp. 138–230.
   6.1. F. Crespo, K. F. Foreman, G. LeRoy, R. Gert, O. Volcker, and R. Eriksson of CIGRE Working Group 33.02, "Part I, Switching Overvoltages in Three-Phase Systems," *ELECTRA*, May 1979, pp. 138–157.
   6.2. A. Pigini, L. Thione, R. Cortina, K. H. Weck, C. Menemenlis, G. N. Alexandrov, and Yu. A. Gerasimov of CIGRE Working Group 33.03, "Part II—Switching Impulse Strength of Phase-to-Phase External Insulation," *ELECTRA*, May 1979, pp. 158–181.
   6.3. K. H. Weck and G. Carrara of CIGRE Working Group 33.06, "Part III—Design and Testing of Phase-to-Phase External Insulation," *ELECTRA*, May 1979, pp. 182–210.
   6.4. A. Pigini, E. Gabagnati, B. Hutzler, J. P. Riu, H. Studinger, K. H. Weck, G. N. Alexandrov, and A. Fisher of CIGRE Task Force 33.03.03, "Part IV—The Influence of Non Standard Conditions on the Switching Impulse Strength of Phase-to-Phase Insulation," *ELECTRA*, May 1979, pp. 211–230.
7. R. Cortina, P. Nicolini, A. Pigini, and L. Thione, "Space Occupation of EHV and UHV Transmission Lines as Affected by the Switching Impulse Strength of Phase–Phase Insulation," Stockholm: CIGRE Symposium on Transmission Lines and the Environment, 1981.
8. R. Cortina, M. Sforzini, and A. Taschini, "Strength Characteristics of Air Gaps Subjected to Interphase Switching Surges," *IEEE Trans. on PA&S*, Mar. 1976, pp. 448–452.
9. A. R. Hileman, "Phase–Phase Switching Overvoltage Insulation Coordination," CIGRE Working Group paper, CIGRE SC33-85(WG06)61WD, 1985.
10. T. Udo, "Minimum Phase-to-Phase Electrical Clearances for Substations Based on Switching Surges and Lightning Surges," *IEEE Trans. on PA&S*, Aug. 1966, pp. 838–845.

12 PROBLEMS

1. Using the simplified method with $\rho_{p+} = 1.00$, calculate the phase–phase strike distance for a 625 tower, 500 kV line (1 pu = 449 kV) at an altitude of 1000 meters for a phase–phase SSFOR of 1.0 flashover per 100 breaker operations for the following conditions:

Stress

For phase–ground: $f(V^+)$ is Gaussian having the parameters

$$E_2^+ = 1.8 \, \text{pu} \qquad \frac{\sigma^+}{E_2^+} = 0.07 \qquad \gamma^+ = 0.88 \tag{59}$$

For phase–phase: $f(V_p)$ is normal having the parameters

$$E_{2p} = 2.9 \, \text{pu} \qquad \frac{\sigma_p}{E_{2p}} = 0.10 \qquad \gamma_p = 0.88 \tag{60}$$

Phase–Phase SOVs, Transmission Lines

Strength:

$$K_{\text{GP}} = 1.26 \quad K_{\text{L}} = 0.68 \quad \frac{\sigma_{\text{FP}}}{\text{CFO}_0} = 0.02 \quad (61)$$

2. For a 500 kV line (1 pu = 449 kV) the phase–phase SOV distribution of V_Z approximates an extreme value positive skew distribution with the parameters $E_{2Z} = 2.8$ pu and $\beta_Z/E_{2Z} = 0.10$. The phase-to-ground SOV distribution is also an extreme value positive skew with $E_2^+ = 1.8$ pu and $\beta/E_2^+ = 0.10$. The voltage profiles are $\gamma^+ = \gamma_Z = 0.90$. The phase–phase strength parameters are $K_{\text{L}} = 0.68$ $K_{\text{GP}} = 1.26$, and $\sigma_{\text{FP}}/\text{CFO}_0 = 0.02$. The phase–ground strength parameters are $k_g = 1.2$ (no decrease for wet conditions) and $\sigma_f/\text{CFO} = 0.05$. The line is composed of 500 towers. Estimate the phase–phase and phase–ground strike distance for a phase–phase and phase–ground SSFOR of 1 flashover per 100 breaker operations. Assume sea level conditions, i.e., the altitude is zero.

3. For a 230 kV line (242 kV maximum, 1.0 pu = 198 kV), estimate the phase–phase strike distance for a SSFOR of 1.0/100. The line has 625 towers and is at sea level. Assume that $K_{\text{GP}} = 1.30$ and $K_{\text{L}} = 1.00$. Assume that the correlation coefficient is 1.00. Also for a Gaussian SOV distribution

$$E_{2p} = 4.7 \text{ pu} \quad \frac{\sigma_p}{E_{2p}} = 0.10 \quad \gamma_p = 0.90$$

$$E_2^+ = 3.0 \text{ pu} \quad \frac{\sigma^+}{E_2^+} = 0.10 \quad \gamma^+ = 0.90 \quad (62)$$

$$\frac{\sigma_{\text{FP}}}{\text{CFO}_0} = 0.02$$

5
Switching Overvoltages, Substations

1 INTRODUCTION

In Chapters 3 and 4 switching overvoltage insulation coordination for transmission lines was presented. The objective of this chapter is to present switching overvoltage insulation coordination methods for substations. That is, the purpose is to estimate both the phase–ground and the phase–phase BSLs for substation equipment and phase–ground and phase–phase clearances. In performing this task, the methods developed in Chapters 3 and 4 and the insulation strength data presented in Chapters 2 and 4 are used. Although the basic methods of Chapters 3 and 4 are valid, there are some differences when adapting these to station insulation. It is assumed that the reader is familiar with the methods of Chapters 3 and 4 and therefore the discussion will center on the differences. The insulation strength will be reviewed, but again it is assumed that the reader is familiar with Chapters 2 and 4.

Although the selection of the BSL of the transformer is considered, of primary concern is the insulation coordination of the self-restoring apparatus. Therefore the primary emphasis is on the selection of BSLs for self-restoring insulations and the selection of clearances.

2 PHASE–GROUND INSULATION COORDINATION

As for transmission lines, switching overvoltages only become important for systems whose nominal system voltage is equal to or greater than 345 kV. The differences between the insulation coordination of transmission lines and that of the station lies in

1. Station insulation and line insulation must be coordinated.
2. The number of insulations in parallel: for the station, $n = 5$ to 10.

3. The voltage profile factor γ or E_S/E_R is 1.00 in a station.
4. Not all insulations within a station have equal insulation strengths.
5. The values of K_f and K_G or K_E must be altered, since the number of insulations is small and therefore the value of σ_0 may not be significantly greater than that of σ_{fn}, see Chapter 3, Section 7.2, Eqs. 61–63.
6. Station insulation strengths are described by BSL (for apparatus) or V_3 or CFO (for air clearances).
7. The design value of the SSFOR for the station may be a decade step lower than that of the line.

Each of these points is discussed below.

2.1 Coordination of Station and Line Insulation

As illustrated in Fig. 1, the switching surges at the end of the line are those that impinge on the station insulation—or the station insulation is the insulation at the end of the line. Considering the station insulation as the line-end insulation, if no arresters are used on the line side of the circuit breaker, it is apparent that as a first rule, so as not to degrade the line insulation, the station insulation strength must be equal to or greater than the line insulation strength required for switching surges. That is, the switching impulse strength, i.e., the value of V_3, for the station must equal or exceed the switching impulse strength, the V_3, for the line. Note carefully that the line insulation strength referred to is the switching impulse design value, which because of lightning or contamination may be less than that required for contamination or lightning. As will be demonstrated, usually this criterion necessitates either the increase of the station switching impulse insulation strength or the use of line-end (station-entrance) arresters.

The use of arresters on the line side of the breaker essentially isolates the station from the line. In this case, the station insulation may be selected based on the arrester characteristics without regard for the line insulation strength.

2.2 Number of Insulations in Parallel

As illustrated in Fig. 1, the number of parallel insulations in the station, usually $n = 5$ to 10, is much smaller than for the line, where usually $n = 100$ to 1000. For example, assuming a breaker and a half scheme per Fig. 1, the station insulation

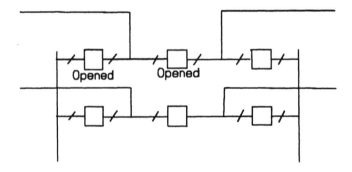

Figure 1 Breaker and a half scheme showing two open breakers.

consists of a single bus support insulator, two disconnecting switches, two opened breakers, and one or more "clearances," and thus $n = 5$ to 10.

2.3 Voltage Profile

As noted from Fig. 1, distances between various pieces of equipment are small. Since the time to crest of the SOVs are large compared with the travel time for these distances, the shape and magnitude of the SOVs are essentially constant throughout the station. Thus the voltage profile factor γ is 1.00, or in other words, the voltage profile need not be considered.

2.4 Unequal Insulation Strengths

The insulation strengths of equipment within the station may not be equal. For example, at 500 kV, only a single BSL of 1300 kV (in the opened position) exists for the circuit breaker. Other equipment in the station, such as the disconnecting switch or the bus support insulators, may have lower BSLs, and the clearances may have lower insulation strengths. Therefore the equation for the SSFOR is altered to

$$\text{SSFOR} = \frac{1}{2} \int_{E_1}^{E_m} f(V) \left[1 - \prod_{i=1}^{n} q_i \right] dV \tag{1}$$

where, as before, $f(V)$ is the probability density function of the SOVs, q_i is the probability of no flashover for each apparatus, and n is the number of insulations in parallel. As before, E_1 is the minimum SOV, usually 1.0 pu of maximum line-ground system voltage, and E_m is the maximum SOV.

As expected, thinking of the phenomenon of the weak link, the SSFOR of the station insulation is to a large extent dictated by the apparatus or clearance having the lowest insulation strength. For example, for $n = 5$, decreasing the BSL by 10% on all insulations increases the SSFOR by about 500% whereas decreasing only one of the insulations by 10% increases the SSFOR by about 250%. Therefore a conservative estimate of the station SSFOR is to assume that all the insulation strengths are equal to the lowest strength. Thus Eq. 1 may be simplified to

$$\text{SSFOR} = \frac{1}{2} \int_{E_1}^{E_m} f(V) [1 - q^n] dV \tag{2}$$

2.5 Estimating the SSFOR and Strike Distance

In Chapter 3, a method was developed to estimate the strike distance by use of the factors K_f and K_G (or K_E). The method was derived assuming that the number of insulations was large, in the order of 100 to 1000, and thus the insulation characteristic curve became so steep that it could be approximated as a single value, the CFO_n. Because the number of insulations in parallel in a station is small, this assumption is no longer completely valid. That is, the value of σ_{fn} may not be insignificant as compared to σ_0 as discussed in Chapter 3, Section 7.2, Eqs. 61–63. To retain the

Table 1 K_f Modified for Stations

n	K_f for $\sigma_f/\text{CFO} = 7\%$
5	0.890
10	0.910

simplified method, a small correction is made in the values of K_f, K_G, and K_E. These revised values are presented in Tables 1 and 2.

In anticipation of the effect of arresters, values of K_G and K_E are provided for small values of σ_0/E_2 and β/E_2. And in anticipation of the phase–phase methods, a value of Z_e for the Gaussian distribution is given in Table 2 (see Table 6 of Chapter 3).

2.6 Station Insulation Strengths

Relationship of Strengths

The insulation strength can be specified by the CFO, by the BSL, or by V_3. The relationship between these quantities is given by the equations

$$V_3 = \text{CFO}\left(1 - 3\frac{\sigma_f}{\text{CFO}}\right) = 0.79\,\text{CFO} \tag{3}$$

$$\text{BSL} = \text{CFO}\left(1 - 1.28\frac{\sigma_f}{\text{CFO}}\right) = 0.91\,\text{CFO} \tag{4}$$

where the numerical values assume a σ_f/CFO of 0.07.

Clearances or Strike Distances

Gap factors for typical gaps or clearances are presented in Table 3, as obtained from Chapter 2. (See Chapter 2 for other values.) The authors of Ref. 1 found that σ_f/CFO ranges from 4 to 9%, averaging 6%. Since the larger values of σ_f/CFO lead to larger strike distances and BSLs, a conservative value of 7% of the CFO is recommended as shown in Table 3. However, σ_f for the tower is 5% of the CFO per Chapter 3. Rain or wet conditions do not significantly decrease the insulation strength of pure air gaps. The conductor-to-lateral-structure gap emulates the condition of a phase conductor passing through a portal tower. The conductor-to-lower-rod represents the tip of an opened disconnecting switch on the lower bus to the conductor on the upper bus. Gap factors for rod-to-plane generally never occur. The

Table 2 K_E and K_G Adjusted from Chapter 3

SSFOR	Z_e	Gaussian: K_G for $\sigma_0/E_2 =$					Extreme value (+), K_E for $\beta/E_2 =$			
		0.01	0.02	0.05	0.10	0.15	0.01	0.02	0.05	0.10
1/10	0.8415	0.99	0.97	0.93	0.86	0.80	0.98	0.95	0.87	0.74
1/100	2.0542	1.07	1.05	1.02	1.00	0.99	1.05	1.03	1.00	0.98
0.5/100	2.3268	1.09	1.07	1.04	1.03	1.03	1.08	1.05	1.03	1.05
1/1000	2.8785	1.13	1.12	1.09	1.09	1.12	1.12	1.10	1.12	1.21

Table 3 Typical Value of Gap Factors k_g for Phase–Ground Insulations

Gap configuration	Range of k_g	Typical value of k_g	σ_f/CFO
Rod–plane	1.00	1.00	0.07
Rod–rod (vertical)	1.25–1.35	1.30	0.07
Rod–rod (horizontal)	1.25–1.45	1.35	0.07
Conductor–lateral structure	1.25–1.40	1.30	0.07
Conductor–lower rod	1.40–1.60	1.50	0.07
Conductor–plane	1.15	1.15	0.07
Post insulators	1.18	1.18	0.07
Tower, center phase	—	0.96 (1.20)	0.05
Tower, outside phase	—	0.96 (1.30)	0.05
Insulator string	—	0.96 (1.30)	0.05

rod–rod, vertical and horizontal, may be used for gaps between small diameter grading rings (more about this later when considering phase–phase). The conductor-to-plane gap does occur between the phase conductor and ground or earth. However, it is never, or seldom, important, since, even though it has a low gap factor, the distance to earth is large as compared to the other clearances. Practically, the lowest gap factor in the substation is 1.3, which normally is conservative.

The CFO is given by the equation

$$\text{CFO}_S = k_g \frac{3400}{1 + (8/S)} \tag{5}$$

where S is the strike distance in meters and the CFO_S in kV denotes the CFO for standard atmospheric conditions.

Apparatus

Apparatus BILs and BSLs for nominal system voltages of 230 kV and higher are presented in Table 4 as obtained from ANSI/IEEE standards (see Chapter 2). Per standards, the BSLs are for wet conditions. Note that, except for transformers, BSLs are specified only for nominal system voltages of 345 kV and above.

Note also that BSL of circuit breakers is greater in the opened position than in the closed position.

Also note that BSLs for disconnecting switches (and bus support insulators) are not provided. However, from chapter 2 for post insulators,

$$\text{BSL}_S = 1.07 \frac{3400}{1 + (8/S)} \tag{6}$$

or using Eq. 4,

$$\text{CFO}_S = 1.18 \frac{3400}{1 + (8/S)} \tag{7}$$

Table 4 Apparatus BIL/BSL, Phase–Ground, kV

System voltage, kV	Transformer BSL	Transformer BIL	Circuit breaker BSL	Circuit breaker BIL	Transformer bushings BSL	Transformer bushings BIL	Disconnect switches BIL
230/242	650	540	900	NA	750	NA	900
	750	620			900	NA	1050
	825	685			1050	NA	
	900	745					
	1050	870					
345/362	900	745	1300	825/900	900	700	1050
	1050	870			1050	825	1175
	1175	975			1175	825	
	1300	1080					
500/550	1300	1080	1800	1175/1300	1300	1050	1500
	1425	1180			1425	1110	1800
	1550	1290			1550	1175	
	1675	1390			1675	1175	
765/800	1800	1500	2050	1425/1550	1800	1360	2050
	1925	1600					
	2050	1700					

For circuit breakers, first value of BSL is for closed breaker, second value is for opened breaker. NA = not available.

where the BSL_S and CFO_S for wet conditions are the BSL and CFO at standard weather conditions. This gap factor of 1.18 is also shown in Table 3. Also from Chapter 2, the BIL is

$$BIL_S = 450S \qquad (8)$$

where S is the insulator height or strike distance in meters and the BIL_S in kV is the BIL for standard atmospheric conditions.

For the ANSI and IEC standard BILs, the strike distance is obtained per Eq. 8, from which the BSL is estimated per Eq. 6 and shown in Table 5. BILs marked as IEC or ANSI only apply to these standards, while the unmarked values apply to both ANSI and IEC. Because no standard BSL values exist in the ANSI standards, these values per Table 5 are assumed to be acceptable estimates both for post insulators, all bus support insulators, and disconnecting switches. For apparatus, the suggested value of σ_f/CFO is also 0.07.

For nonstandard atmospheric conditions, i.e., altitudes greater than sea level, correction procedures require the use of a gap factor. This gap factor may be obtained by use of a strike distance obtained from Eqs. 6 or 8. The required accuracy of this gap factor is not great, and a gap factor of 1.18 per Eq. 7 can be used for all apparatus, i.e., bus support insulators, disconnecting switches, and circuit breakers.

Switching Overvoltages, Substations

Table 5 Assumed Values of BSL for Standard Values of BIL for Disconnecting Switches and Bus Support Insulators

BIL, kV	S, meters	BSL, kV
825 ANSI	1.83	678
850 IEC	1.89	695
900 ANSI	2.00	728
950 IEC	2.11	760
1050	2.33	821
1175	2.61	895
1300	2.89	965
1425	3.17	1032
1550	3.44	1094
1675	3.72	1154
1800	4.00	1213
1925 ANSI	4.28	1268
1950 IEC	4.33	1278
2050 IEC	4.56	1321

2.7 Design Value of SSFOR

The normal design value of SSFOR for the line is 1 flashover/100 breaker operations. Considering that the consequence of failure or flashover within the station is higher than that for the line, the SSFOR of the station is sometimes set at a decade step lower than that of the line. Thus a value of 1/1000 is sometimes used, although it appears that presently the value of 1/100 is more frequently used.

2.8 Equations—Review

$$\delta = e^{-A/8.6} \qquad (9)$$

$$\mathrm{CFO}_A = \delta^m \mathrm{CFO}_S \qquad \mathrm{BSL}_A = \delta^m \mathrm{BSL}_S \qquad (10)$$

$$\mathrm{CFO}_S = k_g \frac{3400}{1 + (8/S)} \qquad (11)$$

$$m = 1.25 G_0 (G_0 - 0.2) \qquad G_0 = \frac{\mathrm{CFO}_S}{500 S} \qquad (12)$$

$$S = \frac{8}{(3400 k_g \delta^m / \mathrm{CFO}_A) - 1} = \frac{8}{(3400 k_g / \mathrm{CFO}_S) - 1} \qquad (13)$$

Because the value of K_f has been modified for stations, when calculating the SSFOR it is better to use Eqs. 82 and 83 of Chapter 3, i.e., for a Gaussian distribution of SOVs,

$$\text{SSFOR} = \frac{1}{2}[1 - F(Z_e)]$$

$$Z_e = 2.054 - \frac{1 - [(V_3/E_2)/K_f]}{\sigma_0/E_2} \tag{14}$$

and for an extreme value positive skew distribution of SOVs

$$\text{SSFOR} = \frac{1}{2}[1 - F(y_e)] = \frac{1}{2}\left[1 - e^{-e^{-y_e}}\right]$$

$$y_e = 3.902 - \frac{1 - [(V_3/E_2)/K_f]}{\beta/E_2} \tag{15}$$

2.9

Example 1. Estimating the SSFOR. Assume that a BSL of 850 kV is used for all apparatus in a 500 kV station. Also assume that the distribution of the SOVs can be approximated by a Gaussian distribution having an E_2 of 808 kV (1.8 pu) and a σ_0/E_2 of 0.10. Also assume a total of 10 parallel insulations or $n = 10$. Also $\sigma_f/\text{CFO} = 0.07$. From Eq. 4, $\text{CFO}_S = 934$ kV, $V_{3S} = 738$ kV, and $V_{3S}/E_2 = 0.9134$. From Table 1, $K_f = 0.91$. Then using Eq. 14, $Z_e = 2.091$. From Table 1 of Chapter 3,

$$\text{SSFOR} = \frac{0.0183}{2} = 0.0091 = 0.91/100 \tag{16}$$

Using a computer program to perform the numerical integration, the SSFOR is 0.86/100.

To continue this example, assume now that the station is at an elevation of 1500 meters, so that the relative air density is $\delta = 0.840$. Also to obtain reasonable values of SSFOR assume a BSL_S of 950 kV. Thus the CFO_S is 1044 kV. For this BSL, the strike distance per Eq. 6 is 2.827 meters and therefore

$$G_0 = \frac{1044}{500(2.827)} = 0.7386 \qquad m = 0.4972 \tag{17}$$

Therefore

$$\text{CFO}_A = 0.840^{0.4972}(1044) = 957 \text{ kV}$$

$$V_{3A} = 756 \text{ kV} \qquad V_{3A}/E_2 = 0.936 \tag{18}$$

Then using Eq. 14

$$Z_e = 2.339 \qquad \text{SSFOR} = \frac{1}{2}(0.0097) = 0.49/100 \tag{19}$$

Using a computer program to perform numerical integration, the SSFOR is 0.50/100.

2.10

Example 2. Estimating the BSL and Clearance. Consider the same station as for Example 1, except that the station insulation is to be specified based on a SSFOR of 1/100. First assume sea level conditions.

From Tables 1 and 2, $K_f = 0.91$ and $K_G = 1.0$, and therefore $V_3/E_2 = 0.91$. Thus $V_3 = 735$ kV, CFO = 931 kV, and the BSL = 847 kV. Thus the required BSL is 847 kV. The next highest BSL, per Table 5, is 895 kV, which is the BSL for a BIL of 1175 kV. Thus 1175 kV BIL is selected for the bus support insulators and the disconnecting switches. The BSL for the circuit breaker is 1300 kV, which is more than adequate. Using Eq. 13 with $k_g = 1.3$, the clearance is 2.13 meters.

Using a computer program, the required BSL is 844 kV and the clearance is 2.12 meters. Therefore the same BIL would be selected.

Before continuing this example to include the effect of altitude, consider the requirement that the V_3 of the station must be equal to or greater than that of the line. For the line assume that $n = 500$ and that $E_S/E_R = 0.9$. Therefore $n_e = 100$ and for a SSFOR of 1/100, V_3/E_2 must equal 0.9693, and V_3 for the line is 783 kV. Since the V_3 for the line, 783 kV, is larger than the V_3 for the station, 735 kV, the insulation in the station must be designed for a V_3 of 783 kV. Unfortunately, this is always true since for the same SSFOR the K_G or K_E for the station and line are approximately equal but the K_f for the line is greater since the number of insulations for the line exceeds that for the station. However, the use of a V_3 of 783 kV for the station results in a SSFOR for the station of about 2.12/1000. To expand, from Tables 1 and 2, the required V_3/E_2 for a SSFOR of 1/1000 is 0.9919, which produces a V_3 of 801 kV, which is greater than the 783 kV as required by the line.

Using a required V_3 of 783 kV results in a required CFO of 991 kV and a required BSL of 902 kV. From Table 5, the next larger BSL is 965 kV, which is the BSL for a BIL of 1300 kV. Thus a BIL of 1300 kV is selected for the bus support insulators and the disconnecting switches. The circuit breaker having a BSL of 1300 kV is again more than adequate.

To determine the required clearance of strike distance, use Eq. 13, with a gap factor of 1.3 and a relative air density of 1.00. Then $S = 2.31$ meters.

Now assume that the station is at 1500 meters. The required CFO, BSL, and V_3 are the same as before except that these values are required at 1500 meters. Thus they are $CFO_A = 991$ kV, $V_{3A} = 783$ kV, and $BSL_A = 902$ kV. A first estimate of the BSL_S can be made by assuming that $m = 0.5$ ($\delta = 0.840$) and therefore that the $BSL_S = 902/\delta^m = 984$ kV. The iterations are shown in Table 6. Equation 6 is used to find S, then the CFO_S is the $BSL_S/0.91$. G_0, m, and δ^m can now be found. Then since $BSL_A = 902$ kV, from Eq. 10, the BSL_S is found. As shown, only a single calculation is necessary, since the iterated BSL_S is 981 kV.

Therefore a BSL_S of 980 is required. Since the circuit breaker BSL of 1300 kV is greater than this, the circuit breaker is more than acceptable. Entering Table 5 with a

Table 6 Iteration to Find the BSL_S for 1500 meters

BSL_S	S, meters	CFO_S	G_0	m	BSL_S
980	2.95	1077	0.730	0.484	981

required BSL of 980 kV, the next highest is 1032 kV, which translates to a BIL of 1425 kV. If a computer program is used, the BSL required is 981 kV.

As noticed, the required circuit breaker BSL is also obtained using the equations for the post insulators. In explanation, the equations for the post insulators are used only to obtain an estimate of the exponent m, and therefore, since m is not a highly sensitive value, the estimate appears justifiable. Also note that an m of 0.5 is a crude estimate and if used above would require a BSL_S of 984 kV. This is not a bad guess.

The remaining task is to determine the strike distance or clearance. This is accomplished in the same manner as in Chapter 3 except that a gap factor of 1.3 is used. From above, a CFO_A of 991 kV is required. As an initial guess, with $m = 0.5$, $S = 2.6$ meters. The final value from the iterations of Table 7 is 2.69 meters. From a computer program, $S = 2.68$ meters, much the same value.

Some final comments on this example: In all except one case, BPA, the BIL used for all station insulation except for transformer is 1800 kV. Thus in most cases the BSL employed is much greater than required. As will be shown in Chapter 13, lightning insulation coordination requirements usually exceed the requirements for switching surges.

2.11 Phase–Ground Insulation Coordination with Arresters

If the required BSL exceeds the circuit breaker BSL, or the required BIL/BSL of the bus support insulators is considered excessive, then line-entrance arresters may be used. As discussed previously, if arresters are used on the line side of the breaker, the requirement that the station insulation strength be equal to or greater than the line insulation strength is eliminated. The arresters essentially isolate the line from the station. As an example, if no closing resistors are used in the breaker, an estimate of the value of E_2 is 2.8 pu or 1257 kV. Using the data of the example, the required BSL at 1500 meters is about 1400 kV, which exceeds the standard BSL of the breaker. Thus in this case, line-entrance arresters are required.

As is discussed in Chapter 11, line-entrance arresters are sometimes used to protect the breaker when it is in the opened position. This is seldom required at EHV levels but may be necessary for lower voltage systems.

Returning to the use of arresters to mitigate the SOVs, assuming that the arrester discharge voltage is less than the maximum SOV, arresters will not only decrease the SOVs within the station but also decrease the SOVs along a portion of the line. Thus both the line and station apparatus insulation strength may be reduced. Recent failures of the closing resistors within the breaker have stimulated the consideration of the use of line-entrance arresters (or line-end arresters), and some utilities have successfully deleted the resistors and added arresters. However, closing resistors provided a superior means of reducing the SOVs along the entire line, whereas arresters only decrease the SOVs within a relatively short distance of the arrester.

Table 7 Finding the Clearance

S, meters	CFO_S	G_0	m	S, meters
2.6	1084	0.834	0.661	2.69
2.69	1112	0.827	0.648	2.68

Returning to the application of arresters, as depicted in Fig. 2A, assume that the arrester discharge voltage E_A can be represented by a single linear line having the equation

$$E_A = E_0 + I_A R_A \tag{20}$$

Also, per the equivalent circuit of Fig. 2B, the switching overvoltage E is

$$E = E_A + I_A Z \tag{21}$$

where Z is the line surge impedance. Thus

$$E_A = K_A E + (1 - K_A) E_0 \tag{22}$$

where

$$I_A = \frac{E - E_0}{Z + R_A} \tag{23}$$

and

$$K_A = \frac{R_A}{Z + R_A} \tag{24}$$

If the original SOV density function or distribution $f(V)$ is Gaussian with a mean, μ_0 and a standard deviation σ_0, then the distribution from E_0 to the maximum SOV, E_{mA}, is also a Gaussian distribution with a mean μ_A and a standard deviation σ_A given by the equations

$$\begin{aligned} \mu_A &= K_A \mu_0 + (1 - K_A) E_0 \\ \sigma_A &= K_A \sigma_0 \\ E_{mA} &= K_A E_m + (1 - K_A) E_0 \end{aligned} \tag{25}$$

From E_1, the minimum SOV, to E_0, the SOV distribution is equal to the original distribution, i.e., μ_0 and σ_0. See Fig. 3.

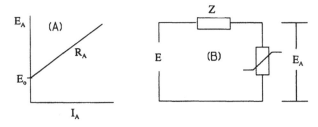

Figure 2 Arrester characteristics. Equivalent circuit.

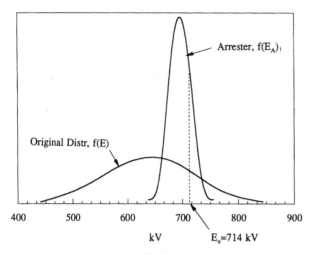

Figure 3 Arresters alter SOV density.

Alternately, if the original SOV distribution is an extreme value positive skew distribution with the parameters u and β, then the modified distribution from E_0 to E_{mA} is also an extreme value with the parameters u_A and β_A given by

$$u_A = K_A u + (1 - K_A)E_0$$
$$\beta_A = K_A \beta \qquad (26)$$
$$E_{mA} = K_A E_m + (1 - K_A)E_0$$

The switching impulse discharge voltage as published in ANSI/IEEE standards (C62.11) [2] is the discharge voltage for the discharge currents of Table 8. Ranges of this discharge voltage are listed in Table 9, and a typical characteristic based on the discharge voltage at 1.0 kA is presented in Table 10. The resultant values of R_A and E_0 are given in Table 11. As can be noted, the value of K_A and therefore σ_A or β_A is small. Also μ_A or u_A is primarily determined by E_0. The following examples illustrate the procedure.

Table 8 SI Discharge Current Used to Obtain the SI Discharge Voltage

Arrester class	Nominal system voltage, kV	SI current, kA
Station	3–150	0.5
	151–325	1.0
	326–900	2.0
Intermediate	3–150	0.5

Switching Overvoltages, Substations

Table 9 SI Discharge Voltage in Per Unit of Crest MCOV

Arrester class	Discharge voltage, kV
Station	1.63–1.86
Intermediate	1.66–1.91

Table 10 Typical Discharge Voltage–Current Characteristics

Discharge current, kA	Discharge voltage per unit of voltage, at 1 kA
0.001	0.832
0.01	0.866
0.10	0.913
0.50	0.970
1.00	1.000
2.00	1.037

Table 11 Arrester Start Voltage and Resistance in Per unit of Voltage at 1 kA per Table 6

Current range, kA	R_A	E_0
0.001–0.01	3.78	0.828
0.01–0.10	0.522	0.862
0.10–0.50	0.143	0.899
0.50–1.00	0.060	0.940
1.00–2.00	0.037	0.963

2.12

Example 3. With Arresters. Consider the same example as used in Example 2 with the altitude at sea level but with the use of a 318-kV MCOV arrester whose SI discharge voltage is 823 kV. From Table 8, this discharge voltage is for a 2.0 kA discharge current, and therefore the discharge voltage at 1.0 kA is 794 kV; see Table 10. The first task is to select the values of R_A and E_0 so that the resultant arrester current at the voltage E_2 matches the range of currents per Table 11. Selecting the range of 0.1 to 0.5 kA, $R_A = 113$ and $E_0 = 714$ kV. For $E = E_2 = 808$ kV and for a line surge impedance of 350 ohms, from Eq. 23, $I_A = 0.203$ kA, which is in the selected range of 0.1 to 0.5 kA. For this example, since $\mu_0 = 642$ kV and $\sigma_0 = 80.8$ kV, $K_A = 0.2441$, $\sigma_A = 19.72$ kV, and $\mu_A = 696.4$ kV. From these values the value of E_2 is reduced to E_{2A} of 737 kV and $\sigma_A/E_{2A} = 0.0268$. The resultant cumulative SOV distribution is shown in Fig. 4 illustrating the decrease in the SOVs.

Using the new values of E_2 and σ_0/E_2 for the arrester, the BSL and the clearance can be determined. From Table 1 for $n = 10$, $K_f = 0.91$. From Table 2 for a σ_0/E_2 of 0.02 and a SSFOR of 1/100, $K_G = 1.05$. Thus the required value of V_3/E_2 is 0.9555,

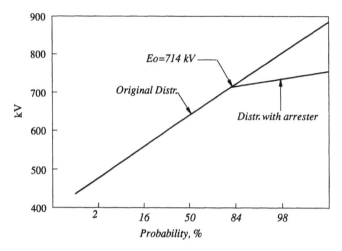

Figure 4 Arresters effect cumulative SOV distribution.

which in turn results in a required V_3 of $0.9555(737) = 704$ kV, a required CFO of 891 kV, and a required BSL of 811 kV assuming sea level conditions. From Table 5, selecting a BSL of 821 kV gives a BIL of 1050 kV. Therefore for the bus supports and the disconnecting switches, a BIL of 1050 kV can be specified. For the breaker, the BSL of 1300 kV is more than adequate. For the conductor-to-tower-leg gap, $k_g = 1.3$ and the required clearance is 2.0 m. Using a PC program results in a BSL of 804 kV and a clearance of 2.00 m.

Table 12 compares the results of the examples plus the use of arresters at 1500 meters. (A PC program gives a BSL of 881 kV and a clearance of 2.33 m.) Interestingly, the use of arresters at 1500 m results in a lower BIL than required at sea level when no arresters are used, 1300 vs. 1175 kV BIL and 2.31 vs. 2.33 m. As mentioned previously, the normal BIL used at a nominal system voltage of 500 kV is 1800 kV. As will be evident in Chapter 13, the BIL required from a lightning standpoint is larger than that required by switching overvoltages.

In this example, E_{2A} was less than E_2, and thus arresters were beneficial. If E_{2A} is equal to or greater than E_2, then the design should be based on the original value of E_2.

Table 12 Comparison of BSLs and Clearances

Altitude, meters	Criterion	Req'd BSL, kV	Selected BIL, kV	Clearance, m
0	w/o considering line	844	1175	2.12
0	Considering line	902	1300	2.31
0	With arresters	904	1050	2.00
1500	Considering line	981	1425	2.68
1500	With arresters	881	1175	2.33

Also note that in Table 2, K_G and K_E vary significantly dependent on the values of σ_0/E_2 or β/E_2 and the design value of SSFOR. For the Gaussian distribution of SOVs, K_G increases as the σ_0/E_2 decreases. Thus the required value of V_3 increases.

2.13 Transformer and Transformer Bushing

Almost universally, an arrester is located immediately adjacent to the transformer, either at the transformer or on the transformer bus, and therefore the arrester switching impulse discharge voltage is compared to the BSL. Since the transformer and the internal insulation of the transformer bushing is non-self-restoring, a deterministic insulation coordination method must be used with a minimum safety margin of 15%. Therefore

$$\text{BSL} = 1.15 E_{\text{SI}} \tag{27}$$

where E_{SI} is the arrester switching impulse discharge voltage as published by the manufacturer for the discharge current as specified in Table 8. Using the previous example, where this voltage E_{SI} was 823 kV, the required BSL is 946 kV. From Table 4, the next highest transformer BSL is 975 kV, which translates to a BIL of 1175 kV. Again using Table 4 for the transformer bushing, the next highest BSL is 1050 kV, which is for a bushing BIL of 1300 kV. Thus the required BILs of the transformer and of the internal insulation of the bushing are 1175 and 1300 kV, respectively. Examining Table 4, it is noted that the minimum BIL used at 500 kV is 1300 kV. Thus, considering only switching overvoltages, a BIL of 1300 kV would be specified.

For the external transformer bushing insulation, the effect of altitude must be considered. In addition, the use of a safety margin is debatable. In selecting the BSL of other self-restoring insulations in the stations, a safety margin was not considered. It appears better to decrease the desired or design SSFOR than to use a safety margin. Thus, for sea level conditions, for the external bushing, Eq. 27 could be used without a margin. Not using a safety margin, the required BSL of the external bushing would be 823 kV, which translates to a BIL of 1175 kV. However, since the internal insulation BIL of the transformer is 1300 kV, the bushing would normally be specified as a 1300 kV BIL.

For higher altitudes, the insulation strength of the external bushing insulation decreases so that the required BSL increases. That is, the required BSL at an altitude A, BSL_A, is

$$\text{BSL}_A = \delta^m (\text{BSL}_S) \tag{28}$$

where BSL_S is the standard value (at sea level). To obtain the value of m, the strike distance across the bushing and the CFO must be known. If the strike distance is not known, it can be estimated from the equation for a rod–plane gap. However, first the CFO is determined by Eq. 4. The strike distance is then obtained from the equation for a rod–plane gap, i.e.,

$$S = \frac{8}{3400/\text{CFO}_S - 1} \tag{29}$$

Table 13 Finding the Bushing BSL

BSL$_S$	CFO$_S$	S	G_0	m	BSL$_S$
997	1096	3.80	0.5763	0.271	992
992	1090	3.78	0.5774	0.272	992

where S is the strike distance in meters. Then the value of m and G_0 are found from the previous equations. To demonstrate, assume as before a required BSL of 946 kV at an altitude of 1500 meters, $\delta = 0.840$, i.e., BSL$_A$ = 946 kV. To start, assume $m = 0.3$, resulting in a BSL$_S$ of $946/0.9165 = 997$ kV. Proceeding as in Table 13 results in a BSL$_S$ of 992 kV.

Normally, the value of m ranges from about 0.3 to 0.5, and to obtain a quick and easy estimate, a conservative value of 0.3 is recommended.

Applying no safety margin, the required BSL is 992 kV. From Table 4, the next highest BSL is 1050 kV, which translates to a BIL of 1300 kV. If a safety margin of 15% is used, the BIL would increase to 1550 kV.

Comparing this to previous results with arresters at 1500 meters, the BSL required for other equipment was only 881 kV, which for a transformer bushing would lead to a BIL of 1300 kV.

3 PHASE–PHASE INSULATION COORDINATION

3.1 Insulation Strength

As explained in Chapter 4, two methods exist for determining the strength of phase–phase insulations. To review,

The Alpha Method

A positive voltage αV_p is applied to one phase and a negative voltage $(1-\alpha)V_p$ is applied to the other phase, where V_p is the phase–phase voltage. With α held constant, the voltage V_p is varied to determine the phase–phase CFO denoted as CFO$_p$. Per this method, α is defined as

$$\alpha = \frac{V^-}{V^+ + V^-} = \frac{V^-}{V_p} = 1 - \frac{V^+}{V_p} \qquad (30)$$

The equation for the CFO$_p$ is

$$\mathrm{CFO}_p = k_{gp} \frac{3400}{1 + (8/S)} \qquad (31)$$

where S is the phase–phase spacing and k_{gp} is the gap factor for this method. This method is employed in the IEC and used with alpha equal to 0.5, i.e. $V^+ = V^-$, to estimate the strike distances (clearances). The phase–phase BSL, BSL$_p$, in IEC standards [3] is defined for an alpha of 0.5 and therefore

$$\text{BSL}_p = \text{CFO}_p\left(1 - 1.28\frac{\sigma_{fp}}{\text{CFO}_p}\right) \quad \text{with } \alpha = 0.5 \tag{32}$$

The V^+-V^- Method

With a constant negative polarity voltage V^- applied to one phase, the positive polarity voltage on the other phase is varied to obtain the positive polarity CFO^+. This CFO^+ may be related to the negative polarity voltage by the equation

$$\text{CFO}^+ = \text{CFO}_0 - K_L V^- \tag{33}$$

where K_L is a constant dependent on the gap configuration and CFO_0 is the CFO when the negative polarity component is zero (one of the two electrodes is grounded) and is given by the equation

$$\text{CFO}_0 = K_{GP}\frac{3400}{1 + (8/S)}$$

$$V_{30} = \text{CFO}_0\left(1 - 3\frac{\sigma_{FP}}{\text{CFO}_0}\right) \tag{34}$$

Because the alpha method was first employed, most if not all test results are available using this method. However, these test results may be transcribed or changed so as to apply to the other method. That is,

$$K_{GP} = k_{gp}[1 - \alpha(1 - K_L)] \tag{35}$$

Using this equation, the gap factors for the two methods are shown in Table 14, as obtained from Chapter 4, for alternate gap configurations. As discussed in Chapter 4, the CFO_0 is the phase–phase CFO when one of the electrodes is grounded. Thus the CFO_0 is similar to the CFO_g^+ since this CFO is also the CFO when one of the electrodes is grounded. Therefore σ_f/CFO or σ_f/CFO_g^+ should be identical to σ_{FG}/CFO_0, and also K_{GP} should be equal to k_g. It is true that the CFO_0 is used differently from CFO_g^+, but the gap factors are measured or determined in the same manner or by the same type of test, i.e., with one electrode grounded. Therefore in Table 14, for the V^+-V^- method, the nomenclature for the gap factor has been changed to k_g and that for the standard deviation has been changed to σ_f/CFO_0. In addition, since σ_{fg}/CFO_p is also equal to σ_f/CFO, this nomenclature has been changed to σ_f/CFO_p. This should make the presentation and use of these variables clearer and simpler.

Referring to Fig. 1, the length of the bus upon which the switching surge impinges is relatively short, less than the span length of the line. Thus the conductor–conductor gap configuration whose length is large is not given in Table 14. Rather the 10-meter conductor–conductor configuration will be used for setting the phase–phase clearance of the bus. The ring–ring or large smooth electrode configuration is applicable to large-diameter grading rings such as used at EHV on bushings. At lower voltages where the grading rings are considerably smaller, the rod–rod configuration is more applicable. Referring to the rod–rod configuration,

Table 14 Gap Factors for Phase–Phase Switching Impulses

	Alpha method			V^+–V^- method		
Gap configuration	α	k_{gp}	σ_f/CFO_p	K_L	k_g	σ_f/CFO_0
Ring–ring or large smooth electrodes [4]	0.33 0.50	1.70 1.80	0.05 0.05	0.70	1.53	0.05
Crossed conductors [4]	0.33 0.50	1.53 1.65	0.05 0.05	0.62	1.34	0.05
Rod–rod [4]	0.33 0.50	1.52 1.62	0.05 0.05	0.67	1.35	0.05
Conductor–conductor, 10 m length [4]	0.33 0.50	1.52 1.62	0.035 0.035	0.67	1.35	0.035
Supported busbar fittings [4]	0.33 0.50	1.40 1.50	0.05 0.05	0.65	1.23	0.05
Asymmetrical geometries, e.g., rod–conductor [4]	0.33 0.50	1.36 1.45	0.05 0.05	0.67	1.21	0.05
Jumper–shield ring and conductor–conductor (short length) [5]	0.33 0.50	1.57 1.68	0.04 0.04	0.66	1.39	0.04
Conductor–conductor (47 m length) [6]	0.33 0.50	1.56 1.66	— —	0.68	1.40	—

note that the gap factor of Table 3, 1.25 to 1.45, with a typical value of 1.35, is practically identical to the gap factor of 1.35 of Table 14. Thus either of these gap factors could be used.

As mentioned in Chapter 4, for small air gaps of less than 2 or 3 meters, i.e., for nominal system voltages less than about 500 kV, K_L is approximately equal to one. If $K_L = 1.00$, the separation of the phase–phase into components is unnecessary. Also $K_L = 1.00$ leads to larger gap spacings or clearances.

Using Eq. 35, the two CFOs can also be related by the equation

$$\text{CFO}_p = \frac{\text{CFO}_0}{1 - \alpha(1 - K_L)} \tag{36}$$

In IEC [3], the phase–phase BSL is defined for $\alpha = 0.5$, which stipulates that half the crest voltage is positive and applied to one terminal and half the crest voltage is negative and applied to the other terminal. Thus $\alpha = 0.5$, and Eq. 36 becomes

$$\text{CFO}_p = \frac{2\,\text{CFO}_0}{1 + K_L}$$
$$\text{BSL}_p = \frac{2\,\text{BSL}_0}{1 + K_L} \tag{37}$$

3.2 SOV Distribution

As discussed in Chapter 4, three SOV voltages exist, V^-, V^+, and V_p. Since these are related by the equation

$$V_p = V^+ + V^- \tag{38}$$

only two of these voltages need be obtained.

Usually, since V_p and V^+ are definable quantities and can be better estimated, these SOV distributions are obtained. As discussed in Chapter 4, the maximum phase–phase SOV does not occur at the same exact time instant as the maximum phase–ground SOV. Therefore theoretically these voltages should be collected at two time instants, the time of maximum phase–phase SOV and the time of maximum positive phase–ground SOV. However, usually the phase–phase voltage is collected at its time instant and the maximum phase–ground voltage is collected at its time instant. Thus the highest phase–phase voltage is considered as being time coincident with the highest phase–ground voltage. This method of collection should result in conservative values of phase–phase BSL and clearance. Further, in most cases, at the higher SOVs that are of primary concern, the maximum phase–phase SOVs occur at approximately the same time instant as the maximum phase–ground SOVs.

To date, sufficient studies have not been made to estimate accurately the phase–phase switching overvoltage. However, per Chapter 4, for a Gaussian distribution, the ratio of E_{2p}/E_2^+ is approximately constant at 1.55 [3].

3.3 Combined Distribution of V_Z

In Chapter 4, the reduced variable V_Z was defined by one of the following equations:

$$\begin{aligned} V_Z &= V^+ + K_L V^- \\ V_Z &= (1 - K_L)V^+ + K_L V_p \end{aligned} \tag{39}$$

Assuming that the distribution of V^+ and V_p are Gaussian, the distribution of V_Z is also Gaussian with the parameters μ_Z and σ_Z given by

$$\begin{aligned} \mu_Z &= (1 - K_L)\mu^+ + K_L \mu_p \\ \sigma_Z &= \sqrt{[(1 - K_L)\sigma^+]^2 + (K_L \sigma_p)^2 + 2\rho_{p+}(1 - K_L)K_L \sigma_p \sigma^+} \end{aligned} \tag{40}$$

If $\rho_{p+} = 1.0$, then

$$\begin{aligned} \sigma_Z &= (1 - K_L)\sigma^+ + K_L \sigma_p \\ E_{2Z} &= (1 - K_L)E_2^+ + K_L E_{2p} \end{aligned} \tag{41}$$

If $K_L = 1.0$,

$$\mu_Z = \mu_p \qquad \sigma_Z = \sigma_p \qquad E_{2Z} = E_{2p} \tag{42}$$

The correlation coefficient ρ_{p+} is positive and normally ranges from about 0.8 to 1.0.

Or considering the second form of Eq. 39, if the distractions of V^- and V^+ are Gaussian, then the distribution of V_Z is also Gaussian with the parameters μ_Z and σ_Z, i.e.,

$$V_Z = V^+ + K_L V^-$$
$$\mu_Z = \mu^+ + K_L \mu^- \tag{43}$$
$$\sigma_Z = \sqrt{(\sigma^+)^2 + (K_L \sigma^-)^2 + 2\rho_{+-} K_L \sigma^+ \sigma^-}$$

If $\rho_{+-} = 1.0$, then

$$\sigma_Z = \sigma^2 + K_L \sigma^-$$
$$E_{2Z} = E_2^+ + K_L E_2^- \tag{44}$$

and with $\rho_{+-} = 1.0$ and $K_L = 1.00$,

$$\mu_Z = \mu^+ + \mu^- \quad \sigma_Z = \sigma^+ + \sigma^- \quad E_{2Z} = E_2^+ + E_2^- \tag{45}$$

3.4 Estimating Method

As in Chapter 4, the method employed to estimate the strike distance or clearance or the SSFOR assumes that

1. All correlation coefficients are equal to 1.00.
2. Only the parameters of the original SOV distribution are used. That is, do not use "reversed parameters".

However, in contrast to the methods of Chapter 4,

3. The number of parallel insulations is considered to be equal to 1, and therefore an improved estimating method can be used that accounts for both σ_f and σ_0, the standard deviation of the Gaussian SOV distribution. As developed in Section 7.2 of Chapter 3, for a Gaussian SOV distribution and a Gaussian strength characteristic,

$$Z_e = \frac{CFO_0 - \mu_Z}{\sqrt{\sigma_f^2 + \sigma_Z^2}} \tag{46}$$

Then the SSFOR is

$$SSFOR = \frac{1}{2}[1 - F(Z_e)] \tag{47}$$

Alternately, if the clearance or BSL is desired, then Z_e is obtained from Eq. 47 or from Table 2, and Eq. 46 is used to find the CFO_0, i.e.,

$$\mathrm{CFO}_0 = \mu_Z + Z_e\sqrt{\sigma_f^2 + \sigma_Z^2} \tag{48}$$

The clearance is then obtained from Eq. 34. To find the BSL, the CFO_0 must be changed to CFO_p using Eq. 37. Then the BSL_p is found from

$$\mathrm{BSL}_p = \mathrm{CFO}_p\left(1 - 1.28\frac{\sigma_f}{\mathrm{CFO}_p}\right) \tag{49}$$

where the σ_f/CFO_p is obtained from Table 14.

As for the previous phase–ground case, the design value of the SSFOR usually ranges from 0.10/100 to 1.0/100, although the de facto standard appears to be 1.0/100.

3.5

Example 4. Estimating the BSL and Clearance. Consider a 500/500 kV system (1 pu = 449 kV) for which the distribution of phase–phase SOVs is approximated by a Gaussian distribution having a E_{2p} of 2.8 pu and a σ_p/E_{2p} of 0.10. In addition, the distribution of the positive SOVs is considered Gaussian with $E_2^+ = 1.8$ pu and $\sigma^+/E_2^+ = 0.10$. As per the above, let the correlation coefficient between the phase–phase and the positive SOVs equal 1.0, and let the number of insulations equal one. Determine the required phase–phase clearance between conductors of 10 m length, i.e., per Table 14, $K_L = 0.67$, $k_g = 1.35$, and $\sigma_f/\mathrm{CFO}_0 = \sigma_f/\mathrm{CFO}_p = 0.035$. Then, designing for an SSFOR of 1/100,

$$\mu_p = 2.225 \quad \sigma_p = 0.280 \quad \mu^+ = 1.430 \quad \sigma^+ = 0.180 \tag{50}$$

$$\mu_Z = 1.963 \quad \sigma_Z = 0.247 \quad Z_e = 2.0542 \tag{51}$$

$$\mathrm{CFO}_0 = 1.963 + 2.0542\sqrt{0.247^2 + \sigma_f^2} \tag{52}$$

Since the value of σ_f is unknown, a short iteration is necessary, see Table 15. Therefore $\mathrm{CFO}_0 = 2.50$ per unit or 1122.5 kV. Using Eq. 34, $S_p = 2.59$ meters. Using a computer program, $S_p = 2.59$ m.

Table 15 Finding the CFO_0

σ_f	CFO_0	σ_f
0.0	2.470	0.0865
0.0865	2.500	0.0875
0.0875	2.500	0.0875

Repeat this example with $k_g = 1.53$, $K_L = 0.70$ and $\sigma_f/\text{CFO}_0 = 0.05$, which are the factors for a ring–ring gap representing the spacing between the circuit breakers. Then

$$\mu_Z = 1.9865 \qquad \sigma_Z = 0.250 \qquad E_{2Z} = 2.50 \qquad \sigma_Z/E_{23Z} = 0.100 \qquad (53)$$

$$\text{CFO}_0 = 1.9865 + 2.0542\sqrt{0.250^2 + \sigma_f^2} = 1151\,\text{kV} \qquad \sigma_f = 0.128 \qquad (54)$$

Using Eq. 34, the clearance is $S_p = 2.227$ meters. To find the phase–phase BSL, BSL_p, use Eqs. 37 and 49, i.e.,

$$\text{CFO}_p = \frac{2(1151)}{1.7} = 1354\,\text{kV} \qquad \text{BSL}_p = 0.936(1354) = 1267\,\text{kV} \qquad (55)$$

Using a computer program, $S_p = 2.27\,\text{m}$ and $\text{BSL}_p = 1268\,\text{kV}$.

Thus, for the standard test, a positive voltage of 633.5 kV is applied to one phase and a negative voltage of 633.5 kV to the other phase.

This example assumes sea level conditions. For altitudes greater than zero, the strike distance increases and may be calculated using the same methods as are used for the phase–phase ground SOVs.

3.6 Arresters

Provided the SOVs are sufficiently high so that at least one arrester operates, the phase–phase SOVs are reduced, thus reducing the required clearance and BSL. The procedure, in this case, is first to determine if one or more arresters operate. Then the SOV distribution parameters are calculated for one or both arresters. In contrast to the previous phase–phase calculations, for this case the positive and *negative* SOVs are used, and for conservatism the correlation coefficient between the positive and negative SOVs is set to 1.00.

If both arresters operate, then the SSFORs for the original and for the reversed parameters are equal. Therefore to design for a SSFOR of 1.0/100, the input design SSFOR is 0.5/100. Similarly, for a design of a SSFOR of 0.1/100, an SSFOR of 0.05/100 should be used. An example should clarify the procedure.

3.7

Example 5. With Arresters. In Example 4, the positive polarity and phase–phase SOVs are given. However, if arresters are used, the distribution of the negative SOVs are required. The parameters of the negative SOV distribution can be obtained from the equations

$$\begin{aligned}\mu^- &= \mu_p - \mu^+ \\ \sigma^- &= \sqrt{(\sigma_p)^2 + (\sigma^+)^2 - 2\rho_{p+}\sigma_p\sigma^-}\end{aligned} \qquad (56)$$

Switching Overvoltages, Substations

where for an assumed value of $\rho_{p+} = 1.00$,

$$\sigma^- = |\sigma_p - \sigma^+| \tag{57}$$

Using the results of Example 4 and Eqs. 56 and 57, $\mu^- = 0.795\,\text{pu}$, $\sigma^- = 0.100$, and $E_2^- = 1.00\,\text{pu}$ or 449 kV. Assuming the use of a 318 kV MCOV arrester with a switching impulse discharge voltage of 823 kV at 2 kA, per Example 3, the arrester will modify the SOVs on the positive phase but not on the negative phase. Using $K_L = 0.700$, $k_g = 1.53$, $\sigma_f/CFO_0 = \sigma_f/CFO_p = 0.05$ (ring-ring gap), and a line surge impedance of 350 ohms, we obtain, for the positive phase, from Example 3,

$$R_A = 113\,\text{ohms} \quad E_0 = 714\,\text{kV} \quad K_A = 0.2441 \quad \mu_A = 696\,\text{kV} \tag{58}$$
$$\sigma_A = 19.7\,\text{kV} \quad E_{2A} = 737\,\text{kV} \quad \sigma_A/E_{2A} = 0.0268$$

For the negative phase we obtain

$$E_2^- = 1.00\,\text{pu} \quad \sigma^- = 0.10\,\text{pu} = 45\,\text{kV} \quad \mu^- = 0.795\,\text{pu} = 357\,\text{kV} \tag{59}$$

And the combined Z distribution (see Eqs. 43–44) is

$$\mu_Z = 946\,kV \quad \sigma_Z = 51.2\,kV \quad E_{2Z} = 1051\,\text{kV} \quad \sigma_Z/E_{2Z} = 0.049 \tag{60}$$

For a SSFOR of 1.0/100, $Z_e = 2.0542$. Then

$$CFO_0 = 946 + 2.0542\sqrt{51.2^2 + \sigma_f^2} \tag{61}$$

and

$$\sigma_f = 55.0\,\text{kV} \quad CFO_0 = 1100\,\text{kV} \quad CFO_p = 1294\,\text{kV} \tag{62}$$

$$BSL_p = 1211\,\text{kV} \quad S_p = 2.15\,\text{meters} \tag{63}$$

Using a computer program, $S_p = 2.14\,\text{m}$ and $BSL_p = 1210\,\text{kV}$.

The calculated values should be compared to an S of 2.82 meters and a BSL_p of 1267 kV when arresters are not used.

Frequently, to obtain a conservative answer, arresters are assumed to operate on both phases. That is, both the positive and negative phases are assumed to have identical values of E_2 and σ_0, i.e., E_{2A} and σ_A. For this case, a SSFOR of 0.5/100 is used, and $Z_e = 2.3268$. The values are

$$\begin{array}{lll} \mu_Z = 1183\text{ kV} & \sigma_Z = 33.5\text{ kV} & E_{2Z} = 1253\text{ kV} \\ \sigma_Z/E_{2Z} = 0.027 & \sigma_f = 68\text{ kV} & CFO_0 = 1360\text{ kV} \\ CFO_p = 1600\text{ kV} & S = 2.83\text{ m} & BSL_p = 1497\text{ kV} \end{array} \tag{64}$$

Using a computer program, $S_p = 2.83\,\text{m}$ and $BSL_p = 1495\,\text{kV}$.

4 CFO AND K_L AS A FUNCTION OF S_p/h

As discussed in Chapter 4, K_L appears to be a function of S_p/h. In many cases, the CFO_0 is also a function of S_p/h. Thus iteration is required to determine the value of S_p. To demonstrate, consider the problem of estimating the phase–phase clearance between a rod–rod gap. To expand the complexity, let the altitude of the substation be 1000 meters. Then assume the SOV data of Example 4. The gap factor for the rod–rod gap from Chapter 2 and the equation for K_L from Chapter 4 are

$$K_L = 1 - .08 \frac{S_p}{h}$$
$$k_g = 1.35 - \left(\frac{S_p}{h} - 0.5\right) \tag{65}$$

To be complete, the required CFO_0 at the altitude of the substation, CFO_{0A}, is

$$CFO_{0A} = \mu_Z + 2.0542\sqrt{\sigma_Z^2 + \sigma_f^2} \tag{66}$$

and given the CFO_{0A}, the phase–phase clearance is given by the equation

$$S_p = \frac{8}{\dfrac{3400\,k_g\delta^m}{CFO_{0A}} - 1} \tag{67}$$

Also, at sea level,

$$CFO_{0S} = k_g \frac{3400}{1+(8/S_p)} \tag{68}$$

Letting $h = 7$ meters and starting with an $S_p = 2.6$ meters, the iteration proceeds as in Table 16. Therefore, $S_p = 2.73$ meters. Note that from the first to the final values, S_p and S_p/h increase by 5%, k_g decreases by 1.3%, and S_p increases by 5%. Thus the initial guess is fairly good and only 1 or 2 iterations become necessary.

However, not all gap factors are formulated with the parameter S_p/h. Therefore, at present, a single value of k_g must be used. In contrast, if desired a value of K_L dependent on S_p/h may be used.

Table 16 Finding S_p

S_p	S_p/h	K_L	k_g	μ_Z	σ_Z	CFO_{0A}	CFO_{0S}	G_0	m	S_p
2.6	0.371	0.703	1.479	1.989	0.250	1152	1233	0.949	0.888	2.73
2.73	0.390	0.688	1.460	1.977	0.249	1147	1263	0.925	0.839	2.73

5 GENERAL GAP FACTORS IN SUBSTATIONS

As a general guide, typical gap factors for use in a substation are presented in Fig. 5. As presented in this chapter, some of these gap factors are to be used with the phase–phase CFO_0 and some are for the CFO to ground. Not shown are the rod–plane gap factor ($k_g = 1.00$), which is virtually never used, and the conductor-to-lower-rod gap factor, which is used between a conductor on the upper bus to the tip of an opened disconnecting switch on the lower bus ($k_g = 1.50$). Again, these are typical but should only be used as a guide.

6 COMPARISON WITH IEC

6.1 IEC Methods: Phase–Ground [3]

For *internal* insulations, e.g. transformer, protected by an arrester, the IEC equation used to determine the BSL is:

$$BSL = K_S K_{cd} E_{SI}$$

where K_S is a safety factor for internal insulations of 1.15, E_{SI} is the arrester switching impulse discharge voltage and K_{cd} is a factor to account for the low value of σ_A as discussed in section 2.11 and 2.12. In the example in IEC 71.2, $E_{SI} = 1300$ and $K_{cd} = 1.03$. Thus the required BSL is 1540 kV and the selected standard BSL is 1550 kV

For *external* insulations, the IEC equation for the BSL is:

$$BSL = K_{cs} K_a K_S E_2 \tag{70}$$

K_a is the altitude correction factor. For an altitude of 1000 m used in the IEC example, the relative air density per IEC is 0.885 and using a m of 0.6, δ^m is 0.929 and the inverse is about 1.07 which is K_a and K_S is a safety factor of 1.05 recommended for external insulations, E_2 is identical to that defined in this and previous chapters, 2% of the SOVs equal or exceed E_2.

The factor K_{cs} is determined based on the design SSFOR. That is:

$$K_{cs} = \frac{BSL}{E_2} \tag{71}$$

In the IEC example, $K_{cs} = 1.15$ which is stated to produce a SSFOR of 1/10,000. For the IEC example, for a maximum system voltage of 765-kV, $E_2 = 1200$ kV (1.92 pu). Thus the required BSL becomes 1550 kV.

The phase–ground clearance is not calculated in the IEC example. Rather, the clearance is obtained from a standard table, i.e. Table 18. Thus the clearance for a conductor-structure gap ($k_g = 1.30$) is 4.9 m and the clearance for a rod-structure gap ($k_g = 1.10$) is 6.4 m.

Using a computer program, for the BSL and strike distances determined by the IEC method, the SSFOR is less than 0.5/100,000. Using a computer program, for a SSFOR of 1/1000, the BSL is 1325 kV for $n = 1$ and 1426 kV for $n = 10$. For this same SSFOR, for $k_g = 1.30$, the strike distance is 4.0 m for $n = 1$ and 4.5 m for $n = 10$.

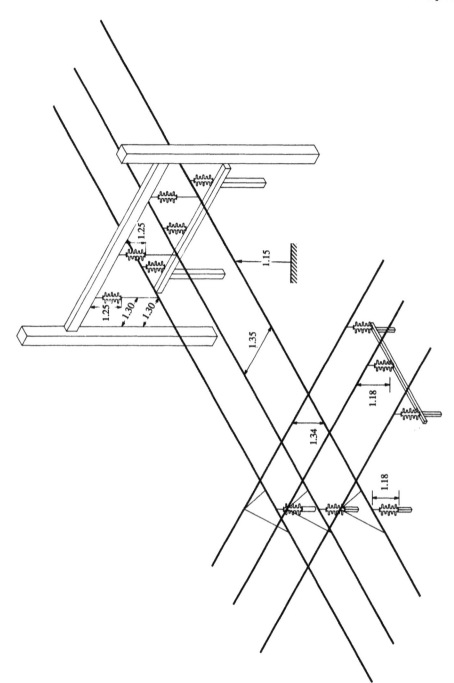

Figure 5 Typical gap factors in a substation.

Switching Overvoltages, Substations

Using the above results to evaluate the IEC method, the IEC method is very conservative resulting in BSLs and strike distances that are excessive.

6.2 IEC Method – Phase–Phase [3]

Again, through some method, the values of E_2 for the phase to ground SOVs, E_2^+, and for the phase–phase SOVs, E_{2p}, are obtained. Per IEC, the BSL_p for 3-phase, external insulation is found from the equation:

$$BSL_p = 2K_{cs}K_aK_S(F_2E_2^+ + F_1E_{2p}) \qquad (72)$$

where

$$F_1 = \frac{1}{2-\sqrt{2}}\left[1-\frac{\sqrt{1+K_L^2}}{1+K_L}\right] \quad F_2 = \frac{1}{2-\sqrt{2}}\left[2\frac{\sqrt{1+K_L^2}}{1+K_L}-\sqrt{2}\right] \qquad (73)$$

To derive a similar equation using the equation from this chapter, i.e.,

$$E_{2Z} = (1 - K_L)E_2^+ + K_L E_{2p} \qquad (74)$$

$$BSL_p = \frac{2\,BSL_0}{1+K_L} \qquad (75)$$

Also, let the design criterion be:

$$\frac{BSL_0}{E_{2Z}} = K_{cs} \qquad (76)$$

Combining these equations and inserting K_a and K_S:

$$BSL_p = \frac{2K_{cs}K_aK_S}{1+K_L}\left[(1-K_L)E_2^+ + K_L E_{2p}\right] \qquad (77)$$

As noted, this equation differs from the IEC equation. To illustrate the difference, let $E_2^+ = 1200$ kV (1.92 pu), $E_{2p} = 2040$ kV (3.27 pu), $K_L = 0.6$, $K_{cs} = 1.15$, $K_a = 1.07$ (altitude correction for 1000 m) and $K_S = 1.05$ (safety factor for external insulation). These values are identical values to those given in IEC as an example for a system having a maximum system voltage of 765 kV. Using the IEC equation, the $BSL_p = 2670$ kV while using the above equation the $BSL_p = 2752$ kV, a difference of about 3%. Examining the standard BSL_p, the largest value is 2480 kV [7]. Thus, for either method, standard BSL_p are not adequate. To be noted is that $K_{cs} = 1.15$, which provides an SSFOR of 1/1000 to 1/10,000, is multiplied by a safety factor, K_S, of 1.05. This is equivalent to increasing the value K_{cs} to 1.21, thus decreasing the design value of the SSFOR. In this chapter, K_S is not used, i.e. $K_S = 1.00$. Using this value, the BSL_p is 2620 kV, which is about the same value as given by the IEC equation.

The clearance per IEC is determined using the same equations as before plus the equation:

$$BSL_0 = (1 - 1.28\sigma_f/CFO_0)CFO_0 \tag{78}$$

The objective is to determine the CFO_0. Therefore, the equation is:

$$CFO_0 = \frac{K_{cs}K_aK_S(1+K_L)(F_2E_2^+ + F_1E_{2p})}{1 - 1.28\sigma_f/CFO_0} \tag{79}$$

The equation derived from this chapter is:

$$CFO_0 = \frac{K_{cs}K_aK_S[(1-K_L)E_2^+ + K_LE_{2p}]}{1 - 1.28\sigma_f/CFO_0} \tag{80}$$

Using the same value of the parameters as previously employed, the $CFO_0 = 2314\,kV$. The clearance is determined by use of the equation:

$$CFO_0 = 1080\,k_g\ln(0.46\,S_p + 1) \tag{81}$$

In IEC the gap factor for a conductor–conductor gap is chosen as 1.62 and therefore the clearance or strike distance, S_p, is 6.0 m. Also for a rod to conductor gap with a gap factor of 1.45, $S_p = 7.4$ m. These gap factors are presented in Table 14 for the α method for $\alpha = 0.5$. However, these gap factors should be applied to CFO_p. For the $V^+ - V^-$ method, i.e. CFO_0, Table 14 shows a gap factor of 1.35 for the conductor–conductor gap and 1.21 for the rod–conductor gap. Using these gap factors the clearances are 8.46 m and 10.60 m, respectively.

Using equation 79, the $CFO_0 = 2385\,kV$. Then with the gap factors of 1.35 and 1.21, $S_p = 8.98$ and 11.31 m. If the safety factor of 1.05 is not used, $CFO_0 = 2271\,kV$ and $S_p = 8.15$ and 10.18 m.

Using the computer program to calculate the $SSFOR_p$, for a BSL_p of 2670 kV as found using the IEC equations, the $SSFOR_p$ is 0.8/100,000 and for the strike distances of 8.46 m and 10.60 m, the $SSFOR_p$s are 0.1/100,000 and 0.8/100,000. If a design $SSFOR_p$ of 1/100 is used, the resultant BSL_p is 2170 kV and the strike distances are 5.51 m and 6.79 m. For a $SSFOR_p$ of 1/1000, the BSL_p becomes 2359 kV and the strike distances are 6.33 m and 8.02 m. These values, even for a $SSFOR_p$ of 1/1000 are significantly lower than those calculated using the IEC method.

In evaluation of the IEC method, the method makes some of the same assumptions as made in this chapter. That is, (1) $\rho_{p+} = 1.0$ and (2) no "reversed parameter". However, the IEC method is very conservative leading to excessive insulation levels and strike distances.

6.3 IEC Clearances [3, 7]

The BSLs, phase–ground and phase–phase, taken from IEC Publication 71 [7], are shown in Table 17. Note that the ratio of BSLs, phase–phase to phase–ground, increase as the system voltage increases. This reflects the thought that at higher voltages, some type of control of phase–ground SOVs occurs—such as closing resistors in the circuit breaker. However, controlling the phase–ground SOVs does not have as much effect on the phase–phase SOVs. Thus the phase–phase to phase–ground BSL ratio tends to increase slightly. Also note that the IEC specifies the

Table 17 IEC BIL/BSLs

Max. system voltage, kV	Phase–ground BSL, BSL_g, kV	Ratio BSL_p/BSL_g	BIL, kV
300	750	1.50	850 or 950
	850	1.50	950 or 1050
362	850	1.50	950 or 1050
	950	1.50	1050 or 1175
420	850	1.60	1050 or 1175
	950	1.50	1175 or 1300
	1050	1.50	1300 or 1425
550	950	1.70	1175 or 1300
	1050	1.60	1300 or 1425
	1175	1.50	1425 or 1550
800	1300	1.70	1675 or 1800
	1425	1.70	1800 or 1950
	1550	1.60	1950 or 2100

Source: From IEC Publication 71 [7].

BSLs and BILs for each system voltage, whereas ANSI/IEEE standards leave this to the system designer. In addition, as discussed in Chapter 1, the standard values of BIL and BSL differ in the IEC and ANSI/IEEE standards.

The phase–ground and phase–phase clearances recommended by IEC are presented in Tables 18 and 19. These are shown as a function of the phase–ground or phase–phase BSLs. The phase–ground clearances are calculated using the older Paris–Cortina equation for BSLs up to and including 1550 kV. As discussed in Chapter 1, this equation is only valid for one wave front, whereas the presently

Table 18 IEC Phase–Ground Clearances for Switching Overvoltages

$$\text{CFO} = 500 k_g S^{0.6} \qquad \sigma_f/\text{CFO} = 0.06$$

Phase–ground BSL BSL_g, kV	Clearance, meters Conductor–structure, $k_g = 1.30$	Clearance, meters Rod–structure, $k_g = 1.10$
750	1.6	1.9
850	1.8	2.4
950	2.2	2.9
1050	2.6	3.4
1175	3.1	4.1
1300	3.6	4.8
1425	4.2	5.6
1550	4.9	6.4

Source: IEC Publication 71 [3].

Table 19 IEC Phase–Phase Clearances for Switching Overvoltages

$$\text{CFO}_p = k_{gp} \frac{3400}{1+(8.S)} \qquad \alpha = 0.5 \qquad \sigma_{fp}/\text{CFO}_p = 0.06$$

Phase–ground BSL, BSL_g, kV	Phase–phase BSL, BSL_p, kV	Ratio BSL_p/BSL_g	Clearance, meters conductor–conductor $k_{gp} = 1.62$	Clearance, meters Rod–conductor $k_{gp} = 1.45$
750	1125	1.5	2.3	2.6
850	1275	1.5	2.6	3.1
850	1360	1.6	2.9	3.4
950	1425	1.5	3.1	3.6
950	1615	1.7	3.7	4.3
1050	1575	1.5	3.6	4.2
1050	1680	1.6	3.9	4.6
1175	1763	1.5	4.2	5.0
1300	2210	1.7	6.1	7.4
1425	2423	1.7	7.2	9.0
1550	2480	1.6	7.6	9.4

Source: IEC Publication 71 [3].

used Gallet–LeRoy equation applies for the critical wave front. The reason for the use of the older equation is simply that these same values appeared in earlier editions of the standard and were used with apparent success. Therefore the changing of the clearances would be detrimental to standardization. However, when adding the 1675 and 1800 kV BSLs, the newer Gallet–LeRoy equation was used.

For phase-to-ground clearances, the gap factor 1.10 for the rod–structure is the worst electrode configuration normally encountered in a station, while the gap factor of 1.30 for the conductor–structure is applicable to a wide range of gap configurations encountered in a station; see Table 3.

The phase–phase clearances, presented in Table 19, are correlated to the phase–phase BSLs and are shown for two electrode configurations. The unsymmetrical rod–structure gap is the worst electrode configuration normally encountered while the conductor–conductor gap (for 10 meters) is a normal type of configuration and further may be conservatively applied to other types of electrode configurations; see Table 15.

Although presenting the clearances as a function of the BSL is convenient and makes for easy calculation, it may lead to large conservative errors in application. The BIL/BSL of post insulators or of equipment is based not only on switching surges but also on lightning and contamination. In addition, BILs and BSLs are standardized numbers that sometimes greatly exceed those required. Clearances should be based on the actual system conditions as illustrated in this chapter. That is, the clearances as listed in these tables can be used if the BSL is the *required* BSL and not the BSL actually selected or used. Phase–phase BSLs have not been established in ANSI/IEEE standards, nor do recommended phase-ground or phase–phase clearances exist.

7 REFERENCES

1. A Pigini, L. Thione, R. Cortina, K. H. Weck, C. Menemenlis, G. N. Alexandrov, Yu A. Gerasimov, CIGRE Working Group 33.03, "Part II – Switching Impulse Strength of Phase-to-Phase External Insulation," *ELECTRA*, May 1979, pp. 138–157.
2. ANSI/IEEE C62.11, "Standard for Metal Oxide Surge Arresters for Alternating Current Power Circuits."
3. IEC Standard 71-2, "Insulation Coordination, Part 2, Application Guide," 1996.
4. G. Gallet, B. Hutzler, and J. Riu, "Analysis of the Switching Impulse Strength of Phase-to-Phase Air Gaps," *IEEE Trans. on PA&S*, Mar./Apr. 1978, pp. 485–494.
5. M. Miyake, Y. Watanabe, and E. Ohasaki, "Effects of Parameters on the Phase-to-Phase Flashover Characteristics of UHV Transmission lines," *IEEE Trans. on Power Delivery*, Oct. 1987, pp. 1285–1291.
6. I. S. Grant and A. S. Paulson, "Phase-to-Phase Switching Surge Design," EPRI EL-3147, Project 1492, June 1983.
7. IEC Publication 71-1, "Insulation Coordination, Part 1, Definitions, Principles, and Rules," 1993.

8 PROBLEMS

1. Specify the phase–ground and phase–phase BSLs and clearances for a 500 kV station for the following conditions:

   1. Positive polarity SOV, Gaussian, $E_2^+ = 810\,\text{kV}$, $\sigma^+/E_2^+ = 0.10$.
   2. Phase–phase SOV, Gaussian $E_{2p} = 1215\,\text{kV}$, $\sigma_p/E_{2p} = 0.10$.
   3. All correlation coefficients $= 1.0$.
   4. Substation altitude $= 1000$ meters.
   5. Phase–ground SI Line $\text{CFO}_S = 1000\,\text{kV}$ (at sea level), strike distance $= 2.8\,\text{m}$, $\sigma_f/\text{CFO} = 0.05$.
   6. Design for an $\text{SSFOR} = 1/100$ for both phase–ground and phase–phase.
   7. For phase–ground clearance—for tower leg to conductor, $k_g = 1.30$, $\sigma_f/\text{CFO} = 0.07$, $n = 10$.
   8. For phase–ground BSL use $\sigma_f/\text{CFO} = 0.07$, $n = 10$.
   9. For phase–phase bus clearance (conductor–conductor), $K_L = 0.67$, $k_g = 1.35$, $\sigma_f/\text{CFO}_0 = \sigma_f/\text{CRO}_p = 0.035$, $n = 1$.
   10. For phase–phase (ring–ring) BSL, $K_L = 0.70$, $k_g = 1.53$, $\sigma_f/\text{CFO}_0 = \sigma_f/\text{CFO}_p = 0.05$, $n = 1$.
   11. $\gamma^+ = \gamma_Z = 1.00$

6
The Lightning Flash

1 INTRODUCTION—GENERAL BACKGROUND [1-3]

In this world, there are 2000 thunderstorms in progress at any time resulting in 100 lightning flashes to ground per second—8 million per day. Lightning causes about 100 deaths and 250 injuries in the United States per year, more deaths than from any other weather-related phenomenon, be it hurricanes, tornadoes, or floods. Even in a plane one is not immune to lightning, a plane being struck about once per 5000 flying hours.

Considering that lightning activity was several decade steps greater during the early existence of mankind, it is not difficult to imagine our ancestors, the cave men, cowering in fear in their caves. People did not know how to protect themselves or their property against lightning. But although it must have been a terrifying sight, it did give them fire for light and heat. Since people could not understand this fire from heaven, they established gods who had dominion over lightning. The Norse god Thor hurled lightning bolts down from heaven, and in Greek and Roman mythology, Zeus and Jupiter had similar powers.

As mankind evolved and became more educated, people tried to generate theories or explanations and to devise protection methods. For many centuries in Europe and England it was common practice to ring church bells during a lightning storm. The poor and uneducated people believed that this dispersed evil spirits, while the more educated knew that it caused a vibration in air that broke up the continuity of the lightning path. Eventually this practice was banned by various authorities such as Charlemagne because of the high fatality rate among bell ringers. A book published in Munich in 1784, entitled *A Proof that the Ringing of Bells During Thunderstorms May Be More Dangerous than Useful*, showed that in a 33-year period, 386 church steeples were hit by lightning, killing 103 bell ringers at the rope.

During this period of time, other property damages and disasters have been recorded. The bell tower of St. Mark's in Venice stands about 100 meters high. From 1388 to 1762, it was damaged six times and destroyed three times.

After gunpowder and artillery came into use in the eighteenth century, it was necessary to store large quantities of gunpowder. In many cases church vaults were used. The combination of a tall steeple and a basement full of gunpowder sometimes produced disastrous results. In 1769, 100 tons of gunpowder exploded in a church in France, killing 3000 people and destroying one-sixth of the surrounding city. In 1856, 4000 people were killed when gunpowder in the vaults of St. Jean on the Island of Rhodes was ignited by lightning. Similar lightning-caused explosions occurred in military storage depots. In 1782, lightning ignited 400 barrels of powder belonging to the East Indian Company at Fort Malaga in Sumatra.

Ships at sea were hit as well. A survey of British naval ships showed that in 16 years (1799–1815) there were 150 cases of damage. In addition, one ship, the 44-gun *Resistance* was destroyed by a lightning flash in 1798.

In 1746, Benjamin Franklin began experimenting and studying lightning and electricity by use of the Leyden jar. He observed that there were many similarities between sparks in the Leyden jar and lightning, e.g., color, smell, tortuosity. Prior to his observations there were other investigators who also theorized that electricity and lightning must be the same, but unlike these theorists, Franklin devised an experiment to prove his conjecture.

In July 1750, he wrote to his friend Peter Collinson in London describing his proposed sentry box experiment. Collinson acted as Franklin's agent in submitting his letters for publication to the *Philosophical Transactions of the Royal Society*. The sentry box experiment consisted of placing an iron pointed rod 6 to 9 meters (20 or 30 feet) above a small sentry box. The bottom of the rod was connected to a conductive platform that was insulated from the ground. During a storm, the investigator was to stand on the platform, grasp the rod, and with the other hand extended determine whether sparks came off his fingers! Franklin stated that if "any danger to the man be apprehended, he should ground the mast or rod."

Before Franklin could perform such an experiment, D'Alibard in France, in May 1752, set up the sentry box experiment and observed sparks from his hand, proving that thunderheads did contain electricity. Other investigators repeated the experiment with similar results, except in one case where an experimenter, G. W. Richman, in 1752 in Russia, was killed by a direct stroke to the rod.

In about June of 1752, Franklin devised a better experiment, the kite. He reasoned that a longer rod could be formed by a conducting kite string. This was attached to a key and to a silk insulating handkerchief which he held. Interestingly, Franklin appeared a little dubious about the experiment and therefore only permitted his son to be a witness. As he approached the key with a ring on his finger, sparks were emitted. Interestingly, Franklin delayed publicizing this experiment until 1788. It is also believed that Franklin did not know of D'Alibard's earlier sentry box experiment until after he flew his kite.

Franklin did not stop at simply theorizing. Even in his 1749 letter (before the sentry box and kite experiments), which was published in May 1750, he invented the lightning rod. His first theory was that a sharp pointed rod on a house or steeple would give off sparks, thereby discharging the cloud. However in a letter in 1755, Franklin also stated that if it did not discharge the cloud, it would guide the lightning to it instead of to the house. Today we know that his first observation was incorrect, the second correct.

Controversy abounded as to whether the rod should be pointed or blunt. Should it have a round ball on top? And, of course, there were skeptics. But eventually the rod was used with virtually total success.

From this time until the early 1900s, little additional knowledge was gained. The reason appears clear: the rod did its job. It protected people from lightning and there was no need for further research. However, in the early 1900s, electric utility systems were begun and lightning became the primary source of trouble. Lines and equipment were damaged. Some utilities, in hopes of saving their systems, simply shutdown the entire system during a storm, grounding their lines. So now there was a problem and a need, and investigations began. Engineers studied the phenomena of lightning, the mechanism, the characteristics. They devised theories and applied these to develop lightning protection methods.

By about 1950, the general thought existed that the then present knowledge of lightning phenomena and lightning protection was essentially complete. Little more could be gained by further study. Lightning investigations and studies came to a virtual standstill. During this decade, American Gas and Electric (now AEP), as part of the Ohio Valley Electric Corporation, was asked to supply power to a nuclear fusion plant. To supply this large load, the next step in voltage was required, i.e., 345 kV. Thus double circuit 345-kV lines were built, and the engineering world stamped these as virtually lightning-proof lines, perhaps as much as 0.5 flashovers per 100 mile-year. However, the engineering world was shocked to find that field data indicated a performance of over seven flashovers per 100 mile-year, an error of about 15 : 1. Again, a maelstrom of activity occurred. Investigations were made into all facets of the problem. Lightning phenomena was reinvestigated. Laboratory studies were made to investigate the strength of insulation and to develop new theories. Theoretical studies were started into the methods of calculation of flashover rates. As a result of this new activity, much additional knowledge was obtained, and the problem was solved. There will be more about this later, in Chapters 7 and 10. Here suffice it to say that today our knowledge is vastly improved. But unlike before, it is realized that even today our knowledge is lacking and should further improve in coming years.

With this as a background, the primary purposes of this chapter are: to discuss the lightning stroke mechanism with the objective of developing a simplified model of the last step of the lightning stroke, and to examine the important characteristics of lightning necessary for studying and developing lightning protection methods.

2 THE STROKE MECHANISM

In discussing the mechanism of the lightning stroke, we must constantly realize that our primary interest is in the last step of the stroke, where (if a personality is ascribed to the stroke) it decides where it will terminate. We are interested in the formation of the thunderhead and in the mechanism that starts a stroke proceeding toward the ground, but again our primary goal is to obtain an understanding so that we can develop a mathematical model of the last step of the lightning stroke. With this in mind, let us briefly look at a thunderhead immediately preceding the start of a "stepped leader." By some method not conclusively known, a charge separation takes place as illustrated in Fig. 1. The lower portion of the cloud is negatively charged; the upper portion is positively charged. Also, positive charges build up on the ground beneath the cloud. In the lower portion of the cloud, there may also exist a small positively charged pocket. Temperatures within the cloud may reach −20°C, and wind speeds greater than 100 miles per hour have been recorded. Thunderheads as high as 60,000 feet have been recorded, although the average height is about 30,000 to 40,000 feet. The base of the cloud on nonmountainous terrain is at an elevation of about 5,000 feet. As more charge separation takes place, and the potential between charge centers increases, a point is reached when air breakdown occurs. This breakdown or arc formation is thought to occur initially between the negatively charge region and the lower positively charged pocket or from the major negative and positive charges. Following this event, a sufficient voltage gradient occurs at the edge of the cloud, and air breakdown begins from cloud to ground and the stepped leader moves toward ground.

Figure 2 illustrates the general phenomenon. As its name implies, the stepped leader moves toward earth in halting steps of about 50 meters (Fig. 2a). After each step, the stepped leader pauses, then proceeds along one or more paths. The time for each step is about 50 μs near the base of the cloud but decreases to about 13 μs as it approaches the earth. The velocity of the stepped leader is relatively slow, about 0.10% of the speed of light. The leader is not visible to the naked eye and contains a current 50 to 200 amperes.

As this stepped or downward leader nears the earth (Fig. 2b), an upward leader (or return stroke) is initiated that meets the downward leader. This upward leader travels upward toward the cloud (Fig. 2C) at a velocity of between 10 to 30% that of light. It is highly visible to the naked eye. The current brought to earth by this upward channel may exceed 200 kA but has a median value of about 33 kA. The temperature of this channel exceeds 50,000°F, about five times the temperature of the surface of the sun. The rapid increase in temperature to this high value creates shock waves that we hear as thunder. The total length of the downward leader or the upward channel averages about 5 to 6 km (3 to 4 miles).

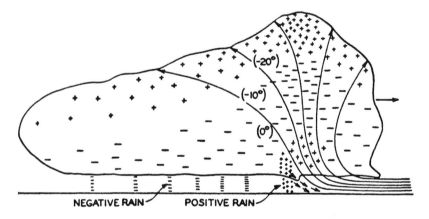

Figure 1 Charge distribution in a thunderhead.

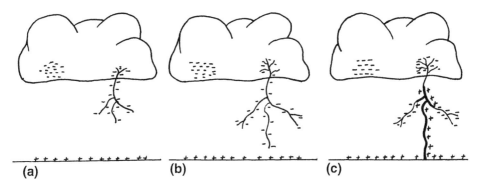

Figure 2 The first stroke. (a) Stepped leader starts. (b) Stepped leader reaches ground. (c) Upward channel moves toward cloud.

The above describes the mechanism of the first stroke of a lightning flash—and a flash may be composed of up to 54 strokes although the average is three strokes per flash. (Multiple strokes of the flash may be frequently seen by the naked eye. That is, the noticeable pulsations of a flash are caused by subsequent strokes of the flash, i.e., one can count the pulsations.) Figure 3 illustrates this mechanism. After a time between about 10 to 100 ms, a second leader, called a dart leader, again starts downward from the cloud, Fig. 3b. To initiate this leader, another portion of the charge in the cloud is discharged. This dart leader, as its name implies, has no pronounced steps but proceeds in a direct manner toward ground. Its velocity of about 1% that of light is much greater than that of the stepped leader since it follows the ionized path forged by the stepped leader. As the head of the dart leader nears the earth, an upward channel is again drawn from earth to meet it, and again a current is discharged to earth, although this current is usually only about 40% that of the first stroke. Other charge centers in the cloud may send other dart leaders from cloud to ground, thus initiating another stroke of the flash, and so on.

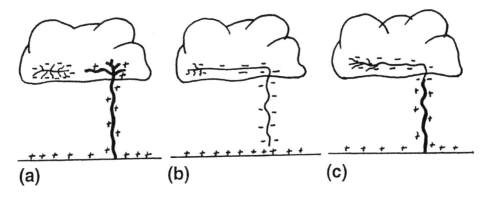

Figure 3 A second stroke. (a) Upward channel of first stroke reaches cloud. (b) Dart leader progresses to ground. (c) Upward channel begins.

To understand more fully the mechanism of the first stroke, examine more closely the stepped or downward leader. As shown in Fig. 4, the downward leader is composed of two parts: a thin highly conductive core or channel and a negative space charge that precedes and surrounds the channel. The diameter of the channel is about 2 mm and has a voltage drop of about 50 kV/m. The charge from the cloud is lowered by the progress of the leader and is distributed in space laterally by corona streamers. Figure 4 depicts the downward leader at the instant during the stepping process at which the space charge has expanded to its maximum extent and the leader is ready for the next step. The potential of the leader in Fig. 4 is about

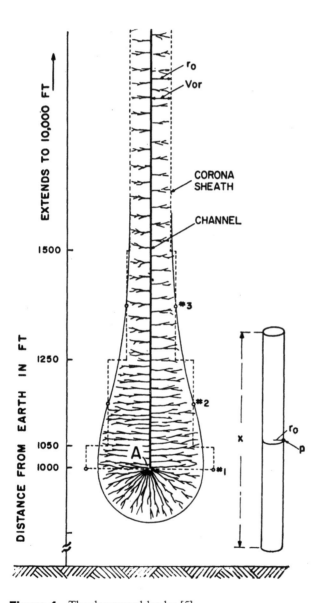

Figure 4 The downward leader [5].

50,000 kV. The stepping process consists of the rapid extension of the channel to about the edge of the corona sheath. At this point, the leader pauses, corona expands around it, and again, a step takes place. This process proceeds from the cloud, where the step interval is about 50 µs, to near the earth, where the step interval decreases to about 13 µs or less. Indeed, as an approximation, we can say that as the leader nears the earth and is about to make its final step, it is moving at essentially a constant velocity.

Figure 5 illustrates the important stages of the last step of the stroke as the downward leader nears the earth, takes its last step, and in so doing sacrifices itself to the high-velocity, high-current return stroke. Assume that the potential of the downward leader is 50,000 kV and that the leader is approaching the earth with a velocity of 1 ft/µs. The leader is shown approaching a 100 ft mast. In A, a corona discharge, positive polarity, has been initiated from the mast. The distance between the top of the mast, point "a", and the tip of the downward leader, point "b," is 275 feet. In B, the leader has traveled downward another 5 feet and the corona envelopes meet. At B, time '0," "the point of discrimination," is reached, and the downward leader determines that it will strike the top of the mast rather than the earth. That is, from laboratory measurements on rod–rod gaps, negative polarity, the breakdown gradient has been determined to be about 605 kV/m or 185 kV/ft. Dividing the potential of the downward leader, 50,000 kV, by the breakdown gradient, 185 kV/ft, results in a critical distance of 270 ft, which is the distance between points "a" and "b" in Fig. 5B. This distance is the distance at which the phenomenon of the last step occurs. It is the point of discrimination, and the distance is known as the "striking distance." The breakdown process is similar to the breakdown of rod–rod gaps in a laboratory. In C, the channels are shown growing downward from the downward leader and upward from the mast. They travel toward each other at ever-increasing velocities, producing ever-increasing values of current until at F the channels meet and crest current is attained. The single channel in G and H continues its upper movement, tapping as it goes the charge previously deposited by the downward leader and draining this charge to earth. Interesting! The current is produced by the charge in the corona sheath of the downward leader and not directly by the

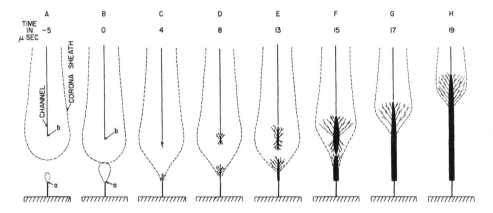

Figure 5 Stages in the development of the upward channel [5].

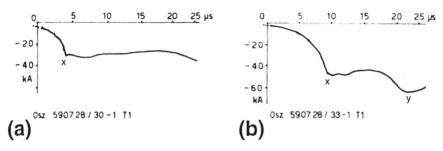

Figure 6 First stroke currents recorded by Berger.

charge in the cloud. Also crest current is attained when the channels meet in F, i.e., there is a time to current crest of 15 μs.

Per the previous discussion, the discharge process is compared to that of rod–rod gap in the laboratory. To achieve breakdown of the gap in the laboratory, the voltage is steadily increased, while in the lightning stroke the voltage is constant and the "gap" is steadily decreased until breakdown occurs.

In the above description of the last step, the two channels moving together at ever-increasing velocity are said to produce ever-increasing values of current since the current is the product of the velocity and the charge density. Assuming a constant linear charge density, the plot of both velocity and current as a function of time would have a concave upward shape. Indeed this is the shape of the front of the current through a rod–rod gap, and as shown by Berger's measurements it is also the shape of the front of the current for the first stroke. Two of Berger's oscillograms are shown in Fig. 6. In Fig. 6a, crest current is attained at point "x", which shows a predicted discontinuity. After this point the current is essentially constant or decreasing. However, in Fig. 6b, although a discontinuity at point "x" exists, crest current is attained at "y." The previous theoretical explanation using Fig. 5 assumes a downward leader without branches. If branches exist, as they will for a first stroke, the charge in these branches is also drained to earth and therefore these additional charges tend to increase the current after the channels meet at point "x."

As shown by Wagner [4], the surge impedance of the stroke for the example of Fig. 5 is about 920 ohms. This surge impedance is a function of the height and the velocity of the return stroke but is between 900 and 2000 ohms, i.e., 900 ohms for a stroke current of 50 kA and 2000 ohms for a stroke current of 10 kA. However, the conservative assumption that the stroke is a constant current source is almost universally used, i.e., the surge impedance of the stroke is infinite.

Before proceeding to the development of an analytical or geometric model of the last step, first dwell on the different types of flashes and their characteristics.

3 TYPES OF LIGHTNING FLASHES

The simplified description of the last step of the first stroke as presented in the previous section is that proposed by C. F. Wagner [5] and assumes a negative downward stroke or flash. This type of flash is the predominant type to open ground or to structures of moderate height, i.e., up to about 100 meters. However, three other types of flashes are possible as defined by Berger [6]. The four types are illustrated in Fig. 7. The name associated with each type is (1) the polarity of the charge in the cloud from which the leader is initiated or to which the leader propagates and (2) the direction of the leader. Note that the polarity portion of the name also denotes the polarity of the resultant current to ground.

The first type of flash, the negative downward flash, predominates for structures having heights of less than about 100 meters. Approximately 85 to 95% of the flashes to these structures are negative downward. The median current is about 33 kA

The negative upward flash was first observed at the Empire State Building in New York City (23 flashes per year). These predominate for high structures. For example, Berger's 70- and 80-meter masts, located atop 650-m Mt. San Salvatore in Switzerland, were struck by 1196 flashes in 11 years. Of these, 75% were negative upward and only about 11% were negative downward. (The remaining were classified as positive upward flashes.) The negative upward flash has a median current of less than 25 kA.

The third type of flash as denoted by Berger is the positive upward flash and is also known as the "Super Flash". About 14% of the flashes recorded by Berger were of this type. Current magnitudes are about 1.2 to 2.2 times that of the negative downward flash, and the action integral, the integral of the current squared with respect to time, is significantly larger than that of the negative downward flash. That is, the tail or time to half value is significantly larger. Positive flashes generally have only one stroke per flash and generally occur at the beginning or at the end of a storm and occur over the ocean. They may also be the predominant flash type during the winter season. Typically, only 2 to 10% of total flashes are positive polarity.

As to the positive downward flash, there exists no comprehensive source of data. Indeed Berger originally analyzed the positive flashes as downward, but in his subsequent analysis they were classified as upward. Thus the positive flash may be upward or downward. There is no clear separation.

In conclusion, about 85 to 95% of the flashes to structures having heights less than about 100 meters on flat or rolling terrain are negative downward. The other 5 to 15% are either negative upward or positive. Thus, from a transmission or substation viewpoint, except for mountainous terrain or very high river crossing towers, the negative downward flash is of primary concern.

4 PARAMETERS OF THE FLASH

To the electric utility engineer, the parameters of the flash that are of primary interest are

1. The crest current for the first and subsequent strokes
2. The waveshape of these currents
3. Any correlation between the parameters
4. The number of strokes per flash
5. Flash incidence rates: the ground flash density, flashes per square km-year, symbolized by N_g.

Also of some interest may be the charge lowered by the flash and perhaps the integral of the current squared, frequently called the "action integral."

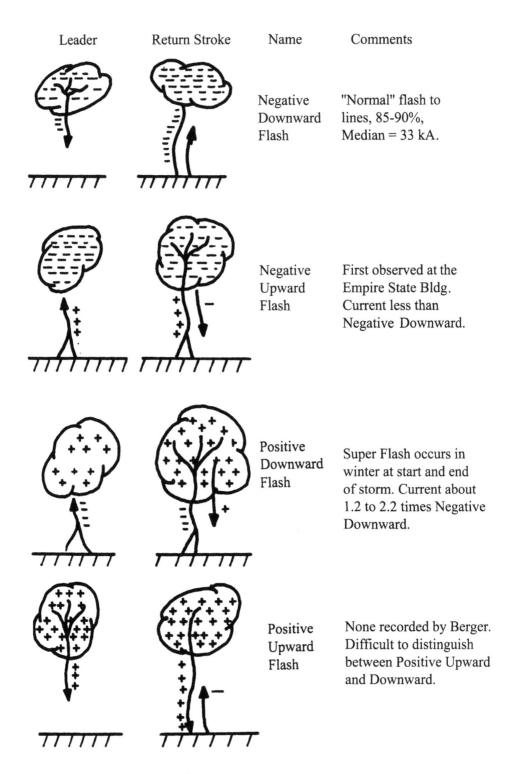

Figure 7 Types of lightning flashes.

The first three listed parameters, as we know them today, are to a very large extent based on the measurements of Berger. Berger's masts, 70 and 80 meters high, were mounted atop Mt. San Salvatore (Switzerland), which is 650 meters above Lake Lugano [6, 7]. As stated previously, although 75% of the 1196 flashes measured were negative upward, about 11% or 125 flashes were negative downward. When it is realized that Berger made oscillographic recordings of the currents in both the first and the subsequent strokes of the flash, making all waveshape parameters and their correlation available, it can be readily noted that these 125 records represent the best and most extensive set of data available to the industry to date.

5 BERGER'S DATA

Berger's data were first analyzed by Berger et al. [6] in 1975 and reexamined by Anderson and Eriksson in 1980 [7]. Further discussion concerning these parameters is presented in the CIGRE Working Group report in 1991 [8]. The statistical distribution of all the parameters of the flash can be approximated by the lognormal distribution whose probability density function is of the form

$$f(x) = \frac{1}{\sqrt{2\pi}\beta x} e^{-\frac{1}{2}\left[\frac{\ln(x/M)}{\beta}\right]^2} \tag{1}$$

where M is the median and β is the log standard deviation. The parameters M and β obtained from Berger's data are presented in Table 1, and the correlation coefficients between parameters are presented in Table 2. The correlation coefficients within brackets should be considered invalid and not be used, i.e., assume a zero correlation coefficient. For subsequent strokes, the correlation coefficients between the initial current and the other parameter were not given. However, it is assumed that they are equal to those for the final current.

The median is the 50/50 statistic. That is, 50% of the observations are above this value, 50% below. Or the probability that the value of the parameter is above or below the median is 0.50. The mean or average value of the parameter can be obtained from the two parameters of the distribution, i.e.,

$$\mu = M e^{\frac{\beta^2}{2}} \tag{2}$$

Thus, for example, the mean or average value of the tail is about 92 μs.

The definitions of the parameters are graphically illustrated in Fig. 8. For example, the parameter $S_{30/90}$ is the steepness of the front as measured by a linear line drawn through the 30% and 90% points. The parameter $t_{30/90}$ is the time to crest measured in the same manner as for a lightning impulse voltage (see Chapter 1). The initial crest current is the first crest as discussed in reference to Fig. 6b at point "x." The final current is that per point "y" of Fig. 6b. The parameter S_{10} is the steepness at the 10% point and was presented for use in distribution lines where the arrester may spark over at this current/voltage level. The parameter S_m is the maximum steepness of the front, which occurs at the crest of the surge. The time t_m is denoted as the minimum equivalent front and is a derived characteristic. That is, this parameter was not directly obtained from the oscillographic data but from the maximum

Table 1 Parameters for the Log Normal Distribution from Berger's Data, Negative Downward Flashes

	First stroke		Subsequent strokes	
Parameter	Median, M	β, log std. deviation	Median, M	β, log std. deviation
Front, µs				
$t_{10/90}$	5.63	0.576	0.75	0.921
$t_{30/90}$	3.83	0.553	0.67	1.013
t_m = min eq. front[a]	1.28	0.611	0.308	0.708
t'_m = min eq. front[b]	1.14	0.578	0.296	0.708
Steepness, kA/µs				
$S_{10/90}$	5.0	0.645	15.4	0.944
$S_{30/90}$	7.2	0.622	20.1	0.967
S_{10}	2.6	0.921	18.9	1.404
S_m	24.3	0.599	39.9	0.852
Crest, kA				
I_I, initial	27.7	0.461	11.8	0.530
I_F, final	31.1	0.484	12.3	0.530
Initial/final	0.9	0.230	0.9	0.207
Charge, C	4.65	0.882	0.938	0.882
Tail, µs	77.5	0.577	30.2	0.933
$I^2\,dt$, $(kA)^2\,s$	0.057	1.373	0.0055	1.366
Inter-stroke interval, ms		1st to 2nd stroke, $M = 45$ ms		
		2d stroke onward, $M = 35$ ms		
		$\beta = 1.066$ for both		
Flash duration, ms, excluding single stroke flashes		$M = 200$		
		$\beta = 0.69$		

[a] t_m is the minimum equivalent front and is derived from I_F and S_m; see text.
[b] t'_m is the minimum equivalent front and is derived from I_I and S_m; see text.

steepness and the final current as follows. The minimum equivalent front is defined either as

$$t_m = \frac{I_F}{S_m} \qquad (3)$$

or as

$$t'_m = \frac{I_I}{S_m} \qquad (4)$$

Since the distribution of I_I, I_F, and S_m are lognormal, the distributions of t_m and t'_m are also log normal with the parameters (for t_m)

The Lightning Flash

Table 2 Correlation Coefficients from Berger's Data, Negative Downward Flashes

Crest current, kA	Front, μs				Steepness, kA/μs			
	t_m (1)	t'_m [a]	t_{10}	$t_{30/90}$	S_m	S_{10}	$S_{10/90}$	$S_{30/90}$
First stroke								
I_I, initial	—	0.355	0.40	0.47	0.43	(0.12)	0.30	(0.19)
I_F, final	0.42	—	0.33	0.45	0.38	(0.06)	(0.20)	(0.17)
Subsequent strokes								
I_F, final	0.075[a]	—	(0.15)	(0)	0.56	(0.05)	0.31	0.23
I_I, initial (assumed)	—	0.075	0	0	0.56	0	0.31	0.23

Parentheses denote that coefficients are below critical values at 5% level of significance.
[a] derived statistics.

$$M_{t_m} = \frac{M_{I_F}}{M_{S_m}} \tag{5}$$

$$B_{t_m} = \sqrt{(\beta_{I_F})^2 + (\beta_{S_m})^2 - 2\rho_{I_F S_m} \beta_{I_F} \beta_{S_m}}$$

The correlation coefficients between t_m and I_F and between t'_m and I_I are also listed in Table 2. They are determined by the equation

$$\rho_{t_m I_F} = \frac{\beta_{I_F} - \rho_{S_m I_F} \beta_{S_m}}{\beta_{t_m}} \tag{6}$$

From these data other conditional distributions can also be derived. The parameters of some of these conditional distributions, which may be useful, are presented in Table 3. The conditional density function is of the form

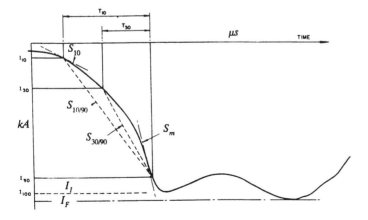

Figure 8 Definition of front and steepness.

Table 3 Derived Parameters of Conditional Lognormal Distributions from Berger's Data, Negative Downward Flashes

	First strokes		Subsequent strokes	
Parameter	Median, M	β, log std. dev.	Median, M	β, log std. dev.
$I_F\|S_m$	$11.7S_m^{0.31}$	0.448	$3.41S_m^{0.35}$	0.439
$S_m\|I_F$	$4.83I_F^{0.47}$	0.554	$4.17I_F^{0.90}$	0.706
$t_m\|I_F$	$0.207I_F^{0.53}$	0.554	$0.24I_F^{0.10}$	0.706
$I_F\|t_m$	$28.7t_m^{0.33}$	0.439	$13.1t_m^{0.056}$	0.529
$t'_m\|I_I$	$0.26I_I^{0.44}$	0.448	$0.23I_I^{0.10}$	0.706
$I_I\|t'_m$	$26.7t_m^{0.28}$	0.432	$12.6t_m^{0.056}$	0.529
$t_{30/90}\|I_F$	$0.636I_F^{0.51}$	0.494	0.67	1.013
$I_F\|t_{30/90}$	$19.5t_{30/90}^{0.39}$	0.432	12.3	0.530

$$f(y|x) = \frac{1}{\sqrt{2\pi}\beta_{y|x}y} e^{-\frac{1}{2}z^2}$$

$$Z = \frac{\ln(y/M_{y|x})}{\beta_{y|x}} \quad (7)$$

The parameters are given by the equations

$$\ln M_{y|x} = \ln M_y + \frac{\beta_y}{\beta_x}\rho_{xy}[\ln x - \ln M_x]$$

$$\beta_{y|x} = \beta_y\sqrt{1 - \rho_{xy}^2} \quad (8)$$

The conditional median can also be placed in the form

$$M_{y|x} = Ax^a \quad (9)$$

where

$$a = \frac{\beta_y}{\beta_x}\rho_{xy}$$

$$\ln A = \ln M_y - a(\ln M_x) \quad (10)$$

For example, consider the conditional of I_F given $t_{30/90}$, and assume that it is desired to obtain the probability that I_F is greater than 50 kA given that $t_{30/90}$ is equal to 2 µs. Thus

$$M_{I_F|t_{30/90}} = 25.55 \text{ kA} \quad \beta_{I_F|t_{30/90}} = 0.432 \quad (11)$$

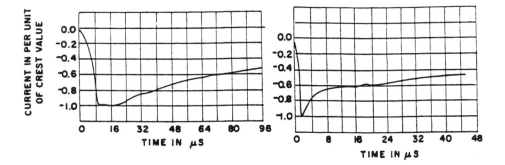

Figure 9 Average wave shapes of negative downward strokes [8].

Thus

$$Z = \frac{\ln(50/25.55)}{0.432} = 1.5539 \qquad (12)$$

From tables for the normal distribution, or from calculators,

$$P[I_F > 50\,\text{kA})|(t_f = 2\,\mu\text{s})] = 0.0601 \qquad (13)$$

The average waveshape of the first and subsequent stroke currents as developed by CIGRE are shown in Fig. 9. Note again the pronounced concave upward front of the first stroke, less pronounced in the subsequent stroke. Also note the increased steepness of the front of the subsequent stroke current. As shown in Table 1, the steepness of the subsequent stroke currents is about 65% larger than that of the first. The median crest current of the subsequent stroke is only about 40% that of the first stroke. No correlation could be obtained between the first and subsequent stroke currents, and therefore they are considered statistically independent. However, for low values of the first stroke current, subsequent stroke current may be larger than the first stroke current. For first stroke currents less than about 20 kA, approximately 12% of the flashes have subsequent stroke current that exceeds that of the first. The maximum subsequent stroke current observed was 80 kA.

Table 1 also contains the parameters for the interstroke interval and the total flash duration. These will be used later to evaluate the need for protection of an open breaker in a substation.

6 THE SEARCH FOR THE CREST CURRENT DISTRIBUTION

Early investigators, aiming at determining the crest current distribution, used magnetic links that responded to the maximum or crest current. In 1950, an AIEE Working Group [9] analyzed these magnetic link records and published the crest current distribution as presented in Fig. 10. Since the first stroke of the flash normally contains the highest crest current, this curve is thus ascribable to the first stroke of the flash. However, the curve does include both upward negative flashes and positive flashes, to the extent that they occur. This curve can be approximated by

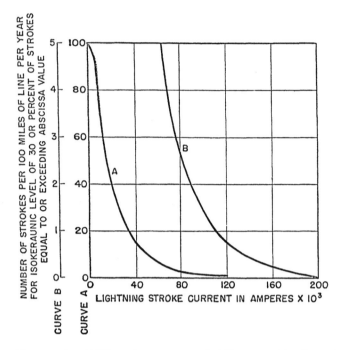

Figure 10 AIEE stroke current probability curve [9]. (Copyright IEEE, 1950.)

a lognormal distribution having a median of 15 kA and a log standard deviation of 0.98 as listed in Table 4. This AIEE Working Group also selected a 4 μs front as being the best representative value for the time to crest.

In the late 1950s, this crest current distribution was challenged on the basis that the magnetic links on tower legs could not be added to obtain the total stroke current. J. G. Anderson [10] reanalyzed these data and proposed an alternate distribution as listed in Table 4 having a median of 46.5 kA. The log standard deviation is 0.71 for currents above the median and 0.41 for currents below the median. Anderson also proposed that a wavefront distribution be used having a median of 1.57 μs and a log standard deviation of 0.60. (This is remarkably similar to the distribution of t_m of Table 1, i.e., $M = 1.28$, $\beta = 0.611$.)

Both Spzor's [11] and Popolansky's [12] investigations using magnetic links on chimneys indicated median crest currents of about 25 to 30 kA. Eriksson's initial 11 measurements on a 60 meter mast in South Africa indicated a median of 41 kA.

At this point in time, there appears little doubt that the median crest current of the first downward negative stroke to chimneys and masts—and to transmission lines—was larger than that given by the AIEE distribution. In 1972, Popolansky [13] derived a new global summary and proposed a median value of 25 kA and a β of 0.90.

However, there still existed a doubt concerning the use of chimney and mast data for transmission lines. Popolansky [12] reanalyzed his data on chimneys and segregated the measurements into groups for different chimney heights. He found that the median current decreased as chimney height increased. (As is shown later,

Table 4 Summary of Measurements of Crest Current and Steepness

Source	Object	Crest Current, kA		Steepness, kA/μs	
		M	β	M	β
AIEE WG [9]	Lines	15	0.98	front = 4 μs	
Anderson et al. [10]	$I < 46.5$ kA	46.5	0.71	$M = 1.57$ μs	
	$I > 46.5$ kA	46.5	0.41	$\beta = 0.60$	
Spzor [11]	Chimneys	25	0.97		
Popolansky [12]	Lines, 20 m	30.5	0.58		
	Chimneys, 60 m	24.5	0.82	5.5	1.85
Popolansky [13]	Summarizing	25	0.90		
Anderson and Eriksson [7]	Summarizing < 60 m	31	0.69		
Cianos and Pierce	First, neg down	20	0.92		
	Sub. neg. down	10	0.92		
CIGRE [8]	First, neg. down	34	0.74	24.3	0.60
	Sub. neg. down	12.3	0.53	39.9	0.85

^a β = 0.64 for UHV lines, 0.80 for EHV lines, and 1.00 for others.
^b β = 22/y, y = conductor height, 0.6 < β < 0.9.
^c β = 0.36 + 0.17 ln(43 − h); if h > 40, then h = 40.
^d For masts, Mousa uses an A of 8.8.
^e γ = 444/(462 − h) for h > 18 m; γ = 1 for h < 18 m.

this is in contradiction to that predicted by use of the geometric model of the last step of the lightning stroke, i.e., increasing the height increases the median.)

At this time, Berger's data provided the answer. All link data for negative polarity must contain both downward and upward strokes. Since Berger's data indicated that negative upward flashes have lower currents, the decrease as noted by Popolansky could be possible if the number of negative upward flashes increased with tower height—and at this time it was known that the number of downward flashes increased with structure height. In 1977, Eriksson reanalyzed existing data to show that (1) the number of flashes to a structure increases dramatically for structure heights about 60 meters, (2) upward flashes appear to occur to structures whose heights exceed about 100 meters, and (3) the median current of downward flashes to structures of heights less than about 60 meters appeared to be approximately constant. Thus he postulated that practically only negative downward flashes would occur to structures of heights of 60 meters or less. Eriksson therefore argued that the current distributions as obtained from chimneys and masts could also be used for transmission lines.

Using only data for negative polarity currents measured on structures having heights less than 60 meters resulted in a median value of 34 kA and a β of 0.737 [7]. A total of 383 observations were used as follows:

1. Czechoslovakia, chimney data, $n = 123$
2. Australia, 230-kV lines, $n = 18$
3. Poland, chimneys, $n = 3$

4. USA, 345-kV lines, $n = 44$
5. Sweden, lines, $n = 14$
6. South Africa, research mast, $n = 11$
7. Switzerland, Berger's masts, $n = 125$

Subsequently, in the recent CIGRE working group report [8], 25 additional values were added from South Africa to boost the total to 408 values. These new added values did not significantly alter the distribution, so that the median remained at 34 kA with a β of 0.737. The minimum current measured was 3 kA, and five observations exceeded 100 kA.

This final distribution can be more accurately represented in a lognormal piecewise manner. Using two piecewise representations, as shown in Fig. 11 (lognormal probability paper), the parameters of the distributions are given in Table 5 and represent the present distribution as suggested by CIGRE. To be carefully noted is that the median current for currents less than 20 kA is *not* 61.1 kA. These values are simply the parameters of the lognormal distribution.

An alternate representation based on Popolansky's data developed by J. G. Anderson [14] and used by the IEEE Working Group [14, 15] is

$$P(I) = \frac{1}{1 + \left(\frac{I}{31}\right)^{2.6}} \tag{14}$$

which is also shown in Fig. 11. $P(I)$ is the probability that the current is equal to or greater than a current I. The IEEE curve generally agrees with the distribution suggested by CIGRE except at the important ends of the distribution. Because the CIGRE distribution is based on the latest data available and better represents the actual data, it is deemed superior and will be used in other chapters in this book.

Using Berger's correlation coefficients with the current distribution of Table 5 results in the distribution of t_m and in the conditional distributions of Table 6. Because the currents are defined by parameters in the two current regions, t_m and the conditional distributions must also be defined within these two regions.

6.1 Negative Upward Flashes

From Berger's data [7], the median current for the upward first negative strokes is about 0.25 kA with a β of 1.29, while for subsequent strokes the median is about 10 kA with a β of about 0.71. However the measurements of Garbagnati of Italy indicate a median of about 25 kA. Thus these negative upward flashes have currents less than 75% of those for the negative downward flashes.

Table 5 Suggested First Stroke Current Distribution

Parameter	$I_F < 20$ kA	$I_F > 20$ kA
M, median	61.1	33.3
β, log std. dev.	1.33	0.605

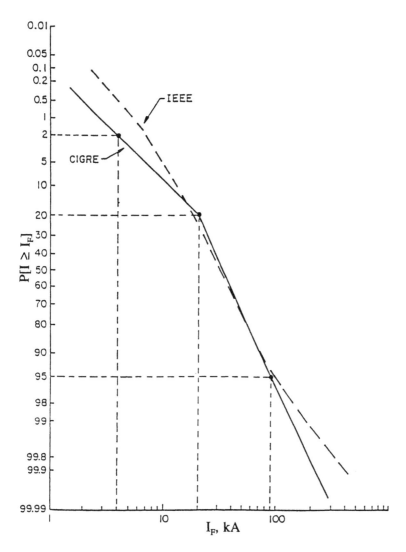

Figure 11 CIGRE and IEEE stroke current probability curves, first stroke. Negative downward flash [8].

6.2 Positive Flashes

From Berger's data [8], the median for all positive flashes is about 40 kA with a β of about 0.98. Comparing this to the negative downward flash we find that (1) 10% of the negative downward flashes have currents greater than 67 kA and 5% have currents greater than 90 kA; (2) 10% of the positive flashes have currents greater than 127 kA and 5% have currents greater than 200 kA. Thus the ratio of the currents for a positive flash to those for a negative flash varies from 1.2 for the median to 2.2 for the 5% level.

Table 6 Parameters of the Log Normal Distribution Using CIGRE Current Distribution of Table 5, First Negative Downward Strokes

	Application range			
	$3 \leq I \leq 20\,\text{kA}$		$I \geq 20\,\text{kA}$	
Parameter	M, median	β, log std. dev.	M, median	β, log std. dev.
Derived statistics				
$t_m = I_F/S_m$	2.51	1.23	1.37	0.670
Conditional distr.				
$S_m\|I_F$	$12.0 I_F^{0.171}$	0.554	$6.50 I_F^{0.376}$	0.554
$t_m\|I_F$	$0.0834 I_F^{0.828}$	0.554	$0.154 I_F^{0.624}$	0.554
$I_F\|t_m$	$25.1 t_m^{0.962}$	0.597	$28.4 t_m^{0.508}$	0.500
$t_{30/90}\|I_F$	$1.77 I_F^{0.188}$	0.494	$0.906 I_F^{0.411}$	0.494
$I_F\|t_{30/90}$	$14.4 t_{30/90}^{1.08}$	1.184	$17.2 t_{30/90}^{0.492}$	0.540
Derived correlation coefficients				
$\rho(t_m, I_F)^*$	0.893		0.563	

The number of positive flashes as a percent of the total number of flashes ranges from about 2.5% [16] to 10% [8] if only flashes over land are considered. Most positive flashes occur over oceans, during winter, and at the beginning and end of a storm.

7 MULTIPLE STROKES (STROKE MULTIPLICITY)

Based on 6000 flash records from different regions of the world, the distribution of the number of strokes per flash for downward negative flashes per Anderson and Eriksson [7] is give in Table 7. The median of the distribution is 2 and the mean or average value is 3. As stated before, almost universally, there is only one stroke per flash for positive polarity flashes.

Table 7 Number of Strokes per Flash, Negative Downward Flash

Strokes/flash	Probability	Cumulative, equal to or greater than
1	0.45	1.00
2	0.14	0.55
3	0.09	0.41
4	0.08	0.32
5	0.07	0.24
6	0.04	0.17
7	0.03	0.13
8	0.02	0.10
9	0.02	0.08
> 10	0.06	0.06

8 LIGHTNING INCIDENCE

The primary objective in obtaining the flash incidence rate is to determine the number of flashes per year that terminate on transmission lines or on substations. As will be shown later, this can be obtained by use of the geometric model of the last step of the lightning stroke or by regression type equations. The fundamental quantity required for this and other calculations is the ground flash density denoted as N_g and given in units of flashes per square km-year. Obviously, the best method of obtaining N_g is by direct measurement of this quantity. However, if reliable data for N_g are not available, then some approximation is necessary to convert the number of thunderstorm days (the keraunic level) as collected by weather bureaus to ground flash density.

Direct measurements of the ground flash density can be made by CIGRE flash counters, which have an observation range of 15 to 20 km, or by more recently developed systems, which extend this range to about 300 to 400 km. These recent systems are either (1) gated wideband direction-finding (DF) systems or (2) time of arrival (TOA) systems.

Prior to these recent lightning location systems, the ground flash density as obtained by the CIGRE 10 kHz counter was used in many European countries and in South Africa to obtain the ground flash density and to compare it to the number of storm days per year so as to arrive at an approximate equation for the ground flash density. The general form of the regression equation is

$$N_g = k T_d^a \tag{15}$$

where T_d is the number of thunderstorm days per year, the keraunic level. Many values of the constants k and a have been proposed. For example, in England, Stringfellow [17] analyzed English data and obtained $k = 0.0026$ and $a = 2$, while in Sweden, Muller-Hillebrand's [18] analysis gave $k = 0.0046$ and $a = 2.0$. Within the USA, J. G. Anderson et al. [10] suggested that $a = 1$ and $k = 0.12$, while Young et al. [19] used $a = 1$ and $k = 0.177$. In the former USSR, Kolokolov and Pavolova [20] suggested the equation

$$N_g = 0.036 T_d^{1.3} \tag{16}$$

The best data to date comes from investigation in South Africa. Eriksson [21] proposed the equation

$$N_g = 0.04 T_d^{1.25} \tag{17}$$

which essentially agrees with that proposed in the USSR. This equation has been accepted by both CIGRE and IEEE. This equation provides the average or mean value of N_g. The standard deviation is approximately 32% of the mean. There are approximately two thunderstorm hours per day.

Contour maps giving the number of thunderstorm days per year or the keraunic level are general available from weather bureaus. Examples of these maps are provided in Fig. 12 [22] for the USA, Fig. 13 for Canada [23], and in Fig. 14 [24] for the world. The world's highest incidence rate occurs in Java, 223 storm days per year.

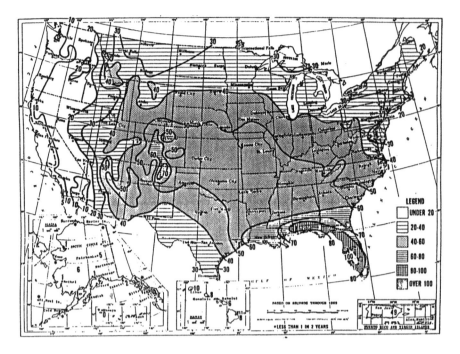

Figure 12 Annual frequency of thunderstorm days in the USA [22].

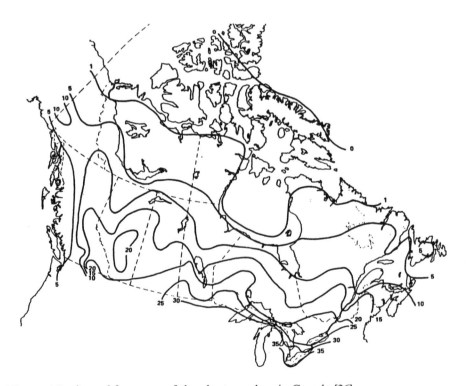

Figure 13 Annual frequency of thunderstorm days in Canada [26].

The Lightning Flash

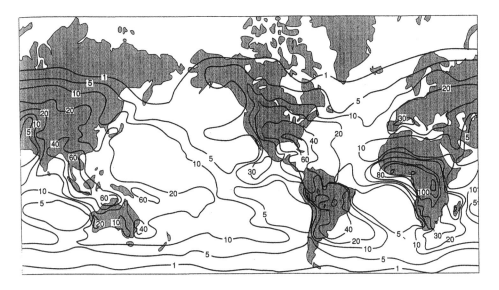

Figure 14 Annual frequency of thunderstorm days in the world [24].

As for the ground flash density, the number of thunderstorm days provided by these maps is the average or mean value. The standard deviation [25] averages about 19% of the mean. For low values of T_d, the standard deviation increases, e.g., 55% for $T_d = 4$, and for high values of T_d the standard deviation, e.g., 14% for $T_d = 70$.

Maps are also available that provide an estimate of the yearly duration of thunderstorms, i.e., thunderstorm hours; see Fig. 15 [26]. The relationship between thunderstorm hours per year T_h and ground flash density as suggested by CIGRE [8] is

$$N_g = 0.05 T_h \tag{18}$$

Figure 16 [26] is a map showing contours of ground flash density based on the authors' (MacGorman et al.) evaluation using thunderstorm hours.

Although it is claimed by some investigators that the use of thunderstorm hours provides a better estimate of the ground flash density, considering the variability of the observations, this conjecture is questioned, and the use of thunderstorm days to estimate N_g is recommended.

In many countries, improved wide-range lightning location or detection systems have been installed [8, 27]. Within the USA, in 1994, there were 125 DF antenna stations, which provided coverage for the entire continental USA. This system was initiated by Richard Orville of the State University of New York (Albany) [16] who constructed eight DF antenna stations. At this stage, EPRI contracted with this university to provide additional stations, and 72 were added to cover the area east of the Mississippi River. Other stations owned by the National Severe Storm Center and the Bureau of Land Management were added to boost the total to 125 and provide coverage of the USA. The system was operated through satellite communication at Albany. In 1991, the system was taken over by GeoMet Data Services,

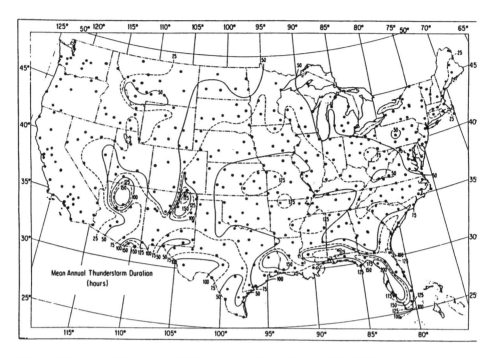

Figure 15 Annual frequency of thunderstorm hours in the USA [26].

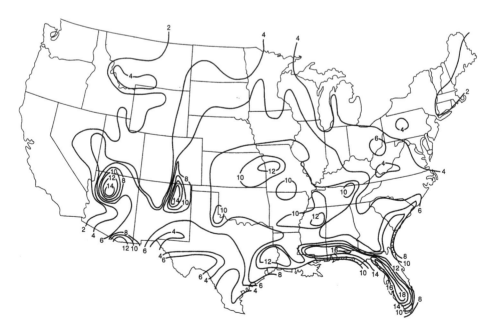

Figure 16 Annual ground flash density in USA obtained from thunderstorm hours [26].

Inc. located in Tucson, AZ. At this location, GeoMet (new name, Global Atmospherics) operates the National Lightning Detection Network.

As noted previously, this USA National Lightning Detection Network employed the DF system for detecting the location of flashes. The other system, the Lightning Position and Tracking System, employed the TOA system, which several utilities have installed. The location accuracy (location of the flash) in either system is within about 2 km. The detection efficiency of the system is in the range of 70%, that is, they record about 70% of the actual flashes. A detailed examination of these systems is provided in CIGRE Technical Bulletin 94 [27], which is highly recommended.

In 1995, the manufacturers of the TOA system, Atmospheric Research Systems, Inc., and the manufacturers of the DF system, Lightning Location and Protection, joined together and are now members of the Dynatech group of companies. Today, the DF stations have been changed so that about 60% of them are now either TOA systems or a combined DF and TOA. These systems are operated by Global Atmospherics in Tucson, AZ.

The data from this national USA system is available on a cost basis to all. The data have been installed in a recent EPRI computer workstation. Other than the single IEEE paper on this network [16], there are a few EPRI reports that present portions of the data [28].

Global Atmospherics has made the five-year ground flash density map available, which is shown in Fig. 17. Except for the original East Coast system of Orville, the data collection period has not been sufficiently long to provide an accurate estimate of the ground flash density. (About 11 years of data are needed.)

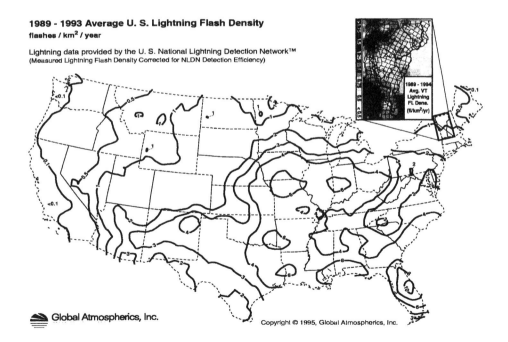

Figure 17 Average five-year, 1989–1993, ground flash density by Global Atmospherics.

Investigators assigned to the USA system also provide estimates of the distributions of crest current through measurements of the maximum field strength by use of the equation

$$I = \frac{2\pi D}{\mu_0 v} \beta \qquad (19)$$

where β is the crest magnetic radiation field, D is the distance to the flash, μ_0 is the magnetic permeability of free space, and v is the velocity of the return stroke in per unit of that of light. Errors in this equation can occur from the measurement of the field strength and distance, and from errors in the assumed velocity of the return stroke. To obviate some of these errors, the investigators have equated the median value of the maximum field strength for all flashes in the USA to the median value from Berger, i.e., 31.1 kA. Thus the measurements are not independent.

These systems are also used to provide an on-time estimate of the location of flashes within the utility system. In general, provided the system has been properly calibrated, the location can be identified to within about 2 km.

Presently available maps of N_g indicate that for 30 thunderstorm days per year, the ground flash density ranges between 1 and 2. Use of the CIGRE and IEEE equation results in a ground flash density of 2.8. As a final comment, use of older equations such as those suggested by Young and Anderson would result in ground flash densities of 3 to 5.3 for 30 thunderstorm days, whereas the above more recent equations or the actual data suggest values that are significantly smaller. Some interesting observations have been made in some of the articles published by authors who are associated with the USA network. For example, Orville [29] produced a map for the East Coast system showing that the median current varies throughout the country. Examining Fig. 18, some correlation exists with the ground flash density,

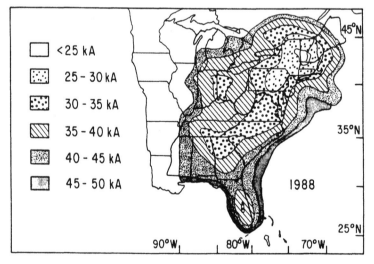

Figure 18 Variation of crest current with latitude [29]. (Reprinted with permission from *Nature*, 1990.)

The Lightning Flash

but note that the same median current occurs in Florida and in Pennsylvania, where the ground flash densities are about 10 and 1, respectively.

Other authors investigated the multiplicity of flashes [30]. Examining 46 flashes having multiple strokes, they found that about 20% of the flashes had subsequent stroke currents that exceeded those of the first stroke. This should be compared to Berger's data reported by CIGRE [8], where it was found that only for first strokes having currents under 20 kA does this occur. That is, 12% of the flashes having first stroke currents under 20 kA have subsequent stroke currents greater than that of the first.

9 GEOMETRIC MODEL OF THE LAST STEP OF THE LIGHTNING STROKE

In 1961, Wagner [5] proposed the simplified model of the last step as previously described. In this explanation the downward leader progressed toward ground until it reached a "point of discrimination." Assuming a critical breakdown gradient of 605 kV/m, and a stroke potential of 50,000 kV, this point was reached when the distance between the core of the downward leader and the top of the tower was $50{,}000/605 = 83.3$ meters or 270 feet. That is, the "striking distance" was about 83 meters. Assuming that the breakdown gradient of 605 kV/m is a good estimate, an equation or method is needed to estimate the potential of the downward leader.

Previous to Wagner's development, Lundholm [31] developed a relationship between the stroke current and the velocity of the return stroke. Later, Rusck [32] slightly modified the relationship to

$$v = \frac{1}{\sqrt{1 + 500/I}} \qquad (20)$$

where v is the velocity of the return stroke in per unit of that of light and I is the crest current in kA. Wagner [33] developed a similar relationship. Both these relationships are shown in Fig. 19.

Thus, if the stroke current is known, the return stroke velocity can be found, and from this velocity the potential of the downward leader can be estimated. Per Wagner, the potential V in MV is

$$V = 120 \frac{v}{1 - 2.2v^2} \qquad (21)$$

where v is the velocity of the return stroke in per unit of that of light. Therefore, all is in place to estimate the striking distance:

1. Given the stroke current, the velocity is estimated from Fig. 19.
2. From the velocity, the potential can be calculated.
3. From the leader potential, the striking distance r is found from $r = V/G$, where G is the breakdown gradient of 605 kV/m.

Using these steps, a single striking distance is obtained. However, in general the striking distance to a conductor or to a top of a tower differs from the striking distance to earth. This appears somewhat obvious, since the breakdown gradient for

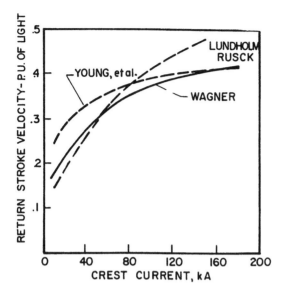

Figure 19 Velocity of return stroke as a function of current.

a rod–plane gap (core of downward leader to ground) differs from the breakdown gradient for a rod–rod (downward leader to top of tower). Thus in general, two striking distances exist, one to the phase conductor or ground wires r_c, and one to the earth or ground r_g.

The resultant geometric model of the last step of the lightning stroke assuming a single overhead ground wire is shown in Fig. 20. The construction of this model is as follows:

1. For a specific current I, calculate the striking distance r_g and r_c.
2. Draw a line parallel to the ground at a distance r_g from the ground.

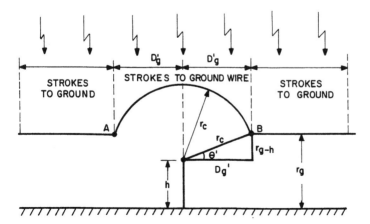

Figure 20 Geometric model for a single ground wire.

3. With a compass centered at the tower top, draw an arc of radius r_c until it intersects the parallel lines drawn in 2, above.

Any stroke that arrives between A and B will terminate on the ground wire, and any stroke that arrives to the left of A or to the right of B will terminate to ground. Thus, given this specific crest current, the number of strokes that terminate on the ground wire $N(G)$ is

$$N(G)|I = 2N_g L D'_g \qquad (22)$$

while L is the length of the line. That is, the area that collects the strokes is $2D'_g$ times the length of the line L. Multiplying this by the ground flash density gives the number of strokes. The probability that this current will occur is $f(I)\,dI$, so that the incremental number of strokes of current I is

$$dN(G) = 2N_g L D'_g f(I)\,dI \qquad (23)$$

and the total number of strokes that terminate on the ground wire is

$$N(G) = 2N_g L \int_3^\infty D'_g f(I)\,dI \qquad (24)$$

Per Fig. 20, the distance D'_g is

$$D'_g = \sqrt{r_c^2 - (r_g - h)^2} \qquad (25)$$

or

$$\theta' = \sin^{-1}\frac{r_g - h}{r_c}$$
$$D'_g = r_c \cos\theta' \qquad (26)$$

As noted, the integration is taken over all possible values of stroke current. The lower integration limit of 3 kA recognizes that there must exist a lower limit for the stroke current, i.e., there cannot be a stroke with zero current. Since for the CIGRE distribution, the lowest stroke current measured is 3 kA, this value has been selected as a reasonable lower limit. However, other investigators believe that values such as 1 or 2 kA are more reasonable, so that limits of zero are sometimes used. However, the use of this value between 0 and 3 kA has little effect on the number of strokes to the ground wire.

The cumulative distribution function of currents that terminate on the ground wire I_G is

$$F(I_G) = \frac{2N_g L}{N(G)} \int_3^{I_G} D'_g f(I)\,dI \qquad (27)$$

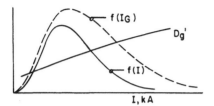

Figure 21 Obtaining the density $f(I_G)$.

from which the probability density function can be found, i.e.,

$$f(I_G) = \frac{2N_g L}{N(G)} D'_g f(I) \tag{28}$$

which may be visualized by examining Fig. 21.

In the above equations, the value of D'_g is only valid for $r_g > h$. If r_g is less than h, for the assumption of vertical strokes, D'_g is equal to r_c.

For the case of two overhead ground wires separated by a distance S_g (Fig. 22), the development is similar. The number of strokes that terminate on the ground wire is

$$N(G) = N_g L \left[\int_3^\infty (2D'_g + S_g) f(I) \, dI \right] \tag{29}$$

or

$$N(G) = 2 N_g L \int_3^\infty D'_g f(I) \, dI + N_g L S_g \tag{30}$$

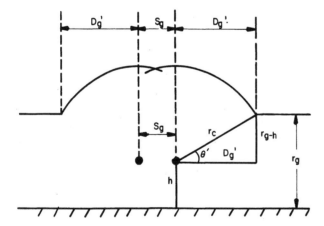

Figure 22 Geometric model for two ground wires.

And the cumulative and density functions are

$$F(I_G) = \frac{2N_gL}{N(G)}\left[\int_3^{I_G} D'_g f(I)\,dI + \frac{S_g}{2} F(I)\right]$$
$$f(I_G) = \frac{2N_gL}{N(G)}\left[D'_g + \frac{S_g}{4}\right]f(I)$$
(31)

10 STRIKING DISTANCE EQUATIONS

Wagner's development of the last step of the lightning flash resulted in the following equation for the striking distance:

$$r = 14.2 I^{0.424} \tag{32}$$

This applied both to the phase conductor/ground wires and to the earth or ground. Young [19] accepted this model and applied it to calculation of the number of shielding failure flashovers. He also presented curves of suggested shielding angle as a function of height. In arriving at these curves Young found that Wagner's equations required some modification so that calculations would agree with field performance. The modifications were

1. The breakdown gradient to ground or to the conductors for heights less than 18 meters was assumed at 605 kV/m, Wagner's value. However, the author reasoned that the last step to towers of higher heights would be similar to the breakdown of rod–rod gaps. Therefore he modified the breakdown gradient for ground wire heights greater than 18 meters to

$$\begin{aligned}G &= 605\,\text{kV/m} \quad h \leq 18\,\text{m}\\ G &= 605\left[\frac{462-h}{444}\right] \quad h > 18\,\text{m}\end{aligned} \tag{33}$$

2. The velocity–current was modified per Fig. 19.

The resultant striking distance equations are

$$\begin{aligned} r_g &= 27 I^{0.32}\\ r_c &= \gamma r_g\\ \gamma &= 1.00 \quad \text{for} \quad h \leq 18\,\text{m}\\ \gamma &= \frac{444}{462-h} \quad \text{for} \quad h > 18\,\text{m}\end{aligned} \tag{34}$$

Following the development of these equations, Armstrong and Whitehead [34] suggested some modifications. First, they reasoned that the breakdown voltage between the downward leader and the ground wires, phase conductors, or ground should be governed by the switching impulse breakdown characteristics rather than

Table 8 Expressions for the Striking Distance $r = AI^b$

	r_g to earth or ground		r_c to phase conductors and ground wires	
Source	A	b	A	b
Wagner [5]	14.2	0.42	14.2	0.42
Young [19]	27.0	0.32	γr_g [e]	0.32
Armstrong and Whitehead [34]	6.0	0.80	6.7	0.80
Brown and Whitehead [35]	6.4	0.75	7.1	0.75
Love [36]	10.0	0.65	10.0	0.65
Anderson and IEEE-1985 [14, 15]	βr_c [a]	0.65	8.0	0.65
IEEE-1991 T&D Committee	βr_c [b]	0.65	8.0	0.65
IEEE-1992 [37] T&D Committee	βr_c [c]	0.65	10.0	0.65
Mousa and IEEE-1995 [38, 39] Substations Committee[d]	8.0	0.65	8.0	0.65
Eriksson [40]			To phase conductor: $r_c = 0.67 y^{0.6} I^{0.74}$	
			To ground wire: $r_s = 0.67 h^{0.6} I^{0.74}$	
			To earth: none	

[a] $\beta = 0.64$ for UHV lines, 0.80 for EHV lines, and 1.00 for others.
[b] $\beta = 22/y$, y = conductor height, $0.6 < \beta < 0.9$.
[c] $\beta = 0.36 + 0.17 \ln(43-h)$, if $h > 40$, then $h = 40$.
[d] For masts, Mousa uses an A of 8.8.
[e] $\gamma = 444/(462-h)$ for $h > 18$ m, $\gamma = 1$ for $h < 18$ m.

by the lightning impulse characteristics, and therefore they employed Paris's data for negative polarity switching impulse, the lower rod being 3 meters above the grounded plane. Whitehead and his coauthors and students [35, 36] later produced other modifications as listed in Table 8. In addition, the IEEE Working Group has accepted various equations, the latest of which are those of 1992 [37]. The CIGRE Working Group employed the Brown–Whitehead equations, since these produced reasonable lower limits of shielding failure flashover rates. Figure 23 compares

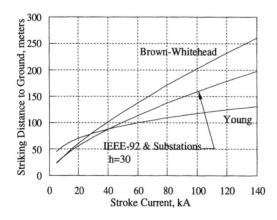

Figure 23 Comparison of striking distance equations.

The Lightning Flash

some of these equations. Further discussion concerning these equations is left to Chapter 7.

Referring again to the expression in Table 8, note that in many cases the primary striking distance is that to the conductors r_c and the striking distance to ground or earth r_g is some functions of r_c. Thus if two overhead ground wires are of unequal height, as is possible in a station, then the striking distance to ground is different for each of these ground wires. Also, as in Chapter 15, when the effect of trees is considered, these types of equations cannot be used. This use of r_g as a function of r_c is not correct. It has been challenged by A. M. Mousa in his discussion of Ref. 15. At the present time the IEEE Lightning Working Group is attempting to correct this error.

11 ERIKSSON'S MODIFIED GEOMETRIC MODEL

As noted from Table 8, the striking distances attributed to Eriksson vary significantly from those of other investigators. For Eriksson's method, see Fig. 24.

1. Striking distances are given for the phase conductor and the overhead ground or shield wire, r_c and r_s, respectively. There exists no striking distance to earth or ground. It is postulated that any downward leader that does not result in a stroke terminating on the ground wire will strike ground. Once past the ground wire striking distance, the downward leader will not change directions and move upward to the ground wire. Thus the stroke to ground or earth is taken as a default condition.
2. The striking distance equations are functions of the height of the ground wire or phase conductor.

Referring to Fig. 24, a horizontal line is first constructed at the height of the ground wire. Next, an arc of radius r_s is constructed with a center at the ground wire and continued until the horizontal is reached. Downward leaders that reach the position between points A and B will result in a stroke terminating on the ground wire. Otherwise, the stroke will terminate to ground. Thus the value of D'_g is

$$D'_g = r_s = 0.67 h^{0.6} I^{0.74} \tag{35}$$

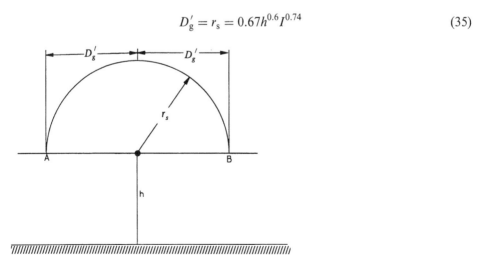

Figure 24 Eriksson's modified model for strokes to ground wire [21, 40].

and the number of strokes to the ground wire is

$$N(G) = 2N_g L \int_3^\infty D'_g f(I)\, dI \qquad (36)$$

which is the same as Eq. 23. Also for two ground wires, Eq. 30 is applicable. That is, the only change is in the equation for D'_g.

12 NUMBER OF FLASHES TO THE GROUND WIRE

12.1 From Field Data

In 1967, Eriksson [21] analyzed field data on the number of strokes to freestanding structures, e.g., masts, chimneys, and power lines, to arrive at an equation to estimate the number of strokes to a line. First, from extensive data on freestanding structures, he found that the number of strokes could be estimated by the power law equation

$$N(G) = 2.4 \times 10^{-5} h^{2.05} \qquad \text{for } N_g = 1 \qquad (37)$$

where h is the height of the structure in meters and h is between 20 and 500 meters. The number of flashes/year as described by Eq. 37 is shown by the dotted line curve in Fig. 25 and is compared to curves presented by L. Dellera and E. Garbagnati in a discussion of Eriksson's paper [40]. As noted, these discussers indicate that the Eriksson's curve, which is assumed to apply to flat terrain, may underestimate the

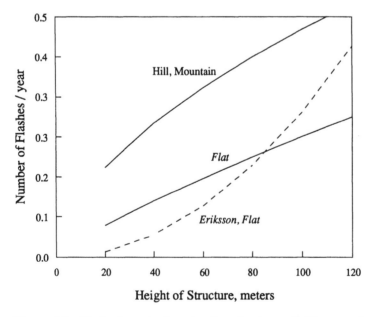

Figure 25 Flashes/year to free-standing structures. Solid curves by L. Dellera and G. Garbagnati in discussion of Ref. 40. Dashed curve by Eriksson.

number of flashes for low structure heights and overestimate the number of flashes for high structure heights. Or, more probably, it may be that Eriksson's curve is for a combination of flat and mountainous terrain.

However, since upward strokes occur for high structures, for heights above about 100 meters this equation is not totally valid for the important negative downward stroke. Eriksson then presented an equation for the number of upward strokes in per unit of the total number of strokes $N(G^-)$:

$$N(G^-) = 1.26 \times 10^{-4} h^{1.48} \quad \text{for } N_g = 1 \tag{38}$$

Using this equation to modify Eq. 37 for only downward strokes, he arrived at the equations

$$N(G) = N_g \pi R_A^2 \times 10^{-6} \tag{39}$$

$$R_A = 16.6 h^{0.55} \tag{40}$$

where N_g is in flashes/km^2-year and R_A is called the attractive radius of the structure in meters. Eriksson then analytically derived the following equation for the attractive radius for heights from 10 to 100 meters.

$$R_A = 0.84 I^{0.74} h^{0.6} \tag{41}$$

For an approximate median current of 35 kA, this equation becomes

$$R_A = 12 h^{0.6} \tag{42}$$

which compares favorably with Eq. 40. Considering both Eq. 40 and 42, Eriksson suggested that

$$R_A = 14 h^{0.6} \tag{43}$$

Although there exist a significant amount of data for freestanding structures, only limited data are available for lines. using five reliable sources and comparing these with the attractive radius of Eq. 43, Eriksson concluded that Eq. 43 was applicable to lines. Thus

$$N(G) = \frac{N_g(2R_A + S_g)}{10} = \frac{N_g(28 h^{0.6} + S_g)}{10} \tag{44}$$

where R_A and S_g, the ground wire separations (Fig. 22), are in meters and $N(G)$ is in strokes per 100 km-years. Although Eriksson notes that the use of h as the tower height yields a better estimate of $N(G)$ than the average line height, in a companion paper [40] he recommends that h be the average ground wire height, i.e., the tower height minus 2/3 of the sag.

12.2 Comparison, Field Data vs. Geometric Model

Using numerical integration to solve the integral equation for the number of flashes per 100 km-year for an N_g of 2.8, which is equivalent to 30 thunderstorm days per the CIGRE equation, a comparison of the results is shown in Fig. 26. These results are compared to Eriksson's suggested Eq. 44 and to the data from Wagner et al. [41]. The suggested Eq. 44 matches the Wagner et al. data very closely. However, the curves obtained by use of the geometric model show far fewer flashes. Interestingly, Eriksson's formulation provides the best comparison but is less than Wagner et al. by about 40%.

This failure of the geometric model to compare favorably with the field data is disturbing and casts doubt on the geometric model. In the original model as presented by Young et al. [19], the resultant curve of number of flashes closely matched the Wagner et al. data. In partial explanation, Young used the old AIEE stroke current distribution ($M = 15$ kA, $\beta = 0.98$). Using the Wagner et al. data of Fig. 26, a conductor having a height of 100 feet collected 100 strokes per 100 mile-years in an area having 30 thunderstorm days per year, resulting in a ground flash density of 13.6 flashes/100 mile-years or 5.3 flashes/100 km-years. He assumed that the ground flash density was a linear function of T_d and therefore

$$N_g = \left(\frac{5.3}{30}\right) T_d = 0.177 T_d \tag{45}$$

Young also modified Wagner's velocity vs. current equation. Thus Young carefully developed his striking distance equations and other data so as to produce an acceptable match to (1) the then available field data on the number of strokes to a conductor and (2) the shielding failure flashover rate of existing lines.

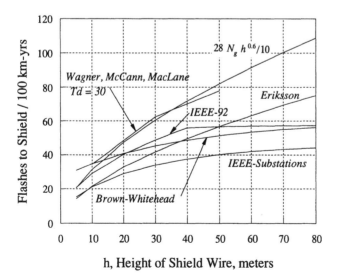

Figure 26 Comparison of number of flashes to a single overhead ground wire with $N_g = 2.8$ equivalent to a $T_d = 30$.

Following Young's development, the CIGRE current distribution was substituted for the AIEE distribution, and the value of N_g as a function of T_d was changed. These alterations produced the nonconformity as shown in Fig. 26. Thus it remains a challenge to produce a new conformity.

Although the geometric model does not result in an acceptable match of field data, it is used today as the most appropriate model of the last step of the stroke. In Chapter 7, it is used to assess the shielding failure flashover rate and to estimate the required shielding angle. This calculation does not require the input of the number of flashes to the overhead ground wire. However, the estimation of the backflash rate as is considered in Chapter 10 does require this value. For this case, Eq. 44 is recommended to estimate the number of flashes to the line. For further conservatism, the height in Eq. 44 will be taken as the height at the tower. That is, to estimate the number of flashes to the line,

$$N(G) = \frac{N_g(28h_T^{0.6} + S_g)}{10} \qquad \text{flashes/100 km-year} \qquad (46)$$

where h_T is the height of the tower. Both h_T and S_g are in meters, and N_g is in flashes per km^2-year.

13 INCREASE IN CURRENT MAGNITUDE WITH HEIGHT

The geometric model predicts an increase in the median value of the stroke current as the height of the line increases. This is illustrated in Fig. 22 where $f(I_G)$ is compared to $f(I)$. More definitively, Fig. 27 shows D'_g as a function of the stroke current. Both curves show an increase as the current increases. Thus more high-magnitude strokes were collected than low-magnitude strokes. Therefore the median of the current terminating at the tower should be greater.

Thus assuming the geometric model to be applicable, the currents collected by Berger, those collected on chimneys, those collected on towers, etc. should have

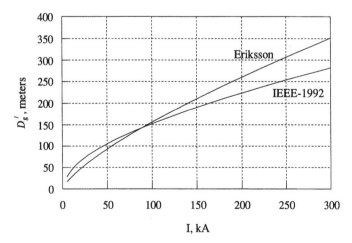

Figure 27 D'_g for $h = 30$ meters.

median currents greater than those to the earth or ground. Thus, conceptually, a ground level current distribution is needed for use with the geometric model. The first investigation of this phenomenon was performed by Sargent [42]. As a result of his analysis, he suggested a ground level distribution having a median of 12.5 kA and a log standard deviation of 0.72. More recently, Mousa and Srivastava [43] suggested a ground level current distribution having a median of 24 kA and a log standard deviation of 0.72, which has been accepted by the IEEE Substations Committee [39].

To investigate these concepts, assume that Berger's distribution having a median of 31.1 kA and a log standard deviation of 0.484 is valid for a conductor height of 30 meters. Using the IEEE-1992 equations, the required ground level distribution (which is lognormal) must have a median of 27.4 kA and a log standard deviation of 0.49. A plot of the median current as a function of conductor height is shown in Fig. 28 and reveals that the median increases sharply from 27.4 to 30.3 for a height increase from 0 to about 1 meter. However, from 1 meter to 60 meters, the median current increases only to 32 kA. Thus the values of 31.1 and 0.484 are valid for all practical conductor heights.

Consider now the CIGRE distribution. Since it is piecewise lognormal, the ground level distribution is also piecewise lognormal. Assuming that the CIGRE distribution is valid for a conductor height of 30 meters, the required parameters of the ground level distribution are presented in Table 9 for alternate striking distance equations.

A plot of the median current as a function of height using the IEEE-1992 equations is shown in Fig. 29. The lower curve for $I > 20\,kA$ is similar in shape to that of Fig. 28, indicating that the parameters 33.3 and 0.605 are valid for all practical conductor heights. However, the upper curve for $I < 20\,kA$ is much different in shape. The median of about 56 kA appears valid up to a height of 20 meters. Above about 40, a median of about 70 kA could be used. Of course, the values for 30

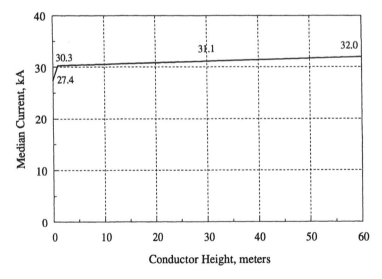

Figure 28 Effect of conductor height on medium stroke current assuming Berger's values at 30 meters.

Table 9 Parameters of the Ground-Level Current Distribution to Obtain the CIGRE Distribution at 30 m

Striking distance equations	For $I < 20$ kA median/beta	For $I > 20$ kA median/beta
IEEE-1992	31/1.52	25.65/0.625
IEEE-1991	29.4/1.48	25.2/0.630
Brown and Whitehead	28.55/1.58	25.3/0.630
Love	38.0/1.58	27.0/0.630
Mousa	348/1.58	26.8/0.630
Young	31.0/1.52	30.0/0.610

meters of 61.1 kA and beta of 1.33 still appear justified as average values for the full range of heights.

Returning to the consideration of the number of flashes collected by conductors, Fig. 30 shows the number of flashes collected for $N_g = 2.8$. The "Original CIGRE" distribution is the ground level distribution with the parameters of median/log standard deviation of 61.1 kA/1.33 and 33.3 kA/0.605, and the ground level distribution is that derived using the IEEE-1992 striking distance equations, "Ground Level CIGRE." The parameters of the two current distributions are 31 kA/1.52 and 25.65 kA/0.625. Assuming the CIGRE distribution to be valid at ground level provides a more conservative estimate of the number of flashes. However, as developed in Chapter 7, the number of flashes that terminate on the phase conductor, that is, the shielding failure flashover rate, increases slightly when the ground level distribution is used.

As a final note, recently, alternate approaches to the geometric model using the leader progression model have been proposed by Dellera and Garbagnati [4] and by Rizk [45]. Although the geometric model has proven successful in determining the

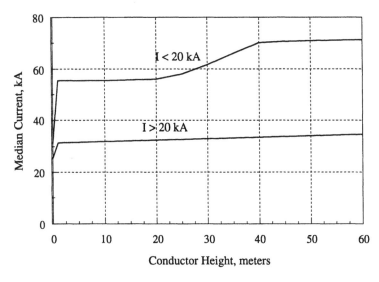

Figure 29 Effect of conductor height on median stroke current assuming CIGRE values at 30 meters.

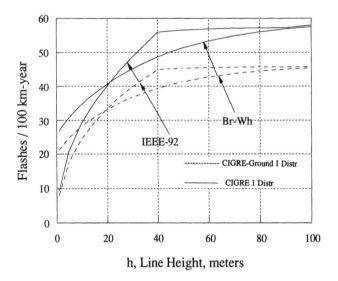

Figure 30 Comparison using the original CIGRE current distribution and the CIGRE ground-level current distribution, $N_g = 2.8$.

proper shielding angle, these methods are expected to offer considerable improvement and solve the aforementioned problems when using the geometric model. At present, these models have not been sufficiently simplified for general use and therefore the geometric model is the method primarily used.

Presently, a CIGRE working group is studying this problem with respect to lightning protection of structures, which is presently standardized in IEC1024 [48]. A report from this task force is expected to be published in *ELECTRA* in 1997 [46].

14 HIGH ALTITUDE

There exist some indications that the median current at high altitude is less than at sea level. A 1941 paper [47] concluded that the median current at 10,000 feet (3,000 meters) is only 10 kA, also that at high altitudes, 64% of the flashes are negative and 36% positive. The authors further postulated that there may be no lightning strokes above an altitude of 18,000 feet. The reasoning behind these observations appears to be that the lines or towers are at times within the cloud.

15 SUMMARY

1. The model of the last step of the lightning stroke consists of development of striking distances from the downward leader to earth, r_g, and from the downward leader to the phase conductor and ground wires, r_c.

2. Eriksson's modified geometric model assumes a striking distance to the ground wire and to the phase conductor but no striking distance to ground. His striking distances are functions of both current and height.

3. The suggested formulations of the striking distance equations are

1. IEEE-1992, as suggested by the IEEE Working Group:

$$r_c = 10.0I^{0.65} \quad r_g = \beta r_c$$
$$\beta = 0.36 + 0.17\ln(43-h) \quad (47)$$
$$\text{for } h > 40, \text{ set } h = 40$$

2. Brown–Whitehead, as suggested by the CIGRE Working Group:

$$r_g = 6.4I^{0.75} \quad r_c = 7.1I^{0.75} \quad (48)$$

4. For tower heights less than about 100 meters, approximately 85 to 95% of the flashes are negative downward flashes. The remaining are negative upward and positive upward and downward.

5. Negative upward flashes are predominant for high towers. These flashes generally have lower currents than for the negative downward flash.

6. Positive flashes are considered to have 1.2 to 2.2 times the currents in the negative downward flash. They tend to predominate in the winter. They also occur at the start and end of a storm.

7. Except for the current, the parameters of the flash are based on Berger's measurements.

8. The statistical distribution of the flash parameters are approximated by a lognormal distribution with the parameters M, the median, and β, the log standard deviation. The mean or average value of the parameter is given by the equation

$$\mu = Me^{\frac{\beta^2}{2}} \quad (49)$$

9. The use of the CIGRE distribution of the first stroke of the negative downward flash is suggested, i.e.,

Suggested First Stroke Current Distribution

Parameter	$I_F < 20\,\text{kA}$	$I_F > 20\,\text{kA}$
M, median	61.1	33.3
β, log std. dev.	1.33	0.605

10. Other parameters of the first stroke of the negative downward flash are provided in Table 1. Of primary importance are

Parameter	M, median	β, log std. dev.
t_m, min. eq. front, µs	1.28	0.611
$t_{30/90}$, 30–90% front, µs	3.83	0.553
S_m, max. steepness, kA/µs	24.3	0.599
t_T, tail, µs	77.5	0.577

11. Two versions of the conditional distributions of the flash parameters are presented in Tables 3 and 6. The conditional distributions using all of Berger's data including his first stroke current distribution are presented in Table 3. The conditional distributions using Berger's correlation coefficients and the CIGRE first stroke current distribution are presented in Table 6. Either of these are equally variable options. However, the distributions of Table 3 are easier to use and appear more "reasonable." Therefore, the conditional distributions of Table 3 are recommended. The parameters of the conditional distributions of primary interest, taken from Table 3, are

Conditional distribution	M, median	β log std. dev.
$I_F|t_m$	$28.7 t_m^{0.33}$	0.439
$t_m|I_F$	$0.21 I_F^{0.53}$	0.554
$I_F|S_m$	$11.7 S_m^{0.31}$	0.448
$S_m|I_F$	$4.83 I_F^{0.47}$	0.554

12. 45% of negative downward flashes have one stroke per flash. The mean is three per flash; the median is two per flash. See Table 7.

13. If available and considered reliable, the directly measured ground flash density should be used. Otherwise the thunderstorm days per year T_d available from weather bureaus may be used to estimate the ground flash density N_g in flashes/km^2-years.

$$N_g = 0.04 T_d^{1.25} \tag{50}$$

14. The number of strokes to a line may be estimated by the equation

$$N(G) = \frac{N_g(28 h_T^{0.6} + S_g)}{10} \tag{51}$$

where h_T is the height of the ground wire at the tower in meters, S_g is the horizontal distance between the ground wires in meters, and $N(G)$ is in strokes or flashes per 100 km-years.

16 REFERENCES

1. H. Prinz, "Lightning in History," Chapter 1 of *Lightning* (R. H. Golde, ed.) London: Academic Press, 1977.
2. B. Dibner, "Benjamin Franklin," Chapter 2 of *Lightning* (R. H. Golde, ed.) London: Academic Press, 1977.
3. B. F. J. Schonland, *The Flight of Thunderbolts*, Oxford: Clarendon Press.
4. C. F. Wagner and A. R. Hileman, "Surge Impedance and Its Application to the Lightning Stroke," *AIEE Trans* on PA&S, 80(3), 1961, pp. 1011–1020.
5. C. F. Wagner and A. R. Hileman, "The Lightning Stroke-II," *AIEE Trans. on PA&S*, Oct. 1961, pp. 622–642.
6. K. Berger, R. B. Anderson, and H. Kroninger, "Parameters of the Lightning Stroke," *ELECTRA*, 41, Jul. 1975, pp. 23–37.

7. R. B. Anderson and A. J. Eriksson, "Lightning Parameters for Engineering Application," *ELECTRA* 69, Mar. 1980, pp. 65–102.
8. CIGRE Working Group 33.01, "Guide to Procedures for Estimating the Lightning Performance of Transmission Lines," Technical Brochure No. 63, Oct. 1991.
9. AIEE Committee Report, "A Method of Estimating the Lightning Performance of Transmission Lines," *AIEE Trans.* 69(3), 1950, pp. 1187–1196.
10. J. G. Anderson, J. G. Fisher, and E. F. Magnusson, "Estimating Lightning Performance of Transmission Lines," Chapter 8 of *EHV Transmission Line Reference Book*, New York: Edison Electric Institute, 1968.
11. S. Spzor, "Comparison of Polish Versus American Lightning Records," *IEEE Trans. on PA&S*, May 1969, pp. 646–652.
12. F. Popolansky, "Preliminary Report on Lightning Observations on High Objects in CSSR," 33–76 (WG 01) 24 IWD, and "Frequency Distribution of Lightning Current Amplitudes to Lower Objects," 33–79 (WG 01) 5 IWD.
13. F. Popolansky, "Frequency Distribution of Lightning Currents," *ELECTRA* 22, May 1972, pp. 139–147.
14. J. G. Anderson, 'Lightning Performance of Transmission Lines," Chapter 12 of *Transmission Line Reference Book*, Palo Alto, CA: Electric Power Research Institute.
15. IEEE Working Group on Lightning Performance of Transmission Lines, "A Simplified Method for Estimating the Lightning Performance of Transmission Lines," *IEEE Trans. on PA&S*, Apr. 1985, pp. 919–932.
16. R. E. Orville and H. Songster, "The East Coast Lightning Detection Network," *IEEE Trans. on Power Delivery*, Jul. 1987, pp. 899–907.
17. M. F. Stringfellow, "Lightning Incidence in the United Kingdom," IEE Conf. Publ. 108, "Lightning and the Distribution system," 1974, pp. 30–40.
18. D. Muller-Hillebrand, "Experiments with Lightning Ground Flash Counters," *Eltenika*, pp. 59–68.
19. F. S. Young, J. M. Clayton, and A. R. Hileman, "Shielding of Transmission Lines," *AIEE Trans.*, 61S, 1951, pp. 132–154.
20. R. H. Golde, ed. *Lightning*, 2 vols., New York: Academic Press.
21. A. J. Eriksson, "The Incidence of Lightning Strikes to Transmission Lines," *IEEE Trans. on Power Delivery*, Jul. 1987, pp. 859–870.
22. J. L. Baldwin, "Climates of the United states," US Dept of Commerce, Washington D.C., 1973, p. 81.
23. D. W. Bodle, A. J. Ghazi, and R. L. Woodside, "Characteristics of the Electrical Environment," Toronto: University of Toronto Press.
24. World Distribution of Thunderstorm Days," World Meteorological Organization, Geneva, Switzerland, WMO No. 21, 1056.
25. A. R. Hileman, "Weather and Its Effects on Air Insulation Specifications," *IEEE Trans. on PA&S* Oct. 1984, pp. 3104–3116.
26. D. R. MacGorman, M. W. Maier and W. D. Rust, "Lightning Strike Density for the Contiguous United States from Thunderstorm Duration Records," NUREG/CR-3759, Washington, D.C., 1984.
27. CIGRE Task Force 33.01.02, 'Lightning Characteristics Relevant for Electrical Engineer: Assessment of Sensing, Recording and Mapping Requirements in the Light of Present Technological Advancements," Technical Bulletin 94, 1995.
28. EPRI Ground Flash Density Reports by R. E. Orville, R. W. Henderson and R. B. Pyle are EL-4729, Aug. 1985, EL-5667, Feb. 1988, EL-6413, Aug. 1989. Also EPRI TR-103603, 1988–1991 Data, Dec. 1993.
29. R. E. Orville, "Peak Current Variations of Lightning Return Strokes as a Function of Latitude," *Nature*, Jan. 111, 1990, pp. 149–151.

30. R. Thottappillil, V. A. Rakor, M. A. Uman, W. A. Beasley, M. J. Master, and D. V. Shelukhin, "Lightning Subsequent Stroke Electric Field Peak Greater Than the First Stroke Peak and Multiple Ground Terminations," *Journal of Geophysical Research*, May 20, 1992, pp. 7503–7509.
31. R. Lindholm, "Induced Overvoltages on Transmission Lines and Their Bearing on the Lightning Performance of Medium Voltage Networks," Diploic Goteborg, Gothenburg, Sweden, 1955. Also see *Trans. of Chalmers Univ. of Technology* (Gothenberg, Sweden), 188, 1957, pp. 1–17.
32. Sune Rusck, "Induced Lightning Over-Voltages on Power-Transmission Lines with Special Reference to Over-Voltage Protection of Low-Voltage Networks. *Trans. of Royal Institute of Technology* (Stockholm, Sweden) 120, 1958.
33. C. F. Wagner, "The Relation Between Stroke Current and the Velocity of the Return Stroke," *IEEE Trans. on PA&S*, Oct. 1963, pp. 609–617.
34. H. R. Armstrong and E. R. Whitehead, "Field and Analytical Studies of Transmission Line Shielding," *IEEE Trans. on PA&S*, 1969, pp. 617–626.
35. G. E. Brown and E. R. Whitehead, "Field and Analytical Studies of Transmission Line Shielding-II," *IEEE Trans. on PA&S*, 1969, pp. 617–626.
36. E. R. Love, "Improvements on the Lightning Stroke Modeling and Application to Design of EHV and UHV Transmission Lines," M.Sc. thesis, University of Colorado, 1973.
37. IEEE Working Group, "Estimating the Lightning Performance of Transmission Lines II—updates to Analytical Models," *IEEE Trans. on Power Delivery*, Jul. 1993, pp. 1254–1267.
38. A. M. Mousa and K. D. Srivastava, "A Revised Electrogeometric Model for the Termination of Lightning Strokes on Grounded Objects," Proceedings of International Aerospace Conference on Lightning and Static Electricity, Oklahoma City, Apr. 1988, pp. 342–352.
39. IEEE Standard 998-1996, "Guide for Direct Stroke Shielding of Substations."
40. A. J. Eriksson, "An Improved Electrogeometric Model for Transmission Line Shielding Analysis," *IEEE Trans. on Power Delivery*, Jul. 1987, 871–886.
41. C. F. Wagner, G. D. McCann, and G. L. MacLane, "Shielding of Transmission Lines," *AIEE Trans.* 60, 1941, pp. 313–328.
42. M. A. Sargent, "The Frequency Distribution of Current Magnitudes of Lightning Strokes to Tall Structures," *IEEE Trans. on PS&S*, 1972, pp. 2224–2229.
43. A. M. Mousa and K. D. Srivastava, "The Implications of the Electrogeometric Model Regarding Effect of Height of Structure of the Median Amplitude of Collected Strokes," *IEEE Trans. on Power Delivery*, Apr. 1989, pp. 1450–1460.
44. L. Dellera and E. Garbagnati, "Lightning Stroke Simulation by Means of the Leader progression Model," Parts I and II, *IEEE Trans. on Power Delivery*, Oct. 1990, pp. 2009–2029.
45. F. A. M. Rizk, "Modeling of Transmission Line Exposure to Direct Strokes," *IEEE Trans on Power Delivery*, Oct. 1990, 1983–1997.
46. CIGRE Working Group 33.01, Task Force 03, "Lightning Exposure of Structures and Interception Efficiency of Air Terminals," Draft, Jan. 1996.
47. L. M. Robertson, W. W. Lewis, and C. M. Foust, "Lightning Investigation at High Altitudes in Colorado," *AIEE Trans.*, Dec. 1941.
48. IEC 1024-1, "Protection of Structures Against Lightning," Part 1, General Principles, 1990.

The Lightning Flash

17 PROBLEMS

1. A test line is constructed by installing a single conductor on top of 120-foot high poles. From test results, an investigator finds that 100 strokes per 100 mile-year terminate on the line. Given that the stroke current probability density function is

$$f(I) = \frac{1}{50}\left[1 - \frac{I-2}{100}\right] \quad 2 \leq I \leq 102 \, \text{kA}$$

and that the striking distance to ground and to the conductor are given by the equations

$$r_g = 27 I^{0.32} \quad r_c = 28.2 I^{0.32}$$

find

(1) The ground flash density N_g in flashes/km^2-year.
(2) The cumulative distribution of currents to the 120-foot conductor. Plot this as an equal to or greater than curve.
(3) The number of strokes per 100 km-year to a 40-foot high conductor. Assume that

$$r_c = r_g = 27 I^{0.32}$$

(4) The cumulative distribution of currents to a 40-foot high conductor.

2. Using Table 3, find the probability that, for the first stroke of a negative downward flash,

(1) The time to crest t_m is greater than 0.701 µs for a stroke current I_F of 10 kA.
(2) The time to crest t_m is greater than 1.33 µs for a stroke current I_F of 33.3 kA.
(3) The time to crest t_m is greater than 2.95 µs for a stroke current I_F of 150 kA.
(4) Repeat the above calculations assuming that t and I are not correlated.

3. A 120 foot mast is erected in an area having an N_g as determined in problem 1.

(1) Determine the number of strokes/year to the mast.
(2) Plot the cumulative distribution of the currents terminating on the mast.

4. The attractive radius R_A is a distance such that the number of strokes to the ground wire or mast may be calculated using the equations

$$\text{for lines} \quad N(G) = 2 N_g R_A L \quad \text{for masts} \quad N(G) = N_g \pi R_A^2$$

For the lines and mast problems 1 and 3, find R_A and the ratio R_A/h, where h is the height of the line or mast.

5. A station having the dimensions of 300 × 300 meters is to be constructed in an area having an average ground flash density of 5 flashes/km^2-year with a standard deviation of 25% of this average. This ground flash density is the average for an area

30×30 km. Find the average ground flash density and its standard deviation in percent of the average for the station.

6. Find the probability that the minimum equivalent front t_m of the first stroke of a negative downward flash is greater than 1.28 μs given that the current I_F is (1) 15 kA, (2) 31.1 kA, and (3) 200 kA. Use Table 3.

7. Find the probability that the maximum steepness S_m of a subsequent stroke (negative downward flash) is greater than 39.9 kA/μs given that the final current I_f is 12.3 kA. Also find this probability if the steepness and the final current are statistically independent.

8. Assume the following parameters for log normal distributions: for the stroke current, $M_I = 33.3$ kA, $\beta_I = 0.605$; for the maximum steepness, $M_S = 24.3$, $\beta_S = 0.599$.

   1. Assume no correlation between the stroke current and the steepness. Calculate the probability of exceeding a current of 80 kA and an S of less than 20 kA/μs, i.e., $P[(I > 80), (S < 20)]$.
   2. Same as the above but assume a correlation coefficient such that the conditional of Table 3 is valid.
   3. Same as in 1 but the correlation coefficient is 1.00.

7
Shielding of Transmission Lines

1 INTRODUCTION

In the preceding chapter, the geometric model of the last step of the lightning stroke was introduced and used to determine the number of flashes to the shield wires. The purpose of these shield or overhead ground wires is to act as collectors of the flashes and insofar as possible to prevent flashes from terminating on the phase conductors and causing a flashover. However, in the practical case, flashes cannot be totally prevented from reaching the conductor, unless the phase conductor is completely surrounded by shield wires. In addition, it may be uneconomical to shield the conductor so that no flashovers occur. Therefore, the goal should be to locate the shield wires so that a specific number of flashes result in flashover. For example, the goal could be to shield the line so that the shielding failure flashover rate, the SFFOR, is 0.05 flashes per 100 km-year.

2 BACKGROUND

Following the development of the simplified model of the last step of the lightning stroke by Wagner [1, 2], Young [3] developed the geometric model (GM) with the primary and indeed sole purpose of showing that shielding angles should be decreased as tower height increases. Prior to this investigation, shielding angles of about 30° were used with success on all lines for which tower heights were in the range of 80 feet (24 meters) [4, 5]. For example, a Philadelphia Electric 230-kV line with 80-foot towers had a shielding angle of 35° and a flashover rate of 1.4/100 mile-years, and an Ontario Hydro 230-kV line with 80-foot towers had a shielding angle of 32° and a flashover rate of 0.33/100 mile-years. A Pennsylvania Water & Power

230-kV line with 80-foot towers and a shielding angle of 28° had a flashover rate of zero.

The impetus for both Wagner's and Young's studies was the poor performance, 7.2 flashovers/100 miles-years, of the AG&E – OVEC 345-kV double circuit line, which had a tower height of 150 feet (46 meters) and employed a shielding angle of 33° [6]. The result of Young's study is shown in Fig. 1, where the recommended shielding angle is plotted versus tower height. Note that Young's recommendation resulted in a shielding angle of about 12° for the 345-KV tower while maintaining the previous recommendation of 30 to 35° for the then normal tower heights of 80 to 90 feet (24 to 27 meters). Subsequently the 345-kV tower was redesigned with a 12° shielding angle. The line performance was reduced to under 1.0/100 mile-years, but shielding failure flashovers now occurred to the middle phase since the shielding angle to the middle phase exceeded that for the top phase. That is, because of icing, the middle phase conductor was horizontally displaced further than the top or bottom phase conductors. Thus another lesson was learned: check the shielding angle to the middle phase for a vertical phase configuration.

Subsequently, Armstrong and Whitehead [7] and Brown and Whitehead [8] further developed the GM. The breakdown gradient was modified, a stroke angle distribution was added (Young assumed vertical strokes), and their calculations were compared to the results of the Pathfinder experiment, which produced data showing shielding failures on instrumented lines [9, 10]. Their recommendation is shown in Fig. 2, where the *average* shielding angle is plotted against the *average* shield wire height. Young's recommendation as shown in Fig. 2 uses the height at the tower. To be noted is that if Young's curve were plotted as average height, the two curves would compare favorably. The Pathfinder data are also shown, plotted as average height and average angle. Each data point represents one or more shielding failure

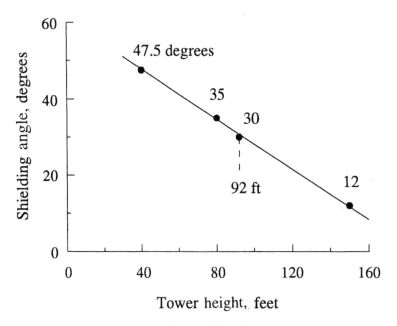

Figure 1 Young's recommended shielding angles.

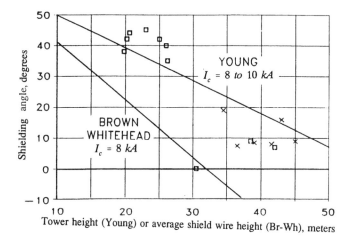

Figure 2 Recommended shielding angles by Young and by Brown–Whitehead compared to Pathfinder data. Squares for critical currents of 4 kA, crosses for critical currents of 8.5 kA.

flashovers. Comparing the data points to the two curves indicates a good agreement between field data and the recommended angles.

To be emphasized further is that the primary purpose of the GM is to show the sensitivity of the shielding angle with tower or line height.

Following these initial studies, further investigations were undertaken to improve, extend, and expand the GM [11–27]. The primary problem was that Young [3] calibrated the GM using the AIEE stroke current distribution which had a median of 15 kA and a log standard deviation of 0.98. As discussed in Chapter 6, this calibration consisted of matching the GM calculations with the then known number of flashes to alternate height lines. The change to Berger's or the CIGRE distribution produced problems in this calibration [14, 15]. As concluded in Chapter 5, the CIGRE distribution will continue to be used. It is the proper distribution when considering flashes to the ground wires or towers. However, it is debatable whether this distribution should be used in calculating shielding failures. As reported in Chapter 6, a current distribution to ground that is developed from the CIGRE distribution may be more appropriate, and using this ground level distribution results in higher shielding failure rates, or requires smaller design angles. This will be discussed further in this chapter.

The seven striking distance equations that are considered in this chapter are presented in Table 1 as obtained from Chapter 6.

Recently, alternate approaches, using leader progression model concepts, have been proposed. Eriksson's approach led to a modified GM [21, 22]. His formulation of the striking distance equations, presented in Table 1, as taken from Chapter 6, provide a significant height sensitivity. Further, the stroke terminating to earth is treated as a default condition, and thus a striking distance equation to earth is not required. The Dellera–Garbagnati [23] approach requires a significant calculation effort, although curves are provided to obtain a quick estimate. The Rizk [16] approach results in two simple sets of curves describing the perfect shielding angle.

Although the geometric model has proven successful in determining the proper shielding angle, the aforementioned methods based on an improved theory of the last

Table 1 Expressions for the Striking Distance $r = AI^b$

Source	r_g to earth or ground		r_c to phase conductors and ground wires	
	A	b	A	b
Young	27.0	0.32	γr_g [d]	0.32
Brown–Whitehead	6.4	0.75	7.1	0.75
Love	10.0	0.65	10.0	0.65
IEEE-1991 T&D Committee	βr_c [a]	0.65	8.0	0.65
IEEE-1992 T&D Committee	βr_c [b]	0.65	10.0	0.65
Mousa and IEEE-1995 Substations Committee[c]	8.0	0.65	8.0	0.65
Eriksson [40]			To phase conductor: $r_c = 0.67 y^{0.6} I^{0.74}$	
			To ground wire: $r_s = 0.67 h^{0.6} I^{0.74}$	
			To earth: none	

[a] $\beta = 22/y$, y = phase conductor height, $0.6 < \beta < 0.9$.
[b] $\beta = 0.36 + 0.17 \ln(43-h)$; if $h > 40$, then $h = 40$.
[c] For masts, Mousa uses $A = 8.8$.
[d] $\gamma = 444/(462 - h)$ for $h > 18\,\text{m}$; $\gamma = 1$ for $h < 18\,\text{m}$.

step of the lightning stroke has recently been proposed. Today, these methods are not in general use, primarily because of the complexity of the calculation. To be expected is that these methods will be simplified until they can be easily employed. For now the geometric model is the primary tool.

Therefore the geometric model will be presented first, following by a brief look at these newer methods.

3 THE GEOMETRIC MODEL

3.1 Basic Concept

Consider the general concept as depicted in Fig. 3. For a *specific value of stroke current*, arcs of radii r_c are drawn from the phase conductors and from the shield wires. In addition, a horizontal line a distance r_g from the earth's surface is constructed. The intersections of these arcs and the intersection of the arcs with the horizontal line are marked A, B, and C. Downward leaders that reach the arc between A and B will terminate on the phase conductor. Those that reach the arc between B and C will terminate on the shield wires, and those that terminate beyond A will terminate to ground or earth.

Assuming only vertical strokes, the distances D_c and D_g are defined in Fig. 3 and are the exposure distance for the phase conductors and shield wires, respectively. Therefore for the specific value of current for which the arcs of Fig. 3 are drawn, the number of strokes that terminate on the phase conductor, or the shielding failure rate SFR, is the area formed by D_c and the length of the line L times the ground flash density, i.e.,

$$\text{SFR}|I = 2N_g L D_c \qquad (1)$$

Shielding of Transmission Lines

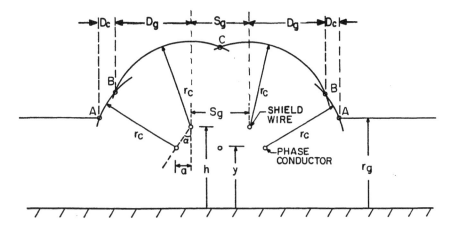

Figure 3 The geometric model, definitions of angles and distances.

The probability of occurrence of this current is $f(I)\,dI$ so that the incremental failure rate $d(\text{SFR})$ is

$$d(\text{SFR}) = 2N_g L D_c f(I)\,dI \tag{2}$$

and the SFR for all currents is

$$\text{SFR} = 2N_g L \int_3^{I_m} D_c f(I)\,dI \tag{3}$$

As noted, the integration limits are 3 kA and I_m, where I_m is the maximum current at and above which no strokes will terminate on the phase conductor. To explain I_m, consider Fig. 4, where the diagram per Fig. 3 is repeated for higher and higher currents. As the current increases, D_c decreases until a point is reached at which all three striking distances meet and D_c becomes zero. This point is defined by the current I_m. As in Chapter 6, the 3 kA lower limit merely recognizes that the first stroke cannot have zero current—that it must have some lower limit. Since the lowest value of current in the CIGRE data is 3 kA, this was selected as the minimum value. However, other investigators believe that values such as 1 or 2 kA are more reasonable and therefore a lower limit of zero current is sometimes used.

Above I_m, as illustrated in Fig. 5, the exposure distance for the shield wires becomes D'_g as defined previously in Chapter 6. Thus the number of strokes or flashes to the shield wires, $N(G)$, is

$$N(G) = N_g L \left[\int_3^{I_m} (2D_g + S_g) f(I)\,dI + \int_{I_m}^{\infty} (2D'_g + S_g) f(I)\,dI \right] \tag{4}$$

Or since S_g is a constant,

$$N(G) = 2N_g L \left[\int_3^{I_m} D_g f(I)\,dI + \int_{I_m}^{\infty} D'_g f(I)\,dI \right] + N_g L S_g \tag{5}$$

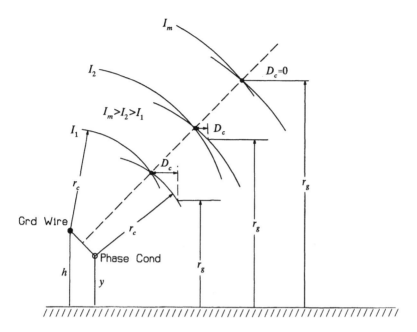

Figure 4 Definition of I_m where $D_c = 0$.

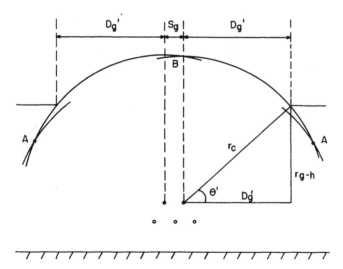

Figure 5 Definition of D_g' for $I > I_c$.

3.2 Distances D_c and D_g

Figure 6 shows one side of the shield wire–phase conductor diagram of Fig. 3. The angle between the two radii r_c is defined as β and is

$$2\beta = \sin^{-1}\frac{c}{2r_c} = \sin^{-1}\frac{\sqrt{a^2+(h-y)^2}}{2r_c} = \sin^{-1}\frac{(h-y)\sqrt{1+\tan^2\alpha}}{2r_c} \quad (6)$$

The angles θ and α are

$$\theta = \sin^{-1}\frac{r_g - y}{r_c} \qquad \alpha = \tan^{-1}\frac{a}{h-y} \quad (7)$$

From this figure,

$$\begin{aligned} D_c &= r_c[\cos\theta - \cos(\alpha+\beta)] \\ D_g &= r_c\cos(\alpha-\beta) \end{aligned} \quad (8)$$

If r_g is less than or equal to y, set θ to zero in Eq. 8.

3.3 The Maximum Shielding Failure Current I_m

Figure 7 depicts the situation where all striking distances coincide at a single point, where I_m is defined. From this diagram, the value of r_g at I_m or r_{gm} is found by first finding the value of a as

$$a = \sqrt{r_c^2 - (r_g - h)^2} - \sqrt{r_c^2 - (r_g - y)^2} \quad (9)$$

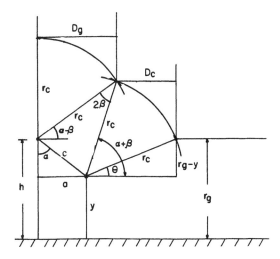

Figure 6 Expanded view of Fig. 3.

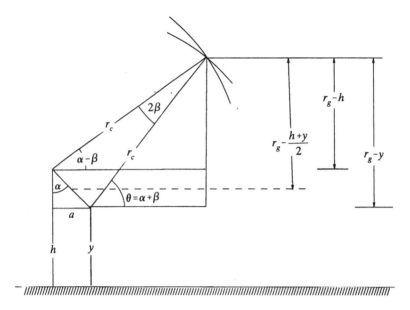

Figure 7 Finding I_m and the perfect shielding angle.

And thus

$$r_{gm} = \frac{h+y}{2k_0}\left[1 + \sqrt{1 - k_0\left[1 + \left(\frac{a}{h+y}\right)^2\right]}\,\right] \tag{10}$$

$$k_0 = 1 - \gamma^2 \sin^2 \alpha \qquad \gamma = \frac{r_c}{r_g}$$

Also from this figure

$$\sin \alpha = \frac{r_{gm} - \dfrac{h+y}{2}}{\sqrt{r_{cm}^2 - \dfrac{c^2}{4}}} \tag{11}$$

This may also be used to derive Fig. 10. Usually,

$$r_{cm}^2 \gg \frac{c^2}{4} \tag{12}$$

and therefore as a good approximation,

$$r_{gm} = \frac{(h+y)/2}{1 - \gamma \sin \alpha} \tag{13}$$

Shielding of Transmission Lines

which is far easier to handle and recommended for use.

r_{rgm} and I_m are related by

$$I_m = \left[\frac{r_{gm}}{A}\right]^{\frac{1}{b}} \tag{14}$$

where the striking distance is of the form

$$r_g = AI^b \tag{15}$$

3.4 Shielding Failure Flashover Rate SFFOR

The SFR is the number of strokes that terminate on the phase conductor. Not all of these will result in flashover. However, if the voltage produced by a stroke to the conductor exceeds the CFO, flashover occurs. Thus the SFR includes both the strokes that cause flashover and those that do not. To determine the flashover rate, note that per Fig. 8 the voltage on the conductor and across the line insulation E is

$$E = I\frac{Z_c}{2} \tag{16}$$

where Z_c is the surge impedance of the phase conductor.

If the voltage E is set to the CFO, negative polarity, then the critical current, at and above which flashover occurs, is

$$I_c = \frac{2(\text{CFO})}{Z_c} \tag{17}$$

The impulse waveshape produced by the stroke is the same as that of the stroke current. Although the time to half value of this surge exceeds that of the standard lightning impulse, and thus the CFO for this surge would be less than the standard lightning impulse CFO, the CFO employed is usually assumed as the standard CFO, negative polarity, which from Chapter 1 is 605 kV/m times the strike distance S.

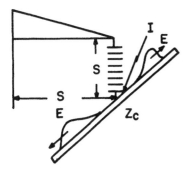

Figure 8 For flashover, voltage must be greater than the CFO.

Revising the equation for the SFR to obtain the SFFOR,

$$\text{SFFOR} = 2N_g L \int_{Ic}^{I_m} D_c f(I) \, dI \qquad (18)$$

3.5 The "Perfect" Shielding Angle

Defining "perfect" shielding as a SFFOR of zero, it is noted from Eq. 18 that this can be achieved by setting I_c to I_m. Therefore Fig. 7 can be reused to develop the equations for perfect shielding angle α_p. Three forms of the equation can be found. First, the easiest one, from the diagram, Fig. 7

$$\alpha_p - \beta = \sin^{-1} \frac{r_g - h}{r_c}$$
$$\alpha_p + \beta = \sin^{-1} \frac{r_g - y}{r_c} \qquad (19)$$

Adding these equations results in

$$\alpha_p = \frac{1}{2} \left[\sin^{-1} \frac{r_g - h}{r_c} + \sin^{-1} \frac{r_g - y}{r_c} \right] \qquad (20)$$

Another way is to determine first the horizontal distance a for perfect shielding, a_p.

$$a_p = \sqrt{r_c^2 - (r_g - h)^2} - \sqrt{r_c^2 - (r_g - y)^2} \qquad (21)$$

Then the perfect angle can be found by

$$\alpha_p = \tan^{-1} \frac{a_p}{h - y} \qquad (22)$$

And yet another way,

$$\alpha_p = \sin^{-1} \frac{r_g - (h + y)/2}{\sqrt{r_c^2 - (c^2/4)}} \qquad (23)$$

Since $c/2 \ll r_c$, an approximate equation is

$$\alpha_p = \sin^{-1} \frac{r_g - (h + y)/2}{r_c} \qquad (24)$$

Note that for small values of the angles,

$$\sin^{-1} X = X \qquad (25)$$

where X is in radians. Therefore, approximately, from Eq. 20,

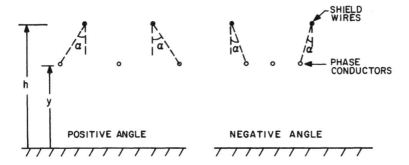

Figure 9 Definition of shielding angles.

$$\alpha_p = \frac{r_g - (h+y)/2}{r_c} = \frac{r_g}{r_c} - \frac{1}{r_c}\left(\frac{h+y}{2}\right) \qquad (26)$$

indicating that approximately α_p is a linear function of $(h+y)/2$, i.e., the average height of the shield wires and phase conductor.

3.6 Stroke Angle

Before attempting an analysis of the alternate striking distance equations, the stroke angle should be considered. In Young's original derivation of the geometric model, only vertical strokes were considered. That is, the downward leader was assumed to be perpendicular to the line, and the previously developed equations apply for these vertical strokes. Later, Whitehead and his associates developed the concept that the downward leader could approach the line from any direction and further suggested the probability density function

$$f(\psi) = k \cos^2 \psi \qquad (27)$$

where the ψ is the angle to the vertical axis and varies between $\pm\pi/2$. This assumption adds a considerable degree of complexity to the calculation of SFFOR and only increases the SFFOR by about 10 to 29%. Note that if $r_g < (h-y)/2$, the shielding angle is negative per Eq. 20 (see the next section). Considering the distribution of the stroke angle, as a limit, near-horizontal strokes could occur.

The definition of a negative shielding angle is illustrated in Fig. 9.

4 ERIKSSON'S MODIFIED MODEL

The previous presentation applies to the normal geometric model. In contrast, Eriksson's [21] modified model must be considered separately since it (1) does not consider a striking distance to ground and (2) assumes all angles of the stroke are equally likely.

4.1 The SFFOR

Eriksson's modified geometric model was introduced in Chapter 6. To reiterate, there are two striking distances, r_s to the shield wire and r_c to the phase conductor, whose equations are

$$r_s = 0.67 I^{0.74} h^{0.6}$$
$$r_c = 0.67 I^{0.74} y^{0.6}$$
(28)

The geometric diagram is shown in Fig. 10. As shown, the striking distance to ground does not exist; it is a default condition. That is, any downward leader that does not meet the arc described by r_c will terminate to ground. Thus all stroke angles are considered, and all are considered equally likely. However, the downward leader is not permitted to travel below the height of the phase conductor and then travel upward to the phase conductor. The exposure of the phase conductor is specified by the arc D_c and therefore

$$D_c = r_c \theta$$
(29)

where the angle θ is in radians. Therefore, as before, the SFR and the SFFOR are

$$\text{SFR} = 2 N_g L \int_3^{I_m} D_c f(I) \, dI$$
(30)

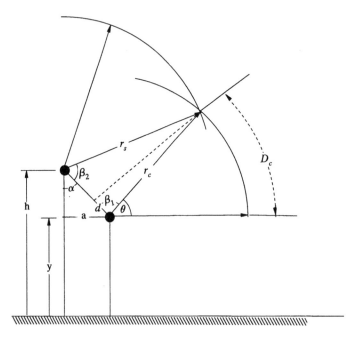

Figure 10 Eriksson's modified geometric model.

Shielding of Transmission Lines

$$\text{SFFOR} = 2N_g L \int_{I_c}^{I_M} D_c f(I) dI \tag{31}$$

From Fig. 10, D_c is found from the equation

$$r_s^2 - (c-d)^2 = r_c^2 - d^2$$
$$d = \frac{r_c^2 - r_s^2 + c^2}{2c} \tag{32}$$

Therefore

$$\beta_1 = \cos^{-1}\frac{d}{r_c}$$
$$\beta_2 = \cos^{-1}\frac{c-d}{r_s} \tag{33}$$
$$\theta = \alpha - \beta_1 + \frac{\pi}{2}$$

Or to be complete,

$$D_c = \left(\alpha - \beta_1 + \frac{\pi}{2}\right) r_c \tag{34}$$

where the angles are in radians.

4.2 Perfect Shielding and Maximum Shielding Failure Current

The perfect horizontal distance a_p and the perfect shielding angle α_p can be obtained from Fig. 11:

$$a_p = \sqrt{r_s^2 - (h-y)^2} - r_c$$
$$\alpha_p = \tan^{-1}\frac{a_p}{h-y} \tag{35}$$

For a specific value of a, the maximum current I_m can also be obtained from Fig. 11. The maximum value of r_c, r_{cm} is

$$r_{cm} = \frac{1}{\gamma^2 - 1}\left[a + \sqrt{a^2 + c^2(\gamma^2 - 1)}\right]$$
$$\gamma = \frac{r_s}{r_c} = \left(\frac{h}{y}\right)^{0.6} \tag{36}$$

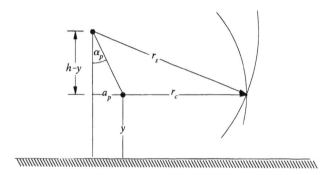

Figure 11 The perfect angle and I_m for Eriksson's modified model.

and thus

$$I_m = \left[\frac{r_{cm}}{0.67 y^{0.60}}\right]^{\frac{1}{0.74}} \tag{37}$$

5 SENSITIVITY ANALYSIS—PERFECT SHIELDING

The perfect shielding angles, which are calculated using the seven major formulations of striking distance as obtained from Table 1, are shown in Figs. 12 and 13.

In contrast to the other striking distance formulations, Eriksson's formulation indicates an almost constant shielding angle independent of the average height of the shield wire and phase conductor. This is in variance with the data of Figs. 1 and 2, and therefore Eriksson's results will not be considered further.

Except for Young's formulation, which only permits vertical strokes, the other striking distance equations result in severe negative angles for higher average heights. For a critical current of 5 kA, negative shielding angles occur for average heights above about 20 meters. For a critical current of 10 kA, negative angles occur above an average height of about 28 meters. Again, the severe negative angles required do not agree with the data shown in Figs. 1 and 2.

Further, if it is desired to design for this perfect angle, then, for example, for an average height of 30 meters, for a 5-kA critical current, angles of from about $-35°$ to $+19°$ could be obtained from these curves. If only the Brown–Whitehead, IEEE-1992, and Substations formulations are considered, the spread decreases to about $-18°$ to $-13°$, which again does not match the data of Figs. 1 and 2. The most obvious source of error appears to be the assumption that severe stroke angles, up to $90°$ from the vertical (a horizontal stroke) can occur. If stroke angles were more limited, these angles would increase. To simplify the use of these striking distance equations and in an attempt to rectify the severe assumptions of a horizontal stroke, both IEEE and CIGRE have changed the stroke angle assumption to that of Young, that is, only vertical strokes are considered.

But some further explanation is required as to how these negative angles occur. Two conditions are required for negative angles to occur. First, as explained in the

Shielding of Transmission Lines

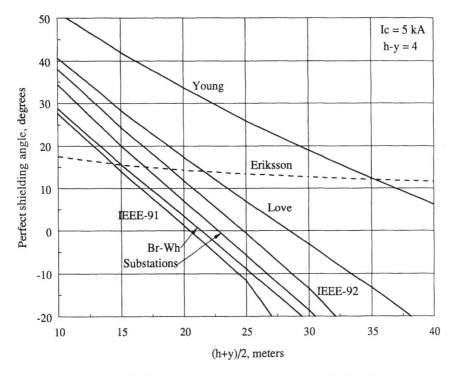

Figure 12 Perfect shielding angles using the geometric model, $I_c = 5\,\text{kA}$.

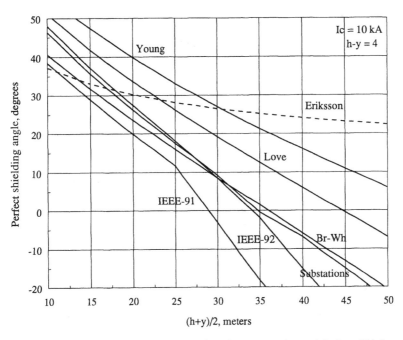

Figure 13 Perfect shielding angles using the geometric model, $I_c = 10\,\text{kA}$.

previous section, except for Young's formulation, near horizontal strokes are possible, and second, the conductor height y must be equal to or greater than the striking distance to ground, i.e., $y \geq r_g$. To explain this later reason, consider Fig. 14. As is normal, the task of the engineer is to locate the shield wires or wire given the location of the conductors. To perform this in a graphical manner, the striking distances are calculated and a horizontal line is drawn at a distance of r_g from the earth. Arcs of radii r_c are then drawn with centers at the conductors, and an intersection point A is obtained. Now with a compass at point A, an arc is drawn from the conductor upward; see the solid line in Fig. 14a. These two arcs define the location of the ground wires, assuming r_c is constant. The shield wires can be located anywhere to the outside of these arcs. In Fig. 14b the point of intersection for these two arcs is the location for a single shield wire.

The solid line of Fig. 14a assumes that r_c is constant with height of the shield wires. Because r_c is a function of the height of the shield wires, the dotted line curves of Fig. 14a apply.

Using the same critical current, Figs. 14c and 14d illustrate the arcs for a constant r_c when the conductor height is increased. Figure 14c is drawn for the condition of $y < r_g$ but $h > r_g$. To be noted is that depending on the location of the shield wire, negative angles can occur and further, that in this case, two shield wires are required. For greater line heights, where y is greater than or equal to r_g, Fig. 14d shows that negative angles are required.

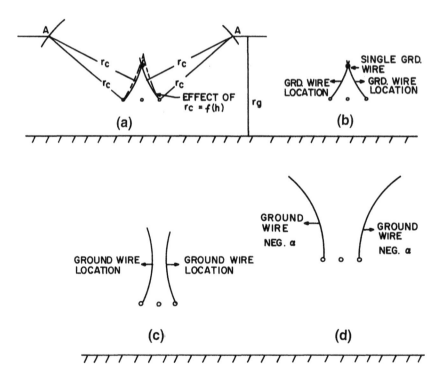

Figure 14 The reason for negative shielding angles.

Referring back to Figs. 12 and 13, since only vertical strokes are considered by Young, his minimum shielding angle is zero.

6 EFFECT OF GROUND LEVEL CURRENT DISTRIBUTION

The use of a ground level stroke current distribution instead of the usual CIGRE distribution increases the SFFOR and thus will decrease the required shielding angle for a desired SFFOR. To illustrate this effect assume a desired SFFOR of 0.05 flashovers/100 km-years. The resultant required shielding angles are shown in Table 2. If the ground level CIGRE distribution per Chapter 6 is used, the required shielding angles decrease by only about 8%. Because of this small decrease and all the other variabilities of selecting the shielding angle, the use of the original CIGRE distribution will be maintained.

7 SELECTION OF THE SHIELDING ANGLE BASED ON SFFOR

The primary aim in the selection of the number and location of shield wires is to provide a means of intercepting the lightning flash and to reduce the shielding failure rate to an acceptable level, fully realizing that a SFFOR of zero is virtually impossible. In the past, the design shielding angle was frequently selected on the basis of perfect shielding. While this may be proper for areas of very high ground flash densities, in areas where N_g is 1 to 4 this restriction to a perfect angle may severely handicap an economical design. Thus one shield wire may be adequate for areas of low ground flash density, whereas two shield wires are required in areas of higher lightning activity. Therefore, a design based on a nonzero value of SFFOR is suggested. The actual design value of SFFOR must be the prerogative of the designer so as to permit economical designs. For lines serving critical loads, a design value of 0.05/100 km-years may be suitable. However, in Europe, design values as high as 2.0 have been reported. In general, a design value of 0.05/100 km-years is recommended.

An additional benefit of selecting a nonzero SFFOR is that the large deviations of shielding angles between alternate methods as shown in Figs. 12 and 13 are greatly reduced. Further reductions in these deviations result if only vertical strokes are considered.

To examine the second premise, that of considering only vertical strokes, the SFFOR as a function of the shielding angle using the four major formulations of the striking distances are presented in Fig. 15. (These formulations, taken from Chapter 6, are shown in Table 1.) the shield wire and phase conductor heights are held

Table 2 Effect of Current Distribution on the Shielding Angle SFFOR = 0.05/100 km-years, $h = 32\,m$, $y = 28\,m$, $I_c = 10\,kA$, $N_g = 4$

Equations	CIGRE	CIGRE-ground	Ratio
Brown–Whitehead	19.2	17.6	0.92
IEEE-1992	17.7	16.3	0.92
Substations	20.1	18.4	0.92

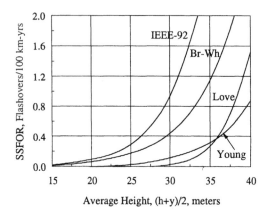

Figure 15 Shielding angles assuming vertical strokes using the geometric model, $h = 30\,\text{m}$, $y = 28\,\text{m}$, $N_g = 4$, $I_c = 10\,\text{kA}$.

constant and the critical current is set at 10 kA. Note first that the curves have an initial slow upward trend followed by a quick change to a very steep increase. Thus the conclusion is that the angle should be selected within the slow upward trend region, and some conservatism is justifiable.

The four formulations of striking distance give a wide variation in shielding angle. For example, for 0.2 flashovers per 100 km-year, Fig. 15 shows that the shielding angle could be 24, 26, 33, or 34 degrees depending on the striking distance equations used. Love's and Young's equations essentially give the same result, but the Brown–Whitehead equations are more conservative and the IEEE-1992 equations more conservative still.

To illustrate the point that these curves further converge if a nonzero value of SFFOR is selected as a design value, Fig. 16 presents the results for a SFFOR of 0.05/100 km-years. Note that all strike distance formulations provide reasonable values of shielding angle. Again, Love's and Young's curves are almost identical,

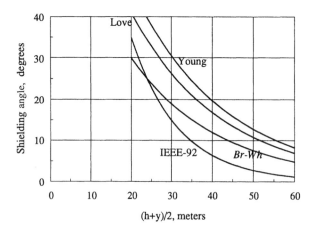

Figure 16 Shielding angles for a SFFOR of 0.05/100 km-years, $h - y = 4\,\text{m}$, $I_c = 10\,\text{kA}$.

while the Brown–Whitehead curve is more conservative. The sharp move to small angles displayed by the IEEE curve is a result of the restricted use of the value of beta as given in Table 1.

As a further point, Young's curves give the shielding angle at the *tower*, while the other curves are for the average shielding angle along the span, and thus the shielding angle at the tower would be greater. Thus to compare these curves critically, Young's curve should be moved to the left, or the other curves should be moved to the right. This factor would result in an even better comparison of the shielding angles.

For general design, it is recommended that the design value of SFFOR be set at 0.05 flashovers per 100 km-year, that either the Brown–Whitehead or the IEEE-1992 equations be employed, and that only vertical strokes be assumed. These equations provide a reasonable conservative limit to shielding angle. Again, these angles should be considered as the average shielding angle along the span. That is, they are the shielding angles at the tower height minus 2/3 of the sag. If more conservatism is desired, the shielding angle may be considered as that at the tower.

To assist in selection of the shielding angle, the curves of Figs. 17 and 18 have been prepared for the Brown–Whitehead and the IEEE-1992 formulations of the striking distance, for vertical strokes and for a SFFOR of 0.05/100 km-years. As an illustration of the use of Figs. 17 and 18, assume a design value of SFFOR of 0.05/100 km-years in an area having a ground flash density of 10. For an average phase conductor and shield wire height of 30 meters and a critical current of 5 kA, a shielding angle of about 8° is obtained from either figure. For these same conditions but for a ground flash density of 1, a shielding of 17° to 18° results.

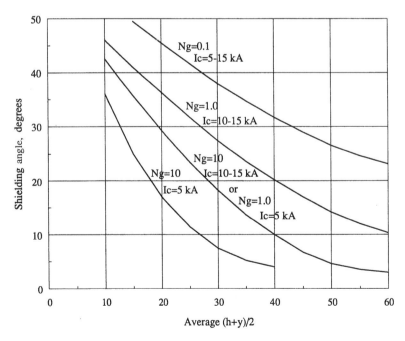

Figure 17 Shielding angles for a SFFOR of 0.05/100 km-years, Brown–Whitehead equations, vertical strokes.

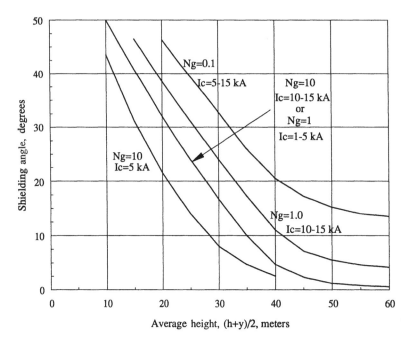

Figure 18 Shielding angles for a SFFOR of 0.05/100 km-years. IEEE-1992 equations, vertical strokes.

8 VARIABLES OF DESIGN

8.1 Terrain–Hillside Effects

The calculated shielding angles assume flat or rolling terrain. For towers located on hillsides, the average shielding angle is that obtained from Figs. 17 and 18 minus the hillside angle. To explain, consider Fig. 19 where the hillside angle is θ_G. The line constructed at a distance r_g from the earth is now parallel to the hill. Thus the horizontal distance for the perfect shielding angle is

$$a' = \sqrt{r_c^2 - (r_g - h')^2} - \sqrt{r_c^2 - (r_g - y')^2} \tag{38}$$

where

$$a' = (a + b\tan\theta_G) \qquad h' = (h - a\tan\theta_G) \qquad y' = ya^1\cos\theta_G \tag{39}$$

These equations may be solved for a', but as a slightly conservative but excellent approximation, the shielding angle is the previous angle calculated without hillside effects minus the hillside angle. For example, if the shielding angle selected for level terrain were 30° and the hillside angle 15°, the shielding angle for the hillside location would be 15°.

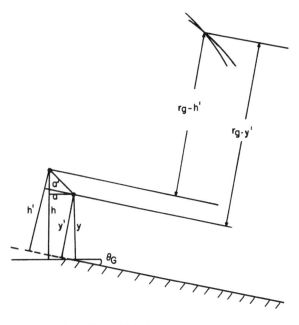

Figure 19 Effect of hillsides.

8.2 Terrain—Trees Along the Right-of-Way

Trees, structures, etc., along the line right-of-way are beneficial in that they increase the effective earth plane or decrease the height of the line, as illustrated in Fig. 20. The striking distance to earth is from the structure or treetops. Thus, for this case, larger angles can be used. Of interest is that these structures or trees bring remote strokes closer to the line and thus increase the probability of flashover caused by induced voltage from nearby strokes. This increase is most dramatic for low-voltage lines where the insulation strength is low. This effect is considered in Chapter 15.

8.3 Terrain—Hilltops

Towers located on hilltops are especially vulnerable, since more strokes are collected by these towers. The hilltop site is even more vulnerable to the backflash, since the tower footing resistance tends to be larger than normal (see Chapter 10). As has been

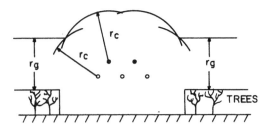

Figure 20 Effect of trees.

noted by the industry, the performance of a few towers or line sections frequently determines the total performance of the line; thus the name "rogue" towers has come into the colloquial language.

8.4 The Center Phase

This discussion has been centered on shielding the outside phase of a line. While shielding failures to the center phase may occur in extremely rare cases, they are predicted to be and are essentially zero. For all practical cases of tower design, the shielding failures to the center phase should be considered nonexistent. This factor of greatly improved shielding to objects between shield wires or masts is employed in the design of station shielding, a subject to be considered in the following Chapter 8.

With all the above factors, the final selection of the shielding angle at the tower must be a matter of judgment, based on the experience of the designer and on the performance of other lines within the utility system.

8.5 An Example

An example of the selection of the shielding angle occurred in the design of the Allegheny Power System 500-kV line [28]. From the expected span length distribution of Fig. 21, the tower heights were established assuming level terrain. Using these tower heights, a required distribution of "perfect' shielding angles was determined as shown in Fig. 22. Per this figure, a 25° angle provides "perfect" shielding for 50% of the line, while a 17° angle provides "perfect", shielding for 75% of the line. In practice, the towers will be located on hillsides and hilltops. Their height will

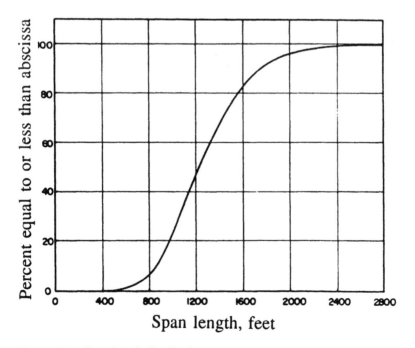

Figure 21 Span length distribution.

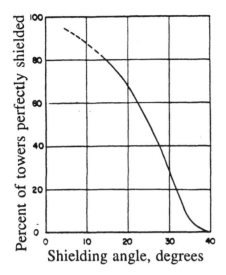

Figure 22 Percent of towers "perfectly" shielded.

decrease below those assumed in Fig. 21, but their lightning exposure may increase. The cost to decrease the shielding angle from 20° to 15° was estimated at $60 per mile (1966 cost), and an additional $90 per mile would be required to decrease this angle to 10°. Mechanical design limited further reduction to 5° or less. From this analysis, a 15° angle was selected. This line has a lightning performance of less than 0.6 flashovers/100 km-year. The majority of flashovers appear to occur from the back-flash.

9 A SIMPLIFIED METHOD OF CALCULATING THE SFFOR

The equations developed must be solved by numerical integration. An excellent approximation to calculate the SFFOR was suggested by J. G. Anderson [26]. Observing that the value of D_c when $I = I_m$ is zero, Anderson suggested that the average value of D_c over the interval from I_c to I_m is the half of the value of D_c at $I = I_c$. More formally, let D_{cc} equal the value of D_c at I_c. then, since D_{cc} is assumed constant, it may be taken outside the integral, i.e.,

$$\begin{aligned}
\text{SFFOR} &= 2N_g L \frac{D_{cc}}{2} \int_{I_c}^{I_m} f(I)\, dI \\
&= 2N_g L \frac{D_{cc}}{2} P(I_m \geq I \geq I_c) \\
&= 2N_g L \frac{D_{cc}}{2} [F(I_m) - F(I_c)] \\
&= 2N_g L \frac{D_{cc}}{2} [Q(I_c) - Q(I_m)]
\end{aligned} \quad (40)$$

where

$$Q(I) = 1 - F(I) \tag{41}$$

If a normal distribution table is not available, an approximation to the CIGRE cumulative distribution is

Range of current I, kA	Approximate equation
3 to 20	$Q = 1 - 0.31 e^{-\frac{Z^2}{1.6}}$
20 to 60	$Q = 0.50 - 0.35 Z$
60 to 200	$Q = 0.278 e^{-\frac{Z^2}{1.7}}$

where Z is, as before,

$$Z = \frac{\ln I - \ln M_I}{\beta_I} = \frac{\ln(I/M_I)}{\beta_I} \tag{42}$$

From Chapter 6, the median and log standard deviation for the CIGRE distribution are

Current range, kA	Media, M_I	Beta, β_I
3 to 20	61.1	1.33
Greater than 20	33.3	0.605

Two examples may help

Example 1. Two shield wires are located at an average height of 30 meters, and the conductors are at 26 meters. The shielding angle is 25° and the critical current is 10 kA. Also the ground flash density is 4. Using the Brown–Whitehead equations, the approximate Eq. 13 and Eq. 15,

$$r_{gm} = 52.72 \quad I_m = 16.63 \, \text{kA} \tag{43}$$

At I_c (Eqs. 6, 7, and 8),

$$r_g = 36.0 \quad r_c = 39.9 \quad \beta = 3.17 \quad \theta = 14.5 \quad D_c = 3.459$$
$$Z_c = \frac{\ln(10/61.1)}{1.33} = -1.361 \quad Q_c = 0.9026 \tag{44}$$

At I_m,

$$Z_m = \frac{\ln(16.65/61.1)}{1.33} = -0.9782 \quad Q_m = 0.8295 \tag{45}$$

Therefore

$$\text{SFFOR} = 2(4)(100)(3.459/2)(1/1000)(0.0731) = 0.10/100 \, \text{km-years} \tag{46}$$

A computer program gives 0.12/100 km-years.

Example 2. Same as example 1 except that the shielding angle is 35°.

$$r_{gm} = 77.1 \qquad I_m = 27.6 \qquad D_c = 7.41$$

$$Z_c = -1.361 \qquad Z_m = \frac{\ln(27.6/33.3)}{0.605} = -0.3103 \tag{47}$$

$$Q_c = 0.9026 \qquad Q_m = 0.6095 \qquad \text{SFFOR} = 0.87/100 \text{ km-years}$$

And the computer gives 0.86/100. the second example shows that because the maximum current is over 20 kA, a different median and log standard deviation must be used.

10 SUBSEQUENT STROKES—A PROBLEM

To this point in the presentation, only the first stroke of the flash has been considered. It has been assumed that subsequent strokes of the flash will have magnitudes less than that of the first and that therefore if the first stroke terminates on the phase conductor and does not cause a flashover, subsequent strokes will also not cause a flashover. This premise, although used by all investigators, may not be correct. Although this was not stated in Chapter 6, no correlation exists between the first stroke current and currents of subsequent strokes. However, it has been noted that the subsequent strokes *generally* have lower currents.

It currents of subsequent strokes are independent of the first stroke current, then a probability does exist that subsequent strokes of a flash may cause a flashover even though the first stroke of the flash does not. Thus, even if a line is "perfectly" shielded, i.e. has a SFFOR of zero, a subsequent stroke could cause flashover. To expand the thought, assume that the critical current is 10 kA and that the line is perfectly shielded. Under this assumption, if the first stroke current is 10 kA or less, no flashover occurs. Now, let us assume that a first stroke current of 8 kA terminates on the conductor. No flashover occurs. Then a subsequent stroke occurs having a magnitude of 15 kA. Now, flashover occurs!

As source data, Table 7 of Chapter 6 presents the probability of subsequent strokes. Also, subsequent stroke currents have a median of 12.3 kA and a log standard deviation of 0.530. To develop the equation for the total SFFOR including subsequent strokes, let the probability of n strokes per flash per Table 7 of Chapter 6 be denoted by P_n. Now consider only one stroke per flash. The SFFOR is

$$\text{SFFOR} = P_1(\text{SFFOR}_1) \tag{48}$$

where the SFFOR$_1$ is that calculated previously when subsequent strokes were not considered.

The SFFOR for two strokes per flash is of the form

$$\text{No. of flashovers} = P[2 \text{ strokes/flash}] \{\text{No. of FO's on 1st}$$
$$+ P(\text{FO on 2d}) \text{ (No. of no FO's on 1st)}\} \tag{49}$$

or

$$\text{SFFOR} = P_2\{\text{SFFOR}_1 + pN_1\} = P_2\{\text{SFFOR}_1 + N_1(1-q)\} \qquad (50)$$

where N_1 is the number of first strokes that do not result in flashover and q is the probability of no flashover on a subsequent stroke, i.e.,

$$N_1 = 2N_g L \int_3^{I_c} D_c f(I)\,dI = \text{SFR}_1 - \text{SFFOR}_1$$
$$q = \int_0^{I_c} f(I_s)\,dI_s \qquad (51)$$

where SFR_1 is the SFR as calculated previously for the first stroke. I_s is the subsequent stroke current, and $f(I_s)$ is the probability density function.

The SFFOR for three strokes per flash is somewhat similar:

No. of Flashovers = P[3 strokes/flash] {No. of flashovers on 1st

+ P (of at least one flashover on 1st or 2d)

(no. of no flashovers on 1st)} $\qquad (52)$

or

$$\text{SFFOR} = P_3\{\text{SFFOR}_1 + N_1(1-q^2)\} \qquad (53)$$

The value of $1-q^2$ is the probability that at least one of the subsequent strokes results in a flashover.

Continuing to set down the equations for 4, 5, 6, etc., strokes per flash and then adding these to obtain the total SFFOR, SFFOR_T, results in

$$\text{SFFOR}_T = \text{SSFOR}_1 + 2N_g L P_s \int_3^{I_c} D_c f(I)\,dI$$
$$= \text{SFFOR}_1 + P_s(\text{SFR}_1 - \text{SFFOR}_1) \qquad (54)$$

where

$$P_s = \sum_{n=1}^{\infty} P_n[1-q^n] \qquad (55)$$

As an approximation, P_s is 0.50 and therefore the total SFFOR is approximately

$$\text{SFFOR}_T \approx \text{SFFOR}_1 + (0.50)(\text{SFR}_1 - \text{SFFOR}_1) = 0.50(\text{SSFOR}_1 + \text{SFR}_1) \qquad (56)$$

As an example, consider a line having a shielding angle of 30° and a ground wire height of 32 meters. If the critical current is 10 kA, the SFFOR_1 is 0.9441 flashovers/100 km-years and the number of strokes that do not cause flashover is 0.3413/

Shielding of Transmission Lines

100 km-years resulting in a SFR_1 of 1.2854/100 km-years. The value of q in the previous equations is 0.345. Thus the total shielding failure flashover rate is 1.107, i.e., 0.166 is added by subsequent strokes. If the approximation of Eq. 56 is used, the total SFFOR is 1.115. Thus the total shielding failure flashover rate is between the original SFFOR and the SFR.

If the critical current is changed to 5 kA, the $SFFOR_1$ is 12.16, the SFR_1 is 12.83, the number of strokes that do not cause flashover is 0.67, and q is 0.0446. Therefore the total SFFOR is 12.53, which is approximately the same result as given by Eq. 56. Again the total SFFOR is between the original SFFOR and the SFR.

At the present time, it is recommended that the effect of the subsequent strokes not be considered. However, it should be kept in mind that flashovers can occur from subsequent strokes.

11 DENSITIES AND CDF

To complete the discussion of the geometric model, the density and cumulative distribution function are listed below.

11.1 SFFOR Currents

Let the shielding failure flashover currents equal I_f,

$$F(I_f) = \frac{2N_g L}{\text{SFFOR}} \int_3^{I_f} D_c f(I)\, dI \quad \text{for } I_c \leq I_f \leq I_m$$

$$f(I_f) = \frac{2N_g L}{\text{SFFOR}} D_c f(I)$$
(57)

11.2 SFR Currents

Let the shielding failure currents equal I_s.

$$F(I_s) = \frac{2N_g L}{\text{SFR}} \int_3^{I_s} D_c f(I)\, dI \quad \text{for } 3 \leq I_s \leq I_m$$

$$f(I_s) = \frac{2N_g L}{\text{SFR}} D_c f(I)$$
(58)

11.3 Currents to Shield Wires

Let these currents to the shield wires equal I_G. For currents between 3 and I_m,

$$F(I_G) = \frac{N_g L}{N(G)} \left[2\int_3^{I_G} D_g f(I)\, dI + S_g \int_3^{I_G} f(I)\, dI \right]$$

$$f(I_G) = \frac{N_g L}{N(G)} [2 D_g f(I) + S_g f(I)]$$
(59)

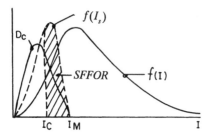

Figure 23 Densities of the shielding failure current.

For currents greater than I_m,

$$F(I_G) = \frac{N_g L}{N(G)} \left[2 \int_3^{I_G} D_g f(I) dI + 2 \int_{I_G}^{\infty} D'_g f(I) dI + S_g \int_3^{I_G} f(I) dI \right]$$

$$f(I_G) = \frac{N_g L}{N(G)} [2D'_g f(I) + S_g f(I)]$$

(60)

The equation for the density of $f(I_g)$ is illustrated in Fig. 23. The crosshatched area is the SFFOR.

12 COMPARISON OF RESULTS OF GM WITH LPM METHODS

As mentioned in the introduction, other methods exist from which the shielding angle may be obtained. In addition to Eriksson's modified GM [23], these methods are principally (1) Rizk [16] and (2) Dellera–Garbagnati [23], both of which employed a leader progression model for the downward leader. The purpose of this section is to compare the results of these alternate methods.

The results of Eriksson and Rizk, Fig. 24, indicate a significant agreement. Since the Dellera–Garbagnati approach does not permit the determination of the perfect angle, the results are shown for a SFFOR/N_g of 0.0125 and 0.05 for an N_g of 4 in Fig. 25. When compared to the results of Fig. 24, significant deviations are apparent.

Assuming vertical strokes, a further comparison of results for a SFFOR/N_g ratio of 0.01 is shown in Tables 3 and 4. The dispersion of results for the GM is significantly reduced from those of Figs. 12 and 13. However, the results shown in Table 4 for the leader progression model indicate a significant dispersion. In particular, the results from Dellera and Garbagnati produce negative angles—for these cases all stroke angles are considered.

13 PRESENTLY USED SHIELDING ANGLES

The suggestion of altering the shielding angle as a function of the ground flash density is not a new idea. Utilities have been doing this for years. At 500 kV, Bonneville Power Administration has used one shield wire for low ground flash density regions and two shield wires for higher ground flash density areas.

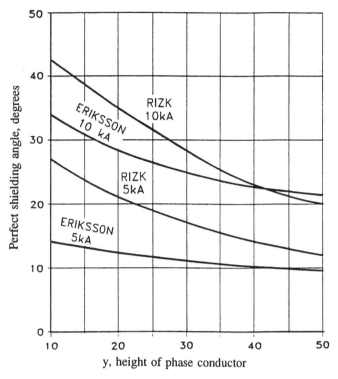

Figure 24 Perfect shielding angles as determined by Eriksson's and Rizk's methods for $h - y = 4$ [21, 16, 29].

The lowering of the shielding angle with increased tower height has now been entirely accepted. Some verification of this is contained in the CIGRE survey [18]. Also the data contained in Ref. 19, although not giving the tower height, may be helpful. As shown in Table 5, higher towers have lower shielding angles, and higher angles are used in areas of lower lightning intensity.

14 CONCLUSIONS

1. The geometric model of the last step of the lightning stroke was developed to show that the shielding angle should be decreased as the tower height increases.

2. The exact value of the shielding angle is dependent on the striking distance equation. Five of these formulations are presently in use: (1) Young, (2) Brown–Whitehead (used by CIGRE), (3) Love, (4) IEEE-1992 (from the T&D Committee), and (5) IEEE-1995 (from the Substations Committee).

3. New methods using the leader progression model have been proposed that promise to improve the shielding model. However, these methods are not at the stage of supplanting the geometric model.

4. The geometric model should be employed to estimate the shielding angle based on a nonzero value of SFFOR. In general, a SFFOR of 0.05 flashovers/ 100 km-years is recommended.

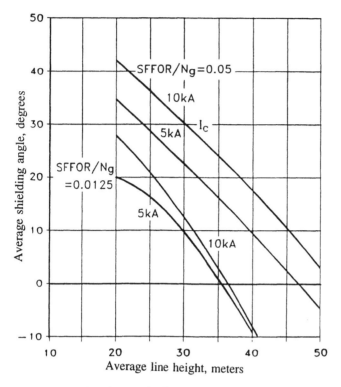

Figure 25 Shielding angles for Dellera–Garbagnati method, $N_g = 4$ [23, 29].

5. If designs are based on a nonzero SFFOR, the shielding angle will be a function of the ground flash density.

6. Either the Brown–Whitehead equations or the IEEE-1992 striking distance equations, using the assumption of vertical strokes are recommended for general use. The angles so determined are the angles at the average height of the line. In equation form,

$$\bar{h} = h_t - \frac{2}{3} \text{GW sag}$$

$$\bar{y} = y_t - \frac{2}{3} \text{ Ph. cond. sag}$$

(61)

Table 3 Comparison of Shielding Angles using the GM for a SFFOR/N_g Ratio of 0.01 for Critical Currents of 5 and 10 kA, $(h - y) = 4$

$(h+y)/2$ meters	Brown–Whitehead		Young		IEEE-1991		Eriksson		IEEE-1992	
	5 kA	10 kA	5 kA	10 kA	5 kA	10 kA	5 kA	10 kA	5 kA	10 kA
20	20	30	39	43	18	27	37	45	24	33
30	14	18	26	31	10	10	33	41	11	17
40	11	11	16	21	5	5	31	38	4	5
50	10	10	9	13	4	4	29	36	2	2

Table 4 Comparison of Shielding Angles Using Leader Progression Model Concepts for a SFFOR/N_g Ratio of 0.01 for Critical Currents of 5 and 10 kA

	Eriksson (1)		Dellera–Garbagnati[a]		Rizk[b]	
$(h+y)/2$ meters	5 kA	10 kA	5 kA	10 kA	5 kA	10 kA
20	37	45	21	28	21	43
30	33	41	9	13	17	35
40	31	38	−9	−8	14	29
50	29	36	—[c]	—[c]	11	20

[a]For $h - y = 4$ m. [b]For perfect shielding. [c]Values could not be determined.

Table 5 Shielding Angles Used for 345- and 500-kV Lines in the USA, Lightning Tripout Rate Less Than 0.6/100 km-year

Nominal system voltage, kV	Type	Thunderstorm days/year, range	Shielding angle Range	Shielding angle Average
345	Single circuit	22 to 50	0 to 33	22
345	Double circuit	20 to 60	0 to 30	11
500	Single circuit	30 to 110	−9 to 20	12
500	Single circuit	2 to 25	15 to 30	23

where $\bar{h}$ and $\bar{y}$ are the average heights of the ground wire and phase conductor, respectively, and h_t and y_t are the heights at the tower.

Setting the calculated shielding angle at the average height results in a larger angle at the tower.

7. If the currents of subsequent strokes are statistically independent of the currents of the first stroke, subsequent strokes will increase the SFFOR to 0.5 (SFFOR + SFR). Since the relationship or correlation between the first and the subsequent strokes is questionable, it is recommended that the effect of subsequent strokes be neglected.

15 REFERENCES

1. C. F. Wagner and A. R. Hileman, "The Lightning Stroke—II," *IEEE Trans. on PA&S*, Oct. 1961, pp. 622–642.
2. C. F. Wagner, "The Relation Between Stroke Current and the Velocity of the Return Stroke," *IEEE Trans. on PA&S*, Oct. 1963, pp. 609–617.
3. F. S. Young, J. M. Clayton, and A. R. Hileman, "Shielding of Transmission Lines," *IEEE Trans. on PA&S*, S82, 1963, pp. 132–154.
4. C. F. Wagner, G. D. McCann, and G. L. MacLane, "Shielding of Transmission Lines," *AIEE Trans.*, 60, 1941, pp. 318–328.
5. C. F. Wagner, G. D. McCann, and C. M. Lear, "Shielding of Substations," *AIEE Trans.* 61, 1942, pp. 96–100.
6. W. S. Price, S C. Bartlett, and E. S. Zobel, "Lightning and Corona Performance of 330-kV Lines of the American Gas and Electric and the Ohio Valley Electric Corporation Systems," *AIEE Trans.*, 75(3), Aug. 1956, pp. 583–597.

7. H. R. Armstrong and E. R. Whitehead, "Field and Analytical Studies of Transmission Line Shielding," *IEEE Trans. on PA&S*. Jan. 1968, pp. 270–281.
8. G. W. Brown and E. R. Whitehead, "Field and Analytical Studies of Transmission Line Shielding—*IEEE Trans. on PA&S*, 1969, pp. 617–626.
9. H. R. Armstrong and E. R. Whitehead, "A Lightning Stroke Pathfinder," *IEEE Trans. on PA&S*, 1964, pp. 1223–1227.
10. E. R. Whitehead, "Mechanism of Lightning Flashover Research Project," Final Report of the Edison Electric Institute, Publication No. 72-900, Edison Electric Institute, New York, NY.
11. G. W. Brown, "Lightning Performance—I, Shielding Failures Simplified," *IEEE Trans. on PA&S*, Jan./Feb. 1978, pp. 33–38.
12. J. R. Currie, L. Ah Choy, and M. Darrenzia, "Monte Carlo Determination of the Frequency of Lightning Strokes and Shielding Failures," *IEEE Trans. on PA&S*, Sep./Oct. 1971, pp. 2305–2312.
13. E. R. Love, "Improvements on the Lightning Stroke Modelling and Application to Design of EHV and UHV Transmission Lines," M.Sc. thesis, University of Colorado, 1973.
14. M. A. Sargent, "The Frequency Distribution of Current Magnitudes of Lightning Strokes to Tall Structures," *IEEE Trans. on PA&S*, 1972, pp. 2224–2229.
15. A. M. Mousa and K. D. Srivastava, "The Implications of the Electrogeometric Model Regarding Effect of Height of Structure on the Median Amplitude of Collected Strokes," *IEEE Trans. on Power Delivery*, Apr. 1989, pp. 1450–1460.
16. F. A. M. Rizk, "Modeling of Transmission Line Exposure to Direct Strokes," *IEEE Trans. on Power Delivery*, Oct. 1990, pp. 1983–1997.
17. IEEE Working Group on Lightning Performance on Transmission Lines, "A Simplified Method for Estimating Lightning Performance of a Transmission Line," *IEEE Trans. on PA&S*, 1985, pp. 919–932.
18. E. R. Whitehead, "CIGRE Survey of the Lightning Performance of Extra High-Voltage Transmission Lines," *ELECTRA*, Mar. 1974, pp. 63–89.
19. F. J. Ellert, S. A. Miske, and C. J. Truax, "EHV and UHV Transmission Systems," Chapter 2 of *Transmission Line Reference Book*, Palo Alto, CA: Electric Power Research Institute.
21. A. J. Eriksson, "An Improved Electrogeometric Model for Transmission Line Shielding analysis," *IEEE Trans. on Power Delivery*, Jul. 1987, pp. 871–886.
22. A. J. Eriksson, "Lightning and Tall Structures," *Trans. SAIEE*, 69(8), Aug. 1978.
23. L. Dellera and E. Garbagnati, "Lightning Stroke Simulation by Means of the Leader Progression Model, Parts I and II," *IEEE Trans. on Power Delivery*, presented at the IEEE/PES Summer Meeting, Jul. 1989.
24. IEEE Working Group, "Estimating Lightning Performance of Transmission Lines II, Updates to Analytical Models," *IEEE Trans. on Power Delivery*, Jul. 1993, pp. 1254–1267.
25. IEEE Substations Committee, "IEEE Guide for Direct Stroke Shielding of Substations," IEEE Std. 988–1996.
26. J. G. Anderson, "Lightning Performance of Transmission Lines," Chapter 12 of *Transmission Line Reference Book*, Palo Alto, CA: Electric Power Research Institute.
27. A. M. Mousa and K. D. Srivastava, "A Revised electrogeometric Model for the Termination of Lightning Strokes on Grounded Objects," Proceedings of International Aerospace Conference on Lightning and Static Electricity, Oklahoma City, OK.
28. A. R. Hileman, W. C. Guyker, H. M. Smith, and G. E. Grosser, Jr, "Line Insulation Design for APS 500 kV System," *IEEE Trans. on PA&S*, Aug. 1967, pp. 987–994.

29. CIGRE WG 33.01, "Guide to Procedures for Estimating the Lightning Performance of Transmission Lines," Technical Bulletin 63, Oct. 1991.

16 PROBLEMS

1. Determine the SFFOR for the 500 kV line whose dimensions are shown below. Use (A) the Brown–Whitehead and (B) Young's striking distance equations. Assume all vertical strokes. Also calculate the perfect shielding angle. Do this by the simplified method as presented in this chapter.

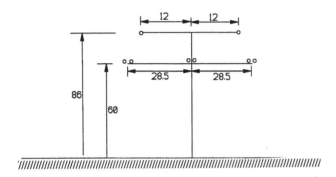

Figure 26 500-kV line, distances in feet.

Subconductor diameter = 1.65 in.
Subconductor spacing = 18 in.
Shield wire diameter = 0.5 in.
Conductor surge impedance = 355 Ω
Ground flash density = 5.0 flashes/km^2-year

Phase conductor sag = 21 ft
Ground wire sag = 14 ft
Span length = 1000 ft
Minimum strike distance = 11.2 ft

2. Based on a SFFOR of 0.08 flashovers/100 mile-year, determine the location and number of overhead ground wires for the 500-kV line of problem 1 assuming only the location of the phase conductors and that the ground wire height is given below. Use the Brown–Whitehead equations for vertical strokes. Use the regular CIGRE current distribution, assume the minimum current is 3 kA, and do not consider subsequent strokes. Use the computer program ALPD.

1. Let $T_d = 30$ thunderstorm days/year and a ground wire height of 86 feet.
2. Let $T_d = 4$ thunderstorm days/year and a ground wire height of 86 feet.
3. Let $T_d = 30$ thunderstorm days/year and a ground wire height of 100 feet.

3. Based on a SFFOR of 0.08 flashovers/100 mile-year, determine the location and number of ground wires for a 230 kV double circuit line as shown in Fig. 27. Assume a ground wire height of 110 feet. Use the Brown–Whitehead equations for vertical strokes. Assume (1) that $T_d = 30$ and (2) that $T_d = 5$. Use the regular CIGRE current distribution, assume that the minimum current is 3 kA, and do not consider subsequent strokes. Use the computer program ALPD.

Conductor diameter = 1.65 in.
Shield wire diameter = 0.5 in.

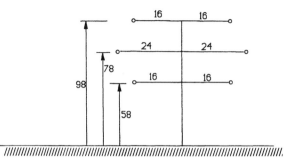

Figure 27 230-kV line, distances in feet.

Conductor surge impedance, phase A = 477 Ω
  phase B = 463 Ω
  phase C = 446 Ω
Minimum strike distance = 5.8 ft (13 insulators)
Phase conductor sag = 21 ft
Ground wire sag = 14 ft
Span length = 1000 ft

4. Estimate the SFFOR for each of the lines of problem 3 of Chapter 2. Assume that the phase conductor surge impedances are 465, 455, and 472 ohms for the 34.5-, 69-, and 115-kV lines, respectively. Use the Brown–Whitehead and the IEEE-1992 striking distance equations. Assume a ground flash density of 5 flashes/km^2-year. Use the regular CIGRE current distribution, assume that the minimum current is 3 kA, and do not consider subsequent strokes. Use the computer program SFFOR.

5. Estimate the SFFOR for the 115-kV line of problem 4 of Chapter 2. Assume that the phase conductor surge impedance is 373 ohms. Use the Brown–Whitehead and the IEEE-1992 striking distance equations. Assume a ground flash density of 5 flashes/km^2-year. Use the regular CIGRE current distribution, assume that the minimum current is 3 kA, and do not consider subsequent strokes. Use the computer program SFFOR.

6. A single circuit 500 kV line has two overhead ground wires and horizontal disposed phase conductors as in problem 1. The average heights of the ground wires and phase conductors are 28 and 18 meters, respectively. The shielding angle is 33°, the CFO is 2000 kV, and the conductor surge impedance is 400 ohms. Assume a ground flash density of 10 flashes/km^2-year. Use the regular CIGRE current distribution, assume that the minimum current is 3 kA, and do not consider subsequent strokes. Using the Brown–Whitehead equations, estimate the SFFOR using the simplified method and check this using the computer program SFFOR.

7. Using Eriksson's modified GM and assuming vertical strokes, derive the equation for D_c. Also derive the equation for the "perfect" angle.

8
Shielding of Substations

1 INTRODUCTION

In Chapter 7 the shielding of transmission lines was considered. The emphasis was on shielding the phase conductor, which is normally outboard from the shield wire, i.e., there was a positive shielding angle. It was mentioned that the center phase, that phase between the shield wires, need not be considered, since it would be more than adequately shielded. In this chapter this condition is paramount. As we shall see, the best method to provide shielding to the bus and equipment is to locate the shield wires or masts so that they enclose the objects to be protected.

The basis of design for substation shielding is somewhat different from that for lines. While the same concept of designing to a specific SFFOR is valid for buses in the substation, the design based on a SFFOR for specific pieces of equipment is difficult. For this reason and for simplicity, the design is approached on the basis of a design current.

Yet another difference in substation shielding is that either or both shield wires and masts may be used, the decision being that of the designer.

In this chapter, the basis of design is discussed first; showing the differences and similarities to the shielding of lines. Then the equations are given. (In presenting this chapter to students, skipping the theoretical development and using only the application material contained in the summary creates more interest.)

2 BASIS OF DESIGN

Ideally, the basis of design should be approached, as in Chapter 7, from a SFFOR standpoint. However, this is only practical for the station bus. For equipment, which is defined in three dimensions, the calculation of the SFFOR is theoretically possible but is complex and cumbersome so that, in all cases, the design is based on a specific design current, which in turn is derived on the basis of a SFFOR or a MTBF, the mean time between a shielding failure flashover.

To begin, assume that only the high-level bus is to be considered. Since the consequence of failure in a station is greater than that in a line, the equivalent design SFFOR is usually set to a lower value. Assume that the design SFFOR is 0.01 flashovers/100 km-years. Assuming that the high-level bus length L is 0.5 km, the MTBF is

$$\text{MTBF} = \frac{1}{L(\text{SSFOR})} = \frac{100}{0.5(0.01)} = 20{,}000 \text{ years} \tag{1}$$

Or, approaching this problem from a design MTBF, the first task is to select a desired MTBF. Assume that a MTBF of 1000 years is desired and that the length of the high-level bus is 0.5 km. Then the equivalent SFFOR is

$$\text{SFFOR} = \frac{1}{L(\text{MTBF})} = \frac{100}{0.5(1000)} = 0.2 \text{ flashover/100 km-years} \tag{2}$$

(As noted, the equivalent design SFFOR exceeds that recommended for the lines, i.e., 0.05/100 km-years.)

If the bus conductors were on the outboard side of the shield wires, the shielding angle could be obtained in the identical manner as for the lines, as given in Chapter 7. To simplify this procedure and also make it applicable for the equipment and the bus, a design current is necessary. This design current has the same definition as the critical current of Chapter 7 but is usually somewhat larger. Per Chapter 7, the critical current is

$$I_c = \frac{2(\text{CFO})}{Z} \approx \frac{2(\text{BIL})}{Z} \tag{3}$$

To arrive at a design current I_d, first consider a SFFOR of 0.05 flashovers/100 km-years, which is usually greater than that for the substation. Using the calculation methods of Chapter 7, the ratio of the maximum current to the critical current, I_m/I_c, for the Substation Committee's formulation of striking distance is shown in Fig. 1 as a function of the ground flash density N_g. Note that the maximum current is the equivalent critical current if the line were designed for perfect shielding. Thus the maximum may be considered as the design current.

Figure 1 also shows the regression equations for the two curves. Combining these equations results in

$$I_d = (1.27 + 0.72 e^{-N_g/4}) I_c \tag{4}$$

Before using this equation, consider the value of N_g. The value of the ground flash density is usually derived over a considerable area. Assume that this area is 30×30 km, an area of 900 km². From Chapter 6, the standard deviation of this mean value of N_g is 32% of N_g. Assuming an average value of N_g of 5, in this area the mean number of flashes μ is 4500 flashes per year, and the standard

Shielding of Substations

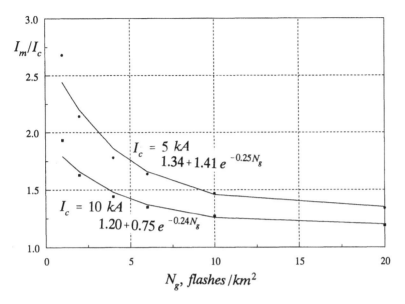

Figure 1 SFFOR = 0.05 flashovers/100 km-years, substation equations.

deviation σ is 1125 flashes per year. Now consider the area of a substation of 300 × 300 meters, or 0.09 km². Within this smaller area the mean μ_s and standard deviation of the number of flashes σ_s are

$$\mu_s = \frac{\mu}{n} = \frac{4,500}{10,000} = 0.45 \text{ flashes/year}$$

$$\sigma_s = \frac{\sigma}{\sqrt{n}} = 11.25 \text{ flashes/year} \tag{5}$$

where n is the ratio of the areas. The standard deviation within the substation area is greatly increased, and the value of σ_s in per unit of the mean has increased from 0.32 to 25. This is as expected and is observable in practice, since a flash can terminate on the station shielding several times in one year and not at all for several years. The primary conclusion is that the design value of N_g should be greater than the average value. As a suggestion, assume that the design value of N_g is at least twice the average value.

Returning now to the selection of the design current, Table 1 lists the critical currents and the design currents as obtained using Eq. 4 and assuming that N_g is double the average value. The last column of this table shows suggested integer values of the design current. For design currents less than 3 kA, the suggested value is 3 kA, since this has been used as the minimum value of stroke current.

Further, for practical designs, the design current values in Table 1 may be further reduced to 5 kA for system voltages below 230 kV and to 10 kA for system voltages at and above 230 kV.

To complicate the issue further, arresters within the substation will decrease the surge voltage and thus the design current. For example, in a 345-kV station using an

Table 1 Selection of the Design Current

System nominal voltage, kV	I_c, kA	N_g	T_d	I_d from Eq. 4	I_d, suggested
34.5	1	2.8	30	1.6	3
		10.0	83	1.3	6
69	2	2.8	30	3.3	3
		10.0	83	2.7	3
115	3	2.8	30	4.9	5
		10.0	83	4.0	4
138	3	2.8	30	4.9	5
		10.0	83	4.0	4
230	5	2.8	30	8.1	8
		10.0	83	6.6	7
345	6	2.8	30	9.8	10
		10.0	83	8.0	8
500	10	2.8	30	16.3	16
		10.0	83	13.3	13
765	12	2.8	30	19.5	20
		10.0	83	15.9	16

arrester rated 209 kV MCOV, the 10 kA discharge current is about 600 kV. This is in contrast to the usual BIL of 1050 or 1300 kV used at 345 kV. Although the 600 kV would only occur at the arrester location, voltages lower than the 1050 or 1300 kV would usually occur at other locations.

Thus the suggested values of 5 kA for system voltages of below 230 kV and 10 kA for system voltages equal to or greater than 230 kV appear conservative.

To complete this section, a correction to the number of flashes to a substation should be revised from the example above, since the substation collects more flashes than its physical area times the ground flash density. The substation exposed area A_S is

$$A_S = (W + 2R_A)(L + 2R_A) \tag{6}$$

where W and L are the physical length and width of the station and per Chapter 6

$$R_A = 16h^{0.6} \tag{7}$$

where h is the height of the substation.

3 STRIKING DISTANCE EQUATIONS

A significant revision of the striking distance equations is required. For transmission lines, the heights of the shield wire and the phase conductors are not significantly different, so that the striking distances to the shield wire and to the phase conductors were assumed to be equal. However, for station shielding, this assumption may not be true. That is, the height of the object to be protected may range from near ground level to near the height of the shield wire. thus the striking distance to the object to be

Shielding of Substations

protected should be equal to the striking distance to ground for objects of low height but may increase to a value equal to the striking distance to the shield wire for heights approximating the shield wire height. Therefore, three striking distances should be considered: (1) the striking distance to the shield wire or shielding mast, r_s, (2) the striking distance to the object to be protected, r_c, and (3) the striking distance to ground, r_g. To simplify

$$r_s = \gamma_s r_g \qquad r_c = \gamma_c r_g \qquad (8)$$

For Young's formulation, only a slight adjustment is required, and for Love's or for the IEEE-1995 equation, no modification is necessary, since all striking distances are equal. However, for the Brown–Whitehead equations and the IEEE-1992 equations, a significant change is required. In summary, for station shielding, we have the following.

3.1 Young's Equations

$$r_g = 27 I^{0.32}$$
$$\gamma_s = \frac{444}{462 - h} \quad \text{for } h \geq 18\,\text{m} \quad \text{otherwise } \gamma_s = 1 \qquad (9)$$
$$\gamma_c = \frac{444}{462 - y} \quad \text{for } y \geq 18\,\text{m} \quad \text{otherwise } \gamma_c = 1$$

3.2 Love's Equations

$$r_g = 10 I^{0.65}$$
$$\gamma_s = \gamma_c = 1 \qquad (10)$$

3.3 Brown–Whitehead—CIGRE Equations

$$r_g = 6.4 I^{0.75}$$
$$\gamma_s = 1 + \frac{h - 18}{108} \quad \text{for } h \geq 18\,\text{m} \quad \text{otherwise } \gamma_s = 1 \qquad (11)$$
$$\gamma_c = 1 + \frac{y - 18}{108} \quad \text{for } y \geq 18\,\text{m} \quad \text{otherwise } \gamma_c = 1$$

3.4 IEEE-1992—IEEE T&D Committee Equations

$$r_g = 9.0 I^{0.65}$$

$$\gamma_s = \frac{1}{0.36 + 0.17 \ln(43 - h)} \quad \gamma_c = \frac{1}{0.36 + 0.17 \ln(43 - y)} \quad (12)$$

for $h \geq 30$, set $h = 30$ for $y \geq 30$, set $y = 30$

3.5 IEEE-1995—IEEE Substations Committee Equations

$$r_g = 8 I^{0.65}$$
$$\gamma_s = \gamma_c = 1 \quad (13)$$

where, as before, I is the magnitude of the stroke current in kA, r_g, r_s and r_c are distances in meters, h is the height of the shield wire or shielding mast, and y is the height of the object to be protected. Both h and y are in meters.

4 SHIELDING USING SHIELD WIRES

The shielding zone offered by two shield wires is shown in Figs. 2 and 3. Figure 2 shows the zones when the shield wires are remote from one another. To construct these zones, draw an arc centered at the top of the shield wire of radius r_s until it intersects the striking distance to ground r_g. With the center at this intersection, draw an arc from the top of the shield wire until it intersects ground. This represents the protective zone, and any object that is under this arc or in this zone is protected. This is the identical zone as would be shown for transmission lines. That is, the value of a per Fig. 3 is the same as in Eq. 9 of Chapter 7, i.e.,

$$a = \sqrt{r_c^2 - (r_g - h)^2} - \sqrt{r_c^2 - (r_g - y)^2} \quad (14)$$

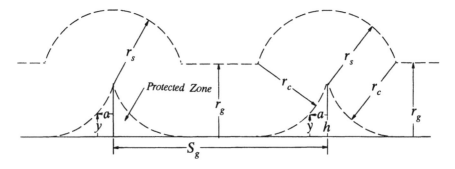

Figure 2 Protected zone with remote shield wires.

Shielding of Substations

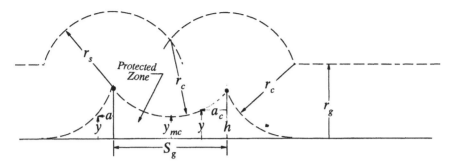

Figure 3 Protected zone improves with decreased separation.

As the shield wires are brought closer together per Fig. 3, the two arcs from the shield wires meet at a location that is above the horizontal line for the striking distance to ground r_g. Thus a stroke or downward leader approaching the system between the shield wires will reach the arcs from the shield wires before reaching the horizontal line for the striking distance to ground. In other words, all strokes will terminate on the shield wires and none will reach ground. Now the protective zone is described by a single arc of radius r_c drawn from the intersection of the arcs from the shield wires as shown in Fig. 3. From these sketches the improvement in the protective zone is apparent.

This protective zone between the shield wires can be described by the distances a_c or R_{PC}, which from Fig. 4 can be calculated by the equations

$$R_{PC} = \sqrt{r_c^2 - \left(h - y + \sqrt{r_s^2 - R_c^2}\right)^2} \tag{15}$$

$$a_c = R_c - R_{PC}$$

where R_c is half the horizontal distance between the shield wires S_g.

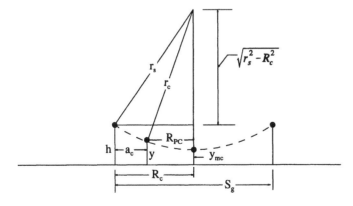

Figure 4 Between the shield wires.

The minimum protective height y_{mc} as shown in Fig. 3 occurs half way between the two shield wires and can be obtained by setting R_{PC} equal to zero in Eq. 15 or directly from Fig. 4.

$$y_{mc} = h - r_c + \sqrt{r_s^2 - R_c^2} \tag{16}$$

The protective zone outward from the shield wires can be described by Eq. 14, which can be conveniently separated into two components, a_0 and R_{PO}, as shown in Fig. 5:

$$\begin{aligned} R_{PO} &= \sqrt{r_c^2 - (r_g - y)^2} \\ a_0 &= \sqrt{r_s^2 - (r_g - h)^2} \end{aligned} \tag{17}$$

and therefore

$$a = a_0 - R_{PO} \tag{18}$$

Since for $y = 0$ and $r_c = r_g$, $R_{PO} = 0$, and therefore a_0 is the distance a for $y = 0$, and the protective height is zero at a_0.

Equation 17 can be used to determine the required height of the shield wire for a specific value of y_{mc}. As an example, assume that the equipment height y is 12 meters. The equipment is located between two shield wires that are separated by 60 meters, i.e., $R_c = 30$ meters and $y_{mc} = 12$ meters. The design current is 10 kA, and Young's striking distance equations are used.

Rearranging Eq. 16,

$$h = y_{mc} + r_c - \sqrt{r_s^2 - R_c^2} \tag{19}$$

To find the required height of the shield wire, the height must be iterated, since r_s is dependent on h. Since the equipment height is 12 meters, $r_c = r_g = 56.4$ meters. Starting with $h = 18$, Table 2 shows the iteration process, which results in a shield

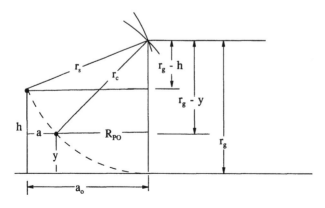

Figure 5 Outward from shield wire.

Shielding of Substations

Table 2 Iteration to Find Height of Two Shield Wires

h	r_s	h
18	56.41	20.64
20.65	56.75	20.24
20.25	56.70	20.30
20.30	56.70	20.29

wire height of about 20.3 meters. Figure 6 shows the resultant plan and profile views of the protective zone.

If more than two shield wires are present, the shielding zones can be found by sequential use of the above equations for two shield wires. The protective zone for a single shield wire can be described by Eqs. 17 and 18.

5 SHIELDING USING MASTS

5.1 One Mast

The protective zone for a single or isolated mast can be described by the same equations as used for a single shield wire, i.e.,

$$a = a_0 - R_{PO}$$
$$a_0 = \sqrt{r_s^2 - (r_g - h)^2} \qquad (20)$$
$$R_{PO} = \sqrt{r_c^2 - (r_g - y)^2}$$

However in this case the quantities a and a_0 are radii of circles as shown in the plan view of the protective zone in Fig. 7. Again, as noted, a_0 is the radius for a protective

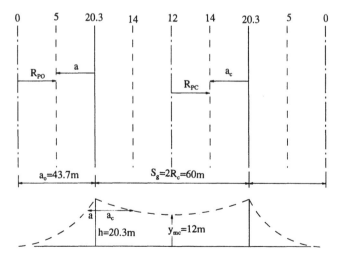

Figure 6 Example: plan and profile view of shielding zones for two shield wires.

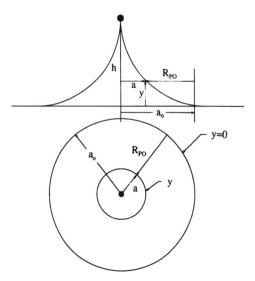

Figure 7 Protective zone for a single mast.

height of zero, $y = 0$. Again, from these equations, the required mast height can be determined from any height and location of the object to be protected.

5.2 Two Masts

The major problem in the analysis of two or more masts is simply that it is a three-dimensional problem. The visualization of the problem is difficult, and so is the task of sketching properly the three-dimensional figure. Nevertheless consider Fig. 8, which attempts to illustrate that above each mast, a portion of space is described by the radius r_s from the top of the mast. Each of these portions of a sphere is ended or terminated by the striking distance to ground r_g. If the two masts are sufficiently

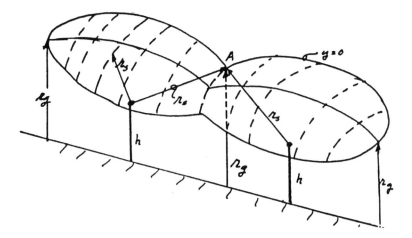

Figure 8 Three-dimensional view of shielding above two masts.

Shielding of Substations

close to one another, the two spheres will intersect. The important aspects of the problem is that unlike the case of two shield wires, the effect of earth is still present.

The critical intersection, location A, is illustrated in Fig. 8, where the two radii r_s from the masts meet r_g. This point can be described as the condition where $y = 0$. Figure 9 presents a plan view of this case. The two circles of radius a_0 meet midway between the masts

The sketch of Fig. 10 also shows the critical intersection A along with the striking distances.

From this diagram,

$$a_0 = \sqrt{r_s^2 - (r_g - h)^2} \qquad (21)$$

which is identical to a_0 for the isolated mast and for the outward side of the shield wires. If $h > r_g$, then $a_0 = r_s$. Also from Fig. 9,

$$d = \sqrt{a_0^2 - R_c^2} \qquad (22)$$

where R_c is half the distance between the masts. Combining the last two equations,

$$h = r_g - \sqrt{r_s^2 - R_c^2 - d^2} = r_g - \sqrt{r_s^2 - a_0^2} \qquad (23)$$

Of more importance is the protective radius R_{PO}, which from Fig. 10 is

$$R_{PO} = \sqrt{r_c^2 - (r_g - y)^2} \qquad (24)$$

Note that this describes an arc centered at point A in Fig. 10. If $y > r_g$, then $R_{PO} = r_c$. The minimum protected height y_{m2}, which occurs midway between the masts, is obtained from Eq. 24 by setting $R_{PO} = d$. Then, also using Eq. 22,

$$y_{m2} = r_g - \sqrt{r_c^2 - a_0^2 + R_c^2} \qquad (25)$$

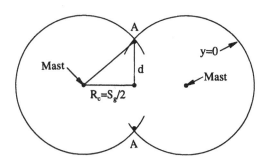

Figure 9 Plan view of Fig. 8.

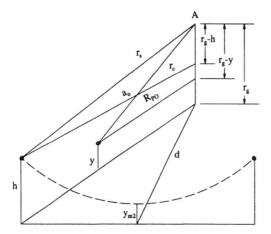

Figure 10 Between two masts.

The resultant protective zone for two masts is illustrated in Fig. 11. Note that to the right or left of the masts, the protective zone is identical to that for an isolated mast, and that only between the masts is the protective zone altered. The protective zone is described by lines of constant height y. Along these isoprotected height lines, an object is protected if its height is equal to or less than y.

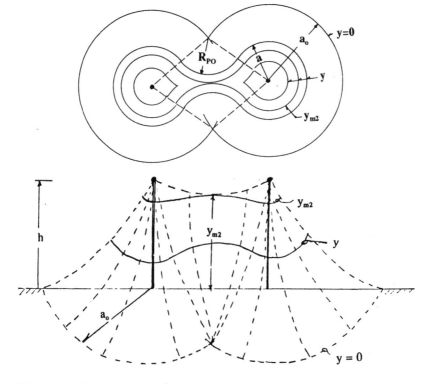

Figure 11 Protective zone for two masts.

Shielding of Substations

The protected zone is described by circles or arcs of circles. The procedure is

1. Draw circles of radius a_0 from each mast. These circles describe the limit of the protective zone and are for $y = 0$.
2. Using the intersection A as a center, construct arcs having a radius R_{PO}.
3. Using the masts as center points, construct arcs or partial circles of radius a or by noting that

$$a = a_0 - R_{PO} \qquad (26)$$

4. The intersection of the arcs of the above steps 2 and 3 complete the protected zone for the height y.

Note that for protected heights less than y_{m2}, separate lines of constant protected height y occur around each mast. To illustrate by example, Fig. 12 shows the protective zones for $I_c = 10\,\text{kA}$, $h = 30$ meters, and $R_c = 30$ meters using Young's equations.

Although the shielding diagrams are simple and interesting to draw, for most problems it is unnecessary to construct them. Using only a knowledge of their construction most engineering problems can be solved. As an example, consider the station of Fig. 13a, where the equipment to be protected is shown in the cross-hatched area having a height of 6 meters. Assume that the separation between the

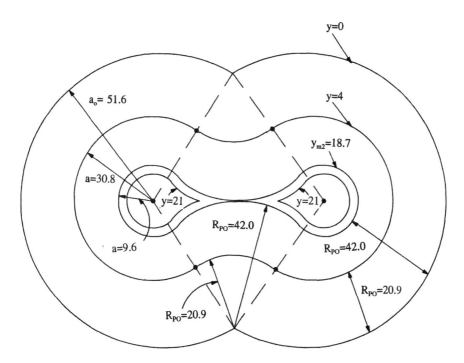

Figure 12 Example for two masts.

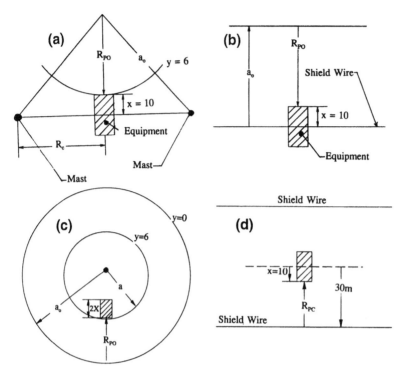

Figure 13 Example with $y = 6$ m, (a) Two masts. (b) One shield wire. (c) One mast. (d) Two shield wires.

two masts is 60 meters. The problem is to find the required height of the masts. Using a design current of 10 kA and Young's equations, the governing equations are

$$r_g = r_c = 56.4 \text{ m}$$
$$R_{PO} = \sqrt{r_c^2 - (r_g - y)^2} = 25.3 \text{ m} \tag{27}$$

From the diagram of Fig. 13a,

$$a_0 = \sqrt{(R_{PO} + x)^2 + R_c^2} = 46.33 \text{ m} \tag{28}$$

Then from Eq. 21,

$$h = r_g - \sqrt{r_s^2 - a_0^2} \tag{29}$$

which, as before, requires iteration to obtain the height. As shown in Table 3, the resultant required height is about 23 meters.

Next, to continue this example, consider protecting the equipment with a single shield wire as shown in Fig. 13b. The shield wire is located directly over the equipment. The value of R_{PO} remains the same at 25.3 meters. However, a_0 is now

Shielding of Substations

Table 3 Iteration to Find h for Two-Mast Case

h	γ_s	r_s	h
24.4	1.015	57.2	22.8
22.8	1.011	57.0	23.2
23.2	1.012	57.1	23.1

$R_{PO} + 10 = 35.3$ meters. To find the required height, the same equation as before, Eq. 29, is used. If the required height is less than 18 meters, no iteration is necessary. For this case, the required height is only 12.4 meters.

Again, continuing the example, use a single mast per Fig. 13c. For this case, set a at 30 meters; thus a_0 is $R_{PO} + 30$ or 55.3 meters. Now use Eq. 29 to find h. Because h is greater than 18 meters, iteration is necessary and is similar to that in Table 3. Performing the iteration, h is 36.3 meters.

As a last case, use two shield wires per Fig. 13d. Assume the shield wire separation is 60 meters or R_c is 30 meters. From Fig. 13d, R_{PC} is $30 - 10 = 20$ meters. From Eq. 15,

$$h = y + \sqrt{r_c^2 - R_{PC}^2} - \sqrt{r_s^2 - R_c^2} \tag{30}$$

Substituting in this equation, h is 11 meters.

These examples should demonstrate that most problems can be solved by the knowledge of the shielding diagram and a few basic equations. That is, the complete shielding diagram need not be drawn.

5.3 Three Masts

Assuming masts of equal height at the corners of a triangle as in Fig. 14, the three striking distances r_s from each mast intersect in the middle of the triangle so that the horizontal distance to each mast is R_c. The value of R_c illustrated in Fig. 15 is

$$R_c = \frac{ac}{2h_b} = \frac{a}{2\sin\alpha} \tag{31}$$

Note that in Fig. 14, as for the case of shielding between two shield wires, the striking distance to ground r_g is not involved, since the distance from the intersection point to ground is greater than r_g.

From Figs. 16 and 17,

$$y_{mc} = h - r_c + \sqrt{r_s^2 - R_c^2}$$

$$R_{PC} = \sqrt{r_c^2 - \left(h - y + \sqrt{r_s^2 - R_c^2}\right)^2} \tag{32}$$

Note that R_{PC} is the radius of a circle centered at the midpoint of the triangle described by R_c. Figure 18 is a crude sketch of the protected zone within a triangle.

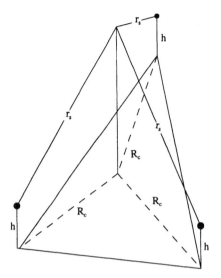

Figure 14 R_c is distance to center point.

As noted, a minimum height of the protected zone is y_{mc} centered at R_c. Circles of radius R_{PC} are drawn from this point with $y_2 > y_1 > y_{mc}$, etc.

While Fig. 18 illustrates the protective zone within the triangle, it is incorrect in that it does not show the influence of ground, or r_g, outside the triangle. Consider the case of three masts located on corners of an equilateral triangle, each side equal to 60 meters. That is, $R_c = 34.64$ meters. Let $h = 30$ meters and let the design current be 10 kA. using Young's equations, Fig. 19 shows the equiheight protective lines for this situation. Both the equations for three masts and those for two masts are used. Note the significant intrusion of the two-mast lines into the area of the triangle.

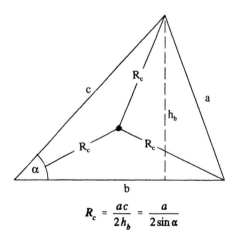

$$R_c = \frac{ac}{2h_b} = \frac{a}{2\sin\alpha}$$

Figure 15 Calculating R_c.

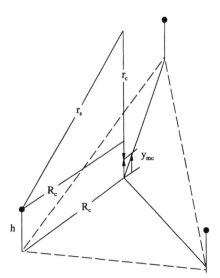

Figure 16 Finding y_{mc}.

For this case, at the center of the triangle, the minimum protective height y_{mc} is 19.9 meters. The protective height y increases away from this point until a critical point of $y_c = 21.5$ meters is reached, as is shown by the solid line. This is the point at which the earth (or the one- and two-mast equations) becomes effective, i.e., there is a point of discontinuity. From this point outward, the protective height decreases as the two mast equations rule. This characteristic is shown in Fig. 20. The value of X and its direction are shown in Fig. 19.

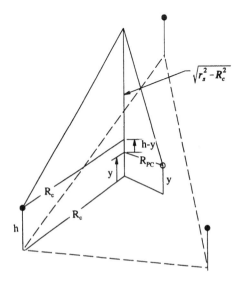

Figure 17 In center of three masts.

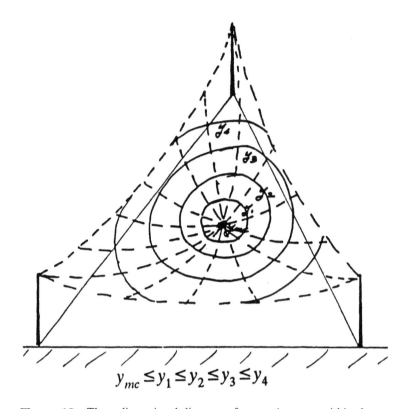

Figure 18 Three-dimensional diagram of protective zone within three masts.

As an example of providing protection using three masts, consider the case of three masts set at the corners of an equilateral triangle having sides of 30 meters, as shown in Fig. 21. The equipment to be protected has a height of 6 meters. Locating the masts so that the equipment is at the center, $y = y_{mc}$ meters. Also, $R_c = 17.3$ meters. Assume that the design current is 10 kA and that Young's equations are used. Then, from Eq. 32,

$$h = y_{mc} + r_c - \sqrt{r_s^2 - R_c^2} \qquad (33)$$

Thus $h = 8.7$ meters. As a point of interest, the shielding angle is about 81°.

The value of $y = y_c$, as described by the solid line in Fig. 19, at which the equations for two masts must be used, can be determined by the following equation obtained from Fig. 22:

$$R_{PC} + R_{PO} = \sqrt{R_c^2 - \left(\frac{S_g}{2}\right)^2} + \sqrt{a_0^2 - \left(\frac{S_g}{2}\right)^2} \qquad (34)$$

For the example of Fig. 19

Shielding of Substations

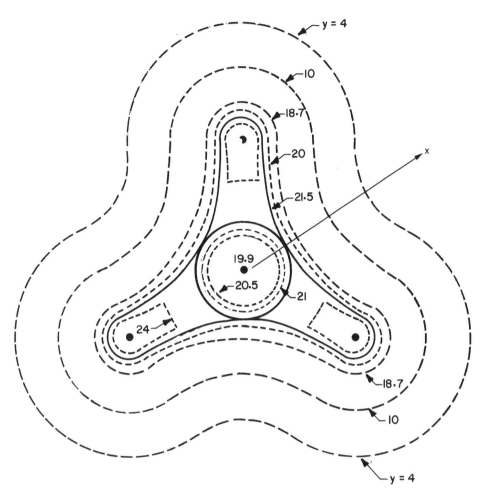

Figure 19 Plan view of protective zone for three masts.

$$R_{PO} + R_{PC} = 59.32 \tag{35}$$

Iterating on y, at $y_c = 21.5$, $R_{PO} = 44.88$ and $R_{PC} = 14.44$.

5.4 More than Three Masts

The three-mast triangular case represents the basic or fundamental mast arrangement. If more than three masts exist, they can be treated as a series of three masts. For example, consider the four masts of Fig. 23. These can be considered as two sets of three masts, each set having its own value of R_c, i.e., R_{c1} and R_{c2}.

However, since four-mast arrangements such as those of Fig. 24 frequently occur, these will be considered as a special case. Note that in these cases $R_{c1} = R_{c2}$. That is, only one value of R_c exists. The four-mast rectangular case is shown in Figs. 25 and 26. From these figures,

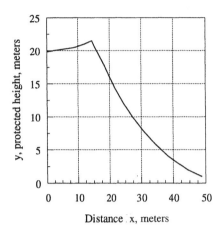

Figure 20 Protective height as a function of X of Fig. 19.

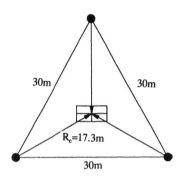

Figure 21 Example for three masts.

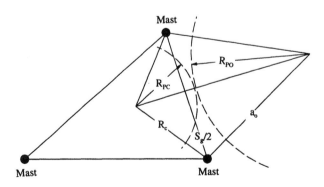

Figure 22 Finding the critical value of y where y for two-mast equations equals y for three-mast equations.

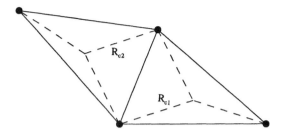

Figure 23 Four masts can be separated into two three-mast cases.

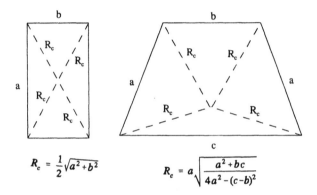

$$R_c = \frac{1}{2}\sqrt{a^2+b^2} \qquad R_c = a\sqrt{\frac{a^2+bc}{4a^2-(c-b)^2}}$$

Figure 24 Special four-mast cases.

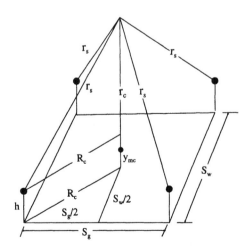

Figure 25 Four-mast case showing y_{mc}.

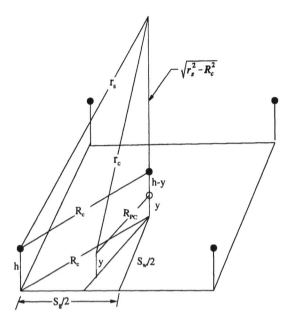

Figure 26 Determining the protective zone inside the four masts.

$$y_{mc} = h - r_c + \sqrt{r_s^2 - R_c^2}$$

$$R_{PC} = \sqrt{r_c^2 - \left(h - y + \sqrt{r_s^2 - R_c^2}\right)^2} \qquad (36)$$

$$R_c = \frac{1}{2}\sqrt{S_g^2 + S_w^2}$$

Figure 27 is a three-dimensional sketch of the protective zone only considering the inside of the masts. However, as for the three-mast case, the effect of earth, or the region outside the rectangle, influences or protrudes into this space. Again, the four-mast and two-mast equations must be used. Figure 28 is a diagram showing this effect for masts arranged in a square with sides of 30 meters, giving an R_c of 21.21 meters. Using Young's equations with $I_c = 10\,\text{kA}$, for $h = 30\,\text{m}$, y_c is 20.5 m and y_{mc} is 13.1 m.

5.5 Special Cases

Equations can also be developed for masts of unequal height or for two shield wires of unequal height. In addition, other special cases, such as for shield wires that are not parallel, can be considered. However, seldom are detailed equations for these special cases necessary. Virtually all these cases can be analyzed by use of the equations presented for equal height shield wires and masts. To illustrate, consider the case of two masts of unequal height, the plan view of the shielding zone shown in Fig. 29. As is intuitive, two circles having alternate values of a_0 are drawn to describe

Shielding of Substations

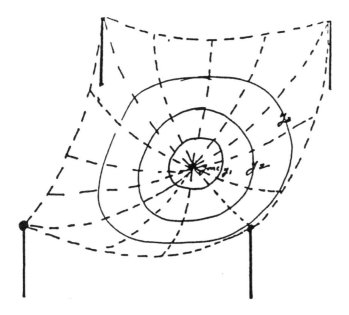

Figure 27 Three-dimensional sketch of the four-mast case.

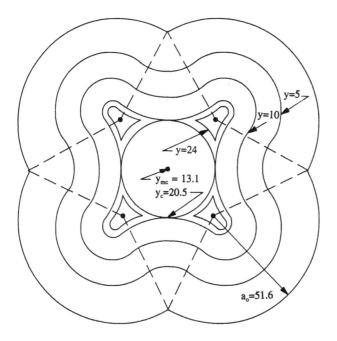

Figure 28 Example showing protective zone for four masts.

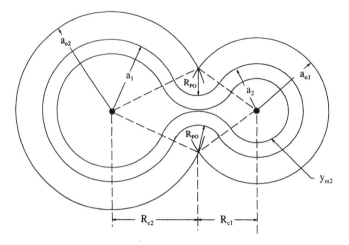

Figure 29 Two masts of unequal height.

the $y = 0$ location. Since R_{PO} is independent of the mast height, these arcs are constant for each mast. However, values of a are different for values of y. Per Fig. 29 use either R_{c1} or R_{c2} and a_{02} to find y_{m2}.

Similarly, a shield wire supported by masts of unequal height has the same equivalent diagram; see Fig. 30. R_{PO} is constant, but a_{01} and a_{02} differ.

In some substations, two of the shield wires of the transmission line are continued over the station to a single support. Figure 31 illustrates this case for shield wires of a constant height of 30 meters with an included angle of 45° using Young's equations for 10 kA. To find the protective height contours, first construct a line to divide the angle per Fig. 31. Next, calculate a_0 and set this value perpendicular to the shield wires at a distance L along the shield wire of

$$L = \frac{a_0}{\tan(\alpha/2)} \quad (37)$$

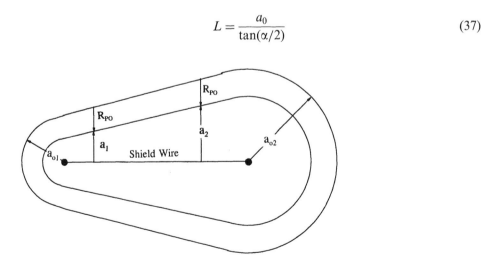

Figure 30 One shield wire on supports of unequal height.

Shielding of Substations

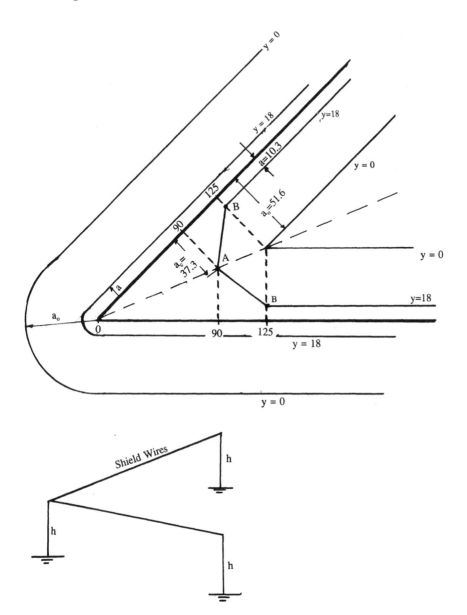

Figure 31 Two shield wires in a V-configuration.

where α is the angle between the shield wires. Next calculate a value of a_c for the height y, setting R_{PO} equal to zero, i.e.,

$$a_c = \sqrt{r_s^2 - (r_c - h + y)^2} \tag{38}$$

Set this value perpendicular to the shield wires at L, where L is calculated using a_c in Eq. 38. Now calculate a for this protective height and mark this distance along the line that was used to construct a_0. Construct the contour by joining A and B. Continue a line parallel to the shield wires at a distance a. The protective height contour lines are also drawn outside of the shield wires. If shield wire heights vary, the equiprotected lines will be skewed. Figures 30 and 31 also illustrate that at the end of the shield wires, the protected height contours are defined by the mast equations.

Another interesting example is the case of a mast adjacent to a shield wire. This can be solved by considering that the shield wire is an infinite series of masts with zero spacing between them. The solution is shown in Fig. 32, where one value of y is considered. An example is shown in Fig. 33 using the IEEE substation equations with the height of the shield wire and mast set at 30 meters and the separation between the mast and shield wire set at 6 meters.

Similarly, other cases can be either solved, by adapting the equal height equations, or approximated.

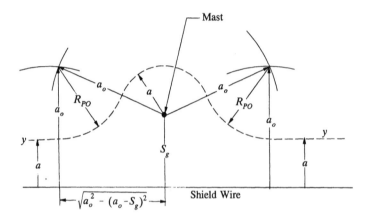

Figure 32 One mast and one shield wire.

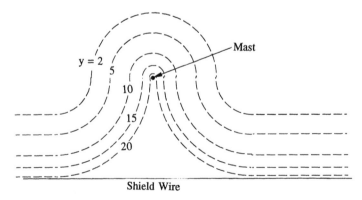

Figure 33 Example of mast–shield wire combination, $h = 30$, $S_g = 6$ meters.

Shielding of Substations

6 RECONSIDERING THE SFFOR FOR SUBSTATIONS

In introducing station shielding, the calculation of the SFFOR or designing for a SFFOR was deemed so complex that design for a specific design current was used as an alternate method. With the development of shielding protective diagrams, this problem can now be reassessed. First consider a single shield wire with an adjacent object to be protected, as displayed in Fig. 34. From this diagram, the SFFOR is

$$\text{SFFOR} = N_g L \int_3^\infty D_c f(I) \, dI \tag{39}$$

where L is the length or width of the object to be protected, i.e., the distance into the figure. Thus for this case, the SFFOR is calculated in the same manner as for the transmission line. However, now think of two shield wires with an object between them, as in Fig. 35. As noted, the equation for the SFFOR remains the same, but the development of the equation for D_c now becomes more complex. Now conceive of a similar diagram for the three- and four-mast cases, and the calculation methodology becomes hopelessly complex. When the overall accuracy of the method is also considered, it becomes quickly apparent that the simplified method of designing for a specific current is entirely adequate.

7 MASTS OR SHIELD WIRES

As portrayed by the shielding diagrams, shield wires provide superior protection. However, there exists a concern among some utilities that breakage of the shield wires may occur resulting in faults within the station. In opposition to this concern, there exists no documented evidence of breakage of the shield wires. This concern is to the advantage of the shielding masts, and in addition there is the advantage of low cost.

Within the USA, masts are normally employed for stations with small areas such as those servicing lower voltage distribution systems. For large high-voltage stations, almost universally, shield wires are used in the USA. In other countries, masts are used for high-voltage stations, e.g., the 800 kV station in Hungary.

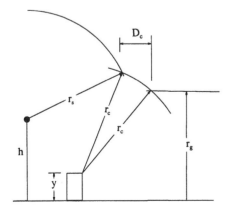

Figure 34 SFFOR for a single shield wire.

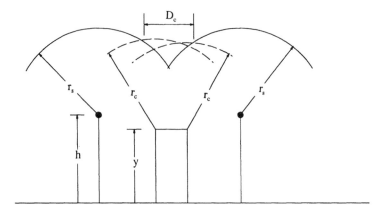

Figure 35 Considering two masts.

8 BACKGROUND

In 1942, Wagner et al. [1] published a paper titled "Shielding of Substations" in which they presented results from laboratory tests. The tests performed attempted to model the downward leader from the cloud by using a rod placed vertically above a model shield wire or mast. The authors then produced curves showing the percentage of the strokes that terminated on a protected object. These curves were then used to design the shielding system for substations. Although the basis of the curves produced was incorrect, these curves continued in use into the 1950s and even today are sometimes used by designers.

With the advent of the geometric model of the lightning flash, new impetus was generated to revise the method though not until 1979 did Ralph H. Lee devise the so-called rolling ball theory [2]. Assuming as for Love's equations that all the striking distances are equal, the rolling ball has a radius equal to the striking distance. The ball could then be rolled around and on top of the station. Any object that the ball did not contact was protected for the stroke current represented by the ball. This theory is correct provided that all striking distances are equal. If otherwise, the ball is rather a variable-diameter ball.

The idea of the rolling ball is useful in visualizing the protected areas and the protection contour. Indeed, in the development of the equations and thoughts in this chapter, a foam ball was used to roll among doles fixed into a pegboard.

The shielding diagrams produced by Wagner et al. as obtained from Fig. 11 of their paper are shown in Fig. 36. As noted, remarkably, they are not unlike those shown in this chapter.

9 CONCLUSIONS

1. Recommended striking distance equations are (1) IEEE Substations Committee 1995, (2) IEEE T&D Committee 1992, and (3) Brown–Whitehead as used by CIGRE.

Shielding of Substations

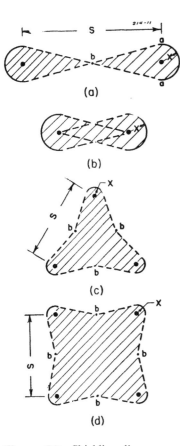

Figure 36 Shielding diagrams as shown in Ref. 1. (Copyright IEEE, 1942.)

2. Either masts or shield wires can be used to shield substations. Shield wires are generally used for large-area stations, whereas masts are in normal use for low-voltage, small-area substations. Shield wires usually provide better protection.

3. For the best protection, surround the object to be protected by the shielding system.

4. Three or more masts are superior to one or two masts.

5. Usually, the entire shielding diagram need not be drawn. Knowing the method of construction is sufficient to determine the shielding required.

10 A REVIEW

As noted in this chapter, only a few equations are necessary to develop shielding patterns for shield wires and masts. In addition, fortuitously, the same equations apply to shield wires and masts. Following are the equations.

10.1 Design Current

The suggested design current is given in Table 1. In general, suggested values are 5 kA for nominal system voltages below 230 kV and 10 kA for system voltages at and above 230 kV. The design current is a type of reliability index, so that for highly important substations the design current should be reduced below that of Table 1. The design current is greater than the critical current, which is

$$I_c = \frac{2(\text{CFO})}{Z_c} \qquad (40)$$

where the CFO is the negative polarity lightning impulse CFO, i.e., 605 kV/m

10.2 Striking Distance Equations

Striking distances: $r_g =$ to ground; $r_c =$ to object to be protected; $r_s =$ to ground wire or mast. Heights: $y =$ height of object to be protected; $h =$ height of ground wire or mast. The general striking distance equations are of the form

$$r_c = \gamma_c r_g \qquad r_s = \gamma_s r_g \qquad (41)$$

The striking distance equations are provided in Section 3.

10.3 General Equations

$$a_0 = \sqrt{r_s^2 - (r_g - h)^2} \qquad (42)$$

$$R_{po} = \sqrt{r_c^2 - (r_g - y)^2} \qquad (43)$$

$$a = a_0 - R_{po} \qquad (44)$$

$$a_c = R_c - R_{pc} \qquad (45)$$

$$R_{pc} = \sqrt{r_c^2 - \left[h - y + \sqrt{r_s^2 - R_c^2}\right]^2} \qquad (46)$$

$$y_{mc} = h - r_c + \sqrt{r_s^2 - R_c^2} \qquad (47)$$

$$y_{m2} = r_g - \sqrt{r_c^2 - a_0^2 + R_c^2} \qquad (48)$$

For two masts, R_c is half the distance between the masts, i.e., $R_c = S_g/2$.

10.4 R_c, Center Point of Masts

For three-masts, with distances between the masts denoted as S_1, S_2 and S_3,

$$R_c = \frac{S_2 S_3}{2 H_b}$$

$$H_b = \frac{2}{S_1}\sqrt{S(S-S_1)(S-S_2)(S-S_3)} \qquad (49)$$

$$S = \frac{1}{2}(S_1 + S_2 + S_3)$$

10.5 To Find $y = y_c$ Where $R_{PC} = R_{PO}$

Solve iteratively:

$$R_{pc} + R_{po} = \sqrt{R_c^2 - (S_g/2)^2} + \sqrt{a_0^2 - (S_g/2)^2} \qquad (50)$$

where S_g is the distance between the two masts.

10.6 To Determine Whether a Three-Mast Case Exists

For the three-mast case to exist, a_0 must be greater than R_c, and the center of the masts must be within the area described by lines joining the masts, i.e., R_c must be equal to or less than Z_n, where,

$$Z_n = \sqrt{H_{bn}^2 + \left[\frac{S_{max}}{2} - x\right]^2}$$

$$x = \sqrt{S_{min}^2 - H_{bn}^2}$$

$$H_{bn} = \frac{2}{S_{max}}\sqrt{S(S-S_1)(S-S_2)(S-S_3)} \qquad (51)$$

$$S = \frac{1}{2}(S_1 + S_2 + S_3)$$

where S_1, S_2, and S_3 are the distances between masts and S_{max} and S_{min} are the maximum and minimum of these distances.

10.7 Definition of R_c

R_c = distance to center point between masts (Table 4).

Table 4 Some Values of R_c

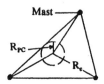

Configuration	R_c	Drawing
Two shield wires	$\dfrac{S_g}{2}$	
Triangles	$\dfrac{ac}{2h_b}$	
	$\dfrac{a}{\sqrt{3}}$	
	$\dfrac{a^2}{\sqrt{4a^2 - b^2}}$	
Square	$\dfrac{a}{\sqrt{2}}$	
Rectangle	$\dfrac{1}{2}\sqrt{a^2 + b^2}$	
Trapezoid	$a\sqrt{\dfrac{a^2 - bc}{4a^2 - (c-b)^2}}$	

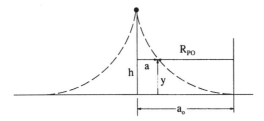

Figure 37 Plan view of shielding diagram for one shield wire.

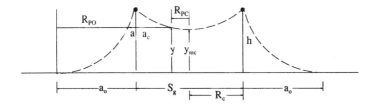

Figure 38 Plan view of shielding diagram for two shield wires.

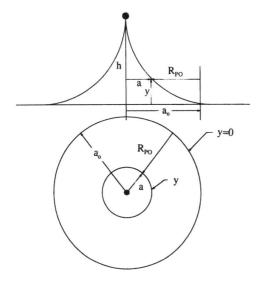

Figure 39 Plan and profile views of shielding diagram for one mast.

11 REFERENCES

1. C. F. Wagner, G. D. McCann, and C. M. Lear, "Shielding of Substations," *AIEE Trans.*, 61, 1942, pp. 96–100.
2. R. H. Lee, "Lightning Protection of Buildings," *IEEE Trans. on Industry and Applications*, May/Jun. 1979, pp. 236–240.
3. IEEE Substations Committee, WG E5, "Guide to Direct Lightning Shielding of Substations," IEEE Std. 988-1996.
4. A. M. Mousa, "Shielding of HV and EHV Substations," *IEEE Trans. on PA&S*, 1976, pp. 1303–1310.

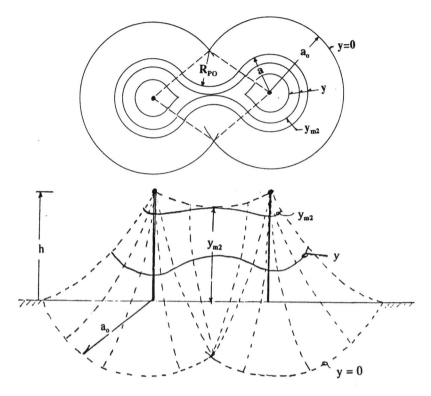

Figure 40 Plan and profile views of shielding diagram for two masts.

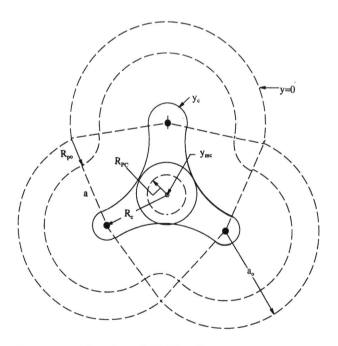

Figure 41 Plan view of shielding diagram for three masts.

Shielding of Substations

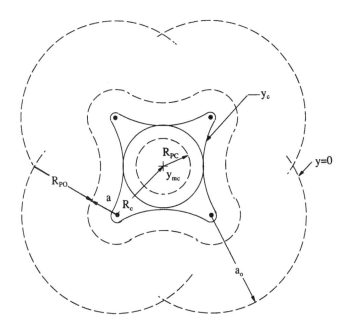

Figure 42 Plan view of shielding diagram for four masts.

12 PROBLEMS

1. Centered within a 30 × 30 meter station shown in Fig. 43 is a group of equipment having a height of 10 meters. The equipment area is 10 × 10 meters. Using Young's equations with a design current of 10 kA, determine the height of 2, 3, and 4 masts with the requirement that the mast must be located along or outside the borders of the station, i.e., outside the 30 × 30 meter area.

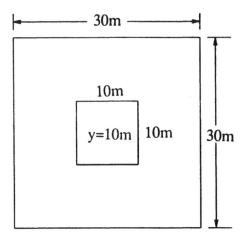

Figure 43 Diagram for Problem 1.

2. Centered in a 50 × 50 meter station shown in Fig. 44 is a group of equipment having a height of 10 meters and an area of 6 × 60 meters. Design a shielding system using (A) two shield wires and also (B) four masts. That is, determine the required heights of the shield wires and masts. Use Young's equations with a design current of 10 kA. Masts and shield wires must be outside the 50 × 100 meter area.

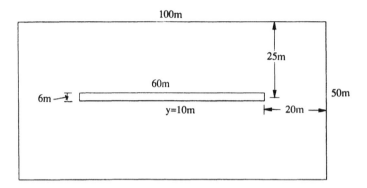

Figure 44 Diagram for Problem 2.

3. A 138/12 kV station is shown in Fig. 45. It is proposed to shield the station by using shield wires strung from the two dead-end towers to a single pole. The two shield wires are indicated by dashed lines. The shield wire height is 50 feet (15.24 m). The maximum height of the bus work, circuit breakers, potential transformers, and power transformer is 24 feet (7.32 m). The maximum height of the 12 kV switchgear is 14 feet (3.66 m), and the maximum height of the control building is 10 feet (3.05 m). Neglect the enclosing substation fence. Assume a critical or design current of 5 kA and use Young's equations.

Equipment	x, ft	y, ft	height, ft
Control house	9	46	10
End of 12 kV bus	45	34	14
Center of 12 kV bus	111	34	14
End of 138 kV bus	49	122	24
Center of 138 kV bus	111	122	24
Shield wire supports			
At towers	70	190	50
	152	190	50
Support at bottom	111	0	50

1. Check the proposed scheme for shielding of all equipment, including the control house.
2. If the proposed scheme is not adequately shielded, relocate the shield wires but retain the shield wire terminals at the dead-end towers.
3. Propose an all mast design.

Shielding of Substations

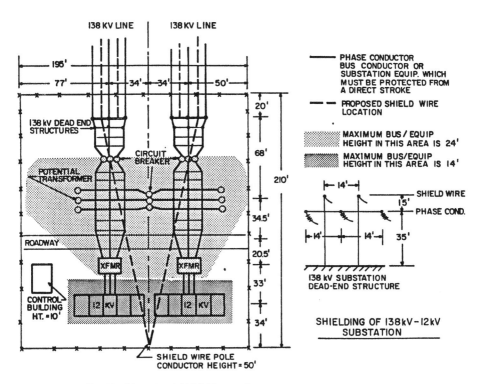

Figure 45 For Problem 3, A 138/12kv station.

As a basis of conformity, assume that the equipment and present ground wire supports are at the x–y coordinates in the table and that the station is 105×210 feet, i.e., $x = 195$ and $y = 210$ feet.

4. Draw or sketch the shielding diagram for three masts whose heights are 30 meters and which are spaced 50, 50, and 64 for protective heights of 4, 9, and 25 meters. Also draw the protective height diagram for the critical protective height y_c and indicate the minimum protective height y_{mc}. Use the Brown–Whitehead equations with a design current of 8 kA. Use the appropriate computer program to obtain the values for the drawing.

5. Draw or sketch the shielding diagram for three masts whose heights are 30 meters and which are spaced 30, 30, and 42.426 meters (a right triangle) for protective heights of 12 and 22 meters. Also draw the protective height diagram for the critical protective height y_c and indicate the minimum protective height y_{mc}. Use the Brown–Whitehead equations with a design current of 8 kA. Use the appropriate computer program to obtain the values for the drawing.

6. The plan and profile views of a typical 34.5/12.5 kV substation are shown in Figs. 46 and 47. Three masts are used for shielding. Specify the mast height assuming (A) a critical current of 5 kA and (B) a critical current of 10 kA. Use the Brown–Whitehead equations. Also specify an improved location of the masts.

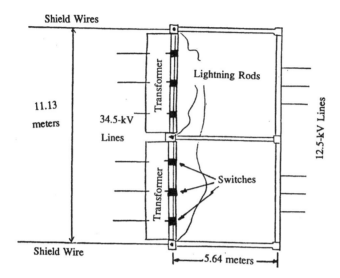

Figure 46 For Problem 6, a plan view of a 34.5/12 kv substation.

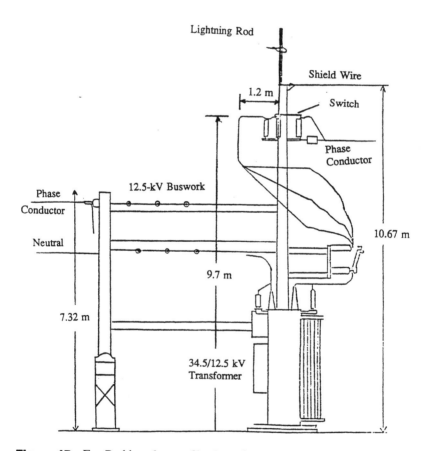

Figure 47 For Problem 6, a profile view of a 34.5/12 kV substation.

9
A Review of Traveling Waves

1 CONCEPT: SURGE IMPEDANCE

Any transient disturbance, such as a lightning stroke terminating on a phase conductor or the closing or opening of a circuit breaker, can be analyzed by use of traveling waves. Normally, this subject is approached by first noting that a transmission line is a distributed parameter network composed of a series inductance and resistance and shunt capacitance and resistance. Partial differential equations are then written and solved for the voltage and current, and normally the series and shunt resistances are neglected. This seemingly high theoretical development does not represent the most sophisticated approach, nor does it in many cases provide a needed insight into the approximations required to view transient phenomena as traveling waves. The superior method begins with Maxwell's equations and uses retarded potential equations.

However, both these methods can be circumvented for the normal presentation by simply stating that a lightning stroke to a conductor or the closing of a breaker produces traveling waves of voltage e and current i that are related by a surge impedance Z equal to e/i that travels along the conductor at the speed of light c as portrayed by Fig. 1.

The surge impedance Z is purely resistive; therefore, e and i have the same shape. Only truly distributed parameter circuits such as a transmission line, a cable, or a SF_6 bus can "possess" a surge impedance. The surge impedance and velocity of propagation can be obtained from the inductance and capacitance, i.e.,

$$Z = \sqrt{\frac{L}{C}} \qquad v = \frac{1}{\sqrt{LC}} \qquad (1)$$

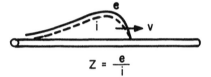

Figure 1 e and i are related by Z.

from which the following useful equations can be derived:

$$L = \frac{Z}{v} \qquad C = \frac{1}{Zv} \tag{2}$$

where L is the inductance and C is the capacitance per unit length.

1.1 Overhead Lines

For a single conductor having a radius r located at a height h above ground (Fig. 2) and assuming an earth of zero resistivity, the inductance and capacitance are

$$L = 0.20 \ln \frac{2h}{r} \quad \mu\text{H/m}$$

$$C = \frac{10^{-3}}{18 \ln \frac{2h}{r}} \quad \mu\text{F/m} \tag{3}$$

Therefore

$$Z = 60 \ln \frac{2h}{r} \text{ ohms} \qquad v = 30 \text{ m/}\mu\text{s} \quad \text{or} \quad \approx 1000 \text{ ft/}\mu\text{s} \tag{4}$$

As noted, in this case, the velocity is equal to the speed of light. The surge impedance of a single conductor varies in a narrow band between about 400 and 500 ohms.

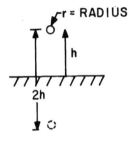

Figure 2 A single overhead conductor.

1.2 Cables

Equation 4 for the velocity is a special case of the general phenomenon that the velocity of propagation varies inversely as the square root of the permittivity of the medium. For cables, see Fig. 3, the permittivity k varies from about 2.4 to 4.0. The surge impedance and velocity of propagation are

$$Z = \frac{60}{\sqrt{k}} \ln \frac{r_2}{r_1} \qquad v = \frac{300}{\sqrt{k}} \text{ m/}\mu\text{s} \quad \text{or} \quad \frac{1000}{\sqrt{k}} \text{ ft/}\mu\text{s} \qquad (5)$$

The surge impedance of a cable varies from about 30 to 60 ohms and the velocity of propagation is about 1/3 to 1/2 the speed of light.

1.3 SF$_6$ Cables

SF$_6$ has a permittivity of about 1 and therefore the velocity of propagation is equal to that of light. For all SF$_6$ designed cables up to the UHV level, the ratio of r_2 to r_1 is constant and therefore the surge impedance is constant at between 60 and 65 ohms.

1.4 An Example

Assume that $Z = 400$ ohms and $v = 300$ m/μs. Therefore from Eq. 2, $L = 1.33\,\mu\text{H/m}$ or $0.4\,\mu\text{H/ft}$ and $C = 8.33 \times 10^{-6}\,\mu\text{F/m}$ or $8.33\,\text{pF/m}$ or $2.5 \times 10^{-6}\,\mu\text{F/ft}$ or $2.5\,\text{pF/ft}$.

1.5 The Micro-System

Note that in these calculations, a consistent set of units should be used. For work in this transient area, the micro-system set of units is suggested, i.e.,

L in μH/m or μH/ft
C in μF/m or μF/ft
v in m/μs or ft/μs
Z in ohms

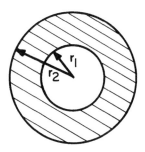

Figure 3 A cable.

2 BEHAVIOR OF WAVES AT A POINT OF DISCONTINUITY

2.1 General

When voltage and current waves traveling on a transmission line reach a point of discontinuity, i.e., a change in circuit impedance, voltage and current waves are "reflected" backward toward their origin, and voltage and current waves are "transmitted" onward. To develop the concept, first consider that the line is terminated in some generalized impedance Z_k per Fig. 4.

Consider the circuit of Fig. 4 with a conductor surge impedance of Z and the impedance Z_k, which may be resistive, capacitive, inductive, or any combination of these. The original or forward waves are denoted as e and i. The resultant voltage across the impedance Z_k and the current through this impedance, known as the transmitted voltage and current, are called the double prime quantities e'' and i''. The voltage and current reflected backward from the discontinuity are called the single prime quantities e' and i'. In general, so that the sum of all currents is zero at A in Fig. 4, reflected and transmitted waves must exist. The equations used for solution are generally divided into normal equations, which describe the traveling waves, and boundary equations, which specify the conditions necessary at the point of discontinuity. Following are these equations:

$$\begin{array}{ll} \text{Normal equations} & \text{Boundary equations} \\ e = iZ & i'' = i - i' \\ e' = i'Z & e'' = e + e' \\ e'' = i''Z_k & \end{array} \qquad (6)$$

Therefore to find e'' in terms of e

$$e'' = e + e' = e + i'Z = e + (i - i'')Z$$
$$= 2e - i''Z = 2e - \frac{Z}{Z_k}e'' \qquad (7)$$

Therefore

$$e'' = \frac{2Z_k}{Z + Z_k}e \qquad (8)$$

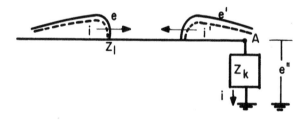

Figure 4 General circuit to develop reflection and transmitted equations.

A Review of Traveling Waves

Also

$$i'' = \frac{e''}{Z_k} = \frac{2e}{Z + Z_k} = \frac{2Z}{Z + Z_k} i$$

$$e' = e'' - e = \frac{Z_k - Z}{Z + Z_k} e \quad (9)$$

$$i' = \frac{e'}{Z} = \frac{Z_k - Z}{Z + Z_k} i$$

First let $Z_k = 0$, a short circuit. Then

$$\begin{aligned} e'' &= 0 & i'' &= 2i \\ e' &= -e & i' &= -i \end{aligned} \quad (10)$$

Next, let $Z_k =$ infinity, an open circuit. Then

$$\begin{aligned} e'' &= 2e & i'' &= 0 \\ e' &= e & i' &= i \end{aligned} \quad (11)$$

That is, the voltage doubles at an open circuit; the current doubles at a short circuit.

2.2 The Use of Thevenin's Theorem

Thevenin's theorem can be applied to the circuit of Fig. 5 to obtain the voltage across the impedance Z_k. First open the circuit at the point of discontinuity and calculate the opened circuit voltage. From Eq. 11, this open circuit voltage is equal to $2e$. Next, find the impedance of the circuit by "standing" at the open circuit point and looking backward. Then we find the impedance as simply Z. Therefore Thevenin's equivalent circuit is as shown in Fig. 6, which can be used to calculate

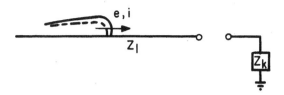

Figure 5 Developing Thevenin's circuit.

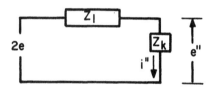

Figure 6 Thevenin's equivalent circuit.

the voltage e'' and the current i''. The voltage e' and the current i' can be found by noting that $e' = e'' - e$ and $i' = i - i''$.

2.3 Z_k Is a Capacitor

Assume that Z_k is a capacitor C_k and that e is a unit step function or square wave of magnitude E. Then

$$e'' = \frac{2/C_k s}{Z + (1/C_k s)} \frac{E}{s} = \frac{2E}{ZC_k} \frac{1}{s[s + (1/ZC_k)]} \tag{12}$$

where s is the Laplace operator. Then

$$\begin{aligned} e'' &= 2E\left[1 - e^{-\frac{t}{ZC_k}}\right] \\ e' &= E\left[1 - 2e^{-\frac{t}{ZC_k}}\right] \end{aligned} \tag{13}$$

2.4 A Special Case: Two Conductors

For this case, the impedance Z_k is another line or a cable. Thus Z_k is a pure resistance and the voltage e'' is the voltage transmitted onto this line or cable per Fig. 7. The Thevenin circuit remains as per Fig. 6, and thus the equations developed previously apply. For example, for $Z_k = 30$ and $Z_1 = 400$, $e'' = (0.14)e$.

2.5 The n-Line Station

The special and important case for application in station insulation coordination consists of a single line to which several other lines are connected, as in an n-line station. Let there be n lines, one incoming line and $(n-1)$ outgoing lines per Fig. 8. Let the lines have surge impedances Z_1, Z_2, Z_3, etc. Thus assuming a total of four lines,

$$Z_k = \frac{Z_2 Z_3 Z_4}{Z_2 Z_3 + Z_2 Z_4 + Z_3 Z_4} \tag{14}$$

The Thevenin circuit of Fig. 9 applies, and as seen, the transmitted voltages on each of the outgoing lines will be equal to e'', or as before,

$$e'' = \frac{2Z_k}{Z_1 + Z_k} e \tag{15}$$

Figure 7 Case of two conductors.

A Review of Traveling Waves

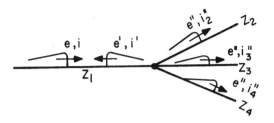

Figure 8 An *n*-line station with unequal surge impedances.

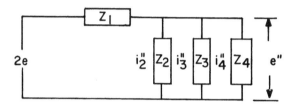

Figure 9 Thevenin's circuit of Fig. 8.

However, the currents on each of the lines will differ, since each line has a different surge impedance; see Fig. 9. Thus

$$i_2'' = \frac{e''}{Z_2} \qquad i_3'' = \frac{e''}{Z_3} \qquad i_4'' = \frac{e''}{Z_4} \tag{16}$$

As a further example of use, consider a transmission switching station having a total of *n* lines, each of which has a surge impedance of Z. Per Fig. 10, assume that a surge voltage and current, e and i, travel in toward the station. Now assume that the buses and equipment within the station do not act as points of discontinuity. (They do, but for now, for the sake of approximation, neglect this effect.) Therefore the Thevenin circuit is as shown in Fig. 11 and the voltage transmitted outward on each line, and more importantly the voltage at the station, e'', is

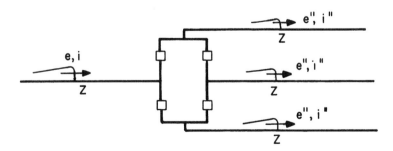

Figure 10 An *n*-line station; all Zs are equal.

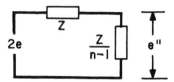

Figure 11 Thevenin's circuit of Fig. 10.

$$e'' = \frac{2[Z/(n-1)]}{Z+[Z/(n-1)]}e = \frac{2e}{n} \qquad (17)$$

For example, if $n = 4$, then $e'' = \frac{1}{2}e$.

2.6 Line–Cable Junction

As another example, consider Fig. 12, where a surge is arriving at the line–cable junction. The objective is to find the voltage at the line–cable junction, which is the same as the voltage transmitted onto the cable. Assuming a cable surge impedance of 50 ohms and a line surge impedance of 400 ohms, per Fig. 13, the voltage at the line–cable junction is

$$e'' = \frac{100}{450}e = \frac{2}{9}e \qquad (18)$$

and the reflected surge voltage is

$$e' = e'' - e = -\frac{7}{9}e \qquad (19)$$

Thus the low impedance of the cable reduces the incoming surge voltage to a low value.

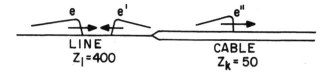

Figure 12 Voltages at a line–cable junction.

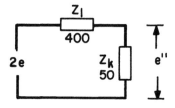

Figure 13 Thevenin's circuit of Fig. 12.

3 A CALCULATION METHOD: THE LATTICE DIAGRAM

As discussed in the previous section, when waves impinge on any point of discontinuity, they produce both reflected and transmitted waves. In the normal problem there exist many points of discontinuity, and the problem is to find a methodology to permit keeping track of all the reflections. Such a method is the lattice diagram as suggested by Bewley [1]. Although today complex problems are solved efficiently by digital transient programs such as the EMTP or the ATP, lattice diagrams are still in use for obtaining general equation solutions to simplified problems and also to provide a check on computer programs. To illustrate the use of the lattice diagram, consider the problem shown in Fig. 12. If the cable has an infinite length, the voltage is as calculated. However, for a finite length cable, surges are reflected from the end of the cable and arrive back at the cable–line junction and usually tend to increase the voltage at this point. Now consider that the cable is of finite length as shown in Fig. 14. And to add general interest, also assume that the end of the cable is terminated in a resistor R.

The first task is to calculate the reflection and transmission coefficients and place them on the "sign" posts as in Fig. 14. That is, for example, for a surge e traveling on the cable toward the line, the sign post tells us that the voltage transmitted onto the line is δe and that the surge reflected back onto the cable is ϕe. The coefficients are determined using Eqs. 8 and 9. Thus

$$\beta = \frac{2Z_2}{Z_1 + Z_2} \qquad \gamma = \frac{Z_2 - Z_1}{Z_1 + Z_2}$$

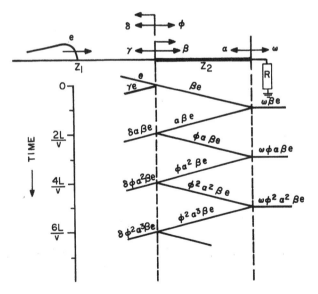

Figure 14 Lattice diagram for line–cable circuit.

$$\omega = \frac{2R}{R+Z_2} \qquad \alpha = \frac{R-Z_2}{R+Z_2}$$
$$\delta = \frac{2Z_1}{Z_1+Z_2} \qquad \phi = \frac{Z_1-Z_2}{Z_1+Z_2} \qquad (20)$$

The lattice diagram is constructed with time increasing downward. Starting with the time $t = 0$, t is defined as the time that the surge e reaches the line–cable junction; lines are drawn diagonally downward with a constant slope as illustrated. A value of the surge voltages, represented as a combination of coefficients, is placed on each of these lines. In addition, the total voltage at the end of the cable is shown on a horizontal line. To clarify, the voltage e arriving at the line–cable junction transmits a surge of βe and reflects a surge of γe. The transmitted surge travels to the end of the cable and reflects a surge of $\alpha\beta e$. The voltage produced at the end of the cable is $\omega\beta e$. Now this reflected surge travels back to the line–cable junction and transmits a voltage $\delta\alpha\beta e$ and reflects a voltage $\phi\alpha\beta e$, etc.

The voltage at any location or point on the cable can be calculated by adding those voltages that arrive at the selected location. However, they must be added with respect to their time of arrival at this location. For example, the voltage at the line–cable junction, e_T, is

$$e_T = e\big[\beta + \delta\alpha\beta(t-2T) + \delta\phi\alpha^2\beta(t-4T) + \delta\phi^2\alpha^3\beta(t-6T)\ldots\big] \qquad (21)$$

where T is the time required for a surge to travel one length of the cable L, i.e., the cable travel time is

$$T = \frac{L}{v} \qquad (22)$$

The term $(t - 2T)$ denotes that this surge is delayed from time zero by one travel time. That is, it is only used in calculating the voltage at and after $t = 2T$.

Note the first term of Eq. 21, βe. This could have been taken as $(1+\gamma)e$, or that the second term $\delta\alpha\beta e$ could have been taken as $(\alpha\beta + \phi\alpha\beta)e$, depending on which side of the dotted line is traversed. There is no difference, since $\beta = 1 + \gamma$ and $\delta\alpha\beta = \alpha\beta + \phi\alpha\beta$. The above equation can be somewhat simplified to

$$e_T = \beta e\big[1 + \delta\alpha(t-2T) + \delta\phi\alpha^2(t-4T) + \delta\phi^2\alpha^3(t-6T)\big] \qquad (23)$$

For a numerical example, let

$$\begin{aligned} Z_1 &= 400 & Z_2 &= 30 & R &= 10 \\ \alpha &= -0.5 & \beta &= 0.14 & \delta &= 1.86 \\ \omega &= 0.5 & \gamma &= -0.86 & \phi &= 0.86 \end{aligned} \qquad (24)$$

and let e be defined by a linear front and an infinite tail. Let the time to crest of this surge equal 4 µs and the travel time T be 1 µs. Then

$$e_T = e[0.14 - 0.13(t-2) + 0.056(t-4) - 0.0241(t-6) + 0.0104(t-8) + \cdots] \qquad (25)$$

A Review of Traveling Waves

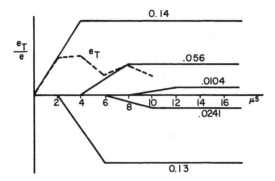

Figure 15 Adding the voltages at the line–cable junction.

These surges and their addition to determine the shape and magnitude of e_T are shown in Fig. 15. The maximum voltage occurs at 4 µs and is $0.075e$. As a point of interest, the voltage at $t = $ infinity is

$$E_T(t = \infty) = \beta e[1 + \delta\alpha(1 + \phi\alpha + \phi^2\alpha^2 + \phi^3\alpha^3 + \cdots]$$
$$= \beta e\left[1 + \frac{\delta\alpha}{1 - \phi\alpha}\right] \quad (26)$$
$$= \frac{2R}{R + Z_1} e = 0.049e$$

which shows that at time equal infinity, the reflections have eliminated or "wiped out" the cable, since the above equation is equivalent to a surge traveling on a line of surge impedance Z_1 terminated in a resistor R.

4 STROKE TO TOWER

Assume that the first stroke of a flash terminates at the top of a tower as illustrated in Fig. 16. A voltage e is produced at the top of the tower, creating a traveling wave that travels down the tower and out on the overhead ground wires. The voltage e is the product of the stroke current I and the combined impedance of the tower and the ground wires, i.e.

$$e = \frac{Z_T(Z_g/2)}{Z_T + (Z_g/2)} I = \frac{Z_T Z_g}{Z_g + 2Z_T} I \quad (27)$$

where Z_g is the ground wire surge impedance and Z_T is the surge impedance of the tower. As a good approximation that will be further examined later, let

$$Z_T = \frac{Z_g}{2} \quad (28)$$

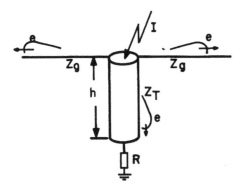

Figure 16 Stroke to tower.

Therefore, Eq. 27 becomes

$$e = \frac{Z_T}{2}I = \frac{Z_g}{4}I \tag{29}$$

Let the tower travel time be T_T and assume that the waveshape of the voltage e is defined as a linear rising front and an infinite or constant tail. Let the time to crest equal t_f. To calculate the voltage at the top of the tower, or at any point along the tower, the lattice diagram of Fig. 17 is used. Another point on the tower, A, is also

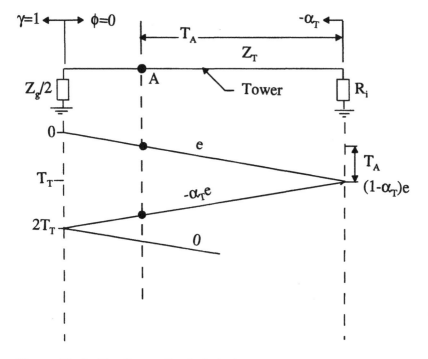

Figure 17 Lattice diagram for stroke to tower.

A Review of Traveling Waves

shown, and the voltage at this location will be considered later. As noted, the lattice diagram is very simple, since per Eq. 28 there is no reflection at the top of the tower. The reflection and transmission coefficients are

$$\alpha_T = \frac{Z_T - R_i}{Z_T + R_i} \approx \frac{Z_g - 2R_i}{Z_g + 2R_i} \qquad \gamma = \frac{2Z_g}{Z_g + 2Z_T} \qquad \phi = \frac{Z_g - 2Z_t}{Z_g + 2Z_T} \tag{30}$$

Note that the reflection coefficient α_T in Eq. 30 must be used as a negative value in Fig. 17. The voltage at the tower top is illustrated in Fig. 18. Three voltage magnitudes are of interest, V_{TT}, the crest voltage, V_T, the voltage at the tower top prior to any reflections from the footing resistance, and V_F, the final voltage. The equation for V_T is

$$V_T = \frac{2T_T}{t_f} e = Z_T \frac{T_T}{t_f} I = L_T \frac{I}{t_f} \tag{31}$$

where L_T is the total inductance of the tower and $L_T = Z_T T_T$. The factor I/t_f is the rate of rise or the steepness of the front and is frequently denoted as S_i or more simply dI/dt. Thus V_T is very simply the voltage drop caused by the tower, $L\, dI/dt$.

The voltage V_{TT} is

$$\begin{aligned} V_{TT} &= e - \alpha_T e \frac{t_f - 2T_T}{t_f} = (1 - \alpha_T)e + \alpha_T e \frac{2T_T}{t_f} \\ &= \left[\frac{R_i Z_g}{Z_g + 2T_i} + \alpha_T Z_T \frac{T_T}{t_f}\right] I = \left[R_e + \alpha_T Z_T \frac{T_T}{t_f}\right] I \\ &= K_{TT} I \end{aligned} \tag{32}$$

where

$$R_e = \frac{R_i Z_g}{Z_g + 2R_i}$$

$$K_{TT} = R_e + \alpha_T Z_T \frac{Z_t}{t_f} \tag{33}$$

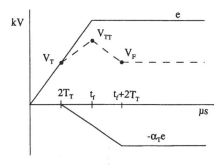

Figure 18 Voltage at tower top.

The final voltage V_F is

$$V_F = (1 - \alpha_T)e = R_e I \tag{34}$$

The voltage across the footing resistance, V_R, and the current through this resistance, I_R, are

$$V_R = \frac{R_i Z_T}{Z_T + R_i} \qquad I_R = \frac{Z_T}{Z_T + R_i} I = \frac{R_e}{R_i} I \tag{35}$$

For $R_i \ll Z_T$, which is the normal case, $\alpha_T < 1$ and V_{TT} is the initial tower component of voltage multiplied by α_T plus the final voltage V_F. The final voltage in per unit of the stroke current is the footing resistance in parallel with half the ground wire surge impedance. And if $Z_g \gg R_i$, then the final voltage is simply IR_i.

An example: Let $Z_g = 350$ ohms, $Z_T = 200$ ohms, $R_i = 20$ ohms, $h = 30$ meters, and $t_f = 2\,\mu s$. Then $R_e = 17.95$ and

$$\frac{V_T}{I} = 10 \text{ ohms}$$

$$\frac{V_{TT}}{I} = K_{TT} = 17.95 + 8.18 = 26.13 \text{ ohms} \tag{36}$$

$$\frac{V_F}{I} = 17.95 \text{ ohms} \qquad \frac{V_R}{I} = 17.95 \text{ ohms} \qquad \frac{I_R}{I} = 0.898$$

First note that even for a low tower footing resistance of 20 ohms, the footing resistance component is dominant, about 78% of V_{TT}. However, the tower component is 22% of V_{TT}. Also note that the current through the footing resistance is about 90% of the stroke current; little current travels out the ground wires.

In the above example, the tower surge impedance was *not* equal to half of the ground wire surge impedance although the equations were derived using this assumption. That is, even though the assumption is made in the derivation, any value of Z_T can be used to obtain an approximate answer. The accuracy of using these equations for any value of Z_T was examined. For practical values of tower surge impedance and for t_f greater than 1 μs, the error is about 5%. The error increases for smaller values of the wave front, but these smaller values of wave front are improbable. Considering, as will be discussed later, that Z_T is a time varying quantity, and considering the inaccuracy in establishing an equivalent constant value of Z_T, the use of these approximate equations appears justified.

Returning to Figs. 16 and 17, the voltage at point A on the tower is also of interest and will be used subsequently in Chapter 10. The equations for this voltage V_{TA} are the same as for the voltage at the top of the tower, provided T_A is substituted for T_T, i.e.,

$$V_{TA} = \left[R_e + \alpha_T Z_T \frac{T_A}{t_f} \right] I = K_{TA} I \tag{37}$$

A Review of Traveling Waves

where

$$K_{TA} = R_e + \alpha_T Z_T \frac{T_A}{t_f} \tag{38}$$

5 STROKE TO TOWER: EFFECT OF REFLECTION FROM ADJACENT TOWERS

5.1 Reduction of the Crest Voltage at the Struck Tower

In the preceding section, simplified equations were developed to determine the crest voltage at the tower top and at any point along the tower. However, in so doing it was assumed that the ground wires were infinitely long. That is, reflections from adjacent towers were neglected. To be expected is that these reflections will reduce the voltages at the struck tower. The lattice diagram of Fig. 19 considers one adjacent tower on each side of the struck tower. As noted, the line is "folded" at the struck tower so that the surge impedance of the ground wires is now half of Z_g.

In anticipation of Chapter 10, where these developments are used, it is assumed that the footing resistance of the struck tower, R_i, is different from that of the adjacent tower, R_0. That is, as will be discussed in Chapter 10, the large current flowing through the footing resistance of the struck tower results in a decreased resistance. However, as will be shown, the current flowing through the footing resistance of the adjacent towers is only a few percent of that of the struck tower, and therefore this footing resistance will remain at approximately the measured or low current value.

As before, to simplify, let $Z_T = Z_g/2$ so that

$$\alpha_T = \frac{Z_T - R_i}{Z_T + R_i} \approx \frac{Z_g - 2R_i}{Z_g + 2R_i} \tag{39}$$

Also let

$$\alpha_R = \frac{Z_g}{Z_g + 2R_0} \tag{40}$$

Defining T_T as the tower travel time and T_s as the span travel time, from the lattice diagram, the voltage at the top of the struck tower, e_{TT}, is

$$\begin{aligned}e_{TT} = [1 - \alpha_T(t - 2T_T)]e\{&[1 - \alpha_R(t - 2T_s) + \alpha_R\alpha_T[t - (2T_s + 2T_T)] \\ &- \alpha_R^2\alpha_T[t - (4T_s + 2T_T)] + (\alpha_R\alpha_T)^2[1 - (4T_s + 4T_T)] \\ &- \alpha_R^3\alpha_T[t - (6T_s + 6T_T)] + (\alpha_R\alpha_T)^3[t - (6T_s + 8T_T)] - \cdots\}\end{aligned} \tag{41}$$

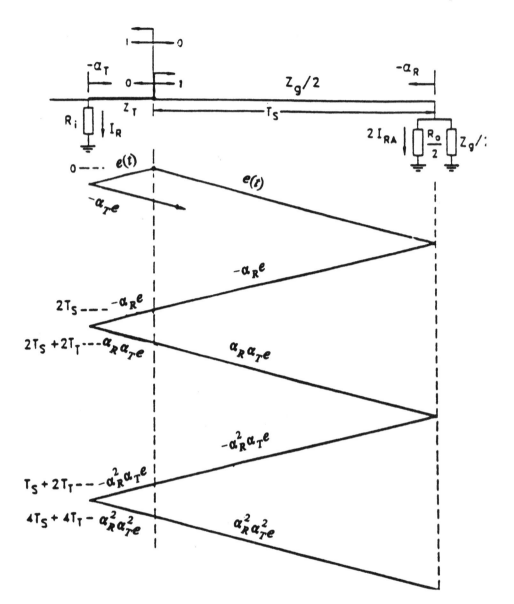

Figure 19 Lattice diagram; effect of adjacent towers on crest voltage.

Neglecting the travel time of the struck tower for reflections,

$$e_{\mathrm{TT}} = [1 - \alpha_{\mathrm{T}}(t - 2T_{\mathrm{T}})]e\{1 - \alpha_{\mathrm{R}}(1 - \alpha_{\mathrm{T}})[(t - 2T_{\mathrm{s}}) + \alpha_{\mathrm{R}}\alpha_{\mathrm{T}}(t - 4T_{\mathrm{s}}) + (\alpha_{\mathrm{R}}\alpha_{\mathrm{T}})^2(t - 6T_{\mathrm{s}}) - \cdots]\} \quad (42)$$

The quantity

$$[1 - \alpha_{\mathrm{T}}(t - 2T_{\mathrm{T}})]e \quad (43)$$

is the value of e_{TT} developed previously when reflections from adjacent towers were neglected. Therefore this quantity can be replaced by K_{TT}.

For purposes of simplification and approximation, assume that the voltage described by $(K_{TT}I)$ has a linearly rising front. Then the crest voltage becomes

$$V_{TT} = K_{SP}K_{TT}I \qquad (44)$$

where

$$K_{SP} = 1 - \alpha_R(1-\alpha_T)\left[\left(1-\frac{2T_s}{t_f}\right) + \alpha_R\alpha_T\left(1-\frac{4T_s}{t_f}\right) + (\alpha_R\alpha_T)^2\left(1-\frac{6T_s}{t_f}\right) + \cdots\right] \qquad (45)$$

The terms of this equation are only valid where the term $(1 - nT_s/t_f)$ is positive. That is, if $T_s = 0.5\,\mu s$ and $t_f = 4\,\mu s$, then four reflections are considered, i.e., $n = 8$.

To clarify, assume $t_f = 6\,\mu s$, $T_s = 1\,\mu s$, $Z_g = 300$ ohms, $Z_T = 150$ ohms, $R_0 = 40$ ohms, $R_i = 20$ ohms. then $K_{TT} = 19.56$, $\alpha_T = 09.7647$, $\alpha_R = 0.7985$, and $K_{SP} = 0.8388$. Therefore

$$V_{TT} = 0.8388(19.56)I = 16.41I \qquad (46)$$

For this case the crest tower top voltage is decreased by about 20%.

Reflections from other towers can further reduce the crest voltage, providing they arrive before crest voltage is attained at the struck tower. For example, the first reflection from the second tower, arriving at $4T_s$, is equal to

$$\Delta e_{TT} = -\alpha_R(1-\alpha_T)(1-\alpha_R)^2 \qquad (47)$$

For practical values of the variables, this reflection decreases the tower top voltage by less than 1%. Therefore to approximate the crest voltage at the struck tower, only the first adjacent towers need be considered.

The factor K_{SP} also applies to V_{TA} and V_F. That is

$$V_{TA} = K_{SP}K_{TA}I \qquad V_F = K_{SP}R_e I \qquad (48)$$

5.2 Reduction of the Tail

For the stroke to the tower, the tail of the stroke current was assumed infinite, i.e., the crest current was held constant, and therefore the tails of the tower voltages were also infinite. Even though reflections from adjacent towers do not decrease the crest voltages at the struck tower, they will decrease the tail or time to half value. To assess the magnitude of this decrease, the surge impedance and length of the shield wire is replaced by its equivalent inductance, the tower is neglected, and additional inductive–resistive pi-sections are added to represent the entire line. For an infinite line, the final voltage approaches zero, and, as may be noted from this network, the method of achieving zero voltage is through time constants consisting of the inductance and various combinations of R_0 and R_i. However, the tail or voltage e_R for times equal to or greater than $t_f + 2T_T$ can be approximated by a single time constant τ such that

$$e_R(t) = V_F e^{-\frac{t-(t_f+2T_T)}{\tau}} \tag{49}$$

To evaluate the apparent or approximate value τ, a step function of current having a magnitude of 1.0 per unit was injected into the inductance–resistance network. The time to decrease to 0.607 per unit was obtained and multiplied by 2 to obtain the apparent time constant. Figure 20 shows the variation of τ as a function of R_i using the ratio R_0/R_i as a parameter. Expected ratios of R_0/R_i vary between about 2 and 5, and as shown in the figure, for ratios of 2 to 5, the equation given by the dotted line curve is conservative. That is, the time constant of the tail can be conservatively estimated by the equation

$$\tau = \frac{Z_g}{R_i} T_s \tag{50}$$

To be noted is that this equation is simply an L/R time constant where the inductance L is the inductance of a span, i.e.,

$$L = \frac{Z_g}{c}(\text{span length}) = Z_g T_s \tag{51}$$

The current through the footing resistance of the adjacent tower can be obtained from the lattice diagram of Fig. 19. First, the surge voltage at the adjacent tower, e_A, is

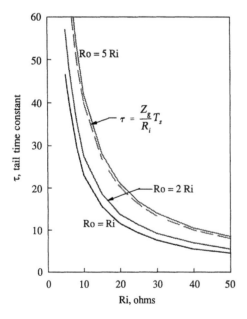

Figure 20 Apparent time constant of tail, $Z_g = 400$ ohms, $T_S = 1$ μs.

A Review of Traveling Waves

$$e_A = (1-\alpha_T)(1-\alpha_R)e\left[1+\alpha_R\alpha_T\left(1-\frac{2T_s}{t_f}\right)+(\alpha_R\alpha_T)^2\left(1-\frac{4T_s}{t_f}\right)+\cdots\right] \quad (52)$$

where the voltage s is $(I_L Z_g/2)$ and I_L, the current flowing out on the shield wires, is

$$I_L = \frac{2R_i}{Z_g + 2R_i} I \quad (53)$$

Combining these equations and remembering that the result is twice this current, and using the equation for the current through the footing resistance of the struck tower, the current through the footing of the adjacent tower, I_{RA}, is

$$I_{RA} = \frac{8R_i^2}{(Z_g+2R_0)(Z_g+2R_i)} I_R \left[1+\alpha_R\alpha_T\left(1-\frac{2T_s}{t_f}\right)+(\alpha_R\alpha_T)^2\left(1-\frac{4T_s}{t_f}\right)+\cdots\right] \quad (54)$$

Using the same value of the parameters as before, Section 5.1, then

$$I_{RA} = 0.065 I_R \quad (55)$$

Lower values of current occur for lower values of the time to crest, higher values for shorter span lengths. In general, the current through the footing of the adjacent tower will be in the range of 4 to 8% of the current in the struck tower. Therefore it appears justified and conservative to maintain the footing resistance of adjacent towers at their measured or low current values, i.e., R_0.

6 ARRESTERS

6.1 General: Separation Distance

To demonstrate further the use of traveling wave theory, consider the effect of arresters in limiting the surge voltage at locations remote from the arrester. In this presentation, the arrester is considered as an "ideal" or constant-voltage arrester. That is, the assumption is that the arrester maintains a constant voltage E_A that is independent of the current being discharged by the arrester. Further assumed is that the arrester appears as an opened circuit until this voltage is reached. At that time, the arrester appears as a short circuit, since the voltage is held constant. Of course this is untrue, but the assumption does produce some useful equations. Now consider the circuit of Fig. 21, which shows an arrester located in front of a transformer, which is here represented by an opened circuit and located behind a piece of equipment, generically denoted as a breaker. As a word of caution, the transformer should not be modeled as an opened circuit but as a capacitance to ground whose capacitance varies from about 1 to 6 nF. A value of about 2 nF is frequently used for the transformer. The effect of the capacitance is to increase the voltage at the transformer and increase the current through the arrester. These effects will be considered in Chapter 13.

Continuing the development, the lattice diagram is shown in Fig. 22. Assume that a surge having a front steepness of S and an unlimited crest voltage travels in

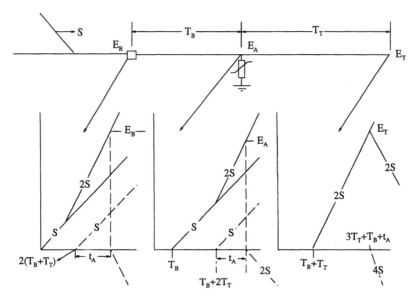

Figure 21 Surge voltage at remote location from arrester is a function of separation distance.

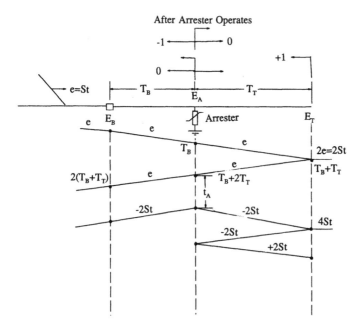

Figure 22 Lattice diagram: arrester protection.

A Review of Traveling Waves

toward the breaker–arrester–transformer. The travel time between the arrester and the breaker is defined as T_B, and the travel time between the arrester and the transformer is defined as T_T. The surge arrives at the breaker at time zero, at the arresters at time T_B, and at the transformer at time $T_B + T_T$. Because of the opened circuit, the steepness at the transformer doubles, and a reflected surge having a steepness S travels back toward the arrester and breaker. Upon arrival at these other two locations, the steepness doubles. This situation continues with no further reflections until the arrester operates. Assume that the arrester operates at time t_A after the reflection from the transformer. Therefore the voltage at the arrester can be described by the equation

$$E_A = 2ST_T + 2St_A \tag{56}$$

and the voltages E_T and E_B are

$$E_T = 2S(2T_T + t_A)$$
$$E_B = 2S(T_B + T_T + t_A) \tag{57}$$

Substituting for t_A from Eq. 56, we obtain

$$E_T = E_A + 2ST_T$$
$$E_B = E_A + 2ST_B \tag{58}$$

These same equations could be derived by separately considering the transformer and the breaker. As noted by the equations, at both locations the voltage is increased by twice the steepness multiplied by the travel time between the arrester and the equipment. The maximum voltage at the transformer is $2E_A$, which occurs at

$$\frac{ST_T}{E_A} = 0.5 \tag{59}$$

The maximum voltage at the breaker is the crest magnitude of the incoming surge plus half of the arrester voltage as shown in Fig. 23, where t_f is the time to crest of the incoming surge. The maximum voltage is

for $(T_A + 2T_B) \geq t_f$

$$E_A = 2St_A$$
$$E_B = E + St_A = E + \frac{E_A}{2} \tag{60}$$

6.2 Voltage at Transformer for an *n*-Line Station

Now consider an *n*-line station and determine the voltage at the opened circuit which is here called the transformer. The surge voltage with steepness S approaches the arrester and is reduced to αS after passing the arrester. The transmitted and reflected coefficients for a surge approaching the arrester from the right, as used in Fig. 24, are

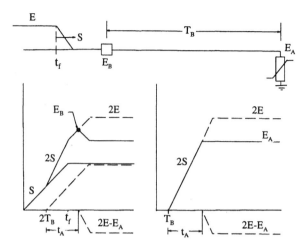

Figure 23 Maximum breaker voltage for a single-line station.

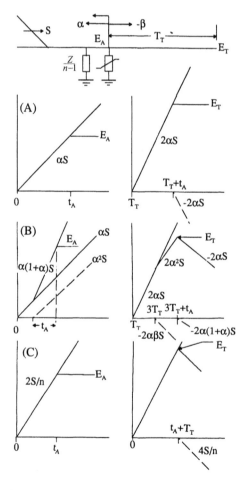

Figure 24 For an n-line station.

A Review of Traveling Waves

$$\alpha = \frac{2}{n+1} \qquad \beta = \frac{n-1}{n+1} \qquad \alpha = 1 - \beta \qquad (61)$$

To obtain the maximum voltage at the arrester, assume that the arrester operation occurs before a reflection returns from the opened end of the line, as depicted in Fig. 24A. The arrester voltage and the voltage at the opened end are

$$E_A = \alpha S t_A$$
$$E_T = 2\alpha S t_A = 2E_A \qquad (62)$$

And the maximum voltage at the transformer is twice the arrester voltage.

Next, consider that the arrester operates after the first reflection from the transformer as shown in Fig. 24B where reflections are shown by the dotted lines. The voltages at the arrester and transformer are

$$E_A = 2\alpha S T_T + \alpha(1 + \alpha) S t_A$$
$$E_T = 4\alpha S T_T + 2\alpha^2 S t_A \qquad (63)$$

from which the equation is obtained

$$E_T = \frac{4}{n+3}(E_A + 2ST_T) \qquad (64)$$

To confirm the solutions, the circuit of Fig. 24 was set up on the ATP, the results of which are shown in Fig. 25.

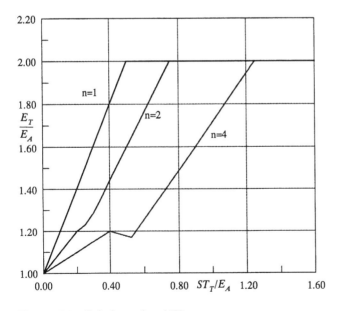

Figure 25 Solution using ATP.

Equating Eqs. 62 and 64 shows that the maximum voltage occurs where

$$\frac{ST_T}{E_A} = \frac{n+1}{4} \tag{65}$$

which agrees with Eq. 59.

Next, for small values of T_T, as shown in Fig. 24C, the voltage steepness at the arrester and the transformer are equal at $2S/n$. The arrester operates at time t_A, and the voltage at the transformer occurs at $t_A + T_T$, resulting in the equations

$$E_A = \frac{2S}{n} t_A$$
$$E_T = \frac{2S}{n}(t_A + T_T) \tag{66}$$
$$= E_A + \frac{2S}{n} T_T$$

Equating Eqs. 64 and 66 shows that the two equations intersect at $ST_T/E_A = n/6$, at which point $E_T/E_A = 4/3$.

Figure 26 illustrates the general solution.

6.3 E_B for an *n*-Line Station

The voltage behind the arrester for an *n*-line station is independent of the number of lines. Using Fig. 27, where

$$\alpha = \frac{2}{n} \quad \beta = \frac{n-2}{n}$$
$$\alpha = 1 - \beta \tag{67}$$

the voltage at the breaker is

$$E_B = E_A + 2ST_B \tag{68}$$

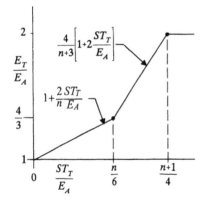

Figure 26 General solution for transformer voltage.

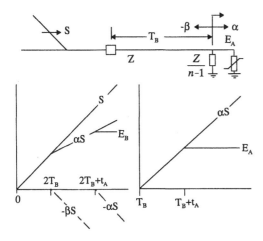

Figure 27 Voltage at break are for n lines.

6.4 Arrester Lead Length

Returning to Fig. 22, an arrester lead length with a travel time of T_A should be added from the bus to the top of the arrester. As shown in Fig. 28a, the effect of the lead length on the voltage at the breaker is obvious, in that the travel times T_B and T_A should be added and therefore

$$E_B = E_A + 2S(T_B + T_A) \tag{69}$$

The effect of the arrester lead length on the voltage at the transformer is not as obvious as seen from Fig. 28b. To obtain an approximation of the effect, assume that the separation distance to the transformer is infinite or T_T is infinite. Then the circuit of Fig. 28B can be reduced to that of Fig. 29. Using the lattice diagram of Fig. 29, the voltages E_J and E_T are shown in Figs. 30a and b. As noted, approximately, both

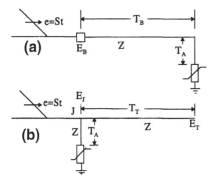

Figure 28 Circuits when considering arrester lead length: (a) for breaker; (b) for transformer.

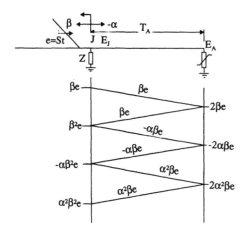

Figure 29 Lattice diagram for Fig. 28b.

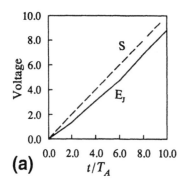

(a)

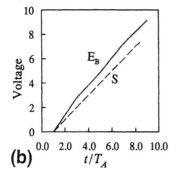

(b)

Figure 30 Comparison of actual voltage with steepness S.

A Review of Traveling Waves

voltages increase at the same steepness S, that of the incoming surge. Assuming both voltages have a steepness S, the voltage at the arrester is

$$E_A = St_A \tag{70}$$

and the voltage at the junction J is

$$\begin{aligned} E_J &= 2ST_A + St_A \\ &= E_A + 2ST_A \end{aligned} \tag{71}$$

and thus the lead length adds a voltage $2ST_A$ to both the voltage at the breaker and the voltage at the transformer or

$$\begin{aligned} E_B &= E_A + 2S(T_B + T_A) \\ E_T &= E_A + 2S(T_T + T_A) \end{aligned} \tag{72}$$

Similarly for the n-line case, the travel time T_A should be added to the travel times T_T and T_B.

6.5 Power Frequency Voltage

As will be presented in Chapter 11, the usually incoming surge to a substation rides atop an opposite polarity power frequency voltage V_{PF} as illustrated in Fig. 31, where E is the surge voltage.

The method used to handle this situation is first to determine the surge voltages at the transformer and breaker using the surge voltage E. Afterwards add (subtract) the power frequency voltage V_{PF}.

To formalize, let

E_t = voltage to ground at the transformer
E_T = surge voltage at the transformer
E_b = voltage to ground at the breaker
E_B = surge voltage at the breaker

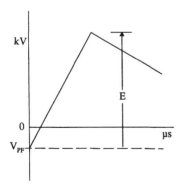

Figure 31 Incoming surge rides on top of opposite polarity power frequency voltage.

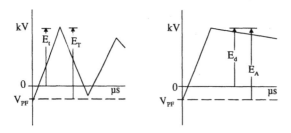

Figure 32 Examples of definitions; see Eqs. 73.

E_j = voltage to ground at the junction J
E_J = surge voltage at the junction J
E_d = voltage to ground at the arrester, i.e., the discharge voltage
E_A = surge voltage at the arrester

As illustrated in Fig. 32, the equations are

$$E_t = E_T - V_{PF}$$
$$E_b = E_B - V_{PF}$$
$$E_j = E_J - V_{PF}$$
$$E_d = E_A - V_{PF}$$

(73)

Thus to calculate the voltage at the transformer, E_T, use the steepness and magnitude of the incoming *surge voltage*. To obtain the voltage to ground, E_t, subtract V_{PF}. For example, if the surge voltage at transformer is 1000 kV and the opposite polarity power voltage is 300 kV, the voltage to ground, that is, the voltage across the insulation, is 700 kV, which should be compared to the transformer insulation strength.

There should be no confusion with the designations of E_A and E_d. In previous equations, the arrester voltage was denoted as E_A. That is, since the power frequency voltage was not considered, E_A is both the surge voltage and the voltage to ground at the arrester. For the single-line case with no transformer capacitance, the power frequency voltage has no effect on the voltage to ground. That is,

$$E_T = E_A + 2S(T_A + T_T) \tag{74}$$

Substituting

$$E_t + V_{PF} = E_d + V_{PF} + 2S(T_A + T_T) \tag{75}$$

or

$$E_t = E_d + 2S(T_A + T_T) \tag{76}$$

and thus since the term V_{PF} on each side of the equation cancels, the power frequency voltage has no effect on the voltage at the transformer. The canceling of the power frequency voltage as above may also be true for others of the equations developed. However, in general, each case must be considered separately. For example, for an n-line station, the power frequency voltage has no effect for one of the equations for the voltage at the transformer or for the voltage at the breaker. However, for other equations, the power frequency voltage must be considered.

6.6 Arrester Current

Figure 33a shows the circuit for the single-line case. Note that the voltage across the arrester is shown as E_d. Per Fig. 32, the voltage to ground is defined as E_d. This is the discharge voltage, and since the arrester current is only a function of the voltage across the arrester, the voltage E_d must be used. The source is twice the surge voltage E minus the power frequency voltage. The arrester I_A is therefore

$$I_A = \frac{2E - E_d - V_{PF}}{Z} = \frac{2E - E_A}{Z} \tag{77}$$

For an n-line station, the circuit of Fig. 33b applies, which using Thevenin's theorem can be reduced to that of Fig. 33C. Thus, in general

$$I_A = \frac{2(E/n) - E_d - V_{PF}}{(Z/N)} = \frac{2(E/n) - E_A}{(Z/N)} \tag{78}$$

The arrester discharge voltage E_d is a function of the arrester discharge current I_A, and therefore these values are not independent. An illustration of the discharge voltage–current characteristic is shown by the solid line of Fig. 34. Through any two currents I_{A1} and I_{A2}, the characteristic can be approximated as a straight line per the dotted line of Fig. 34. Therefore the arrester discharge voltage is

$$E_d = E_0 + I_A R_A \tag{79}$$

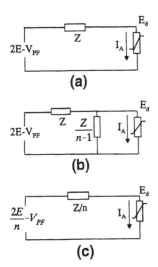

Figure 33 Circuits to obtain the arrester current: (a) single line; (b) n lines; (c) Z/n lines.

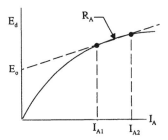

Figure 34 Arrester voltage–current characteristic.

With this modification, the arrester current becomes

$$I_A = \frac{(2E/n) - E_0 - V_{PF}}{(Z/n) + R_A} \tag{80}$$

To illustrate by example, assume a single line, $n = 1$, and that $Z = 400$ ohms, $E = 1500\,\text{kV}$, and $V_{PF} = 90\,\text{kV}$. Assume that the arrester characteristics for an 84-kV MCOV station class arrester are given per Table 1. The values of E_0 and R_A are also given. Note first that the maximum current is $2E/Z$ or $3000/400 = 7.5\,\text{kA}$, and therefore as a first step assume that the current is below $7.5\,\text{kA}$, or is between 5 and 10 kA. Using these characteristics from Table 1, the current is

$$I_A = \frac{3000 - 203 - 90}{404} = 6.70\,\text{kA} \tag{81}$$

which is between 5 and 10 kA. If the answer is not between 5 and 10 kA, then a revised iteration must be done. For example, if the calculated current is 4 kA, then the calculation should be repeated using the 3 to 5 kA values from Table 1. Continuing, the arrester discharge voltage is

$$E_d = 203 + (6.70)(4) = 230\,\text{kV} \tag{82}$$

The calculation of arrester current and voltage will be considered further in Chapter 12, where the effect of the transformer capacitance is shown to increase the arrester current by about 60%.

Table 1 Arrester Characteristics

Arrester current, kA	R_A, ohms	E_0, kV
3 → 5	5.0 →	198
5 → 10	4.0 →	203
10 → 15	3.4 →	209
15 → 20	2.2 →	227

7 MULTIPLE CONDUCTORS

7.1 General

Figures 35 and 36 show two conductors separated by a distance d_{12} with unequal heights above ground. The radii of the conductors may also be different, r_1 and r_2 for conductors 1 and 2. Z_1 and Z_2 are the surge impedances of each conductor, or better, the self-surge impedances of the conductors, that surge impedance in absence of the other conductor. The self-surge impedances, as before, are

$$Z_1 = 60 \ln \frac{2h_1}{r_1}$$
$$Z_2 = 60 \ln \frac{2h_2}{r_2} \tag{83}$$

The mutual surge impedance between the conductors is defined as

$$Z_{12} = 60 \ln \frac{D_{12}}{d_{12}} \tag{84}$$

The traveling wave equations are similar to those for a single conductor except a voltage is induced by a current in the other conductor. Thus

$$e_1 = i_1 Z_1 + i_2 Z_{12}$$
$$e_2 = i_1 Z_{12} + i_2 Z_2 \tag{85}$$

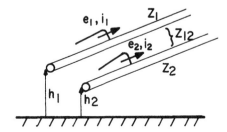

Figure 35 Definition, two conductors with mutual impedance.

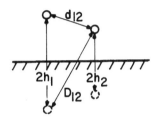

Figure 36 Distance definitions.

Solving for the currents,

$$i_1 = \frac{Z_2 e_1 - Z_{12} e_2}{D}$$
$$i_2 = \frac{Z_1 e_2 - Z_{12} e_1}{D} \tag{86}$$

where

$$D = Z_1 Z_2 - Z_{12}^2 \tag{87}$$

7.2 Equivalent Surge Impedance

In many cases, an equivalent surge impedance or a combined surge impedance of two or more conductors is desired. For example, the surge impedance of two overhead ground wires is needed for calculation of the tower top voltage. Thus first consider the case of two conductors having equal surge impedances Z. The surge voltage e on each of the conductors is also equal as is illustrated in Fig. 37. The combined surge impedance of the two conductors, Z_e, is the surge voltage divided by the total current i_T. That is,

$$Z_e = \frac{e}{i_T} = \frac{e}{i_1 + i_2} \tag{88}$$

But since the self-surge impedances and surge voltages on each of the conductors are equal, the currents in each conductor are equal and from Eq. 86,

$$i = i_1 = i_2 = \frac{Z - Z_{12}}{D} e \tag{89}$$

Thus

$$Z_e = \frac{e}{2i} = \frac{Z^2 - Z_{12}^2}{2(Z - Z_{12})} = \frac{Z + Z_{12}}{2} \tag{90}$$

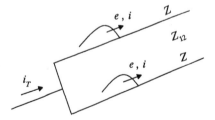

Figure 37 Combined surge impedance of two conductors.

A Review of Traveling Waves

In general, for n conductors, if $Z = Z_1 = Z_2 = Z_3$ etc., and $Z_m = Z_{12} = Z_{13} = Z_{23}$ etc., then

$$Z_n = \frac{Z + (n-1)Z_m}{n} \quad (91)$$

If the self-surge impedances are not all equal and the mutual surge impedances are not at all equal, the above equation is still valid to within about 2% if Z and Z_m are defined as

Z = the average self-surge impedance of all conductors
Z_m = the average mutual surge impedance of all conductors

7.3 The Coupling Factor

If a traveling wave voltage and current are impressed on only one conductor, a voltage will be induced or coupled to the other conductor. Referring to Fig. 38, the coupling factor C is defined as

$$C = \frac{e_2}{e_1} \quad (92)$$

Since $i_2 = 0$, the equations are

$$\begin{aligned} e_1 &= i_1 Z_1 \\ e_2 &= i_1 Z_{12} \\ e_2 &= \frac{Z_{12}}{Z_1} e_1 \end{aligned} \quad (93)$$

and the coupling factor is

$$C = \frac{Z_{12}}{Z_1} \quad (94)$$

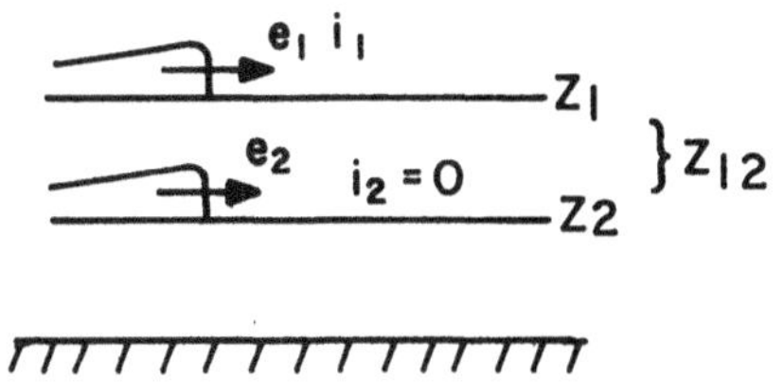

Figure 38 The coupling factor e_2/e_1.

7.4 Coupling Factor Between Two Ground Wires and One Phase Conductor

In this practical case, the coupling factor between the two overhead ground wires and a single phase conductor is needed. The surge impedances are defined in Fig. 39, where the subscript c refers to the phase conductor. Let the voltage on the ground wires be e. Remembering that the current in the phase conductor is zero, then

$$e = e_1 = i_1 Z_1 + i_2 Z_{12}$$
$$e = e_2 = i_1 Z_{12} + i_2 Z_2 \qquad (95)$$
$$e_c = i_1 Z_{1c} + i_2 Z_{2c}$$

where Z_{1c} and Z_{2c} are the mutual surge impedances between the ground wires and the phase conductor. Letting $Z = Z_1 = Z_2$, then

$$i_1 = i_2 = \frac{e}{Z + Z_{12}} \qquad (96)$$

Therefore

$$\begin{aligned} e_c &= \frac{Z_{1c} + Z_{2c}}{Z + Z_{12}} e \\ &= \frac{(Z_{1c} + Z_{2c})/2}{(Z + Z_{12})/2} \\ &= \frac{(Z_{1c} + Z_{2c})/2}{Z_e} \\ &= \frac{\text{average mutual surge impedance}}{\text{combined self-surge impedance}} \end{aligned} \qquad (97)$$

where the combined surge impedance of the two overhead ground wires is Z_e per Eq. 91.

As noted, the coupling factor is simply the average mutual surge impedance divided by the combined surge impedance of the two ground wires.

$$C = \frac{\text{Average mutual surge impedance}}{\text{equivalent ground wire surge impedance}} \qquad (98)$$

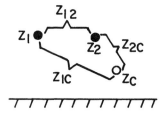

Figure 39 Coupling factor: two ground wires, one conductor.

8 MULTIPLE CONDUCTORS: POINTS OF DISCONTINUITY

As with single conductors, traveling waves on multiple conductors produce reflected and transmitted waves at points of discontinuity. To illustrate, Fig. 40 shows e_1, i_1 and e_2, i_2 arriving at a point of discontinuity. The normal and boundary equations are

$$
\begin{array}{ll}
\text{Normal equations} & \text{Boundary equations} \\
e_1 = i_1 Z_1 + i_2 Z_{12} & e_1'' = e_1 + e_1' \\
e_2 = i_1 Z_{12} + i_2 Z_2 & e_2'' = e_2 + e_2' \\
e_1' = i_1' Z_1 + i_2' Z_{12} & i_1'' = i_1 - i_1' \\
e_2' = i_1' Z_{12} + i_2' Z_2 & i_2'' = i_2 - i_2' \\
e_1'' = i_1'' Z_3 + i_2'' Z_{34} & \\
e_2'' = i_1'' Z_{34} + i_2'' Z_4 &
\end{array}
\qquad (99)
$$

8.1

Example. Per Fig. 41, a voltage and current are injected into the top conductor only. A coupled voltage appears on the bottom conductor (no current). The problem is to calculate e_1'' and e_2'. Therefore

$$
\begin{array}{ll}
e_1 = i_1 Z_1 & e_1'' = e_1 + e_1' \\
e_2 = i_1 Z_{12} & e_2'' = e_2 + e_2' \\
e_1' = i_1' Z_1 & i_1'' = i_1 - i_1' \\
e_2' = i_1' Z_{12} & i_2'' = i_2 - i_2' = 0 \\
e_1'' = i_1'' Z_1 & i_2 = 0
\end{array}
\qquad (100)
$$

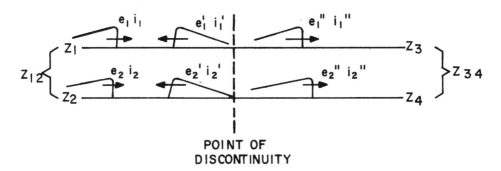

Figure 40 Surges arrive at a discontinuity.

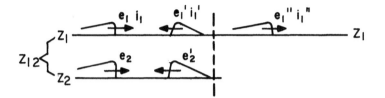

Figure 41 Example.

From which

$$i_1'' = i_1 - i_1'$$
$$\frac{e_1''}{Z_1} = \frac{e_1}{Z_1} - \frac{e_1'}{Z_1} \qquad (101)$$
$$e_1'' = e_1 - e_1'$$

and therefore

$$\begin{array}{ccc} e_1' = 0 & e_1'' = e_1 & e_2' = 0 \\ e_2'' = e_2 & i_1'' = i_1 & i_2' = 0 \end{array} \qquad (102)$$

as would be expected.

8.2

Another Example. The case described by Fig. 42 will be used in future chapters. In general, a surge on the phase conductor e_2 and a surge on the ground wire e_1 approach a tower that is grounded through a resistor R. The problem is to determine the value of e_2'' and e_1''. The normal and boundary equations are

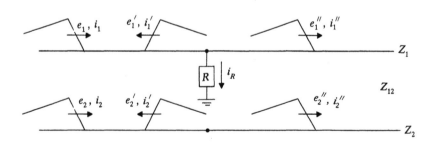

Figure 42 Surges approach a tower.

A Review of Traveling Waves

$$\begin{aligned} e_1 &= i_1 Z_1 + i_2 Z_{12} & e_1'' &= e_1 + e_1' = i_R R \\ e_2 &= i_1 Z_{12} + i_2 Z_2 & e_2'' &= e_2 + e_2' \\ e_1' &= i_1' Z_1 + i_2' Z_{12} & i_1'' + i_R &= i_1 - i_1' \\ e_2' &= i_1' Z_{12} + i_2' Z_2 & i_2'' &= i_2 - i_2' \\ e_1'' &= i_1'' Z_1 + i_2'' Z_{12} \\ e_2'' &= i_1'' Z_{12} + i_2'' Z_2 \end{aligned} \qquad (103)$$

Solving for the currents,

$$i_1 = \frac{1}{D}[Z_2 e_1 - Z_{12} e_2) \qquad i_2 = \frac{1}{D}[Z_1 e_2 - Z_{12} e_1] \\ i_1' = \frac{1}{D}[Z_2 e_1' - Z_{12} e_2'] \qquad i_2' = \frac{1}{D}[Z_1 e_2' - Z_{12} e_1'] \qquad (104)$$

where

$$D = Z_1 Z_2 - Z_{12}^2 \qquad (105)$$

Then

$$\begin{aligned} e_1'' &= e_1 + e_1' = e_1 + i_1' Z_1 + i_2' Z_{12} = i_R R \\ e_1 &+ (i_1 - i_1'' - i_R) Z_1 + (i_2 - i_2'') Z_{12} = i_R R \\ 2e_1 &= i_R (Z_1 + 2R) \end{aligned} \qquad (106)$$

Therefore

$$i_R = \frac{2e_1}{Z_1 + 2R} \qquad (107)$$

then

$$e_1'' = i_R R = \frac{2R}{Z_1 + 2R} e_1 \qquad (108)$$

then

$$e_1' = e_1'' - e_1 = -\frac{Z_1}{Z_1 + 2R} e_1 \qquad (109)$$

Next, find e_2'':

$$e_2'' = e_2 + e_2' = e_2 + i_1' Z_{12} + i_2' Z_2$$
$$e_2'' = e_2 + (i_1 - i_1'' - i_R) Z_{12} + (i_2 - i_2'') Z_2 \tag{110}$$
$$e_2'' = e_2 - \frac{Z_{12}}{Z_1 + 2R} e_1$$

and therefore

$$e_2' = e_2'' - e_2 = -\frac{Z_{12}}{Z_1 + 2R} e_1 \tag{111}$$

To find the currents, use Eqs. 104 and the previous equations.

$$i_2'' = i_2 - i_2' = \frac{1}{D}(Z_1 e_2 - Z_{12} e_1) = i_2 \tag{112}$$
$$i_2' = 0$$

Also

$$i_1'' = i_1 - i_1' - i_R = i_1 - \frac{e_1}{Z_1 + 2R} \tag{113}$$
$$i_1' = -\frac{e_1}{Z_1 + 2R}$$

The process of reduction of the surge voltages continues as each tower is passed. that is the voltage e_1'' becomes e_1 and e_2'' becomes e_2 at the next tower, and so on. The equations that show this progression are

$$e_1'' = \left(\frac{2R}{Z_1 + 2R}\right)^n e_1 \tag{114}$$

$$e_2'' = e_2 - \left(\frac{Z_{12}}{Z_1 + 2R}\right) \sum_{m=0}^{m=n-1} \left(\frac{2R}{Z_1 + 2R}\right)^m e_1 \tag{115}$$

where n is the number of towers. As n approaches infinity,

$$e_1'' = 0$$
$$e_2'' = e_2 - \frac{Z_{12}}{Z_1} e_1 = e_2 - C e_1 \tag{116}$$

where, as before, C is the coupling factor. Figure 43 summarizes the results.

A Review of Traveling Waves

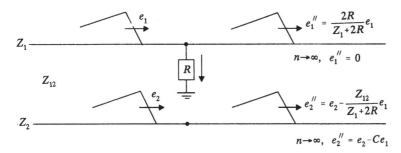

Figure 43 Solution to Fig. 42.

8.3 Extending the Example

The previous example can be extended to consider the case where a stroke terminates on the tower or ground wire resulting in a flashover to the phase conductor. Thus, after flashover, the voltages on the ground wire and phase conductor are equal, that is

$$e = e_1 = e_2 \tag{117}$$

and therefore, for an infinite number of towers, Eq. 116 becomes

$$e_2'' = (1 - C)e \tag{118}$$

The progression to the value of Eq. 118 is slow. However, the value per Eq. 118 is a reasonable approximation and will be used in estimating the crest of the surge that appears at the substation entrance resulting from a backflash.

9 TOWER SURGE IMPEDANCE

To understand the use and concept of a tower surge impedance, we must briefly return to our college classes where we studied electric fields. First, consider the circuit of Fig. 44a, where a conductor at height h above ground is energized by a switch connected to a battery. Upon switching, waves of current i and charge q are produced that travel down the conductor and produce electric fields. Voltage or potential is a measured or calculated quantity and is the line integral of the electric field. If the field is integrated along the dotted line, the voltage or potential V is obtained. The surge impedance V/i is a function of time and is illustrated in Fig. 44b. The surge impedance that we have calculated, $60 \ln 2h/r$, is attained at a time equal to $2h/c$, where c is the velocity of light. That is, the field must travel down to ground and be reflected to the conductor before the conductor knows that a ground exists.

Now, consider Fig. 44c where a 1000-kV surge is traveling on a conductor. The problem is to determine the voltage between points A and B. Usually, we would immediately answer 1000-kV, since the voltage at point B is zero (assuming a perfect earth). But that answer, although correct, assumes that the integration of the electric field is along path 1. However, equally valid is the integration along path 2, which for a perfect conductor (no resistance) is zero!! Other values of potential may also be

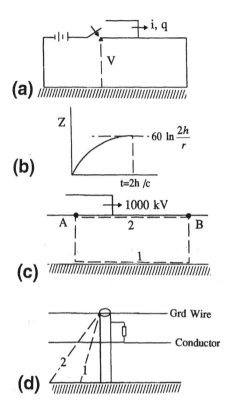

Figure 44 Explaining the concept of a surge impedance.

obtained and are equally valid. Thus we see that the potential is a function of the path of integration of the electric field.

Now take the next step and consider a transmission tower as shown in Fig. 44d. Assume that a lightning stroke terminates at the top of the tower and produces waves of current and charge that travel down the tower and out on the ground wires. As before, these waves of current and charge produce electric fields. Let us now consider the potential or voltage at the top of the tower. If a voltage divider is located along path 1, a voltage will be measured. If the divider is moved and located along path 2, a different voltage is measured. But now integrate the field along the tower. Assuming a perfect conductor, the voltage is zero. That is, the tower top potential is zero.

Yet we know that if lightning terminates on the tower, a voltage is generated across the insulators. The answer lies in the last statement. When the voltage across the insulator or the voltage from the tower to the phase conductor is calculated using field theory, a term in the equation appears that is denoted as the tower surge impedance [2], i.e.,

$$Z_T = 60 \ln \sqrt{2} \frac{ct}{r} \tag{119}$$

A Review of Traveling Waves

where r is the radius of the tower assumed here to be a cylinder. The equation is valid from $t = 0$ to $t = 2h/c$. Thus the "tower" surge impedance is time varying, having a shape similar to that of Fig. 44b. At $t = 2h/c$, the surge impedance reaches a maximum value of

$$Z_T = 60 \ln \sqrt{2} \frac{2h}{r} \tag{120}$$

This maximum value of Z_T was suggested for use by the authors of Ref. 2.

Later, Sargent and Darveniza [3] suggested a modified form of this expression based on the average surge impedance over the time $2h/c$, i.e., for a cylinder,

$$Z_T = 60 \left[\ln \sqrt{2} \frac{2h}{r} - 1 \right] \tag{121}$$

For a cone, Fig. 45,

$$Z_T = 60 \ln \frac{\sqrt{2}}{\sin \theta} \tag{122}$$

where a cone may perhaps represent a double circuit tower.

All the above expressions are approximations since the surge impedance is a time varying quantity. Fortunately, as will be shown in Chapter 10, the tower surge impedance is not a sensitive parameter in estimating the backflash rate, and therefore Eq. 121 for the cylinder is suggested.

In some cases, such as the wood-pole, H-frame, two downleads are brought down, one alongside each pole. In this case, r is the radius of the downlead. The self-surge impedance of one downlead, Z_T is estimated per Eq. 121. The mutual surge impedance Z_m between the downleads may be approximated as

$$Z_m = 60 \left[\ln \sqrt{2} \frac{2h}{D} - 1 \right] \tag{123}$$

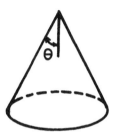

Figure 45 Definition, cone tower.

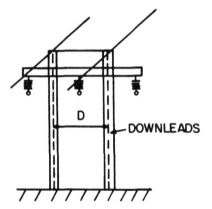

Figure 46 Wood-pole with downleads.

where, per Fig. 46, D is the separation distance between the poles. Then the total surge impedance Z'_T is

$$Z'_T = \frac{Z_T + Z_m}{2} \qquad (124)$$

10 EFFECT OF CORONA ON TRAVELING WAVES

10.1 Introduction

As the voltage on a conductor is increased, a threshold voltage is reached, above which streamers emanate from the conductor, thus increasing the radius of the conductor. As suggested by Boehne [4] in 1937, this streamer formation could be viewed as an increase in conductor radius, which therefore increases the capacitance to ground. Since the inductance of the conductor remains constant, this increase in capacitance results in a decrease in the velocity of propagation and a decrease in conductor surge impedance. As illustrated in Fig. 47, this decrease in velocity results in a distortion of the surge voltage. That is, the wave front is pushed back so that the

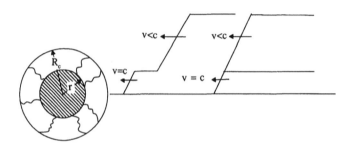

Figure 47 Modeling corona by an increase in actual radius, r to R_c.

steepness of the surge is deceased, and depending on the tail of the initial surge, the crest voltage is also decreased.

The decrease in steepness is of importance in determining the steepness of the surge, which arrives at the station as a result of either a backflash or a shielding failure on the line. While it is true that other phenomena, such as earth resistivity effects, also cause attenuation and distortion of traveling waves, for surges that emanate close to the station, corona is the dominant effect. The effect of corona in decreasing the steepness will be employed in Chapter 11 to determine the incoming surge steepness and magnitude.

The reduction in surge impedance will be used in assessment of the backflash rate of a line in Chapter 10. As will be shown, since the mutual surge impedance is unaffected by corona, the coupling factor is increased, which tends to decrease the backflash rate.

Thus the objective is to assess the shape of the traveling wave as it enters the station and to estimate the decrease in surge impedance with its attendant increase in coupling factor.

The most thorough investigation of the effects of corona began in the early 1950s with tests on a 7170 foot (2185-m) Tidd line [5]. This test line was primarily constructed for the investigation of power frequency corona loss and radio noise, which were expected to be the primary design criteria for the next voltage level of 345 kV. However, the most important data came from the experiments on this test line that investigated the effects of corona on traveling waves. Following these results, reported by Wagner et al. [5]. Wagner and Lloyd [6] continued their investigation by performing laboratory experiments that justified their data analysis of the transmission line tests and first brought a theoretical basis to the effect of corona. Indeed both these papers [5, 6] are truly classic and are recommended to anyone pursuing an investigation of the effects of corona.

The latest analysis of the effects of corona is contained in the CIGRE Technical Brochure 63 [7]. The analysis made in this brochure comes from many Internal Working Documents of Working Group 33.01 written by K.-H. Weck and from investigations made by C. Gary and associates, the results of which are contained in another CIGRE Technical Brochure [8].

10.2 Attenuation and Distortion

General

To investigate the effect of corona on traveling waves, surges were applied to a 7170-foot (2185-m) test line that was terminated in its surge impedance. The surge voltages were measured at intervals along the line. A sample of these results is shown in Fig. 48. Three conductors were employed; a 0.927 inch and a 2.00 inch ACSR and a 1.65 inch HH segmented conductor. As can be seen, below the corona inception voltage, the front of the wave suffered little distortion. However, above the corona inception voltage, the front of the surge was pushed back, and the degree of this push back is dependent on the distance traveled. Note that corona inception voltage decreases as the travel distance increases. This corona inception voltage is a statistical quantity and is also a function of the voltage steepness. Thus as the surge progresses further in its travel, the minimum corona inception voltage is attained.

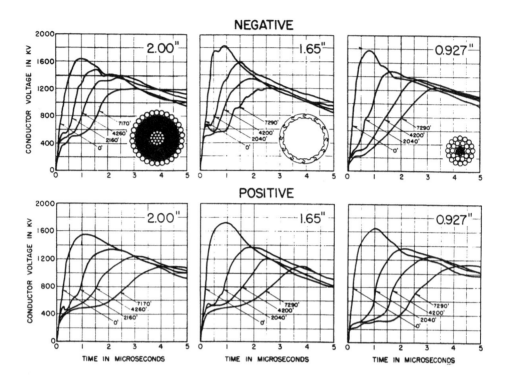

Figure 48 Sample of data from Tidd tests [5, 6]. (Copyright IEEE, 1954.)

In analyzing the data, the authors found that the data could be succinctly presented in form of the amount of pushback of the front per unit of travel or $\Delta T/d$. The data presented in this form are shown in Fig. 49 where the curves are the average value of $\Delta T/d$. The method of use is illustrated in Fig. 50. For any instantaneous voltage on the wave front, a value of $\Delta T/d$ is read from the curves of Fig. 49 as illustrated in Fig. 50A. This value is multiplied by the distance traveled to obtain a time ΔT, which is then used in Fig. 50B. This procedure continues until the curve or front of the voltage e_c meets the tail of the original surge e. At that point, the crest of the surge e_c is attained and the voltage follows the remaining tail of the surge. Thus not only is the front pushed back but also the crest magnitude is decreased, provided that the tail of the surge is of short duration.

Following these tests on transmission lines, Wagner and Lloyd performed tests on conductors in a high-voltage laboratory and obtained oscillograms of the charge vs. voltage (a q–e curve) as illustrated in Fig. 51. Since the capacitance is dq/de, to be noted is that as the voltage increases, the capacitance increases until the crest voltage is attained. The dotted line indicates the natural capacitance of the conductor, C_n, and an additional dotted line represents an increased capacitance, $C_n + \Delta C$. After the crest of the voltage, the q–e curve indicates that the capacitance returns to the natural capacitance of the conductor. The identical shape of curve was obtained regardless of the wave front. If the crest voltage of the surge was increased, the same q–e curve was obtained up to the voltage level of the other surge as shown in Fig. 52. Thus the authors showed that the q–e curve is only dependent on the

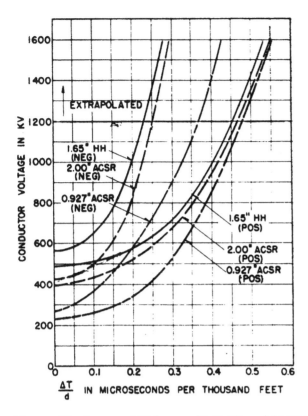

Figure 49 Average $\Delta T/d$ curves [5, 6]. (Copyright IEEE, 1954.)

instantaneous value of the applied voltage. Again, to emphasize, corona only affects the surge on the front or when the voltage is increasing. There is no effect on the tail of the wave, or when the voltage is decreasing.

Thus the authors' equation for the capacitance of the conductor was

$$C = C_n + \Delta C \tag{125}$$

where ΔC is only applicable above the corona start voltage V_i. To derive the equation for $\Delta T/d$, now denoted as $\Delta T_T/d$, consider Fig. 53, where the originating surge

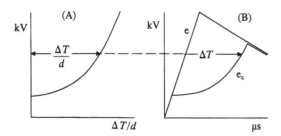

Figure 50 Illustration of use of $\Delta T/d$ curves.

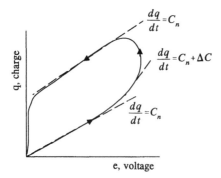

Figure 51 Illustration of q–e curves and capacitances.

e_0 and the surge e_c after traveling a distance d are shown. The voltage at distance d arrives at t_0, which is equal to d/c, where c is the velocity of light. If no corona is present, the voltage above the corona start voltage V_i is shown by the dotted curve. The time delay between this voltage and the actual voltage is $\Delta T_T/d$, which is

$$\frac{\Delta T_T}{d} = \frac{1}{v} - \frac{1}{c} \tag{126}$$

where v is the velocity of propagation above the corona inception voltage and c is the velocity of light. Since the velocity of propagation is the inverse of the square root of the inductance L times the capacitance, and the inductance is not altered by corona,

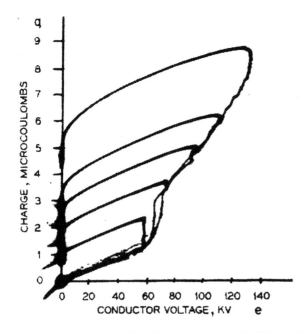

Figure 52 q–e nest for alternate voltages [6]. (Copyright IEEE, 1955.)

A Review of Traveling Waves

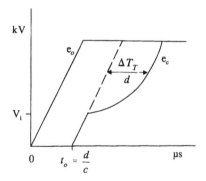

Figure 53 Illustration used to derive the equations for $\Delta T_T/d$.

$$v = [L(C_n + \Delta C)]^{-\frac{1}{2}} = (LC_n)^{-\frac{1}{2}}\left(1 + \frac{\Delta C}{C_n}\right)^{-\frac{1}{2}} \quad (127)$$

Therefore

$$\frac{\Delta T_T}{d} = \sqrt{LC_n}\left[\sqrt{1 + \frac{\Delta C}{C_n}} - 1\right] \quad (128)$$

Approximating the value of the square root

$$\frac{\Delta T_T}{d} \approx \frac{1}{2}Z_0 \Delta C \quad (129)$$

where Z_0 is the natural or noncorona surge impedance, i.e.,

$$Z_0 = \sqrt{\frac{L}{C_n}} \quad (130)$$

Thus $\Delta T_T/d$ is only a function of the ratio of the increase in capacitance to the natural capacitance or a function of the surge impedance and the added capacitance above corona. More importantly, the assumptions made in presenting the field data in the form of Fig. 49 are proven and are justified. If the added capacitance is simply an addition of a single fixed capacitance C_1, as illustrated in Fig. 54a, then the L-section of the line would be as shown in Fig. 54B.

Subsequently, several authors have suggested equations to represent the added capacitance above the corona inception voltage.

Weck. From examining the Wagner–Lloyd data and from other experiments, Weck [9, 10] suggested that the capacitance ΔC be modeled by the equation

$$\Delta C = C_{iw} + k_w(e - V_i) \quad (131)$$

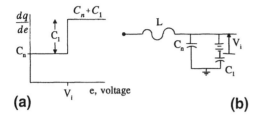

Figure 54 If $\Delta C = C_1$.

as illustrated in Fig. 55A, where C_{iw} is the abrupt increase in capacitance when the voltage e is equal to the corona inception voltage, and k_w is a constant.

CIGRE. In Technical Bulletin 63, CIGRE [7] uses the equation

$$\Delta C = C_{ic} + k_c(2e - V_i) = C_{ic} + k_c V_i + 2k_c(e - V_i) \tag{132}$$

as illustrated in Fig. 55B. The initial jump in capacitance is now $C_{ic} + k_c(V_i)$, and k_c is a constant.

Cary. Cary [8] had earlier suggested that a power-law equation be used, i.e.,

$$\Delta C = C_n E_i B \left(\frac{e}{V_i}\right)^{B-1} \tag{133}$$

Since Cary was later one of the authors of the corona portion of Technical Bulletin 63, it is assumed that he has accepted the above simpler CIGRE equation. Therefore Eq. 133 will not be considered further. Table 2 is presented to assist in further understanding the difference between the CIGRE and the Weck equations. From the analysis provided in Table 2 or from Eqs. 131 and 132,

$$\begin{aligned} C_{iw} &= C_{ic} + k_c V_i \\ k_w &= 2k_c \end{aligned} \tag{134}$$

Thus, the two equations are essentially the same.

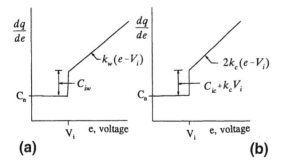

Figure 55 (a) Weck's and (b) CIGRE's interpretations of q–e curves into equations for capacitances.

A Review of Traveling Waves

Table 2 Comparison of the CIGRE and Weck Equations

Voltage	Capacitance	
	Weck	CIGRE
$e < E_i$	C_n	C_n
$e = E_i$	$C_n + C_{iw}$	$C_n + C_{ic} + k_c V_i$
$e > E_i$	$C_n + C_{iw} + k_w(e - V_i)$	$C_n + C_{ic} + k_c E_i + 2k_c(e - V_i)$
		or
		$C_n + C_{ic} + k_c(2e - V_i)$

To estimate the values of the constants in Table 2, consider Fig. 56. The surge voltage e at some location on the line is assumed to have a linearly rising front having a steepness of S_0 and an infinite or flat tail. At a distance d or a time d/c, the surge voltage e_c below corona is identical to that of the original surge. When the surge voltage is equal to the corona inception voltage, the front is pushed back by a time ΔT_0, or in general by $\Delta T_0/d$. Above the corona inception voltage, the front is pushed back by an additional time ΔT_c or by $\Delta T_c/d$. The total incremental time $\Delta T_T/d$ is the sum of the two $\Delta T/d$s. The steepness of the surge above the corona inception voltage is defined as S.

At the Corona Inception Voltage

To derive the equations, the CIGRE equation for capacitance is used. Following this derivation, the equivalent equation using Weck's formulation of capacitance is given. For $e = V_i$, using the capacitance from Table 2, the value of $\Delta T_0/d$ is

$$\frac{\Delta T_0}{d} = \frac{1}{v_0} - \frac{1}{c} = \sqrt{L(C_n + C_{ic} + k_c V_i)} - \sqrt{LC_n}$$

$$= \sqrt{LC_n}\left[\sqrt{1 + \frac{C_{ic} + k_c V_i}{C_n}} - 1\right] \tag{135}$$

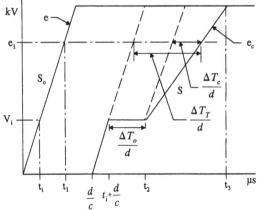

Figure 56 Illustration to derive Equations for $\Delta T_0/d$ and $\Delta T_c/d$.

As an approximation,

$$\frac{\Delta T_0}{d} \approx \sqrt{LC_n}\left[1 + \frac{1}{2}\frac{C_{ic} + k_c V_i}{C_n} - 1\right]$$

$$\approx \frac{1}{2} Z_0 [C_{ic} + k_c V_i] \qquad (136)$$

Using Weck's formulation, Eq. 136 becomes

$$\frac{\Delta T_0}{d} = \frac{1}{2} Z_0 C_{iw} \qquad (137)$$

Above Corona Start Voltage

Again using the CIGRE formulation and referring to Fig. 56, at a voltage e_1,

$$\frac{\Delta T_c}{d} = \sqrt{L(C_n + C_{ic} + k_c(2e_1 - V_i))} - \sqrt{L(C_n + C_{ic} + k_c V_i)}$$

$$= \sqrt{LC_n}\left[\sqrt{1 + \frac{C_{ic} + k_c(2e_1 - V_i)}{C_n}} - \sqrt{1 + \frac{C_{ic} + k_c V_i}{C_n}}\right] \qquad (138)$$

Again approximating,

$$\frac{\Delta T_c}{d} \approx \sqrt{LC_n}\left[1 + \frac{1}{2}\frac{C_{ic} + k_c(2e_1 - V_i)}{C_n} - 1 - \frac{1}{2}\frac{C_{ic} + k_c V_i}{C_n}\right]$$

$$\approx Z_0 k_c (e_1 - V_i) \qquad (139)$$

If the Weck formulation is used,

$$\frac{\Delta T_c}{d} = \frac{Z_0 k_w}{2}(e_1 - V_i) \qquad (14)$$

Returning to the CIGRE equation, to find the steepness S of the surge note first that the steepness of e, denoted as S_0, is

$$S_0 = \frac{e_1 - V_i}{t_1 - t_i} \qquad (141)$$

and that the steepness of e_c, denoted as S, is

$$S = \frac{e_1 - V_i}{t_3 - t_2} = \frac{e_1 - V_i}{t_1 - t_i + \Delta t_c} = \frac{S_0}{1 + (\Delta T_c S_0 / e_1 - V_i)} = \frac{S_0}{1 + k_c Z_0 S_0 d} \qquad (142)$$

Setting

$$K_c = \frac{1}{k_c Z_0} \qquad (143)$$

Then

$$S = \frac{S_0}{1 + (S_0 d/K_c)} \quad (144)$$

Note that

$$\frac{1}{S} - \frac{1}{S_0} = \frac{d}{K_c} \quad (145)$$

and if $S_0 = $ infinity,

$$S = \frac{K_c}{d} \quad (146)$$

For Weck's equation, Eqs. 144, 145, and 146 are valid. However, the value of K_c changes to

$$K_c = \frac{2}{k_w Z_0} \quad (147)$$

Value of Parameters

From the tests reported from Wagner, Gross, and Lloyd as provided in Fig. 49, the values of the parameters C_i, k, and K_c may be estimated as shown in Table 3. Since the incoming surge to a station will be of negative polarity, only these values are shown. Other estimates of these parameters are made in CIGRE Technical Bulletin 63 as shown in Table 4.

Another estimate can be obtained from one of the figures from the CIGRE report. The reported k_w is about 1.8×10^{-3} pF/kV-m which is a k_c of 0.9×10^{-3} pF/kV-m, about half of the value reported in Table 4.

In IEC Publication 71-2 [11], other values of K_c are provided by Weck. From these values, the values of k_c can be obtained with an assumed value of Z_0 as shown in Table 5.

As noted, there appears to be a considerable difference in values of K_c and k_c even in the same publication [7]. For a single conductor, the value k_c varies from 0.9 to 3.3×10^{-3}, and K_c varies from 667 to 2432 kV-km/μs. Since the proposed IEC Publication is considered to contain the most recent data, it is recommended that

Table 3 Values of Parameters from Analysis of Fig. 3 for Negative Polarity

Conductor diameter inches/mm	E_i, kV	$\Delta t_0/d$, μs/km	K_c, kV-km/μs	C_{iw}, pF/m	$k_w = 2k_c$ pF/kV-m	C_{ic} pF/m
0.927/23.5	270	0.472	1132	1.92	3.6×10^{-3}	1.43
1.65/41.9	570	0.379	1531	1.66	2.9×10^{-3}	0.85
2.00/50.8	420	0.492	2432	2.21	1.9×10^{-3}	1.82

Table 4 Values of Parameters per CIGRE, Negative Polarity

Conductor diameter, inches/mm	Assumed Z_0	k_c, pF/kV-m	K_c, kV-km/μs
0.63/16	450	1.6×10^{-3}	1390
1.22/31	450	1.8×10^{-3}	1230
$2 \times 1.22/2 \times 31$	350	1.6×10^{-3}	1786

these data be used. In addition, rounded values of K_c are listed in the last column of Table 5, which will be used in Chapter 11.

Corona Inception Voltage

The corona inception voltage for a single conductor can be estimated from the equation

$$V_i = \frac{Z_0 r E_0}{60} \quad (148)$$

where E_0 is the critical gradient usually in kV/cm, r is the conductor radius in cm, and Z_0 is the natural or noncorona surge impedance. The critical gradient per Ref. 7 is

$$E_0 = 23\left(1 + \frac{1.22}{d^{0.37}}\right) \text{ kV/cm} \quad \text{CIGRE} \quad (149)$$

Another form of this equation from Skilling and Dykes [12, 13] is:

$$E_0 = 23\delta^{0.67}\left(1 + \frac{0.3}{\sqrt{r}}\right) \text{ kV/cm} \quad \text{Skilling–Dykes} \quad (150)$$

where δ is the relative air density. For bundle conductors, Skilling and Dykes provide an equation, for the equivalent radius r_{eq}, which should be substituted into Eq. 150 for the conductor radius, where

$$r_{eq} = \frac{nr}{1 + 2(n-1)\sin\frac{\pi}{n}\frac{r}{A}} \approx \frac{nr}{1 + 2(n-1)\frac{\pi}{n}\frac{r}{A}} \approx nr \quad (151)$$

Table 5 Values Proposed in IEC Publication 71-2, Negative Polarity

Number of subconductors	Proposed K_c, kV-km/μs	Assumed Z_0	k_c, pF/kV-m	Suggested K_c
1	667	450	3.3×10^{-3}	700
2	1000	350	2.9×10^{-3}	1000
3 or 4	1667	320	1.9×10^{-3}	1700
6 or 8	2500	300	1.3×10^{-3}	2500

where A is the subconductor spacing in cm and n is the number of subconductors. Approximately, the equivalent radius is equal to n times the conductor radius. The surge impedance Z_0 becomes the surge impedance of the bundle conductors; see problem 4 of this chapter; i.e.,

$$Z_0 = \sqrt[n]{rA\left(\sin\frac{\pi}{n}\right)^{-1} \prod_{i=1}^{i=n-1} \sin\frac{i\pi}{n}} \tag{152}$$

As a check on these equations, the corona inception voltage for the conductors tested per Fig. 49 is calculated in Table 6 using both the CIGRE and the Skilling–Dykes equations and compared with the actual values of V_i. As noted, the Skilling-Dykes appear to provide somewhat better values.

10.3 Surge Impedance

Since the capacitance increases above the corona inception voltage, it would be expected that the surge impedance would decrease. Letting Z_c be the surge impedance above corona, then

$$Z_c = \sqrt{\frac{L}{C_n + \Delta C}} = Z_0 \sqrt{\frac{C_n}{C_n + \Delta C}} \tag{153}$$

and thus it is evident that under corona conditions, the surge impedance decreases. Note that since ΔC is a function of voltage, the decrease of the surge impedance is also a function of voltage. This corona surge impedance is only to be used on the front of the surge, where the voltage is increasing. On the tail of the surge, where the voltage is decreasing, the surge impedance returns to its noncorona value. To gain a further insight on the corona surge impedance, recall from Section 1.1,

$$L = 0.2 \ln \frac{2h}{r} \qquad C = \frac{10^{-3}}{18 \ln \frac{2h}{r}} \tag{154}$$

Assume that the increase in capacitance can be simulated by an increase in radius of the conductor. That is, the ionization surrounding the conductor expands until a critical gradient is reached. Let this corona radius be R_c. Then the capacitance is

Table 6 Comparison of Corona Inception Voltage

Conductor diameter, inches/mm	E_0, Skilling–Dykes, kV/cm	E_0, CIGRE, kV/cm	V_i, Skilling–Dykes, kV	V_i CIGRE, kV	V_i Actual, kV	Z_0
2.0/50.8	27.3	38.4	515	723	420	445
1.65/4.2	27.8	39.5	443	630	570	456
0.927/23.5	29.4	43.5	283	420	270	491

$$C_n + \Delta C = \frac{10^{-3}}{18 \ln \frac{2h}{R_c}} \tag{155}$$

Therefore Z_c is

$$Z_c = \sqrt{0.2 \ln \frac{2h}{r} 18 \ln \frac{2h}{R_c} 10^3} = \sqrt{60 \ln \frac{2h}{r} 60 \ln \frac{2h}{R_c}} = \sqrt{Z_0 Z'_c} \tag{156}$$

where

$$Z'_c = 60 \ln \frac{2h}{R_c} \tag{157}$$

To obtain an estimate of the corona radius

$$Z'_c = \frac{Z_c^2}{Z_0}$$

$$\ln R_c = \ln 2h - \frac{Z_c^2}{60 Z_0} \tag{158}$$

Anderson [14] also provides an estimate of the corona radius. He assumes a critical corona gradient E_0 of 15 kV/cm and iteratively solves the equation

$$R_c \ln \frac{2h}{R_c} = \frac{e}{E_0} \tag{159}$$

Assuming a single conductor line having $C_n = 4.7 \, \text{pF/m}$, $Z_0 = 477$, $r = 12.7 \, \text{mm}$, $V_i = 350 \, \text{kV}$, $k_w = 6$ and $3 \times 10^{-3} \, \text{pF/kV-m}$ and $C_{iw} = 1.5 \, \text{pF.m}$, the corona radius and corona surge impedance are calculated for alternate voltages and compared to those using Anderson's approach in Table 7. The height of the conductor is 18 meters. As noted, the CIGRE method results in larger corona radii and lower surge impedances.

10.4 Coupling Factor

The coupling factor C_0 under noncorona conditions is

$$C_0 = \frac{Z_m}{Z_0} \tag{160}$$

Under corona conditions, the surge impedance Z_0 is lowered to Z_c, but the mutual surge impedance, being the log of the ratio of two distances, is unchanged. Therefore the corona coupling factor becomes

A Review of Traveling Waves

Table 7 Comparison of Corona Radius and Surge Impedance

Voltage, kV	R_c, meters		Z_c, ohms	
	CIGRE-Weck $k_w = 6/3 \times 10^{-3}$	Anderson	CIGRE/Weck, $k_w = 6/3 \times 10^{-3}$	Anderson
200	0.0127	0.017	$Z_0 = 477$	468
350	0.049/0.049	0.034	435/435	446
1000	0.36/0.16	0.12	363/394	405
2000	1.58/0.52	0.28	299/348	373
3000	3.36/1.10	0.46	261/316	353

$$C_c = \frac{Z_m}{Z_c} = \frac{Z_m}{Z_0}\frac{Z_0}{Z_c} = C_0\frac{Z_0}{Z_c} = C_0\sqrt{1+\frac{\Delta C}{C_n}} \qquad (161)$$

and therefore the coupling factor increases under corona conditions and is a function of the voltage. Using the results of Table 7 for CIGRE, the increase in coupling factor provided by the ratio of C_c/C_0 is shown in Table 8. As will be shown in Chapter 10, the voltage across the line insulation is not a direct function of the coupling factor but directly a function of 1-coupling factor. Therefore the last column of Table 8 shows the ratio of 1-coupling factors for a C_0 of 0.30. As can be seen, the effect of the corona coupling factor is diminished. For example, for 2000 kV, the effect diminishes from 1.60 to 1.34.

As for the surge impedance, the corona coupling factor is only applicable when the voltage is increasing—on the front of the surge. When the voltage is decreasing, the coupling factor returns to the noncorona value.

As a final comment, Wagner and Lloyd in an unpublished report evaluated the corona reduced surge impedance and the corona-increased coupling factor from their field tests [15]. They show that the equations presented here provided good estimates of these parameters.

Table 8 Increase in Coupling Factor

Voltage, kV	Z_0, ohms $k_w = 6/3 \times 10^{-3}$	Ratio C_c/C_0, $k_w = 6/3 \times 10^{-3}$	Ratio $(1-C_0)/(1-C_c)$, $k_w = 6/3 \times 10^{-3}$
200	477/477	1.00/1.00	1.00/1.00
350	435/435	1.10/1.10	1.04/1.04
1000	363/394	1.31/1.21	1.15/1.10
2000	299/348	1.60/1.37	1.34/1.19
3000	261/316	1.83/1.50	1.55/1.27

11 REFERENCES

1. L. V. Bewley, *Traveling Waves on Transmission Systems*, John Wiley, 1951.
2. C. F. Wagner and A. R. Hileman, "A New Approach to the Calculation of the Lightning Performance of Transmission Lines, III—A Simplified Method: Stroke to Tower, *IEEE Trans. on PA&S*, Oct. 1960, pp. 589–603.
3. M. A. Sargent and M. Darveniza, "Tower Surge Impedance," *IEEE Trans. on PA&S*, May 1969, pp. 680–687.
4. E. W. Boehne, Discussion, *AIEE Trans.*, vol. 50, Jun. 1931, pp. 558–559.
5. C. F. Wagner, I. W. Gross, and B. L. Lloyd, "High-Voltage Impulse Tests on Transmission Lines," *AIEE Trans.*, Apr. 1954, pp. 196–210.
6. C. F. Wagner and B. L. Lloyd, "Effects of Corona on Traveling Waves," *AIEE Trans.*, Oct. 1955, pp. 858–872.
7. CIGRE Technical Bulletin 63, "Guide to Procedures for Estimating the Lightning Performance on Transmission Lines," Oct. 1991.
8. C. Gary, CIGRE Technical Brochure 55, "Distortion and Attenuation of Traveling Waves Caused by Transient Corona," 1990.
9. H. J. Koster and K. H. Weck, "Attenuation of Traveling Waves by Impulse Corona," CIGRE 33.01, IWD 21, 1981. Also "Distortion of Lightning Overvoltages by Corona," CIGRE 33.01, IWD 5A, 1984. Also H. J. Koster and K. H. Weck, "The Effect of Corona on Lightning Surges on Transmission Lines," CIGRE 33.01, IWD, Aug. 1973. Also K. H. Weck, "Impulse Corona on Conductors," CIGRE 33.01, IWD 20, 1974.
10. A. R. Hileman and K. H. Weck, "Practical Methods for GIS Insulation Coordination," Part 1 of "Insulation Coordination and Testing of GIS," IWD 5, CIGRE SC33 Colloquium, Edinburgh, Jun. 1983.
11. IEC 71-2, "Insulation Coordination—Part 2: Application Guide," 1996.
12. H. H. Skilling and P. K. Dykes, "Distortion of Traveling Waves by Corona," *AIEE Trans,.*, pt. III, 1954, vol. 73, pp. 196–210.
13. W. Disendorf, *Insulation Coordination in High Voltage Electric Systems*, New York: Crane, Russak, 1974.
14. J. G. Anderson, "Lightning Performance of Transmission Lines," Chapter 12 of *Transmission Line Reference Book*, Palo Alto, CA: Electric Power Research Institute, 1982.
15. C. F. Wagner and B. L. Lloyd, "Corona Effects on Traveling Waves Determined by Field and Laboratory Tests," Unpublished report, circa 1956.

12 PROBLEMS

1. Figure 57 provides the dimensions of a tower for a 500 kV line. The ground wire diameter is 3/8 inch and each of the subconductors of the phase has a diameter of 1.68 inches. Calculate

   1. The equivalent surge impedance of the ground wires.
   2. The surge impedance of each of the phase conductors.
   3. Coupling factors to each of the phase conductors.
   4. Assume that phase A has flashed over, thus becoming a ground wire. Now recalculate the coupling factors.

A Review of Traveling Waves

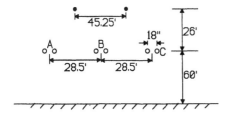

Figure 57 A 500-kV tower of Problem 1.

2. Figure 58 provides the dimensions of a tower for a 230 kV line. The ground wire diameter is 1/4 inch and the phase conductor diameter is 1.65 inches. Calculate

1. The equivalent surge impedance of the ground wires.
2. The surge impedances of the phase conductors.
3. The coupling factors to each of the phases.
4. Assume that phase C has flashed over, thus becoming a ground wire. Now recalculate the coupling factors.

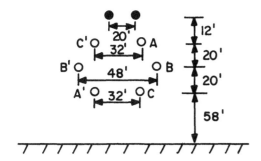

Figure 58 A 230-kV tower of Problem 2.

3. Figure 59 depicts the case of a stroke to the conductor but with no flashover to the shield wire. Find e_2'' in terms of e_2.

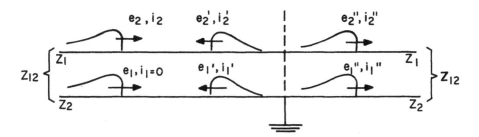

Figure 59 Stroke to phase conductor, no flashover.

4. Develop the general equation for the surge impedance of a bundle conductor having n subconductors arranged in a circle, each subconductor is separated a distance d from the subconductor on each side. Let h equal the height of the bundle above ground and let r_c equal the radius of the subconductors. See Fig. 60.

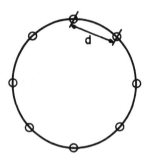

Figure 60 An n-conductor bundle of Problem 4.

5. Tests are to be performed on a three-phase transmission line to determine the attenuation and distortion of lightning impulses caused by corona. The test engineer plans to apply impulses to one, two, or all three conductors simultaneously. He would like to terminate the end of the transmission line so that no reflections occur. He wants this termination to be such that it need not be changed if a one-, two-, or three-phase surge is applied. He knows that if he were testing only a single conductor, the termination would be a resistor whose resistance was equal to the conductor surge impedance.

Find this termination. But before you attempt this for all three phases, determine the termination if the transmission line were composed of only two phases.

6. As depicted by Fig. 61, a lightning flash can terminate on both or only one of two shield wires at the midspan. At the tower, the shield wires are joined or shorted together. To show the difference in voltage at the tower location, determine the voltage e'' for both cases, assuming a stroke current I. Neglect the tower and do not consider multiple reflections along the shield wire.

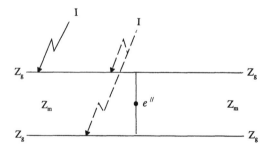

Figure 61 Lightning flash to one or two ground wires.

7. As shown in Fig. 62, a 10 kA stroke terminates on a conductor having a surge impedance Z_1. Below this conductor is another conductor have a surge impedance Z_2. Surge voltages e_1 and e_2 are created and travel to a pole or tower. At this point the lower conductor is grounded through a resistor R. Find i_R, e_1'', e_2'', and the differential voltages e_I and e_I''.

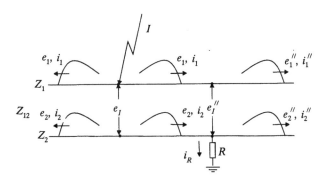

Figure 62 Problem 7.

8. Figure 63 is intended to represents either (1) a shielding failure on a transmission line or (2) a stroke to the phase conductor on a distribution line (lower conductor is the neutral). Assume that a 12 kA stroke terminates on the phase conductor. Also assume that the surge impedances of the ground wire or neutral and the surge impedance of the phase conductor are equal at 400 ohms and that the mutual surge impedance is 120 ohms. Set the grounding resistance at 40 ohms. Assume that the arrester is a constant-voltage arrester whose discharge voltage is 400 kV. Find the currents through the grounding resistance and the current discharged by the arrester. Also find the voltages e_1 and e_2. Assuming that the arrester discharge voltage has a duration of 200 μs and that the current discharged by the arrester has a tail time constant of 133 μs, calculate the energy discharged by the arrester.

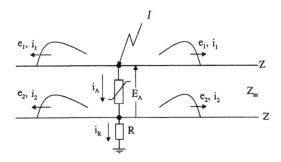

Figure 63 Problem 8.

9. Figure 64 is intended to represent a stroke to the overhead ground wire of a transmission line in which an arrester is installed between the ground wire and the phase conductor. The arrester is a constant-voltage arrester having a discharge voltage of 400 kV. The tower footing resistance is 40 ohms. The surge impedances of the overhead ground wire and the phase conductor are equal at 400 ohms, and the mutual surge impedance is 120 ohms. For a stroke of 100 kA, find the current through the footing resistance and the current discharged by the arrester. Also find the voltages e_1 and e_2.

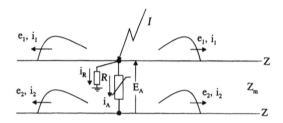

Figure 64 Problem 9.

10. The voltages e_1 and e_2 of Fig. 64, problem 9, travel onward to an adjacent tower as depicted in Fig. 65. Using the same assumptions as for problem 9, calculate the currents through the footing resistance and the arrester and find the voltages e_1'' and e_2''.

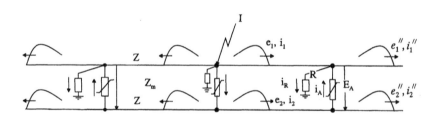

Figure 65 Problem 10.

11. Determine the surge impedances and coupling factors for the lines of problem 3, Chapter 2. Also add an underslung ground wire at heights of 23, 31, and 38 feet on the poles for 34.5-, 69-, and 115-kV towers, respectively, and repeat the calculations. use the program SRGKON.

12. Determine the surge impedances and coupling factors for the line of problem 4, Chapter 2. Use the program SRGKON.

10
The Backflash

1 INTRODUCTION AND REVIEW

To this point, the overhead ground wires or shield wires have been located so as to minimize the number of lightning strokes that terminate on the phase conductors. The remaining and the vast majority of strokes and flashes now terminate on the overhead ground wires. A stroke that so terminates forces currents to flow down the tower and out on the ground wires. Thus voltages are built up across the line insulation. If these voltages equal or exceed the line CFO, flashover occurs. This event is called a backflash. The origin of the word backflash is interesting. In the laboratory, an impulse is normally applied to the conductor and flashover occurs from the conductor to ground. For the backflash, the highest voltage is on the tower rather than on the conductor and flashover appears to occur from the tower or ground to the conductor. The flashover is backwards from that in the laboratory, thus the term backflash.

In Chapter 9, the voltages that occur for a stroke to the tower were derived. Additionally, the beneficial effect of the adjacent towers in reducing these voltages was discussed and an additional equation was formulated. As a review, Fig. 1 shows the location of the crest voltages and illustrates the wave shapes. From Chapter 9,

$$\begin{aligned} V_{TT} &= (K_{sp}K_{TT})I \\ V_{TA} &= (K_{sp}K_{TA})I \\ V_F &= R_e I \end{aligned} \quad (1)$$

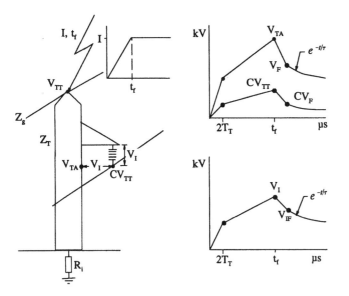

Figure 1 Surge voltages at the tower and across the insulation.

and the current through the footing resistance is

$$I_R = \frac{R_e}{R_i} I \qquad (2)$$

where

$$K_{TT} = R_e + \alpha_T Z_T \frac{T_T}{t_f}$$

$$K_{TA} = R_e + \alpha_T Z_T \frac{T_A}{t_f}$$

$$K_{SP} = 1 - \alpha_R(1 - \alpha_T)\left[\left(1 - 2\frac{T_s}{t_f}\right) + \alpha_R \alpha_T\left(1 - 4\frac{T_s}{t_f}\right) + (\alpha_R \alpha_T)^2\left(1 - 6\frac{T_s}{t_f}\right) + \cdots\right] \qquad (3)$$

For these equations:

$$R_e = \frac{Z_g R_i}{Z_g + 2R_i} \quad \alpha_T = \frac{Z_T - R_i}{Z_T + R_i} \approx \frac{Z_g - 2R_i}{Z_g + 2R_i} \quad \alpha_R = \frac{Z_g}{Z_g + 2R_i} \qquad (4)$$

Also, from Chapter 9, the tail of the voltages can be conservatively approximated by a time constant τ:

$$\tau = \frac{Z_g}{R_i} T_s \qquad (5)$$

That is, the equation for the tail of the surge is

$$e_{TT} = V_F e^{-(t-t_f)/\tau} \tag{6}$$

To be complete, the definitions of the variables (see also Fig. 1) are

- t_f = time to crest of the stroke current, μs
- C = coupling factor
- Z_T = surge impedance of the tower, ohms
- Z_g = surge impedance of the ground wires, ohms
- T_T = tower travel time, μs
- T_A = tower travel time to any location on the tower A, μs
- T_S = travel time of a span, μs
- I = stroke current, kA
- I_R = current through footing of struck tower, kA
- R_0 = measured or low-current footing resistance, ohms
- R_i = impulse or high-current footing resistance, ohms
- τ = time constant of tail, μs

2 A FIRST ESTIMATE OF THE BACKFLASH RATE

To provide a first estimate of the backflash rate, the BFR, examine Fig. 1. The surge voltage on the ground wires produces a surge voltage on the phase conductor equal to the coupling factor C times the voltage on the ground wires, or CV_{TT}. Also note that the voltage V_{TA} is located on the tower opposite the phase conductor. Therefore the crest voltage across the insulation V_I is

$$V_I = I[K_{TA} - CK_{TT}]K_{SP} \tag{7}$$

Also note that the crest voltage V_{IF} across the insulation caused by the footing resistance is

$$V_{IF} = (1 - C)R_e I \tag{8}$$

This is also the voltage if the tower component of voltage is neglected. As shown in Ref. 1 by the use of field theory, the voltage across the air gap is equal to the voltage across the insulator string.

For a flashover to occur, the voltage across the insulator V_I, must be equal to or greater than the CFO of the insulation. This CFO will differ from the CFO for a 1.2/50-μs impulse, since the waveshape is significantly different. Therefore call this CFO_{NS} or the nonstandard CFO. Replacing V_I of Eq. 7 with the CFO_{NS}, the current obtained is the critical current I_c at and above which flashover occurs, i.e.,

$$I_c = \frac{CFO_{NS}}{(K_{TA} - CK_{TT})K_{SP}} \tag{9}$$

Since K_{TT} is in many cases approximately equal to K_{TA}, then approximately,

$$I_c = \frac{\mathrm{CFO_{NS}}}{(1-C)K_{TT}K_{SP}} \tag{10}$$

as was assumed in Chapter 9 when discussing corona.

The probability of a flashover is the probability that the stroke current I equals or exceeds the critical current I_c, or

$$\mathrm{Prob}(I \geq I_c) = P(I_c) = \int_{I_c}^{\infty} f(I)\,dI \tag{11}$$

The backflash rate BFR is this probability times the number of strokes, N_L, that terminate on the ground wire, or

$$\mathrm{BFR} = N_L P(I_c) \tag{12}$$

where from Chapter 6,

$$N_L = N_g \frac{(28h^{0.6} + S_g)}{10} \tag{13}$$

where h is the tower height (meters), S_g is the horizontal distance between the ground wires (meters), and N_g is the ground flash density (flashes/km²-year), and therefore N_L is in units of flashes per 100 km-year. Thus the BFR is in terms of flashovers per 100 km-years.

Equation 12 thus represents the simple equation to estimate the BFR. However, there are effects that to this point have not been considered, such as

1. Strokes to Span: The equations developed for the voltage across the insulation are based on a stroke to the tower. Since strokes can terminate at any point on the ground wire, the effect of strokes terminating along the span must be considered.

2. Footing Resistance: As discussed previously, the footing resistance to be employed in the above equations for the struck tower is the impulse resistance R_i and not the measured resistance R_0. That is, when a high current flows through the soil, breakdown or flashover of the soil particles occurs, which essentially increases the dimensions of the ground rod or the footing, which in turn decreases the resistance. Therefore some method is necessary to estimate this impulse resistance.

3. Number of Phases and Power Frequency Voltage: To this point only one phase conductor has been assumed. Consideration should be given to the effect of more than a single phase line. This effect is tied to the effect of the power frequency voltage. That is, in addition to the surge voltage that occurs across the line insulation, a power frequency voltage exists. As will be shown, the magnitude of this power frequency voltage depends on the time instant of the stroke with respect to the power frequency voltage and may add or subtract from the voltage across the insulator. Since at any one time instant, the power frequency voltage differs for each of the phases, alternate phases may flash over and thus the BFR is affected—it will be increased.

4. $\mathrm{CFO_{NS}}$: As discussed previously, the waveshape of the voltage across the line insulation significantly differs form the standard 1.2/50 µs impulse upon which the standard CFO is based. A method is needed to estimate this nonstandard CFO.

5. t_f, the Time to Crest of the Stroke Current: As noted from the equations, the crest voltages are a function of the time to crest. So what value of t_f should be used? From Chapter 6, the time to crest is a probabilistic value, and further, it is dependent on the stroke current, i.e., the time to crest is conditioned on the magnitude of the stroke current, a conditional probability density function.

6. Corona: In Chapter 9, it was shown that the effect of corona was to decrease the ground wire surge impedance, thus increasing the coupling factor, which per the above equations appears to decrease the BFR. However, in mitigation of this apparent decrease in the BFR, corona effects only occur on the front of the surge voltage and have no effect on the tail of the surge.

Thus the job is to determine these effects and, if required, to modify the equations presented. This will create some complications and make the estimate of the BFR more difficult. In fact, the inclusion of all these effects complicates the estimation to the point that the calculation can only be effectively performed using a computer program. However, in some cases, many of these effects can be ignored, which fortunately leads to some simplified equations that can be done by hand calculation.

Now consider the above effects in the order given.

3 EFFECT OF STROKES WITHIN THE SPAN

This subject is considered in detail in Appendix 1 of this chapter. Therefore only a summary is given here.

A stroke terminating on the shield wire within the span produces voltage across the air insulation between the shield wire and the phase conductor and also across the air–porcelain insulation at the tower. Although the voltage across the span insulation exceeds that across the tower insulation, the span insulation strength exceeds that of the tower. Thus dependent on the relative voltages and insulation strengths, flashover can occur either across the span or across tower insulations.

3.1 Flashovers Within the Span

Considering a stroke to the shield wire, as defined in Figs. 2 and 3, the voltage at the stroke terminating point attempts to reach a crest voltage of $Z_g I/2$. However, reflections from adjacent towers reduce this voltage, provided t_f is greater than $2(T_S - T_{ST})$. The maximum voltage occurs at the stroke terminating point, and voltages decrease as the distance from the stroke terminating point increases, reaching a minimum at the tower.

To obtain an approximation of the expected number of span flashovers as opposed to tower flashovers, assume that the waveshapes of all the voltages are identical so that the nonstandard critical flashover voltage, CFO_{NS}, is a linear function of the gap spacing. For a typical 500-kV line, the minimum strike distance at the tower is 3.35 m, while the shield wire to phase conductor spacing varies from 9.2 m at the tower to 11.6 m at midspan; see Fig. 2. Thus the ratio of insulation strength is 3.5. For a stroke terminating at $T_{ST}/T_S = 0.20$, as defined in Fig. 3, and for $t_f = 2.0\,\mu\text{s}$ and $R_i = 20$ ohms, the ratio of the voltages at midspan to the voltage at the tower is 2.4. Thus, for this case, flashover would occur at the tower. If all stroke-terminating

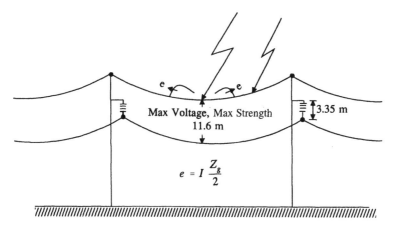

Figure 2 For strokes within the span, the maximum voltage occurs within the span. However, the maximum strength is also within the span.

points are considered, for $t_f = 2.0\,\mu s$, approximately 16% of the strokes result in span flashover. For $t_f = 4.0\,\mu s$, span flashover is reduced to about 2%.

Another phenomenon further reduces the probability of span flashover. At high overvoltages, predischarge currents flow from the shield wire to the phase conductor, producing a voltage on the phase conductor that decreases the voltage across the span insulation [2, 3]. Although no quantitative calculation will be made, suffice it to note that this phenomenon inhibits flashover.

Thus, considering both the example calculations and the predischarge current phenomenon, although flashovers within the span are possible, they appear to be insignificant compared to flashovers at the tower.

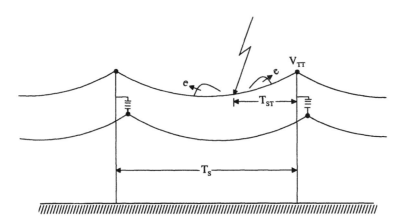

Figure 3 Definition of the stroke-terminating point.

3.2 Flashovers at the Tower Caused by Strokes to the Shield Wire

For a stroke terminating within the span, the crest voltage at the tower, in terms of K_{TT}, is
If $2T_T \leq t_f \leq 2(T_T - T_{ST})$, then

$$V_{TT} = IK_{TT} \tag{14}$$

where

$$K_{TT} = R_e + \alpha_T Z_T \frac{T_T}{t_f} \tag{15}$$

If $2(T_S - T_{ST}) \leq t_f \leq 2T_S$, then the crest voltage occurs either at t_f or at $2(T_S - T_{ST})$, dependent on the value of the tower footing resistance, i.e., if the crest voltage occurs at t_f, then

$$K_{TT} = R_e \frac{2(T_S - T_{TS})}{t_f} + \alpha_T Z_T \frac{T_T}{t_f} \tag{16}$$

or if the crest voltage occurs at $2(T_S - T_{ST})$, then

$$K_{TT} = \frac{R_e}{Z_g + 2R_i} \left[2R_i + Z_g + \frac{2(T_T - T_{ST})}{t_f} + \frac{4\alpha_T R_i}{R_e} Z_T \frac{T_T}{t_f} \right] \tag{17}$$

As noted by comparing these equations to Eq. 15, the voltage resulting from a stroke within the span equals that produced by a stroke to the tower only when $t_f \leq 2(T_S - T_{ST})$. Therefore the voltage produced at the tower by a stroke within the span is equal to or less than that produced by a stroke to the tower.

Thus (1) flashovers within the span can be neglected, and (2) voltages produced at the tower for strokes terminating within the span produce lower voltages than voltages produced by a stroke to the tower. Therefore if the BFR is based on only strokes to the tower, the BFR will be significantly greater than if strokes within the span are considered. Since it is desirable to base the BFR on strokes terminating at the tower, some adjustment must be made to this BFR. In Appendix 1, this is analyzed, and the conclusion is that the BFR can be based on the BFR for strokes to the tower provided than this BFR is multiplied by 0.6. That is, Eq. 12 should be modified to

$$\text{BFR} = 0.6 N_L P(I_C) \tag{18}$$

4 IMPULSE RESISTANCE OF GROUND ELECTRODES

The purpose of this section is to present simplified equations to estimate the impulse or high-current resistance of concentrated grounds and to attempt to examine the impulse resistance of counterpoises. A more through analysis of concentrated grounds is in Appendix 2.

4.1 Concentrated Grounds—Ground Rods

Concentrated grounds are defined as ground rods or counterpoises within about 15 meters of the base of a tower.

High magnitudes of lightning current, flowing through the ground resistance, decrease the resistance significantly below the measured low-current values. Although this has been known for many years, most lightning performance estimating methods, while they acknowledge this fact, have not provided a means of estimating the impulse resistance, primarily because of the lack of data and the lack of an adequate simplified calculation procedure. Within the CIGRE Working Group 33.01. Weck [4] analyzed measured impulse resistance data to arrive at a simplified method. It is the purpose of this section to present this simplified method as presented in CIGRE Technical Brochure No. 63 [5].

For high currents, representative of lightning, when the gradient exceeds a critical gradient E_0, breakdown of soil occurs. That is, as the current increases, streamers are generated that evaporate the soil moisture, which in turn produces arcs. Within the streamer and arcing zones, the resistivity decreases from its original value, and as a limit approaches zero and becomes a perfect conductor. This soil breakdown can be viewed as increasing the diameter and length of the rod as shown in Fig. 4, which shows the initial limit or area. As the ionization increases, the shape of the zone becomes more spherical, as illustrated in Fig. 4, which also shows the final limit or area. Figure 5 illustrates that the hemispherical distribution occurs for multiple ground rods.

Assuming a hemispherical electrode of radius r_0 per Fig. 6, the breakdown of soil begins when the gradient at the hemisphere surface exceeds the critical gradient E_0. The current required to achieve this gradient is denoted as I_g and is determined by the equation

$$I_g = \frac{1}{2\pi} \frac{\rho E_0}{R_0^2} \tag{19}$$

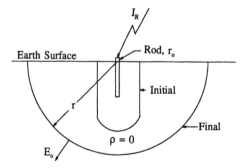

Figure 4 At high currents, rod becomes a hemisphere.

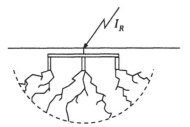

Figure 5 Multiple rods act as a hemisphere.

As before, the low-current or measured resistance is R_0 and the soil resistivity is ρ (ohm-meters); E_0 is approximated at 400 kV/m.

For currents greater than I_g, breakdown of soil continues and expands, reaching a radius r. Within this area described by r, the soil resistivity is considered zero, the soil being a perfect conductor. Thus the resistance under high currents is simply the resistance of a hemisphere of radius r. Therefore, the resistance R_i becomes

$$R_i = \frac{R_0}{\sqrt{I_R/I_g}} \qquad (20)$$

The plot of the resistance as a function of the current is illustrated in Fig. 7. The low-current resistance R_0 is maintained until the current exceeds, I_g, after which the resistance is given by Eq. 20.

Returning to a single rod, since the dimensions of the rod permit the gradient E_0 to be achieved essentially instantaneously, the decrease in resistance also occurs instantaneously. However, this decrease is not rapid until the streamer and arcing zones approximate a hemisphere. The plot of the resistance is shown in Fig. 8, where the change from an increased dimensioned rod to a hemisphere occurs at a current of I_g. For a rod or rods—or any concentrated grounds—this characteristic can be approximated by the equation

$$R_i = \frac{R_0}{\sqrt{1 + (I_R/I_g)}} \qquad (21)$$

where I_g is given by Eq. 19.

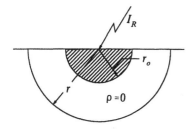

Figure 6 The hemisphere electrode.

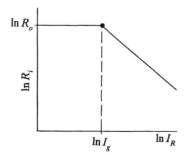

Figure 7 Impulse resistance of hemisphere.

As an example, let $E_0 = 400 \, \text{kV/m}$, $R_0 = 40$ ohms, $I_R = 100 \, \text{kA}$, and $\rho = 800$ ohm-meters. Then I_g is 31.8 kA and $R_i = 19.7$ ohms, about a 50% reduction. The value of R_0 can be estimated by equations presented in Appendix 3.

4.2 Counterpoises

Counterpoises are horizontal conductors buried in the earth at a depth of about 1 meter and connected to the base of the tower. The term counterpoise was coined because counterpoises were believed to be effective because of their capacitive coupling to the phase conductors. However, it was found that this coupling was only in the range of 3 to 10% and that the effectiveness was a result of the decreased grounding impedance provided by the counterpoises.

In about 1934, tests were performed on counterpoises, and analytical investigations were made. Unfortunately, these tests used currents of less than 100 amps and thus did not consider the decrease in resistance caused by high currents. No additional tests have been made to date to consider less this fundamental condition. Therefore, this presentation will begin with the results of these previous tests followed by a brief examination of possible methods to take into account the effect of high currents in reducing the ground impedance.

Figure 9 illustrates the low-current phenomenon. Voltage and current waves traveling down the tower impinge on the combination of the concentrated ground R_i and the counterpoises, resulting in waves of current e_c and voltage i_c traveling out along the counterpoises at about 1/3 the speed of light. Symbolically, these waves of

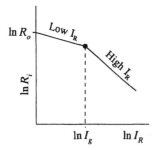

Figure 8 Impulse resistance of concentrated grounds.

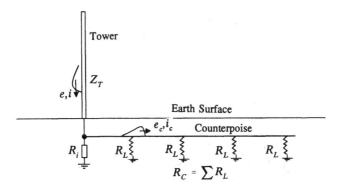

Figure 9 Waves of current and voltage travel outward on the counterpoise and decrease the footing resistance as a function of time.

current and voltage meet and reflect from the leakage resistance along the counterpoises and thus with a time delay decrease the total tower footing resistance. The major components of the counterpoise impedance are

1. Initially, at time zero, the counterpoise appears as a surge impedance Z_C of about 120 to 220 ohms. Usually a value of 150 ohms is assumed.
2. At a time equal to twice the travel time T_C of the counterpoise, the impedance is reduced to the total leakage resistance of the counterpoise, R_C.

Bewley [6] represented these components by an equivalent circuit for a single counterpoise as shown in Fig. 10. The response of this circuit to a rectangular current wave is

$$Z'_c = \frac{e}{i} = R_C + (Z_C - R_C)e^{-t/\tau_c} \tag{22}$$

where

$$\tau_c = \frac{L}{Z_C - R_C} \tag{23}$$

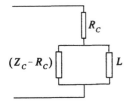

Figure 10 Bewley's [6] equivalent circuit of a single-counterpoise.

To determine the inductance L, Bewley set the value of L such that the transient was 95% complete in two travel times, i.e., $2T_c$, where T_c is equal to the counterpoise length divided by the velocity of propagation of 1/3 that of light. Thus

$$L = \frac{2}{3} T_C (Z_C - R_C) \qquad (24)$$

To illustrate the effect of counterpoise length and the number of counterpoises emanating from the tower, Bewley assumed a 300-m counterpoise having a total leakage resistance R_C of 10 ohms and a surge impedance of 150 ohms. The transient impedance of this counterpoise per Eq. 22 is shown as $N = 1$ in Fig. 11. This 300 m length was then broken into two counterpoises of 150 m length ($N = 2$), into three counterpoises of 100 m length ($N = 3$), and into four counterpoises of 75 m length ($N = 4$). The advantages of breaking the 300-m length into several counterpoises are (1) a reduced surge impedance and (2) a shorter time constant to the final leakage resistance. Thus the lesson is that improved performance is realized when a single length is broken into several smaller lengths. Bewley further points out that the leakage impedance should be significantly less than the surge impedance.

As stated previously, these observations were concluded from tests using low currents. They must obviously be modified when high currents are involved. To date, there exists no agreement as to the necessary modifications, nor are there any high-current test results. Although the modifications must await test results, one suggestion is to subdivide the counterpoises into 30-m sections or lengths and to apply the equations for concentrated electrodes, taking into account the travel times between the sections. This suggestion has not yet been implemented in any method of calculating the BFR.

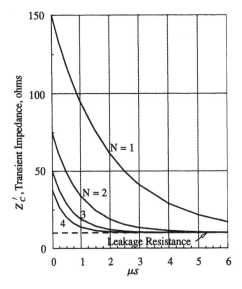

Figure 11 Effect of length and number of counterpoises [6].

4.3 Grounding Considerations

Obviously, the BFR is reduced by decreasing the footing resistance. Some methods of tower grounding are shown in Fig. 12. A so-called butt-wrap around a wooden pole is one form, in some cases a steel plate is used at the bottom of the pole. For augured footing for a steel tower, the lower figure shows a footing used at 500 kV. A cage of reinforcing bars is lowered into the augured footing and attached to the leg stub angle. In this case, because of the fear of breakage of the concrete, the hole was lined with tar paper and an additional copper wire was placed outside of the tar paper ending in a coil at the bottom of the footing. Following this placement, the concrete was poured.

If the normal tower grounding does not result in the desired value of footing resistance, supplemental ground in the form of driven rods or counterpoises can be used. Since the counterpoise may be viewed as a horizontal ground rod (or the ground rod as a vertical counterpoise), in general, for the same lengths, both methods of improving grounding should provide the same result. Since the counterpoise can be used in lengths greater than a ground rod can be driven, ground rods are generally employed for low-resistivity soils while counterpoises are generally used in high-resistivity soils.

Soil resistivity and thus footing resistance depends largely on the water content of the soil and the resistivity of the water; these values will vary considerably with weather conditions. In addition, soil resistivities in localized areas may greatly exceed these average values. For thunderstorms that follow an extended dry season, the footing resistance will be high until the moisture penetrates the soil. Thus for these conditions, flashovers are more probable at the beginning of a storm.

In the following subsections ground-rod and counterpoise resistances are discussed assuming a constant value of soil resistivity. However, rarely is the soil resistivity constant with depth or location. Thus the equations and curves are primarily used for planning a grounding installation. The actual number and length (depth) of the rods or counterpoises must be decided during installation and field

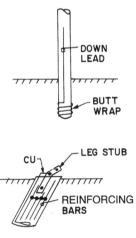

Figure 12 Methods of tower grounding.

measurement. Frequently, the installation crew is given an objective as to (1) the target value of footing resistance and (2) the maximum number of rods or maximum number and length of counterpoises to install.

Ground Rods. As shown in Appendix 3, for a constant resistivity soil, the driven depth of rods should be from 2 to 6 meters. However, since soil resistivity is seldom constant with depth, rods are frequently driven to greater depths. Multiple rods decrease the resistance. However, mutual effects exist between the rods, and the benefit decreases as more rods are added. Thus three or five rods, spaced about 3 meters or more apart, is normally the limit. The diameter of the rod is not important; any rod diameter that is mechanically suitable is acceptable from an electrical viewpoint.

Counterpoises. Per Appendix 3, for a constant soil resistivity, counterpoise length should be limited to about 50 meters. Additional counterpoises decrease the resistance, but spacings should be in the range of about 10 meters. Some typical arrangements of counterpoises are shown in Fig. 13, where the counterpoise is brought to the edge of the right-of-way to decrease any mutual effects. The number of parallel counterpoises, on each side of the tower, should be limited to about three. The depth of burial is usually set so that a farmer's plow will not contact or disturb the counterpoise, a depth of about 1 meter. The counterpoise wire is normally copperweld, #2AWG, although steel has been used successfully. Use of aluminium is not recommended, since this material will vanish in a few years.

Effect of Soil Ionization on Spacing. Using the equations from Appendix 2, the final ionized diameter D for the ground rod can be approximated by the equations for sphere electrodes, i.e.,

$$D = \sqrt{\frac{2\rho I_R}{\pi E_g}} \qquad (25)$$

Thus ionized diameters can range from 5 to 10 meters. To obtain the maximum effectiveness of parallel rods, spacings should be increased to approximately 5 meters.

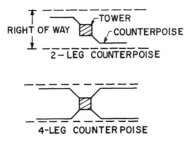

Figure 13 Counterpoise locations within the right-of-way.

5 EFFECT OF POWER FREQUENCY VOLTAGE AND NUMBER OF PHASES

(See Appendix 4 for a more thorough analysis of this effect.) To this point in the development, only a single phase has been considered with a coupling factor C from the ground wires to this conductor. Now consider a three-phase line with coupling factors C_A, C_B, and C_C as illustrated in Fig. 14. Also the voltages on the tower will be different for each of the phases. That is, there are now three voltages, V_{TA}, V_{TB} and V_{TC}. Therefore the *surge voltages* across the line insulation for phases A, B, and C, namely V_{IA}, V_{IB}, and V_{IC}, are now given by the equations

$$V_{IA} = (K_{TA} - C_A K_{TT})K_{SP}I$$
$$V_{IB} = (K_{TB} - C_B K_{TT})K_{SP}I \qquad (26)$$
$$V_{IC} = (K_{TC} - C_C K_{TT})K_{SP}I$$

Next, consider the power frequency voltage. Letting the crest line-neutral power frequency voltage equal V_{LN}, then

$$V_{IA} = (K_{TA} - C_A K_{TT})K_{SP}I + V_{LN}\sin\omega t$$
$$V_{IB} = (K_{TB} - C_B K_{TT})K_{SP}I + V_{LN}\sin(\omega t - 120°) \qquad (27)$$
$$V_{IC} = (K_{TC} - C_C K_{TT})K_{SP}I + V_{LN}\sin(\omega t + 120°)$$

Assuming that all CFOs for all phases are equal and setting these voltages across the insulation to this CFO, i.e., CFO_{NS}, the currents become the critical currents for each phase, i.e.,

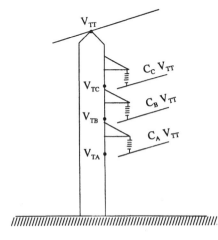

Figure 14 Three-phase line with different coupling factors and tower voltages.

$$I_{CA} = \frac{CFO_{NS} - V_{LN}\sin\omega t}{(K_{TA} - C_A K_{TT})K_{SP}}$$

$$I_{CB} = \frac{CFO_{NS} - V_{LN}(\sin\omega t - 120°)}{(K_{TB} - C_B K_{TT})K_{SP}} \tag{28}$$

$$I_{CC} = \frac{CFO_{NS} - V_{LN}(\sin\omega t + 120°)}{(K_{TC} - C_C K_{TT})K_{SP}}$$

Therefore, the calculation of the BFR becomes more complex, since there is more than a single critical current. Obviously, the smallest of these critical currents is controlling. That is, if one phase flashes over, this is counted as a line flashover.

Per Eq. 28, the critical current for each phase depends on the phase angle of the power frequency voltage at the instant that the stroke terminates on the overhead ground wire and also is dependent on the coupling factor and the voltage on the tower opposite the phase conductor. Assuming a double circuit tower with a vertical phase configuration, K_{TC} is less than K_{TB}, which is less than K_{TA}, thus indicating that I_{CA} should be the smallest. But C_C is less than C_B, which is less than C_A, thus indicating that I_{CC} should be the smallest. Thus all these factors should be considered.

To illustrate with a detailed calculation, a 115-kV, single-circuit, horizontal configured line is considered to have the following characteristics:

Ground flash density = 6.0 flashes/km²/yr   Nominal system voltage = 115 kV
Surge impedances: Ground wire, 339 ohms; Tower, 170 ohms
Coupling factors: phase A/B/C = 0.331/0.386/0.331
Heights: Ground wires 57 ft; All phase conductors, 46 ft
Horizontal separation of ground wires: 12.5 ft
Span length: 750 ft   CFO: Footing resistance, $R_0 = 20$ ohms
Soil resistivity: 400 ohm-meters

The power frequency voltage was considered by calculating the critical current and BFR for each of the phases for instantaneous power frequency voltages determined for each of twelve 30° steps. The results are shown in Table 1. The first column gives the angle of the power frequency voltage for phase A. The other columns give the critical currents and BFRs. The phase that flashes over is the phase with the lowest critical current or with the highest BFR. To obtain the total BFR, the maximum BFRs at each time instant, those in the last column, are added and divided by the number of time steps. The total BFR is therefore 2.279/12 = 0.190 flashovers/100 km-year. The number of flashovers on A, B, and C phases are 5.5, 1.0, and 5.5, respectively. Dividing these by 12 results in the conclusion that 45.8% occur to phase A and 45.8% occur to phase C. Also, 8.3% occur to phase B, the middle phase. Note for 30°, the critical current and BFR for phases A and C are equal. Therefore each phase is assigned 1/2 flashover. This same example was used in Appendix 4 using thirty-six 10° steps. The resultant BFR is the same, but 13.9% of the flashover occurred to phase B, the remainder being divided between phases A and C. Thus there is a significant number of flashovers that occur to the B phase even though the coupling factor is higher than that for phases A or C, that is, that power frequency voltage overcomes the deficiency.

Table 1 115-kV Single-Circuit Line

	I_C critical current, kA			BFR, FO/100 km-yrs			Flashover phase	Line BFR
ωt, degrees	A	B	C	A	B	C		
0	188	243	166	.119	.029	.225	C	.225
30	175	247	175	.172	.026	.172	A&C	.172
60	166	243	188	.225	.029	.119	A	.225
90	163	230	202	.249	.039	.083	A	.249
120	166	216	211	.225	.058	.064	A	.225
150	175	201	215	.172	.086	.058	A	.172
180	188	190	212	.119	.114	.064	A	.119
210	202	186	202	.083	.127	.083	B	.127
240	212	190	188	.064	.114	.119	C	.119
270	215	201	175	.058	.086	.172	C	.172
300	212	216	166	.064	.058	.225	C	.225
330	202	231	163	.083	.039	.249	C	.249

This procedure to calculate the probability, although exact, is not desirable except for computer use, since it is difficult and laborious. More desirable is some approximate procedure that is simple to use and that will result in reasonable accuracy. This approximation and procedure, developed in Appendix 4, uses the following equation for the critical current.

$$I_C = \frac{\text{CFO}_{NS} - K_{PF} V_{LN}}{K_{SP}(K_{TA} - C_A K_{TT})} = \frac{\text{CFO}_{NS} - V_{PF}}{K_{SP}(K_{TA} - C_A K_{TT})} \qquad (29)$$

where from Appendix 4,

1. K_{PF} is the power frequency factor and varies with the phase configuration. For a vertical phase configuration, representative of double circuit towers, K_{PF} varies from about 0.25 to 0.55 dependent on the ratio of the nominal system voltage to the CFO. The average and recommended value is $K_{PF} = 0.40$. For a horizontal phase configuration, representative of a single circuit tower, K_{PF} varies from about 0.65 to 0.76. The average and recommended value is $K_{PF} = 0.70$. If unsure as to the value of K_{PF}, set $K_{PF} = 0.70$.
2. C_A is the lowest coupling factor.
3. K_{TA} is set to the same phase that is used for the coupling factor per 2.

As a reminder.

$$V_{PF} = K_{PF} V_{LN}$$

$$V_{LN} = \frac{\sqrt{2}}{\sqrt{3}} V_{L-L} \qquad (30)$$

where V_{L-L} is the nominal system voltage. For example, for a nominal system voltage of 230 kV, V_{LN} is 188 kV.

6 FINDING THE NONSTANDARD CFO, CFO$_{NS}$

The waveshape of the voltage across the tower insulation, shown by the solid line of Fig. 15, is composed of a power frequency voltage V_{PF}, a voltage produced by the tower footing resistance V_{IF}, and a voltage produced by the tower ΔV, i.e.,

$$\begin{aligned} \Delta V &= V_I - (1-C)V_F \\ &= K_{SP}[K_{TA} - CK_{TT}]I - K_{SP}(1-C)R_e I \\ &= \frac{\alpha_T Z_T}{t_f}(T_A - CT_T)K_{SP}I \end{aligned} \quad (31)$$

$$V_{IF} = V_F(1-C)I = K_{SP}R_e(1-C)I$$

The decreasing voltage on the tail of the surge voltage can be described by a time constant τ. A further simplified approximation of this voltage is shown by the dotted line.

As can be observed from Fig. 15, the waveshape of this voltage is far from the standard lightning impulse, a 1.2/50 µs wave. As discussed in Chapter 1, all data on the lightning impulse insulation strength assume the standard lightning impulse waveshape, and thus some method must be used to estimate the CFO for the waveshape of Fig. 15 and placed in terms of the CFO for the standard 1.2/50 µs waveshape.

Several methods of estimating the CFO of nonstandard waveshape impulses are in use today. In general, these methods can be divided into those that attempt to model directly the breakdown process and those that are derived from the breakdown process. The method considered in this section and used to establish the nonstandard CFO, CFO$_{NS}$, is the method that directly models the breakdown pro-

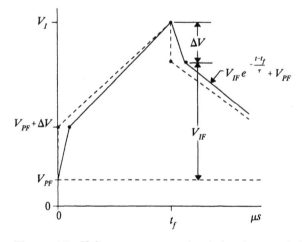

Figure 15 Voltage across tower insulation (— actual shape; --- approximation problems).

cess, called the leader progression model or LPM. The other methods are discussed in an appendix to Chapter 13. To gain a more complete understanding of the LPM method, first consider the breakdown process as illustrated in Fig. 16. Consider a gap with a spacing d upon which is applied an impulse voltage. Following the bridging of the gap by the streamers, the leader begins its progress across the gap when the voltage gradient exceeds a voltage gradient of E_0. As the leader proceeds, the voltage across the gap increases, and the distance from the leader tip to the ground electrode decreases, thus increasing the voltage gradient across the unbridged gap, distance x of Fig. 16. Because of the increase in voltage gradient, the velocity of the leader v increases. As this process continues, the velocity continues to increase until the leader reaches the ground electrode, at which time gap breakdown occurs.

Models of the LPM consist of a single equation for the velocity of propagation of the leader. Many equations have been proposed; a summary of these is contained in Ref. 7. The equation selected by CIGRE Working Group 33.01 for use in CIGRE Technical Bulletin 63 [5] for analysis of the voltage shown in Fig. 15 is

$$v(t) = k_L e(t) \left[\frac{e(t)}{x} - E_0 \right] \quad (32)$$

where $v(t)$ is the leader velocity, $e(t)$ is the voltage as described by Fig. 15, E_0 is the critical leader inception gradient, x is the distance of the unbridged gap, and k_L is a constant. The value of E_0 is primarily dependent on the gap configuration or, in more practical terms, on the critical flashover voltage gradient for the standard

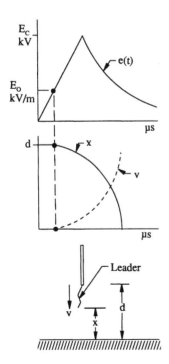

Figure 16 Breakdown process.

lightning impulse, whereas the value of k_L is primarily dependent on the upward curvature of the time-lag curve for the standard lightning impulse.

The calculation procedure consists of determining the velocity at each time instant, finding the extension of the leader for this time instant, determining the total leader length, and subtracting this from the gap spacing to find a new value of x. This process is continued until the leader bridges the gap. This method is called the leader progression model or LPM.

Although the LPM can be directly used within a computer program, a better method is to derive a regression equation from the LPM for the voltage waveshape as depicted in Fig. 15. To accomplish this task, the following insulation strength is assumed as obtained from Chapter 2 for the standard lightning impulse waveshape: (1) CFO gradient $= 560\,\text{kV/m}$; (2) breakdown voltage for a $3\,\mu s$ chopped wave $= 1.38$ times the CFO. Further, the $1.2/50\,\mu s$ waveshape is assumed to be given by a double exponential as

$$e(t) = AV_c(e^{-\alpha t} - e^{-\beta t}) \tag{33}$$

where the constants selected are

$$\alpha = 0.0146591 \qquad \beta = 2.46893 \qquad A = 1.03725 \tag{34}$$

and V_c is the crest voltage. The values of the constants of Eq. 32 are

$$\begin{aligned} E_0 &= 535.0\,\text{kV/m} \\ k_L &= 7.875 \times 10^{-7} \end{aligned} \tag{35}$$

for x in meters and $e(t)$ in kV. The resultant time-lag curve displays a $2\,\mu s$ chopped wave voltage of 1.66 times the standard CFO, which checks the value given in Chapter 2.

Using Eqs. 32 to 35 and employing regression analysis, the critical flashover voltage of the nonstandard surge voltage of Fig. 15, CFO_{NS}, was found to be best approximated by the equation

$$\frac{\text{CFO}_{NS}}{\text{CFO}} = \left[0.977 + \frac{2.82}{\tau}\right]\left[1 + \frac{\Delta V}{V_{IF}}\right]\left[1 - 0.2\left(1 + \frac{\Delta V}{V_{IF}}\right)\frac{V_{PF}}{\text{CFO}}\right]$$
$$\left[1 - 0.09\left(1 + \frac{10}{\tau}\right)\frac{\Delta V}{V_{IF}}\right]\exp\left[-\frac{\Delta V}{V_{IF}}\frac{t_f}{13}\right] \tag{36}$$

This equation was developed for values of τ between 10 and $100\,\mu s$, for values of $\Delta V/V_{IF}$ between 0 and 1.0, and for values of t_f between 0.5 and $5\,\mu s$.

Note that if the tower or the tower component of voltage ΔV is neglected, Eq. 36 becomes

$$\frac{\text{CFO}_{NS}}{\text{CFO}} = \left[0.977 + \frac{2.82}{\tau}\right]\left[1 - 0.2\frac{V_{PF}}{\text{CFO}}\right] \tag{37}$$

7 TIME TO CREST OF THE STROKE CURRENT, t_f

Before examining the effect of the time to crest, consider first that to this point, the shape of the front is assumed linear, whereas, per Chapter 6, the front is actually concave upward. If a concave upward front is assumed, the tower top voltage waveshape for a 100 kA stroke is shown by the solid line in Fig. 17. This shape of the concave upward front was developed using a minimum equivalent front of 2.7 µs and a front defined by the 30%/90% points of 6.0 µs. The dotted line curves of Fig. 17 show the tower top voltages for linearly rising fronts of 2.7 and 6.0 µs. As noted, the actual crest voltage is between the two dotted curves and that for the minimum equivalent front exceeds the actual voltage by about 9%. Thus the use of the minimum equivalent front is conservative and therefore will be used.

The equations for K_{TT} and K_I show that the voltage across the insulation increases as the time to crest of the stroke current decreases. This is caused by the tower component of voltage, ΔV. Thus the critical current increases as the time to crest increases. As an example, Fig. 18 shows this variation for a typical 230-kV and a typical 500-kV line. Thus, theoretically, all fronts should be considered. To do this, first the equation for the BFR should be changed from

$$\text{BFR} = 0.6 N_L P(I_c) \qquad (38)$$

to a conditional BFR, that is, a BFR given or assuming a specific time to crest, i.e.,

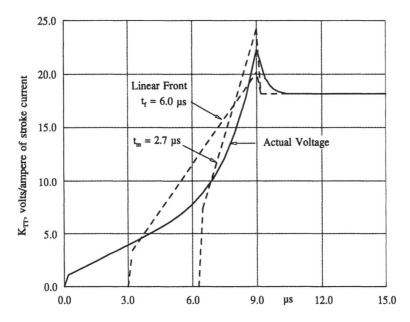

Figure 17 Comparison of tower top voltages.

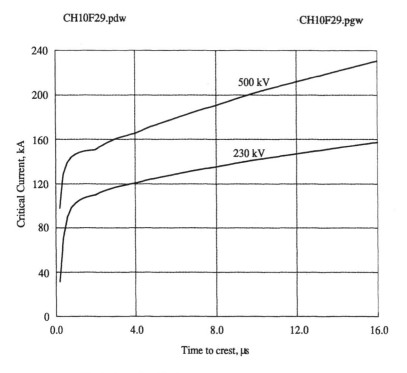

Figure 18 Variation of critical currents.

$$\text{BFR}|t_f = 0.6N_L \int_{I_c}^{\infty} f(I|t_f)\,dI$$
$$= 0.6N_L P(I_c|t_f) \tag{39}$$

Then, to consider all fronts, this equation must be integrated for all times to crest or

$$\text{BFR} = \int_0^{\infty} (\text{BFR}|t_f) f(t_f)\,dt_f \tag{40}$$

or in form of the double integral

$$\text{BFR} = 0.6N_L \int_0^{\infty} \int_{I_c}^{\infty} f(I|t_f) f(t_f)\,dI\,dt_f \tag{41}$$

where $f(I|t_f)$ is the conditional distribution of I given t_f. From Chapter 6, the parameters of this distribution are

$$M_{t_f|I} = 0.207 I^{0.53} \qquad \beta = 0.554 \tag{42}$$

The Backflash

where M is the median and beta is the log standard deviation. The distribution of the minimum equivalent time to crest is also given in Chapter 6 with the parameters

$$M_{t_f} = 1.28 \qquad \beta_{t_f} = 0.611 \tag{43}$$

The solution to these equations is illustrated in Fig. 19. First, using the curve of I_c vs. t_f, the probability that I_c is exceeded for any given value of t_f is calculated. This is the BFR|t_f. Now locate this value times $f(t_f)$ on the t_f axis. Continue to do this until a curve is established. The area under this curve times 0.6 is the total BFR.

Naturally, the objective is to simplify this procedure by using only a single value of the time to crest such that this value results in the same BFR as the complete procedure of considering all times to crest. This can be accomplished if the time to crest is coordinated with the median value of the time to crest for a value of the critical current, i.e., from Eq. 42,

$$t_f = 0.207 I_c^{0.53} \tag{44}$$

This is an iterative type of solution. To show this, consider the 230-kV line used in Fig. 18. The total BFR considering all times to crest is 1.15 flashovers per 100 km-years. This BFR can also be obtained using a constant time to crest of 2.6 μs for which the critical current is 116 kA. That is, using Eq. 44, t_f for a current of 116 kA is 2.6 μs. For the 500-kV line, the total BFR considering all times to crest is 0.21 flashovers/100 km-years. This BFR can also be obtained using a constant time to crest of 3.09 μs per Eq. 44, for a critical current of 164 kA.

Thus the conclusion is that the BFR can be calculated using only one value of time to crest, which is given by Eq. 44.

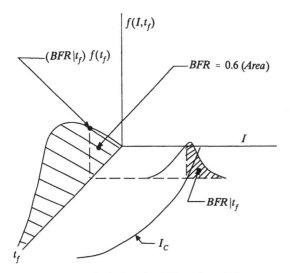

Figure 19 Calculating the BFR using all fronts.

8 EFFECT OF CORONA

As presented in Chapter 9, above the corona inception voltage and on the front of the surge, the ground wire surge impedance decreases, thus increasing the coupling factor. For example, depending on the assumptions, the ground wire surge impedance decreases by about 30 to 60% and thus the coupling factor increases by these same percentages. However, the critical current increases as $(1 - C)$ decreases. The decrease of $(1 - C)$ is about half of the increase of C.

To illustrate the effect of corona, consider Fig. 20 and assume that the noncorona coupling factor, C_0 is 0.30 while the corona coupling factor is 0.50. Assume that V_{TA} and V_{TT} are 10.0 and V_F is 6.0. In Figs. 20A and B, corona is not considered, so that the voltage V_I is 7.0 while the voltage V_{IF} is 4.2, thus giving a ΔV of 2.8. Now, consider corona. In Fig. 20C, the coupled voltage on the conductor has a crest of 5.0, but since there is no effect of corona on the tail, the voltage drops to $0.3(6) = 1.8$ as before. Thus the crest voltage V_I is decreased to 5.0, but the voltage V_{IF} is the same as before; see Fig. 20D. Thus ΔV has decreased to 0.8, demonstrating that the important portion of the surge, the tail, has not been altered.

Therefore corona is not expected to have a predominant effect on the BFR—although this will be confirmed later with a sensitivity study. Thus, with a little conservatism, corona can be neglected.

9 CALCULATING THE BFR—THE CIGRE METHOD

The method as developed to this point is the CIGRE method [5]. The BFR can be calculated by hand. However, in actual practice, as with most methods [8–10], the calculations are sufficiently complicated that a computer is necessary to eliminate inaccuracies and reduce boredom.

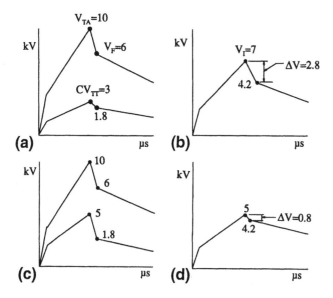

Figure 20 Effect of corona:. (a) and (b) without corona; (c) and (d) with corona.

The Backflash

The calculation of the BFR is an iterative process. Two DO loops are required, the outer loop on the time to crest of the stroke current and the inner loop on the impulse resistance, per Fig. 21. First, select the time to crest: for 115- to 230-kV lines, a front of 2.5 µs is appropriate; for 345 kV or above, a 4.0-µs front is suggested. Next, assume a value of R_i equal to about 50% of R_0 and solve for I_c. Then calculate I_C and I_R. Find R_i from I_R. If R_i is not within the desired degree of accuracy of the initially assumed value of R_i, iterate on R_i. When the value of R_i is satisfactory, calculate the median front for the value of I_c. If this front does not match the assumed front, iterate. Finally, calculate the BFR. The remaining objective is to determine whether this procedure can be simplified and if so, what the limitations are.

10 A SIMPLIFIED METHOD

If the tower component of voltage can be neglected, the calculation of the BFR is greatly simplified, since the time to crest of the stroke current is no longer a parameter. That is, the outer loop of Fig. 21 can be eliminated and a hand calculation method is now truly available. Neglecting the tower component of voltage has been suggested from Bewley [6] to the present [5]. The limitation of this method is considered in the next section. For the present, assume that in some cases this is viable. To clarify, following are the pertinent equations:

$$I_c = \frac{\text{CFO}_{NS} - V_{PF}}{R_e(1-C)} \qquad R_e = \frac{R_i Z_g}{Z_g + 2R_i} \tag{45}$$

$$\tau = \frac{Z_g}{R_i} T_S \qquad \text{CFO}_{NS} = \left(0.977 + \frac{2.82}{\tau}\right)\left(1 - 0.2\frac{V_{PF}}{\text{CFO}}\right)\text{CFO} \tag{46}$$

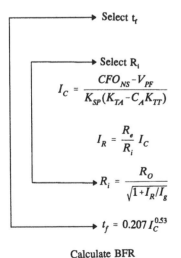

Figure 21 Flow diagram to calculate the BFR.

$$I_R = \frac{R_e}{R_i} I_c \qquad R_i = \frac{R_0}{\sqrt{1 + (I_R/I_g)}} \qquad (47)$$

$$I_g = \frac{1}{2\pi} \frac{E_0 \rho}{R_0^2} \qquad E_0 = 400 \text{ kV/m} \qquad (48)$$

$$N_L = (28 h^{0.6} + S_g) \frac{N_g}{10} \qquad \text{BFR} = 0.6 N_L P(I > I_c) \qquad (49)$$

An Example. Consider a 230-kV single-circuit line, $Z_g = 400$, CFO = 960 kV, $C = 0.30$, $R_0 = 50$ ohms, $\rho = 1000$ ohm-meters, $V_{PF} = K_{PF}$, $V_{LN} = 0.70(188) = 131$ kV, ground wire height = 30 meters, phase conductor height = 24 meters, span length = 300 meters, $N_g = 4$, spacing between ground wires $S_g = 5$ meters. Therefore

$$I_g = 25.5 \text{ kA} \qquad N_L = 88.2 \text{ flashes}/100 \text{ km-years} \qquad \frac{V_{PF}}{\text{CFO}} = 0.136$$

and iterating,

$P(I > I_c) = 0.235 \quad \text{BFR} = 0.6 \, (0.235) \, (88.2) = 12.4 \text{ flashovers}/100 \text{ km-years}$
Note that $\text{BFR}/N_L = 0.14$, i.e. 14% of the strokes result in flashover.
By computer program: CIGRE method: 13.2 flashovers/100 km-yr, a 17% error.

11 A SENSITIVITY ANALYSIS

The objective of this section is to answer several questions such as

1. Is the regression equation for the CFO_{NS} valid? How does it compare to the direct use of the leader progression model?
2. What are the limitations of the simplified method? How accurate is it when compared to the CIGRE method?
3. What is the effect of corona? Can it be neglected?
4. How important is the decrease in footing resistance from R_0 to R_i?
5. What is the effect of one vs. two ground wires?
6. Can underbuilt ground wires significantly decrease the BFR?
7. What about counterpoises?
8. What BFRs are expected for medium-voltage 34.5-kV lines?
9. What BFRs are expected for distribution lines? Can distribution lines be protected?
10. How does the IEEE method compare to the CIGRE method?

Table 2 An Example

R_i	R_e	τ	CFO_{NS}	I_c	I_R	R_i
25	22.2	18	1059	59.6	53.0	28.5
28.5	24.9	16	1077	54.2	47.4	29.6
29.6	25.8	15.5	1115	49.0	42.6	30.5
30.5	26.5	15.1	1087	51.6	44.8	30.1

The Backflash

To perform this analysis and answer these questions the high-voltage lines of Fig. 22, whose characteristics are given in Table 3, are used. The span length of these lines is 300 meters and the CFO is 1200 kV. The ground flash density is assumed as four flashes/km^2-year.

11.1 The CFO$_{NS}$ Regression Equation—How Good Is It?

The leader progression model was used to develop the regression equation for the critical flashover voltage for the nonstandard waveshape of surge voltage appearing across the line insulation. Using a 230-kV double-circuit line with tower heights of 35 and 70 meters and two ground wires, the BFR is calculated using the regression equation (solid lines) and also by computer calculation using the full leader progression model. The results are presented in Fig. 23. The dotted line for a tower height of 70 meters coincides with the solid line except for resistances of 5 to 15 ohms. The conclusions is that the LPM equation provides an excellent approximation of the LPM.

11.2 CIGRE Method vs. the Simplified Method

Figure 24 compares the CIGRE method with the simplified method. As expected, the comparison appears acceptable for the line with tower heights of 35 meters, but for tower heights of 70 meters the simplified method is inadequate. Of course since a computer program is available, the CIGRE method is always the proper tool.

11.3 Effect of Corona

Using the CIGRE method with and without corona, the BFR for the 230-kV double-circuit line with two ground wires is shown in Fig. 25. Since the tower component of voltage is greater for the 70 meter tower, the effect of corona is greater. The conclusion is that conservative values are obtained when neglecting corona. For high towers, the corona effect should be included in the calculations.

If now the CIGRE method with the effect of corona is compared to the simplified method as in Fig. 26, the comparison is excellent for the 35 meter tower; but for the 70

Table 3 Characteristics of Lines, Distances in Meters; See Fig. 22

System voltage, kV	h	y_A	y_B	y_C	S_g	S_a	S_b	Z_g	Z_T	C_A	C_B	C_c
Double-circuit lines												
230	35	29.5	24.1	18.7	5	8.5	11.0	379	190	.350	.248	.183
[a]230	35	29.5	24.1	18.7	0	8.5	11.0	600	190	.223	.158	.116
230	70	64.5	59.1	53.1	5	8.5	11.0	421	210	.420	.335	.283
[b]230	35	29.5	24.1	18.7	5	8.5	11.0	239	190	.441	.347	.307
Single-circuit lines												
230	20	15.6	15.6	15.6	6	5.5	5.5	340	170	.264	.301	.264
500	25	17.0	17.0	15.6	14	9.0	9.0	329	165	.232	.253	.232

[a] Single ground wire. [b] Underbuilt ground wire at $h = 12$ m at center of tower.

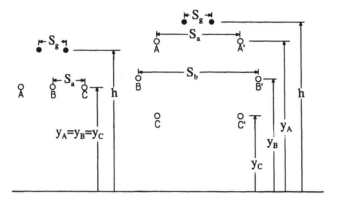

Figure 22 Definition of dimensions for HV towers; see Table 3.

meter tower, the simplified method shows a BFR that is too low. Thus the simplified method should only be used for towers whose heights are below about 50 meters.

11.4 Effect of Decrease of Resistance from R_0 vs. R_i

Using the CIGRE method, the BFR of the single-circuit 230-kV is shown in Fig. 27 as a function of R_0 with the ratio ρ/R_0 as a parameter. To illustrate the dramatic effect of the decrease of resistance with current, a curve labeled $R_i = R_0$ for which the footing resistance is not decreased is also presented. The curve labeled $\rho/R_0 = 20$ represents a reasonable value for concentrated grounds and is used for further illustrations of sensitivities. As shown in Fig. 28, for $\rho/R_0 = 20$ of Fig. 27, the stroke current times to crest vary between 6 and 2 μs since for low values of resistance, the critical current is large and thus the time to crest would be high. Figure 29, for the same case, shows that the impulse resistance is about 50% of the low-current value.

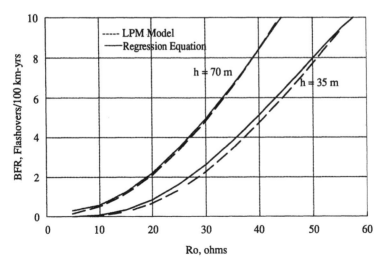

Figure 23 Comparison: LPM model and regression equation for CFO_{NS} for 230-kV double-circuit tower with two ground wires.

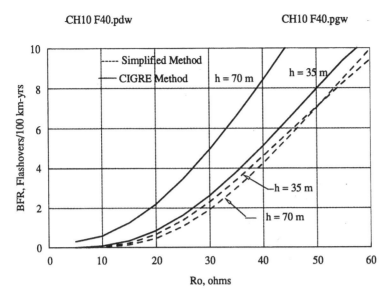

Figure 24 Comparison of BFRs for CIGRE method and simplified method, 230-kV double-circuit towers with two ground wires.

Figure 29 also shows that $\Delta V/V_{IF}$ ranges from about 0.4 to 0.1, which coordinates with the time to crest of Fig. 28. Also shown is the effect of changing the span length from 300 meters. At 600 meters, the BFR increases by 60%.

11.5 One vs. Two Shield Wires

For some applications, where the cost of two shield wires is not economically and technically justified, or where there is low ground flash density, a single shield wire can be used. This single wire increases the value of R_e, decreases the coupling factors,

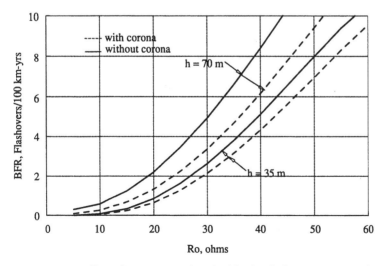

Figure 25 Effect of corona, 230-kV double-circuit lines, two ground.

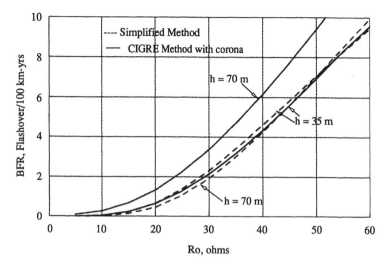

Figure 26 Comparison: CIGRE method with corona and simplified method, 230 kV double-circuit lines.

and thus increases the BFR. To illustrate, the curves of Fig. 30 have been constructed to compare one and two shield wires for a 230 kV double-circuit line and two shield wires for a single-circuit 230-kV line. Using one shield wire on the double-circuit line essentially doubles the BFR as compared to the two-shield-wire case.

11.6 Underbuilt Shield or Ground Wire

A ground wire located below the phase conductors cannot truthfully be called a shield wire, since it has no shielding function. Rather, its function is to increase the coupling factor to the lower phases, those phases that are most likely to flash over. For example, for the 230-kV double-circuit, two-ground-wire line with a shield wire

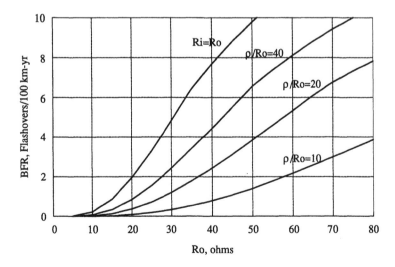

Figure 27 Effect of decrease to high-current footing resistance.

The Backflash

CH10F37.PDW   Ch10 F37B.pgw

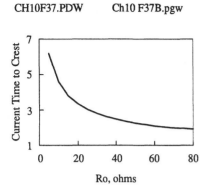

Figure 28 For $\rho/R_0 = 20$ of Fig. 27.

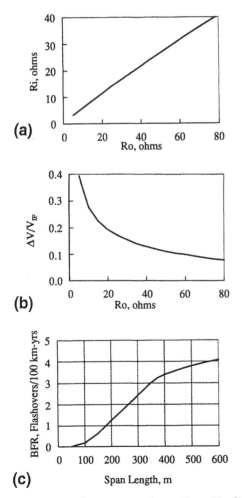

Figure 29 Some more using $\rho/R_0 = 30$ of Fig. 27.

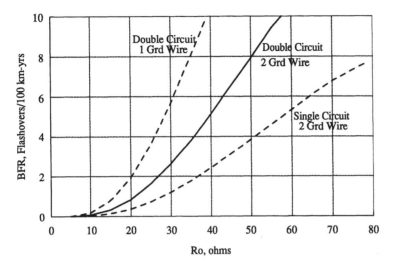

Figure 30 Two shield wires for the 230-kV double-circuit line with $h = 35$ m decrease the BFR, $\rho/R_0 = 20$.

height of 35 meters and coupling factors to the top, middle, and bottom phase of 0.350, 0.248, and 0.183, respectively, installing a ground wire at 12 meters above ground at the center of the tower increases these coupling factors to 0.441, 0.347, and 0.307, respectively. Thus all coupling factors are increased and are more uniform. Figure 31 shows the dramatic decrease in BFR for this case.

Viewing this result would indicate that all utilities should immediately use this remedial action. However, this is not the case. The reason appears to be that sagging of this ground wire is troublesome when considering the various sags of the phase

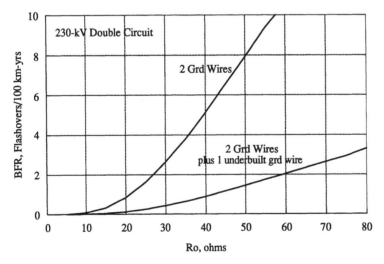

Figure 31 An underbuilt ground wire decreases the BFR, 230-kV double-circuit, $h = 35$ m, $\rho/R_0 = 20$.

conductors under various loading and atmospheric conditions. One hopes that, in the future, this problem can be overcome so that an experiment can occur.

11.7 Effect of Counterpoises

As discussed earlier, the true model of a counterpoise has to date not been thoroughly developed. If the counterpoise is modeled in the identical manner as the concentrated ground footing, then Fig. 32 applies. Figure 32 has been constructed for a concentrated footing resistance R_0 of 100 ohms and a soil resistivity of 2000 ohm-meters, which results in a BFR of 17.7 flashovers per 100 km-year. The effect of the BFR by adding one or two counterpoise on one side of the tower or two counterpoise on each side of the tower (4 counterpoise) is shown.

11.8 BFRs for 34.5 kV Lines

As an example, the 34.5-kV line of problem 3 of Chapter 2 is considered. This line is analyzed in problem 11 of Chapter 9. The dimensions in feet are shown in Fig. 33, and parameters are given in Table 4. The BFR as a function of R_0 is shown in Fig. 34 including the use of an underbuilt ground wire per Fig. 33. The simplified method provides a good approximation to the CIGRE Method. Corona does significantly affect the BFRs and an underbuilt ground wire, again, gives astounding results. Overall, the performance of a 34.5-kV line is in the range of from about 3 to 8 flashovers/100 km-years for 20 to 40 ohms of tower footing resistance, respectively. As for the high-voltage lines, the assumed ground flash density is 4.0 flashes/km^2-year. For other values, the BFRs are proportional. That is, for Florida, where the ground flash density approaches 10, the BFRs would increase to about 7.5 to 20. However, since low tower footing resistances can be obtained in Florida, for an R_0 of 10, the BFR would be about 1.5 flashovers/100 km-year, a favorable value. In conclusion, medium-voltage lines can be constructed to have a reasonable performance.

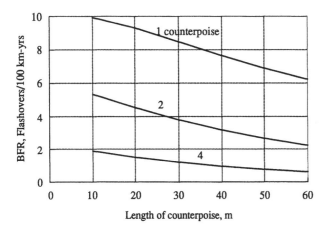

Figure 32 Effect of counterpoise, 230-kV double circuit, two ground wires, $h = 35$ m.

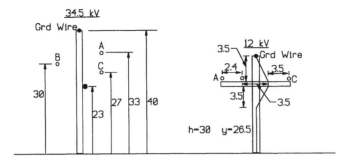

Figure 33 Dimensions in feet of a 34.5-kV and a 12-kV line; see Table 4.

Table 4 Characteristics of Lines, Distances in Feet; See Fig. 33

System voltage	Z_g	Z_r	C_A	C_B	C_C	CFO	span
34.5	511	557	.375	.219	.189	640	300
*34.5	286	557	.469	.391	.406	640	300
12	520	580	.271	.368	.254	300	100

* Underbuilt ground wire.

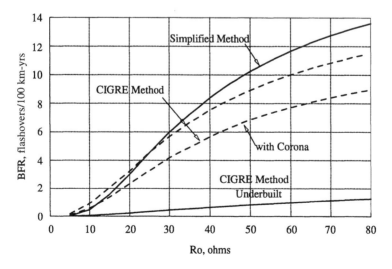

Figure 34 Performance of a 34.5-kV line.

11.9 The Performance of Distribution Lines

Because of the low insulation strength of distribution lines, virtually all strokes that terminate on the phase conductors of a distribution line will result in flashover. For example, assume that the surge impedance of the phase conductor is 500 ohms and that the CFO is 300 kV. Then a stroke having a current of 1.2 kA or greater will cause a flashover. However, in some cases, shield wires may provide some measure of protection from these direct strokes. Consider the distribution line of Fig. 33 and Table 4, the performance of which is shown in Fig. 35 for $\rho/R_0 = 20$ and $N_g = 4$. For footing resistances of 20 to 40 ohms, the BFR ranges from 6 to 7.5 flashovers/100 km-years. Note that the number of strokes that terminate on the line is $N_L = 42.3$ strokes/100 km-years. Thus only 14 to 18% of the strokes that terminate on the ground wire result in flashover. If, however, the CFO is less 300 kV of Table 4, the BFR increases as shown in Fig. 36 for $R_0 = 20$ ohms. The primary reason for the relatively low BFRs of the distribution line is the short span length and the relatively low height of the line. The short span length reduces the voltage across the insulation, and the low heights decrease the number of strokes that terminate on the line.

Distribution lines are also subjected to overvoltages induced by strokes that terminate adjacent to the line. In fact, most of the flashovers that occur are caused by these induced voltages. This is discussed further and the flashover rate is estimated in Chapter 15. The recent guide by the IEEE Working Group on distribution lines [11] is an excellent source of information.

11.10 Comparison: CIGRE vs. IEEE Methods

Another recognized and frequently used method for estimating the BFR is the IEEE Method [9, 10, 12]. This method is essentially that formulated by Anderson [8] in 1982. The IEEE Working Group accepted this method but subsequently changed some of the assumptions in an attempt to improve the predictions of the BFR. For

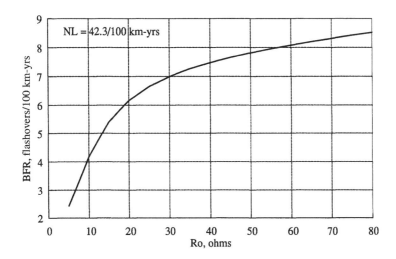

Figure 35 Performance of the 12-kV line of Fig. 32 for $N_g = 4$.

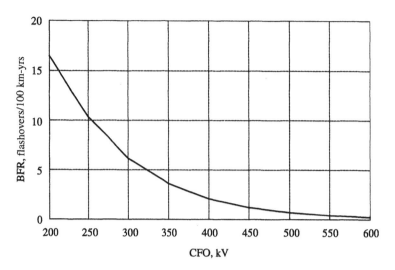

Figure 36 Performance of 12 kV if CFO is altered from assumed 300 kV.

an identical set of parameters, the calculated voltage across the insulation for these methods agrees to within 0.4 to 4.4%. Thus the differences in the methods are to be found in the assumptions of the parameters.

However, before discussing these parameters, a comparison of the predicted performance is presented in Figs 37 and 38. The curves of Fig. 37 are constructed for the 230-kV single-circuit line having a height of 20 meters, and the curves of Fig. 38 are constructed for the 230-kV double-circuit, two-ground-wire line having a height of 70-meters. The IEEE method does not consider the reduction of the footing

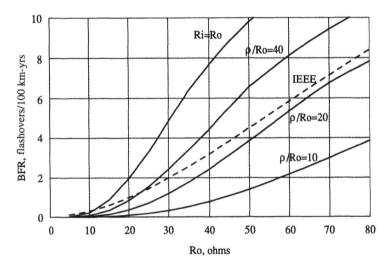

Figure 37 Comparison of CIGRE and IEEE methods for 230-kV single-circuit tower, two ground wires, $h = 20$ m.

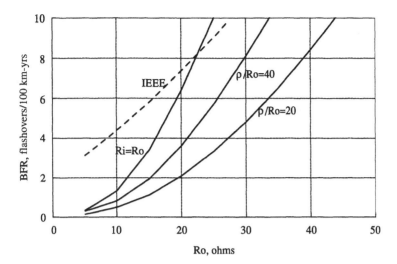

Figure 38 Comparison of CIGRE and IEEE methods for 230-kV double-circuit tower, $h = 70$ m.

resistance caused by high currents, and thus the BFRs are provided by a single curve. For the single-circuit line, the IEEE method compares favorably with the CIGRE method for $\rho/R_0 = 20$. Since the value of ρ/R_0 is considered an average value, the conclusion is that the IEEE and CIGRE methods should provide similar results for average conditions.

However, for the higher double-circuit line the comparison is not as good. The IEEE method appears to estimate much larger BFRs per Fig. 38. The differences in the methods are

1. High current footing resistances: The IEEE method does reduce the footing resistance caused by high currents.

2. Corona: The IEEE method assumes that the effect of corona occurs for the entire voltage waveshape, whereas the CIGRE method neglects all effects of corona. As discussed in this chapter, corona effects only occur when the voltage is increasing on the front of the voltage wave. Thus the IEEE method overestimates the effect of corona. In contrast, the CIGRE method should include the effects of corona. Thus both methods need improvement.

3. Stroke current time to crest: The IEEE method assumes that the stroke current time to crest is 2 µs for all stroke currents. This assumption becomes important for higher voltage lines for which the critical current is high—and therefore the time to crest would be much greater than 2 µs. Also, for low-voltage lines with low critical currents, the time to crest should be lower than 2 µs.

4. CFO_{NS}: The IEEE method employs the time lag curve for the standard lightning impulse waveshape to estimate the CFO_{NS}. The voltage across the insulation is examined and compared to the insulation strength at two time instants:

(a) At 2 μs, at which the crest voltage across the insulation occurs, which is compared to the 2-μs point of the time lag curve, which is calculated using a CFO gradient of 822 kV/m. Actually, the following equation for the time lag curve is used where $V(t)$ is the breakdown voltage.

$$V(t) = \left(400 + \frac{710}{t^{0.75}}\right) S \tag{50}$$

where t is the time to breakdown or time to flashover in μs and S is the strike distance in meters. This equation is considered valid for time up to 16 μs, at which the CFO gradient is 489 kV/m.

(b) At 6 μs if the span travel time is less than 1 μs and at twice the span travel time if the span travel time is equal to or greater than 1 μs. The voltage across the insulation is compared at this time instant to the values using Eq. 50. For example at 6 μs, $V(t)$ of Eq. 50 is 585 kV/m, while if the span travel time is 2 μs, $V(t)$ is 651 kV/m.

5. Probability Distribution of Stroke Current: In the IEEE method the stroke current distribution is provided by the simple equation

$$P(I > I_c) = \frac{1}{1 + (I_c/31)^{2.6}} \tag{51}$$

This equation is an approximation of the stroke current probability distribution presented in Chapter 6. However, Eq. 51 overestimates the probability for currents greater than 70 kA, and thus the IEEE method will produce significantly higher BFRs for critical currents higher than about 120 kA. Oppositely, Eq. 51 underestimates the probability for currents in the shielding failure zone, i.e., below about 10 kA. See Fig. 11 of Chapter 6.

6. Power Frequency Voltages: In the IEEE method, the variation of the magnitude of the power frequency voltages at the instant of stroke termination is handled by calculating the BFR for each 10° of phase angle as is shown in Table 3 of Appendix 4. As discussed previously, the CIGRE method approximates this by subtracting $K_P V_{LN}$ from the CFO. The IEEE method is superior in this regard, and the CIGRE method could be improved by this method. Of course, this can only be accomplished in a computer program.

7. Some Very Minor Differences: The velocity of propagation along the span and down the tower are assumed as 90% and 85% of the speed of light, respectively.

In conclusion to this comparison, it appears remarkable that both methods should provide a good comparison for low height lines—just to prove that various assumptions appear to cancel each other in this inexact science.

The IEEE method is contained in a computer program, FLASH 1.7, or FLSH17, and is available to IEEE members. The CIGRE method is contained in the computer program BFR which is included with this book. It is available to all.

12 COMMENTS

12.1 The Power Frequency Voltage and Multiphase Flashovers

It was previously recommended that K_{PF} be set to 0.70 for a horizontal configuration of phase conductors and to 0.40 for a vertical configuration of phase conductors, and that the lowest value of coupling factor should be used. The resulting BFR is an excellent approximation of the actual BFR when all phases are considered. The BFR so calculated is the total BFR. While it is true that the phase with the lowest coupling factor has the highest probability of flashover, it is not true that this is the only phase that will flash over. For example, for a horizontal configuration of phases, usually the coupling factors to the two outside phases are equal, and therefore the probabilities of flashover for each of these phases are equal. Even though the coupling factor to the center phase is higher than the coupling factors to the outside phases, because of the effect of the power frequency voltage, the probability of flashover is not zero. As shown in Appendix 4, for a typical line, flashovers will be divided into about 43% to each of the outside phases and the remaining 14% to the middle phase.

The total BFR of the line is the sum of single-phase flashovers, double-phase flashovers and three-phase flashovers. That is, the critical current calculated per the equations presented is the minimum critical current to cause a flashover. Stroke currents equal to or above this critical current will produce either (1) a single-phase flashover, (2) a two-phase flashover, or (3) a three-phase flashover. A single-phase flashover results in the phase conductor becoming a ground wire. Because of this, the coupling factors to the other two phases increase. Thus a new critical current can be calculated, which will be the critical current for either a two-phase flashover or a three-phase flashover. Next, when the flashover occurs to the second phase, this now becomes a ground wire and the coupling factor increases to the remaining phase. Again, a new critical current is calculated which is the critical current for a three-phase flashover. From these critical currents, the BFRs can be calculated and subtracted to obtain the single-phase, the two-phase, and the three-phase BFR.

As would be expected, on a single-circuit line, the three-phase BFR is extremely low, normally being less than 1% of the total BFR.

12.2 DC Lines

The BFR for DC lines can also be determined, but in this case the value of $K_{PF}V_{LN}$ is equal to the crest pole-to-ground voltage.

12.3 Double-Circuit Flashover Rates

The BFR as calculated by the preceding equations is in terms of flashovers per 100 km of line route. Per Section 12.1, for a single-circuit line, flashovers may involve one or more phases. For a double-circuit line, flashovers may involve one or more phases and one or more circuits.

For a double-circuit line having a vertical phase configuration, flashover to one of the lower phases is usually most probable. Assuming that this phase flashes over, the most probable phase to flash over next is usually the lower phase of the other circuit. Estimates of the double-circuit flashover rate can be made by use of the previous equations, modified to include the decreased value of Z_g and the increased value of the coupling factor. To amplify, Z_g is recalculated to include the flashed over phase, since it is now a shield wire. Also, new values of coupling factors must be calculated, since the mutual surge impedance to the phase conductors have been altered.

With the changed values of coupling factors and ground wire surge impedance, the previous equations can be used to determine the double-circuit flashover rate. This type of calculation is required in the problems at the end of the chapter.

To illustrate the results of the calculation, for a 230-kV double-circuit line having a critical flashover voltage of 1200 kV, an R_0 of 30 ohms, and an earth resistivity of 600 ohm-meters, the total flashover rate is estimated as 3.00 flashovers/100 km-years, and the double-circuit flashover rate is 0.60/100 km-years. Thus 2.40 backflashes only involve one circuit, and 20% of the total backflashes involve both circuits.

To decrease the double-circuit BFR, differential insulation has been used with varying degrees of success. For example, for the same parameters as before, except assuming that one circuit has a critical flashover voltage of 1400 kV with the other remaining at 1200 kV, the total BFR remains at 3.00 while the double-circuit BFR is reduced to 0.18, that is, only 6% are double-circuit flashovers. However, since for this vertical configuration, space within the tower must be capable of employing a sufficient umber of insulators to obtain a critical flashover voltage of 1400 kV, this improvement should be compared to a tower having 1400 kV on both circuits. For this case, the total BFR becomes 1.19 and the double-circuit BFR remains at 0.18, so now 15% are double-circuit flashovers; see Table 5. While it appears true that differential insulation decreases the percentage of double-circuit flashovers, the use of increased insulation on both circuits results in improved performance for both the double-circuit BFR and the total BFR and is therefore preferred.

The calculation method as used above has been substantially improved by Sargent and Darveniza [13, 14]. These authors show the same tendency as illustrated above and in addition suggest designs that may substantially improve the double-circuit BFR. For a horizontal phase configuration with one circuit below the other, the most probable phases to flash over are those of the lower circuit. That is, flashover to the lower circuit essentially provides shield wires that completely encase the upper circuit, thus improving coupling factors to all phases of the upper circuit and inhibiting flashovers to this upper circuit.

As an interesting adjunct to the above observation, in many countries it is normal to install one or more lower voltage circuits below the high-voltage circuit on the same tower. This will improve the performance of the high-voltage circuit,

Table 5 BFRs, FO.100 km-years, for Alternate CFOs of a Double-Circuit Line

CFOs 1st circuit/ 2nd circuit	Total BFR	Double-circuit BFR	Percent double-circuit BFR
1200/1200	3.00	0.60	20
1200/1400	3.00	0.18	6
1400/1400	1.19	0.18	15

since flashovers will first occur to the lower voltage circuits, thus creating improved coupling to the high-voltage circuit. Oppositely, the performance of the lower voltage circuits will be degraded since more strokes will be collected by the higher towers. Thus such an arrangement is great for the high voltage line but bad for the lower voltage lines.

12.4 Tower Surge Impedance

The tower surge impedance can be determined by use of equations from Chapter 9. However, because the tower component of voltage is normally not of major importance, and considering that it is a time varying quantity that acts as a type of transfer function to obtain the voltage across the insulation, an approximation appears valid. For two shield wire lines, a value equal to 0.5 times the shield wire surge impedance is suggested. In general a value between 150 and 200 ohms can be used. For a single downlead on a wood-pole line, the tower surge impedance increases to 550 to 600 ohms.

12.5 Alternate Flashover Paths and Their Critical Flashover Voltage

The standard lightning impulse critical flashover voltage, called the CFO, is the lowest value considering all possible flashover paths. In most cases, except for wood-pole designs, only two flashover paths need be considered: to the tower side and across the insulator strings. For a tower using a vertical or I-string, the insulator string length is normally limiting. The value of this insulation strength, in terms of the CFO gradient, can be considered constant at 560 kV/m (positive polarity) for either air strike distances or insulator lengths.

For wood-pole towers, more than two flashover paths are normally possible. Therefore the CFO of all flashover paths must be calculated and the lowest CFO used. See the problems of Chapter 2.

The insulation of wood-pole lines usually consists of wood and porcelain in series. The wood increases the insulation strength and may provide an arc deionization for distribution lines, which means that the flashover is self-extinguishing. That is, the breaker tripout rate is less than the flashover rate. These effects have been investigated by several authors [15, 16] with varying results. Chapter 2 contains a method of evaluating the insulation strength. A summary including results of new investigations is contained in an excellent book by Darveniza [17]. In general, as discussed in Chapter 2, if the length of wood should be equal to twice the insulator length, the CFO is the CFO of the insulator plus 100 kV/m (300 kV/ft) times the length of the wood. If the length of the wood is greater than twice the length of the insulators, then the CFO is the CFO of the wood alone, about 300 kV/m (90 kV/ft). If the length of wood is less than twice the insulator length, the CFO added by wood is small, i.e., the CFO is the CFO of the insulator plus about 40 kV/m (10 kV/ft) times the length of wood.

Insulators not constrained from movement, i.e., vertical insulator strings, can be moved closer to the tower by the action of wind, thus decreasing the strike distance to the tower and decreasing the critical flashover voltage. This effect has been studied [20], and while it does increase the BFR, the increase is minor and can be neglected.

12.6 Flashover vs. Outage

Depending on the power frequency voltage gradient across wood, flashover may not result in a breaker trip, i.e., an outage. As discussed by Darveniza [17], the power frequency arc within or on the surface of the wood creates an arc voltage that decreases the current flow and may extinguish the arc before the breaker operates. The primary variable in establishing the probability of an outage is the power frequency voltage gradient across the wood. Darveniza's probability curve can be mathematically represented by a cumulative Weibull distribution function, that is

$$p = 1 - e^{-(\frac{G}{26})} \quad \text{for the crossarm}$$
$$p = 1 - e^{-(\frac{G}{56})^{1.7}} \quad \text{for the pole} \tag{52}$$

where p is the probability of an outage and G is the power frequency line to ground voltage gradient in kVrms/meter. These probability of outage curves are shown in Fig. 40. Assuming a 34.5-kV line and a 4-foot wood crossarm, $G = 16\,\text{kV/m}$, which results in a probability of an outage of 0.46. Thus the BFRs as calculated should be multiplied by 0.46 to obtain the outage rate. The use of this probability also applies to unshielded lines.

Even for air or porcelain insulations, all flashovers do not result in an outage, i.e., a breaker trip. That is, depending on the phase angle of the power frequency voltage at the instant of the surge voltage, the flashover may not result in an outage. This outage-to-flashover ratio is in the range of 0.85, which can be applied to the BFR as calculated in this chapter. However, this ratio was not used, since it is of minor importance compared to other factors.

Interestingly, Darveniza used this property of wood to construct a "wooden" arrester that operates in a manner similar to that of the old Protector Tube [38].

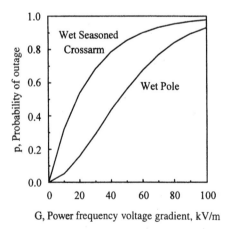

Figure 39 Probability of an outage for wood.

The Backflash 415

Darveniza's book [17] is recommended for further study and to all those using wood-pole lines.

Before leaving this subject, it is well to point out that flashovers along wet wood crossarms frequently expel wooden splinters. These and natural weather conditions may weaken the crossarm to the extent that maintenance personnel are in jeopardy and therefore replacement is necessary. Today, some utilities replace the crossarms with steel I-beams that are attached to the pole at an increased height of 1 foot, thus permitting two extra insulators to be added while maintaining the clearance to ground. However, even adding two extra insulators does not compensate for the loss of the insulation strength of wood, and thus these lines are predicted to have an increased flashover rate.

12.7 Distribution of Footing Resistance

Usually, towers of a line do not possess the identical tower footing resistance. In this case it is essential to consider all possible values of footing resistance. That is, using an average value frequently results in a BFR that is smaller than actual, since higher footing resistances result in a disproportional increase in BFR. As an example, Fig. 39 shows the distribution of footing resistance along the line via the dotted line. The BFRs in percent of the total BFR for each of the line sections is also shown in Fig. 39, via the solid line. Note that 25% of the line has an R_0 greater than 100 ohms and that the BFR for these sections is 61% of the total BFR. The total BFR is 4.98 flashovers/100 km-years. The average footing resistance is 65 ohms, which if used to calculate the BFR would give 6.09 flashovers/100 km-years, an error of over 20%.

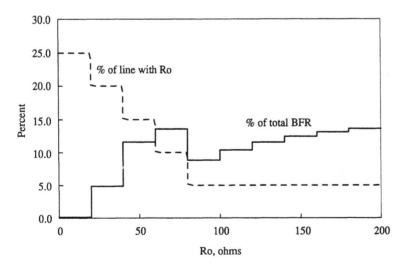

Figure 40 Footing resistance and BFR distribution along a line, 230-kV single-circuit tower, two ground wires, $\rho/R_0 = 20$.

13 LINE DESIGN AND METHODS TO IMPROVE PERFORMANCE

13.1 Designing the Line

The lightning design of the line usually begins with the specification of the desired or design BFR. To achieve this desired BFR, the primary variables under the control of the design engineer are

1. Insulation length, i.e., the length of the insulator string and the strike distance
2. Footing resistance, i.e., the type and extent of supplemental grounding
3. The number of shield wires, i.e., one or two or possibly an underbuilt ground wire
4. The use of surge arresters
5. And perhaps, if a double-circuit line is used, the insulation specification and phase arrangement

All other variables of design, the span length, the ground flash density, the tower height, the line route, etc. must be accepted as input conditions.

13.2 BFR Design Value

No universal agreement exists as to the recommended design value of the BFR, since this is (1) a function of economics and the total utility system design and (2) the voltage level being considered.

Considering the first item, customer reliability requirements are becoming more stringent, and in some locations the lightning tripout design values have decreased. The total lightning tripout rate, the sum of the shielding failure flashover rate and the BFR, combined with the probability of a successful reclose operation, should be considered.

As for the second item, normally the BFR design objective changes with the voltage level. For the highest voltage level of a system, since a greater degree of reliability is required, the design value of the BFR would normally be less than for lower voltages. For example, if 500-kV transmission is the highest voltage of a utility system, the BFR design goal may be 0.6 flashovers per 100 km-years (1.0 per 100 mile-years), whereas for the next lower voltage, 230 kV, the design goal may be 1.2 per 100 km-years (2.0 per 100 mile-years). But for a system where 230 kV is the highest voltage, the design goal for this voltage may be 0.6 per 100 km-years.

13.3 Rogue Tower

Towers located on hilltops not only are exposed to a larger than normal number of strokes but also usually have footing resistances larger than normal. The BFR of these towers can be expected to be considerably higher than those for the rest of the line, and thus the term rogue tower has come into the jargon of the industry to describe these towers. These towers produce a disproportional increase in BFR, and therefore consideration should be given to improved grounding or application of arresters.

13.4 Use of Surge Arresters

For cases where acceptable BFRs are not attainable, surge arresters can be placed across the line insulation [21, 23], thus preventing flashovers. Although line arresters have been primarily employed on lines with overhead shield wires, they are also used on lower voltage distribution lines and some higher voltage lines without shield wires. Application of these line arresters is the subject of Chapter 14.

The primary application problem is the arrester energy and the current discharged through the arrester. For shielded lines, the energy and current discharged through the arrester are in general within the arrester capability. For strokes to the shield wire, the majority of the stroke current is discharged through the footing resistance, even for high footing resistances. Strokes to the phase conductor are limited in magnitude to the maximum shield failure current, which for the usual line is between 5 and 15 kA. These shielding failure currents plus currents from subsequent strokes produce arrester energies that normally exceed the energy caused by strokes to the shield wire and thus represent the primary application criteria. However, in general, the energy discharged through the arrester is within the energy capability of the arrester.

Arresters, applied to lines without a shield wire, are in a highly hostile lightning environment, since all magnitudes of stroke currents can terminate on the phase conductor. For this situation, the application of line arresters is critical (see Chapter 14).

For distribution lines whose insulation strength is very low, CFOs of about 100 to 300 kV, the primary cause of flashovers is the induced voltage from a stroke that terminates close to the line. This induced voltage with its magnitude and probability of occurrence is the subject of Chapter 15. The arrester energy discharged is usually within the arrester's capability, and arresters on these lines will successfully eliminate these flashovers.

Another excellent use of line arresters is to apply arresters to the three phases of one circuit of a double-circuit line, thus eliminating double-circuit flashovers. This method is extensively used in Japan even at 500 kV, where over 6000 arresters have been applied.

In many utilities, the use of line arresters is considered the last resort when all other methods to reduce the BFR have failed. In general, if the soil conditions permit the installation of counterpoises, the use of this method to improve the BFR is more economical than that of arresters. However, when soil conditions (e.g., rock) do not permit the installation of counterpoises, line arresters are an excellent alternative.

In summary, candidate situations for arrester applications are

1. Lines with shield wires in areas of high soil resistivity, where because of rock formation, etc., it is impossible to install counterpoises
2. River-crossing towers
3. Lines serving critical loads where loss of service must be kept to an absolute minimum regardless of cost
4. Lower voltage lines for which shield wires are not effective in decreasing the number of flashovers.

13.5 Comparison of CIGRE method with Actual Performance

In general, the comparison of actual performance of lines to that calculated using the CIGRE method has been satisfactory, providing estimates that are within about 20 to 30% of the field performance. For example, the lightning performance of a 115-kV line of Carolina Power and Light averaged 4.2 flashovers/100 km-years. The calculated performance was 3.8 flashovers/100 km-years.

14 CONCLUSIONS

1. Although flashovers within the span are possible, they appear to be insignificant compared to flashovers at the tower. Therefore flashovers within the span can be neglected.
2. To account for flashovers at the tower caused by strokes terminating within the span, the BFR as calculated for a stroke to the tower should be multiplied by 0.6.
3. Simple equations were developed to estimate the decrease in the footing resistance caused by high surge currents.
4. The effect of the power frequency voltage and the number of phases can be approximated by a factor, $K_{PF}V_{LN}$, that is subtracted from the CFO_{NS}. The lowest coupling factor is then used to calculate the critical current.
5. A regression equation was developed from the leader progression model to estimate the CFO_{NS}. The use of this equation was compared to the use of the full leader progression model. The conclusion was that the regression equation was an accurate approximation of the leader progression model.
6. The use of the statistical distribution of time to crest of the stroke current can be replaced by a single equivalent front whose value is approximately equal to the median value of time to crest for the specific critical current.
7. Because corona only acts on the wave front of the voltage, it only marginally decreases the BFR. The effect of corona is neglected in the CIGRE method.
8. Equations were developed to estimate the BFR that included the tower component of voltage; their use is called the CIGRE method. This method is sufficiently complex so that the use of a computer program is suggested. A sensitivity analysis indicated that the calculation of the BFR can be further simplified by neglecting the tower component of voltage for tower heights of less than about 50 m. This simplified method is amenable to hand calculations.
9. The effects of corona, neglected in the CIGRE method, decrease the BFR. Although they may be neglected to give conservative results, they should be added to the CIGRE method.
10. Of primary importance is the decrease of footing resistance caused by high surge currents.
11. Two ground wires in contrast to a single ground wire significantly decrease the BFR.
12. An underbuilt ground wire dramatically decreases the BFR.
13. Medium voltage lines can be constructed to have an acceptable BFR.

14. Depending on the CFO, a ground wire installed on a distribution line can decrease the flashover rate caused by strokes to the line.
15. Comparison of the CIGRE and IEEE methods showed a good comparison for low-height towers. For high towers, the IEEE method produced larger BFRs than did the CIGRE method.

15 BACKGROUND

Usually the background of a subject is presented within the introduction to the subject. However, since this is primarily a teaching text, the background is presented here so as not to interfere with the presentation. However, assuredly, the authors here referred to have made significant contributions to this subject.

1. In 1950, measurements of currents on transmission line towers were assembled into an AIEE statistical distribution having a median of 15 kA and a log standard deviation of 0.98. Using this distribution and assuming a 2-μs linear front of the stroke current, Harder and Clayton [24] produced curves to estimate the BFR. This was followed by an AIEE Committee Report [19] in which the authors changed the front assumption to 4 μs, since the 2-μs front resulted in more midspan flashovers than were justified by field data.

2. Field theory was invoked to provide a detailed analysis of the shield wire–tower–phase conductor system [25–28]. Lundholm et al [27] developed the "loop-voltage" method to determine potential differences across the line insulation. Wagner and Hileman [1] applied this method to show that traveling wave theory could be used to approximate the results of field theory, provided that the artifice of a surge impedance be employed to represent the tower. They further calculated the voltage produced across the insulation caused by the charge in the channel above the tower and suggested that this component may exceed that produced by the current and charge injected into the tower.

3. Wagner and Hileman [2, 3] also investigated predischarge currents, those which are precursors of the breakdown process, and noted that the current shape is similar to that of the lightning stroke current and that these currents could inhibit flashovers; further, that these predischarge currents may be responsible for the apparent lack of midspan flashovers.

4. Fisher et al. used reduced scale models of the system (called nanosecond models) to obtain the response of the tower–shield wire system [29]. Monte Carlo methods were then used by Anderson to determine the BFR [30]. In this study, Anderson employed new distributions for both the current and the time to crest. The stroke current distribution was piecewise lognormal, having a median of 46.5 kA and log standard deviations of 0.71 for currents below the median and 0.41 for currents above the median. The distribution of time to crest had a median of 1.57 μs and a log standard deviation of 0.60.

5. During the following years, new methods of estimating the BFRs were devised by several investigators taking into account the results of the above investigations but with the overriding intention to produce methods that would agree with field experience. In 1964, Clayton and Young [18] reformed the previous estimating method [24] using the AIEE current distribution but employing a relationship

between the crest current and the wavefronts of 2, 4, and 6 μs. Anderson et al. [31] and Anderson [32] furthered their earlier work and produced a comparative method that employed the results of reduced-scale model tests and the current and time to crest distributions as mentioned in the preceding paragraph. Sargent and Darveniza [13] developed a method to estimate both the single- and the double-circuit BFRs and applied these to evaluate alternate designs [14]. For this method they also used the AIEE current distribution and the crest current–time to crest relationship of Clayton and Young.

6. During this period, the results of more extensive measurements of lightning parameters were becoming available. Spzor [33] showed that Polish records indicated median currents of 30 kA, Popolansky's [34] records showed median currents of 25 kA, and Berger et al. [35, 36] reported results that show a median of 31 kA and a log standard deviation of 0.46. The use of these results, as shown by Ah Choy and Darveniza [37], produced BFRs that exceeded those previously calculated by over 200%.

7. Ah Choy and Darveniza [37] analyzed the effect of the charge in the stroke channel as first discussed by Wagner. Assuming a finite length of an upward streamer, as opposed to zero length, decreased the voltage across the insulator by up to 100%. However, the voltage still represented 17 to 28% of the insulator strength. A discussion of this paper by Giudice and Piparo showed results of their investigation, which indicated further reductions in voltage. Therefore this component of voltage has been neglected in all estimating methods.

8. In 1982, with the recognition of the revised stroke current distribution, Anderson [8] produced a new estimating method using a constant 2-μs front. This method with minor modifications was adopted by the IEEE working group and published as the IEEE method [9].

As noted in this brief background, significant advancements have been made both in the theory of calculation of the BFR and in the practical application of this theory to produce estimating methods. To a significant degree, these developments were a result of attempts to explain the high flashover rate of the AG&E OVEC 345-kV double-circuit line. However, of equal importance, the maelstrom of activity was a result of the new lightning crest current distributions, which required reformation of estimating methods.

16 REFERENCES

1. C. F. Wagner and A. R. Hileman, "A New Approach to the Calculation of the Lightning Performance of Transmission Lines III—A Simplified Method—Stroke to Tower," *AIEE Trans. on PA&S*, 79 (3), 1960, pp. 589–603.
2. C. F. Wagner and A. R. Hileman, "Effect of Predischarge Currents upon Line Performance," *AIEE Transactions on PA&S*, 1963, pp. 117–128.
3. C. F. Wagner and A. R. Hileman, "Predischarge Current Characteristics of Parallel Electrode Gaps," *IEEE Transactions of PA&S*, 83, 1964, pp. 1236–1242.
4. K. H. Weck, "The Current Dependence of Tower Footing Resistance," CIGRE 33-88(WG01), 14 IWD, 1988 and 33-89(WG01), 7 IWD, 1989.

5. CIGRE Working Group 33.01, "Guide to Procedures for Estimating the Lightning Performance of Transmission Lines," Technical Brochure 63, 1991.
6. L. V. Bewley, *Traveling Waves on Transmission Lines*, New York: John Wiley, 1951.
7. A. Pigini, G. Rizzi, E. Garbagnati, A. Porrino, G. Baldo, and G. Pesavento, "Performance of Large Gaps under Lightning Overvoltages: Experimental Studies and Analysis of Accuracy of Predetermination Methods," *IEEE Trans. on Power Delivery*, Apr. 1989, pp. 1379–1392.
8. J. G. Anderson, "Lightning Performance of Transmission Lines," Chapter 12 of *Transmission Line Reference Book*, Palo Alto, CA: Electric Power Research Institute, 1982.
9. IEEE Working Group on the Lightning Performance of Transmission Lines, "A Simplified Method for Estimating the Lightning Performance of Transmission Lines," *IEEE Transactions on PA&S*, Apr. 1985, pp. 919–932.
10. IEEE Working Group on Lightning Performance of Transmission Lines, "Estimating Lighting Performance of Transmission Lines II, Updates to Analytical Models," *IEEE Trans. on Power Delivery*, Jul. 1993, pp. 1254–1267.
11. Working Group on the Lightning Performance of Distribution Lines, "Guide for Improving the Lightning Performance of Electric Power Overhead Distribution Lines," IEEE Std, 1410, 1997.
12. Working Group on the Lightning Performance of Transmission Lines, "Guide for Improving the Lightning Performance of Transmission Lines," IEEE Std, 1243, 1997.
13. M. A. Sargent and M. Darveniza, "The Calculation of Double Circuit Outage Rate of Transmission Lines," *IEEE Trans. on PA&S*, 1969.
14. M. A. Sargent and M. Darveniza, "Lightning Performance of Double Circuit Transmission Lines," *IEEE Trans. on PA&S*, May/Jun. 1970, pp. 913–925.
15. J. T. Lusigan and C. J. Miller, "What Wood May Add to Primary Insulation for Withstanding Lightning," *AIEE Transactions*, 74, 1955, pp. 534–540.
16. J. M. Clayton and D. F. Shankle, "Insulation Characteristics of Wood and Suspension Insulators," *AIEE Trans.*, 74(3), 1955, pp. 1305–1312.
17. M. Darveniza, *Electrical Properties of Wood and Line Design*, St. Lucia, Queensland, Australia: Univ. of Queensland Press, 1980.
18. J. M. Clayton and F. S. Young, "Estimating Lightning Performance of Transmission Lines," *IEEE Trans. on PA&S*, 83(3), 1964, pp. 1103–1110.
19. AIEE Committee Report, "A Method of Estimating the Lightning Performance of Transmission Lines," *AIEE Trans.*, 69(2), 1950, pp. 1187–1196.
20. A. R. Hileman, "Weather and Its Effect on Air Insulation Specifications," *IEEE Trans. on PA&S*, Oct. 1984, pp. 3104–3116.
21. R. E. Koch, J. A. Timoshenko, J. G. Anderson, and C. H. Shih, "Design of Zinc Oxide Transmission Line Arresters for Application on 138-kV Towers," *IEEE Trans. on PA&S*, Oct. 1985, pp. 2675–2680.
22. C. H. Shih, R. M. Hayes, D. K. Nichols, R. E. Koch, J. A. Timoshenko, and J. G. Anderson, "Application of Special Arresters on 138-kV Lines of the Appalachian Power Company," *IEEE Trans. on PA&S.*, Oct. 1985, pp. 2857–2863.
23. E. J. Los, "Transmission Line Lightning Design with Surge Suppressors at Towers," *IEEE Trans. PA&S*, 99 (2), 1980, pp. 720–728.
24. E. L. Harder and J. M. Clayton, "Transmission Line Design and Performance Based On Direct Lightning Strokes," *AIEE Trans.*, 68 (1), 1949, pp. 439–449.
25. C. F. Wagner, "A New Approach to the Calculation of the Lightning Performance of Transmission Lines," *AIEE Trans. on PA&S.*, 75, Dec. 1956, pp. 1233–1256.
26. C. F. Wagner and A. R. Hileman, "A New Approach to the Calculation of the Lightning Performance of Transmission Lines—II," *AIEE Trans. on PA&S*, 78, 1959, pp. 996–1020.

27. R. Lundholm, R. B. Finn, and W. S. Price, "Calculation of Transmission Line Lightning Voltages by Field Concepts," *AIEE Trans. PA&S*, Feb. 1958, pp. 1271–1283.
28. C. F. Wagner and A. R. Hileman, "Surge Impedance and Its Application to the Lightning Stroke," *AIEE Trans. on PA&S*, 80 (3), 1961, pp. 1011–1020.
29. F. A. Fisher, J. G. Anderson, and J. H. Hagenguth, "Determination of Transmission Line Performance by Geometrical Models," *IEEE Trans. PA&S*, 78, 1960, pp. 1725–1736.
30. J. G. Anderson, "Monte Carlo Computer Calculation of Transmission Line Lightning Performance," *IEEE Trans. on PA&S*, Aug. 1961, pp. 414–420.
31. J. G. Anderson, F. A. Fisher, and E. F. Magnusson, "Calculation of Lightning Performance of EHV Lines," Chapter 8 of *EHV Transmission Line Reference Book*, New York: Edison Electric Institute, 1968.
32. J. G. Anderson, "Lightning Performance of EHV-UHV Lines," *Transmission Line Reference Book—345 kV and Above*, Palo Alto, CA: Electric Power Research Institute, 1975.
33. S. Szpor, "Comparison of Polish Versus American Lightning Records," *IEEE Transactions on PA&S*, May 1969, pp. 646–652.
34. F. Popolansky, "Frequency Distribution of Amplitudes of Lightning Currents," *ELECTRA*, No. 22, May 1972, pp. 139–147.
35. K. Berger, R. B. Anderson, and H. Kroninger, "Parameters of Lightning Flashes," *ELECTRA*, No. 41, Jul. 1975, pp. 23–37.
36. R. B. Anderson and A. J. Eriksson, "Lightning Parameters for Engineering Application," *ELECTRA2*, No. 69, Mar. 1980, pp. 65–102.
37. Liew Ah Choy and M. Darveniza, "A Sensitivity Analysis of Lightning Performance Calculations for Transmission Lines," *IEEE Trans. on PA&S*, 1971, pp. 1443–1551.
38. E. L. Harder and J. M. Clayton, "Line Design Based upon Direct Strokes," Chapter 17 of *Electrical Transmission and Distribution Book*, Westinghouse Electric Corporation, 1950.

17 PROBLEMS

1. Using the simplified method (hand calculation), determine the BFR for the 500 kV line of problem 1 of Chapter 9. Assume a value of R_0 of 40 ohms in a soil of 800 ohm-meters resistivity. Also assume that the span length is 300 meters and $N_g = 4$ flashes/km^2-year. Use $K_{PF} = 0.70$. The standard CFO is 1700 kV. Using the computer program for the CIGRE method, repeat this calculation. Also using the CIGRE method, determine the BFR for a two- or three-phase flashover. From this find the single-phase flashover rate.

2. Using the simplified method (hand calculation), determine the BFR for the 230 kV, double-circuit line of problem 2 of Chapter 9. Assume an $R_0 = 40$ ohms and a soil resistivity of 800 ohm-meters. The standard CFO is 1200 kV, $N_g = 4$, and the span length is 200 meters. Use a K_{PF} of 0.40. Using the computer program for the CIGRE method, repeat this calculation and also calculate the double-circuit flashover rate. From this find the single-circuit flashover rate.

3. Using the CIGRE method, for the line of the above problem 2, for the distribution of footing resistances in the table, calculate the BFR. The computer program may be used.

The Backflash

R_0, ohms	Soil resistivity, ohm-m	Percent of line
15	200	20
45	500	40
85	800	30
150	1200	10

4. For the 69-kV line of problem 3, Chapter 2, draw a curve of the BFR as a function of the measured footing resistance R_0 for R_0 of 10 to 50 ohms. Assume a ρ/R_0 of 10 and 20. Assume a ground flash density of 5.0 flashes/km^2-year. Use the CIGRE and IEEE methods, i.e., the computer programs. Let $K_{PF} = 0.40$ and $Z_T = 572$ ohms.

Add an underbuilt ground wire beneath the bottom phase conductor per problem 11 of Chapter 9 and calculate the BFR versus R_0 as before. Use $\rho/R_0 = 20$ and the CIGRE method, i.e., the computer program.

5. For the 115 kV line of problem 3, Chapter 2, draw a curve of the BFR as a function of the measured footing resistance R_0 for R_0 of 10 to 50 ohms. Assume a ρ/R_0 of 10 and 20. Assume a ground flash density of 5.0 flashes/km^2-year. Use the CIGRE and IEEE methods, i.e., the computer programs. Let $K_{PF} = 0.40$ and $Z_T = 590$ ohms.

Add an underbuilt ground wire beneath the bottom phase conductor per problem 11 of Chapter 9 and calculate the BFR versus R_0 as before. Use $\rho/R_0 = 20$ and the CIGRE method, i.e., the computer program.

6. Using the CIGRE and IEEE methods for the line of problem 4, Chapter 2, draw a curve of the BFR as a function of the measured footing resistance R_0 for R_0 of 10 to 80 ohms. Assume a ρ/R_0 of 10 and 20. Also assume a ground flash density of 5.0 flashes/km^2-year. For $\rho/R_0 = 20$, repeat the calculations using the CIGRE method with corona. Let $Z_T = 365$ ohms and $K_{PF} = 0.70$.

7. Select any line of your choice and select the design BFR. What insulation level, CFO, and footing resistance would you specify? Is supplemental grounding required?

10—Appendix 1
Effect of Strokes Within the Span

A stroke terminating on the shield wire within the span produces voltages across the air insulation between the shield wire and the phase conductor and also across the air–porcelain insulation at the tower. Although the voltage across the span insulation exceeds that across the tower insulation, the span insulation exceeds that of the tower. Thus dependent on the relative voltages and insulation strengths, flashover can occur either across the span or across tower insulations.

1 FLASHOVERS WITHIN THE SPAN

Considering a stroke to the shield wire, as defined in Fig. 1, the voltage at the stroke terminating point attempts to reach a crest voltage of $Z_g I/2$. However, reflections from adjacent towers reduce this voltage, provided t_f is greater than $2(T_S - T_{ST})$. The maximum voltage occurs at the stroke terminating point, and voltages decrease as the distance from the stroke terminating point increases, reaching a minimum at the tower. To illustrate, Fig. 2 shows the voltage at the midspan and at a location defined as $T_S/5$ from the tower for a stroke terminating at midspan. This decrease in voltage is better illustrated by the curves of Fig. 3, where the parameter is the stroke terminating point defined by T_{ST}.

To obtain an approximation of the expected number of span flashovers as opposed to tower flashovers, assume that the waveshapes of all the voltages are identical so that the nonstandard critical flashover voltage CFO_{NS} is a linear function of the gap spacing. For a typical 500-kV line, the minimum strike distance at the tower is 3.35 m, while the shield wire to phase conductor spacing varies from 9.2 m at the tower to 11.6 m at midspan. Thus the ratio of insulation strength is 3.5. For a stroke terminating at $T_{ST}/T_S = 0.20$, and for $t_f = 2.0\,\mu s$ and $R_i = 20$ ohms, the ratio of the voltages at midspan to the voltage at the tower is 2.4. Thus, for this case,

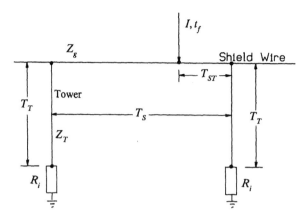

Figure 1 Definitions of variables, stroke within span.

flashover would occur at the tower. If all stroke terminating points are considered, for $t_f = 2.0\,\mu s$, approximately 16% of the strokes result in span flashover. For $t_f = 4.0\,\mu s$, span flashover is reduced to about 2%.

Another phenomenon further reduces the probability of span flashover. At high overvoltages, predischarge currents flow from the shield wire to the phase conductor producing a voltage on the phase conductor that decreases the voltage across the span insulation [2, 3]. Although no quantitative calculation will be made, suffice it to note that this phenomenon inhibits flashover. Thus, considering both the example calcula-

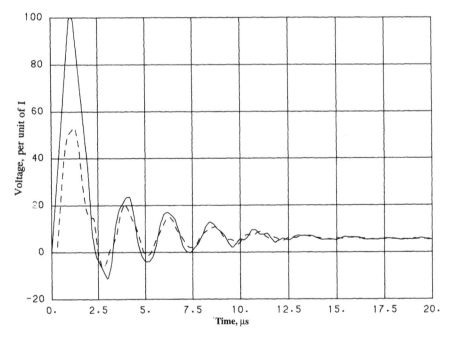

Figure 2 Comparison of surge voltages for stroke terminating at midspan, $t_f = 2\,\mu s$, $R_i = 20$ ohms. — voltage at midspan; --- voltage at $T_{ST} = T_S/5$.

Effect of Strokes Within the Span

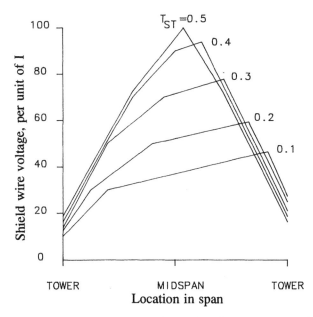

Figure 3 Voltages on shield wire for stroke terminating points defined by T_{ST}, $R_i = 20$ ohms, $T_S = 1$ μs.

tions and the predischarge current phenomena, although flashovers within the span are possible, they appear to be insignificant compared to flashovers at the tower.

2 FLASHOVERS AT THE TOWER CAUSED BY STROKES TO THE SHIELD WIRE

For a stroke terminating within the span, the crest voltage at the tower, in terms of K_{TT}, is, if $2T_T \leq t_f \leq 2(T_T - T_{ST})$,

$$V_{TT} = IK_{TT} \tag{1}$$

where

$$K_{TT} = R_e + \alpha_T Z_T \frac{T_T}{t_f} \tag{2}$$

If $2(T_S - T_{ST}) \leq t_f \leq 2T_S$, then the crest voltage occurs either at t_f or at $2(T_S - T_{ST})$, dependent on the value of the tower footing resistance, i.e., if the crest voltage occurs at t_f, then

$$K_{TT} = R_e \frac{2(T_S - T_{TS})}{t_f} + \alpha_T Z_T \frac{T_T}{t_f} \tag{3}$$

or if the crest voltage occurs at $2(T_S - T_{ST})$, then

$$K_{TT} = \frac{R_e}{Z_g + 2R_i} \left[2R_i + Z_g \frac{2(T_T - T_{ST})}{t_f} + \frac{4\alpha_T R_i}{R_e} Z_T \frac{T_T}{t_f} \right] \tag{4}$$

As noted by comparing these equations to Eq. 1, the voltage resulting from a stroke within the span equals that produced by a stroke to the tower only when $t_f \leq 2(T_S - T_{ST})$. Therefore the voltage produced at the tower by a stroke within the span is equal to or less than that produced by a stroke to the tower. Figure 4 shows a

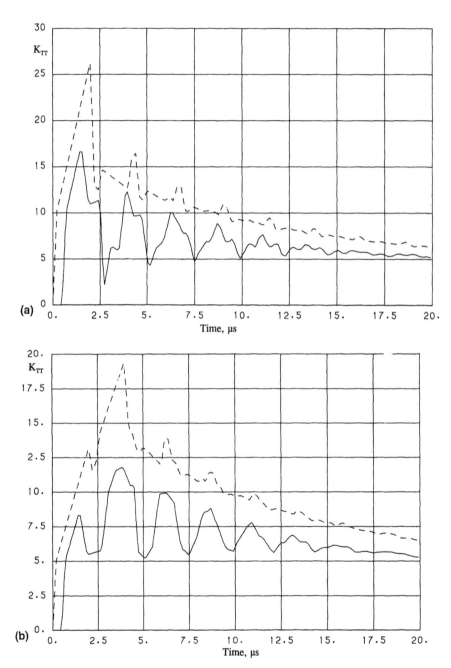

Figure 4 Comparison of surge voltages at the tower from --- stroke to tower, — stroke to midspan; $R_i = 20$ ohms, $T_S = 1\,\mu s$. (a) $t_f = 2\,\mu s$; (b) $t_f = 4\,\mu s$.

comparison of these voltages, and as noted, for these cases, the voltage produced by a stroke to midspan is approximately 60% of the voltage produced by a stroke to the tower.

The crest voltage at the tower in terms of K_{TT} for $R_i = 20$ ohms, for alternate stroke terminating points, and for alternate wave fronts, is presented in Fig. 5. Values on the curves are ratios of the K_{TT} for strokes to the midspan to the K_{TT} for strokes to the tower. This ratio initially decreases as the front t_f increases but then gradually increases as the front increases further. This ratio, called the K_{TT} ratio, is plotted as a function of t_f/T_S in Fig. 6, for $t_f > 2T_S$. The range of this ratio is relatively narrow, from about 0.58 to 0.77.

Assuming a K_{TT} ratio of 0.7, the effect of strokes to the span on the backflash rate BFR can be estimated. Knowing the K_{TT} ratio and the critical current for strokes to the tower, I_c, the critical current for strokes along the span can be obtained if the CFO_{NS} for all voltages is considered equal, since I_c is approximately

$$I_c = \frac{CFO_{NS}}{K_{TT}(1 - C)} \quad (5)$$

Using this approach, the ratio K_S of the total BFR, considering all possible stroke terminating points to the BFR when only considering only strokes to the tower, can be obtained and is shown by the solid line curve of Fig. 7 as a function of I_c. This assumes that the number of strokes to each incremental length of the shield wire is constant. If the number of strokes is assumed to vary with the height of the shield

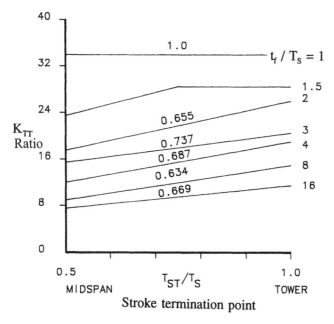

Figure 5 Voltages at tower as a function of stroke terminating point for various fronts. Values on curve are K_{TT} ratios of the K_{TT} at midspan to the K_{TT} at the tower; $R_i = 20$ ohms.

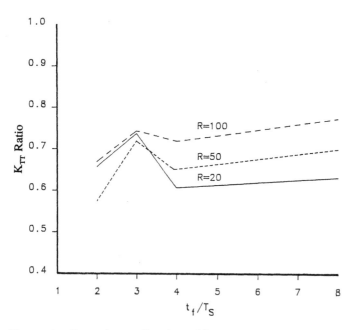

Figure 6 K_{TT} ratio as a function of front.

wire, the curve decreases, an estimate being provided by the dotted line in Fig. 7. The resultant ratio K_S or span factor ranges from 0.63 to 50 kA to 0.42 for 200 kA and thus is a function of the system operating voltage. However, for purposes of estimating the BFR, a single value of 0.6 is suggested.

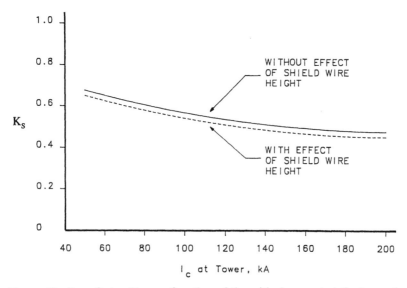

Figure 7 Span factor K_A as a function of the critical current at the tower I_C.

Thus, in conclusion,

1. For strokes within the span, although flashovers can occur within the span, they are insignificant to flashovers that occur at the tower and therefore can be neglected.
2. Strokes within the span cause flashovers at the tower.
3. Strokes within the span produce voltages at the tower that are usually less than those produced by strokes to the tower.
4. The BFR considering all stroke terminating points is equal to about 60% of the BFR if only strokes to the tower are considered.

Therefore, if only the strokes to the tower are considered, the BFR must be modified to

$$\text{BFR} = 0.6 N_\text{L} P(I_\text{c}) \tag{6}$$

3 REFERENCES

1. E. K. Saraoja, "Lightning Earths," chapter 18 of *Lightning* (R. H. Golde, ed.), Academic Press, 1977.
2. H. B. Dwight, "Calculation of Resistances to Ground," *AIEE Trans.*, Dec. 1936, pp. 1319–328.
3. E. D. Sunde, *Earth Conduction Effects in Transmission Systems*, New York: Dover, 1968.

10—Appendix 2
Impulse Resistance of Ground Electrodes

High magnitudes of lightning current, flowing through the ground resistance, decreases the resistance significantly below the measured low-current values. Although this has been known for many years, most lightning performance estimating methods, while acknowledging this fact, have not provided means of estimating the impulse resistance, primarily because of the lack of data and the lack of an adequate simplified calculation procedure. Within the CIGRE Working Group 33.01, Popolansky's reports [1] on developments of a similitude relationship by Korsuntcev [2] sparked new interest and was directly responsible for suggestions by other authors as to methods to formulate mathematically the similitude relationship [3–4]. Subsequently, Weck [5] analyzed measured impulse resistance data by Berger [6] to arrive at a simplified method.

The purpose of this appendix is to present K. H. Weck's simplified method of estimating the impulse resistance of concentrated ground electrodes, i.e., tower footings and ground rods, as presented in CIGRE Technical Brochure No. 63 [7].

1 GROUND ROD

The low-current, low-frequency resistance R_0 of a single ground rod of length L and radius r_0 driven in soil having a resistivity of ρ, is

$$R_0 = \frac{\rho}{2\pi L}\left[\ln\frac{4L}{r_0} - 1\right] \quad (1)$$

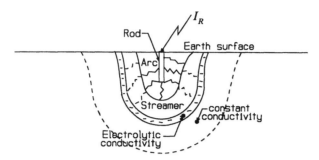

Figure 1 Impulse breakdown of soil surrounding a rod electrode.

The current density J at a distance r from the rod and the voltage gradient E_0 for an injected current I_R are

$$J = \frac{I_R}{2\pi rL} \qquad E = \rho J = \frac{\rho I_R}{2\pi rL} \tag{2}$$

For high currents, representative of lightning, when the gradient exceeds a critical gradient E_0, breakdown of soil occurs. The process is illustrated in Fig. 1. As the current increases, streamers are generated that evaporate the soil moisture, which in turn produces arcs. Thus within the streamer and arc zones, the resistivity decreases from its original value and as a limit approaches zero for a perfect conductor. This soil breakdown can be viewed as increasing the diameter and length of the rod. Indeed, most investigators represent this process with the simplified model of Fig. 2, where the streamer and arc zones are modeled as an ionization zone having zero resistivity (the electrolytic zone is small and is neglected). The ionization zone is described by the critical field strength E_0 at which the radius is equal to r. As the ionization increases, the shape of the zone becomes more spherical, as illustrated in Fig. 3 for multiple electrodes. Thus at high currents, the ground rod can be simply modeled as a hemisphere electrode. Therefore the hemisphere electrode is studied first.

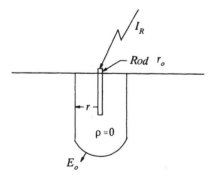

Figure 2 Simplified model of ionization zone.

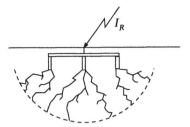

Figure 3 Multiple rods act as a hemisphere.

2 THE HEMISPHERE ELECTRODE

For the hemisphere electrode of Fig. 4 having a radius r_0 the equations are

$$R_0 = \frac{\rho}{2\pi r_0}$$

$$J = \frac{I_R}{2\pi r^2} \qquad E = \frac{\rho I_R}{2\pi r^2} \tag{3}$$

Assuming the soil resistivity is zero within the ionization zone simply means that the perfectly conducting hemisphere radius has expanded to a radius r as defined by setting $E = E_0$ in Eq. 3. Then, replacing r_0 with this new radius and denoting this impulse resistance as R_i results in the equation

$$R_i = \sqrt{\frac{E_0 \rho}{2\pi I_R}} \tag{4}$$

or taking the log of both sides,

$$\ln R_i = \frac{1}{2}\left[\ln \frac{E_0 \rho}{2\pi} - \ln I_R\right] \tag{5}$$

and thus the impulse resistance is inversely proportional to the reciprocal of the square root of the current, or on log-to-log paper as in Fig. 5, the impulse resistance vs. current is a straight line.

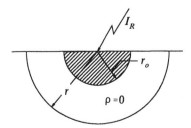

Figure 4 The hemisphere electrode.

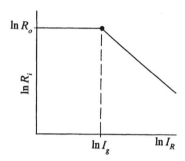

Figure 5 Impulse resistance of hemisphere.

However, this relationship does not exist until there is sufficient current to produce the critical gradient E_0 at the surface of the sphere. To determine this current I_g, set $r = r_0$, $E = E_0$, and $I_R = I_g$ in Eq. 3. Then

$$I_g = \frac{2\pi r^2 E_0}{\rho} = \frac{1}{2\pi} \frac{\rho E_0}{R_0^2} \tag{6}$$

Then substituting Eq. 6 into Eq. 4, we obtain

$$R_i = R_0 \sqrt{\frac{I_g}{I_R}} \tag{7}$$

For example, let $r_0 = 1$ m, $E_0 = 400$ kV/m, and $\rho = 200$ ohm-meters. Then I_g is 12.5 kA and the relationship per Eq. 7 does not occur until 12.5 kA or higher, as illustrated in Fig. 5. Letting $I_R = 50$ kA and $R_0 = 50$ ohms, then $R_i = 35$ ohms, a 30% reduction.

Rearranging Eq. 4 and letting D equal the diameter of the hemisphere,

$$\frac{R_i D}{\rho} = \sqrt{\frac{E_0 D^2}{2\pi \rho I_R}} \tag{8}$$

and letting

$$\Pi_1 = \frac{R_i D}{\rho} \qquad \Pi_2 = \frac{\rho I_R}{E_0 D^2} \tag{9}$$

results in

$$\ln \Pi_1 = \frac{1}{2} \ln \frac{1}{2} \pi - \ln \Pi_2 \tag{10}$$

The two variables Π_1 and Π_2 are dimensionless and form the similitude relationship, and will be discussed later.

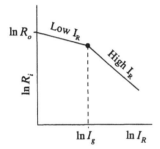

Figure 6 At high currents, rod becomes a hemisphere.

3 THE ROD ELECTRODE

As for the hemisphere, replace r_0 with r for $E = E_0$. The result is

$$R_i = \frac{\rho}{2\pi L}\left[\ln\frac{8\pi L^2 E_0}{\rho} - 1 - \ln I\right] \qquad (11)$$

And thus R_i is a function of the log of I. As the current increases, a point is reached where, as stated before, the ionized zone is approximately spherical and therefore R_i decreases as for a hemisphere as illustrated in Fig. 6.

Because of the small geometry of the rod, the onset current to obtain ionization is small and can be neglected. Now, rearranging Eq. 11,

$$\Pi_1 = \frac{1}{2\pi}[\ln 8\pi - 1 - \ln \Pi_2] \qquad (12)$$

where in this case the length L is used in place of the diameter D of Eq. 10. Thus in both cases, data can be plotted in terms of Π_1 and Π_2 as illustrated in Fig. 7. This figure obtained from ref. 1 shows Korsuntcev's similitude relationship. The data

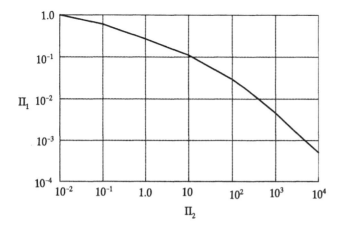

Figure 7 Korsuntcev's similitude relationship of the generalized impulse resistance [1].

points from which the curve is derived have been omitted. The data are primarily from rods and hemispheres.

4 THE SIMPLIFICATION

Weck's simplification resulted from his review and investigation of tests performed by Berger [6]. In summary, for the impulse resistance for a hemisphere, Eq. 7 can be rearranged as

$$R_i = \frac{R_0}{\sqrt{I_R/I_g}} \tag{13}$$

As discussed previously, Eq. 13 is applicable for currents above I_g. The desire is to adapt this equation for ground rods noting that for high currents, rods act as spheres. Therefore it is only the initial portion of the function that needs modification, and Weck has suggested

$$R_i = \frac{R_0}{\sqrt{1 + (I_R/I_g)}} \tag{14}$$

where I_g is given by Eq. 6.

The final comparison of this equation vs. the data is shown in Fig. 8 by the solid line. The dotted line is that obtained by using Eq. 6 for spheres. The curve was drawn for an E_0 of 400 kV/m, which is suggested for general use, for both rods and tower footings.

A comparison of results with those from Ryabkova and Mishkin [8] and those from Liew and Darveniza [9] is shown in Fig. 9.

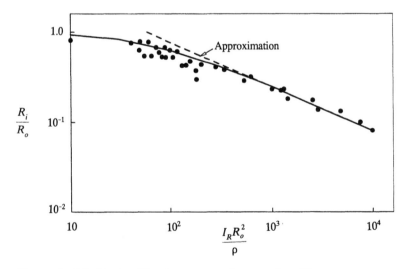

Figure 8 Weck's simplification compared to test data [7].

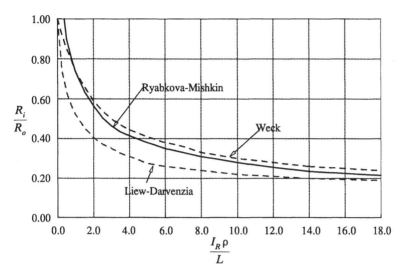

Figure 9 Comparison of results [7].

5 RESULTS AND OTHER INVESTIGATIONS

1. Liew and Darveniza (1974) [9] developed a time- and current-dependent algorithm. Of significance is that their development included two time constants, τ_1, the ionization time constant, and τ_2, the deionization time constant. From their test data, τ_1 is 1.5 to 2.0 μs and τ_2 is 0.5 to 4.6 μs. These time constants are not considered in the Weck equations. The use of these time constants represents a refinement that is normally not justified.

2. Oettle (1988) [3] used Popolansky's data and added her own. Regression analysis resulted in the following equation for the normalized curve.

$$\Pi_1 = 0.22 \Pi_2^{-0.29} \tag{15}$$

Oettle suggests an E_0 value of 1000 kV/m.

3. Chisholm and Janischowsky (1989) [4] used Popolansky's data, and for the practical range of Π_2 between 0.3 and 10 they suggested the equations

$$\Pi_1 = 0.26 \Pi_2^{-0.33} \qquad E_0 = 241 \rho^{0.215} \tag{16}$$

The equation for E_0 was obtained from a draft of Oettle's paper presented at the CIGRE Working Group but deleted in Oettle's published paper. Of significance is that the authors proposed an equation to estimate the footing inductance. With T_T as the tower travel time and t_f as the time to crest of the stroke current, the footing inductance L_f is

$$L_f = 60 T_T \ln \frac{t_f}{T_t} \tag{17}$$

If the values of time are in μs, then the inductance is in μH. The inductance per this equation results in an inductance approximately equal to that of the tower, i.e., the tower height is doubled.

6 REFERENCES

1. F. Popolansky, "Generalization of Model Measurements Results of Impulse Characteristics of Concentrated Earths," CIGRE sc 33-80 (WG01), IWD, Aug. 1980; "Determination of Impulse Characteristics of Concentrated electrodes," CIGRE SC 33-86 (WG-1), IWD, 1986.
2. Korsuntcev, "Application of the Theory of Similitude to the Calculation of Concentrated Earth Electrodes," *Electrichestvo*, No. 5, 31, 1958.
3. E. E. Oettle, "A New General Estimating Curve for Predicting the Impulse Impedance of Concentrated Earth Electrodes," *IEEE Trans. on Power Delivery*, 1987.
4. W. A. Chisholm and W. Janischewskyj, "Lightning Surge Response of Ground electrodes," *IEEE Trans. on Power Delivery*, Apr. 1989.
5. K. H. Weck, "The Current Dependence of Tower Footing Resistance," CIGRE 33-88 (WG 01), 14 IWD, 1988 and 33-89 (WG01), 7 IWD, 1989.
6. K. Berger, "Das Verhalten von Erdungen unter hohen Stossströmen," *Bull. Assoc. Suisse Elek. SEV* 37, 197, 1946.
7. CIGRE Working Group 33.01 (Lightning), "Guide to procedures for Estimating the Lightning Performance of Transmission Lines," CIGRE Technical Brochure 63, Oct. 1991.
8. Ryabkova and Miskin, "Impulse Characteristics of Earthings for Transmission Line Towers," *Electrichestvo*, No. 8, 1976, pp. 62–70.
9. A. C. Liew and M. Darveniza, "Dynamic Model of Impulse Characteristics of Concentrated Earths," *IEE Proc.*, 1974, pp. 123–135.

10—Appendix 3
Estimating the Measured Footing Resistance

Following are the equations for counterpoises and grounds as obtained from the work of H. B. Dwight and E. D. Sunde [1–3].
  Definitions

- r = conductor or ground rod radius
- d = burial depth of counterpoise
- L = counterpoise or ground rod length
- n = number of counterpoises or ground rods in parallel
- ρ = soil resistivity
- R_0 = resistance of a single counterpoise or ground rod
- R_n = resistance of n counterpoises or n ground rods in parallel
- a_{ij} = horizontal distance between counterpoises i and j or between ground rods i and j.
- R_{ij} = mutual resistance between ground rods i and j or between counterpoises i and j.

1 COUNTERPOISES

$$a'_{ij} = \sqrt{a_{ij}^2 + (2d)^2} \tag{1}$$

$$f(x) = \ln\frac{L + \sqrt{x^2 + L^2}}{x} + \frac{x - \sqrt{x^2 + L^2}}{L} \tag{2}$$

1.1 Single Counterpoise

$$R_0 = \frac{\rho}{2\pi L}\left[f(r) + f(2d)\right] \tag{3}$$

and if $L \ggg d$ and r,

$$R_0 = \frac{\rho}{\pi L}\left[\ln\frac{2L}{\sqrt{2rd}} - 1\right] \tag{4}$$

1.2 Two or More Counterpoises

$$R_n = \frac{1}{n}R_0 + \frac{2}{n^2}\sum_{i=1}^{n-1}\sum_{j=i+1}^{n} R_{ij} \tag{5}$$

where

$$R_{ij} = \frac{\rho}{2\pi L}\left[f(a_{ij}) + f(a'_{ij})\right] \tag{6}$$

and if $L \ggg d$, r and a_{ij}

$$R_n = \frac{\rho}{\pi L}\left[\ln\frac{2L}{a_{eq}} - 1\right] \tag{7}$$

$$a_{eq} = \left[(2rd)^n \sum_{i=1}^{n-1}\sum_{j=n+1}^{n}(a_{ij})^2(a'_{ij})^2\right]^{\frac{1}{2n^2}} \tag{8}$$

2 GROUND RODS

$$f(x) = \ln\frac{2L + \sqrt{x^2 + (2L)^2}}{x} + \frac{x - \sqrt{x^2 + (2L)^2}}{2L} \tag{9}$$

2.1 One Ground Rod

$$R_0 = \frac{\rho}{2\pi L}f(r) \tag{10}$$

and if $L \ggg r$,

$$R_0 = \frac{\rho}{2\pi L}\left[\ln\frac{4L}{r} - 1\right] \tag{11}$$

2.2 Two or More Ground Rods

$$R_n = \frac{1}{n}R_0 + \frac{2}{n^2}\sum_{i=1}^{n-1}\sum_{j=i+1}^{n} R_{ij} \tag{12}$$

where

$$R_{ij} = \frac{\rho}{2\pi L}f(a_{ij}) \tag{13}$$

If rods are in a circle of diameter D then

$$a_{ij} = D\sin\left[(j-i)\frac{\pi}{n}\right] \tag{14}$$

and approximately, if the spacing between adjacent rods is greater than the length of the rod,

$$R_n = \frac{1}{n}\frac{\rho}{2\pi L}\left[\ln\frac{4L}{a} - 1 + \frac{L}{D}\sum_{i=1}^{n-1}\frac{1}{\sin(i\pi/n)}\right] \tag{15}$$

3 EXAMPLES

Example 1. *Three Counterpoises.*

$$R_3 = \frac{1}{3} + \frac{2}{9}(R_{12} + R_{13} + R_{23}) \tag{16}$$

and approximately,

$$R_3 = \frac{\rho}{\pi L}\left[\ln\frac{2L}{a_{eq}}\right] \tag{17}$$

where

$$a_{eq} = \left[(2ad)^3(a_{12})^2(a'_{12})^2(a_{13})^2(a'_{13})^2(a_{23})^2(a'_{23})^2\right]^{\frac{1}{16}} \tag{18}$$

Example 2. *Three Ground Rods in a Circle.*

$$R_3 = \frac{1}{3}R_0 + \frac{2}{9}(R_{12} + R_{13} + R_{23}) \tag{19}$$

and approximately,

$$R_3 = \frac{1}{3}\frac{\rho}{2\pi L}\left[\ln\frac{4L}{r} - 1 + \frac{2L}{D\sin(\pi/3)}\right] \tag{20}$$

4 COMMENTS

1. Equations 5 and 12, for two or more rods or counterpoises, were developed using the assumption that currents are identical in each rod or counterpoise. They also assume that the lengths of the rods or counterpoises are equal.

2. The assumption of equal currents in each counterpoise was examined for three counterpoises. The error is less than 0.5%.

3. In practical situations, frequently, multiple ground rods of unequal length are driven. In this case, mutual resistances must be calculated to account for these unequal lengths. For this case, the distance to be employed between any two rods is the minimum length of the two rods.

5 SENSITIVITY

Using the previous equations, the sensitivity of the parameters can be studied. For the following cases the ground rod diameter and the counterpoise diameter is 13 mm. From the equations, the diameter has only an insignificant effect.

1. A ground rod can be thought of as a vertical counterpoise, and a counterpoise can be thought of as a horizontal ground rod. Thus the resistance of the ground rod as a function of depth and the resistance of a counterpoise as a function of length should be approximately equal. Figure 1 illustrates this thought. As noted, the resistance of the rod and counterpoise are approximately equal.

2. Figure 2 shows the effect of driven depth of a single rod on the resistance for $\rho = 200$ ohm-meters and a rod diameter of 13 mm. As noted, beyond a depth of about 6 meters, the decrease is minimal. However, this assumes a constant soil resistivity with depth. In some cases, the soil resistivity decreases with depth so that long depth rods are beneficial.

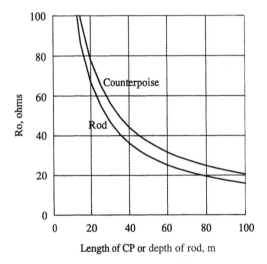

Figure 1 Comparison of a single ground rod and counterpoise, $\rho = 1000$ ohm-meters, counterpoise depth = 1 meter.

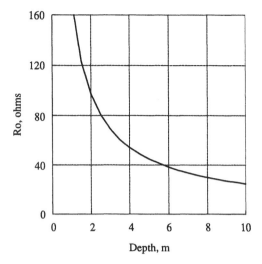

Figure 2 Resistance of a single ground rod, $\rho = 200$ ohm-meters.

3. Figure 3 illustrates the effect of multiple rods and the spacing between rods, i.e., the mutual effect. Rods in parallel significantly decrease the resistance. However, as shown, increasing the number beyond about four shows a small improvement. The dotted curve shows the resistance if the mutual effects are not considered. Thus spacings in a circle of about 5 meters are recommended.
4. The solid line curve of Fig. 4 is constructed for two counterpoises of 50-meter length. For this case the mutual effects between counterpoises is significant for spacings below about 20-meters. usually counterpoises are separated by the width of the line right-of-way, and thus 20-meters is fairly easy to obtain except for low-voltage lines and perhaps in mountainous conditions.

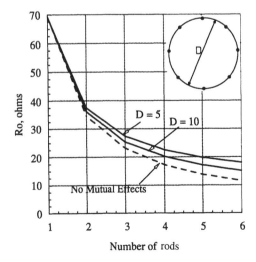

Figure 3 Effect of number of ground rods, $\rho = 200$ ohm-meters, depth = 3 meters.

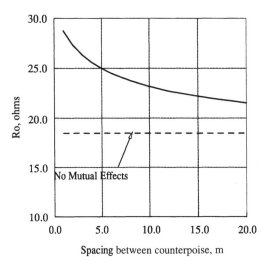

Figure 4 Effect of spacing between two 50 m counterpoises at a depth of 1 m, $\rho = 1000$ ohm-meters.

6 REFERENCES

1. E. K. Saraoja, "Lightning Earths," Chapter 18 of *Lightning* (R. H. Golde, ed.), Academic Press, 1977.
2. H. B. Dwight, "Calculation of Resistances to Ground," *AIEE Trans.*, Dec. 1936, pp. 1319–1328.
3. E. D. Sunde, *Earth Conduction Effects in Transmission Systems,* New York: Dover, 1968.

10—Appendix 4
Effect of Power Frequency Voltage and Number of Phases

For clarity, in this presentation the span factor K_{SP} is neglected, i.e., assumed to be 1.0. Later, in the conclusions, Section 6, it is again included. To this point in the development, only a single phase has been considered with a coupling factor C from the ground wires to the conductor. Now consider a three-phase line with coupling factors C_A, C_B, and C_C as illustrated in Fig. 1. Also the voltages on the tower will be different for each of the phases. That is, there are now three voltages, V_{TA}, V_{TB}, and V_{TC}. Therefore the *surge voltages* across the line insulation for phases A, B, and C, V_{IA}, V_{IB}, and V_{IC}, are now given by the abbreviated equations

$$V_{IA} = K_{IA}I$$
$$V_{IB} = K_{IB}I \quad (1)$$
$$V_{IC} = K_{IC}I$$

where

$$K_{IA} = K_{TA} - C_A K_{TT}$$
$$K_{IB} = K_{TB} - C_B K_{TT} \quad (2)$$
$$K_{IC} = K_{TC} - C_C K_{TT}$$

To determine the critical current I_C, the CFO_{NS} is substituted for these voltages across the insulation, and the highest value of K_{IA}, K_{IB}, and K_{IC} is used. Letting

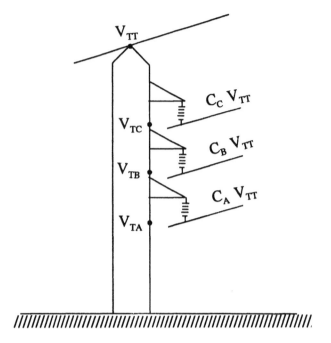

Figure 1 Three-phase line with different coupling factors and tower voltages.

this highest value be denoted as K_{IH},

$$I_C = \frac{CFO_{NS}}{K_{IH}} \tag{3}$$

Now consider the actual case where $K_{IA} \neq K_{IB} \neq K_{IC}$ and the power frequency voltage is considered. Letting the crest line-neutral power frequency voltage equal V_{LN}, then

$$\begin{aligned} V_{IA} &= K_{IA}I + V_{LN} \sin \omega t \\ V_{IB} &= K_{IB}I + V_{LN} \sin(\omega t - 120°) \\ V_{IC} &= K_{IC}I + V_{LN} \sin(\omega t + 120°) \end{aligned} \tag{4}$$

These voltages are illustrated in Fig. 2 and show that the maximum voltage across the insulation occurs at alternate times on phases A, B, and C. Note that the designation of the phases differs from that in the main text. That is, the phase that has the maximum voltage is dependent both on the surge voltage and on the phase angle ωt. For a portion of the time, the maximum voltage occurs across phase A insulation. But even though the phase B and phase C surge voltages across the insulation are less than the surge voltage across phase A, the maximum voltage occurs across phases B and C a portion of the time. Thus flashover will occur primarily across phase A, but flashovers will also occur across phases B and C. The fraction of the time that the maximum voltage occurs across each of the phases is a function of the

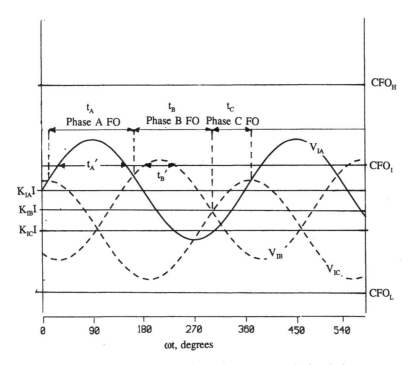

Figure 2 General diagram showing voltages across the insulation.

relative magnitudes of K_{IA}, K_{IB}, and K_{IC}. The probability of flashover is a function of relative magnitude of these K_I's and the CFO_{NS}. To explain, assume that the CFO_{NS} is equal to CFO_H, as depicted in Fig. 2. Then, since all V_I's are less than CFO_H, no flashovers will occur. Now, assume that the CFO_{NS} is equal to CFO_L, where CFO_L is less than any of the V_I's. Then flashover will occur on one of the phases. The portion of the time that flashover will occur to phase A is $t_A/(t_A + t_B + t_C)$, to phase B, $t_B/(t_A + t_B + t_C)$, and in a like manner for phase C. However, now consider that the CFO_{NS} is equal to CFO_1. Now flashover occurs to phase A during time t'_A, which is less than t_A, to phase B during time t'_B, which is less than t_B, and never to phase C.

To determine the values of t_A, t_B, and t_C, the points of intersection of V_{IA}, V_{IB} and V_{IC} must be found. This can be done by use of Eq. 4. For example, to find θ_{AB} of Fig. 3, equate V_{IA} to V_{IB}. The resultant equations are

$$\theta_{AB} = 150° + \sin^{-1}\frac{(K_{IA} - K_{IB})I}{\sqrt{3}V_{LN}}$$

$$\theta_{BC} = 270° + \sin^{-1}\frac{(K_{IB} - K_{IC})I}{\sqrt{3}V_{LN}} \qquad (5)$$

$$\theta_{AC} = 360° + \theta_{CA}$$

$$\theta_{CA} = 30° + \sin^{-1}\frac{(K_{IA} - K_{IC})I}{\sqrt{3}V_{LN}}$$

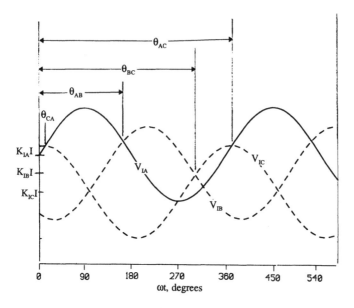

Figure 3 Defining the angles θ.

As an example, let $K_{IA} = 18$, $K_{IB} = 17$, and $K_{IC} = 16$. Also let $I = 100\,\text{kA}$ and $V_{LN} = 400\,\text{kV}$. Then

$$\theta_{AB} = 158.3° \qquad \theta_{BC} = 278.3° \qquad \theta_{AC} = 373.22° \qquad \theta_{CA} = 13.22° \tag{6}$$

Then the portion of the flashovers, in percent, that will occur on phases A, B, and C given the CFO_{NS} as depicted by CFO_L per Fig. 2 is

$$\text{FO}_A = \frac{\theta_{AB} - \theta_{CA}}{360} 100 = 40.3\%$$

$$\text{FO}_B = \frac{\theta_{BC} - \theta_{AB}}{360} 100 = 33.3\% \tag{7}$$

$$\text{FO}_C = \frac{\theta_{AC} - \theta_{BC}}{360} 100 = 26.4\%$$

where FO_A is the percentage of the flashovers that occur across phase A, etc.

Now consider the problem that the CFO_{NS} is given and the objective is to calculate the BFR. To perform this task, the critical current must first be found. The dilemma is shown in Figs. 4 and 5. As the current increases, V_{IA}, V_{IB}, and V_{IC} increase, but no flashovers occur until the current is sufficiently large so that CFO_{NS} and the maximum value of V_{IA} coincide; see Fig. 4. The critical current required for this event, denoted as I_{CL}, is

$$V_{IA} = K_{IA} I_{CL} + V_{LN} = \text{CFO}_{NS} \tag{8}$$

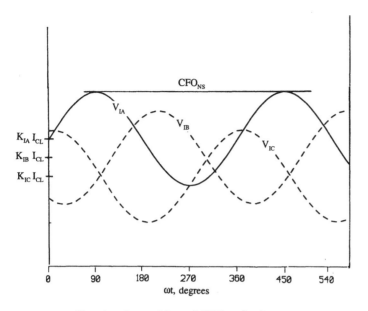

Figure 4 Showing the position of CFO_{NS} for I_{CL}.

Therefore

$$I_{CL} = \frac{CFO_{NS} - V_{LN}}{K_{IA}} \qquad (9)$$

In general, to find I_{CL}, use the largest of the K_I's.

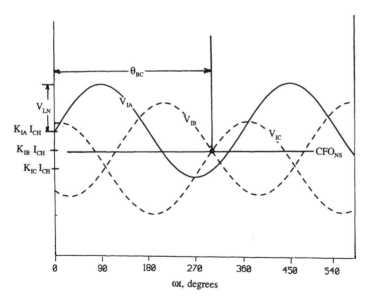

Figure 5 Position of CFO_{NS} for I_{CH}.

As the current increases further, phase A continues to flash over. Then, with further increase in current, phases B or C may flash over. Finally, phase A, B, or C may flash over as the current increases further.

However, for currents greater than some current I_{CH}, flashover is a certainty. At and above this current, flashover will occur and may occur on phase A, B, or C. This critical point is illustrated in Fig. 5 and occurs when the two lower voltages V_{IB} and V_{IC} are simultaneously equal to the CFO_{NS}, i.e.,

$$CFO_{NS} = V_{LN} \sin(\theta_{BC} - 120°) + K_{IB}I_{CH}$$
$$CFO_{NS} = V_{LN} \sin(\theta_{BC} + 120°) + K_{IC}I_{CH} \qquad (10)$$

Adding these two equations results in

$$I_{CH} = \frac{2CFO_{NS} + V_{LN} \sin \theta_{BC}}{K_{IB} + K_{IC}} \qquad (11)$$

However,

$$\theta_{BC} = 270° + \sin^{-1} \frac{(K_{IB} - K_{IC})I_{CH}}{\sqrt{3}V_{LN}} \qquad (12)$$

and Eqs. 10 and 11 must be iterated to obtain I_{CH}. Assuming again that $K_{IA} = 18$, $K_{IB} = 17$, $K_{IC} = 16$, $V_{PF} = 400\,kV$, and $CFO_{NS} = 2000\,kV$, then

$$\theta_{BC} = 279.06° \qquad I_{CL} = 88.89\,kA \qquad I_{CH} = 109.24\,kA \qquad (13)$$

Thus below 88.89 kA, no flashovers occur, and above 109.24 kA, flashover of the *line* has a probability of 100%. Between 88.89 kA and 109.24 kA, flashover of the line has a probability of between 0 and 100%. To demonstrate, Table 1 presents the voltages V_I assuming a current of 100 kA for various values of ωt.

Per Table 1, for $\omega t = 30°$, $V_{IA} = 2000\,kV$, which is equal to CFO_{NS}, and thus a flashover occurs on phase A. Continuing, there are five flashovers on phase A, three on phase B, and one on phase C. There are also three cases for which no flashover occurs. In total, nine flashovers out of twelve occur for this case of $I = 100\,kA$. Therefore the probability of flashover is 9/12. This estimate could be improved by increasing the number of time steps for the 360° cycle. However, assuming that the probability of line flashover, P_I, is determined in this manner, then the total probability of flashover is

$$P(FO) = \sum_{I_{CL}}^{I_{CH}} [P_I f(I) \Delta I] + P(I > I_{CH}) \qquad (14)$$

where $f(I)$ is the probability density function of the stroke current. Note that as before, for currents above I_{CH}, the $P(FO)$ is calculated as before, i.e., $P(FO) = P(I > I_{CH})$.

This procedure to calculate the probability, although exact, is not desirable except for computer use, since it is difficult and laborious. More desirable is some

Table 1 Probability of Flashover for $I = 100$ kA, $\text{CFO}_{\text{NS}} = 2000$ kV, $V_{\text{LN}} = 400$ kV, $K_{\text{IA}} = 18$, $K_{\text{IB}} = 17$, $K_{\text{IC}} = 16$

	Power freq. voltage			V_{I}				
ωt	A	B	C	A	B	C	P(FO)	FO phase
0	0	−346	346	1800	1354	1946	0	0
30	200	−400	200	2000	1300	1800	1	A
60	346	−346	0	2146	1354	1600	1	A
90	400	−200	−200	2200	1599	1400	1	A
120	346	0	−346	2146	1700	1254	1	A
150	200	200	−400	2000	1900	1200	1	A
180	0	346	−346	1800	2046	1254	1	B
210	−200	400	−200	1600	2100	1400	1	B
240	−346	346	0	1454	2046	1600	1	B
270	−400	200	200	1400	1900	1800	0	0
300	−346	0	346	1454	1700	1946	0	0
330	−200	−200	400	1600	1500	2000	1	C

approximate procedure that is simple to use and that will result in reasonable accuracy. To gain an insight, consider three situations: (1) the surge voltages across the insulation of phases A, B, and C are equal, i.e., $K_{\text{IA}} = K_{\text{IB}} = K_{\text{IC}}$; (2) the surge voltages across phases A and B are equal and the surge voltage across phase C is so much lower that it need not be considered, i.e., $K_{\text{IA}} = K_{\text{IB}}$ and $K_{\text{IC}} \lll K_{\text{IA}}$; and (3) the surge voltages across phases B and C are so much lower than that across phase A that they need not be considered, i.e., $K_{\text{IB}} = K_{\text{IC}}$ and $K_{\text{IA}} \ggg K_{\text{IB}}$.

1 $K_{\text{IA}} = K_{\text{IB}} = K_{\text{IC}} = K_{\text{I}}$

The voltages across phases A, B, and C are shown in Fig. 6. Using Eq. 5

$$\theta_{\text{AB}} = 150° \quad \theta_{\text{BC}} = 270° \quad \theta_{\text{CA}} = 30° \quad \theta_{\text{AC}} = 360° \tag{15}$$

which are marked on Fig. 6. A horizontal line is drawn in this figure to represent the CFO_{NS}. If I is decreased so that the CFO_{NS} just equals V_{IA}, then again

$$I_{\text{CL}} = \frac{\text{CFO}_{\text{NS}} - V_{\text{LN}}}{K_{\text{I}}} \tag{16}$$

If the current is increased so that the line for the CFO_{NS} is as depicted in Fig. 6, flashovers will occur across phases A, B, and C insulations. That is, the P(FO) given this value of current is $(\omega t_{\text{A}} + \omega t_{\text{B}} + \omega t_{\text{C}})/2\pi$. Or, since $t_{\text{A}} = t_{\text{B}} = t_{\text{C}}$, $P(\text{FO}) = 3\omega t_{\text{A}}/2\pi$. When the current increases further so that the CFO_{NS} intersects the values of θ_{AB}, θ_{BC}, and θ_{CA} (i.e., 30, 150, and 270 degrees), the value of I_{CH} is reached. From Eq. 11,

$$I_{\text{CH}} = \frac{\text{CFO}_{\text{NS}} - 0.5 V_{\text{LN}}}{K_{\text{I}}} \tag{17}$$

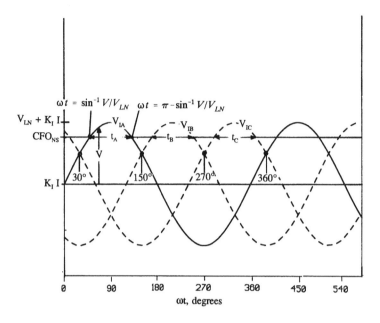

Figure 6 For $V_{IA} = V_{IB} = V_{IC}$.

And for currents above I_{CH}, the $P(FO)$ is 100%. The range of $I_{CH} - I_{CL}$ is

$$I_{CH} - I_{CL} = \frac{0.5 V_{LN}}{K_I} \tag{18}$$

Returning to the calculation of $P(FO)$ for the line, previously the $P(FO)$ was calculated as the probability that $V_{IA} >$ CFO$_{NS}$ or $V_{IB} >$ CFO$_{NS}$ or $V_{IC} >$ CFO$_{NS}$. Since $V_{IA} = V_{IB} = V_{IC}$, then

$$\begin{aligned} P(FO) &= 3P[V_{IA} \geq \text{CFO}_{NS}] \\ &= 3P[(K_{IA}I + V_{LN}\sin\omega t) \geq \text{CFO}_{NS}] \\ &= 3P\left[1 \geq \frac{\text{CFO}_{NS} - V_{LN}\sin\omega t}{K_{IA}}\right] \end{aligned} \tag{19}$$

This equation shows that the critical current is variable, that is, I_C is

$$I_C = \frac{\text{CFO}_{NS} - V_{LN}\sin\omega t}{K_{IA}} \tag{20}$$

which is valid from $\omega t = \pi/2$ where $I = I_{CL}$ to $\omega t = \pi/6$ where $I = I_{CH}$. The objective now is to find a value of K_{PF}, a constant,

$$I_C = \frac{\text{CFO}_{NS} - K_{PF} V_{LN}}{K_{IA}} \tag{21}$$

such that $P(FO) = P(I > I_C)$ results in a good approximation of the true $P(FO)$.

Effect of Power Frequency Voltage and Number of Phases

Once more, examining Fig. 6, the probability that V_I is greater than the CFO_{NS} is

$$P[V_I > CFO_{NS}] = P[(K_I I + V_{LN} \sin \omega t) > CFO_{NS}] = P[V_{LN} \sin \omega t > (CFO_{NS} - K_I I)] \tag{22}$$

Assuming that the interval between I_{CL} and I_{CH} is small so that the stroke current is approximately constant

$$\begin{aligned} P[V_I > CFO_{NS}] &\approx P[V_{LN} \sin \omega t > V] \\ &= \frac{\omega t_A + \omega t_B + \omega t_C}{2\pi} \\ &= \frac{3\omega t_A}{2\pi} \\ &= \frac{3[\pi - 2\sin^{-1}(V/V_{LN})]}{2\pi} \\ &= \frac{3}{2} - \frac{3}{\pi} \sin^{-1} \frac{V}{V_{LN}} \end{aligned} \tag{23}$$

where V is a power frequency voltage as shown in Fig. 7. The cumulative distribution, equal to or less than, is

$$P[\text{PF voltage} \leq V] = F(V/V_{LN}) = -\frac{1}{2} + \frac{3}{\pi} \sin^{-1} \frac{V}{V_{LN}} \tag{24}$$

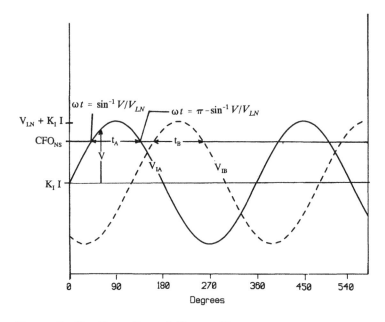

Figure 7 For $V_{IA} = V_{IB}$ and $V_{IC} \lll V_{IA}$.

which is valid between $V/V_{PF} = 0.5$ to 1.0. Differentiating to find the density,

$$f(V/V_{LN}) = \frac{3}{\pi} \frac{1}{\sqrt{1-(V/V_{LN})}} \qquad (25)$$

From this the mean value may be determined as

$$\mu = \frac{3}{\pi} \int_{0.5}^{1.0} \frac{V/V_{LN}}{\sqrt{1-(V/V_{LN})^2}} d(V/V_{LN}) = -\frac{3}{\pi} \left[\sqrt{1-(V/V_{LN})^2}\right]_{0.5}^{1.0} = \frac{3\sqrt{3}}{2\pi} = 0.827 \qquad (26)$$

The mode M can be found from Eq. 24 as 0.866. Therefore as an approximation,

$$I_C = \frac{CFO_{NS} - 0.827 V_{LN}}{K_{IA}} \qquad (27)$$

This approximation of I_C is then used to calculate the probability of flashover of the line as $P(FO) = P(I \geq I_C)$.

2 $K_{IA} = K_{IB}$, $K_{IC} <<< K_{IA}$

Using the same methods as from the previous case (see Fig. 7),

$$I_{CL} = \frac{CFO_{NS} - V_{LN}}{K_{IA}}$$

$$I_{CH} = \frac{CFO_{NS} + 0.5 V_{LN}}{K_{IA}} \qquad (28)$$

$$I_{CH} - I_{CL} = \frac{1.5 V_{LN}}{K_{IA}}$$

Then

$$P[\text{PF voltage} \geq V] = 1 - \frac{2}{\pi} \sin^{-1} V/V_{LN} \qquad \text{for } 0.5 \leq V/V_{LN} \leq 1.0$$

$$= \frac{5}{6} - \frac{1}{\pi} \sin^{-1} V/V_{LN} \qquad \text{for } -0.5 \leq V/V_{LN} < 0.5 \qquad (29)$$

Therefore

$$f(V/V_{LN}) = \frac{2}{\pi\sqrt{1-(V/V_{LN})^2}} \qquad \text{for } 0.5 \leq V/V_{LN} \leq 1.0$$

$$= \frac{1}{\pi\sqrt{1-(V/V_{LN})^2}} \qquad \text{for } -0.5 \leq V/V_{LN} < 0.5 \qquad (30)$$

and

$$\mu = \frac{\sqrt{3}}{\pi} = 0.551 \qquad M = \frac{\sqrt{2}}{2} = 0.707 \qquad (31)$$

Thus, for this case,

$$I_C = \frac{\text{CFO}_{\text{NS}} - 0.551 V_{\text{LN}}}{K_{\text{IA}}} \qquad (32)$$

3 ONLY K_{IA}, THAT IS, $K_{\text{IB}} = K_{\text{IC}}$ AND $K_{\text{IA}} \ggg K_{\text{IB}}$

Using the same methods as previously, the mean and mode for this case are both zero. Thus the critical current is

$$I_C = \frac{\text{CFO}_{\text{NS}}}{K_{\text{IA}}} \qquad (33)$$

The critical currents and their range are

$$I_{\text{CL}} = \frac{\text{CFO}_{\text{NS}} - V_{\text{LN}}}{K_{\text{IA}}}$$

$$I_{\text{CH}} = \frac{\text{CFO}_{\text{NS}} + V_{\text{LN}}}{K_{\text{IA}}} \qquad (34)$$

$$I_{\text{CH}} - I_{\text{CL}} = \frac{2 V_{\text{LN}}}{K_{\text{IA}}}$$

4 VERIFYING THE RESULTS OF SECTIONS 1 AND 3

The conclusions reached in Sections 1 to 3 can be checked by use of a computer program. In this program the BFR is calculated by varying the instantaneous power frequency voltage in steps of 30°, 12 increments. A typical 115-kV single-circuit line is assumed. To simulate the case of Sections 1 to 3, the coupling factors were altered. To show the effect of the nominal system voltage, nominal system voltages of 115, 200, and 300 kV are assumed. The CFO is 1067 kV. The results are shown in Table 2, where the "exact" method is compared against the method for which a *KPF* is used. As noted, the two values of BFR compared most favorably for the lower system voltages and for the case where $V_{\text{IA}} = V_{\text{IB}} = V_{\text{IC}}$. the comparison is worst for higher system voltages and for the case where only one phase is considered. This good or bad comparison is the result of the range of $I_{\text{CH}} - I_{\text{CL}}$, which for the case of two phases considered is three times that when all three phases are considered. The range is largest when only one phase is considered, being four times greater than that when all phases are considered. In addition, the range increases as the ratio of the system voltage to the CFO increases.

5 FINDING K_{PF} FOR PRACTICAL CASES

The value of K_{PF} for three theoretical cases has been determined assuming that the range of $I_{\text{CH}} - I_{\text{CL}}$ is small. In addition, the theoretical cases assume that the other phases not considered add insignificantly to the BFR or to the value of the K_{PF}. In

Table 2 Comparison of "Exact" and Approximate BFRs

Case	Nominal sys. voltage, kV	"Exact" BFR FO/100 km-yrs	Approximate using K_{PF}	
			K_{PF}	BFR
Section 1	115	0.2848	0.827	0.2862
$V_{IA} = V_{IB} = V_{IC}$	200	0.4491	0.827	0.4476
	300	0.7739	0.827	0.7602
Section 2	115	0.2438	0.551	0.2240
$V_{IA} = V_{IB}$	200	0.3570	0.551	0.3150
	300	0.5810	0.551	0.4473
Section 3	115	0.1782	0	0.1566
V_{IA}	200	0.2268	0	0.1566
	300	0.3350	0	0.1156

this section, two practical cases are considered to determine the effect on the value of K_{PF}.

5.1 Horizontal Configuration

For a horizontal configuration of phase conductors on a typical single-circuit line, the coupling factors to the two outside phases are equal and the coupling factor to the center phase is about 17% greater. The V's or K's are approximately $(1 - C)$ times the voltage of the tower top, so that the two of the V's are equal but the third phase has a V_I that is only 9% lower. Thus the center phase cannot be neglected. The value of K_{PF} should be somewhere between 0.551 and 0.827. Using a computer program, K_{PF} was found to be a function of the nominal system voltage V_n divided by the CFO, i.e., V_n/CFO as shown in Fig. 8. The average value of K_{PF} is about 0.70, which can be used as an approximation.

5.2 Vertical Configuration

For a vertical configuration of phase conductors, the V_I or the K_I of the top phase is about 50% greater than that of the bottom phase. The V_I of the middle phase is about 13% greater than that of the bottom phase. Thus it appears that while the top phase could be neglected, the middle phase cannot. Thus K_{PF} should be between 0

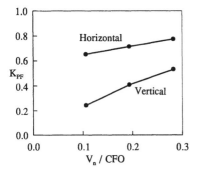

Figure 8 Actual values of K_{PF} for horizontal and vertical phase conductors.

Effect of Power Frequency Voltage and Number of Phases

and 0.551 and closer to 0.551. Per Fig. 8 for a typical line, *KPF* varies from about 0.25 to about 0.55 with an average of about 0.40. Thus for this configuration, K_{PF} may be approximated as 0.40.

5.3 Details of a Calculation

To illustrate a detailed calculation, a 115-kV single-circuit horizontal configured line is considered having the following characteristics:

Grd flash density = 6.0 flashes/km^2/year   Nominal system voltage = 115 kV
Surge impedances: Ground wire, 339 ohms; tower, 170 ohms
Coupling factors: Phases A/B/C = 0.331/0.386/0.331
Heights: Ground wires: 57 ft; all phase conductors, 46 ft
Horizontal separation of ground wires, 12.5 ft   span length, 750 ft
CFO, 1067 kV; footing resistance $R_0 = 20$ ohms; soil resistivity, 400 ohm-meters

The power frequency voltage was considered by calculating the critical current and BFR for each of the phases for instantaneous power frequency voltages determined for each of 36 10° steps. The results are shown in Table 3. The first column gives the angle of the power frequency voltage for phase A. The other columns give the critical currents and BFRs. The phase that flashes over is the phase with the lowest critical current or with the highest BFR. To obtain the total BFR, the maximum BFRs at each time instant, those in the last column, are added and divided by the number of time steps. The total BFR is therefore 6.881/36 = 0.191 flashovers/100 km-year. The number of flashovers on A, B, and C phases are 15.5, 5.0, and 15.5, respectively. Dividing these by 36 results in the conclusion that 43.06% occur to phase A and 43.06% occur to phase C. Also, 13.89% occur to phase B, the middle phase. Note that in Table 3, for 30°, the critical current and BFR for phases A and C are equal. Therefore each phase is assigned 1/2 flashover.

6 CONCLUSION

To conclude this section, adding into the equations the span factor K_{SP}, the BFR can be calculated by first finding the critical current using the equation

$$I_c = \frac{\text{CFO}_{NS} - K_{PF}V_{LN}}{K_{SP}(K_{TA} - C_A K_{TT})} = \frac{\text{CFO}_{NS} - V_{PF}}{K_{SP}(K_{TA} - C_A K_{TT})} \tag{35}$$

where C_A is the lowest value of the coupling factor and K_{TA} is the corresponding value for this phase; V_{PF} is

$$V_{PF} = K_{PF}V_{LN} \tag{36}$$

where V_{LN} is the crest nominal line-to-neutral voltage, i.e.,

$$V_{LN} = \frac{\sqrt{2}}{\sqrt{3}} V_{L-L} \tag{37}$$

where V_{L-L} is the nominal system voltage (line to line). For example, for a nominal system voltage of 230 kV, V_{LN}, is 188 kV.

Table 3 115 kV Single-Circuit Line

ωt degrees	I_C, Critical current, kA			BFR, FO/100 km-yrs			Flashover phase	Line BFR
	A	B	C	A	B	C		
0	188	243	166	.119	.029	.225	C	.225
10	184	246	168	.136	.027	.209	C	.209
20	180	247	172	.153	.027	.191	C	.191
30	175	247	175	.172	.026	.172	A&C	.172
40	171	247	170	.191	.027	.153	A	.191
50	169	246	184	.209	.027	.136	A	.209
60	166	243	188	.225	.029	.119	A	.225
70	164	240	193	.237	.032	.105	A	.237
80	163	236	197	.246	.035	.094	A	.246
90	163	230	202	.249	.039	.083	A	.249
100	163	226	205	.246	.044	.075	A	.246
110	164	220	209	.237	.051	.069	A	.237
120	166	216	211	.225	.058	.064	A	.225
130	169	210	213	.209	.066	.061	A	.209
140	172	205	215	.191	.076	.059	A	.191
150	175	201	215	.172	.086	.058	A	.172
160	180	196	215	.153	.096	.059	A	.153
170	184	193	213	.136	.105	.061	A	.136
180	188	190	212	.119	.114	.064	A	.119
190	193	188	209	.105	.121	.069	B	.121
200	197	187	205	.094	.125	.075	B	.125
210	202	186	202	.083	.127	.083	B	.127
220	205	187	197	.075	.125	.094	B	.125
230	209	188	193	.069	.121	.105	B	.121
240	212	190	188	.064	.114	.119	C	.119
250	213	193	184	.061	.105	.136	C	.136
260	215	196	180	.059	.096	.153	C	.153
270	215	201	175	.058	.086	.172	C	.172
280	215	205	172	.059	.076	.191	C	.191
290	213	210	169	.061	.066	.209	C	.209
300	212	216	166	.064	.058	.225	C	.225
310	209	221	164	.069	.051	.237	C	.237
320	205	226	163	.075	.044	.246	C	.246
330	202	231	163	.083	.039	.249	C	.249
340	197	236	163	.094	.035	.246	C	.246
350	193	240	164	.105	.032	.237	C	.237

The value of K_{PF} is dependent on the phase arrangement

1. For a horizontally configured line, a *KPF* of 0.70 is recommended.
2. For a vertical configuration of phases, a K_{PF} of 0.40 is recommended.
3. If uncertain, the conservative value of 0.70 should be used.

11
The Incoming Surge and Open Breaker Protection

1 DESIGNING FOR AN MTBF

The purpose of this chapter is to provide a method to estimate the magnitude and shape, i.e., the front steepness and the tail time constant, of the surge that arrives at the entrance to the station, which is denoted as the incoming surge. The magnitude and waveshape of this surge is a function of the distance between the station and the stroke-terminating point, the magnitude of the stroke current, and the initiating event—shielding failure or backflash. In turn, the number of surges that arrive at the station is a function of this distance and the BFR or SFR. Since the stroke current and the BFR or SFR are statistical quantities, it is apparent that the magnitude and shape of the surge is a random event and must be considered in probabilistic terms. Thus the incoming surge is statistical, which leads to the concept that the magnitude and shape of the incoming surge may be based on a design rate of the number of surges per year that equal or exceed a specific steepness and magnitude. The reciprocal of the number of surges is the return period or mean time between surges or MTBS in units of years per surge. That is, a surge may be selected so that the probability that its severity is equaled or exceeded is, for example, once in 100 years. If this 100 year surge produces voltages within the station that just equal the insulation strength, the mean time between failure of the equipment, the MTBF, is 100 years. However, this last statement applies only for a single-line station where the MTBS is equal to the MTBF.

For a multiline station, the situation is somewhat more complex, since each line brings a 100-year surge into the station. To clarify, consider a simple multiline station as depicted in Fig. 1 and assume that the desired MTBF is 100 years. If a

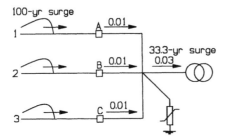

Figure 1 For the transformer, the MTBS is n times the MTBF of the station.

100-year surge, 0.01 surges/year, is sequentially applied to each line, the total number of surges applied to the transformer is 0.03 for an MTBF of 33.3 years. Thus the three lines act as collectors of surges. To compensate and achieve an MTBF of 100 years at the transformer, or for all equipment on the transformer bus, a 300 year surge should be applied to each line, an MTBS of 300 years. However, for the equipment on the other buses, the breaker, the disconnecting switches, the bus support insulators, etc., the 100 year surge is used since, as will be shown in Chapter 13, each line's 100 year surge produces the most severe voltage on the equipment on its bus. So as not to create more confusion, the 100 year surge is denoted as an MTBS, a mean time between surges.

As a general rule, for an n-line station, for the transformer or equipment on the transformer bus, the MTBS should be equal to n times the desired station MTBF, whereas for the breakers or equipment not on the transformer bus, the MTBS is set equal to the MTBF. That is, for this example, a 100 year surge would be used for analysis of the breaker BIL.

Another modification is required to account for contingency switching conditions within a multiline station, where contingency switching refers to $(n-1)$, $(n-2)$, etc. lines in service in an n-line station. As an example, consider again that the desired MTBF is 100 years for the three-line station of Fig. 1. For the contingency of two lines in service, for analysis of the transformer, a 200-year surge would appear to be proper. However, suppose that this condition exists only 15 percent of the time—during thunderstorms, i.e., the probability is 0.15. Then the 200-year surge could be modified to 0.15(200) or 30 years. Continuing the example with assumed probabilities, Table 1 shows the results. If these probabilities are not considered, the MTBS for one line in service is 100 years. Considering these probabilities, the MTBS decreases to only 5 years. However, the MTBS for all lines in service remains high—it only decreases by 20%—but there are mitigating effects that will be considered in Chapter 13.

To complete the example, Table 1 also illustrates the effect for equipment not on the transformer bus. Note that in this case, when the probability of the contingency is not considered, the MTBS remains at 100 years. However, considering these probabilities, the MTBS is considerably reduced. Again, this will be considered in Chapter 13.

The above example assumes a fairly simply substation, one perhaps typical of lower voltage stations. However, at high voltages, the station configuration becomes more complex, and the general rule that the incoming surge should be based on n

The Incoming Surge and Open Breaker Protection

Table 1 Determining the MTBS for a Three-Line Station; Station MTBF = 100 years.

No. of lines in service	Probability of lines in service	MTBS, years, for transformer		MTBS, years, for other equip.	
		Without prob.	With prob.	Without prob.	With prob.
3	0.80	300	240	100	80
2	0.15	200	30	100	15
1	0.05	100	5	100	5

times the MTBF for equipment on the transformer bus is not strictly valid. To explain the concepts more fully, consider the breaker and a half station of Fig. 2. Assume that the station is to be designed for an MTBF of 100 years. Assume that a 100 year surge arrives at the station from each of the lines and that the voltage at the transformer is 1200 kV for the surge arriving on line 1, 1200 kV for the surge arriving on line 2, 1000 kV for the surge arriving on line 3, and 900 kV for a surge arriving on line 4. Thus, twice in 100 years, or once in 50 years, the maximum voltage at the transformer reaches 1200 kV. Therefore to achieve a station MTBF of 100 years, a 200-year surge should be applied from each line. If, alternately, the four voltages were 1200, 1100, 900, and 800 kV, then the use of a 100-year surge on each of the lines would result in a station MTBF of 100 years. Thus, frequently, for more complexly configures stations, the prior determination of the proper MTBS cannot be made. In this case, an initial run or case is required to establish the MTBS.

The incoming surge developed in this chapter is applied to the station, and the voltages at the equipment terminals are calculated and compared to the insulation strength. (Chapter 13 provides a simplified method to estimate these voltages.) Therefore the station insulation is designed on the basis of an MTBF. In this Chapter, however, the task is to determine the shape and magnitude of the incoming surge for a given MTBS.

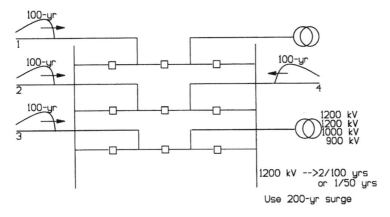

Figure 2 Determining the MTBS for a complex station configuration.

The selection of the MTBF of the station is at the discretion of the designer, but it is normally selected on the basis of the consequence of failure of the insulation and on the expected life of the equipment. To explain, if the life of the equipment is 30 years, it would appear imprudent to select an MTBF of 200 years. For air-insulated stations, an MTBF of between 50 and 100 years has been used [1, 2], whereas in the IEC guide [3], values of 400 and 500 years are used. In contrast, because of the consequence of a failure in a gas-insulated station (longer repair/replacement times), a higher MTBF of 400 years has been used, and MTBFs between 300 and 1000 years have been suggested [4].

As mentioned previously, the incoming surge may be a result of a shielding failure or a backflash. The surge resulting from a shielding failure is limited in magnitude to the maximum shielding failure current times half of the conductor surge impedance. In contrast, the maximum surge resulting from a backflash is only limited by the stroke current. Because of this, and because the BFR is normally much larger than the SFR, the maximum incoming surge resulting from a backflash is usually more severe and in many cases is the only surge considered. However, for completeness, both surge-originating events will be considered.

2 INCOMING SURGE FROM A BACKFLASH

2.1 Review of the Backflash

Using the simplified CIGRE method as presented in Chapter 10, in which the tower component of voltage is neglected, the crest voltages on the tower, V_{TT}, and phase conductor, V_C are illustrated in Fig. 3a. The crest voltage across the insulation, V_I, is shown in Fig. 3b. The defining equations are

$$V_{TT} = R_e I$$
$$V_C = C R_e I - V_{PF} \qquad (1)$$
$$V_I = (1 - C)V_{TT} + V_{PF} = (1 - C)R_e I + V_{PF}$$

where C is the coupling factor, I is the stroke current, and V_{PF} is the power frequency voltage at the instant of stroke termination, i.e.,

$$V_{PF} = K_{PF} V_{LN} \qquad (2)$$

where V_{LN} is the crest line-to-neutral nominal system voltage and K_{PF} is approximately 0.40 for a vertical configuration of phase conductors and about 0.70 for a horizontal configuration. R_e is given by the equation

$$R_e = \frac{R_i Z_g}{Z_g + 2R_i} \qquad (3)$$

where R_i is the impulse or high current resistance and Z_g is the ground wire surge impedance.

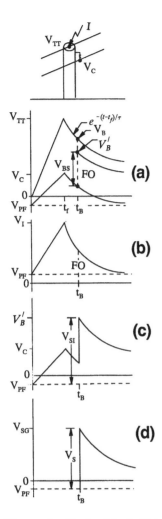

Figure 3 The backflash. Voltages (a) on the tower and conductor, (b) across the insulation, (c) on the conductor at struck point, (d) on the conductor at station without corona.

Also,

$$R_i = \frac{R_0}{\sqrt{1+(I_R/I_g)}}$$
$$I_g = \frac{1}{2\pi}\frac{E_0\rho}{R_0^2} \qquad (4)$$
$$I_R = \frac{R_e}{R_i}I$$

where I_R is the current through the footing resistance, E_0 is the soil breakdown gradient taken as 400 kV/m, I_g is the current required to achieve a gradient of E_0, R_0 is the measured or low-current footing resistance, and ρ is the soil resistivity in ohm-meters.

A backflash occurs if the voltage across the insulation exceeds the critical flashover voltage of the insulation, CFO_{NS}; the resultant critical current I_C at and above which flashover occurs is

$$I_C = \frac{\text{CRO}_{\text{NS}} - V_{\text{PF}}}{(1-C)R_e} \tag{5}$$

where

$$\text{CFO}_{\text{NS}} = \left(0.977 + \frac{2.82}{\tau}\right)\left(1 - 0.2\frac{V_{\text{PF}}}{\text{CFO}}\right)\text{CFO} \tag{6}$$

and

$$\tau = \frac{Z_g}{R_i}T_S \tag{7}$$

where T_S is the span travel time, CFO is the positive polarity CFO for the standard lightning impulse, and CFO_{NS} is the CFO for the nonstandard voltage waveshape that appears across the line insulation. The BFR, in flashovers per 100 km-years, is then found from the equation

$$\text{BFR} = 0.6 N_L P(I \geq I_C) \tag{8}$$

where N_L is the number of flashes to the line per 100 km-years.

2.2 The Incoming Surge

When the stroke current is equal to I_C, flashover "just" occurs, and the time to flashover or breakdown, t_B, is at some point on the tail and at its maximum value. As the stroke current increases, the voltage across the insulation increases, leading to decreased times to flashover.

To develop the theory, let the time to flashover for currents at or above I_C be t_B, as shown in Figs. 3a and 3b. At t_B, the voltage across the insulation is decreased to zero in essentially zero time, and the voltage on the conductor is instantly increased to a voltage V_B' that is slightly lower than the tower voltage V_B. This voltage can be obtained from an analysis of the circuit of Fig. 4, where the closing of the switch represents the flashover. The impedance on the tower side of the switch is R_e and the impedance on the conductor side of the switch is half the conductor surge impedance Z_C. The surge portion of the voltage V_{BS} can be obtained by the cancellation method of analysis. Per Fig. 4a, the voltage across the switch is

$$\Delta V = V_B(1-C) + V_{\text{PF}} \tag{9}$$

To simulate the switch closing, this voltage is applied oppositely across the switch to cancel the voltage. The resultant circuit is shown in Fig. 4b. The surge voltage V_{BS} is

$$V_{\text{BS}} = \frac{Z_c}{Z_c + 2R_e}[V_B(1-C) + V_{\text{PF}}]$$
$$V_{\text{BS}} = V_B\left[\frac{R'}{R_e}(1-C)\right] + \frac{R'}{R_e}V_{\text{PF}} \tag{10}$$

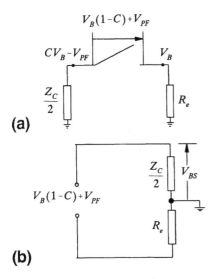

Figure 4 Upon flashover, the surge voltage becomes V_{BS}.

where

$$R' = \frac{R_e Z_c}{Z_c + 2R_e} \tag{11}$$

and the total conductor voltage V'_B is V_{BS} plus the voltage that existed on the conductor before the switch was closed. Therefore as shown in Fig. 3c,

$$V'_B = V_B \left[\frac{R'}{R_e}(1-C) + C \right] - \left[1 - \frac{R'}{R_e} \right] V_{PF} \tag{12}$$

The surge voltage V_{BS} starts at V_{PF}, and therefore the total surge voltage is V'_B plus V_{PF}, which is denoted as V_{SI} per Fig. 3c. That is,

$$V_{SI} = V'_B + V_{PF} = V_B \left[\frac{R'}{R_e}(1-C) + C \right] + \frac{R'}{R_e} V_{PF} \tag{13}$$

As this surge voltage travels in toward the station and progressively passes each adjacent tower, the voltage on the ground wire vanishes. As shown in Chapter 9, if the voltage on the ground wire vanishes, the coupled voltage on the conductor CV_B, vanishes. In addition, the surge voltage V_{SI} must be multiplied by $(1-C)$. Therefore V_S of Fig. 3d becomes

$$V_S = V_B(1-C) \left[\frac{R'}{R_e}(1-C) + C \right] + \frac{R'}{R_e} V_{PF}(1-C) \tag{14}$$

and the total voltage is V_{SG} per Fig. 3d. Since the voltage after crest decreases with the time constant τ, the voltage V_B is

$$V_{\rm B} = V_{\rm TT} e^{-\frac{t_{\rm B}-t_{\rm f}}{\tau}} = IR_{\rm e} e^{-\frac{t_{\rm B}-t_{\rm f}}{\tau}} \qquad (15)$$

where $t_{\rm f}$ is the time to crest of the stroke current; see Fig. 3a. Using Eq. 15, Eq. 14 becomes

$$V_{\rm S} = (1-C)[(1-C)R' + CR_{\rm e}]I e^{-\frac{t_{\rm B}-t_{\rm f}}{\tau}} + \frac{R'}{R_{\rm e}}(1-C)V_{\rm PF} \qquad (16)$$

The remaining task is to estimate the time to breakdown, $t_{\rm B}$. Using the leader progression model, the time to breakdown or flashover can be approximated by the equation

$$\frac{V_{\rm I}}{\rm CFO} = 1.68 e^{-\frac{t_{\rm B}-t_{\rm f}}{B}} \qquad (17)$$

or

$$t_{\rm B} - t_{\rm f} = B \ln \frac{1.68}{V_{\rm I}/{\rm CFO}}$$

where

$$\frac{V_{\rm I}}{\rm CFO_{NS}} \leq \frac{V_{\rm I}}{\rm CFO} \leq 1.68$$

$$B = 15 - \frac{\tau}{13} \qquad (18)$$

That is, the voltage $V_{\rm I}$ must be equal to or greater than the nonstandard critical flashover voltage for flashover to occur, and the 1.68 limitation indicates that the equation is applicable for times greater than the time to crest $t_{\rm f}$. Combining the equations, for $V_{\rm S}$ we obtain

$$V_{\rm S} = (1-C)[R'(1-C) + R_{\rm e}C]I \left[\frac{V_{\rm I}/{\rm CFO}}{1.68}\right]^{\frac{B}{\tau}} + \frac{R'}{R_{\rm e}}(1-C)V_{\rm PF} \qquad (19)$$

or using Eq. 1,

$$V_{\rm S} = (V_{\rm I} - V_{\rm PF})\left[\frac{R'}{R_{\rm e}}(1-C) + C\right]\left[\frac{V_{\rm I}/{\rm CFO}}{1.68}\right]^{\frac{B}{\tau}} + \frac{R'}{R_{\rm e}}(1-C)V_{\rm PF} \qquad (20)$$

To simplify, if $V_{\rm B} = V_{\rm B}'$, then $R' = R_{\rm e}$ and Eq. 20 becomes

$$V_{\rm S} = (V_{\rm I} - V_{\rm PF})\left[\frac{V_{\rm I}/{\rm CFO}}{1.68}\right]^{\frac{B}{\tau}} + (1-C)V_{\rm PF} \qquad (21)$$

Before corona is considered, let us pause and estimate the maximum and minimum surge voltage $V_{\rm BS}$. Assume that $V_{\rm PF}/{\rm CFO} = 0.15$, $C = 0.20$, $R_{\rm i} = 20$ ohms, and $Z_C = Z_g = 350$ ohms. Then $R_e = 17.95$, $R' = 16.28$, and $\tau = 17.5$ (300-m spans). For the minimum, let $V_I = \text{CFO}_{NS}$; then from Eq. 6, $\text{CFO}_{NS} = 1.104$ CFO. From Eq. 18, $B = 13.65$, and from Eq. 17, $(t_B - t_f) = 5.73$. Therefore, from Eq. 20, $V_S = 0.745$ CFO. For the maximum, assume flashover at t_f. Then $V_I = 1.68$ CFO and $V_S = 1.524$ CFO. Therefore the surge voltage varies between about 0.76 and 1.52 times the standard CFO. If Eq. 21 is used, the values are 0.808 CFO and 1.65 CFO, an error of about 8%.

2.3 Effect of Corona

Now to return to the development and consider the effect of corona. As discussed in Chapter 9, as the surge travels in toward the station, corona pushes back the front of the surge and in so doing decreases the front steepness and the crest magnitude. Figure 5a illustrates this effect. Below the corona inception voltage, the surge is attenuated only by ground effects, and thus the vertical front is only slightly modified. However, above corona, the corona effect is predominant and is approximated by the horizontal dotted line of Fig. 5a followed by a linearly rising front having a rate of rise, or steepness, of S. The important part of this characteristic is the steepness, and therefore a further approximation is made per Fig. 5b, i.e., the steepness occurs over the entire front of the surge. The magnitude of this surge is given by the intersection of the steepness S, with the original surge. From Chapter 9, the equation for the steepness is

$$S = \frac{S_0}{1 + (S_0 d / K_c)} \tag{22}$$

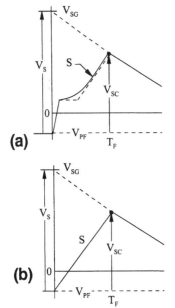

Figure 5 (a) Voltage at station with corona and (b) approximation.

Table 2 K_C in km-kV/μs

Conductor	K_c
Single	700
Two-cond. bundle	1000
Three or four cond. bundle	1700
Six or eight cond. bundle	2500

where S_0 is the steepness of the surge at the struck point, K_c is a constant, and d is the distance traveled in km. Since the steepness of the surge at the struck point is essentially infinity,

$$S = \frac{K_c}{d} \qquad (23)$$

From Chapter 9, suggested values of K_c, the corona constant, are shown in Table 2.
To obtain the crest voltage,

$$V_{SC} = V_S e^{-\frac{T_F}{\tau}} \qquad (24)$$

where

$$T_F = \frac{V_S}{S} \qquad (25)$$

Therefore

$$V_{SC} = V_S e^{-\frac{V_{SC}}{S\tau}} \qquad (26)$$

With less than about 1% error,

$$T_F = \frac{V_S}{S} \qquad (27)$$

and therefore, approximately,

$$V_{SC} = V_S e^{-\frac{V_S}{S\tau}} \qquad (28)$$

Returning now to the estimate of the crest voltage of this incoming surge, assume that $K_S = 1000$, $d = 1$ km, and the CFO = 1200 kV. Then $S = 1000$ kV/μs and the minimum and maximum magnitudes of the incoming surge V_{SC} are 0.71 and 1.37 CFO. Thus use of Eq. 28 only diminishes the incoming surge by 5 to 11%.

2.4 Estimating the Incoming Surge for an MTBS

The above equations provide a method for estimating the steepness and crest voltage of the incoming surge for a given stroke current. The value of this current is a

The Incoming Surge and Open Breaker Protection

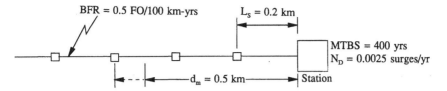

Figure 6 The steepness of the incoming surge is dependent on the stroke location.

function of the degree of reliability desired for the station, the MTBF, which in turn is a function of the MTBS or the return period of the surge. To develop the concept and the equations, consider Fig. 6. It is desired to determine the incoming surge for an MTBS of 400 years, or 400 years/surge. The reciprocal of the MTBS, denoted as N_D, is 0.0025 surges per year. Further assume that the BFR for the line is 0.5 flashovers per 100 km-years. The span length is L_S in km. Per Eq. 23, the maximum steepness of the incoming surge occurs for a stroke-termination point that is closest to the station. The minimum distance from the station, d_m, is where the number of flashovers to the line in distance d_m is equal to the number of surges, N_D, or

$$d_m = \frac{N_D}{\text{BFR}} = \frac{1}{\text{BFR(MTBS)}} = \frac{100}{0.5(400)} = 0.5 \text{ km} \qquad (29)$$

which in this example is 0.5 km. If a span length is now assumed as 200 m, this location is midway between the second and the third tower. Since, in development of the calculation technique for the BFR, it was shown that midspan flashovers are extremely unlikely, flashovers can only occur at the towers. Therefore this minimum distance is increased to the third tower location or to 0.6 km. From Eq. 23, assuming a K_c of 700 km-kV/μs, the steepness is 1167 kV/μs. This is the maximum steepness that can occur for the design or assumed value of the MTBS. For other more distant towers, the steepness can also be calculated by Eq. 23. For example, at the fourth and fifth towers, the steepnesses are 875 and 700 kV/μs.

Knowing the distance d_m or the tower from which the maximum steepness occurs now permits the calculation of the crest voltage of the incoming surge. First consider that the number of surges arriving at the station, N_S, is the number of flashovers within a given distance, and that this number of flashovers must equal N_D. For example, the number of surges occurring from the station to the third tower is the BFR times the distance to the third tower or 0.003 surges. However, the design criterion states that only 0.0025 are permitted. Therefore to obtain a lesser number of surges, the surges must be of larger magnitude than those generated by the minimum stroke current to cause flashover, I_C. To determine this voltage, an iterative approach is needed, as illustrated in Fig. 7. Here the steepness of the surge arriving at the station from tower 1 is S_1, from tower 2, S_2, and from tower 3, S_3; and $S_1 > S_2 > S_3$. Assume that the crest surge voltage arriving at the station has a crest magnitude of V_{SC}. At towers 1, 2, and 3 the voltage V_S is calculated, and as illustrated, because the steepness at the station is less for a stroke to the third tower, the required value V_S to obtain the voltage V_{SC} at the station is larger than for the first tower. At each of the towers, the stroke current required to obtain the voltages

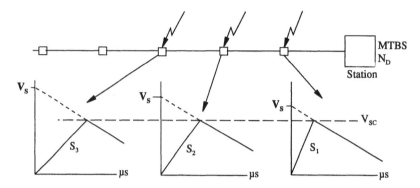

Figure 7 Determining the crest voltage of the incoming surge.

V_S can be calculated. From this the number of surges originating from each tower is obtained. If the sum of these surges does not equal N_D, the value of V_{SC} is increased or decreased until an acceptable accuracy is achieved.

To obviate this lengthy procedure, some approximations may be made. First assume that all values of V_S are the same. The stroke current magnitude can then be directly obtained. First note that the number of surges within a distance d_m is

$$N_S = 0.6 N_L d_m P(I > I_S) \tag{30}$$

where N_L is the number of flashes and I_S is the stroke current, which must be equal to or greater than I_C. Also note that this equation is identical to that for the BFR if $I_S = I_C$ and $d_m = 100$ km. Setting N_S to N_D we obtain

$$P(I > I_S) = \frac{N_D}{0.6 d_m N_L} = \frac{1}{0.6 d_m N_L \text{MTBS}} \tag{31}$$

from which the reduced variate Z can be obtained—from which the current I_S is found. For example, let the MTBS = 400 years, $d_m = 0.6$ km, and $N_L = 50$ flashes per 100 km-year. Then $P(I > I_S) = 0.0139$, $Z = 2.20$, and $I_S = 126$ kA. The voltage is determined from the previous equations.

2.5 An Example

An Example. Consider a 230 kV line having a BFR = 2.0 flashovers/100 km-years, $N_L = 61.5$ flashes/100 km-years, $Z_C = Z_g = 350$, $R_0 = 20$ ohms, $\rho = 400$ ohm-meters, $I_g = 63.7$ kA, CFO = 1040 kV, $C = 0.20$, $V_{PF} = 75$ kV, and span length = 300 m. Therefore $I_g = 63.7$ kA. Assuming a desired MTBS of 100 years, then the minimum distance d_m is 0.5 km. The next tower is at 0.6 km, so this distance is increased to 0.6 km. For a steepness constant K_c of 700, the maximum steepness is 1167 kV/μs. For a flashover at the third tower, $d = 0.9$, and the steepness at the station decreases to 778 kV/μs and for a flashover at the fourth tower to 583 kV/μs.

For a flashover at the second tower, $P(I > I_S) = 0.04517$, $Z = 1.695$ and $I_S = 92.86$ kA. A convenient equation to determine the current is

$$I_S = M_I e^{\beta Z} = 33.3 e^{1.025} = 92.86 \text{ kA} \tag{32}$$

First the value of R_i is determined. Assuming 10.0 ohms, then

R_i	R_e	I_R	R_i
10	9.46	87.9	13.0
13.0	12.1	86.4	13.0

For $R_i = 13.0$ ohms, $\tau = 26.9\,\mu s$, $R' = 11.32$ ohms, and $V_I = 974\,kV$, from Eq. 1. Then from Eq. 18, $B = 12.9\,\mu s$, from Eq. 20, $V_S = 700\,kV$, and from Eq. 28, $V_{SC} = 685\,kV$. Thus the incoming surge from a flashover to the first tower for an MTBS of 100 years is characterized by (1) a steepness of $1167\,kV/\mu s$, (2) a crest voltage of 685 kV (0.66 CFO), and (3) a tail described by a time constant of 26.9 μs. This surge voltage rides on top of an opposite polarity power frequency voltage of 75 kV. That is, the voltage to ground is 610 kV.

For a flashover at the third tower, the procedure is the same except that $d = 0.9$, which results in a surge voltage at the station of 760 kV. The steepness is 778 kV/μs. For a flashover at the fourth tower, the steepness at the station decreases to 583 kV/μs and the voltage increases to about 809 kV. Table 3 summarizes the results and compares these results with the more exact method described earlier as obtained from a computer program. As noted, the difference in crest voltage between these two methods ranges from about 0.4% at the second tower to 9% at the eleventh tower. Interestingly each of these surges was calculated for the same MTBS. Therefore each of these surges is equally likely. They all produce surges equal to the MTBS criterion. The maximum steepness occurs for the closest flashover, as expected, but the minimum crest voltage is tied to this maximum steepness. The maximum crest voltage occurs for towers more distant. Note that the crest voltage for a flashover at the second tower is only 73% of the CFO, and even for the fourth tower, the voltage is only 78% of the CFO of the line. Only for a flashover at a very distant tower does the voltage approach the CFO.

Table 3 Incoming Surges per the Example, Section 2.5

Tower number	Steepness kV/μs	Simplified method		"Exact method"	
		Voltage, kV	Voltage, % of CFO	Voltage, kV	Voltage, % of CFO
2	1167	685	66	689	66
3	778	760	73	771	74
4	583	809	78	825	79
5	467	843	81	867	83
6	389	868	83	901	86
7	333	885	85	927	89
8	292	898	86	949	91
9	259	906	87	967	92
10	233	912	88	983	94
11	212	915	88	997	95

2.6 Flashover at Adjacent Towers

As shown by Table 3, the crest voltage of the incoming surge increases for flashovers at distant towers. These surge voltages impinge on the insulation of adjacent towers and may cause flashovers at these towers. For example, assume that the calculated voltage from a distant tower is 1400 kV. Subtracting the power frequency voltage, the voltage across the line insulation is about 1325 kV. The estimated CFO (negative polarity) for this surge voltage waveshape is about 1125 kV. Thus flashover at the adjacent tower occurs, as illustrated in Fig. 8. This chopped wave surge proceeds to the station, and the front is further pushed back by corona. Thus, because of this chop, the voltage that appears at the station entrance is decreased and is far less severe than a smaller magnitude surge that has not been chopped by a flashover at the adjacent tower. The practical result is that the line insulation limits the maximum surge voltage that appears at the station entrance. Although the limiting surge voltage is a function of the time constant of the tail, an estimated limit to the surge voltage is between about 1.25 and 1.35 per unit of the positive polarity CFO of the line insulation. Thus for the above example, surge voltages are limited to between 1300 and 1400 kV.

2.7 Selecting the Incoming Surge

Since each of the surge voltages per the above calculation are equally probable, each of these surge voltages should theoretically be applied to the station. However, since the steepness of the surge is of primary importance in determining the voltage across the terminals of the station equipment, the surge with the highest steepness is most frequently chosen. In some cases, the surge with the highest steepness has a crest voltage such that the surge voltage at the station would be less than the equipment BIL. For example, the calculated surge voltage for the highest steepness in the previous example is 685 kV. Assuming that this doubles at the station, and subtracting the power frequency voltage, results in voltage to ground of 1295 kV. If the station equipment BIL were equal to or greater than 1295 kV, this surge would be less dangerous than the surge with a smaller steepness but a larger magnitude. For the example, for the 230-kV system, the equipment BIL is usually 900 kV, and therefore the surge with the highest steepness is the most severe—and if a single surge is desired for testing the station, this surge would be selected. In conclusion, the first surge calculated, the one with the highest steepness, is usually selected.

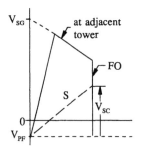

Figure 8 Flashover at adjacent tower decreases the incoming surge voltage.

The Incoming Surge and Open Breaker Protection

Table 4 Reducing the BFR on the First Three Towers

Tower number	BFR, FO/100 km-years	Span length, meters	BFR per tower	Sum of BFRs
1	1.0	300	0.003	0.003
2	1.0	300	0.003	0.006
3	1.0	300	0.003	0.009
4	2.0	300	0.006	0.015
5	2.0	300	0.006	0.021

2.8 Ameliorating Measures

The results of the above example are somewhat typical in that the incoming surge has a steepness of about 1000 kV/μs and a crest voltage equal to about 70% of the line CFO. However, if the BFR of the line were higher—for example, 4.0/100 km-years—or the MTBS were doubled to 200 years, then a surge arriving from the first tower would have a steepness of about 2333 kV/μs. Thus the severity of the surge is a function of both the BFR and the design criteria. To reduce the severity of the surge, the obvious method is to decrease the BFR, that is, increase the line insulation level, or more practically, decrease the footing resistance. This decrease of footing resistance only needs to be made for towers adjacent to the station. In general, decreasing the footing resistance for towers within about 1 km or about 1/2 mile of the station is quite beneficial. Using the previous example, but now assuming that the footing resistance of the first three towers has been decreased to 10 ohms, results in a BFR of 1.0 flashover/100 km-years for these towers. The BFR per tower, which is the span length limes the BFR, and the sum of these BFRs is shown in Table 4. Rearranging Eq. 29,

$$\sum_{1}^{n} L_S(\text{BFR}) \geq \frac{1}{\text{MTBS}} \tag{33}$$

Note that if the BFR is constant, the sum of L_S is equal to d_m. This equation permits the determination of d_m. That is, d_m occurs when the sum in the last column of Table 4 equals or exceeds the value of 1/MTBS or 0.01. This occurs at the fourth tower,

Table 5 Incoming Surges per Table 4

Tower number	Steepness, kV/μs	Simplified method		"Exact method"	
		Voltage, kV	Voltage, % of CFO	Voltage, kV	Voltage, % of CFO
4	583	809	78	669	64
5	467	843	81	743	71
6	389	868	83	793	76
7	333	885	85	837	80
8	292	898	86	869	83

and therefore d_m is 4(0.3) or 1.2 km. Therefore the maximum steepness of the incoming surge has been reduced to $700/1.2 = 583\,\text{kV}/\mu\text{s}$. An approximation of the crest magnitude is obtained as before. That is, begin by using Eq. 31 and proceed as before. Using this method of estimation produces very conservative values, as shown in Table 5, compared to the more exact method by use of a computer program.

The example is designed to illustrate the benefit of reducing the BFR on the first few towers from the station, in this example for about 1 km or 0.5 miles. Compared to the previous example, the steepness has been reduced from 1167 to $583\,\text{kV}/\mu\text{s}$, a 2 : 1 reduction.

3 INCOMING SURGE FROM A SHIELDING FAILURE

3.1 Voltage at the Struck Point

In Chapter 7, shielding failures were segregated into those that caused flashover and those that did not. In designing the shielding system for lines, only those which resulted in flashover were considered, the SFFOR. However, even those shielding failures that do not result in flashover produce a surge voltage on the phase conductor. Therefore, from the viewpoint of determining an incoming surge, the total shielding failure rate or SFR is of importance.

The voltage on the phase conductor at the struck point is simply the shielding failure current multiplied by half the conductor surge impedance. The maximum surge voltage is caused by the maximum shielding failure current I_m, and the minimum surge is the minimum current, 3 kA, multiplied by half the conductor surge impedance.

The distribution of shielding failure currents can be determined by use of the equations of Chapter 7. Typical distributions of this current for maximum shielding failure currents of 5 and 10 kA are presented in Fig. 9. As can be noted, the curves are fairly linear except for currents close to the maximum. Thus the probability of obtaining a current or a resultant voltage close to the maximum is remote. As an approximation, these shielding failure current distributions can be taken as a uniform distribution, and therefore the cumulative distribution is shown by the dotted

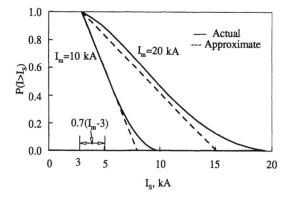

Figure 9 The distribution of shielding failure currents.

lines of Fig. 9. The difference between the maximum and minimum current per this distribution is equal to about 70% of the difference between the maximum current and 3 kA. Therefore, defining the surge voltage at the struck point as V_S, the probability equation is

$$P(V \geq V_S) = \frac{V_M - V_S}{V_M - V_1}$$

where

$$V_M = [0.7(I_m - 3) + 3]\frac{Z_c}{2} \tag{34}$$

$$V_1 = 3\frac{Z_c}{2}$$

The waveshape of the surge at the struck point is a duplicate of the stroke current waveshape. Therefore unlike the backflash case, for which the steepness at the struck point is essentially infinite, a noninfinite steepness exists. The stroke current steepness is given by the conditional distribution of the steepness given the current, and the tail or time to half value is provided by the distribution of the time to half value since no correlation exists between it and other parameters of the stroke current, i.e.,

$$\begin{aligned} M_{S|I} &= 4.83 I^{0.47} & \beta_{S|I} &= 0.554 \\ M_{t_T} &= 77.5 & \beta_{t_T} &= 0.577 \end{aligned} \tag{35}$$

The median current steepness ranges from about 8 kA/μs at 3 kA to about 14 kA/μs at 10 kA. To simplify, the conditional distribution is considered independent of the crest current, and as an approximation,

$$M_S = 11 \qquad \beta_S = 0.554 \tag{36}$$

The voltage steepness, denoted as S_v is simply the current steepness multiplied by half the conductor surge impedance, and therefore the parameters of the distribution of S_v, the voltage steepness, are

$$M_{S_v} = 11\frac{Z_c}{2} = 5.5 Z_c \qquad \beta_{S_v} = 0.544 \tag{37}$$

The median time to half value of the current is 77.5 μs, and the mean or average is about 92 μs. This mean value is normally used to describe the tail, i.e., no statistical distribution—and is translated into a time constant of 133 μs.

Thus the surge at the struck point can be characterized (1) by a voltage magnitude that ranges from V_1 to V_M per Eq. 34 and (2) by a steepness per Eq. 37 and by a tail described by a time constant of 133 μs.

3.2 Voltage at the Station

As the surge voltage at the struck point travels towards the station, corona pushes back the wave front, decreasing the steepness. However, since the surge at the struck point has a noninfinite steepness, the steepness at the station is less than for the backflash event. For a steepness S at the station, the steepness at the struck point, S_0, is, from Eq. 22,

$$S_0 = \frac{S}{1 - \frac{dS}{K_c}} \tag{38}$$

where d is the distance traveled in km and K_c is the corona constant of Table 2.

Because the time to half value of the surge at the struck point is large, corona does not appreciably decrease the crest value of the voltage surge in its travel toward the station. Only the soil resistivity is effective, and for short travel distances, which are of primary interest, this attenuation is minor and can be neglected. Therefore the surge arrives at the station with approximately the same crest voltage as at the struck point, but with a reduced steepness as determined by Eq. 22.

3.3 Estimating the Incoming Surge for an MTBS

The above equations provide a method of estimating the incoming surge for a given shielding failure current. But, as for the backflash case, the incoming surge is desired for a specific MTBS. Per Fig. 10, consider a line having an SFR of 0.534 shielding failures per 100 km-years and that the desired MTBS is 400 years or $N_D = 0.0025$ surges per year. The maximum shielding failure current is 21.13 kA, while the critical current is 5.43 kA, resulting in an SFFOR of 0.476 flashover/100 km-years. The phase conductor surge impedance is 477 ohms, and therefore $V_M = 3742$ kV and $V_1 = 716$ kV. The CFO is 1296 kV. Also assume that $K_c = 700$.

The process of calculation is embodied in setting the number of surges transmitted to the station, N_S, equal to N_D, where

$$N_S = (\text{SFR})L_S \sum_1^N P(S_0)P(V_S) \tag{39}$$

where N is the most distant tower that needs to be considered, $P(S_0)$ is the probability of exceeding the steepness S_0 at the struck point, and $P(V_S)$ is the probability

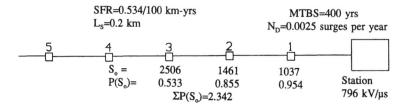

Figure 10 For shielding failure, determining the steepness for the minimum crest voltage.

of exceeding the crest surge V_S at the struck point. Since the crest surge voltage is considered identical at all towers, the equation may be simplified to

$$N_S = (\text{SFR})L_S P(V_S) \sum_1^N P(S_0) \qquad (40)$$

Equating this to N_D results in

$$\sum_1^N P(S_0) = \frac{N_D}{L_S(\text{SFR})P(V_S)} = \frac{1}{\text{MTBS}(L_S)(\text{SFR})P(V_S)} \qquad (41)$$

Thus the problem is to determine the sum of the probabilities of steepness to the most distant tower being considered. First note that from Eq. 29 used for the backflash,

$$d_m = \frac{N_D}{\text{SFR}} = \frac{1}{\text{MTBS}(\text{SFR})} = 0.468\,\text{km} \qquad (42)$$

and assuming 200-m spans, as for the backflash, this distance is increased to 0.6 km and $N = 3$. Reasoning as for the backflash, the number of shielding failures or thus the number of surges within this distance is the distance times the SFR or 0.00320. Assuming that the crest surge voltage is the minimum voltage, V_1, then the $P(V_1) = 1.0$, i.e., all the surges will equal or exceed the minimum crest voltage. Therefore, from Eq. 41, the sum of the probabilities of the steepness at the struck point must equal 2.34 for the three towers. Figure 10 illustrates the selection of this steepness. The procedure is (1) to select the steepness at the station, (2) to calculate the steepness at each of the three towers using Eq. 38, and (3) to calculate the probability of exceeding this steepness using Eq. 37. If this does not equal the target value of 2.34, another steepness is selected and the process repeated. These steps were performed for his sample problem, and the results are shown in Fig. 10. Note that the sum of the steepness probabilities is 2.342 and thus the incoming surge has a steepness of 796 kV/μs and a crest magnitude of 716 kV. Note that the resulting steepness is about equal to the steepness of an infinitely steep surge originating from the next tower, tower 4, i.e.,

$$S = \frac{K_c}{d + L_S} \qquad (43)$$

or 875 kV. This is somewhat logical, since if the fourth tower were considered in Fig. 10, the steepness at the fourth tower would be infinity and its probability of being exceeded would be zero. Thus the number of surges emanating from tower 4 would be zero. Also note that the sum of the steepness probabilities is approximately equal to 3, the number of towers considered.

Next, consider a total of four towers and set the steepness at the station at a value as though an infinite surge arrived from the fifth tower, i.e., 700 kV/μs. Per Fig. 11, the sum of the steepness probabilities is 2.97. Rearranging Eq. 41,

$$P(V_S) = \frac{1}{\text{MTBS}(L_S)(\text{SFR})\sum_1^N P(S_0)} \qquad (44)$$

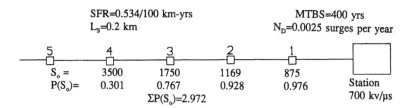

Figure 11 Determining the steepness probabilities for shielding failures to four towers.

and the probability of exceeding V_S must be 0.7876. From Eq. 34, $V_S = 1359\,\text{kV}$. If the sum of the steepness probabilities were 3, equal to 1 minus the number of towers considered, then V_S would be 1380 kV, which appears to be an acceptable estimate of the voltage and is considered adequate for estimating purposes. That is, Eq. 44 becomes

$$P(V_S) = \frac{1}{\text{MTBS}(d)(\text{SFR})} \qquad (45)$$

Equating this to Eq. 34,

$$V_S = V_M - \frac{1}{\text{MTBS}(d)(\text{SFR})}(V_M - V_1) \qquad (46)$$

Thus the procedure can be simplified to

1. For the first surge, determine the distance d by the equation

$$d = \frac{1}{(\text{SFR})(\text{MTBS})} \qquad (47)$$

   and extend this distance to the next tower, i.e., an integer number of spans. Determine the steepness by the equation

$$S = \frac{K_c}{d + L_S} \qquad (48)$$

   The crest voltage is V_1, which is 3 kA times the conductor surge impedance divided by 2.

$$V_1 = 3\frac{Z_c}{2} \qquad (49)$$

2. For the next surge,

$$S = \frac{K_c}{d + 2L_S} \qquad (50)$$

   and the crest voltage is

$$V_S = V_M - \frac{1}{\text{MTBS}(d)(\text{SFR})}(V_M - V_1) \qquad (51)$$

3. For other surges, continue with the process by adding L_S to d in the denominator of Eqs. 50 and 51.

| | Estimate | | Computer program | | | |
| | Before flashover | | Before flashover | | After flashover | |
Number of towers	Steepness	Crest, kV	Steepness	Crest, kV	Steepness	Crest, kV
3	796/875	716	692	716	716	No FO
4	700	1380	681	1935	1935	6.095
5	583	1971	581	2413	1492	2.568
6	500	2325	502	2707	902	1.797
7	437	2561	441	2917	654	1.484

Continuing the example, the steepness and crest voltage of the incoming surges are as in the table. These estimated values are compared to results from a computer program what considers the actual distribution of shielding failure currents and uses Eq. 35 to obtain the median steepness given the current. Although the steepnesses compare well with those estimated, the estimated crest voltages are significantly lower. To be emphasized is that these crest voltages are those assuming no flashover at adjacent towers. Consider the fifth tower and examine Fig. 12. At the station, the steepness is 581 kV/μs, and the crest voltage assuming no flashover is 2413 kV. At the struck tower, using Eq. 38, the steepness is 3418 kv/μs and the crest voltage remains at 2413 kV. At the next adjacent tower, tower 4, the steepness is 1729 kV/μs and the crest voltage is 2413 kV. Since this voltage is above the CFO of 1296 kV, flashover occurs, and an estimate of the time to flashover is 2.568 μs. The surge continues its travel into the station, and the steepness decreases to 581 kV/μs. Multiplying this by the time to breakdown results in a crest magnitude of only 1492 kV. The crest voltage has decreased to 62% of its original value. This crest voltage, although higher than the CFO, only has a short duration, about 2.6 μs. Thus this surge is far less severe than a surge having a full wave.

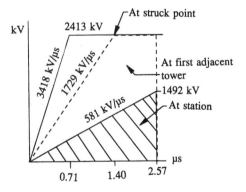

Figure 12 The incoming surge decreased by flashover for five towers.

Therefore the incoming surge to be employed is one without a flashover: either the first surge with the maximum steepness or a higher magnitude surge but without a flashover. As for the backflash, the crest voltage of the incoming surge is limited by the line insulation. As an approximation, this voltage limit is between 1.35 and 1.45 per unit of the line CFO.

This example was used to illustrate all the facets of determining the incoming surge. However, it is not typical. Normally, the SFFOR of a line is zero or near zero, and thus few shielding failure flashovers occur. Assume that a line SFR is 0.10 shielding failures per 100 km-years and that the critical and maximum currents are equal at 10 kA. Assume further that the CFO of the line is 2000 kV and that the phase conductor surge impedance is 350 ohms. If the larger or desired MTBS is 500 years, assuming 300-m spans, V_M is 1383 kV and V_1 is 525 kV. Assuming 300 m spans, the first distance to be considered is $d = 2$ km. For a K_c of 1000, the maximum steepness at the station is 435 kV, and the crest voltage of this surge is 525 kV. The maximum crest voltage of the incoming surge is 1383 kV, well below the CFO. However, because this surge emanates from a considerable distance, the crest will be decreased by effect of soil resistivity. Again, this is the usual case, but it is somewhat uninteresting from the perspective of the development of the incoming surge caused by shielding failures. As noted in the development, (1) the CFO to be employed is the negative polarity CFO, and (2) the power frequency voltage is not considered, since the average power frequency voltage is zero for a shielding failure.

4 SIMPLIFIED AND QUICK ESTIMATE

The methods and equations developed to this point provide an estimate of the magnitude and shape of the incoming surge. Some further simplification can be made, since normally the primary and most sensitive parameter in determining the protection of station equipment is the steepness. Per Eq. 29, the minimum distance to the stroke-terminating point is

$$d_m = \frac{1}{(\text{BFR})(\text{MTBS})} \tag{52}$$

As before, this distance should be extended to the next tower so as to be a multiple of the span length. Letting this extended distance also be denoted as d_m, the steepness of the incoming surge is

$$S = \frac{K_c}{d_m} \tag{53}$$

The crest voltage of the surge is generally between about 85 and 100% of the positive polarity CFO of the line. Historically, the crest has been taken as 1.2 times the CFO [5]. Thus very conservatively the crest voltage can be assumed as 1.2 times the positive polarity CFO of the line.

The tail of the surge can be described by a time constant per Eq. 7, repeated here:

$$\tau = \frac{Z_g}{R_i} T_S \tag{54}$$

Table 6 Typical Values of Steepness

System voltage, kV	S_L, meters	BFR FO/100, km-yrs	MTBS, years	K_c, km-kV/μs	d_m, km	S, kV/μs
500	300	0.6	200	1000	0.9	1110
345	300	0.6	200	1000	0.9	1110
230	300	1.0	150	700	0.6	1170
138	250	3.0	100	700	0.5	1400
69	100	5.0	100	700	0.4	3500

where Z_g is the ground wire surge impedance, R_i is the tower footing impulse resistance, and T_S is the travel time of one span length. Z_g ranges from about 350 for lines with two ground wires to about 500 ohms for lines with one ground wire, R_i is approximately 1/2 of the measured footing resistance, and T_S ranges from about 0.20 for lower voltage lines to over 1.0 for high-voltage lines. Assuming a measured footing resistance of 20 ohms, the tail time constant varies from about 10 to 35 μs, and an average value is about 20 μs.

The value of the opposite polarity power frequency voltage is normally assumed equal to that used when calculating the backflash rate. A conservative value is 70% of the crest line-to-neutral nominal system voltage.

To gain some insight into the parameters of the incoming surge, Table 6 shows some typical steepnesses and Table 7 presents steepness from Table 6 along with the crest voltage E, the power frequency voltage V_{PF}, and the total voltage, which is $(E - V_{PF})$. Also shown is a conservative estimate of the tail time constant. Note that as the system voltage decreases, the steepness increases, although the crest voltage and tail time constant decrease.

5 APPLYING THE SURGE TO THE STATION

The general equivalent circuit used to apply the incoming surge to the station is shown in Fig. 13a. It consists of a voltage source, an impedance Z_i, a line having a surge impedance Z, having a length equal to the minimum distance to the struck point d_m. The impedance Z_i is the impedance as seen by the surge on the return trip from the station. Therefore it consists of the parallel combination of the footing

Table 7 Typical Values of Incoming Surge

System voltage, kV	CFO, kV	S, kV/μs	E, kV	V_{PF}	Total voltage, kV	Tail time constant, μs
500	1900	1110	2300	285	2015	35
345	1500	1110	1800	195	1605	35
230	1000	1170	1200	130	1070	35
138	750	1400	900	75	820	29
69	450	3500	540	40	500	15

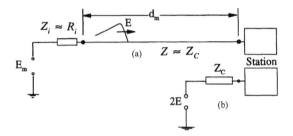

Figure 13 (a) General equivalent circuit; (b) reduced circuit.

resistance R_i, the conductor surge impedance Z_c, and half the ground wire surge impedance, Z_g. Approximately, it is equal to the impulse value of the footing resistance R_i. In equation form

$$Z_i = \frac{R_i Z_c Z_g}{Z_g Z_c + R_i(Z_g + 2Z_c)} \approx R_i \qquad (55)$$

The voltage source consists of the surge voltage E_m of such a value as to produce the incoming surge voltage E on the line. Thus the circuit not only models the initial voltage applied to the station but also models reflections from the struck point, which may increase the current through the arresters and increase the voltage within the station.

The surge voltage E_m is

$$E_m = \frac{Z + Z_i}{Z} E \approx \frac{Z + R_i}{Z} E \qquad (56)$$

Per Chapter 9, the surge impedance Z is

$$Z = Z_c - C^2 Z_g \approx Z_c \qquad (57)$$

where Z_c is the phase conductor surge impedance, Z_g is the ground wire surge impedance, and C is the coupling factor. As noted per Eq. 57, the value of Z is approximately equal to Z_C.

This general circuit is frequently reduced to that of Fig. 13b and is valid for a time equal to twice the travel time to the struck point. For example, if d_m is equal to 900 meters, then the reduced equivalent circuit of Fig. 13b is valid for only 6 μs, since after this time reflections from the struck point will alter the voltage. If only the crest voltage within the station is of interest and this crest voltage occurs before twice the travel time to the struck point, then the circuit of Fig. 13b may be used. However, if the crest voltage occurs after twice the travel time or if the surge is to be evaluated as to its severity, then the circuit of Fig. 13a should be used. Since the reduced equivalent circuit requires less computer time and is therefore desirable, an initial comparison could be made using both circuits. If the reduced circuit provides essentially identical results, then it could be used for further studies. In general, the circuit of Fig. 13b is used.

The Incoming Surge and Open Breaker Protection

As a final note, the source consists only of the surge voltage. The effect of the power frequency voltage on the arrester current and discharge voltage can be accurately simulated by adding the power frequency voltage to each of the arrester discharge voltages. The voltage to ground in the station is then the surge voltage minus the power frequency voltage.

6 THE EFFECT OF THE TOWER

In developing the characteristics of the incoming surge, the tower component of voltage has been neglected. It is proper to question this neglect. As shown in Chapter 10, the tower component of voltage adds a short-duration spike of voltage at the crest. Assuming for the moment that flashover occurs at the crest, Fig. 14 shows the voltage on the phase conductor. The crosshatched tower component of voltage has a duration following flashover of $2h/c$, where h is the tower height and c is the velocity of light: thus $2h/c$ is about 0.15 to 0.30 µs. The effect of corona is also shown in Fig. 14 illustrating that the pushback of the front carries the time to crest of the incoming surge beyond the voltage spike. Thus the tower component of voltage has essentially no effect on the characteristics of the incoming surge.

7 THE INCOMING SURGE FROM SUBSEQUENT STROKES

Although it is not specifically stated, the first stroke of the flash was employed to develop the characteristics of the incoming surge, since the first stroke leads to the most severe surge. However, subsequent strokes of the flash produce voltages that may be dangerous to a circuit breaker that is in the opened position or in the process of opening. Thus the protection of the opened breaker is of importance. In this section, the characteristics of the incoming surge caused by a subsequent stroke are presented, and in the next section, the results of this section are used to develop the concept of open breaker protection.

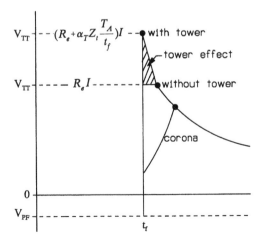

Figure 14 The tower component of voltage can be neglected.

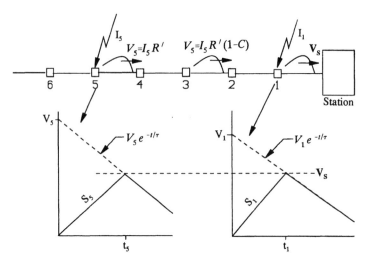

Figure 15 Incoming surge caused by a subsequent stroke.

With the objective of determining the MTBS of a surge having a specific magnitude consider Fig. 15, where it is desired to estimate the MTBS for a 580-kV surge, i.e., $V_S = 580\,\text{kV}$. Subsequent strokes have magnitudes of I_1 to I_n terminate on each of the n towers, producing phase conductor voltages of V_1 to V_n. These voltages travel in toward the station and are modified by corona. Upon reaching the station, their steepnesses are decreased to S_1 to S_n and their magnitudes are decreased to V_S. Thus the voltage is

$$V_n = V_S e^{+\frac{t_f}{\tau}} \tag{58}$$

where

$$t_f = \frac{V_S}{S_n} \qquad S_n = \frac{K_c}{d} = \frac{K_c}{n(S_L)} \tag{59}$$

where n is the number of the towers from the station, d is the distance from the station in km, and S_L is the span length in km. Therefore

$$V_n = V_S e^{+\frac{V_S n S_L}{K_c \tau}} \tag{60}$$

The stroke currents I_n required to obtain the crest voltages V_n can be determined from the equation

$$I_n = \frac{V_n}{R'(1 - C)} \tag{61}$$

where, as before,

$$R' = \frac{Z_C R_e}{Z_C + 2R_e} \qquad R_e = \frac{Z_g R_i}{Z_g + 2R_i} \tag{62}$$

The Incoming Surge and Open Breaker Protection

and

$$I_R = \frac{R'}{R_i} I_n \qquad I_g = \frac{1}{2\pi} \frac{E_0 \rho}{R_0^2} \qquad R_i = \frac{1}{\sqrt{1 + I_R/I_g}} \tag{63}$$

where I_R is the current discharged by the footing resistance, R_0 is the low-current footing resistance, R_i is the impulse or high-current resistance, and I_g is the current required to obtain a soil ionization gradient of $E_0 = 400\,\text{kV/m}$. The value of the time constant τ is found from the equation

$$\tau = \frac{1}{2}\frac{Z_g}{R_i} T_S \tag{64}$$

As noted, this equation is the same as before except that a $\frac{1}{2}$ has been added. The justification is that from Chapter 6, the times to half value of subsequent stroke currents (median = 30.5 µs) are much smaller than the time to half value of the first stroke current (median = 77.5 µs). Thus the tail time constant for voltages produced by subsequent strokes should be smaller and the $\frac{1}{2}$ represents an estimate.

The number of surges from each of the towers is the BFR times the span length times the probability that the subsequent current is greater than I_n. From Chapter 6, the parameters of the subsequent stroke current I_{SUB} are $M = 12.3\,\text{kA}$, $\beta = 0.53$.

The total number of surges having magnitudes equal to or exceeding V_S is

$$N_S = \text{BFR}(S_L) \sum_1^\infty P(I_n \geq I_{SUB}) \tag{65}$$

The MTBS is then

$$\text{MTBS} = \frac{1}{N_S} \tag{66}$$

To demonstrate the procedure, consider a line having a BFR of 3.0/100 km-years. Also let $C = 0.30$, $R_0 = 30$ ohms, $\rho = 600$ ohm-meters, $Z_g = Z_C = 400$ ohms, and span length = 300 m. Then $I_g = 42.44\,\text{kA}$. The first step is to iterate to find R_i. This has been done and the results are shown in Table 8. Adding the last column, we obtain

$$\begin{aligned} N_S &= \frac{3}{100}(0.3)0.02951 = 0.000266\,\text{surges/year} \\ &= 3765\,\text{years} \end{aligned} \tag{67}$$

Using a computer program, the MTBS = 3769 years, a 1% error. The time constant of the tail ranges between 9 and 11 µs.

Table 8 Example: Calculating N_S.

n	V_n	R_i	R_e	R'	I_n	Z	$P(I_n > I_S)$
1	596	21.6	19.5	17.8	47.9	2.57	$5.14 * 10^{-3}$
2	612	21.5	19.4	17.7	49.5	2.63	$4.30 * 10^{-3}$
3	628	21.3	19.2	17.6	51.1	2.69	$3.58 * 10^{-3}$
4	644	21.1	19.1	17.4	52.9	2.75	$2.97 * 10^{-3}$
5	661	20.9	18.9	17.3	54.6	2.81	$2.46 * 10^{-3}$
6	678	20.7	18.8	17.2	56.4	2.87	$2.04 * 10^{-3}$
7	694	20.6	18.7	17.1	58.2	2.93	$1.69 * 10^{-3}$
8	711	20.4	18.5	16.9	60.0	2.99	$1.40 * 10^{-3}$
9	728	20.2	18.4	16.8	61.9	3.05	$1.15 * 10^{-3}$
10	745	20.0	18.2	16.7	63.8	3.11	$0.95 * 10^{-3}$
11	762	19.9	18.1	16.6	65.7	3.16	$0.79 * 10^{-3}$
12	778	19.7	17.9	16.4	67.7	3.22	$0.65 * 10^{-3}$
13	796	19.5	17.8	16.3	69.7	3.27	$0.54 * 10^{-3}$
14	813	19.3	17.6	16.2	71.7	3.33	$0.44 * 10^{-3}$
15	830	19.2	17.5	16.1	73.7	3.38	$0.36 * 10^{-3}$
16	847	19.0	17.4	16.0	75.8	3.43	$0.30 * 10^{-3}$
17	865	18.8	17.2	15.9	77.9	3.48	$0.25 * 10^{-3}$
18	882	18.7	17.1	15.7	80.1	3.53	$0.20 * 10^{-3}$
19	899	18.5	16.9	15.6	82.2	3.59	$0.17 * 10^{-3}$
20	916	18.4	16.8	15.5	84.4	3.64	$0.14 * 10^{-3}$

8 PROTECTION OF THE OPEN BREAKER

In air-insulated stations, arresters are first located immediately adjacent to the transformer. These arresters are frequently the only arresters necessary to protect all the apparatus within the station. That is, with all breakers closed, arresters within the station provide protection to all apparatus. However, with the breaker opened, no protection exists for the line side of the breaker, and a lightning surge may cause flashover of the breaker.

This opened breaker condition may occur when a breaker is standing opened on the line. Normal operating practice, however, dictates that when the breaker is opened for a prolonged period, the breaker disconnecting switches are also opened, so that no surges would impinge on the breaker and any flashover would normally occur to ground at the disconnecting switch. Departures from this practice are rare, but if an opened breaker with a closed disconnect condition normally exists, some form of protection should be used.

Another, more likely, condition that places a breaker in a vulnerable position is caused by a subsequent stroke of the lightning flash. To explain, consider the following event. Lightning terminates on the shield wire or phase conductor of the line, resulting in a flashover. The surge travels into the station, but because the breaker is closed, arresters within the station provide protection to the breaker. Next, the relay senses the fault, the breakers open (in about 50 ms), and in about 300 ms the breaker recloses. However, a lightning flash is composed of one or more strokes, and a subsequent stroke may occur between the 30 and 300 ms and produce a surge voltage that exceeds the breaker insulation strength—when the breaker is opened and unprotected.

Three alternate solutions to this problem have been proposed: (1) apply arresters on the line side of all breakers, (2) apply rod gaps on the line side, or (3) do nothing. All three methods have be used, and all have been "successful" [2, 6, 7, 8]. The first method is best but expensive. The second method is far less expensive but less effective. In addition, it is sometimes difficult to set the gap spacing. The gap spacing should be large enough so that the arresters within the station can prevent flashover when the breaker is closed but small enough to protect the breaker when it is opened. The last method is simply accepting the probabilities of a flashover.

The purpose of this section is to estimate the probability of exceeding the breaker insulation strength when the breaker is in the open position, that is, estimate the MTBF. First, consider the breaker insulation strength. From the previous section, the tail time constant is between 9 and 11 μs. From Chapter 2, the CFO_{NS} for a 9 to 11 μs tail time constant is from 1.29 to 1.23 times the standard CFO. Also from Chapter 2, a 2 μs chopped wave withstand test at 1.29 BIL is applied to the breaker. Thus it appears reasonable to use this test for the strength of the breaker.

Considering that an incoming surge would essentially double at the open breaker, the incoming surge should be limited to 1.29 BIL/2. For example, consider a 230-kV breaker having a BIL of 900 kV with a 2-μs chopped wave test of 1160 kV. Then the incoming surge should be limited to 580 kV. Per the example in the preceding section, the number of surges having a magnitude of 580 kV or greater is 0.000266 surges/year.

The next step is to estimate the probability that the breaker is open when these incoming surges arrive. From Chapter 6, the interstroke interval between the first and second stroke has a median of 45 ms and between the other strokes has a median of 35 ms. Both these distributions have a β of 1.066. Also provided in Chapter 6 is the distribution of multiple strokes. For example, 55% of the flashes have two or more strokes per flash, while 41% have three or more, etc. Calculating the probability that the breaker is open involves determining the probability that the breaker is open on the second stroke but not on the other strokes, the probability that the breaker is closed on the second stroke, opened on the third, but closed on the others, etc. The probability that the breaker is open on any of the subsequent strokes, P_0 can be estimated as

$$P_0 = P(t_0 < t < t_c) \tag{68}$$

where t_0 and t_c are the opening and closing times respectively. Thus the total probability is

$$P_T = N_S P_0 = P(t_0 < t < t_c) \text{BFRS}_L \sum_1^n P(I_n \geq I_{SUB}) \tag{69}$$

Setting $t_0 = 50$ ms and $t_c = 300$ ms, $P_0 = 0.423$. From the preceding section, $N_S = 2.66 \times 10^{-4}$ and therefore $P_T = 1.125 \times 10^{-4}$ surges/year. The MTBF is the reciprocal of this value or MTBF = 8,887 years. Performing this calculation by a computer program, $P_0 = 0.417$ and the MTBF = 8,832 years, an error of 1%. To reiterate, once in 8,832 years, the surge voltage at the terminals of the open breaker will have a magnitude equal to or exceeding the insulation strength of the breaker. Although it

Table 9 MTBFs

Nominal sys. voltage, kV	BIL, kV	BFR, FO/100 km-yrs	K_C	Span length, meters	MTBF, years
500	1800	0.6	1000	300	390×10^6
345	1300	1.0	1000	300	3×10^6
230	900	2.0	700	200	29,000
115	550	3.0	700	200	470
69	350	5.0	700	150	35
34.5	150	10.0	700	150	3

initially appears that no protection of the breaker is required, consider that there may be many 230-kV breakers in a station and many 230-kV stations. For example, if there are a total of 100 breakers installed, then the MTBF would be reduced to 883 years.

Using a computer program, the MTBFs for 34.5-to-500-kV systems are shown in Table 9 for $C = 0.25$, $R_0 = 20$, $\rho = 400$, $t_0 = 50$, $t_c = 300$ ms, and $Z_g = Z_C = 400$ ohms. As indicated by the equation, the MTBF is a function of the BFR and the BIL. As system voltage decreases, the BIL decreases and the BFR increases, both of which act to decrease the MTBF. As seen from Table 9, some type of protection appears required for breakers at system voltage of 69 kV and below, whereas protection may not be required for breakers at system voltages of 345 kV or higher.

9 COMPARISON WITH OTHER SOURCES

9.1 IEEE/AIEE Papers

In 1960, Clayton and Young [1] studied the protection of multiline stations. They pointed out that multiline stations have both a beneficial attribute and detrimental effect. Beneficially, multiple lines reduce the surge voltage and steepness (Chapter 9), but detrimentally, since each of the lines gathers surges to be brought into the station, the voltage magnitude and steepness of the incoming surge must be increased to maintain the desired MTFB. In equation form, they employed the same equation as given in this chapter, i.e.,

$$d_m = \frac{1}{n(\text{BFR})(\text{MTBF})} \quad (70)$$

where n is the number of lines. From d_m, the steepness was determined from Tidd data [9]. In contrast to the developments in this chapter, they applied this concept to both equipment on the transformer bus and equipment on other buses. In addition, consideration was not given to the contingency or line-out conditions. In general, for air-insulated stations, the authors used an MTBF of 50 years, although 30 years was also employed.

In 1954–1955, J. K. Dillard et al. and H. R. Armstrong et al. [10, 11] analyzed the protection of 120-kV and 24-kV stations of the Detroit Edison Company. These authors based the incoming surge on a stroke current that would not result in flashover at an adjacent tower. For the 120-kV station, the distance to the stroke-termi-

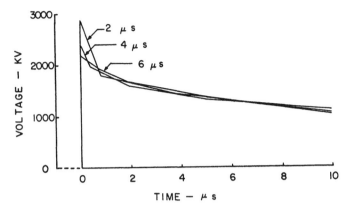

Figure 16 Phase conductor voltage reduced only by effect of decrease in voltage on ground wire [2].

nating point was 1500 m, which resulted in a surge magnitude of 900 kV, which is 1.38 CFO. The steepness was 450 kV/μs, and the tail time constant was 17 μs. Thus a value of K_C of 675 is apparent. For the 24-kV station, the distance to the stroke-terminating point was 1600 m, which resulted in a surge voltage magnitude of 870 kV, which is 93% of the CFO. The steepness was 460 kV/μs, and the tail time constant was 12 μs. Thus the apparent value of K_C is about 740. Interestingly, the authors considered that a shielding failure flashover generates a chopped-wave surge, which because of its short tail can be treated as the same waveshape as a surge from a backflash.

In 1967, a paper was presented that detailed the insulation coordination of 500-kV stations of the Allegheny Power Company [2]. The incoming surges were calculated for a 0.5-, a 1.0-, and a 2.0-mile surge. The authors first determined the phase conductor voltage assuming that flashover occurred at the crest of the voltage, i.e., at t_f. These voltages, decreased by the effect of zero voltage on the ground wire, are shown in Fig. 16. The effect of corona was then considered to arrive at the incoming surges shown in Fig. 17. As shown in Fig. 17 and also in Section 5, the magnitude of the incoming surge is unaffected by the crest of the phase conductor voltage, since corona attenuation pushed back the front beyond the value of t_f. Further analysis of these incoming surges is given in Table 10. For all lines in service, for the three-line Ft. Martin station, the authors used a 0.5-mile surge (MTBF = 67 years), and for the two-line Yukon station, the authors also used a 0.5-mile surge (MTBF = 100 years). For contingency or line-out conditions, the authors used the 1-mile and 2-mile surges

Table 10 MTBF or Return Period, years

Distance from station, miles/km	One-line station	Two-line station	Three-line station	Four-line station
0.5/0.8	200	100	67	50
1.0/1.6	100	50	33	25
2.0/3.2	50	25	17	13

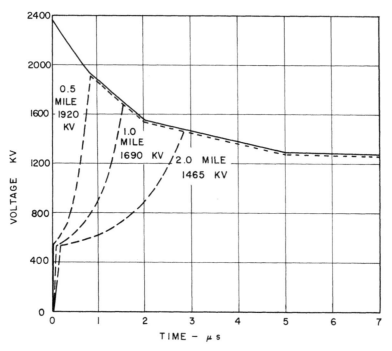

Figure 17 Incoming surges for stroke-termination points 0.5, 1.0, and 2.0 miles from the station [2].

to account for the lower probability of the line-out conditions. From Fig. 17, the average steepness and the surge voltage magnitudes were approximately 2400, 1100, and 500 kV/μs for the 0.5-, 1.0-, and 2-mile surges, respectively. The apparent values of K_C varied from 1700 to 1900, about double that suggested in Chapter 9 and used here.

9.2 CIGRE and IEC

In 1988, Weck and Eriksson [4] presented a paper that described a simplified method to estimate the parameters of the incoming surge. The proposed IEC Application Guide employs the same methods [3]. The authors determine the value of d_m by a different method than that used in this chapter. However, the explanation of this method must await the development in Chapter 13. The authors suggest MTBFs of between 300 and 1000 years for gas-insulated stations and state that for air-insulated stations, MTBFs down to 100 years may still be acceptable.

Backflash. To determine the stroke current for the backflash, they use the equation

$$P(I > I_S) = \frac{1}{d_m N_L \text{MTBS}} \tag{71}$$

which is similar to Eq. 31 except that Eq. 31 has a 0.6 in the denominator. As in Ref. 2, they assume that the breakdown or flashover voltage occurs at the crest of the surge or $t_B = t_f$. The CFO_{NS} is determined by the equation

$$\text{CFO}_{\text{NS}} = \left(1 + \frac{2}{\tau}\right)\text{CFO} \tag{72}$$

The incoming surge voltage magnitude is found from the equation

$$V_{\text{SC}} = I_{\text{S}} R_{\text{i}}(1 - C) \tag{73}$$

To explain, for flashover at the time to crest, V_{SI} of Fig. 3c and V_{S} of Fig. 3d are

$$\begin{aligned} V_{\text{SI}} &= V_{\text{TT}} + V_{\text{PF}} \\ V_{\text{S}} &= (1 - C)(V_{\text{TT}} + V_{\text{PF}}) = (1 - C)I_{\text{S}} R_{\text{e}} - C V_{\text{PF}} \end{aligned} \tag{74}$$

Neglecting the small attenuation caused by corona and setting $V_{\text{PF}} = 0$ and $R_{\text{e}} = R_{\text{i}}$, Eq. 73 results. Note that V_{SC} is equal to V_{I}, the voltage across the line insulation at the struck tower. The authors determine the steepness and tail time constant by the equations

$$S = \frac{K_{\text{C}}}{d_{\text{m}}} \qquad \tau = \frac{Z_{\text{g}}}{R_0} T_{\text{S}} \tag{75}$$

and apply the surge as in Fig. 13a, i.e., through a resistor equal to R_0 to a line of surge impedance Z_{C}. The length of the line is set at 300 m, i.e., not d_{m}.

Using this method for the example in Section 2.5 with $d_{\text{m}} = 0.6$ km, the incoming surge has a steepness of 1667 kV/μs and a tail time constant of 17.5 μs. The surge magnitude is 1110 kV, which is 1.07 times the CFO and only 4% below the CFO_{NS} of 1159 kV. This should be compared to 1167 kV/μs, 685 kV, and 26.9 μs as found in Section 2.5. There is no reason given for the application of the surge at 300 m instead of at d_{m}. To be noted is that their method would naturally lead to a higher crest voltage and that this voltage is still lower than the often used value of 1.2 CFO of 1248 kV.

Shielding Failure. For the shielding failure, the stroke current is determined by the equation

$$P(I > I_{\text{S}}) = P(I > I_{\text{m}}) + \frac{1}{d_{\text{m}} \text{SFR MTBS}} \tag{76}$$

The steepness is found from Eq. 75, and the tail time constant is 200 μs. The crest voltage is found from the equation

$$V_{\text{S}} = I_{\text{S}} \frac{Z_{\text{g}}}{2} \tag{77}$$

In calculating the value of I_{S}, trouble sometimes occurs, since $P(I > I_{\text{S}})$ may be greater than unity. For example, using the value in Section 3.3 and letting $d_{\text{m}} = 0.6$ km gives $P(I > I_{\text{S}}) = 1.55$, and impossible value.

9.3 IEEE Standards

The present and proposed 1998 surge arrester application guide [5] characterizes the incoming surge as (1) a crest voltage of 1.2 times the line CFO and (2) a steepness of 11 kV/μs per kV of the MCOV rating. The steepness was based on the steepness used in testing the silicon carbide gapped arrester. The authors of this guide, the Surge Protective Devices Committee, recognize that the basis of the steepness should not be a function of the arrester but rather a function of the desired MTBF and the BFR of the line. Therefore in future editions of the guide, the steepness will be suggested as proposed in this chapter.

10 SUMMARY

1. In Chapter 13, the station insulation strength will be selected on the basis of a mean time between failures or MTBF. Therefore the surge that arrives at the station entrance, called the incoming surge, is based on this MTBF.

2. For simple stations, to evaluate the equipment on the transformer bus, the incoming surge should be based on n times the MTBF where n is the number of lines. This surge is called the MTBS, the mean time between surges. For example, for an MTBF of 100 years and a three-line station, the equipment on the transformer bus is evaluated with an MTBS of 300 years, i.e., a 300-year surge. The equipment insulation strength on other buses is evaluated with the MTBS = MTBF, e.g., a 100-year surge. Probabilities of contingency (line-out) conditions should be considered and will decrease the MTBS.

3. For more complex station configurations, for the transformer and equipment on the transformer bus, the MTBS should be determined by making preliminary runs or cases.

4. For air-insulated stations, MTBFs of 50 to 100 years have been used, although values of 400 to 500 years are used in the IEC guide. For gas-insulated stations, because of the consequence of failure, larger values are suggested, i.e., 300 to 1000 years is suggested, but 400 years has been used.

5. An incoming surge may result from a backflash or a shielding failure. There is justifiable reason to conclude that only the backflash surge is of importance.

6. The incoming surge can be described by the steepness, crest voltage, and tail time constant. This surge rides atop an opposite polarity power frequency voltage.

7. Incoming surges caused by subsequent strokes of the flash may endanger the breaker when open or during the opening. This usually occurs between the time of opening and that of reclosing. Breakers on low-voltage systems are prone to this event and should be protected (69 kV and below), while higher voltage systems may not require protection. Protection can consist of the installation of arresters or gaps on the line side of the breaker.

11 REFERENCES

1. J. M. Clayton and F. S. Young, "Application of Arresters for Lightning Protection of Multiline Substations," *AIEE Trans. on PA&S*, 1960, pp. 566–575.
2. A. R. Hileman, R. W. Powell, W. A. Richter, and J. M. De Salvo, "Insulation Coordination in APS 500-kV Stations," *IEEE Trans. on PA&S*, 1967, pp. 655–665.

3. IEC Publication 71-2, "Insulation Coordination—Part 2: Application Guide," 1996.
4. K. H. Weck and A. J. Eriksson, "Simplified Procedures for Determining Representative substation Impinging Lightning Overvoltages," CIGRE Paper 33-16, 1988.
5. IEEE Std C62.22-1991, "IEEE Guide for the Application of Metal-Oxide Surge Arresters for Alternating-Current Systems."
6. C. L. Wagner, J. M. Clayton, F. S. Young, and C. L. Rudasill, "Insulation Levels for VEPCO 500-kV Substation Equipment," *IEEE Trans. on PA&S*, March 1964 pp. 235–241.
7. A. R. Hileman, C. L. Wagner, and R. B. Kisner, Jr., "Open Breaker Protection of EHV Systems," *IEEE Trans. on PA&S*, 1969, pp. 105–114.
8. C. L. Wagner, "Insulation Considerations for AC High Voltage Circuit Breakers," IEEE Tutorial Course on Application of Circuit Breakers, 75CH0975-3-PWR, 1975.
9. C. F. Wagner, I. W. Gross, and B. L. Lloyd, "High-Voltage Impulse Tests on Transmission Lines," *AIEE Trans.*, Apr. 1954, pp. 196–210.
10. J. K. Dillard, H. R. Armstrong, and A. R. Hileman, "Lightning Protection in a 120-kV Station—Field and Laboratory Studies," *AIEE Trans. on PA&S*, 1954, pp. 1143–1152.
11. H. R. Armstrong, R. W. Ferguson, A. R. Hileman, "Lightning Protection in a 24-kV Station—Field and Laboratory Studies," *AIEE Trans. on PA&S*, 1955, pp. 1127–1136.

12 PROBLEMS

1. For the 500-kV line of problem 1 of Chapter 9 and problem 1 of Chapter 10, determine the magnitude and steepness of an incoming surge caused by a backflash based on an MTBF of 200 years. Assume a three-line station and find the incoming surge for both the equipment on the transformer bus and the equipment on the other buses. Use a K_c of 1000.

2. For the 230-kV line of problem 2 of Chapter 9 and problem 2 of Chapter 10, determine the magnitude and steepness of an incoming surge caused by a backflash based on an MTBF of 100 years. Assume a three-line station and find the incoming surge for both the equipment on the transformer bus and the equipment on the other buses. Use a K_c of 700.

3. In a single-line 230-kV station, the equipment BIL was based on a maximum steepness of 1600 kV/μs. Find the MTBF of the station. The BFR of the line is 1.5 flashover/100 km-year and the span length is 300 meters. Use a K_c of 700.

12
Metal Oxide Surge Arresters

1 INTRODUCTION

In this chapter the durability or capability characteristics of metal oxide (MO) arresters are presented so that the proper arrester rating can be selected for any set of system requirements. In addition, the protective characteristics of these arresters are presented and used to develop equations to estimate the arrester discharge currents and energies. In the area of durability or capability tests, IEEE [1, 2] and IEC standards [3, 4] differ significantly. Even with these differences, the methods employed to select the proper arrester rating are identical. The tests performed to establish the protective characteristics in these standards are essentially the same. These protective characteristics will be used in other chapters to determine equipment insulation strength.

Primary emphasis is initially placed on the IEEE standards, but this will be followed by a discussion of the IEC standards and a comparison between the two.

Metal oxide surge arresters, first reported in 1971 by Matsuoka [5], were introduced in the USA by Sakshaug et al. [2] in 1977. Because of the concern for the stability and life of the metal oxide, these first station class arresters contained gaps to reduce the normal power frequency voltage placed across the blocks. Subsequently, with improved formulations, the gaps became unnecessary and the present gapless arrester evolved. Later, metal oxide was used for the intermediate and distribution class arresters. Today, all station and intermediate class arresters are metal oxide. However, both metal oxide and silicon carbide (SiC) gapped arresters are still produced in the distribution class.

The gapped station class metal oxide arrester is produced today by both General Electric and Ohio Brass, primarily for use on higher voltage systems. Recently, some manufacturers have produced gapped metal oxide for use on distribution systems.

The advantages of the metal oxide arrester in comparison with the silicon carbide gapped arrester are

1. Simplicity of design, which improves overall quality and decreases moisture ingress
2. Decreased protective characteristics, primarily a result of the elimination of the sparkover gap
3. Increased energy absorption capability

The disadvantage of the metal oxide arrester is one of its chief advantages. Without a gap, the normal power frequency voltage is continually resident across the metal oxide and produces a current of about one milliampere. While this low-magnitude current is not detrimental, higher currents, resulting from excursions of the normal power frequency voltage or from temporary overvoltages (TOV) such as from faults or ferroresonance, produce heating in the metal oxide. If the TOVs are sufficiently large in magnitude and long in duration, temperatures may increase sufficiently so that thermal runaway and failure occur.

The other disadvantage is somewhat facetious in that it occurs with every new product and can be designated as an "educational" item. Because of the gapless design and the resulting sensitivity to power frequency voltages and TOVs, new application rules apply and a new rating system was developed. Replacing the "arrester rating" used for gapped silicon carbide arresters is the MCOV rating, i.e., the maximum continuous operating voltage, a voltage that can be continuously placed across the arrester. In addition, TOVs are now major importance and must be known with a higher degree of accuracy than before so that they can be compared to the TOV capability of the arrester.

1.1 General Characteristics

The voltage–current characteristics of the metal oxide arrester can be divided into three regions [7].

1. In the MCOV region; I is less than 1 mA and is primarily capacitive, thus an I_C. The MCOV of the arrester is selected in this region.
2. In the TOV and switching surge region; I is from 1 mA to about 1000 or 2000 A and is primarily a resistive current, thus an I_R.
3. In the lightning region $I = 1$ to 100 kA. For very large currents, the characteristic approaches a linear relationship with voltage, i.e., becomes a pure resistor.

As shown in Fig. 1 for region 1, as the temperature of the arrester increases, the resistive component of current and thus the power dissipation increases. To prevent thermal instability, i.e., runaway conditions, the power dissipated by the arrester must be transferred to the outside atmosphere through the arrester housing. Voltages that exceed the MCOV rating of the arrester increase arrester temperature, and so the continuous voltage across the arrester must be maintained at or below the MCOV rating, and temporary overvoltages (TOVs) must be limited to within the time limit specific by the manufacturer.

The general characteristic of metal oxide and silicon carbide used in previous arresters is illustrated in Fig. 2a. Because of the sharp turn-on characteristic of the metal oxide characteristic, gaps to isolate the material from the power frequency

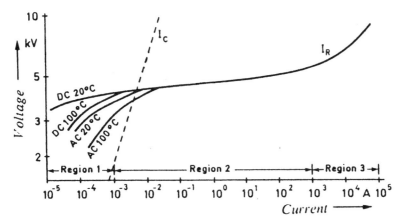

Figure 1 Typical characteristics of a metal oxide arrester disc: diameter = 80 mm, height = 20 mm. (From Ref. 7)

voltage are unnecessary. In contrast, the current drawn by silicon carbide would fail the arrester, and thus gaps are required. For example, assuming that the line-to-ground voltage is 260 kV, Fig. 2a shows that the current drawn by the MO arrester is 1 mA, whereas that drawn by the SiC arrester is 300 A, a current that would destroy the arrester without a gap.

As indicated in Fig. 2a, the SiC characteristic shown can be accurately represented by the power law equation, $I_A = kE_d^\alpha$, where α is about 5.8. E_d and I_A are the discharge voltage and current, respectively. In contrast, as shown in Fig. 2b, if the same power law equation is used for metal oxide, α is variable, reaching a maximum of about 50 in the TOV region and decreasing to about 7 to 10 in the lightning region. Thus, for MO, alpha is primarily used to indicate the "flatness" of the characteristic and should not be employed to model the arrester. However, in some cases, it is convenient to use an alpha within a limited range to assess the arrester's performance, e.g., the TOV capability of the arrester.

1.2 Arrester Classes

Arresters are classified into three primary "durability" or "capability" classes:

1. Station: Used primarily in HV and EHV systems
2. Intermediate: As the name implies, between station and distribution
3. Distribution: Used in distribution systems and further divided into (a) heavy duty, (b) normal duty, and (c) light duty.

In addition, three "specific use" arresters are produced for distribution systems: the riser pole arrester for cables, the dead front arrester for pad-mount transformers, and the liquid immersed arrester used internally in a transformer. These arresters may be of any of the above three classifications. Also arresters are manufactured for use on transmission or distribution lines. They are placed across the line insulation to improve the lightning performance. These later arresters are usually distribution class but may be station or intermediate.

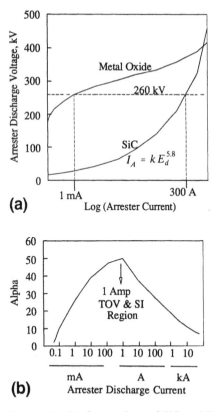

Figure 2 (a) Comparison of SiC and MO characteristics; (b) alpha characteristics of MO.

These three primary classifications may be thought of as good, better, and best varieties. The best, station class, has a superior durability as compared to the good, distribution class. Of course, the best costs more than the better or the good. In general, but not necessarily, the station arrester has better protective characteristics than the intermediate or the distribution arresters.

1.3 Construction types

As mentioned previously, to prevent thermal instability in the initial versions of the MO arresters, gaps were employed. For the quiescent event when normal power frequency is applied to the arrester, either additional MO material is used to decrease the current through the arrester or a decreased voltage is applied to the arrester. The two versions of the gapped MO arrester are shown in Fig. 3a. The first MO arrester, introduced in the USA by General Electric, consisted of two groups of MO elements, one of which was shunted by a gap. The general characteristic is illustrated in Fig. 3b. For normal power frequency voltage, the entire MO material is active. When a lightning or switching surge is applied, the gap sparks over, decreasing the voltage across the arrester and thus producing a lower discharge voltage. For the alternate design by Ohio Brass, a linear component network is placed across the gap (series gap), which is capacitive during normal power frequency voltage. Since the MOA

Metal Oxide Surge Arresters

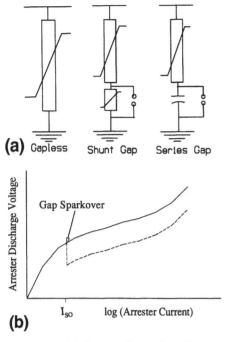

Figure 3 (a) Types of metal oxide arresters; (b) general characteristic of gapped MO arrester.

arrester is capacitive in this region, the network acts as a capacitive divider to decrease the voltage across the MO elements. The network also causes gap sparkover at the higher frequencies of switching and lightning surges at voltages below the power frequency sparkover voltage. Since the introduction of these gapped arresters, metal oxide formulations have been improved to the extent that gaps are unnecessary. However, these arresters are still produced, although primarily for higher voltage systems. In addition, because distribution feeder voltages are at times in excess of the standard "maximum system voltage" and fault TOVs have been larger than realized, gapped metal oxide arresters are being considered for application to these systems.

Since gapped MO arresters are considered as a special application, they will not be considered further in this chapter.

1.4 Voltage Ratings

At present there exist two voltage ratings for the metal oxide arrester:

1. The MCOV or maximum continuous operating voltage (rms)
2. The duty-cycle voltage rating (rms)

As the name implies, the MCOV is the maximum line-to-ground power frequency voltage (rms) that may exist across the arrester—continuously. The duty-cycle voltage rating is equal to the power frequency voltage applied to the arrester during the duty-cycle test (more about this later). For the silicon carbide gapped arrester, the duty-cycle rating was the only rating needed. (Interestingly, the SiC duty-cycle rating

Table 1 IEEE Standard MCOV and Duty-Cycle Ratings of Arresters in kV, rms [1]

MCOV	Duty-cycle	MCOV	Duty-cycle	MCOV	Duty-cycle
2.55	3	42	54	209	258
5.1	6	48	60	212	264
7.65	9	57	72	220	276
8.4	10	70	90	230	288
10.2	12	76	96	235	294
12.7	15	84	108	245	312
15.3	18	98	120	318	396
17.0	21	106	132	335	420
19.5	24	115	144	353	444
22.0	27	131	168	372	468
24.4	30	140	172	392	492
29.0	36	144	180	428	540
31.5	39	152	192	448	564
36.5	45	180	228	462	576
39.0	48	190	240	470	588
				485	612

did not mean that this voltage could be maintained across the arrester continuously.) For the metal oxide arrester, the primary rating and the only rating needed is the MCOV. The duty-cycle rating has been maintained to permit some relationship to the old silicon carbide arrester. (In IEC standards, the duty-cycle rating or simply the voltage rating is defined as the TOV capability at 10 seconds). The standard voltage ratings are presented in Table 1. The ratio of the duty-cycle rating to the MCOV ranges from 1.18 for the lower voltages to 1.26 for the highest voltages.

1.5 Tests—Establishing the MCOV and Protective Characteristics

Tests on arresters can be conveniently divided between those that serve to define the ability of the arrester to exist on the system and protect itself (durability/capability) and those that define the ability to protect the equipment to which it is applied (protective characteristics). The durability/capability tests establish the MCOV rating and the arrester class. These two types of tests will be discussed in the next two sections.

2. DURABILITY/CAPABILITY TESTS

The durability/capability tests, a summary of which is presented in Table 2, establish the arrester voltage rating and arrester class; the primary tests are the discharge withstand tests and the duty-cycle tests. These two tests are preceded by establishing the watt loss of an arrester after 1000 hours at 115°C at the MCOV with excursions to the duty-cycle voltage. This is equivalent to 110 years at 40°C—far beyond the normal life. The ratio of the watt loss after this 1000-hour life to a "new" arrester watt loss at the MCOV is multiplied by the MCOV rating and applied for 30 min to the arrester following these two tests. In this manner the thermal stability is verified,

Table 2 Summary of Durability/Capability Requirements

Arrester class	Max system voltage, kV	Discharge withstand				
		Transmission line			Square wave	4/10 μs
		Voltage, per unit	D_L, line length, km	T, time duration, μs	Amperes, 2000, μs	kA
Station	601–900	2.0	320	2100	—	65
	401–600	2.0	320	2100	—	65
	326–400	2.6	320	2100	—	65
	151–325	2.6	280	1900	—	65
	3–150	2.6	240	1600	—	65
Intermediate	All	2.6	160	1100	—	65
Distribution						
Heavy duty	All	—	—	—	250	100
Normal duty	All	—	—	—	75	65
Light duty	All	—	—	—	75	40

i.e., there is no thermal runaway. The major durability or capability tests prescribed by standards are

1. The high-current, short-duration withstand test
2. The low-current, long-duration withstand test
3. The duty-cycle test
4. "Short-circuit failure" (pressure relief) test
5. The contamination test
6. The temporary overvoltage (TOV) capability test

In addition, manufacturers provide switching surge energy capabilities to define the arrester's ability to discharge the energy in a switching surge. (Tests to establish this capability are presently being considered for standardization.) Adding this capability to the list,

7. Switching surge energy capability

2.1 High-Current, Short-Duration

Two 4/10 μs current impulses must be discharged by the arrester having a crest current of 65 kA for station, intermediate, and distribution normal-duty arresters, 100 kA for distribution heavy-duty arresters, and 40 kA for distribution light duty arresters. Following this test, the MCOV is applied for 30 minutes.

2.2 Low-Current, Long-Duration

For station and intermediate arresters; performed by discharging a pi-section line having a specific length through an arrester, 20 operations required, i.e. three groups of six operations plus one group of two operations, 1 min between groups. See Table

2 for test specifications. For distribution arresters, a 2000 μs square-wave current is discharged through the arrester, 20 operations. Following this test, the MCOV is applied for 30 min.

In Table 2, the voltage to which the line is charged is listed in per unit of the crest line to ground maximum system voltage. This test produces an approximate square-wave current through the arrester having a time duration T in μs of

$$T = \frac{2D_L}{0.3} \qquad (1)$$

where D_L is the line length in km and 0.3 is the velocity of light in km/μs. The energies discharged by the tests of Table 2 are estimated and presented in Table 3. The values are the maximum energies from a single application of either the transmission line discharge, square wave, or a 4/10 μs impulse. The values marked with an asterisk are estimates using the 4/10 μs impulse. The other values are estimates using the transmission line discharge.

2.3 Duty Cycle

While energized at the duty-cycle voltage, an 8/20-μs lightning impulse current is discharged by the arrester, 20 operations, applied once per minute. The crest value of the lightning impulse current is the lightning impulse classifying current of Table 4. Following this test, two lightning impulse currents per Table 4 (except for heavy duty distribution, use 40 kA) are discharged by the arrester without the power frequency voltage. Then MCOV is applied for 30 minutes.

2.4 Pressure Relief "Short-Circuit" Failure

Station and intermediate arresters must have a pressure relief rating. This is designed to show that arrester failure will not be catastrophic. The arrester must vent and all components must lie within a circle whose diameter is twice the height of the arrester plus the arrester diameter. The pressure relief test currents are given in Table 5.

As pressure relief rating is not required for distribution arresters. However, if a fault current withstand is claimed, a test is specified.

Table 3 Estimates of Energy Discharged by Standard Tests of Table 2

Arrester class	Energy, kJ/kV MCOV	Arrester class/max. system voltage, kV	Energy, kJ/kV MCOV
Distribution		Station	
Light duty	*3.0	72.5	*2.3
Normal duty	*4.8	242	*2.3
Heavy duty	*6.7	362	3.9
		550	2.8
		800	3.7
		Intermediate	*3.4
		72.5	

Table 4 Lightning Impulse Classifying Currents

Arrester class	Max. system voltage, kV	Crest current, kV
Station	800	20
	550	15
	< 550	10
Intermediate	All	5
Distribution		
Heavy duty	All	10
Normal duty	All	5
Light duty	All	5

2.5 Contamination

A contamination test is designed to show that the arrester can successfully be applied in contaminated areas. A slurry of bentonite and salt water is poured over the arrester, which is then energized at MCOV for 1 hour. Then MCOV is applied for 30 minutes.

2.6 Switching Impulse Energy

The switching surge energy capability is of importance in selection of the arrester rating. The arrester's capability of discharging the energy contained in a switching surge is partially determined by the low-current, long-duration test. However, usually, higher energies can be safely discharged. Presently, tests to prescribe this energy are not standardized but are being considered for both lightning and switching impulses. In the interim, the switching impulse energy capability is provided by the manufacturers in terms of kJ per kV of MCOV and is given in Tables 6, 7, 8, and 9. This energy is the energy from multiple discharges, distributed over one minute, in which the arrester current is less than a specified magnitude. Thus it becomes apparent that the energy capability depends on the rate at which energy is discharged by the arrester. Some examples from manufacturers' literature are

1. One manufacturer states that the energy from a single arrester operation should not exceed 85% of the energy rating.
2. Two manufacturers state that the energy capability provided is for discharge currents less than specified values.

Table 5 Pressure Relief Test Currents for Station and Intermediate Arresters

	Symmetrical current, A, rms	
Arrester class	High current	Low current
Station	40,000 to 65,000	600 ± 200
Intermediate	16,100	600 ± 200

Table 6 Durability/Capability Characteristics for Station Class Arresters

Mfg	Design	Press relief, kA	MCOV ratings, kV	SI energy, kJ/MCOV	A1, kV/MCOV	B1	A2, kV/MCOV	B2	Ratio	TOV_{10}/MCOV
OB	VLA	10	2.55–22	4.9	1.458	.0213	1.407	.0131	.955	1.326
	VL	65	2.55–39	4.9	1.458	.0213	1.407	.0131	.955	1.326
	VN	93	42–245	8.9	1.458	.0213	1.407	.0131	.955	1.326
GE	XE	65	2.2–39	4.9	1.445	.0203	1.463	.0231	.951	1.311
	XE	65	44–292	8.9	1.445	.0203	1.463	.0231	.951	1.311
Cooper	ATZ	65	2.55–24.4	4.9	1.487	.0185	1.497	.0202	.957	1.364
	ATZ	40	29–31.5	4.9	1.487	.0185	1.497	.0202	.957	1.364
	ATZ	80	34–245	8.9	1.487	.0185	1.497	.0202	.957	1.364
Joslyn	ZS	80	2.55–1.90	6.3	1.617	.0265	1.533	.0157	.957	1.456
	ZSH	80	209–372	9.0	1.617	.0265	1.533	.0157	.957	1.456
ABB	EXLIN-Q	65	2.55–39	4.5	1.491	.0228	1.514	.0263	.953	1.348
	EXLIM-P	80	42–288	7.0	1.491	.0228	1.514	.0263	.953	1.348
	EXLIM-T	65	318–485	10.0	1.491	.0228	1.514	.0263	.953	1.348

Table 7 Durability/Capability Characteristics for Intermediate Class Arresters

Mfg	Design	Press rellief, kA	MCOV ratings, kV	SI energy, kJ/MCOV	A1, kV/MCOV	B1	A2, kV/MCOV	B2	Ratio	TOV_{10}/MCOV
OB	PVI	25	2.55–84	3.4	1.458	.0213	1.407	.0131	.955	1.326
OB	VIA	25	2.55–98	3.4	1.458	.0213	1.407	.0131	.955	1.326
GE	XE	NA	2.2–98	NA	1.436	.0203	1.454	.0231	.951	1.303
Cooper	AZF	NA	2.55–98	NA	1.454	.0214	1.375	.0066	NA	1.315
Joslyn	ZIP	40	2.55–98	3.6	1.436	.0219	1.450	.0195	.972	1.327
Joslyn	ZI	40	2.55–115	4.3	1.617	.0265	1.533	.0157	.957	1.456
ABB	IMX	40	2.55–98	2.6	1.565	.0218	1.590	.0258	.962	1.432

Metal Oxide Surge Arresters

Table 8 Durability/Capability Characteristics for Distribution Class Arresters

Mfg	Type	Design	Fault curr., kA	MCOV ratings, kV	SI energy, kJ/MCOV	A1, kV/MCOV	B1	A2, kV/MCOV	B2	Ratio	TOV_{10}/MCOV
OB	HD	PDV100	20	2.55–29	2.2	1.567	.0216	1.608	.0284	.981	1.463
	ND	PDV65	10	2.55–24.4	1.4	1.567	.0216	1.608	.0284	.981	1.463
	RP	PVR	20	2.55–29	3.4	1.458	.0213	1.407	.0131	.955	1.326
GE	HD	XE	NA	2.55–22	NA	1.548	.0203	1.567	.0231	.951	1.405
	RP	UDIIA	NA	7.65–22	NA	1.408	.0203	1.425	.0231	.951	1.278
	RP	UD-XE	NA	7.65–22	NA	1.436	.0203	1.454	.0231	.951	1.303
Cooper	HD	AZL	NA	2.55–29	NA	1.605	.0232	1.500	.0076	NA	1.445
	HD	AZLP	20	2.55–24.4	NA	1.605	.0232	1.500	.0076	NA	1.445
	ND	AZS	NA	2.55–29	NA	1.605	.0435	1.541	.0341	NA	1.379
	ND	AZSP	NA	2.55–24.4	NA	1.605	.0435	1.541	.0341	NA	1.379
	RP	AZR	NA	7.65–24.4	NA	1.454	.0214	1.375	.0066	NA	1.315
	RP	AZRP	NA	7.7–22	NA	1.454	.0214	1.375	.0066	NA	1.315
Joslyn	HD	ZQ	NA	2.55–22	NA	NA	NA	NA	NA	NA	NA
	HD	ZQP	13	2.55–24.4	NA	NA	NA	NA	NA	NA	NA
	HD	ZR	NA	7.65–22	NA	NA	NA	NA	NA	NA	NA
	RP	ZJ	NA	7.65–22	NA	NA	NA	NA	NA	NA	NA
	RP	ZJP	13	2.55–24.4	NA	NA	NA	NA	NA	NA	NA
	DF	ZE	NA	2.5–22	1.2	1.617	.0265	1.533	.0157	.957	1.456

Table 9 Durability/Capability Characteristics for Line Arresters

Mfg	Design	Fault curr. kA	MCOV ratings, kV	SI energy, kJ/MCOV	A1, kV/MCOV	B1	A2, kV/MCOV	B2	Ratio	TOV_{10}/MCOV
OB	Protec lite	NA	7.65–144	NA	1.567	.0216	1.608	.0284	.981	1.463
Joslyn	ZQPI	12	29–220	NA	NA	NA	NA	NA	NA	NA

3. One manufacturer states that for arresters having MCOV ratings between 2.55 and 39 kV, currents must be below 750 A. For arresters having MCOV ratings between 42 and 245 kV, currents must be below 1200 A.
4. Another manufacturer states that for arresters having MCOV ratings between 2.55 and 39 kV, currents must be below 1000 A. For arresters having MCOV ratings between 42 and 245 kV, currents must be below 1500 A.

The limitation on current magnitude does not affect application on transmission or distribution lines but does affect the application on cables and capacitor banks.

2.7 Temporary Overvoltages

The test results in a curve and table of the TOV capability as a function of time from 0.02 to at least 1000 seconds, and usually this curve is drawn from 0.01 to 10,000 seconds. Figure 4 shows a typical curve for a station class arrester. Two curves are prescribed by standards, one for no prior energy absorption, the other for a prior energy absorption equal to that from two transmission line discharges. In contrast, a minimum TOV capability curve for distribution arresters is shown in Fig. 5 as taken from the IEEE Guide [2].

These curves can be modeled by a power equation

$$\frac{TOV_C}{MCOV} = AT^{-B} \qquad (2)$$

where TOV_C is the TOV capability, A and B are constants, and T is the time duration in seconds. The curves can be divided into two areas:

1. $t = 0.01$ to 100 seconds
2. $t > 100$ to $t = 10,000$ seconds

The values of A and B are presented in Tables 6, 7, 8, and 9 for alternate arrester manufacturers. $A1$ and $B1$ apply to the first time interval and $A2$ and $B2$ apply to the

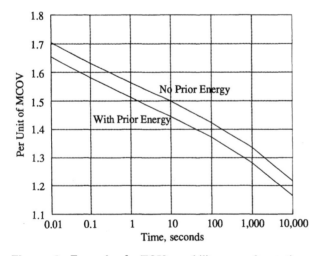

Figure 4 Example of a TOV capability curve for station arresters.

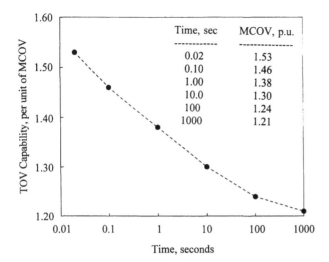

Figure 5 The minimum TOV capability curve for distribution arresters. (From Ref. 2.)

second time interval. ($A1$ is the TOV capability for 1 second.) For times exceeding 100 seconds, the TOV capability is dependent on the heat transfer capability of the arrester and is therefore a function of heat transfer design.

The values of A and B apply to "no prior energy." The ratio listed in these tables is the ratio of the prior energy TOV to the no prior energy TOV. To be conservative, in general, the prior energy TOV should be used for application. The TOV for one second with no prior energy for all arresters classes varies from 1.44 to 1.62 per unit of the MCOV.

The B coefficients can be derived theoretically assuming that alpha is constant in this region. The energy W discharged by the arrester is

$$W = k E_d I_A T \tag{3}$$

where E_d and I_A are the arrester discharge voltage and current, T is the time duration, and k is a constant. Substituting

$$I_A = k_1 E_d^\alpha \tag{4}$$

into Eq. 3,

$$W = k_3 T E_A^{1+\alpha} \tag{5}$$

Then for the first time interval from 0.01 to 100 seconds,

$$\frac{\text{TOV}_C}{\text{MCOV}} = (A1) T^{-\frac{1}{1+\alpha}} = (A1) T^{-B1} \tag{6}$$

If alpha is 50, then $B1$ should be about 0.020. From Tables 6, 7, 8, and 9, $B1$ varies between 0.02 and 0.044, not a bad check.

In IEC [3, 4] the rated voltage of the arrester is defined as the TOV capability at 10 seconds with *prior energy*, i.e., TOV_{10}. Thus

$$\frac{\text{TOV}_{10}}{\text{MCOV}} = (A1)10^{-B1} \tag{7}$$

Using Eq. 2,

$$\text{TOV}_C = \left(\frac{10}{T}\right)^{B1} \text{TOV}_{10} \tag{8}$$

TOV_{10} is also listed in the tables and ranges from 1.30 to 1.46 times MCOV. For cases where no "ratio" exists, the ratio is assumed as 0.95. Note that, since the ratio of the duty-cycle voltage to the MCOV ranges from 1.18 to 1.26, the TOV_{10} is 3 to 24% greater than the duty-cycle voltage.

The minimum TOV characteristics for distribution arresters presented in Fig. 5 may be used if the manufacturer's data is unavailable. The values of A and B are

For time interval of 0.01 to 100 seconds: $A1 = 1.45$, $B1 = 0.0338$
For time interval of 100 to 10,000 seconds: $A2 = 1.30$, $B2 = 0.0106$
The TOV_{10}, assuming a "ratio" of 0.95, is 1.24 MCOV

2.8 General Energy Capability

The energy capability for various waveforms and current magnitudes has been the subject of many investigations [8–10, 21–23]. For these investigations and studies, the authors applied 60-Hz and rectangular pulses in an attempt to identify the energy capability in the TOV, switching surge, and lightning regions. As previously mentioned, tests are only required for currents in the TOV and switching surge regions. With the increased use of arresters for protection of transmission and distribution lines, the energy capability in the lightning region is essential.

The authors of Ref. 8 tested 3 kV (rated voltage) disks having diameters of 62–63 mm and heights of 23–24 mm from three manufacturers. 60 Hz currents of 0.84, 7.4, 67, and 646 A and rectangular or square-wave currents of 4.35 and 35.2 kA were applied to these blocks. The currents were applied until failure occurred; this resulted in the plot of Fig. 6a, where the mean failure energy increases markedly with increased current.

Of most importance, the authors [8] found that (1) the energy capability was variable and (2) the standard deviation ranged from 8 to 23% of the mean, averaging 15%. For example, the average energy capability for 646 A is about 500 J/cm^3, whereas the tests prescribed by standards only produce 200 J/cm^3. Using a 15% standard deviation, the 200 J/cm^3 capability occurs at 4 standard deviations below the mean. The failure mode of the disks was primarily puncture.

Other investigations [21, 22]. show similar probabilistic results. The authors of these reports produced plots of discharge current as a function of energy capability for failure probabilities ranging from 0 to 50%. Analyzing their plot, the standard deviation averages about 13% and the zero probability point is about 4 standard deviations below the mean. Thus the energy capability is probabilistic, a characteristic that while expected has not been confirmed until now. Assuming that the

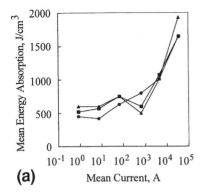

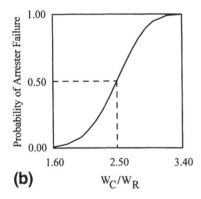

Figure 6 (a) Mean failure energy for alternate currents for disk; (b) energy failure characteristic. (From Ref. 8.)

Weibull cumulative distribution function can be used to model the energy characteristic and setting the discharge energy for the standard tests at the mean minus 4 standard deviations, the characteristic is given by the equations

$$P_{FA} = 1 - 0.5^{\left(\frac{Z}{4}+1\right)^5} \qquad Z = \frac{W_C - \mu}{\sigma} = \frac{(W_C/W_R) - 2.5}{0.375} \qquad (9)$$

where μ is the average or mean energy capability, W_C is the energy capability, and W_R is the rated energy from the standard tests, i.e., that provided by the manufacturer. The equation for Z assumes a standard deviation of 15% of the mean. Figure 6b shows this distribution. To check this equation versus the result in Ref. 8, for a rated energy of 200 J/cm^3, the probability of failure for a mean or average energy of 500 J/cm^3 is 50%, a very fortuitous check. In contrast to the authors of Ref. 8, the authors of [21, 22] state that cracking was the primary failure mode and that energy capability was constant, not dependent on current magnitude or shape.

To clarify these differences, a recent paper [23] presents a detailed examination of the failure modes and the energy capability. The authors examine the three main failure modes, thermal runaway, puncture, and cracking. They found that initially, the energy capability decreases with increasing current but then increases for higher magnitude, short-duration currents, as first pointed out by Sakshaug [9]. The energy capability depends on which of these failure modes is predominant for a current pulse. Each failure mode can be limiting, depending on the disk shape, its electrical uniformity, and the current magnitude.

Another interesting result of the tests performed by the authors of Ref. 8 is that a linear dependence was found between the logarithm of the mean time to failure t and the logarithm of the mean current I_A. Approximately, $I_A = 21/t$.

Although the energy capability for typical lightning impulse waveshapes and for a complete arrester is yet to be shown, there exists sufficient evidence to indicate that the energy capability is probabilistic. Therefore, for the present, Eq. 9 is recommended for use.

3 PROTECTIVE CHARACTERISTICS

The protective characteristics are voltages across the arrester for a specified discharge current magnitude and shape. As specified by standards:

3.1 8/20 µs Discharge Voltage

The manufacturer must tabulate the voltage across the arrester for arrester discharge currents having an 8/20 µs waveshape and crest magnitudes of 1.5, 3.0, 5, 10, and 20 kA. For arresters applicable to 500-kV systems, the 15-kA discharge must also be given. Most manufacturers also provide the 40-kA discharge voltage.

3.2 Front of Wave Impulse Protective Level or the 0.5-µs Discharge Voltage

This is the voltage across the arrester having a time to crest of 0.5 µs when discharging the lightning impulse classifying current of Table 4. This discharge voltage is obtained by using times to crest of the current of 1, 2, and 3 µs and plotting the voltage vs. time to crest of the voltage. The term front of wave protective level, FOW, is a misnomer carried over from previous standards concerning gapped SiC arresters. More properly it should be called the 0.5-µs discharge voltage.

3.3 Switching Impulse Protective Level—SI

This is the voltage across the arrester when discharging a current impulse having a 45- to 60-µs time to crest and a magnitude equal to the switching impulse classifying current per Table 10. It is not required for distribution arresters.

Table 10 Switching Impulse Classifying Currents

Arrester class	Max. system voltage, kV	Crest current, amperes
Station	3–150	500
	151–325	1000
	326–900	2000
Intermediate	3–150	500

Metal Oxide Surge Arresters

Table 11 Range of Protective Characteristics in per unit of crest MCOV [2]

Class	LI current, kA	FOW or 0.5 µs	8/20 µs	SI current, kA	SI
Station	10 to 20 Table 4	2.01–2.48	1.97–2.25	0.5 to 2.0 Table 10	1.64–1.85
Intermediate	5	2.38–2.85	2.28–2.55	0.5	1.71–1.85
Distribution					
Heavy duty	10	2.40–3.75	2.00–3.46	—	—
Normal duty	5	2.90–3.53	2.77–3.32	—	—
Riser pole	10	2.07–3.32	2.65–3.32	—	—

3.4 Protective Characteristics, Discharge Voltages

Table 11 shows the range of discharge voltages or protective characteristics as obtained from the IEEE Application Guide [2], given in per unit of the crest value of the MCOV, i.e., $\sqrt{2}\,MCOV$. The values are given for the lightning and switching impulse classifying currents of Tables 4 and 10.

Because the range of these characteristics is large, and to present realistic values that can be used for application, the protective characteristics of arresters manufactured by the Ohio Brass Company are provided in the Appendix.

3.5 Effect of Time to Crest

As noted from Table 11, the discharge voltage magnitude and time to crest are functions of the time to crest and magnitude of the discharge current. The magnitude increase is shown in Fig. 7 as taken from Ref. 11.

Considering the time to crest of the voltage, an 8-µs current front produces a voltage time to crest of about 7 µs, while at 1-µs current front results in a voltage time to crest of about 0.5 µs. For practical cases, the steepness of the incoming surge per

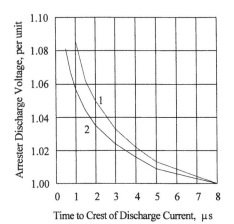

Figure 7 Effect of time to crest of arrester current. 1 = results from Ref. 12, 2 = calculated. (From Ref. 11.)

Chapter 11 is in the range of 500 to 200 kV/µs, and thus the time to crest is much less than 7 µs. Therefore either the protective characteristic should be constructed for the time to crest of the incoming surge, or an arrester model that encompasses the effect of the time to crest should be used (to be considered later). To accomplish this, one manufacturer presents curves of the discharge voltage as a function of the voltage time to crest. However, conservatively, this characteristic can be generated for a 0.5-µs from the available published data. For example, from Appendix 1, the 0.5-µs discharge voltage for a 318-kV MCOV arrester is 1070 kV. From Table 4, since this arrester is to be applied to a 550-kV system, the current employed is 15 kA. The 15-kA, 8/20 µs discharge voltage is 915 kV. Thus the 0.5-µs discharge voltage is 1.17 times the 8/20-µs discharge voltage. Therefore a model of the arrester can be constructed for the 0.5-µs time to crest by multiplying all the 8/20-µs discharge voltages by 1.17.

3.6 Arrester Models

Figure 8 [11] shows a plot of arrester voltage versus arrester current with the arrows indicating the increase in time. Thus the time to crest of the current precedes the time to crest of the voltage, which is also the reason that a current having a front of about 1 µs produces a voltage having a front of about 0.5 µs. A proper model of the arrester should result in a curve such as this and also be sensitive to the effect of the increase in crest voltage with a decrease in current time to crest. Both IEEE and CIGRE have produced models designed to reproduce these desired characteristics [11, 13].

IEEE Model. The IEEE model [13], shown in Fig. 9, is composed of two nonlinear elements separated by a resistance–inductance network where

$$L_1 = 15\frac{d}{n} \qquad R_1 = 65\frac{d}{n} \tag{10}$$

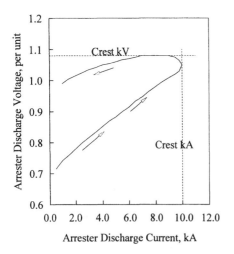

Figure 8 Arrester voltage–current characteristic for a 1/2.5-µs current impulse. (From Ref. 11.)

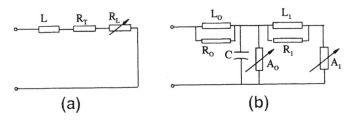

Figure 9 CIGRE and IEEE arrester models. (a) CIGRE. (b) IEEE. (From Refs. 11 and 13.)

where $L1$ is in μH and $R1$ in ohms, d is the height of the arrester in meters and n is the number of parallel columns of MO disks in the arrester. The inductance L_0 in μH, R_0 in ohms, and C, the capacitance of the arrester in pF, are given by the equations

$$L_0 = 0.2\frac{d}{n}$$
$$R_0 = 100\frac{d}{n} \qquad (11)$$
$$C = 100\frac{n}{d}$$

The nonlinear voltage–current characteristics of $A0$ and $A1$ are estimated from the data of Table 12, where the values of voltage are in per unit of the 10-kA, 8/20-μs discharge voltage.

The procedure in constructing the model is to use first the values of the parameters that are stated. Then adjust the values of A_0 and A_1 to obtain a match to the switching impulse discharge voltage for a current time to crest of 45 μs. Then adjust

Table 12 Value of A_0 and A_1 in IEEE Model

Current, kA	Voltage per unit of 10 kA, 8/20 μs for A_0	Voltage per unit of 10 kA, 8/20 μs for A_1
0.01	0.875	—
0.1	0.963	0.769
1	1.050	0.850
2	1.088	0.894
4	1.125	0.925
6	1.138	0.938
8	1.169	0.956
10	1.188	0.969
12	1.206	0.975
14	1.231	0.988
16	1.250	0.994
18	1.281	1.000
20	1.313	1.006

Source: Ref. 13.

the value of L_1 to obtain a match to the published 8/20-μs discharge voltages. The resultant model is stated to be valid for times to crest of the current from 0.5 to 45 μs.

CIGRE Model. The CIGRE model [11], also shown in Fig. 9, has a single non-linear element R_L, which is developed from the 8/20-μs characteristics. R_T is the turn-on resistance and can be obtained from curves [12] or from equations [11]. L is the inductance of the current path through the arrester. This can be represented as an inductance or by a surge impedance and a travel time as follows:

For outdoor arresters: $L = 1\,\mu H/m$ of arrester length of $Z = 300$ ohms; travel time = 3.33 ns/m of arrester length

For GIS arresters: $0.33\,\mu H/m$, $Z = 100$ ohms; travel time = 3.33 ns/m

An example of the use of this model is shown in Ref. 11 by a voltage–current characteristic that results in an excellent match to that of Fig. 8.

Some observations are obvious: (1) The construction of either of these models is complex and requires repeated approximations to obtain the proper value of the parameters. (2) This effort is seldom necessary except for sensitive cases. (3) If the usual model as discussed in Section 3.5 is used, conservative results will be obtained. (4) Examining the voltage–current characteristic of Fig. 8, if the model per Section 3.5 is used, the tail of the voltages at equipment locations will be greater than if the IEEE or CIGRE model is used. This is sometimes detrimental in that larger insulation strength may be required.

4 DETERMINING THE ARRESTER RATING

To determine the arrester rating, the rules are

1. MCOV must be equal to or greater than the maximum line-to-ground system voltage.
2. Switching surge energy discharged by the arrester must be less than the energy capability.
3. The temporary overvoltage across the arrester must be less than the arrester TOV capability.
4. The continuous ambient temperature must be less than 40°C, and the temporary maximum must be less than 105°C.
5. Altitude limit: 1800 m (6000 feet).
6. The pressure relief current must be equal to or greater than the fault current.

4.1 Maximum System Voltage

The first requirement is of most importance. The maximum system voltage is prescribed by standards for each nominal system voltage. However, in some cases the actual maximum may be higher or lower than the standard maximum. For distribution arrester application, the voltage at the customer meter is standardized, but the feeder voltage varies considerably. This factor, while being of relative unimportance when applying silicon carbide gapped arresters, is of major importance for metal oxide arresters.

Metal Oxide Surge Arresters

Example 1. As an example, the maximum system voltage for a 345-kV system is 362 kV. Therefore the maximum phase-to-ground voltage is 209 kV and the MCOV must be equal to or greater than 209 kV. From the standardized list of ratings, a 209-kV MCOV arrester could be used.

4.2 Switching Surge Energy

The switching surge energy discharged by the arrester can be estimated by equations developed using the circuits in Fig. 10. Assuming that a transmission line is charged to a switching surge overvoltage E, the current through the arrester per the equivalent circuit of Fig. 10b is

$$I_A = \frac{E - E_d}{Z} \tag{12}$$

where E_d is the arrester discharge voltage and Z is the surge impedance of the phase conductor. The arrester discharge voltage is represented by a straight line per Fig. 10c, i.e.,

$$E_d = E_0 + I_A R_A \tag{13}$$

Combining these equations, we obtain

$$I_A = \frac{E - E_0}{Z + R_A} \tag{14}$$

Assuming that all the energy is discharged in twice the travel time of the line, T_L, where the travel time is the line length divided by the velocity of light,

$$W = 2 T_L I_A E_d \tag{15}$$

If the current is in kA, the voltage in kV, and the time in ms, then the energy is in kJ.

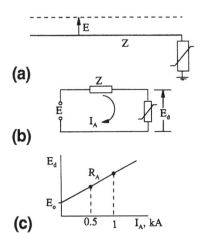

Figure 10 (a) A switching surge E on a line; (b) the equivalent circuit; (c) arrester characteristic.

Table 13 Approximation of SI Discharge Voltage

I_A, kA	E_d, kV
0.5	1.00
1	1.03
2	1.07
5	1.14
10	1.21

The problem with this approach is that the standard tests only provide one value of discharge current and discharge voltage in the switching surge region, so that another value of current–voltage must be estimated. As an approximation, setting the 0.5-kA discharge voltage at 1 per unit, the relative values for alternate currents of Table 13 can be used.

Example 2. Assume that a 900-kV switching surge occurs at the end of a 300-km, 362 kV line where a 209-kV MCOV station class arrester is located. Assume that the conductor surge impedance is 350 ohms and that the arrester SI discharge voltage is 482 kV. From Table 10, the discharge current used to obtain this voltage is 2.0 kA. From Table 13, the discharge voltage at 1.0 kA is 464 kV, R_A becomes 18 ohms, and E_0 is 446 kV. From Eq. 14, the discharge current for a switching surge of 900 kV is 1.234 kA, giving a discharge voltage (Eq. 13) of 468 kV. From Eq. 15 with $T_L = 1.0$ ms, the energy discharged by the arrester is 1155 kJ or 5.53 kJ/kV MCOV. From Table 6, the energy capability ranges from 7 to 9 kJ/kV, and thus a 209-kV MCOV arrester is acceptable even for this long line and high-magnitude surge.

Another more conservative method is to assume that all the energy stored on the line is discharged through the arrester, i.e.,

$$W = \frac{1}{2}CE^2 = \frac{T_L E^2}{2Z} \quad (16)$$

where C is the total capacitance of the line. Use of this equation results in a discharge energy of 1157 kJ, about the same as before.

Another method is to assume that the actual discharge current and voltage through the arrester are the tested values, 2 kA and 482 kV, which gives an energy of 1928 kJ or 9.2 kJ/kV, which is greater than some arrester SI energy ratings.

Example 3. These same methods can be used to estimate the arrester discharge energy for arresters applied to cable circuits. For example, assume that an 84-kV MCOV arrester is applied to a 20-km cable having a surge impedance of 30 ohms. For a velocity of propagation of one-third that of light, the travel time T_L is 0.2 ms. Assume a switching surge of 2.5 pu or 296 kV on this 145-kV cable. The SI discharge voltage of 205 kV is specified for a current of 0.5 kA per Table 10. From Table 13, the 2-kA and 5-kA discharge voltages are 219.35 and 233.70 kV. The R_A becomes 4.783 ohms, and E_0 is 209.8 kV, resulting in a discharge current and voltage of 2.48 kA and 221.7 kV. The discharge energy is 220 kJ or 2.62 kJ/kV MCOV, which, from Tables 6 and 7, is in the range of station and intermediate arresters. Using Eq. 16, the discharge energy is 292 kJ, again a conservative value. However, if the test values of arrester current and voltage are used, the energy is only about 0.5 kJ. Therefore one of the two other methods is suggested.

If the arrester discharge voltage is low compared to the switching overvoltage, the energies determined by use of the first two methods are approximately equal. As a final comment, the SI discharge voltages published by the manufacturer are maximum values. To obtain a more conservative estimate of the energy, the minimum values should be used, which are about 3 to 4% less than the published values. However, the difference between the energies is small. For example, if the switching surge discharge voltage of example 2 is decreased 4%, the energy only increases to 1157 kJ. Thus this refinement is only necessary in critical situations.

4.3 TOVs—Temporary Overvoltages

Temporary overvoltages or TOVs [14, 15] are primarily produced by

1. Faults
2. Load rejections
3. Energization of unloaded lines
4. Resonance

For the usual case, only the first item, faults, needs to be considered. However, the other sources of TOVs should not be totally dismissed, since they may be of primary concern in specific situations.

The TOVs as discussed here are taken from Ref. 14, which should be consulted for detailed explanations. Table 14, showing typical TOVs caused by faults, and Table 15, showing TOVs from other sources, are taken from this reference.

TOVs generated by faults can be quickly calculated or can be obtained from sets of curves. These TOVs, except for resonance grounded systems, are highest at the

Table 14 Typical Fault TOVs [14]

Category	EFF, per unit	COG, %	Duration
Grounded			
Networks, high short circuit	1.2 to 1.4	58 to 80	1 sec
Long radial lines, low short circuit	1.2 to 1.5	58 to 87	1 sec
Partialy or low impedance	1.4 to 1.7	80 to 100	1 sec
Resonant Grounded			
Network or mesh	1.73	100	8 h to 2 days
Long radial lines	1.73 to 1.80	100 to 104	8 h to 2 days
	1.73 to 2.00	100 to 115	Remote from fault
Isolated			
Distribution with overhead lines & industrial with cables	1.73 to 1.80	100 to 104	with fault clearing: lines: 1–2 sec bus: 4 sec w/o fault clearing: 8 h

Table 15 Typical TOVs from Other Than Faults [14]

Category		EFF, per unit	COG, %	Duration
Load rejection				
in a system		1.05	60	> 10 s
arc furnace		1.15	66	> 10 s
generator-transf:	turbo	1.1–1.4	64–80	1 s
	hydro	1.15–1.5	66–85	1 s
Charging, unloaded line: Energization or after load rejection, 200 km				
High short circuit		1.0–1.1	58–64	> 10 s
Low short circuit		1.0–1.2	58–69	> 10 s
Resonance				
Saturation phenomena		2–3?	115–170	0.5–10 s
Coupled circuits		3–5?	170–290	0.5–10 s

fault location. Following are equations that can be used for this purpose. The definitions used are

COG = coefficient of grounding, which is the ratio of the line-to-ground rms voltage on the unfaulted phase to the nominal system *line-to-line* rms voltage, i.e., the prefault voltages

EFF = earth fault factor, which is the ratio of the line-to-ground rms voltage on the unfaulted phase to the normal system *line-to-ground* rms voltage, i.e., the prefault voltage. Note that

$$\text{EFF} = \sqrt{3}\,\text{COG} \qquad (17)$$

SLG Faults. For SLG faults,

$$\text{COG} = \frac{1}{2}\left[\frac{\sqrt{3}K}{2+K} \pm j1\right] \qquad (18)$$

where

$$K = \frac{Z_0 + R_f}{Z_1 + R_f}$$
$$Z_0 = R_0 + jX_0 \qquad (19)$$
$$Z_1 = R_1 + jX_1$$

where the Rs and Xs are the symmetrical component values and R_f is the fault resistance.

DLG Faults. For DLG faults,

$$\text{COG} = \frac{\sqrt{3}K}{K+1} \qquad (20)$$

where in this case

$$K = \frac{Z_0 + 2R_f}{Z_1 + 2R_f} \tag{21}$$

Fault resistance, the use of which is questionable, tends to reduce the COG and the EFF, except in low-resistance systems. For systems having a nominal system voltage equal to or greater than 115 kV, R_0 and R_1 may be assumed as zero. Therefore the previous equations reduce to the following.

SLG Faults. For SLG faults,

$$COG = \frac{\sqrt{K^2 + K + 1}}{K + 2} \tag{22}$$

DLG Faults. For DLG faults,

$$COG = \frac{K}{2K + 1} \tag{23}$$

where

$$K = \frac{X_0}{X_1} \tag{24}$$

An example of the EFFs for $R_1 = R_f = 0$ is shown in Fig. 11. The SLG fault produces the maximum fault voltages for all values of R_0/X_1 except zero, for which the DLG fault voltages are maximum.

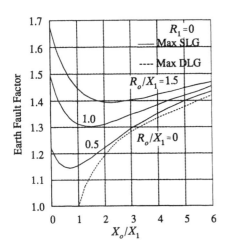

Figure 11 EFFs for $R_1 = R_f = 0$.

Example 4. The system TOV must be equal to or less than the TOV capability of the arrester. As an example, let the system TOV equal 1.4 pu of the rms line-to-neutral voltage or 1.4(209) kV (345/362-kV system) with a TOV time duration of 2 seconds. Assume the use of an OB VN arrester, and from Table 6, $A1 = 1.458$ and $B1 = 0.0213$. Then the arrester capability TOV_C is

$$\text{TOV}_C = 1.458(\text{MCOV})t^{-0.0213} = 1.437\text{MCOV} \qquad (25)$$

Because the system TOV is $1.4(209) = 292.6\,\text{kV}$, the minimum arrester MCOV rating is

$$\text{MCOV} = \frac{292.6}{1.437} = 203.6\,\text{kV} \qquad (26)$$

Because the minimum arrester MCOV is 209 kV from a maximum system voltage standpoint, a 209 kV MCOV would be selected.

4.4 Parallel or Multicolumn Arresters

In some cases, two parallel arresters are installed with the thought that the total current and thus the total energy will be divided between the arresters. This will occur if the arrester voltage–current characteristics are *exactly* equal. However, this is usually not the case, unless the manufacturer is notified before both arresters are purchased. Manufacturing tolerance between arresters may be as little as 3% but is frequently higher. Consider the two arresters having the two voltage–current characteristics of Fig. 12. Assume that the discharge voltage across the arresters is E_2. Therefore the arrester currents are

$$I_{A1} = k_1 E_1^\alpha = k_2 E_2^\alpha$$
$$\frac{k_1}{k_2} = \left(\frac{E_2}{E_1}\right)^\alpha \qquad (27)$$

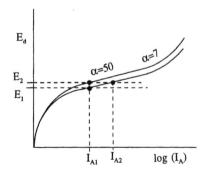

Figure 12 The problem of parallel arresters.

Table 16 Ratio of Arrester Currents

Arrester type	α	E_2/E_1	I_{A2}/I_{A1}
MO	50	1.03	4.4
		1.05	11.5
MO	7	1.03	1.2
		1.05	1.4
SiC	5.8	1.03	1.2
		1.05	1.3

To find the ratio of k_1/k_2, note that

$$I_{A1} = k_2 E_2^\alpha$$
$$I_{A2} = k_1 E_2^\alpha \qquad (28)$$
$$\frac{I_{A2}}{I_{A1}} = \frac{k_1}{k_2}$$

Therefore

$$\frac{I_{A2}}{I_{A1}} = \left(\frac{E_2}{E_1}\right)^\alpha \qquad (29)$$

Table 16 shows the results for the metal oxide and silicon carbide arresters. For the important TOV and switching surge region where α = 50 for the MO, a small difference in the characteristics produces a large increase in differential current. Thus essentially all the current and energy is discharged by one arrester. In the lightning region or for the SiC arrester, the differential current is much less.

The lesson is that if more than a single arrester is required to handle the energy, the manufacturer should be contacted so that a matched set of arresters can be produced. This problem is especially severe for arresters used to protect series capacitors for which over 100 units are sometimes used. The problem also occurs for single arresters constructed of two or more columns. However, in this case, the manufacturer can produce matched sets of arrester columns. Tests to establish the division of current in multicolumn arresters presently exist in the IEC standards and are being considered for IEEE standards.

5 TYPICAL MCOV RATINGS

Tables 17 and 18 contain typical MCOV ratings for HV, EHV and distribution systems.

6 CALCULATING THE DISCHARGE VOLTAGE AND CURRENT FOR LIGHTNING IMPULSES

The purpose of this section is to develop the equations for calculating or estimating the currents discharged by the arrester that in turn produce the arrester voltage. The

Table 17 Typical MCOV Ratings for HV and EHV Systems [2]

Nominal system voltage, kV	Maximum system voltage, kV	Arrester MCOV, kV	
		Grounded neutral circuits	Others
69	72.5	42, 48	57
115	121	70, 76	84
138	145	84	98
161	169	98, 106	115
230	242	140, 144	152
345	362	209, 212	220
400	420	245	
500	550	318	335
765	800	462, 470	485

arrester current and thus the voltage can be easily calculated for an arrester at the end of a line. However, this current is increased by two phenomena. Assuming, as is usually the case, that a transformer is beyond the arrester, reflections from the transformer will increase the arrester current. Further, reflections from the struck point on the line where the surge originates may also increase the arrester current.

Table 18 Typical MCOV Ratings for Distribution Systems [2]

System Voltage, V		Arrester MCOV, kV		
Nominal voltage, V	Max. voltage range, B, V	Four-wire multigrounded neutral wye	Three-wire low-impedance grounded	Three-wire high-impedance grounded
2400	2540			2.55
4160Y/2400	4400Y/2540	2.55	5.1	5.1
4260	4400			5.1
4800	5080			5.1
6900	7260			7.65
8230Y/4800	8800Y/5080	5.1	7.65	
12000Y/6930	12700Y/7330	7.65	10.2	
12470Y/7200	13200Y/7620	7.65 & 8.4	12.7	
13200Y/7620	13970Y/8070	8.4	12.7	
13800Y/7970	14520Y/8380	8.4 & 10.2	12.7	
13800	14520			15.3
20780/12000	22000Y/12700	12.7	17.0	
22860Y/13200	24200Y/13870	15.3	19.5	
23000	24340			24.4
24940Y/14400	26400Y/15240	15.3	22.0	
27600Y/5930	29255Y/16890	17.0	24.4	
34500Y/19920	36510Y/21080	22.0	29.0	

Source: Ref. 2

6.1 Arrester at End of Line

As depicted in Fig. 13a, consider a surge voltage e having a magnitude E and a front steepness S arriving at a surge arrester on a line of surge impedance Z. As defined in Fig. 13b, the surge voltage is traveling atop a negative power frequency voltage V_{PF}. The Thevenin equivalent circuit is shown in Fig. 13c, where $(2E - V_{PF})$ is the applied source voltage. That is, the surge voltage doubles, but the power frequency voltage remains constant. An alternate and more meaningful circuit is shown in Fig. 13d, where the source voltage is double the surge voltage, $2E$, while the voltage across the arrester is E_d, the crest of the arrester voltage plus the power frequency voltage. To clarify, if the power frequency voltage is $-100\,kV$, then a voltage of $+100\,kV$ is added to the crest arrester voltage. In either circuit (Fig. 13c or 13d) the current through the arrester is

$$I_A = \frac{2E - E_d - V_{PF}}{Z} = \frac{2E - E_A}{Z} \tag{30}$$

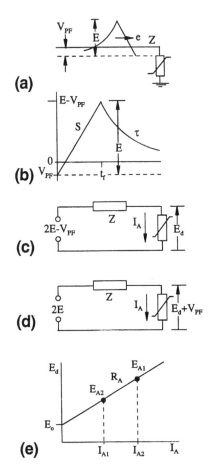

Figure 13 Arrester at end of line.

where

$$E_A = E_d + V_{PF} \tag{31}$$

Now consider that the arrester voltage E_d is a function of the current I_A. Also assume that the arrester voltage between any two values of arrester voltage E_{A1} and E_{A2} can be modeled or approximated by the equation

$$E_d = E_0 + I_A R_A \tag{32}$$

where E_0 and R_A are defined in Fig. 13e, R_A being the apparent arrester resistance. Rewriting the equation obtained from Fig. 13d,

$$2E - E_0 - V_{PF} = 1_A(Z + R_A) \tag{33}$$

so that

$$I_A = \frac{2E - E_0 - V_{PF}}{Z + R_A} \tag{34}$$

Before leaving this equation and the circuits of Fig. 13, consider the solution for the arrester voltage and current when using a transient program such as EMTP or ATP. The circuit and resulting equation shows that to include the power frequency voltage, the power frequency voltage should be added to each value of arrester voltage. That is, in Fig. 13e, the power frequency voltage should be added to the arrester voltages E_{A1} and E_{A2}. After obtaining the solution for the voltage across the arrester, the actual value of the arrester voltage is the measured voltage minus the power frequency voltage.

6.2 For an *n*-Line Station

Next consider an *n*-line station, that is, *n* lines terminate at the station. The surge voltage arrives on one line and thus $n-1$ lines remain. Assume these lines all terminate at the arrester. Thus the equivalent circuit of Fig. 14a results. This circuit can be further reduced to that of Fig. 14b. Therefore the arrester current is

$$I_A = \frac{2E/n - E_A}{Z/n} = \frac{2e/n - E_0 - V_{PF}}{2/n + R_A} \tag{35}$$

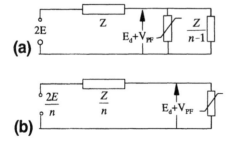

Figure 14 For an *n*-line station.

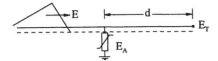

Figure 15 Definitions of quantities.

As shown by this equation and as expected, the arrester current decreases when more than one line is connected to the station.

6.3 Effect of Reflections from Open End-of-Line

Next, again consider that only one line is connected to the station, but now assume that the arrester is distant from the equipment, e.g., a transformer, to be protected. Per Fig. 15, assume that this distance is d meters beyond the arrester, and at present let the equipment be modeled as an opened circuit. Also assume that the distance d is such that the crest arrester voltage E_A is developed at the arrester before reflections return from the equipment, i.e., from the opened end of the line. Also, at this point, to simplify the presentation, assume zero power frequency voltage. At the arrester location, initially before reflections return from the opened end, there exists an additional line, that connected to the equipment, and thus the initial crest arrester current, at point 1 in the lattice diagram of Fig. 16, I_{A1}, is

$$I_{A1} = \frac{2(E - E_A)}{Z} \tag{36}$$

As shown in Fig. 16, a crest voltage E_A is transmitted past the arrester, and at the opened end of the line a voltage E_A is reflected back toward the arrester. Upon arriving at the arrester location, an additional current is pushed through the arrester.

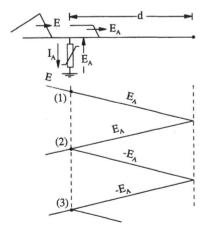

Figure 16 Reflections from open end increase arrester current.

Since the initial current has already overcome the voltage E_A, the additional current I_{A2}, at point 2 in Fig. 16, is

$$I_{A2} = \frac{2E_A}{Z} \tag{37}$$

The total current at this time is obtained by addition of Eqs. 36 and 37:

$$I_A = \frac{2E}{Z} \tag{38}$$

The additional current at point 3 in Fig. 16 cancels the additional current at point 2, and thus the current is oscillatory.

To show this effect, the circuit in the inset of Fig. 17 was set up on the ATP. The arrester voltage was held constant at 620 kV and the crest of the incoming surge E was 1560 kV with a time to crest of 0.5 µs and an infinite tail. Z was 450 ohms. The calculated initial current is 4.18 kA, and the calculated crest current at the first reflection from the opened line is 6.93 kA. Figure 17 shows the oscillogram of current as obtained from the ATP for a distance d of 100 meters. The initial and crest currents are clearly shown.

Of course, this distance of 100 meters is ridiculous in that much shorter distances are necessary to provide adequate protection for the equipment. However, this was done to illustrate the phenomenon. In addition, if a transformer is considered, the model for the transformer is a capacitance to ground that ranges from about 1 to 10 nF. Usually lower values of capacitance in the range of 2 to 4 nF produce the highest voltage at the transformer.

Returning to the circuit of Fig. 17, and using the same incoming surge, let the distance be 5 meters and the capacitance of the transformer at the end of the line be 2 nF. The resulting current is shown in Fig. 18. The initial current is no longer apparent. However, the current reaches a value of 10.5 kA, which is about 50% larger than the crest current with no capacitor.

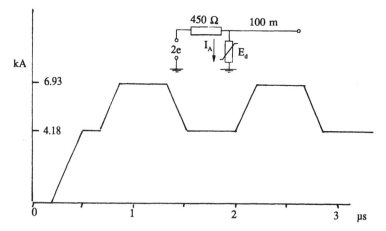

Figure 17 With $C_T = 0$.

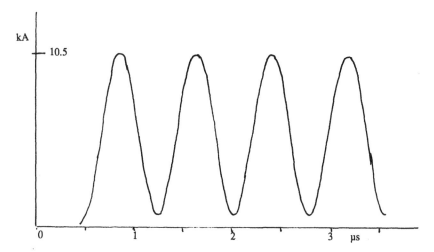

Figure 18 With $C_T = 2\,\text{nF}$.

To investigate this effect more fully, consider the following typical case:

1. Let the incoming surge be defined by a linearly rising front having a steepness S of $1400\,\text{kV}/\mu\text{s}$, a crest voltage E of $1560\,\text{kV}$, and a tail described by a time constant of $14\,\mu\text{s}$.
2. Let the 10-kA discharge voltage be $620\,\text{kV}$, and let the arrester be modeled by its voltage–current characteristics.

For a separation distance between the arrester and transformer of 6 meters, Fig. 19 shows the effect of the capacitance on the arrester current. The solid line curve is for the incoming surge having a tail time constant of $14\,\mu\text{s}$, and the dotted line curve is for an infinite tail. As noted, there is not a significant difference between these curves. Also note that for $C_T = 0$, both curves show a current of $5.6\,\text{kA}$ as given by Eq. 30.

Selecting capacitances of 0 and $4\,\text{nF}$, Fig. 20 shows the effect of the separation distance. From these curves, it is apparent that for a capacitance of 0, i.e., an opened

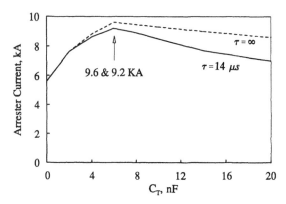

Figure 19 Effect of reflections from transformer; separation distance = 6 meters.

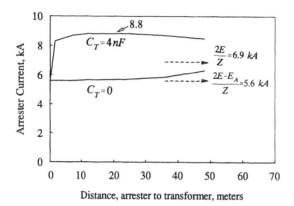

Figure 20 Effect of reflections from transformer.

circuit, Eq. 30 is valid. That is, for short separation distances for which many reflections occur between the arrester and the end of the line before crest current is achieved, the line connecting the arrester to the line end is essentially wiped out. Using this reasoning, the crest current should also be related to this equation when a capacitor is present. Using the results of Fig. 20, when a capacitor is present, i.e., a transformer, then the crest arrester current can be estimated by the equation

$$I_A = 1.6 \left[\frac{2E - E_A}{Z} \right] = 1.6 \left[\frac{2E - E_0 - V_{PF}}{Z + R_A} \right] \quad (39)$$

and by extrapolation for n lines,

$$I_A = 1.6 \left[\frac{2E/n - E_A}{Z/n} \right] = 1.6 \left[\frac{2E/n - E_0 - V_{PF}}{Z/n + R_A} \right] \quad (40)$$

6.4 Effect of Reflections from Struck Point

Reflections from the origin of the incoming surge, that is, the location of the struck point on the line, can also increase the crest arrester current. Consider, as shown in Fig. 21, that a stroke has terminated at or near a tower, causing a flashover and thus producing a surge voltage e on the phase conductor. Neglecting the tower, this voltage e is equal to

$$e = IZ_s$$
$$Z_s = \frac{Z_g Z_c R_i}{Z_g Z_c + 2R_i Z_g + 2R_i Z_c} \quad (41)$$

where as noted the impedance Z_s is the parallel combination of half the ground wire surge impedance Z_g, half the phase conductor surge impedance Z_c, and the impulse resistance of ground R_i. This voltage e travels in toward the station where it meets other lines and a surge arrester. For an n-line station, the other $n-1$ lines can be represented by a resistor to ground having a resistance of $Z_c/(n-1)$. The surge e is

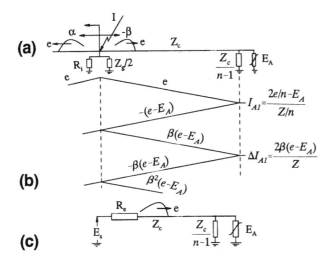

Figure 21 Effect of reflections from struck point.

reflected back toward the struck point, and at the struck point it is again reflected back toward the station. The impedance that this surge meets at the struck point is the parallel combination of half the ground wire surge impedance, the phase conductor surge impedance, and the impulse resistance to ground. Denoting this impedance as R_e,

$$R_e = \frac{Z_g Z_c R_i}{Z_g Z_c + 2 Z_g R_i + Z_c R_i} \tag{42}$$

Thus the circuit of Fig. 21 can be reduced to that of Fig. 21c, where the applied voltage E_s is such as to produce e on the phase conductor or

$$E_s = \frac{Z_c + R_e}{Z_c} e \tag{43}$$

Note that R_s and Z_s are related by

$$Z_s = \frac{R_e Z_c}{R_e + Z_c} \tag{44}$$

The reflection coefficient at the struck point is $-\beta$, where

$$\beta = \frac{Z_c - R_e}{Z_c + R_e} \tag{45}$$

The lattice diagram for this case is also shown in Fig. 21. To simplify the presentation, assume that the power frequency voltage is zero. When the surge e reaches the arrester, an initial current is produced equal to

$$I_{A1} = \frac{2e/n - E_A}{Z_c/n} \tag{46}$$

The voltage reflected back toward the struck point is $e - E_A$ and is reflected back toward the arrester by the reflection coefficient. When this voltage arrives at the arrester, an additional current is produced. Since the first surge voltage overcame the arrester voltage E_A, the additional current ΔI_{A1} is

$$\Delta I_{A1} = \frac{2\beta(e - E_A)}{Z_c} \tag{47}$$

Again, reflections occur at the arrester and at the struck point and produce another additional crest current through the arrester, ΔI_{A2}, of

$$\Delta I_{A2} = \frac{2\beta^2(e - E_A)}{Z_c} \tag{48}$$

The total current through the arrester is the sum of the initial current and all of the additional incremental currents or

$$I_A = I_{Ai} + \sum_{j=1}^{\infty} \Delta I_{Aj} \tag{49}$$

Expanding this equation

$$Z_c I_A = 2e - E_A + 2\beta(e - E_A)\left(t - \frac{2T_s}{\tau}\right) + 2\beta^2(e - E_A)\left(t - \frac{4T_s}{\tau}\right) + \cdots \tag{50}$$

where T_s is the travel time from the struck point to the arrester.

Now assume that e is an infinite rectangular wave having a crest voltage of E. Eq. 50 then becomes

$$Z_c I_A = 2(E - E_A)(1 + \beta + \beta^2 + \beta^3 + \cdots) + (n - 2)E_A \tag{51}$$

or

$$I_A = \frac{2(E - E_A)}{1 - \beta} - (n - 2)E_A = (E - E_A)\frac{Z_c + R_e}{Z_c R_e} - (n - 2)\frac{E_A}{Z_c} \tag{52}$$

Substituting for I, the stroke current,

$$I_A = I - \frac{Z_c + (n - 1)R_e}{Z_c R_e} E_A \tag{53}$$

For $n = 1$

$$I_A = I - \frac{E_A}{R_e} \tag{54}$$

Thus almost all the stroke current is discharged through the arrester. Indeed, if E_A is zero, i.e., a short circuit, all the stroke current is discharged.

The use of an infinite rectangular surge is unrealistic. Per chapter 11, the tail of the surge can be approximated by an exponential voltage decrease having a time constant τ that ranges from about 10 to 20 µs. Therefore, as a better assumption, let the surge be described by a time to crest of zero and a tail described by a time constant τ. Let the crest voltage equal E. Then the equation for the total current becomes

$$Z_c I_A = 2Ee^{-t/\tau} + 2\beta[Ee^{-(t-2T_s)/\tau} - E_A] + 2\beta^2[Ee^{-(t-4T_s)/\tau} - E_A] + \cdots \quad (55)$$

As a crude approximation, let

$$e^x = 1 + x \quad (56)$$

Then the arrester current is

$$I_A = \frac{2(N+1)E}{Z_c}\left[1 - N\frac{T_s}{\tau}\right] - (2N+n)\frac{E_A}{Z_c} \quad (57)$$

where N is the number of reflections from the struck point and therefore $N = 0, 1, 2, 3$, etc. For $N = 0$, the equation becomes I_{A1}.

To determine the crest current, the value of N, N_m, which gives the maximum current must be obtained. To do this, differentiate this equation with respect to the N and set it equal to zero. Then N_m is

$$N_m = \frac{1}{2}\left[\frac{\tau}{T_s}\left(1 - \frac{E_A}{E}\right) - 1\right] \quad (58)$$

where again N_m must be an integer, 0, 1, 2, 3, etc.

To illustrate the effect and quantify the error, a system was set up on the ATP. The incoming surge was assumed to have a zero time to crest with a tail time constant of 14 µs and a crest of 1560 kV. The surge impedance was 450 ohms, and the distance between the struck point and the arrester was 600 meters, i.e., $T_s = 2$ µs. The arrester discharge voltage of the arrester was assumed as 620 kV and is independent of the discharge current, i.e., a constant voltage arrester. The results are shown in Fig. 22 for the number of lines, $n = 1$ and $n = 2$. Note that the value of N is also shown. Using the above equations, the value of N was 1.6, which rounded to the next whole number is 2. Setting $N = 2$, the calculated currents are 8.28 kA and 6.59 kA for $n = 1$ and 2, respectively. Comparing this to the results of Fig. 22 shows that the actual crest current does occur at $N = 2$; the crest values are from 4 to 5% greater. Thus the approximate equations provide a convenient method of quickly estimating the crest current.

6.5 Combining Both Effects

In the previous sections two effects that increase the arrester current were discussed, and equations were developed to estimate the crest current. The question arises as to

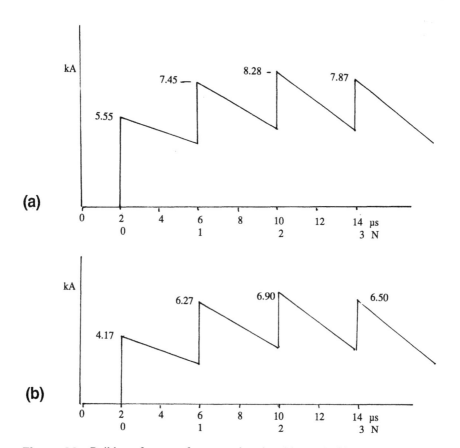

Figure 22 Buildup of current from struck point. (a) $n = 1$; (b) $n = 2$.

the combined effect when both of these effects are present. That is, should the current obtained when considering reflections from the struck point be multiplied by the 1.6 factor of Eq. 39 or 40? Another important question concerns the effect on the voltage at the transformer. Although this latter question or concern should be more properly addressed in Chapter 13, where the insulation coordination of the station will be presented, for sake of convenience, it will be considered here also.

Again the ATP is used to answer these questions. Two situations will be considered for study: first, when the distance from the arrester to the struck point is 600 meters, and second, when this distance is reduced to 150 meters. For both cases

1. The incoming surge $E = 1560\,\text{kV}$, the front steepness $S = 1400\,\text{kV}/\mu\text{s}$, the tail time constant $= 14\,\mu\text{s}$.
2. Arrester: Modeled by the voltage–current characteristics having a 10 kA discharge voltage of 620 kV.
3. Separation distance between arrester and transformer $= 6$ meters.
4. Power frequency voltage $V_{\text{PF}} = 0$.
5. Transformer capacitance $= 4\,\text{nF}$.
6. One-line station, $n = 1$.

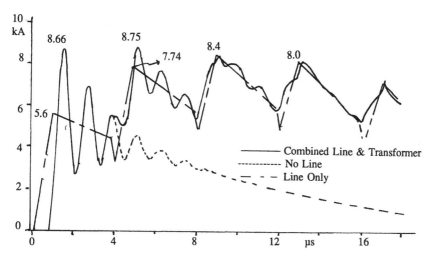

Figure 23 Arrester currents; distance to struck point = 600 meters.

For the 600 meter distance to the struck point, Fig. 23 shows the arrester currents and Fig. 24 presents the transformer voltages. In Fig 23, three traces are shown, (1) for no line between the arrester and the struck point, (2) for the line only, the transformer and 6 meter distance being neglected, and (3) the circuit as originally described, line and transformer represented. As noted for the case of "line only," the arrester current is 5.6 kA initially and reaches a crest of 8.4 kA at about 8 μs. Without the line, because of the capacitor, the current reaches a crest of 8.66 kA, after which it decays sharply. For the full circuit, the initial crest current is 8.66 kA, the same as when no line is considered. However, the current then increases to 8.75 kA at about 6 μs. Note that although the initial current of 5.6 kA increases to 8.66 kA because of the capacitor, the latter current of 7.74 kA does not increase by a similar proportion. In addition, at a later time the currents are equal at 8.4 and 8.0 kA. Thus the two effects may be considered separately. That is, the use of the 1.6 factor does not appear justified when considering the effects of reflections from the struck point.

Now consider the voltages at the transformer, Fig. 24. The crest voltages for the two situations are identical, and there exists only a slight difference in the tails of the voltages. In explanation, the time to crest of the transformer voltage when the line is not considered is 1.26 μs. Thus to affect the crest voltage, the reflection from the struck point must return before this time. However, the first reflection from the struck point returns in 4 μs, i.e., 600 meters to struck point. Therefore it cannot affect the crest voltage at the transformer.

Next, reduce the distance to the struck point to 150 meters so that he reflection from the struck point returns in 1.0 μs, which is less than the time to crest of the voltage at the transformer. For this case, Figs. 25 and 26 apply. The three currents are shown in Fig. 25. As before, when the line to the struck point is set to zero, the crest current is 8.66 kA. If the transformer and distance to it are neglected, the "line only" case, the current increase to a maximum of 16.2 kA. Now combining these effects, i.e., using the complete circuit, results in an initial current of 10.8 kA but a final current of 17.8 kA. As for the preceding case, the final current (16.2 to 17.8 kA)

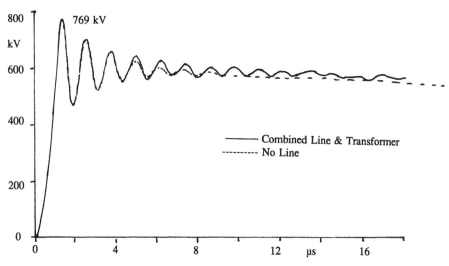

Figure 24 Transformer voltages; distance to struck point = 600 meters.

is not increased as much as the initial current (8.66 to 10.8 kA). Thus, again, the conclusion is that the current determined when considering the effect of reflections should not be multiplied by the 1.6 factor. that is, the currents for each effect should be calculated as independent values and the larger used as the crest current.

As to the transformer voltages as presented in Fig. 26, again the effect is not noticeable. For the combined effect, the crest is 771 kV, and when no line to the struck point is assumed, the crest voltage is 769 kV. As noted, the voltages following the crest are somewhat different. A somewhat larger effect was expected. However, in retrospect, only the first reflection from the struck point at 1.0 µs would be effective in increasing the voltage at the transformer. As shown in Fig. 25, the increase in current is only 13.3 kA, a 23% increase (13.3/10.8). This 23% increase results in only a 1% increase in arrester voltage, and thus only a slight increase in the voltage at the transformer should be expected.

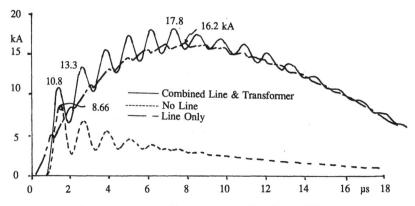

Figure 25 Arrester currents; distance to struck point = 150 meters.

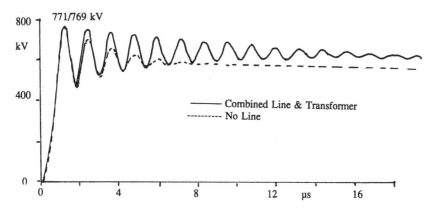

Figure 26 Transformer voltages; distance to struck point = 150 meters.

From these results, it is concluded that when estimating the crest arrester current, the two effects should be considered separately. That is, Eqs. 39 or 40 and Eq. 57 should be used separately, and the maximum of these currents should be considered as the crest arrester current.

Considering the transformer voltage, if the time to crest of the transformer voltage is less than twice the travel distance to the struck point, Eqs. 39 or 40 should be used to calculate the arrester current and thus the arrester voltage. If the time to crest of the voltage at the transformer is greater than twice the travel distance to the struck point, the maximum current as obtained from Eq. 57 should be used. The value of N in Eq. 57 should be such that the number of reflections arrive in time to increase the transformer voltage. This criterion is better stated in mathematical terms as follows:

$$\text{for } t_T \leq 2T_s, \tag{59}$$

$$I_A = 1.6 \frac{2E/n - E_A}{Z/n} = 1.6 \frac{2E/n - E_0 - V_{PF}}{Z/n + R_A} \tag{60}$$

$$\text{for } t_T \geq 2T_s, \tag{61}$$

$$I_A = \frac{2(N+1)E}{Z}\left[1 - N\frac{T_s}{\tau}\right] - (2N = n)\frac{E_A}{Z} \tag{62}$$

where N is an integer less than or equal to the value calculated by the equation

$$N = \frac{t_T}{2T_s} \tag{63}$$

where

t_T = time to crest of the transformer voltage
T_s = travel time between the arrester and the struck point
E = crest of the incoming surge voltage

I_A = crest arrester current
n = number of lines in the station
E_d = arrester discharge voltage
Z = line phase conductor surge impedance
R_A = apparent arrester resistance
τ = tail time constant of the incoming surge
V_{PF} = opposite polarity (to the surge) power frequency voltage

and

$$E_A = E_d + V_{PF}$$
$$E_d = E_0 + I_A R_A \qquad (64)$$

Using Eq. 64, Eq. 62 may be reformed to

$$I_A = \frac{2(N+1)E[1 - NT_s/\tau] - (2N+n)(E_0 + V_{PF})}{Z + (2N+n)R_A} \qquad (65)$$

Example 5. Let $E = 1560\,\text{kV}$, $S = 1400\,\text{kV}/\mu\text{s}$, $Z = 450$ ohms, $n = 1$, $T_s = 0.5\,\mu\text{s}$, $\tau = 14\,\mu\text{s}$, $V_{PF} = 156\,\text{kV}$, and $t_T = 21\,\mu\text{s}$. Let the arrester characteristics be given by the table in which the values of E_0 and R_A have been calculated.

I_A, kA	E_d, kV	R_A, ohms	E_0, kV
5	444		
	→	4.0	424
10	464		
	→	2.6	438
20	490		

As a start, assume that the current is between 5 and 10 kA. Therefore, from the Table, $E_0 = 424\,\text{kV}$ and $R_A = 4.0$ ohms. Since t_T is greater than $2T_s$, use Eq. 65. Then $N = 2.1/1.0 = 2.1$. Decreasing this to the nearest integer, $N = 2$, and using Eq. 65, $I_A = 12.3\,\text{kA}$ and $E_d = 473\,\text{kV}$. Use of Eq. 60 would have produced value of I_A of only about 9.0 kA and an E_d of 460 kV. Note the difference in the discharge voltages, less than 3%.

However, in calculating the current, the values E_0 and R_A were for currents between 5 and 10 kA, and therefore for better accuracy the calculation should be repeated for $E_0 = 438\,\text{kV}$ and $R_A = 2.6$ ohms. Performing this calculation, I_A is 12.4 kA and E_d is 470 kV. There is essentially no change.

6.6 Shielding Failures

The circuit of Fig. 21 is applicable to both backflashes and shielding failure with flashover. However, for shielding failures without flashover, no reflections occur from the stroke-terminating point, and an increase in arrester current does not occur.

7 DISTRIBUTION SYSTEMS

7.1 Selection of the Arrester MCOV

In previous sections, some mention has been made of the unique parameters of distribution systems in selecting the arrester MCOV. The TOV capability has been considered in Table 8, and a few typical values of the system TOV have been shown in Table 14. However, to consider adequately all effects, the distribution system should be considered further. An excellent and extensive guide to the application of distribution arresters is contained in the IEEE Guide [2], and it is not the purpose of this section to repeat this guide. Rather, a general overview is given.

In general, in distribution systems, separate studies are not performed for each arrester location to establish the MCOV. Rather, the arrester MCOV is selected so that this MCOV can be applied anywhere on the system for a similar situation. The rules for selecting the MCOV for distribution systems are identical to those for selecting arresters for any system. However, in distribution systems, the importance or emphasis on each of the rules is somewhat different.

Before considering the alternate distribution systems, the TOV capability as taken from Table 8 and Section 2.7 is

Actual TOV_{10} = 1.28 to 1.46 (MCOV)   Actual TOV_1 = 1.34 to 1.54 (MCOV)
Min. TOV_{10} = 1.24 (MCOV)   Min. TOV_1 = 1.38 (MCOV)

where TOV_{10} is the TOV capability at 10 seconds. The actual values for 1 second were obtained by multiplying the value of $A1$ by the "ratio." Thus all values assume prior energy.

The most important rule in selecting the MCOV is that the MCOV must be equal to or greater than the maximum system line–ground voltage and therefore the maximum voltage must be known. In distribution systems, only the voltage at the customer meter is regulated. In a recent study [16], the unregulated feeder voltage was found to be a maximum of 17% above the nominal system voltage (10% above maximum). The average was 7% above nominal, or only 1% above maximum.

As for higher voltage systems, a knowledge of TOVs is required for distribution systems. However, the magnitudes of TOVs in a distribution system are normally not known to the same degree of accuracy as in higher voltage systems, since the SiC gapped arresters did not require a detailed knowledge of TOVs. That is, the SiC gapped arresters had a 60-Hz sparkover voltage that was 1.2 to 1.3 times the arrester rating (duty cycle). For example, a 9-kV duty-cycle rated SiC arrester could withstand a 60-Hz overvoltage of 10.8 to 11.7 kV for 10 seconds, a TOV of 1.47 to 1.60 p.u. However, a 9-kV duty-cycle/7.65-kV MCOV metal oxide arrester can withstand a minimum of 1.24 p.u. for 10 seconds. Thus the utility engineer did not need to know the TOV with any significant degree of accuracy; the SiC gapped arrester naturally provided a high degree of capability for TOVs. The job of estimating the fault TOVs on distribution systems is not an insignificant job since, for example, for a four-wire multigrounded system, the effect of ground resistance and the size of the neutral conductor can significantly alter the TOVs. And as for high-voltage systems, ferroresonance must also be considered.

Some guidance is given in the following for the alternate distribution systems.

Four-Wire Multigrounded systems. Most distribution systems in North America are in this category. TOVs are low, and the EFF is in the order of 1.25, provided that the grounding resistance is less than 25 ohms and the neutral conductor size is at least 50% of the size of the phase conductor. For an EFF of 1.25, the TOV capability is greater than the fault TOVs, and the minimum arrester MCOV is equal to the maximum system line-to-ground voltage; see Table 18.

Three-Wire, Low-Impedance Grounded systems. The EFF is generally equal to or greater than 1.4. The time duration of the fault could be up to 10 seconds. For example assume the EFF = 1.6 for a 12,700Y/7330 volt system. Assuming that the maximum line-to-ground voltage is 7.33 kV and the TOV capability is 1.24 (MCOV), then the minimum MCOV is

$$\text{MCOV} = \frac{1.6(7.33)}{1.24} = 9.5\,\text{kV} \tag{66}$$

The next highest MCOV rating is 10.2 kV as listed in Table 18. In general, the MCOV ratings in Table 18 are from 1.3 to 2.0 times the maximum line–ground voltage.

Three-Wire, High-Impedance Grounded or Delta-Connected Systems. The EFF for these systems is 1.73, and the time duration can be large, in some cases equal to 8 hours or more. Therefore the time duration is considered infinite, and the MCOV should be equal to the line–line maximum system voltage; see Table 18.

7.2 Arrester Currents

In most cases, distribution lines are unshielded and therefore vulnerable to direct lightning flashes. Thus the lightning surge currents discharged by an arrester are usually greater than for arresters applied within a station. These currents are a function of the ground flash density, the location of the stroke with relation to the arrester location, and the number of arresters in close proximity to each other. In the IEEE Guide [2], a curve is given showing the arrester discharges per year as a function of the arrester current. Table 19, taken from this curve, presents the currents for 0.1 discharges/year (1 in 10 years) as a function of the ground flash density. Two classes of conditions are given: one for rural areas where the line is partially

Table 19 Arrester Discharge Currents for Distribution Systems for 0.1 discharges/year [2]

Ground flash density, flashes/km^2-year	Current, kA, equal to or greater than	
	Rural locations	Suburban locations
1	2	1
5	15	2
10	30	5
20	50	15

Metal Oxide Surge Arresters

shielded by buildings and trees and one for suburban areas where the line is not shielded by these objects. The "coordinating current' is selected from this table or from the surveys of Ref. 2. This current is then used to obtain the arrester discharge voltage, which is employed to select the equipment insulation strength.

8 IEC STANDARDS

The applicable IEC Publications are 99-4, the standard, and 99-5, the application guide. However, frequent reference in 99-5 is made to IEC 99-1. In IEC, arresters are rated or classified primarily by

1. Rated voltage E_R, which is defined as equal to the TOV capability at 10 seconds. Also this rated voltage is used in the duty-cycle test similar to that of IEEE. The rated voltage is similar to the duty-cycle voltage except that in IEEE, this voltage is not defined in terms of the TOV capability.
2. The continuous operating voltage COV, which is the same as the MCOV in IEEE.
3. The nominal discharge current standardized as 1.5, 2.5, 5, 10, or 20 kA.
4. The line discharge class standardized as Class 1, 2, 3, 4, or 5—or none.

The nominal discharge current and line discharge class cannot be selected independently of each other. That is, for a 10-kA nominal discharge current, line discharge classes 1, 2, or 3 may be selected, and for a 20 kA nominal discharge current, line discharge classes 4 and 5 are available. The line discharge class is not given for 1.5-, 2.5-, or 5-A arresters. The combination of the line discharge class and the nominal discharge current determines the low-current long-duration test made on the arrester and thus determines the arrester discharge energy capability.

In contrast to IEEE, a standardized list of arrester rated voltages and MCOV or COV are not provided in IEC. Rather a table of steps of rated voltages are provided per Table 20. No equivalent COVs are given.

In general, the systems to which the arresters are applied are determined by the nominal discharge current, I_n per Table 21. This table also lists approximate values of the COV.

In comparison to IEEE, the IEC standard states that the station arrester approximates the IEC 10 kA arrester and that the intermediate and distribution arresters

Table 20 Steps of Rated Voltages

Range of rated voltage, kV, rms	Steps of rated voltage, kV, rms
<3	Under consideration
3–30	1
30–54	3
54–96	6
96–288	12
288–396	18
396–756	24

Table 21 System Voltages to Which the Arresters are Applied

Nominal discharge current, kA	Rated voltage, kV	Approximate COV, kV
20	360 to 756	285 to 600
10	3 to 360	2.55 to 285
5	132 or less	106 or less
2.5	36 or less	29 or less

approximate the IEC 5 kA arrester.

8.1 Durability/Capability Tests

A summary of the capability requirements is shown in Table 22.

High Current, Short Duration. As part of the operating duty test, two 4/10 μs current impulses are applied to the arresters whose current magnitudes are given in Table 22.

Low Current, Long Duration. For 10- and 20-kA arresters, this test consists of discharging a charged line into the arrester. The parameters of the line are given in Table 22. The charging voltage of the line, the surge impedance, and the time duration of the resultant current impulse are provided in Table 22. This operation is performed 18 times in six groups of three operations (leaving 1 minute between groups).

Duty Cycle. This test is divided into two groups depending on the nominal current and the line discharge class.

Table 22 Summary of the Capability Requirements

Nominal discharge current, kA	Line discharge class	Discharge withstand				
		Transmission line			Square wave	4/10 μs
		Voltage, p.u. of E_R rms	Time duration, μs	Z/E_R	Square wave, amps & μs	kA
20	5	2.4	3200	0.5	—	100
20	4	2.6	2800	0.8	—	100
10	3	2.8	2400	1.3	—	100
10	2	3.2	2000	2.4	—	100
10	1	3.2	2000	4.9	—	100
5	—	—	—	—	75/1000	65
2.5	—	—	—	—	50/500	25

Metal Oxide Surge Arresters

1. High-current operating duty cycle: for 1.5-, 2.5-, and 5.0-kA arresters. Also for 10-kA, Class 1 arresters. First, twenty 8/20-µs impulses are applied (four groups of five), when the arrester is energized with a power frequency voltage equal to 1.2 COV. This is followed by two 4/10-µs impulses having currents per Table 22 (no power frequency voltage). The arrester rated voltage is then applied for 10 seconds followed by the COV for 30 minutes. The test is passed if no thermal runaway occurs.

2. Switching impulse operating duty cycle: for 10-kA, Class 2 and 3, and 20-kA, Class 4 and 5 arresters. First, twenty 8/20-µs impulses are applied (four groups of five); then the arrester is energized with a power frequency voltage equal to 1.2 COV. This is followed by two 4/10-µs 100-kA impulses; see Table 22 (no power frequency voltage). Next, the arrester is subjected to two transmission line discharge operations. The arrester rated voltage is then applied for 10 seconds followed by the COV for 30 minutes. The test is passed if no thermal runaway occurs.

Pressure Relief. This is only required if the arrester is fitted with a pressure relief device. The arrester must vent, and all components must lie within the same circle, as prescribed by the IEEE standard. Pressure relief classes for the 10-kA arrester are 10, 20, 40, 50, 63 and 80 kA, and for the 5-kA arrester, 5 and 16 kA. All arresters also are tested at 800 A.

Pollution. No standard is established at present.

TOV Capability. As a result of these tests, the TOV capability curve is constructed from 0.1 seconds to 20 minutes. The tests differ depending on the nominal current and the line discharge class. The tests on the 10-kA, Class 2 or 3, and 20-kA, Class 4 or 5 arresters are for the "prior duty" category.

1. For 10-kA Class 1 arresters or 5-, 2.5-, or 1.5-kA arresters. A single 4/10-µs current impulse per Table 22 is applied, after which the COV is applied for 30 minutes. Then the TOV-versus-time curve is established.
2. For high lightning duty arresters, the only change from the above test is that three 4/10-µs impulses are applied.
3. For 10-kA, Class 2 or 3, and 20-kA, Class 4 or 5 arresters. Two operations of the transmission line discharge are used and the COV is applied for 30 minutes. This establishes the prior energy, which is then marked on the curve of TOV versus time.

8.2 Protective Characteristics

The tests to establish the protective characteristics of the arrester are essentially identical to those in IEEE. They are

Steep Front Discharge Voltage. A current having a 1-µs front and a crest current equal to the nominal discharge current is applied to the arrester. The resultant discharge voltage has a time to crest of about 0.5 µs and thus is the same as the IEEE front of wave protective level.

Lightning Impulse Discharge Voltage. 8/20-μs currents whose crests are equal to 0.5, 1.0, and 2.0 times the nominal current are applied and the discharge voltage recorded.

Switching Impulse Discharge Voltage. The switching impulse discharge voltage is determined using two current magnitudes per Table 23. The waveshape is described as a front greater than 30 μs and a tail less than 100 μs.

8.3 Selection of the Arrester Ratings

In the IEC Application Guide, 99-5, the ratings to be selected are

1. COV
2. Rated voltage
3. Nominal discharge current
4. Line discharge class
5. Pressure relief class

The COV is selected in the same manner as per IEEE. The rated voltage, which is the TOV capability at 10 seconds, is selected by comparison with the required system TOV. The line discharge class is selected by comparison of the arrester energy capability with the energy discharge required. The nominal discharge current is selected by calculation or estimation of the lightning current discharge by the arrester. The pressure relief class is selected by comparison to the system fault current. Thus the selection of the arrester ratings is virtually identical to that of IEEE. The only difference appears to be the alternate rating method of IEC [17, 18].

8.4 Comparison of IEEE and IEC

While the protective characteristic tests are essentially identical in IEEE and IEC standards, the capability/durability tests are not. Comparisons of the IEEE with the IEC capability tests are hampered by the different rating systems. Recently, IEEE papers have been written on this subject and should be reviewed for better information [19, 20]. Following are some comparisons that may be helpful.

Using the methods of Section 4.2, the energy capability is calculated for the IEC and IEEE discharge withstand tests and shown in Table 24. The comparison is based on a switching impulse discharge voltage of (1) 1.75 crest MCOV for the arresters for 800 kV (2-kA current) and 550 kV (1-kA current) systems and (2) 1.67 crest MCOV for arresters (1-kA current) for 242- and 362-kV systems. The energies calculated are for a single line discharge.

Table 23 Switching Impulse Currents

Classification	Switching impulse current, kA
20 kA, Class 4 or 5	0.5 and 2.0
10 kA, Class 3	0.25 and 1.0
10 kA, Class 1 or 2	0.125 and 0.5

Table 24 Comparison of Energy Discharged by Line Discharge Tests

Max system voltage, kV	MCOV, kV	Duty-cycle voltage, kV	IEC nominal current, kA	IEC line discharge class	IEC energy, kJ/kV MCOV	IEEE Energy, kJ/kV MCOV
800	462	576	20	5	9.3	3.7
	462	576	20	4	5.8	3.7
550	318	396	20	5	9.2	2.8
	318	396	20	4	5.8	2.8
362	209	258	20	4	5.6	3.9
	209	258	10	3	3.9	3.9
242	140	172	10	3	3.9	2.1
	140	172	10	2	2.6	2.1
	140	172	10	1	1.3	2.1

Although not shown in Table 24, comparing the IEC 5-kA arresters with the IEEE heavy-, normal-, and light-duty distribution arresters shows that the IEEE discharge energies are three times the IEC values for heavy-duty and two times the IEC energy for normal- and light-duty.

As noted, for the transmission line discharge, Table 24 indicates that the IEC tests produce larger discharge energies except for two cases, (1) at 362 kV, the 10-kA, Class 2, (2) at 242 kV, the 10-kA, Class 1. Note, however, that the IEC test is composed of 18 discharges, whereas the IEEE test uses 20 discharges. Also, after the line discharge test, IEEE standards specify that the MCOV be applied for 30 minutes, whereas the IEC standards specify that the duty-cycle rating be applied for 10 seconds followed by the MCOV for 30 minutes. Thus it appears that the IEC test is more severe.

As stated previously, the energies per Table 24 are for a single discharge of the line. The total energy discharged by 18 to 20 tests is thus much greater. However, in IEC the discharge energy per Table 24 is compared to the required energy. Interestingly, in the IEC application guide, the energy capability is stated to be at least twice the value given in Table 24. However, this statement is not verifiable.

Comparing the duty-cycle test, the IEC standard has two types of tests, switching impulse and lightning impulse (high current), whereas only one type of test is listed in IEEE. For both types of tests, the IEC test prescribes that the arrester be energized at 1.2 COV, which is less than the duty-cycle voltage or IEC voltage rating. The IEEE test prescribes that the arrester be energized at the duty-cycle rating. As noted following these tests, the IEC switching impulse test requires two line discharge operations. IEEE requires that two 8/20 µs current impulses be applied while energized at MCOV.

The TOV capability tests in IEC and IEEE are virtually identical except for the 10-kA, Class 2 and 3, and the 20-kA, Class 4 and 5. The IEC tests may be viewed as with prior energy, whereas the IEEE tests are with and without prior energy. Again, the IEC rated voltage is the TOV capability at 10 seconds.

9 REFERENCES

1. IEEE Std C62.11-1993, "IEEE Standard for Metal Oxide Surge Arrester for Alternating Current Power Circuits."
2. IEEE Std C62.22-1991 7, "IEEE Guide for the Application of Metal Oxide Surge Arrester for Alternating Current Power Systems."
3. IEC Standard 99-4, 1991-11, "Part 4: Metal Oxide Surge Arresters Without Gaps for A.C. Systems."
4. IEC Standard 99-5, 1996-02, "Part 5: Selection and Application Recommendations."
5. M. Matsuoka, "Nonohmic Properties of Zinc Oxide Ceramics," *Japanese Journal of Applied Physics*, 10, June 1971, p. 46.
6. E. C. Sakshaug, J. S. Kresge, and S. A. Miske, Jr., "A New Concept in Station Arrester Design," *IEEE PA&S*, Mar.
7. A. Schei and K. H. Weck, "Metal Oxide Surge Arresters in AC Systems," *ELECTRA*, Jan. 1990, pp. 101–106.
8. K. G. Ringler, P. Kirby, C. C. Erven, M. V. Lat, and T. A. Malkiewicz, "The Energy Absorption Capability of Varistors used in Station Class Metal Oxide Surge Arresters," *IEEE Trans. on PD*, Jan. 1997, pp. 203–212.
9. E. C. Shakshaug, J. J. Burke, and J. S. Kresge, Jr., "Metal Oxide Arresters on Distribution Systems, Fundamental Considerations," *IEEE Trans. on PD*, Oct. 1989, pp. 1076–1089.
10. P. Kirby, C. C. Erven, and O. Nigol, "Long Term Stability and Energy Discharge Capacity of Metal Oxide Value Elements," *IEEE Trans. on PD*, Oct. 1988, pp. 1656–1165.
11. A. R. Hileman and K. H. Weck, "Protection Performance of Metal Oxide Surge Arresters," *ELECTRA*, Dec. 1990, pp. 133–146.
12. W. Schmidt, J. Meppelink, B. Richter, K. Feser, L. E. Kehl, and D. Qiu, "Behavior of Metal Oxide Surge Arrester Blocks to Fast Transients," *IEEE Trans. on PD*, Jan. 1989, pp. 292–300.
13. IEEE Working Group 3.4.11, "Modeling of Metal Oxide Surge Arrester," *IEEE Trans. on PD*, Jan. 1992, pp. 302–309.
14. J. Elovaara, K. Foreman, A. Schei, and O. Volker, "Temporary Overvoltages and Their Stresses on Metal Oxide Arresters," *ELECTRA*, Jan. 1990, pp. 108–125.
15. N. Menemenlis, M. Ene, J. Balanger, G. Sybille, and L. Snider, "Stresses in Metal Oxide Arresters Due to Temporary harmonic Overvoltages," *ELECTRA*, May 1990, pp. 79–115.
16. J. J. Burke, D. A. Douglas, and D. J. Lawrence, "Distribution Fault Current Analysis," EPRI EL-3085, Project 1209-1.
17. L. Stenstrom, "Selection of Metal Oxide Surge Arrester Characteristics from the Standards," *ELECTRA*, pp. 147–165.
18. B. Backmann and A. Schei, "Performance of Metal Oxide Arresters Under Operating Voltage," *ELECTRA*, Jan. 1990, pp. 107–115.
19. A. Hammy and G. St.-Jean, "Comparison of ANSI, IEC, and CAS Standards Durability Requirements on Station-Type Metal Oxide Surge Arresters for EHV Power Systems," *IEEE Trans. on PD*, Jul. 1992, pp. 1283–1298.
20. J. Osterhout, "Comparison of IEC and U.S. Standards for Metal Oxide Surge Arresters," *IEEE Trans. on PD*, Oct., 1992, pp. 2002–2006.
21. M. L. B. Martinz, L. C. Zanetta Jr, "A Testing Method to Evaluate the Energy Withstanding Capacity of Metal Oxide Resistors for Surge Arresters", *CIGRE SC33 Colloguim*, Toronto, Sept 1997.

22. M. L. B. Martinz, L. C. Zanetta Jr, "Comments on the Energy Withstanding Capacity of Metal Oxide Resistors for Surge Arresters", *CIGRE SC33 Colloguim*, Toronto, Sept 1997.
23. M. Bartkowiak, M. G. Comber, G. D. Mahan, "Failure Modes and Energy Absorption Capability of ZnO Varistors", Paper PE-135-PWRD-0-12-1997, presented at *1998 IEEE/PES Winter Mtg*.

10 PROBLEMS

1. Select a station class arrester for a 500/550-kV substation for the following conditions:

TOV: Only consider faults, EFF = 1.4 p.u. with time duration = 1 second. Use TOV capability with prior energy.
SI Energy: Max. switching overvoltage = 2.2 p.u.
Assume $Z = 350$ ohms, line length = 300 km.
Assume the minimum SI discharge voltage at 2 kA is 1.63 × (crest MCOV).

2. Select a station class arrester for a 230/242-kV substation for the following conditions:

TOV: Only consider faults, EFF = 1.5 p.u. with time duration = 1 second. Use TOV capability with prior energy.
SI Energy: Max. switching overvoltage = 3.1 p.u.
Assume $Z = 400$ ohms, line length = 200 km.
Assume the minimum SI discharge voltage at 1 kA is 1.63 × (crest MCOV).

3. Select a station class arrester for a 69/72-kV substation for the following conditions:

TOV: Only consider faults, EFF = 1.73 p.u. with time duration = 4 seconds. Use TOV capability with prior energy.
SI Energy: Max. switching overvoltage = 3.1 p.u.
Assume $Z = 450$ ohms, line length = 200 km.
Assume the minimum SI discharge voltage at 0.5 kA is 1.63 × (crest MCOV).

4. Calculate the arrester voltage and current using a 318 kV MCOV arrester for a 500/550-kV system for the following conditions: a single line station; an incoming surge of 2200 kV; the 0.5-µs discharge voltage is 1040 kV, $Z = 350$ ohms, $V_{PF} = 338$ kV. Use the derived 0.5-µs discharge voltage characteristics to determine the answer. The 8/20-µs discharge characteristics are as in the table.

I_A, kA	E_A, kV
5	833
10	922
15	948
20	972

12—Appendix 1
Protective Characteristics of Arresters

With permission from the Ohio Brass Company, the following protective characteristics are taken from the Ohio Brass Company catalogs: Section 30, April 1997, for station and intermediate arresters; Section 31, October 1996, for distribution and riser pole arresters.

Table 1 Protective Characteristics of Ohio Brass Polymer-Housed Station Arresters

MCOV, kV rms	Duty-cycle rating, kV rms	0.5-μs discharge voltage, kV at 10 kA	Switching impulse discharge voltage		Discharge voltage, kV for 8/20-μs current					
			Voltage, kV	Current, kA	1.5 kA	3 kA	5 kA	10 kA	20 kA	40 kA
2.55	3	8.4	6	0.5	6.4	6.7	7.1	7.6	8.4	9.6
5.1	6	16.7	11.9	0.5	12.8	13.5	14.1	15.2	16.8	19.1
7.65	9	25	17.8	0.5	19.2	20.2	21.1	22.7	25.1	28.3
8.4	10	27.8	19.8	0.5	21.4	22.5	23.5	25.3	28	31.8
10.2	12	33.3	23.7	0.5	25.6	26.9	28.1	30.3	33.5	38.1
12.7	15	41.7	29.7	0.5	32	33.7	35.2	37.9	42	47.6
15.3	18	50.1	35.6	0.5	38.4	40.4	42.3	45.5	50.4	57.2
17	21	56.3	40.1	0.5	43.2	45.5	47.6	51.2	56.7	64.4
19.5	24	63.9	45.5	0.5	49.1	51.6	54	58.1	64.3	73
22	27	72.9	51.9	0.5	56	58.9	61.6	66.3	73.4	83.3
24.4	30	80.4	57.2	0.5	61.7	64.9	67.9	73.1	80.9	91.9
29	36	95.9	68.3	0.5	73.6	77.4	81	87.2	96.5	109.6
31.5	39	104.2	74.2	0.5	80	84.1	88	94.7	104.8	119
36.5	45	120.9	86.1	0.5	92.8	97.6	102.1	109.9	121.7	138.1
39	48	128.7	91.6	0.5	98.8	103.9	108.7	117	129.5	147.1
42	54	144.4	102.8	0.5	110.9	116.6	122	131.3	145.3	165
48	60	163.5	116.4	0.5	125.5	132	138	148.6	164.5	186.8
57	72	191.8	136.6	0.5	147.3	154.9	162.2	174.4	193.1	219.2
70	90	241.8	172.1	0.5	185.6	195.2	204.2	219.8	243.3	276.3
76	96	257.4	183.2	0.5	197.6	207.8	217.4	234	259	294.1
84	108	288.9	205.6	0.5	221.8	233.2	244	262.6	290.7	330.1
88	108	288.9	205.6	0.5	221.8	233.2	244	262.6	290.7	330.1
98	120	326.9	241.3	1	251	263.9	276.1	297.2	329	373.6
106	132	362.7	267.7	1	278.5	292.8	306.3	329.7	365	414.4
115	144	386.1	285	1	296.5	311.7	326.1	351	388.6	441.2

Table 2 Protective Characteristics of Ohio Brass Porcelain-Housed Station Arresters

MCOV, kV rms	Duty-cycle rating, kV rms	0.5-μs discharge voltage, kV at 10 kA	Switching impulse discharge voltage		Discharge voltage, kV for 8/20-μs current							
			Voltage, kV	Current, kA	1.5 kA	3 kA	5 kA	10 kA	15 kA	20 kA	40 kA
2.55	3	9.1	6.3	0.5	6.9	7.2	7.5	8	8.6	9	10.3
5.1	6	17.9	12.4	0.5	13.6	14.2	14.8	15.8	16.9	17.7	20.3
7.65	9	26.6	18.4	0.5	20.2	21.1	22	23.5	25.1	26.4	30.2
8.4	10	29.3	20.3	0.5	22.2	23.3	24.2	25.9	27.7	29.1	33.3
10.2	12	35.5	24.6	0.5	26.9	28.2	29.4	31.4	33.5	35.2	40.4
12.7	15	44.2	30.6	0.5	33.5	35.1	36.6	39.1	41.8	43.9	50.3
15.3	18	53.3	36.8	0.5	40.4	42.3	44.1	47.1	50.3	52.8	60.6
17	21	59.1	40.9	0.5	44.8	46.9	48.9	52.3	55.8	58.7	67.2
19.5	24	67.8	46.9	0.5	51.4	53.8	56.1	60	64.1	67.3	77.1
22	27	76.5	52.9	0.5	58	60.8	63.3	67.7	72.3	75.9	87
24.4	30	84.9	58.7	0.5	64.3	67.4	70.3	75.1	80.2	84.2	96.5
29	36	101	69.7	0.5	76.4	80	83.4	89.2	95.2	100	115
31.5	39	110	75.8	0.5	83	86.9	90.6	96.9	104	109	125
36.5	45	128	88.3	0.5	96.8	102	106	113	121	127	146
39	48	136	93.8	0.5	103	108	113	120	128	135	155
42	54	135	98	0.5	105	112	115	122	130	136	151
48	60	154	110	0.5	120	127	131	139	149	155	173
57	72	183	131	0.5	142	151	156	165	177	184	205
70	90	223	161	0.5	174	184	190	202	216	226	251
74	90	236	169	0.5	185	195	202	214	229	237	266
76	96	242	175	0.5	190	201	208	220	235	245	274
84	108	267	193	0.5	209	221	229	243	260	271	301
88	108	279	202	0.5	219	232	239	254	272	284	316

Table 2 Protective Characteristics of Ohio Brass Porcelain-Housed Station Arresters (*continued*)

MCOV, kV rms	Duty-cycle rating, kV rms	0.5-μs discharge voltage, kV at 10 kA	Switching impulse discharge voltage		Discharge voltage, kV for 8/20-μs current						
			Voltage, kV	Current, kA	1.5 kA	3 kA	5 kA	10 kA	15 kA	20 kA	40 kA
98	120	311	231	1	244	257	266	283	303	315	351
106	132	340	249	1	264	280	289	306	327	342	381
115	144	368	271	1	287	303	314	332	355	369	413
131	168	418	308	1	326	345	357	379	406	421	470
140	172	446	330	1	348	368	381	404	432	448	502
144	180	458	339	1	359	380	392	417	446	463	517
152	192	483	360	1	379	401	414	440	471	488	546
180	228	571	424	1	447	474	489	520	556	578	645
209	258	665	516	2	522	552	571	604	646	670	752
212	264	675	523	2	527	558	576	613	656	680	760
220	276	700	545	2	547	578	597	635	679	705	788
245	312	778	605	2	609	644	666	708	758	788	878

Table 3 Protective Characteristics of Ohio Brass Polymer-Housed Intermediate Arresters

MCOV, kV rms	Duty-cycle rating, kV rms	0.5 µs discharge voltage, kV at 10 kA	Switching impulse discharge voltage		Discharge voltage, kV for 8/20 µs current					
			Voltage, kV	Current, kA	1.5 kA	3 kA	5 kA	10 kA	20 kA	40 kA
2.55	3	7.9	8.7	6	6.6	7.1	7.4	8.2	9.2	10.7
5.1	6	15.6	17.1	11.8	13	14.1	14.7	16.2	18.1	21.1
7.65	9	23.2	25.5	17.7	19.4	21.1	21.9	24.2	27	31.5
8.4	10	25.6	28.2	19.5	21.4	23.2	24.2	26.7	29.8	34.7
10.2	12	31	34.1	23.6	25.8	28.1	29.3	32.3	36	42
12.7	15	38.6	42.4	29.3	32.2	35	36.4	40.2	44.9	52.3
15.3	18	46.5	51.1	35.5	38.7	42.1	43.9	48.4	54	62.9
17	21	53.8	59.1	40.9	44.8	48.7	50.7	56	62.5	72.8
19.5	24	61.4	67.5	46.7	51.2	55.7	58	64	71.4	83.2
22	27	69.1	76	52.6	57.6	62.6	65.2	72	80.4	93.6
24.4	30	76.8	84.4	58.4	64	69.6	72.5	80	89.3	104
29	36	92.2	101	70.1	76.8	83.5	87	96	107	125
31.5	39	99.8	110	75.9	83.2	90.5	94.2	104	116	135
36.5	45	115.5	127	87.6	96	104	109	120	134	156
39	48	123	135	93.4	102	111	116	128	143	166
42	54	138	152	105	115	125	130	144	161	187
48	60	154	169	117	128	139	145	160	179	208
57	72	184	203	140	154	167	174	192	214	250
70	90	230	253	175	192	209	217	240	268	312
76	96	246	270	187	205	223	232	256	286	333
84	108	276	304	210	230	251	261	288	321	374

Table 4 Protective Characteristics of Ohio Brass Distribution Heavy-Duty Arresters, Polymer-Housed

MCOV, kV rms	Duty-cycle rating, kV rms	0.5 μs discharge voltage, kV at 10 kA	Switching impulse discharge voltage kV at 0.5 kA	Discharge voltage, kV for 8/20 μs current					
				1.5 kA	3 kA	5 kA	10 kA	20 kA	40 kA
2.55	3	12.5	8	9.5	10	10.5	11	13	15.3
5.1	6	25	16	19	20	21	22	26	30.5
7.65	9	34	22.5	24.5	26	27.5	30	35	41
8.4	10	36.5	23.5	26	28	29.5	32	37.5	43.5
10.2	12	43.5	28.2	38	32.9	34.8	38.5	43.8	51.5
12.7	15	54.2	35	38.4	41	43.4	48	54.6	64.2
15.3	18	65	42.1	46	49.1	52	57.5	65.4	76.9
17	21	69.5	44.9	49.2	52.5	55.7	61.5	69.9	82.2
19.5	24	87	56.4	61.6	65.8	69.6	77	87.6	103
22	27	97.7	63.2	69.2	73.9	78.2	86.5	98.4	115.7
24.4	30	108.4	70	76.8	82	86.8	96	109.2	128.4
29	36	130	84.2	92	98.2	104	115	130.8	153.8

Table 5 Protective Characteristics of Ohio Brass Distribution Normal-Duty Arresters, Polymer-Housed

MCOV, kV rms	Duty-cycle rating, kV rms	0.5-μs discharge voltage, kV at 5 kA	Switching impulse discharge voltage kV, at 0.5 kA	Discharge voltage, kV for 8/20-μs current					
				1.5 kA	3 kA	5 kA	10 kA	20 kA	40 kA
2.55	3	12.5	8.5	9.8	10.3	11	12.3	14.3	18.5
5.1	6	25	17	19.5	20.5	22	24.5	28.5	37
7.65	9	33.5	23	26	28	30	33	39	50.5
8.4	10	36	24	27	29.5	31.5	36	41.5	53
10.2	12	50	34	39	41	44	49	57	74
12.7	15	58.5	40	45.5	48.5	52	57.5	67.5	87.5
15.3	18	67	46	52	56	60	66	78	101
17	21	73	49	55	60	64	73	84	107
19.5	24	92	63	71.5	76.5	82	90.5	106.5	138
22	27	100.5	69	78	84	90	99	117	151.5
24.4	30	108	72	81	88.5	94.5	108	124.5	159
29	36	134	92	104	112	120	132	156	202

Table 6 Protective Characteristics of Ohio Brass Polymer-Houses Riser Pole Arresters

MCOV, kV rms	Duty-cycle rating, kV rms	0.5-μs discharge voltage, kV at 10 kA	Switching impulse discharge voltage kV at 0.5 kA	Discharge voltage, kV for 8/20-μs current					
				1.5 kA	3 kA	5 kA	10 kA	20 kA	40 kA
2.55	3	8.7	5.8	6.5	7	7.4	8.1	9	10.6
5.1	6	17.4	11.7	13	14	14.7	16.2	18.1	21.1
7.65	9	25.7	17.5	19.3	21	21.9	24	27	31.6
8.4	10	28.5	19.2	21.2	23	24	26.5	29.8	34.8
10.2	12	34.8	23.3	25.9	28	29.4	32.3	36.2	42.2
12.7	15	43.1	29.1	32.3	35	36.6	40.2	45.1	52.7
15.3	18	51.4	34.9	38.6	41.9	43.8	48	54	63.2
17	21	57.6	38.7	42.8	46.4	48.6	53.6	60.2	70.5
19.5	24	68.8	46.6	51.6	55.9	58.5	64.2	72.1	84.3
22	27	77.1	52.4	57.9	62.9	65.7	72	81	94.8
24.4	30	85.5	57.6	63.5	69	72	79.5	89.4	104.4
29	36	102.8	69.8	77.2	83.8	87.6	96	108	126.6

13
Station Lightning Insulation Coordination

1 INTRODUCTION

With the incoming surge and the arrester rating selected, the process of selecting the BILs of the station equipment can commence. Usually the station type and layout are known, and normally the candidate BILs for the transformer are limited to one to three values. The candidate BILs for the other equipment are even more limited, usually to two values. The circuit breaker BIL is fixed to a single value. These BILs are listed in Table 1, which is reproduced from Chapter 1. Table 2 lists a set of "station BILs" from NEMA Standard SG6. Although these BILs are conservative and in general use, especially for lower voltage stations, they should not be used without additional study. In addition these station BILs apply to apparatus other than the transformer. The overall procedure can be outlined as follows [1, 2].

1. Evaluate the Need for and Type of Opened Circuit Breaker Protection. The need for opened circuit breaker protection is evaluated first, since if arresters are needed, they should be included in the initial study of the station. See Chapter 11.

2. Select the Incoming Surge. The methodology, based on a reliability criterion of an MTBF, is presented in Chapter 11.

3. Select Candidate BILS. Candidate BILs are normally limited to one to three values; see Tables 1 and 2. For circuit breakers, two BSLs are listed for a single BIL for system voltages of 362 kV and higher. The first BSL applies for a closed breaker, while the second BSL applies when the breaker is opened. Two chopped wave tests

Table 1 Transformer and Bushings BILs and BSLs

System nominal/ max. system voltage, kV	Transformers, BIL, kV	Transformers, BSL, kV	Transformer bushings, BIL, kV	Transformer bushings, BSL, kV
1.2/-	30, 45		45	
2.5/-	45, 60		60	
5.0/-	60, 75		75	
8.7/-	75, 95		95	
15.0/-	95, 110		110	
25.0/-	150		150	
34.5/-	200		200	
46/48.3	200, 250		250	
69/72.5	250, 350		350	
115/121	350	280	450	
	*450	375	550	
	550	460		
138/145	450	375	450	
	*550	460	550	
	650	540	650	
161/169	550	460	550	
	*650	540	650	
	750	620	750	
230/242	650	540	650	
	*750	620	750	
	825	685	825	
	900	745	900, 1050	
345/362	900	745	900	700
	*1050	870	1050	825
	1175	975	1175, 1300	825
500/550	1300	1080	1300	1050
	*1425	1180	1425	1110
	1550	1290	1550	1175
	1675	1390	1675	1175
765/800	1800	1500	1800	1360
	1925	1600	1925	—
	2050	1700	2050	—

* Commonly used.
Source: Refs. 3–6

are applied to the breaker: (1) a crest voltage of 1.15 times the BIL chopped at 3 µs and (2) a crest voltage of 1.29 times the BIL chopped at 2 µs. The disconnecting switching BIL across the opened switch is 10% greater than that to ground [7].

4. Evaluate Contingency Conditions. The normal condition of a station during thunderstorm conditions is usually with all lines in service. However, contingency conditions with less than all lines in service may exist. Since these contingencies are normally associated with a low probability of occurrence, consideration of less than all lines in service is seldom required. However, if contingency conditions are likely

Station Lightning Insulation Coordination

Table 2 Insulation Levels for Outdoor Substations and Equipment

Rated max. voltage, kV	NEMA Std, SG6 outdoor substations		Circuit breakers		Disconnect switches	
	BIL, kV	10-s power frequency voltage, kV	BIL, kV	BSL, kV	BIL, kV	BSL, kV, estimate
8.25	95	30	95		95	
15.5	110	45	110		110	
25.8	150	60	150		125	
38.0	200	80	200		150	
					200	
48.3	250	100	250		250	
72.5	350	145	350		350	
121	550	230	550		550	
145	650	275	650		650	
169	750	315	750		750	
242	900	385	900		900	
	1050	455			1050	
362	1050	455	1300	825	1050	820
	1300	525		900	1300	960
550	1550	620	1800	1175	1550	1090
	1800	710		1300	1800	1210
800	2050	830	2050	1425	2050	1320
				1550		

Source: Refs. 8–10.

or deemed important, then the probabilities of the contingency should be evaluated so as to arrive at an appropriate incoming surge. See Chapter 11.

5. Select Arrester Rating and Preliminary Location of Arresters. The methods employed to select the arrester rating are described in Chapter 12. As to the location selected for the initial study, if line entrance arresters are used, they should be placed at the line entrances, and no other arresters should be added. Otherwise, the location selected should be such as to give preference to protection of the transformer. For a simple low-voltage station, this may be on the bus, while for a large breaker and a half scheme, the location is generally near the transformer, e.g., on the transformer bus.

6. Set Up Model on a Digital Transient Program (EMTP or ATP). Generally, only a single-phase model of the station is required. Station buses are modeled as distributed parameter lines described by their surge impedances and lengths. Transformers are modeled by their surge capacitances, which vary between about 1 and 10 nF. If unknown, values of 2 to 4 nF are suggested. Other station equipment can be modeled by their surge capacitances. However, conservatively, except for switching disconnecting switches in a GIS station where high frequencies are expected, this is seldom done. The circuit for applying the incoming surge to the

station is described in Chapter 11. Appendix 1 of this Chapter (13) provides estimates of the transformer surge capacitance and capacitance of other equipment.

7. Surge Voltages/Evaluation—Select BILs and Clearances. With the incoming surge applied to the station, the magnitude and waveshapes of voltages are measured throughout the station, usually at equipment locations and at opened points on the buses. The waveshapes of these voltages do not normally resemble the standard lightning impulse waveshape upon which the BILs or insulation strength are based. Therefore some evaluation method is required to change these surge voltages to equivalent crest voltages for a 1.20/50 μs impulse. For self-restoring insulations, methods employed range from a subjective evaluation to the use of the leader progression model. The leader progression model and the destructive effect method, which is derived from the leader progression model, and their uses are presented in Appendix 2 of this chapter.

For non-self-restoring insulations, e.g., the transformer, only a subjective evaluation is possible. Usually the crest voltage is compared to the 3-μs chopped wave test.

Because the strength of non-self-restoring insulation is specified by a conventional BIL, a safety factor or minimum margin of 15 to 20% is generally recommended. However, for self-restoring insulation, margins are questionable. Preferable to placing a margin on these insulations is the use of a higher MTBF—and thus an increase in the severity of the incoming surge.

Clearances are estimated using the highest equivalent crest voltage of the 1.2/50-μs waveshape divided by a negative polarity CFO gradient. This CFO gradient is a function of the gap configuration and varies from about 540 kV/m to about 750 kV/m, this later value for a rod–plane gap. A value of 605 kV/m, as used previously, is suggested.

For altitudes greater than sea level, the insulation strength decreases as a linear function of the relative air density. Because BILs are defined for the standard sea-level conditions, the BILs and clearances as calculated previously must be divided by the relative air density to obtain the required BILs and clearances.

8. Reevaluation. If the required BILs and clearances are considered excessive, two alternates may be used to decrease these values. Additional arresters can be employed within the station. Alternately, the severity of the incoming surge can be decreased while maintaining the desired MTBF by improving (decreasing) the lightning performance of the towers or lines adjacent to the station. This is practically attained by decreasing the tower footing resistance. For stations whose lines are unshielded, overhead ground wires can be added for a distance from the station as determined by calculation of the minimum distance d_m per Chapter 11. For wood-pole lines without overhead ground wires, consideration should be given to the use of gaps at the first and second tower from the station.

Although this procedure is normally followed for the stations at a new voltage level, because of the experience gained, it may be shortened or circumvented when additional stations are constructed, only examining the significantly different features of the new station.

Station Lightning Insulation Coordination

Usually, distribution or low-voltage stations are treated differently in that these stations are somewhat standardized in layout. Therefore a detailed study for the standardized stations or each of the standardized stations is only required once.

However, in all cases, the recommended procedure is to perform the study using a digital transients program, e.g., the EMTP or the ATP. For small stations, the setup is relatively easy and the voltages are quickly and accurately computed.

However, there are cases for which an approximate, quick, conservative hand-calculation method is desirable. Therefore, in this chapter, instead of proceeding to demonstrate the full method as outlined above, the presentation will be centered on estimating methods that can quickly be used to analyze protection of stations with relatively simple layouts, typically used at lower voltages. These methods can also be used to provide initial estimates of protection requirements and BILs for more complex stations before initiating the full study as outlined above. In addition, such a method can also provide a "sanity" check on computer answers.

As a warning, usually simplified estimating methods are not truly simplified. In a sense, they cannot be, since some degree of accuracy is required. This will become evident in proceeding through this chapter. The presentation plan is first to consider the estimation of the surge voltages. In this regard, some of the work has been already been accomplished in Chapter 9. This will be followed by consideration of the strength, and finally the stress will be compared to the strength to arrive at a BIL and clearances.

2 THE STRESS—CREST VOLTAGE AT EQUIPMENT

The general circuit used to estimate the crest voltages within an n-line station is presented in Fig. 1, in which the crest voltages and travel times are defined. Note that other lines in the station are represented by a resistor connected at the arrester tap point on the bus and that the resistance is equal to the surge impedance of the line divided by $n - 1$. The surge impedance of the phase conductors of the lines may be slightly different from that of the bus. However, this small difference is not of major significance, so it is assumed that the surge impedances of the phase conductor and the bus are equal. For completeness, the travel times are the lengths or distances divided by the velocity of propagation, which for this air-insulated station is $300 \, m/\mu s$ (about $1000 \, ft/\mu s$). Gas-insulated stations will be considered later as a special case. The transformer can be represented by a capacitor to ground. The value of this capacitor can range from only $1 \, nF$ to over $10 \, nF$; see Appendix 1.

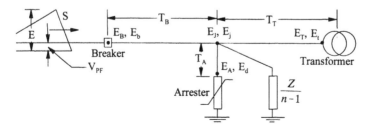

Figure 1 Circuit analyzed.

However, the maximum crest voltage at the transformer is attained for a capacitance of about 4 nF or for 1 nF or less for long separation distances (more later). The transformer in this figure is a generic representation. That is, it may not be present, in which case it is simply the opened end of the line and thus the capacitance of the "transformer" is zero. The voltage behind the arrester denoted by its crest voltage E_B is generally called a breaker. However, it may be any piece of equipment such as a disconnecting switch, a bus support insulator, etc.

The incoming surge voltage is assumed to arrive from a distant struck point so that no reflections from the struck point are considered. This incoming surge has a steepness S and a crest voltage E and a tail that is considered infinite. The surge is riding atop a power frequency voltage V_{PF} that is of opposite polarity to the surge voltage.

As before, the process of calculation is to separate the calculation into (1) the surge voltage and (2) the power frequency voltage. That is, the surge voltage, having a crest of E, is applied to the circuit or station, and the surge voltages throughout the station are calculated. The voltage to ground is then the surge voltage minus the power frequency voltage. To calculate the surge voltage in the station, the arrester discharge voltages must be increased by the power frequency voltage so that the proper current flows through the arrester. Thus the voltages are defined as follows.

E_d = arrester discharge voltage, i.e., voltage to ground
E_A = surge voltage at arrester
E_b = voltage to ground at breaker
E_B = surge voltage at the breaker
E_t = voltage to ground at the transformer
E_T = surge voltage at the transformer
E_j = voltage to ground at the arrester–bus junction
E_J = surge voltage at the arrester–bus junction

In equation form,

$$\begin{aligned} E_A &= E_d + V_{PF} \\ E_B &= E_b + V_{PF} \\ E_T &= E_t + V_{PF} \\ E_J &= E_j + V_{PF} \end{aligned} \quad (1)$$

In each of the following sections, equations will first be developed for no arrester lead length, i.e., $T_A = 0$. Further, the development will proceed by first considering a constant-voltage arrester, that is, an arrester that maintains a constant voltage regardless of the current discharged. This will be followed by considering the actual arrester voltage–current characteristics and the arrester lead length.

2.1 Voltage at the Transformer, Zero Capacitance—An Opened Circuit

The case of zero arrester lead length, $T_A = 0$, and zero capacitance was considered in Chapter 9. The equations are

$$E_T = E_A + \frac{2ST_T}{n} \quad \text{for} \quad \frac{ST_A}{E_A} = 0 \text{ to } \frac{n}{6}$$
$$E_T = \frac{4}{n+3}(E_A + 2ST_T) \quad \text{for} \quad \frac{ST_T}{E_A} = \frac{n}{6} \text{ to } \frac{n+1}{4} \quad (2)$$

Per Eq. 2, at $ST_T/E_A = n/6$,

$$\frac{E_T}{E_A} = 1.33 \quad (3)$$

and at $ST_T/E_A = (n+1)/4$, the above equations show that $E_T = 2E_A$, the maximum voltage at the open circuit.

These equations were developed assuming that the arrester is a constant-voltage arrester. That is, the arrester discharge voltage is constant, independent of the current discharged by the arrester. Also, the tail of the incoming surge was constant, i.e., an infinite time to half value. These two assumptions lead to the conclusion that the equations are conservative.

Effect of Arrester Voltage—Current Characteristics

Using the ATP, the 138 kV case of Table 3 and the discharge voltages of Table 4 were used to develop curves of E_T/E_A as a function of ST_T/E_A, which are shown in Fig. 2 for $n = 1$ and $n = 2$. Comparing these curves to those for the constant-voltage arrester shows that a considerable reduction occurs, more for $n = 1$ than for $n = 2$. Also, the curve for $n = 2$ retains its basic shape showing the break points. Thus the equations for the constant-voltage arrester can be retained but should be reduced by a constant that ranges from about 0.91 for $n = 1$ to 0.97 for $n = 4$. Denoting this constant as K_2 and adding the effect of arrester lead length, the equations become

$$E_T = K_2\left[E_A + \frac{2S}{n}(T_T + T_A)\right] \quad \text{for} \quad \frac{S(T_T + T_A)}{E_A} = 0 \text{ to } \frac{n}{6}$$
$$E_T = \frac{4K_2}{n+3}[E_A + 2S(T + T_A)] \quad \text{for} \quad \frac{S(T_T + T_A)}{E_A} = \frac{n}{6} \text{ to } \frac{n+1}{4} \quad (4)$$

Table 3 Cases Considered, E_A is Voltage for Constant Voltage Arrester; E_{10} is the 10 kA Discharge Voltage

Case	S, kV/µs	E, kV	E_{10}, kV	V_{PF}, kV	E_A, kV
138 kV	1400	1560	464	156	620
500 kV	2000	2300	900	338	1238

Table 4 Assumed Arrester Voltage–Current Characteristics

I_A, discharge current in kA	E_d, discharge voltage in per unit of 10 kA discharge voltage	I_A, discharge current in kA	E_d, discharge voltage in per unit of 10 kA discharge voltage
0.0001	0.7213	1	0.8644
0.001	0.7538	5	0.9577
0.01	0.7701	10	1.0000
0.05	0.7950	20	1.0564
0.10	0.8026	40	1.1161

where K_2 is given in Table 5 (the maximum surge voltage of E_T becomes $2K_2E_A$).

2.2 Voltage at the Transformer and Arrester–Bus Junction with Capacitance

Modeling the transformer with its surge capacitance to ground dramatically increases the voltage at the transformer. While it is possible to develop theoretical equations for this situation, they are extremely complex. Therefore only one simple illustration will be used to show the mechanism of the buildup of the voltage at the transformer.

For a Long Separation Distance

Taking the extreme case for which the distance between the transformer and arrester is sufficiently long that the arrester operates before reflections return to the arrester from the transformer, Fig. 3 illustrates the phenomena assuming $E_A = 600\,\text{kV}$, $S = 1000\,\text{kV/µs}$, $ZC_T = 1.6\,\text{µs}$, and $T_T = 0.6\,\text{µs}$. Per Fig. 3, the voltage $e = St$

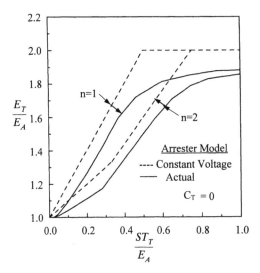

Figure 2 Comparison of voltage at open end for constant voltage arrester and actual arrester.

Table 5 Values of K_2

n number of lines	K_2
1	0.91
2	0.93
3	0.95
4	0.97

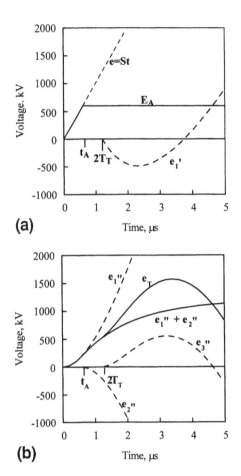

Figure 3 Surge voltages at the (a) arrester and (b) transformer.

arrives at the arrester and is unlimited in magnitude. At a time t_A, the arrester operates and maintains a constant voltage E_A. The initial voltage at the transformer, e_1'', Fig. 3b, is given by the equation

$$e_1'' = 2S[t - ZC(1 - e^{-t/ZC})] \tag{5}$$

Differentiating this equation results in

$$\frac{de_1''}{dt} = 2S(1 - e^{-t/ZC}) \tag{6}$$

which shows that at $t = 0$, the steepness is zero and at t equals infinity, the steepness is equal to $2S$. At time t_A, the arrester reaches a constant voltage E_A, and a negative voltage equal to St arrives at the transformer and produces a voltage e_2'', which is equal to the voltage e_1'' but translated in time by t_A. Thus the voltage at the transformer for times greater than t_A becomes

$$\begin{aligned} e_1'' + e_2'' &= 2S[t - ZC(1 - e^{-t/ZC})] - 2S[(t - t_A) - ZC(1 - e^{-(t-t_A)/ZC})] \\ &= 2S[t_A + ZCe^{-t/ZC}(1 - e^{+t_A/ZC})] \quad \text{for } t > t_A \end{aligned} \tag{7}$$

Note that the maximum voltage at t equals infinity is $2St_A$, which is equal to $2E_A$. Thus as before, per Eq. 2, if nothing else happens, the maximum voltage at the transformer is $2E_A$.

However, the first voltage that arrives at the transformer, e_1'', creates a reflection of e_1' per the equation

$$e_1' = e_1'' - e = S[t - ZC(1 - e^{-t/ZC})] \tag{8}$$

Differentiating,

$$S_1' = \frac{de_1'}{dt} = S[1 - 2e^{-t/ZC}] \tag{9}$$

Setting this equation equal to zero shows that the crest voltage occurs at a time of $0.69ZC$ and is equal to $-0.307SZC$ as illustrated in Fig. 4. Also the initial steepness is $-S$ and the final steepness is $2S$. The steepness at the zero line crossing is $0.59S$. The reflection and initial voltage at the transformer can be visualized by conceiving that initially the capacitor acts as a short circuit, and thus a negative reflection occurs, and finally the capacitor acts as an open circuit, so that a positive reflection results.

To continue, the voltage given by Eq. 8 travels to the arrester and is reflected negatively back to the transformer. This reflection arrives at the transformer at a time of $2T_T$ and the voltage at the transformer caused by this reflection is given by the equation

$$e_3'' = 2S[(3ZC - t) - (3ZC + 2t)e^{-t/ZC}] \tag{10}$$

Again differentiating,

$$\frac{de_3''}{dt} = 2S\left[\left(1 + \frac{2t}{ZC}\right)e^{-t/ZC} - 1\right] \tag{11}$$

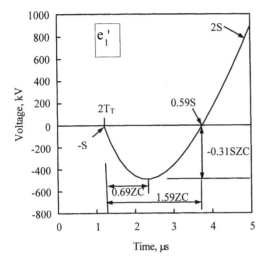

Figure 4 Reflected voltage from the transformer.

Setting this equal to zero shows that the crest voltage is achieved at $t = 1.25ZC + 2T_T = 3.2\,\mu s$ (actually at 3.4 μs because of the shape of the voltage) and the crest voltage is $+0.348ZCS$ or $0.348(1.6)(1000) = 557\,kV$. As time increases, this voltage becomes negative, and at infinity, the steepness is equal to $-2S$.

Thus it is this voltage, the $0.348ZCS$, that adds to the voltage at the transformer, increasing it above that for an open circuit. The maximum voltage is therefore $2E_A + 0.348SZC$ or 1757 kV or $2.9E_A$ for the example. However, this maximum voltage never occurs. First, if the ZC time constant is increased, the voltage $e_1'' + e_2''$ increases more slowly and does not reach $2E_A$ by the time the reflection e_3'' arrives. Next, the incoming voltage does not possess an infinite tail but usually has a short tail described by a 15- to 20-μs constant. In addition, the arrester is not a constant-voltage arrester. All these factors combine so that the maximum voltage at the transformer is in the range of 2.2 to $2.6E_A$. For the example, the $e_1'' + e_2''$ voltage at 3.4 μs is 1026 kV and not $2E_A$. Also, at 3.4 μs, $e_T = 1574\,kV$ or $2.6E_A$. Note that $1026 + 557 = 1583\,kV$, which checks the value of 1574 kV.

Voltage at the Transformer

Since the equations governing the voltage at the transformer are intractable, the alternate method of obtaining a plot of the transformer voltage in terms of E_T/E_A as a function of ST_T/E_A is used [11, 12]. Using the data from this curve, a regression equation can be determined to determine efficiently the voltage at the transformer. The two cases considered are presented in Tables 3 and 4. When the actual arrester is used, the voltage–current characteristic is assumed as per Table 4, where the discharge voltage is in per unit of the 10 kA discharge voltage.

For a 138 kV single-line station, $n = 1$, Fig. 5 shows a comparison of three curves for (1) a constant-voltage arrester with $C_T = 0$, dotted line, (2) a constant-voltage arrester with C_T varied from 1 to 6 nF to obtain the maximum voltage, and (3) an actual arrester with C_T varied from 1 to 6 nF. The maximum value of E_T

Table 6 Constants A and B of Eq. 12

n = number of lines	A	B
1	1.00	0.14
2	0.98	0.16
3	0.84	0.18
4	0.68	0.25

occurred for $C_T = 4\,\text{nF}$ for values of ST_T/E_A below about 0.10. Above this value, the maximum transformer voltage occurred for $C_T = 1$ or $2\,\text{nF}$.

From Fig. 5, the sharp increase in voltage at low values of St_T/E_A is astounding. The maximum value of E_T/E_A was 2.24 at $ST_T/E_A = 1.3$ when the constant-voltage arrester was used for the 138-kV case and for $n = 1$. When the actual arrester is used, the maximum value of E_T/E_A was reduced to 2.13 at $ST_T/E_A = 1.3$. Because the use of the actual arrester characteristics provides a more realistic value of the transformer voltage, the actual arrester characteristic will be used for further analysis.

Figure 6 shows two of the four resultant curves ($n = 1$ and 2) obtained for the two cases of Table 3 using the actual arrester characteristics per Table 4. The dotted line curves are for the case where $C_T = 0$. Regressing the data from the two cases results in the equation

$$\frac{E_T}{E_A} = 1 + \frac{A}{1 + B/K_1} \tag{12}$$

where

$$K_1 = \frac{ST_T}{E_A} \tag{13}$$

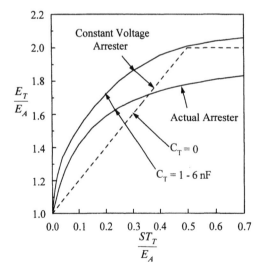

Figure 5 Effect of C_T and arrester model, 138-kV case, $n = 1$.

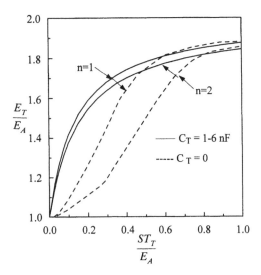

Figure 6 General curves for voltage at transformer.

where the constants A and B are given in Table 6. Considering the arrester lead length, Eq. 13 becomes

$$K_1 = \frac{S(T_T + T_A)}{E_A} \qquad (14)$$

The regression calculation using Eq. 12 showed a correlation coefficient of 0.99 or greater.

Equations 12 and 14 can also be solved for the maximum separation distance given the values of E_T and E_A, i.e.,

$$T_T + T_A = \frac{BE_A}{S} \frac{(E_T/E_A) - 1}{(A+1) - (E_T/E_A)} \qquad (15)$$

As noted from Fig. 6, the curves for C_T from 1 to 6 nF provide higher values of voltage except at high values of the parameter ST_A/E_A, where the curves for with and without C_T are essentially identical.

As an example of the use of the equations, consider the 138-kV case and assume that the arrester discharge voltage E_d is 267 kV, $V_{PF} = 80$ kV, and $S = 1400$ kV/μs. Then the value of E_A is 347 kV. For a separation distance between the arrester and transformer of 12 meters and $T_A = 0.01$, $T_T + T_A = 0.05$ μs. Thus $S(T_T + T_A)/E_A = 0.20173$. For a single-line station, using Eq. 12, $E_T/E_A = 1.5903$ and $E_T = 552$ kV. The voltage to ground at the transformer is $E_t = 552 - 80 = 472$ kV.

As a further example illustrating the use of Eq. 15, assume that the maximum permissible voltage at the transformer is 504 kV, i.e., E_t. Then $E_T = 584$ kV and $E_T/E_A = 1.683$. Using Eq. 15, $T_T + T_A = 0.07476$, and the maximum separation distance is this value times the velocity of propagation (300 m/μs) or 22.4 m.

To illustrate the difference in voltage waveshapes with and without a surge capacitor model, Fig. 7 has been obtained using $C_T = 0$ and 2 nF, $E = 2000$ kV, $S = 2000$ kV/μs, $Z = 400$ ohms, and $V_{PF} = 0$, with the arrester characteristics modeled using Table 4 with a 10-kA discharge voltage of 900 kV. The separation distance between the arrester and the transformer is 20 meters. As shown, the voltage with the 2 nF capacitor is larger and the oscillating frequency is smaller. In evaluating the strength of the non-self-restoring insulation, i.e., the transformer, the time to crest of the voltage is required. To estimate the time to crest of the voltage at the transformer, t_T, the circuit is reduced to an inductance and a capacitance. That is, the total inductance of the line from the arrester to the transformer is $Z(T_T + T_A)$ and the total capacitance of the line is $(T_T + T_A)/Z$. The voltage across the capacitor, E_c, is

$$E_c = V(1 - \cos \omega t) \tag{16}$$

where

$$\omega = \frac{1}{\sqrt{LC}} \tag{17}$$

Therefore the oscillating frequency f_T is

$$f_T = 2\pi\sqrt{(T_A + T_T)(ZC_T + T_A + T_T)} \tag{18}$$

and the period is the inverse of the frequency. The time to crest is 1/2 of the period. Adding the time to crest of the voltage at the arrester, i.e., E_A/S, results in a conservative estimate of the time to crest of

$$t_T = \pi\sqrt{(T_A + T_T)(ZC_T + T_T + T_A)} + \frac{E_A}{S} \tag{19}$$

Testing this equation, for $C_T = 2$ to 4 nF and $T_T = 0.05$ to 0.10 μs, the calculated time to crest exceeds the actual by from 2 to 14%. Thus the equation is somewhat

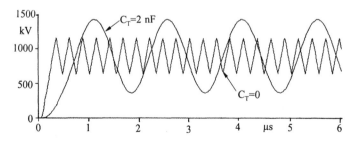

Figure 7 Comparison of voltage waveshapes at the transformer.

inaccurate, but it errs on the proper side. For the case considered in Fig. 7, per Eq. 19, $t_T = 1.21$, whereas from Fig. 7, t_T is about 1.1 µs.

Voltage at the Arrester–Bus Junction

With the transformer modeled as a capacitance, the voltage at the arrester–bus junction can be approximated by the equation

$$E_J = 1 + \frac{A}{1 + B/K_1} \tag{20}$$

where in this case K_1 is

$$K_1 = \frac{ST_A}{E_A} \tag{21}$$

2.3 Voltage Behind the Arrester

Single-Line Station

As shown in Chapter 9, assuming that the incoming surge has a linearly rising front of steepness S and is unlimited in magnitude, for a constant-voltage arrester, the voltage behind the arrester at a location generically denoted as the breaker is

$$E_B = E_A + 2ST_B \tag{22}$$

where E_A is the voltage at the arrester and T_B is the travel time from the breaker to the arrester. Note that E_B is independent of the number of lines. The maximum voltage of E_B is $E + E_A/2$ for $n = 1$ and E if n is greater than 1. However, if the voltage is limited, i.e., has a magnitude of E, and the transformer capacitance C_T is included in the circuit, the voltage at the breaker may decrease depending on the value of the ZC_T time constant and the time to crest of the incoming surge. To show this effect, consider the circuit of Fig. 8, where for purposes of simplification the transformer, i.e., the capacitor, is located at the arrester and there is no arrester lead length. The incoming surge has a magnitude of $E = 1600$ kV and a steepness of $S = 1000$ kV/µs. Also $Z = 400$, $E_A = 300$ kV, and $T_B = 0.5$ µs. Upon arrival at the arrester–transformer combination or the arrester–capacitor combination, the voltage at the arrester is given by the equation

$$e_A = e'' = 2S\left[t - ZC_T\left(1 - e^{-t/ZC_T}\right)\right] \tag{23}$$

As shown in Fig. 8b, this equation is valid up to the time that the arrester "operates" to hold a voltage E_A. Thus assuming a constant-voltage arrester, the arrester voltage is

$$E_A = 2S\left[t_A - ZC_T\left(1 - e^{-t/ZC_T}\right)\right] \tag{24}$$

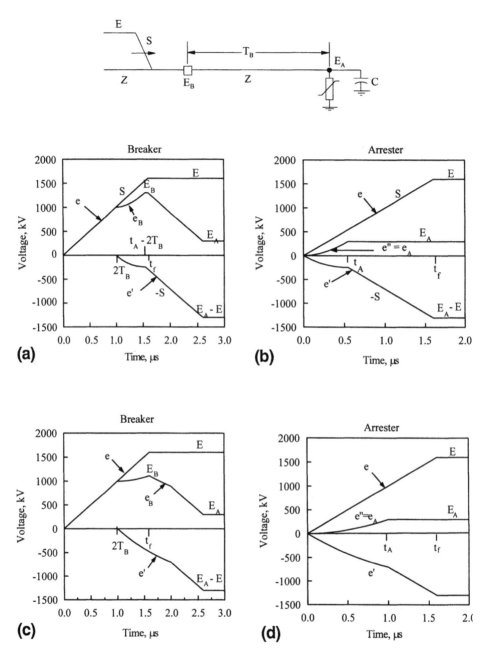

Figure 8 Analysis of voltage at breaker. Voltage at (a) breaker and (b) arrester for $t_A < (t_f - 2T_B)$. Voltages at (c) breaker and (d) arrester for $t_A > (t_f - 2T_B)$.

Station Lightning Insulation Coordination

The voltage reflected back toward the breaker due to the initial surge e is

$$e' = e'' - e = S[t - 2ZC(1 - e^{-t/ZC_T})] \tag{25}$$

Differentiating

$$\frac{de'}{dt} = S[1 - 2e^{-t/ZC_T}] \tag{26}$$

As noted by this equation, the reflected voltage is initially negative, having a steepness of $-S$. The reflected voltage crosses the zero line at about $1.6ZC$ and at $t = $ infinity, the steepness is positive, equal to $2S$. Equations 25 and 26 are valid up to the time that the arrester operates to hold the voltage E_A after which the reflected voltage decreases with a steepness S, since e' is $e_A - St$ and $de'/dt = -S$. At the time to crest of the incoming surge t_T, the voltage is constant at $E_A - E$.

The reflected voltage arrives back at the breaker at time $2T_B$. The sum of this voltage and the original incoming surge is the voltage at the breaker e_B, which has a crest of E_B. Note that the maximum crest voltage is attained at $t_A + 2T_B$, and because the positive steepness of the incoming surge is canceled by the negative steepness of the reflected voltage, the voltage remains constant at E_B to time t_f.

The crest voltage at the breaker, E_B at time $t_A + 2T_B$ is

$$\begin{aligned} E_B &= S(t_A + 2T_B) + S[t_A - 2ZC_T(1 - e^{-t_A/ZC_T})] \\ &= 2T_B S + 2S[t_A - ZC_T(1 - e^{-t_A/ZC_T})] \end{aligned} \tag{27}$$

Substituting Eq. 24 into Eq. 27,

$$E_B = E_A + 2ST_B \quad \text{for } t_A \leq (t_f - 2T_B) \tag{28}$$

which is the same as derived previously when no capacitance was assumed and as noted is valid for a time when the time to reach the voltage E_A, t_A, is less than $t_f - 2T_B$. However, now assume that the capacitance is increased or that the ZC_T time constant is increased so that the time at which the arrester operates, t_A, is greater than $t_f - 2T_B$ as illustrated in Fig. 8d. Now as shown, the crest of the voltage at the breaker, Fig. 8c, occurs at time t_f and is less than that given by Eq. 28. For this condition, the voltage at the breaker, letting $t = t_f - 2T_B$ in Eq. 25 is

$$E_B = E + e' = 2E - 2ST_B - 2ZC_T S\left(1 - e^{-(t_f - 2T_B)/ZC_T}\right) \quad \text{for } t_A \geq (t_f - 2T_B) \tag{29}$$

The minimum voltage for this condition is the arrester voltage E_A. That is, if the use of Eq. 29 results in voltages less than the arrester voltage, the actual voltage is equal to the arrester voltage. The maximum voltage occurs when $t_f = 2T_B$, at which time $E_B = E$. That is, if $2T_B$ is greater than t_f, then $E_B = E$.

For example, let $E = 2300\,\text{kV}$, $E_A = 900\,\text{kV}$, $ZC = 2.4$, $T_B = 0.2$, $V_{PF} = 0$, and $S = 2000\,\text{kV}/\mu\text{s}$. Thus $t_f - 2T_B = 1.5 - 0.4 = 0.75$. To find the value of t_A, Eq. 24 must be iterated. Placing Eq. 24 in a more convenient form,

$$\frac{E_A}{2SZC_T} = \frac{t_A}{ZC_T} - \left(1 - e^{-t_A/ZC_T}\right) \tag{30}$$

For $E_A/2ZCS = 0.938$ per the above data, upon iteration, $t_A/ZC_T = 0.467$ or $t_A = 1.12\,\mu s$. Since t_A is greater than $t_f - 2T_B$, the crest voltage E_B occurs at t_f and is given by Eq. 27. Thus E_B is 1223 kV, whereas the use of Eq. 28 would have resulted in 1700 kV.

Now let $t_A = t_f - 2T_B = 1.12\,\mu s$, the critical value. Therefore $T_B = 0.015\,\mu s$, or the separation distance is 4.5 meters. Using Eqs. 28 or 29 results in the same value of E_B of 960 kV.

Summarizing, Eq. 28 is valid to a time when $t_A = t_f - 2T_B$, after which Eq. 29 applies. Since the iteration to obtain the value of t_A is time consuming, Fig. 9 is provided to give a quick estimate.

More Than One Line

The line development of the theory for more than 1 line is similar. Assuming that $n - 1$ lines are connected at the arrester location, the equation for the voltage at the arrester is

$$E_A = e'' = \frac{2S}{n}\left[t_A - \tau\left(1 - e^{-t_A/\tau}\right)\right] \tag{31}$$

and the reflected voltage is

$$e' = \frac{S}{n}\left[(2-n)t - 2\tau\left(1 - e^{-t/\tau}\right)\right] \tag{32}$$

where

$$\tau = \frac{ZC_T}{n} \tag{33}$$

The voltage at the breaker at time $t_A + 2T_B$ is

$$\begin{aligned}E_B &= S(t_A + 2T_B) + \frac{S}{n}\left[(2-n)t_A - 2\tau\left(1 - e^{-t_A/\tau}\right)\right] \\ &= \frac{2}{n}ST_A - 2ST_B - \frac{2S\tau}{n}\left(1 - e^{-t_A/\tau}\right)\end{aligned} \tag{34}$$

Substituting Eq. 31 into Eq. 34 produces the same equation as Eq. 28, i.e.,

$$E_B = E_A + 2ST_B \quad \text{for } t_A \leq (t_f - 2T_B) \tag{35}$$

If t_A is equal to or greater than $t_f - 2T_B$, then

$$E_B = \frac{2}{n}E - \frac{2-n}{n}(2ST_B) - \frac{2S\tau}{n}\left(1 - e^{-(t_f - 2T_B)/\tau}\right) \quad \text{for } t_A \geq (t_f - 2T_B) \tag{36}$$

Station Lightning Insulation Coordination

By iteration, the time t_A is found from Eq. 31, or rearranging we obtain

$$\frac{nE_A}{2\tau S} = \frac{t_A}{\tau} - \left(1 - e^{-t_A/\tau}\right) \tag{37}$$

Figure 9 can also be used to obtain a quick estimate.

As an example, assume $E = 1600\,\text{kV}$, $S = 1000\,\text{kV}/\mu\text{s}$, $E_d = 300\,\text{kV}$, $V_{PF} = 100\,\text{kV}$, $C_T = 3\,\text{nF}$, $T_B = 0.2\,\mu\text{s}$, and $Z = 400$ ohms ($ZC = 1.2\,\mu\text{s}$), where E_d is the arrester discharge voltage or voltage to ground. The surge voltage at the arrester, E_A, is the discharge voltage plus the power frequency voltage or $400\,\text{kV}$. Thus t_f is $1.6\,\mu\text{s}$ and for $n = 1$, $nE_A/2\tau S = 0.167$, and from Fig. 9, $t_A/ZC_T = 0.64$ or $t_A = 0.768$. Since t_A is less than $t_f - 2T_B$, Eq. 35 applies, and $E_B = 800\,\text{kV}$. Subtracting the power frequency voltage, the surge voltage to ground, E_b, is $700\,\text{kV}$.

As a next step, find the critical distance or the critical value of T_B where both Eq. 35 and Eq. 36 apply. Thus $2T_B = 1.6 - 0.768 = 0.832\,\mu\text{s}$ or the separation distance is 124.8 meters. Using Eq. 35, E_b is $1233 - 100 = 1133\,\text{kV}$. Using Eq. 36, E_b is

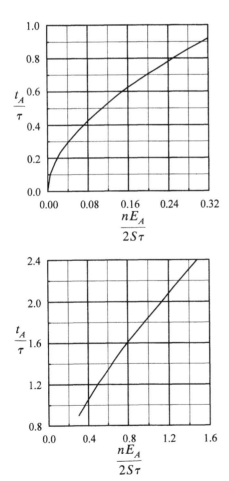

Figure 9 Estimating the value of t_A.

$1233 - 100 = 1133\,\text{kV}$. Now, let $T_B = 0.8\,\mu\text{s}$, so that $t_f - 2T_B = 0$. Thus, using Eq. 36, E_b is $1600 - 100 = 1500\,\text{kV}$. These values, along with an additional point at 180 meters where $E_b = 1220\,\text{kV}$, are plotted in Fig. 10.

Now, develop a similar curve for an additional line connected at the arrester location or for $n = 2$. From Fig. 9, t_A/τ is 1.425, giving a value of t_A of $0.855\,\mu\text{s}$. For the original value of T_B of $0.2\,\mu\text{s}$, $t_f - 2T_B$ of $1.2\,\mu\text{s}$ is less than t_A, and therefore Eq. 35 applies and $E_b = 800 - 100 = 700$, the same as for $n = 1$. The critical value of T_B is $0.3725\,\mu\text{s}$. Using Eq. 35 or Eq. 36, E_b is $1145 - 100 = 1045\,\text{kV}$.

For separation distances of 180 and 240 meters, E_b is $1208\,\text{kV}$ and $1500\,\text{kV}$, respectively. These values of a two-line station are also shown in Fig. 10. As noted from these calculations and from Fig. 10, the voltage at the breaker is somewhat insensitive to the number of lines, but more than one line does produce a small reduction in the voltage. Naturally, the curves must meet at the extreme end, where the maximum voltage is equal to that of the incoming surge voltage of $1600 - 100 = 1500\,\text{kV}$.

To illustrate another limiting case, assume the same parameters as for the previous example except let $C = 20\,\text{nF}$ or $ZC = 8.0\,\mu\text{s}$. Then for $n = 1$, $t_A = 0.235(8) = 1.88\,\mu\text{s}$ and the critical value of T_B is $-0.18\,\mu\text{s}$. This negative value signifies that (1) Eq. 36 must be used and that (2) E_B is equal to E_A until Eq. 36 results in a value greater than E_A, which is at a T_B of approximately $0.08\,\mu\text{s}$ or for a separation distance of 24 meters. Thus the voltage at the breaker is limited to the arrester voltage until the separation distance exceeds 24 meters. This curve for $ZC = 8\,\mu\text{s}$ is also shown in Fig. 10.

Summary with Effect of the Arrester Lead Length

In summary, with the transformer modeled as a capacitor, the voltage at the breaker is

(1) For $C_T \neq 0$ and $t_A \leq [t_f - 2(T_B + T_A)]$,

$$E_B = E_J + 2ST_B \tag{38}$$

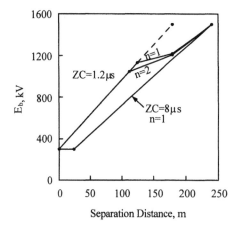

Figure 10 Example of variation of voltage at breaker with separation distance $T_A = 0$.

Station Lightning Insulation Coordination

where E_J is obtained from Eq. 20.

(2) For $C_T \neq 0$ and $t_A > [t_f - 2(T_B + T_A)]$,

$$E_B = \frac{2E}{n} - \frac{2-n}{n}[2S(T_B + T_A)] - \frac{2S\tau}{n}\left\{1 - e^{-[t_f - 2(T_B + T_A)]/\tau}\right\} \quad (39)$$

These formulations provide the best estimates but do not provide a smooth transition from one equation to another. This is discussed and illustrated later in an example.

Effect of Arrester Characteristics—A Special Case [13]

To illustrate the effect of the arrester characteristics on the voltage along a line, consider the circuit of Fig. 11, where a surge e arrives on a line of surge impedance Z, having a crest voltage E, a steepness S, and a tail described by a time constant τ. As shown, assume a single-line station. At the end of the line is an arrester. The surge voltages at the arrester and at a point along the line are shown in Fig. 12. The arrester voltage e_A, having a crest voltage of E_A, is the total voltage at the arrester, and therefore the reflected voltage is

$$e' = e_A - e \quad (40)$$

The equivalent circuit from which the arrester current and voltage can be obtained is shown in Fig. 13, from which we have

$$2e = e_A + i_A Z \quad (41)$$

Combining these equations,

$$e' = \frac{e_A - i_A Z}{2} \quad (42)$$

To determine the crest voltage of the reflected surge, e'_1, take the derivative of this equation and set it equal to zero. The derivative is

$$\frac{de'}{dt} = \frac{1}{2}\frac{di_A}{dt}\left[\frac{de_A}{di_A} - Z\right] \quad (43)$$

Setting this equal to zero shows that either

$$\frac{de_A}{di_A} = Z \quad \text{or} \quad \frac{di_A}{dt} = 0 \quad (44)$$

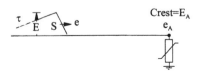

Figure 11 Circuit considered to calculate voltage along the line.

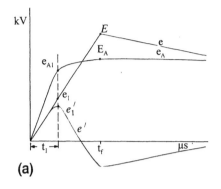

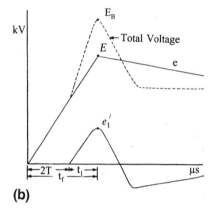

Figure 12 Voltages at (a) arrester and (b) on line [13].

$di_A/dt = 0$ describes the condition at which i_A is a maximum producing a maximum discharge voltage E_A. The other condition occurs prior to this and therefore is the condition where e_1' occurs.

Figure 14 illustrates the arrester discharge voltage–current characteristics and the derivative of e_A with respect to i_A. Denoting this derivative as a resistance R_A, the crest of the reflected voltage occurs when $Z = R_A$. From Eq. 42, this crest voltage is

$$e_1' = \frac{e_{A1} - i_{A1}Z}{2} \qquad (45)$$

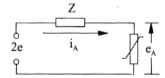

Figure 13 Circuit to calculate arrester current.

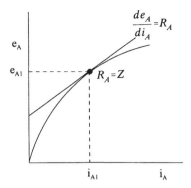

Figure 14 Arrester characteristic with position where $Z = R_A$.

where e_{A1} and i_{A1} are obtained from the arrester voltage–current characteristics; see Fig. 14. At some point along the line, the reflected voltage arrives so than the crest voltage of the incoming surge and the crest voltage of the reflected surge align; see Fig. 12b. Thus the maximum voltage at this point is

$$E_B = E + e_1' \qquad (46)$$

To illustrate by example, let $E = 2300$, $Z = 400$, and the arrester voltage–current characteristics be given by Table 7. Since $Z = 400$, R_A drops below 400 at 0.05 kA. Then $i_A = 0.05\,\text{kA}$ and $e_A = 715.5$. Then

$$e_1' = \frac{715.5 - .05(400)}{2} = 348\,\text{kV} \qquad (47)$$

and the maximum voltage is

$$\text{Max. } E_B = 2300 + 348 = 2648 \qquad (48)$$

Table 7 Assumed Arrester characteristics

Arrester current, kA	Arrester voltage, kV	R_A, ohms
0.01	694.09	
		560.25
0.05	715.50	
		136.25
0.10	722.34	
		61.80
1	777.96	
		20.99
5	861.93	
		7.614
10	900.00	

If the dynamic arrester characteristics are not considered, then the crest arrester discharge must first be determined. For currents between 5 and 10 kA, $R_A = 7.614$ and $E_0 = 824$. Then I_A is

$$I_A = \frac{4600 - 824}{407.6} = 9.26 \text{ kA} \tag{49}$$

and the arrester voltage is 895 kV. Thus the maximum voltage is

$$\text{Max. } E_B = E + \frac{E_A}{2} = 2747 \tag{50}$$

Because of the relatively flat arrester characteristics, this voltage is only about 4% greater than that when the arrester characteristics are considered. However, of interest is that the limitation of the voltage occurs at an arrester of 0.05 kA and not at the maximum arrester current of 9.26 kA.

The conclusion is that for this case, the use of the arrester characteristics is not necessary. Basing the voltage at the breaker on the maximum arrester voltage produced by the maximum arrester current is conservative.

Effect of Arrester Characteristics

Using the 138 kV case of Table 3, it was found that the reduction in the voltage at the breaker was very modest so that the equations need not be modified. This result is as expected from the results of the previous section.

2.4 Estimating the Arrester Discharge Voltage and Current

From Chapter 12, assuming a transformer is located within the circuit being studied, the arrester current can be estimated by the equation

$$I_A = 1.6 \frac{(2E/n) - E_0 - V_{PF}}{(Z/n) + R_A} \tag{51}$$

This equation does not consider the effect of reflections from the struck point; see Chapter 12. However, it provides a good estimate of the arrester current and is recommended for use.

3 INSULATION STRENGTH AND ITS SELECTION

In this section, the insulation for both non-self-restoring insulation, e.g., transformers, and self-restoring insulation are discussed and approximated. This is followed by the presentation of equations to estimate the insulation strength, i.e., the BIL and clearances. However, first the safety factors must be discussed.

3.1 Safety Factors

Before the advent of probability assessment of the insulation coordination of stations, large safety margins were used between the equipment insulation strength or BIL and the surge voltages or stress applied to the equipment. With a definition of an incoming surge in terms of a MTBS and a detailed study of the stations, the safety margins require some reassessment.

Non-Self-Restoring Insulations

For non-self-restoring insulations such as the transformer or the internal insulation of the transformer bushing, there is little doubt that some margin is needed to account for items such as aging and possible insulation degradation as a result of repeated low magnitude surges. Also to be considered is the test method, which basically consists of one application of a full and chopped wave impulse. Also, recently, some concern has been voiced about the effect of power frequency voltage [14]. In a recent paper [15], the authors tested an oil–paper insulation sample having a negative polarity lightning impulse CFO of 177 kV when no power frequency voltage was present. Applying a continuous power frequency voltage and a lightning impulse at various points on the wave, the authors obtained the CFO shown in Fig. 15 for a negative polarity lightning impulse. Of major practical interest is the opposite polarity case. For system voltages of 138 kV and 500 kV, the ratio of the power frequency line to ground voltage to the BIL ranges from about 0.20 to 0.28, which indicates a decrease of the CFO of 16 to 18%. Although these observations and test results need further verification, they indicate that some effect is present.

Margins suggested and used for the transformer insulation range from 10 to 30%. However, the larger margins generally apply to assessment methods that do not consider the actual surge voltage that impinges on the transformer. When the actual voltage at the transformer is considered, the margins used or suggested reduce to about 15 to 20%. The IEC application guide [16] and the IEEE arrester application guide [17] both suggest a margin of 15% for non-self-restoring insulations. A margin of 20% for non-self-restoring insulation is suggested for general use.

Self-Restoring Insulations

For self-restoring insulation, margins of 15% to 20% have been in common use. A margin of 15% is suggested in the IEEE arrester application guide, and a margin of

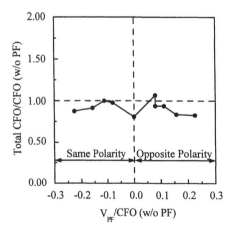

Figure 15 Effect of power frequency voltage on the insulation strength of oil–paper insulation. Negative lightning impulse. (From Ref. 15.) (Copyright IEEE, 1994.)

5% is suggested in the IEC application guide. However, use of a margin and an incoming surge based on probability appear inconsistent. If a higher degree of station reliability is desired, the more proper method is to increase the MTBF. Therefore margins for circuit breakers, disconnecting switches, bus support insulators, and for all self-restoring insulations are questionable. In general, a safety margin should not be used. However, if it is deemed necessary, a small margin of 5% is suggested.

There are some hidden margins. For example, the BIL is specified as the lower value for positive and negative polarity standard lightning impulses. For all practical cases, the lower value is for positive impulses. Since the incoming surge is considered to be of negative polarity, a hidden and nonevaluated margin exists.

3.2 Evaluation of Insulation Strength

General

The waveshapes of the surge voltages throughout the station are functions of (1) the waveshape of the incoming surge, (2) the arrester voltage–current characteristic, and (3) the layout of the station, i.e., the distances between components. In general, these surge voltage waveshapes do not resemble the standard 1.2/50 μs waveshape used to define the BIL. The waveshapes behind the arrester are characterized by an initial spike of voltage followed by a decay to the arrester discharge voltage, while the waveshape of the voltages in front of the arrester have a damped oscillatory shape with a decay to the arrester discharge voltage.

Since the surge voltage waveshapes do not resemble the standard 1.2/50-μs waveshape, some method must be used to estimate the insulation strength when subjected to these nonstandard waveshapes. Two principal methods have been used to accomplish this objective: (1) a subjective evaluation and (2) the use of mathematical methods based on the leader progression model. As shown in Appendix 2, the mathematical models are (1) the direct use of the leader progression model equations and (2) the destructive effect method, which is based on the leader progression model or equations. Ideally, the more exact leader progression model equations should be used. However, this method is only practical on the computer. Some type of simplified method is required for the overall simplified method being developed.

Self-Restoring Insulations

This identical problem was met in Chapter 10 when estimating the line insulation strength for nonstandard waveshapes. In Chapter 10, an equation was developed to estimate the CFO of the nonstandard waveshape, CFO_{NS}. For a surge having an exponential tail with a time constant τ,

$$CFO_{NS} = \left(0.977 + \frac{2.82}{\tau}\right)CFO \qquad (52)$$

From this equation for time constants of 10 and 20 μs, the CFO_{NS} is 1.26 and 1.12 times the CFO, which agrees with the value for 20 μs in Appendix 2. Thus it appears reasonable for voltages behind the arrester that the insulation strength could be set equal to 1.15 times the BIL or set equal to the 3-μs chopped wave test voltage.

From Appendix 2, for an oscillatory waveshape, the CFO_{NS} ranges from 1.13 to 1.20 dependent on the shape of the time-lag curve, i.e., the time-lag curve for apparatus giving the value of 1.13 and the time-lag curve for air-porcelain insulation giving a value of 1.20. These values assume a decay time constant of the oscillations of 20 μs. Thus for surge voltages appearing ahead of the arrester, *for self-restoring insulations*, the CFO_{NS} is also equal to about 1.15 times the CFO or BIL, which again is the 3-μs chopped wave test voltage.

To summarize, for self-restoring apparatus such as the disconnecting switch, the bus support insulators, and the circuit breakers, the insulation strength is approximately equal to 1.15 times the BIL. This 3-μs test voltage, by standards, is applied to the circuit breaker. However, no such test is required for the disconnecting switch or bus support insulators. However, it is reasonable to expect that these insulations will have at least this strength. One more detail remains. For short separation distances, the waveshape of the surge voltage does not possess a significant initial spike of voltage, that is, it is more like a full wave. To guard against this possibility, if the ratio of the crest of the surge voltage to the arrester discharge voltage is less than 1.15, the insulation strength is set equal to the BIL. In equation form,

$$\begin{aligned} \text{BIL} &= \frac{E_d}{\delta} \quad \text{if } \frac{E_b}{E_d} \leq 1.15 \\ \text{BIL} &= \frac{E_b}{1.15\delta} \quad \text{if } \frac{E_b}{E_d} > 1.15 \end{aligned} \quad (53)$$

where E_b is the crest of the voltage to ground, E_d is the arrester discharge voltage, and δ is the relative air density. In this formulation, no safety factor is used, as discussed previously.

In the above equation, the voltage is divided by the relative air density, since for self-restoring or external insulations, the effects of atmospheric conditions must be considered. That is, since the insulation strength is degraded at high altitude, the BIL must be increased to compensate. Per chapter 1, the relative air density is given by the equation

$$\delta = e^{-A/8.6} \quad (54)$$

where A is the altitude in km.

In the IEC application guide [17], the insulation strength for all wave shapes is simply set equal to the BIL. Thus the equation, per IEC using a safety factor of 5%, is

$$\text{BIL} = 1.05 \frac{E_b}{\delta} \quad (55)$$

Non-Self-Restoring Insulation

The above discussion and the resulting estimate of insulation strength only applies to self-restoring insulations. For non-self-restoring or internal insulations as typified by the transformer, a subjective method must be used. That is, even if a complete leader progression model is used to assess the insulation strength of self-restoring insulation, a subjective method must still be used for evaluating the insulation strength of non-self-restoring insulation. The subjective assessment compares the test voltages and their waveshapes to the oscillatory surge that appears at the transformer terminals.

In IEC [17], the insulation strength, independent of the waveshape, is simply set equal to the BIL. But to be noted is that a chopped wave test is not required in international standards. In the USA, historically, the transformer insulation strength has been set equal to the chopped wave test, i.e., 1.10 times the BIL [11, 12]. In an IEEE working group report [18], this single value of strength has been modified to state that it applies only if the time to crest of the voltage at the transformer is less than 3 μs, i.e., the chopping time for the chopped wave test. If the time to crest of the transformer voltage is greater than 3 μs, the strength is set equal to the BIL. In the IEEE arrester application guide [17], the insulation strength of a transformer is based on the time to crest of the voltage at the arrester so as to obviate the calculation of the time to crest at the transformer. If the time to crest of the voltage at the arrester is less than or equal to 2 μs, the strength is set equal to the chopped wave test voltage, i.e., 1.10 times the BIL. If it is greater than 2 μs, the insulation strength is set equal to the BIL. In equation form,

$$\text{if } t_T \leq 3\,\mu s \quad 1.10 \text{BIL} = (\text{SF})E_t \qquad \text{BIL} = \text{SF}\frac{E_t}{1.10}$$

$$\text{if } t_T > 3.0\,\mu s \quad \text{BIL} = (\text{SF})E_t \tag{56}$$

where t_T is the time to crest of the transformer voltage per Eq. 19 and SF is the safety factor.

These criteria do not cover all possible events. A further explanation of the definitions and an addition to cover long-tail voltages is necessary. To explain through example, consider a 230-kV, single-line transformer station where the incoming surge resulting from a backflash has a steepness of 1400 kV/μs, a crest voltage of 1560 kV, and an exponential tail having a time constant of 14 μs. Also let $V_{PF} = 156$ kV. A 140 kV MCOV metal oxide arrester is located a distance L_T in front of the transformer. The arrester lead length is L_A. The transformer surge capacitance is 4 nF except as noted. This circuit was set up on the ATP with the following results.

The Usual Case: $E_t/E_d > 1.10$ and $t_T < 3\,\mu s$. For $L_A = 3$ and $L_T = 9$ meters, Fig. 16 shows the voltages at the transformer, $E_t = 709$ kV, and at the arrester, $E_d = 460$ kV. The test impulse voltages applied to the transformer are shown by the dotted lines. The 3 μs chopped wave test voltage CW, having a crest equal to 1.10 times the full wave test voltage FW, was set equal to the crest of the surge voltage. Figure 16b shows a longer time plot of the voltage and the full wave test voltage. First, from Fig. 16a, since the surge voltage is less than the test voltages, subjective judgment indicates that this application is acceptable. Further, the full wave test voltage is greater than the surge voltage along the tail, as is shown by Fig. 16b. Therefore, as a start, the conclusion is that the crest of the surge voltage E_t

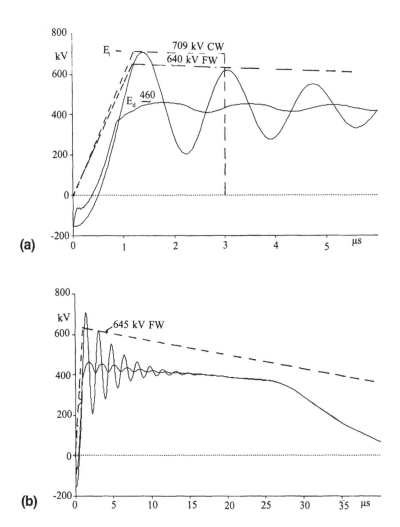

Figure 16 The usual case, $E_t/E_d \geq 1.10$, $t_T < 3\,\mu s$.

should be set equal to the chopped wave test voltage, 1.10 BIL, or the insulation strength is equal to the chopped wave test voltage.

Small Separation Distance: $E_t/E_d \leq 1.10$ and $t_T < 3\,\mu s$. For $L_A = 0.6$ and $L_T = 0.6$ meters, the voltages are shown in Fig. 17, where $E_t = 500\,\text{kV}$ and $E_d = 455\,\text{kV}$. Therefore the ratio of E_t/E_d is 1.10. Again, the 3-μs chopped wave test voltage is set equal to the crest of the surge voltage. The long time plot of Fig. 17b indicates that the tail of the full wave test voltage is slightly below the tail of the surge voltage and thus the coordination appears either marginally acceptable or nonacceptable. If the separation distance or arrester lead length were less than for this case, the crest surge voltage would be less, although the arrester discharge voltage would remain constant. Therefore E_t/E_d would be less than 1.10, and the full wave test voltage would be less than the arrester discharge voltage. To guard

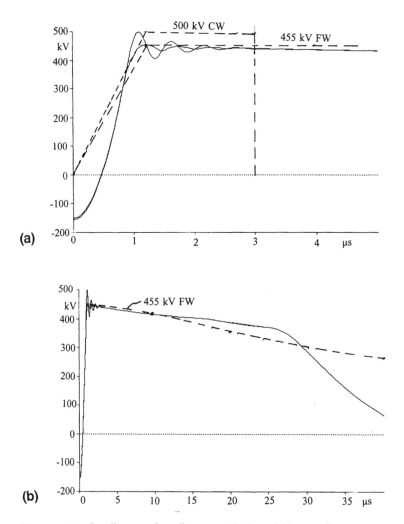

Figure 17 Small separation distances, $E_t/E_d < 1.10$, $t_T < 3\,\mu s$.

against this case, if the voltage ratio E_t/E_d is less than 1.10, the insulation strength should be set equal to the full wave test voltage, i.e., the BIL.

$E_t/E_d > 1,10$ but $t_T > 3\,\mu s$. For $L_A = 3.0$, $L_T = 15$ meters, and $C_T = 20\,\text{nF}$, Fig. 18 shows that $E_t = 579\,\text{kV}$ and $E_d = 454\,\text{kV}$ and thus $E_t/E_d = 1.28$, which meets the previous criterion to set the crest surge voltage equal to the chopped wave test voltage. However, in this case the time to crest of the surge voltage is greater than $3\,\mu s$. This creates a dilemma, since the time to crest of the surge voltage is greater than the chopping time of the test voltage, i.e., $3\,\mu s$. Thus perhaps the chopped wave test voltage should not be used for coordination. Then the only criterion remaining is that the crest surge voltage should be set equal to the full wave test voltage. This coordination is illustrated in Fig. 18. Thus for this case, BIL = (SF)E_t.

Station Lightning Insulation Coordination

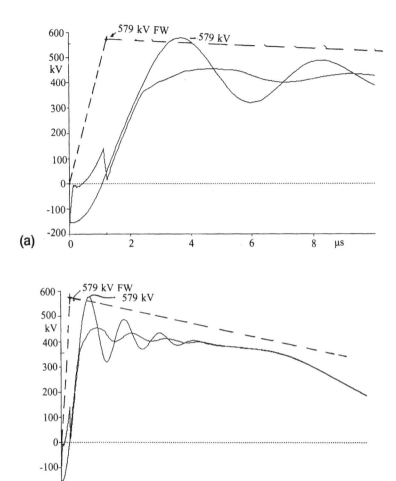

Figure 18 $E_t/E_d > 1.10$, but $t_T > 3\,\mu s$.

Long-Tail Surge, $E_t/E_d > 1.10$, $t_T < 3\,\mu s$. For $L_A = 3.0$, $L_T = 9$ meters. The previous cases assumed an incoming surge with a tail having a time constant of 14 μs. Therefore, this incoming surge was predicated on a backflash. If the incoming surge were due to a shielding failure without flashover, the tail would increase to that of the stroke current. Using the average tail for this event of 92 μs, which translates to a time constant of 133 μs, Fig. 19 shows the results for which $E_t = 711\,kV$, $E_d = 461\,kV$, and $E_t/E_d = 1.54$. Therefore the dotted curves indicating the test voltages are constructed by setting the chopped wave test voltage equal to the crest surge voltage. However, as shown by Fig. 19b, the tail of the full wave test voltage is now significantly below the surge voltage. Thus this does not appear to be a good method of coordination. However, now consider that the switching impulse test voltage SI is about 83% of the full wave test voltage. For a full wave test voltage of 647 kV per Fig. 19a, the SI test voltage would be 536 kV. The switching impulse

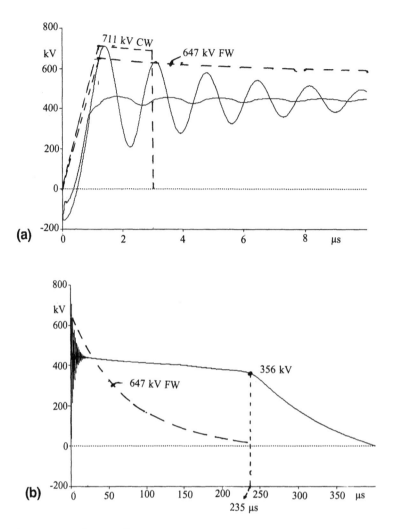

Figure 19 Long-tail surge, $E_t/E_d > 1.10$, $t_T < 3\,\mu s$.

test voltage has a front of greater than 100 μs, a time to zero voltage of greater than 1000 μs, and a time above 90% of the crest of greater than 200 μs. This SI test voltage is about 16% greater than the arrester discharge voltage, and therefore the coordination is acceptable.

Long-Tail Surge, $E_t/E_d \leq 1.10$, $t_T < 3\,\mu s$. For $L_A = L_T = 0.6$ meters, see Fig. 20, where $E_t = 500\,\text{kV}$, $E_d = 455\,\text{kV}$, and $E_t/E_d = 1.10$. As for Fig. 19, the crest surge voltage is set equal to the chopped wave voltage, which indicates a full wave test voltage of 455 kV. This full wave test voltage is compared to the surge voltage in Fig. 20b and indicates that coordination is not attained. However, as before, first check to see if the SI test voltage covers this situation. For this full wave test voltage, the SI test voltage would be 0.83(455) = 378 kV. Since this SI

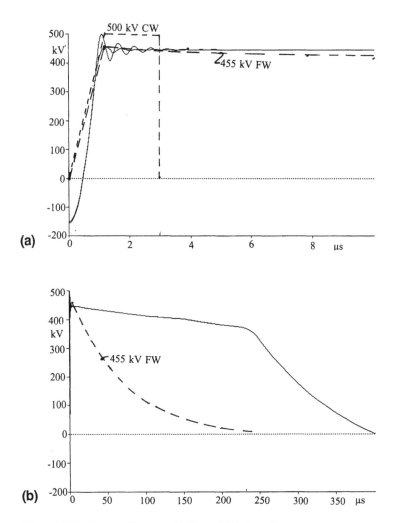

Figure 20 Long-tail surge, $E_t/E_d < 1.10$, $t_T < 3\,\mu s$.

test voltage is less than the arrester voltage, the coordination is not acceptable. To achieve a coordination, the arrester voltage should be set equal to the SI test voltage.

Long-Tail Incoming Surges. To this point, the *added* criterion is that for long-tail incoming surges, the full wave test voltage should also be checked to assure that it is equal to or greater than the SI test voltage. The question now is as to the proper definition of a long tail. From Fig. 19b, the transformer voltage does not significantly decrease until about 235 μs at a voltage of 356 kV, which is approximately the arrester discharge voltage at 10 amperes. The discharge voltage for a 10-A discharge current is equal to about 77% of the 10-kA discharge voltage or $0.77E_{10}$. The time that the arrester voltage significantly decreases, t_c, can be estimated by the equation

$$2Ee^{-\frac{t_c}{\tau}} = 0.77E_{10} + V_{PF}$$

$$t_c = -\tau \ln\left(\frac{0.77E_{10} + V_{PF}}{2E}\right) \quad (57)$$

where E is the crest voltage of the incoming surge and τ is the tail time constant. For this example, for a tail time constant of 133 μs, t_c per Eq. 57 is 240 μs, which represents a good check on the actual value of t_c from Fig. 19b.

The question remains as to when the coordination with the SI test voltage should occur. A proposed method is that if t_c is greater than 50 or 60 μs, the time to half value of the full wave test voltage, the criterion should apply. For this example, this translates into a tail time constant of 28 or 33 μs

A Suggested Criterion. The suggested criterion is as follows where both (1) and (2) should be used and the highest value of BIL accepted. The value of the time to crest of the transformer voltage, t_T, can be obtained from Eq. 19.

1. For an incoming surge with a value of t_c less than 60 μs,

$$\text{BIL} = (\text{SF})E_d \quad \text{if } t_T \leq 3.0 \,\mu\text{s and } E_t/E_d \leq 1.10 \quad (58)$$

$$\text{BIL} = (\text{SF})\frac{E_t}{1.10} \quad \text{if } t_T \leq 3.0 \,\mu\text{s and } E_t/E_d > 1.10 \quad (59)$$

$$\text{BIL} = (\text{SF})E_t \quad \text{if } t_T > 3.0 \,\mu\text{s} \quad (60)$$

2. For an incoming surge having a value of t_c greater than 60 μs or more practically for an incoming surge caused by a shielding failure without a flashover,

$$\text{BIL} = (\text{SF})E_d$$
$$\text{BIL} = \frac{\text{SF}}{0.83}E_d \quad (61)$$

where SF is the safety factor. A value of SF = 1.20 has been suggested.

The criterion concerning the value of t_c essentially translates to a backflash versus a shielding failure. To complete the example, Table 8 shows the voltages and the resultant required BILs using a safety factor of 1.20.

As noted for the cases illustrated by Figs. 19 and 20, only in the case of Fig. 19 is the required BIL affected by the criterion of the long-tail surge. In most cases, to guard against the long-tail surge and to be conservative, the criterion of the long-tail surge per Eq. 61 is universally applied. Besides, it is simpler that way.

A New Transformer Insulation Strength Curve

In 1996, Balma et al. presented a paper [14] in which they suggested that the transformer insulation strength could be represented by a continuous curve. After analysis of the author's curve, the curve was modified and is shown in Fig. 21. In the lightning impulse region, that is, for times between the front of wave test, 0.5 μs, and

Station Lightning Insulation Coordination

Table 8 Results of Examples

Fig. No.	L_A, m	L_T, m	E_t, kV	E_d, kV	E_t/E_d	Tail time constant, μs	t_T, μs	Req'd. BIL (Eq #)	Long tail Req'd. BIL, Eq. 61
16	3	9	709	460	1.54	14	< 3	773 (59)	(665)
17	0.6	0.6	500	454	1.10	14	< 3	545 (58)	(656)
18	3	15	579	454	1.28	14	> 3	695 (60)	(656)
19	3	9	711	461	1.54	133	< 3	776 (59)	667
20	0.6	0.6	500	455	1.10	133	< 3	546 (58)	657

the BIL, 8 μs, the curve is a basic time-lag curve, i.e., the time being the time to breakdown or failure. In the switching impulse and power frequency region, i.e., from times between the switching impulse test, 300 μs, and the induced tests, the times are the time above 90% of crest voltage. The curve is constructed through the following test points.

1. A front of wave (FOW) test of 1.3 to 1.5 BIL at a time of 0.5 μs. This is not a standard test but a test to be specified by the purchaser and agreed upon by the manufacturer.

2. A chopped wave test at 1.10 BIL at a time of 3 μs, a standard test.

3. A full wave test voltage, the BIL plotted at 8 μs, a standard test.

4. A switching impulse test, the BSL equal to 0.83 times the BIL plotted at 300 μs, a standard test.

5. A 1 hour test voltage equal to 1.5 times the maximum line to ground system voltage plotted at 1034 seconds, a standard test. In Fig. 21 this is plotted for a 242-kV maximum system voltage assuming a 750-kV BIL.

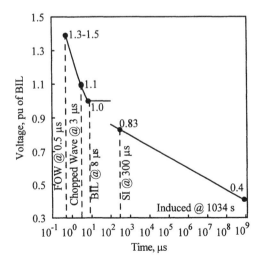

Figure 21 Transformer insulation strength. (From Ref. 14.)

This curve is drawn recognizing alternate modes of stress or failure. In the lightning region, the front of wave and chopped wave tests primarily stress the turn insulation, while the BIL stresses both the ground-wall and turn insulation. In the switching impulse and power frequency region, the tests primarily stress the ground-wall insulation. As noted, the region between the BIL and the BSL is not defined, for it may stress the turn or ground-wall insulation. The curve in the lightning impulse region is unchanged from previous representations [17] except that now the curve is continuous. In the switching impulse and power frequency region, the continuous curve vs. the logarithm of time is new and can be used when examining TOVs. In conclusion, the new curve does not alter the previous presentation in this chapter or the suggested an alternate application criteria. The insulation coordination for the transformer remains the same

Transformer Bushing

The transformer bushing is a special case in that it contains both internal and external insulations. Tests on the bushing as a separate apparatus include a 3-μs chopped wave test at 1.15 times the BIL. Of course, when installed in the transformer, this chopped wave test is decreased to 1.10 times the BIL. Therefore, conservatively, the internal insulation is treated as for the transformer, i.e., the chopped wave test is assumed equal to 1.10 times the BIL. However for the external insulation, the bushing should be treated as the other external or self-restoring insulations and the chopped wave level set at 1.15 time the BIL. As is evident, this dual treatment of the bushing insulations may result in different BILs for the external and internal insulations. If the BIL of the external is lower, then the BIL of the external should be set equal to the BIL of the internal. However, if the opposite is true, i.e., the BIL of the external is greater than the BIL of the internal, then the two different BILs should be accepted. This phenomenon may occur for stations at high altitudes, since the external insulation strength is degraded.

If the BIL of the external is larger that of the internal insulations, the question of testing arises, since the BIL of the external insulation cannot be tested when installed in the transformer. This simply means that the tests on the bushing shell need to be accepted.

Phase–Ground and Phase–Phase Air Clearances

One additional area needs discussion, that of air clearances. The lightning impulse strength of air gaps varies with the type of gap configuration. For positive polarity, the CFO varies from a low of 540 kV/m for a rod–plane gap to about 650 kV/m. The negative polarity strength varies from a low of 540 kV/m to a high of 750 kV/m, this latter value being for a rod–plane gap. Since negative polarity surges are predominant, the suggested value is the same as used previously in Chapters 2 and 3, i.e., 605 kV/m. The 3-μs strength is about 1.38 times the CFO or 835 kV/m. Therefore a maximum value of 835 kV/m can be used. However, for conservatism, the gradient of 605 kV/m is suggested.

Some standards recommend tying the clearance to the bus support insulator BIL, obtaining a clearance by dividing the bus support insulator BIL by a minimum positive polarity gradient of about 500 kV/m. Two margins are obtained by this method. First, the BIL used in this method is the actual BIL used in the station

Station Lightning Insulation Coordination

instead of the required or minimum BIL, and second the 500 kV/m applies to positive polarity. This procedure is not recommended.

Considering phase–phase air clearances, for a line flashover to a single phase, a coupled voltage of the same polarity appears on the other phases. Thus at the struck point, the phase–phase voltage is less than the phase–ground voltage. If these voltages travel a considerable distance in towards the station, a ground mode propagation effect occurs such that the phase–phase voltage is increased. However, because the distances to the struck point are small, the phase–phase voltage seldom exceeds that to ground. Therefore, the phase–ground clearance is considered also to be the phase–phase clearance.

4 STANDARD BILS

In the process of insulation coordination, the standard and available BILs are needed. Tables 9 and 10, produced from Chapter 1, provide a list of the standard BILs from the IEEE standard [18] and from IEC 71-1 [19]. In IEEE, these values are suggested values for use by other equipment standards. In other words, equipment standards may use these values or any others that they deem necessary. However, in general, these values are used. There are exceptions. For any specific type of equipment or type of insulation, there does exist a connection between the BIL and the BSL. For example, for transformers, the BSL is approximately 83% of the BIL. Thus given a standard value of BIL, the BSL may not be a value given in Table 9.

The available BILs and BSLs for a given system voltage are provided in Tables 1 and 2 of this chapter for IEEE and in Tables 7 and 8 of Chapter 1 for IEC [19].

5 APPLICATION OF SIMPLIFIED METHOD

As for all insulation coordination problems, the application of the simplified method consists of comparing the stress to the strength. To illustrate the procedure two situations will be considered. In both cases, the calculations will be performed using transformer surge capacitances of 2 and 4 nF.

Table 9 Standard Values of BIL and BSL per IEEE 1313-1

30	300	825	1925
45	350	900	1050
60	400	975	2175
75	450	1050	2300
95	500	1175	2425
110	550	1300	2550
125	600	1425	2675
150	650	1550	2800
200	700	1675	2925
250	750	1800	3050

Source: Ref. 18

Table 10 Standard Value of BIL and BSL per IEC 71.1

20	325	1300	2550
40	450	1425	2700
60	550	1550	2900
75	650	1675	
95	750	1800	
125	850	1950	
145	950	2100	
170	1050	2250	
250	1175	2400	

Source: Ref. 19.

5.1 Single-Line Station

The Incoming Surge. The 230-kV single-line station of Fig. 22 is to be designed for an MTBF of 100 years. The BFR of the line is 2.0 flashovers/100 km-years, and the span length is 300 meters. Therefore per Chapter 12,

$$d_m = \frac{1}{(2/100)(100)} = 0.5 \, \text{km} \tag{62}$$

Since the span length is 300 m, this distance is increased to 600 meters and the steepness of the incoming surge becomes

$$S = \frac{700}{0.6} = 1167 \, \text{kV}/\mu s \tag{63}$$

The CFO of the line insulation is 1300 kV, and a very conservative estimate of the crest voltage of the incoming surge is 1.2 times the CFO of the line or 1560 kV. Assume that this surge is riding atop an opposite polarity power frequency voltage of 130 kV, i.e., $V_{PF} = 130$ kV.

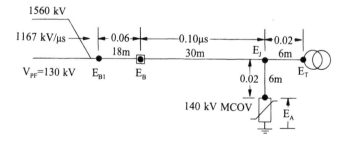

Figure 22 Single-line, 230-kV station, distances in meters, travel times in μs.

Station Lightning Insulation Coordination

Arrester Current and Voltage. The 140 kV MCOV arrester selected has a 0.5 μs discharge voltage of 446 kV at 10 kA. The 8/20-μs discharge voltage at 10 kA is 404 kV, so that a multiplying factor of 1.10 is applied to all 8/20-μs discharge voltages. Therefore the discharge voltage at 5 and 10 kA is 418 and 446 kV, respectively, giving an R_A of 5.6 ohms and an E_0 of 390 kV. for a $Z = 450$ ohms, the arrester current is

$$I_A = 1.6 \frac{2(1560) - 390 - 130}{450 + 5.6} = 9.13 \, \text{kA} \tag{64}$$

and the arrester discharge voltage is

$$E_d = 390 + (9.13)(5.6) = 441 \, \text{kV} \tag{65}$$

and therefore

$$E_A = 441 + 130 = 571 \, \text{kV} \tag{66}$$

Transformer. The surge voltage to ground at the transformer is:

$$t_T = \pi\sqrt{0.04(0.9 + 0.04)} + \frac{571}{1167} = 1.10 \quad \text{for } C_T = 2 \, \text{nF}$$

$$t_T = \pi\sqrt{0.04(1.8 + 0.04)} + \frac{571}{1167} = 1.34 \quad \text{for } C_T = 4 \, \text{nF}$$

$$K_1 = \frac{(1167)(0.04)}{571} = 0.08175 \tag{67}$$

$$\frac{E_T}{E_A} = 1 + \frac{1}{1 + 0.14/0.08175} = 1.3687$$

$$E_T = 781 \, \text{kV} \quad E_t = 781 - 130 = 652 \, \text{kV}$$

Arrester–Bus Junction. The surge and the voltage to ground at the arrester–bus junction are

$$K_1 = \frac{(1167)(0.02)}{571} = 0.04088$$

$$\frac{E_J}{E_A} = 1 + \frac{1}{1 + 0.14/0.04088} = 1.226 \tag{68}$$

$$E_J = 700 \, \text{kV} \quad E_j = 570 \, \text{kV}$$

Circuit Breaker for $C_T = 2\,\text{nF}$.

$$t_f - 2(T_B + t_A) = 1.337 - 2(0.12) = 1.097$$

$$\tau = 0.9\,\mu s \qquad \frac{nE_A}{2S\tau} = \frac{571}{2(1167)(0.9)} = 0.2718 \qquad (69)$$

$$t_A = 0.7563$$

Since t_A is less than $t_f - 2(T_B + T_A)$

$$E_B = 700 + 2(1167)(0.10) = 933\,\text{kV} \qquad E_b = 933 - 130 = 803\,\text{kV} \qquad (70)$$

Circuit Breaker for $C_T = 4\,\text{nF}$.

$$\tau = 1.8\,\mu s \qquad \frac{nE_A}{2S\tau} = \frac{571}{2(1167)(1.8)} = 0.1359 \qquad t_A = 1.0278 \qquad (71)$$

Since t_A is less than $t_f - 2(T_B + T_A)$

$$E_B = 700 + 2(1167)(0.10) = 933\,\text{kV} \qquad E_b = 933 - 130 = 803\,\text{kV} \qquad (72)$$

Station Entrance, $C_T = 2\,\text{nF}$. Let T_{B1} equal the travel time between the arrester-bus junction and the station entrance, i.e., $T_{B1} = 0.16\,\mu s$. Then

$$t_f - 2(T_{B1} + T_A) = 1.337 - 2(0.18) = 0.977$$

$$\tau = 0.9\,\mu s \qquad \frac{nE_A}{2S\tau} = \frac{571}{2(1167)(0.9)} = 0.2718 \qquad t_A = 0.756 \qquad (73)$$

since t_A is less than $t_f - 2(T_b + T_A)$,

$$E_{B1} = 700 + 2(1167)(0.16) = 1073\,\text{kV} \qquad E_{B1} = 1073 - 130 = 943\,\text{kV} \qquad (74)$$

Station Entrance, $C_T = 4\,\text{nF}$.

$$\tau = 1.8\,\mu s \qquad \frac{nE_A}{2S\tau} = \frac{571}{2(1167)(1.8)} = 0.1359 \qquad t_A = 1.023 \qquad (75)$$

Since t_A is greater than $t_f - 2(T_{B1} + T_A)$

$$E_{B1} = 2(1560) - 2(1167)(0.18) - 2(1167)(1.8)\left[1 - e^{-\frac{0.977}{1.8}}\right] = 940\,\text{kV}$$

$$E_{b1} = 940 - 130 = 810\,\text{kV} \qquad (76)$$

To determine the accuracy of these simplified calculations, the ATP was used with the added assumption that the tail time constant of the incoming surge is $14\,\mu s$. A comparison of the results is presented in Table 11. All voltages calculated using the simplified equations are higher than the crest voltages as found from the ATP, i.e., from 1 to 25%. The calculated transformer voltages are only 3 to 6% higher than

Table 11 Comparison of Results for Single-Line Station

	Simplified calculation		ATP	
Voltage	$C_T = 2\,\text{nF}$	$C_T = 4\,\text{nf}$	$C_T = 2\,\text{nF}$	$C_T = 4\,\text{nF}$
E_t	652	652	613	632
E_b	803	803	754	641
E_{b1}	943	810	846	691
E_j	570	570	539	524
E_d	441	441	432	437
I_A	9.1	9.1	7.5	8.3

that from the ATP. The voltages from the ATP are shown in Fig. 23 and illustrate the spike of initial voltage and the decay to the arrester voltage. Note also the oscillatory nature of the transformer voltage.

The process of selecting the BILs is shown in Table 12. The required BILs are first determined, which usually are nonstandard BILs. Next, the next highest standard BILs are selected from Tables 9 or 10. The selected BILs are BILs that exist for the equipment at the system voltage per Tables 1 and 2. The process is detailed below.

Transformer

$$\text{BIL} = 1.20 \frac{652}{1.10} = 711\,\text{kV} \tag{77}$$

From Table 9, the next highest standard BIL is 750 kV, and from Table 1, a transformer BIL of 750 kV is obtainable and therefore is selected.

Transformer Bushing. Both the internal and the external insulation must be considered in evaluating the transformer bushing. The internal bushing is treated in the identical manner as for the transformer. However, when selecting the external bushing BIL, consideration should be given to the altitude of the station, since higher altitude will degrade the BIL. In this case of the single-line station, the assumption is

Table 12 Selection of BILs for Single-Line Station ($C_T = 2\,\text{nF}$)

Equipment	Voltage	Crest, kV	Req'd. BIL, kV	Std. BIL, kV	Selected BIL, kV
Transformer	E_t	652	711	750	750
Transf. bushing,	E_t				
internal		652	711	750	750
external		652	566	600	750
Breaker	E_b	803	698	700	900
Disc switch	E_b	803	698	700	900
Bus insulators	All	943	820	825	900

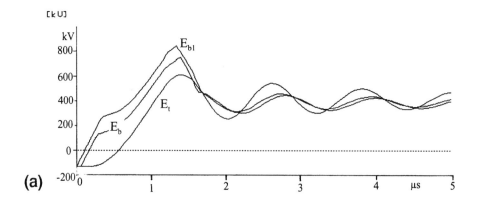

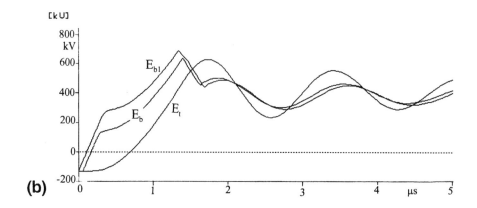

Figure 23 Voltages for single-line station. (a) $C_T = 2\,\text{nF}$. (b) $C_T = 4\,\text{nF}$.

made that the station is at sea level. The required BIL for the external porcelain is therefore

$$\text{BIL} = \frac{652}{1.15} = 566\,\text{kV} \tag{78}$$

The next standard BIL from Table 9 is 600 kV. From Table 1, the minimum available BIL is 650 kV. However, the BIL of the external insulation should be equal to or greater than that of the internal BIL. Therefore the BIL for the external and internal insulation is selected as 750 kV.

If a station is located at a high altitude, the BIL of the external porcelain could possess a higher BIL than the internal insulation. For example, if a station is at an altitude of 1600 meters, the relative air density is

$$\delta = e^{-A/8.6} = 0.830 \tag{79}$$

Station Lightning Insulation Coordination

and the required BIL is then

$$\text{BIL} = \frac{566}{0.830} = 681 \text{ kV} \tag{80}$$

and the standard and selected BIL would be 750 kV.

Circuit Breaker. Since there is only one BIL for the circuit breaker, the selection process is somewhat inconsequential. However, there may be cases where a higher BIL than the standard breaker BIL is required. Since a higher BIL cannot be obtained, the only remedial measure is to decrease the stress at the breaker terminals. The process of finding the required BIL is

$$\text{BIL} = \frac{803}{1.15} = 698 \text{ kV} \tag{81}$$

The next standard BIL is 750 kV, and the selected BIL is the breaker standard BIL (Table 2) of 900 kV. If the station were at 1600 meters, the required BIL would be 841 kV and the breaker standard BIL of 900 kV is still applicable.

Disconnecting Switches. In this sample problem, the disconnecting switches are assumed to be located at the breaker and therefore the BILs are the same as for the breaker.

Bus Support Insulators. The bus support insulators are located through the station, and therefore the surge voltage selected is the maximum found through the station, a value of 964 kV. The required BIL is

$$\text{BIL} = \frac{943}{1.15} = 820 \text{ kV} \tag{82}$$

From Table 9, the next highest BIL is 825 kV, and from Table 2, a 900 kV BIL is selected.

Air Clearances. The highest voltage in the station is 943 kV and therefore the required phase–ground and phase–phase clearance S is

$$S = \frac{943}{605} = 1.56 \text{ meters} \tag{83}$$

5.2 A Two-Line Station

To demonstrate the use of the other equations and to demonstrate consideration of contingency conditions, the two-line station of Fig. 24 is considered. The station is identical to that of Fig. 23 except an additional line has been added.

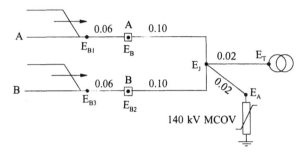

Figure 24 A 230-kV, two-line station.

All Lines in Service

Incoming Surge, for the Transformer and Equipment on Transformer Bus. For the transformer, the MTBS is twice the MTBF or 200 years. As before, the BFR is 2/100 km-years and the span length is 300 m. Therefore

$$d_m = \frac{1}{(2/100)(200)} = 0.25 \text{ km} \tag{84}$$

This distance is increased to one span length so that $d_m = 0.3$ km. Thus

$$S = \frac{700}{0.3} = 2333 \text{ kv/µs} \tag{85}$$

As before, the crest of the incoming surge is conservatively assumed as 1560 kV.

Incoming Surge for Other Equipment Not on Transformer Bus. These equipment BILs are evaluated using a 100-year surge, and thus the steepness S remains at 1167 kV/µs and the crest voltage is 1560 kV.

Arrester Current and Voltage. Using the arrester characteristics as before,

$$I_A = 1.6 \frac{1560 - 390 - 130}{225 + 5.6} = 7.22 \text{ kA} \tag{86}$$

Since the current is between 5 and 10 kA, the calculation is acceptable. The arrester voltages are

$$E_d = 390 + (5.6)(7.22) = 430 \text{ kV} \qquad E_A = 430 + 130 = 560 \text{ kV} \tag{87}$$

Station Lightning Insulation Coordination

Transformer.

$$t_T = \pi\sqrt{0.04(0.9 + 0.04)} + \frac{560}{2333} = 0.849 \quad \text{for } C_T = 2\,\text{nF}$$
$$t_T = \pi\sqrt{0.04(1.8 + 0.04)} + \frac{560}{2333} = 1.092 \quad \text{for } C_T = 4\,\text{nF} \tag{88}$$

$$K_1 = \frac{(2333)(0.04)}{560} = 0.1664 \tag{89}$$

$$\frac{E_T}{E_A} = 1 + \frac{0.98}{1 + 0.16/0.1664} = 1.500$$
$$E_T = 840\,\text{kV} \quad E_t = 840 - 130 = 710\,\text{kV} \tag{90}$$

Arrester–Bus Junction for the Transformer Evaluation.

$$K_1 = \frac{(2333)(0.02)}{560} = 0.0833$$
$$\frac{E_J}{E_A} = 1 + \frac{0.98}{1 + 0.16/0.0833} = 1.336 \tag{91}$$
$$E_J = 748\,\text{kV} \quad E_j = 748 - 130 = 618\,\text{kV}$$

Arrester–Bus Junction for Other Equipment.

$$K_1 = \frac{(1167)(0.02)}{560} = 0.0417$$
$$\frac{E_J}{E_A} = 1 + \frac{0.98}{1 + 0.16/0.0417} = 1.203 \tag{92}$$
$$E_J = 673\,\text{kV} \quad E_j = 673 - 130 = 543\,\text{kV}$$

Circuit Breaker, $C_T = 2\,\text{nF}$

$$t_f - 2(T_B + T_A) = 1.337 - 0.24 = 1.097 \tag{93}$$

$$\tau = 0.45\,\mu\text{s} \quad \frac{nE_A}{2S\tau} = \frac{2(560)}{2(1167)(0.45)} = 1.066$$
$$t_A = 0.864 \quad t_A < t_f - 2(T_B + T_A) \tag{94}$$

$$E_B = E_J + 2ST_B = 673 + 2(1167)(0.10) = 906 \text{ kV}$$
$$E_b = 906 - 130 = 776 \text{ kV} \tag{95}$$

Circuit Breaker, $C_T = 4 \text{ nF}$.

$$t_f - 2(T_B - T_A) = 0.977 \qquad \tau = 0.9 \,\mu\text{s} \qquad \frac{nE_A}{2S\tau} = \frac{2(560)}{2(1167)(0.9)} = 0.533$$
$$t_A = 1.12 \qquad t_A > [t_f - 2(T_B + T_A)] \tag{96}$$
$$E_B = 1560 - (1167)(0.9)\left(1 - e^{-\frac{1.097}{0.9}}\right) = 820 \text{ kV} \qquad E_b = 690 \text{ kV}$$

Station Entrance, $C_T = 2 \text{ nF}$

$$\tau = 0.45 \,\mu\text{s} \qquad [t_f - 2(T_{B1} + T_a)] = 1.337 - 0.36 = 0.977$$
$$\frac{nE_A}{2S\tau} = \frac{2(560)}{2(1167)(0.45)} = 1.066 \qquad t_A = 0.864 \qquad t_A < [t_f - 2(T_{B1} + T_A)] \tag{97}$$

$$E_{B1} = 673 + 2(1167)(0.16) = 1046 \text{ kV} \qquad E_{b1} = 916 \text{ kV} \tag{98}$$

Station Entrance, $C_T = 4 \text{ nF}$

$$\tau = 0.9 \,\mu\text{s} \qquad \frac{nE_A}{2S\tau} = \frac{2(560)}{2(167)(0.9)} = 0.533 \qquad t_A = 1.12 \qquad t_A > [t_f - 2(T_{B1} + T_A)] \tag{99}$$

$$E_{B1} = 1560 - (1167)(0.9)\left(1 - e^{-\frac{0.977}{0.9}}\right) = 864 \text{ kV} \qquad E_{b1} = 734 \text{ kV} \tag{100}$$

Voltages E_{B2} and E_{B3}. The voltages E_{B2} and E_{B3} should be equal to E_J, since the line B does not have any discontinuities.

Comparison with ATP. The voltages calculated by the simplified method and those obtained by use of the ATP are compared in Table 13. All calculated voltages are greater than those from ATP by from 1 to 25%. The calculated transformer voltage is 10 to 21% greater than obtained from the ATP.

Comparison with Single-Line Case. Comparing the results for the two-line case, Table 13, to those for the single-line case, Table 11, shows that the transformer voltage for the two-line case (both the calculated and the ATP results) increased over that for the single-line case by from 4 to 10%. However, except for E_{b1} for

Table 13 Results for Two-Line Station, Surge on Line A

	Simplified calculations		ATP	
Voltage location	$C_T = 2\,nF$	$C_T = 4\,nF$	$C_T = 2\,nF$	$C_T = 4\,nF$
200-year surge				
E_t	710	710	642	586
E_j	618	618	524	496
E_d	430	430	422	425
I_A	7.2	7.2	5.8	6.4
100-year surge				
E_b	776	690	683	560
E_{b1}	916	734	767	627
E_j	543	543	505	489
E_{b2}	543	543	505	489
E_{b3}	543	543	505	489

$C_T = 4\,nF$, the opposite effect occurred for the other voltages. These voltages for the two-line case are about 3% less than those for the single-line case.

Selection of BILs. The selection of the BIL as detailed in Table 14 employs the same methodology as for the single-line case and therefore is not repeated here. Note that in comparison to the single line, the BIL of the transformer has been increased. However, the required and standard BIL of the other equipment has decreased. Since the standard breaker BIL is 900 kV, this value is selected. Similarly, the standard BIL for the bus support insulators is 900 kV, so the selected BIL does not change. Plots of the voltages as obtained from the ATP are presented in Fig. 25.

Clearances. The phase–ground and the phase–phase clearance required is $916/605 = 151$ meters.

Reevaluation of Use of MTBS of 100 years. The BILs of equipment not on the transformer bus was made using a 100 year surge. For this case, the voltage at the disconnecting switch, E_b, was 776 kV. If the surge were placed on line B, the voltage would have been 543 kV. Thus voltages of 776 kV and 543 kV appear once in 100 years, and the voltage of 776 kV is used to determine the BIL. In a similar manner,

Table 14 Selection of BILs for Two-Line Station

Equipment	Voltage	Crest, kV	Req'd BIL, kV	Std. BIL, kV	Selected BIL, kV
Transformer	E_t	710	774	825	825
Transf. bushing,	E_t				
internal		710	774	825	825
external		710	617	650	825
Breaker	E_b	776	675	700	900
Disc switch	E_b	776	675	700	900
Bus insulators	All	916	796	825	900

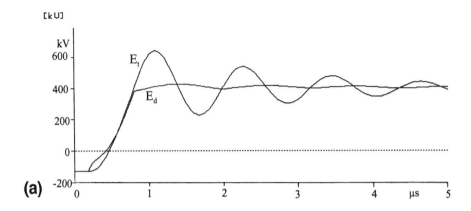

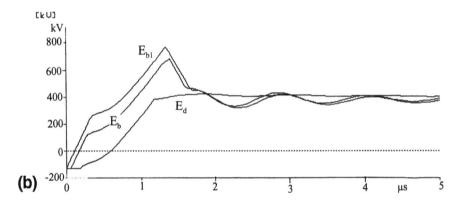

Figure 25 Voltages in two-line station for $C_T = 2\,\text{nF}$ (a) at transformer; (b) at breaker and station entrance; arrester voltage also shown.

the voltages E_{b1} at the disconnecting switch are 916 kV and 543 kV with the BIL based on 916 kV. Thus the use of the 100-year surge is justified.

Contingency Conditions

Now consider the contingency that one line is opened, that is, the disconnects on each side of breaker B are opened as shown in Fig. 26. Assume that the probability of all lines being in service during a thunderstorm is 75% and therefore the probability of only one line being in service is 25%. Thus to maintain the 100-year MTBF, the return period of the surge should be 100 times 0.25 or 25 years. However, with equal probability line A or line B could be out of service. Thus the surge could arrive on line A or on line B. Therefore the transformer BIL should be evaluated on an MTBS of 50 years.

A 50-year surge has a d_m of 1.2 km and therefore a steepness of 483 kV/μs. A 25 year surge has a d_m of 2.1 km and therefore a steepness of 33 kV/μs. Since this is now a single-line station, assuming the crest of the incoming surge remains at

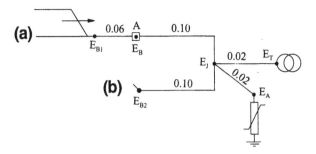

Figure 26 One line out in a two-line station.

1560 kV, the arrester current and voltage are the same as for the single-line case, i.e., $E_d = 441$ kV and $E_A = 571$ kV.

The surge having a steepness of 483 kV produces a transformer voltage E_t of 570 kV, which is significantly less than the voltage of 710 kV for all lines in service.

To evaluate the other equipment BILs, first apply a 25-year surge to Line A ($S = 333$ kV/μs). Since $[t_f(T_B + T_A)] > t_A$, the voltage at J must be determined first.

$$K_1 = \frac{0.02(333)}{471} = 0.01166$$

$$\frac{E_J}{E_A} = \frac{1}{1 + 0.14/0.01166} = 1.077 \tag{101}$$

$$E_J = 615 \text{ kV} \qquad E_j = 485 \text{ kV}$$

Then the voltage E_B is

$$E_B = 615 + 2(333)(0.10) = 682 \text{ kV} \qquad E_b = 552 \text{ kV} \tag{102}$$

The voltage E_{B2} is

$$E_{B2} = 0.91(571 + 2(333)(0.12)) = 592 \text{ kV} \qquad E_{b2} = 462 \text{ kV} \tag{103}$$

Since with equal likelihood, the disconnect switch at A could be opened, now apply the surge to line B with the disconnects at A opened. Therefore the voltage E_b is 462 kV and the voltage E_{b2} is 552 kV. Thus once in 25 years, the two voltages, 552 and 462 kV, appear at the disconnecting switches. Therefore the BIL should be based on a voltage of 552 kV. The required BIL is 480 kV; the standard BIL is only 500 kV. Thus the required standard BIL is much less than for two lines in service. Note that for the case of all lines in service, the calculated voltage E_{b2} for a surge on Line B is 916 kV, about 66% greater than for this contingency case.

Thus the conclusion for this example is that the case of all lines in service is the most critical case and dictates the required BILs. This may not always be true. It depends on the assumed probability of the contingency, which in turn produces the steepness of the incoming surge.

If the voltage at breaker A with B opened and the voltage at B with A opened are equal, the calculation should be repeated with an MTBS of 50 years. For example, assume that these voltages are both equal to 700 kV. Thus two surges of 700 kV occur once in 25 years, or one voltage of 700 kV occurs once in 12.5 years. Therefore the incoming surge should be based on a MTBS of 50 years and the calculations repeated.

5.3 A Nonsymmetrical Station Layout

A 115-kV station [17] of Fig. 27 uses a 84-kV MCOV arrester having a 10-kA discharge voltage of 273 kV located at the end of the bus. In contrast to the previous symmetrical station layouts, incoming surges on lines A, B, and C will result in different voltages at the transformers and circuit breakers (see presentation in Chapter 11, Fig. 2). In addition, since TR2 is more distant from the arrester, the voltage at this transformer should be greater than at TR1. Assuming a MTBF of 100 years, using the ATP, a 100-year surge having a crest voltage of 1080 kV and a steepness of 1000 kV/µs, is applied to each line. Let $V_{PF} = 65$ kV, $Z = 450$ ohms. The crest voltages at TR2 for surges on lines A, B, and C are 394, 417, and 454 kV, respectively for a 4-nF transformer capacitance. Thus, per Chapter 11, a 100-year surge should be used to determine both the transformer and circuit breaker voltages.

A simple method to calculate the voltages applies to both symmetrical and nonsymmetrical layouts and is used in the IEC method [16]. The method consists of selecting the distance as the *maximum distance from the equipment to the closest arrester*. For the transformer, TR2, the distance is 36 meters which includes the arrester ground lead, for the circuit breaker, 39 meters. The calculated voltages at the transformer are within 3% of those using the ATP as shown in Table 15. However, the calculated voltages at the circuit breaker are conservative, exceeding those using the ATP by from 6 to 30%.

Other methods to estimate the transformer and circuit breaker voltages include the IEEE method [17,20,21] and the IEC method [16]. Table 15 presents the results using these methods. As noted the IEEE method overestimates while the IEC method underestimates the voltages.

In general, for nonsymmetrical station layouts, if the voltages at the transformer when the surges are applied to alternate lines are within about 2%, assume a symmetrical layout, i.e. use an incoming surge based on n lines. Otherwise use d_m with $n = 1$.

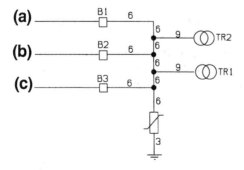

Figure 27 115-kV station, distances in meters.

Table 15 Voltages in Station of Fig. 27

	ATP	Per Chapter 13	IEEE	IEC
at TR2				
$C_T = 2\,nF$	475	461	556	353
$C_T = 4\,nF$	454	461	556	353
at B1				
$C_T = 2\,nF$	431	567		360
$C_T = 4\,nF$	411	444		360

6 GAS-INSULATED STATIONS

Gas-insulated stations (GIS) are used primarily where space is limited, e.g., an urban area, or where adverse atmospheric conditions make their application advantageous, for example, at high altitude or in a highly contaminated area. In the past, some problems in reliability existed. At 500 kV, air-insulated stations were estimated to be from two to three times more reliable. Because of reliability problems at this voltage level, some utilities have constructed a line with an opened disconnecting switch around the station. However, today, for all system voltages, far more reliability exists, and many gas-installations exist throughout the world [22].

Two principal styles exist: a total GIS, in which the transformers are throat-connected to the GIS bus and a partial GIS in which the transformers are connected by an opened bus to an air–gas bushing into the GIS bus. In both cases, the GIS is treated as a single piece of apparatus, and an arrester is located at each line entrance. Normally this arrester is the same type of arrester as is used in the air-insulated station, but in-gas arresters are available. These later type arresters are connected directly to the GIS bus. Their advantage is that the lead length to the arrester and any separation distance between the arrester and the entrance to the GIS is eliminated, thus providing superior protection. Their disadvantage is cost, which ranges from four to five times the cost of a normal arrester. From a technical point of view, these arresters are seldom if ever needed, so that their usage is limited.

The Stress, Lightning. To estimate the surge voltage within the GIS, consider Fig. 28, which shows an incoming line connected to a GIS bus that is open at the end. This open end of the GIS represents an open disconnecting switch. The coefficients of Fig. 28 are

$$\gamma = \frac{2Z_C}{Z + Z_C} \quad \omega = \frac{Z - Z_C}{Z + Z_C} \quad \beta = \frac{2Z}{Z + Z_C} \quad \alpha = \frac{Z - Z_C}{Z + Z_C} \quad (104)$$

First assume that the incoming surge can be represented by a linear rising front that is unlimited in magnitude, i.e., a steepness S. Figure 29a illustrates the development of the voltages at the arrester and the open end, where T is the travel time of the GIS

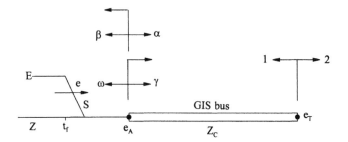

Figure 28 Incoming line connected to GIS.

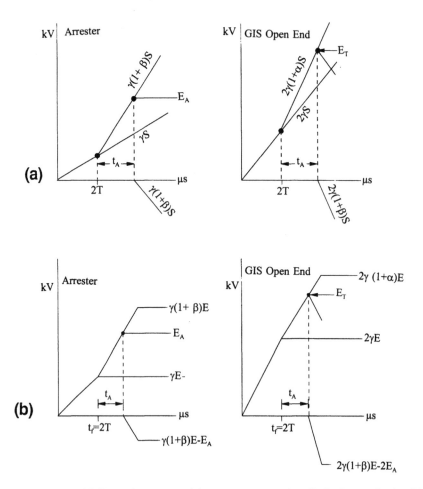

Figure 29 (a) Incoming surge with steepness S and unlimited magnitude; (b) maximum voltage for incoming surge magnitude E.

bus. A constant-voltage arrester is assumed. From this diagram, the arrester voltage E_A is

$$E_A = 2T\gamma S + \gamma(1+\beta)St_A \tag{105}$$

The voltage at the open end of the GIS, E_T, is

$$E_T = 4\gamma ST + 2\gamma(1+\alpha)St_A \tag{106}$$

Combining these equations,

$$\frac{E_T}{E_A} = \frac{2\beta}{1+\beta}\left[1 + \frac{2\gamma ST}{\beta\ E_A}\right] \tag{107}$$

The maximum value of E_T occurs when the arrester voltage E_A is reached before a reflection occurs from the end of the GIS bus. Thus the maximum value is $2E_A$, which per Eq. 107 occurs when

$$\frac{\gamma ST}{E_A} = 0.5 \tag{108}$$

To develop the maximum voltage at the open end for an incoming surge of magnitude E, let the time to crest of this voltage be t_f and let $t_f = 2T$. Then from Fig. 29b,

$$E_A = \gamma E\left[1 + \beta\frac{t_A}{t_f}\right] \tag{109}$$

and

$$E_{\text{Tmax}} = 2\gamma E\left[1 + \alpha\frac{t_A}{t_f}\right] \tag{110}$$

Combining these equations, we find

$$E_{\text{Tmax}} = \frac{2}{\beta}[\alpha E_A + \gamma E] \tag{111}$$

The results of these two equations are shown in Fig. 30, from which the voltage in the GIS can be estimated. The two equations that best approximate the results are given in this figure. The horizontal lines provide the maximum voltages for $Z = 450$ ohms and $Z_C = 60$ ohms.

As an example, consider a 230-kV system. Assume a 140-kV MCOV arrester having an E_d of 446 kV. Also let $E = 1400\,\text{kV}$, $S = 2000\,\text{kV}/\mu\text{s}$, $Z = 450$ ohms, $Z_C = 60$ ohms, and $V_{\text{PF}} = 138\,\text{kV}$. Then $E_A = 584\,\text{kV}$, $\gamma = 0.235$, $\beta = 1.765$, and $\alpha = 0.765$. Also let the length of GIS bus be 12 m or $T = 0.04\,\mu\text{s}$. Then $K_1 = 0.03219$ and $E_T/E_A = 1.1934$. Therefore $E_T = 697\,\text{kV}$ and $E_t = 559\,\text{kV}$. Checking for the maximum voltage, $E_{\text{Tmax}} = 879\,\text{kV}$ and $E_{\text{tmax}} = 741\,\text{kV}$.

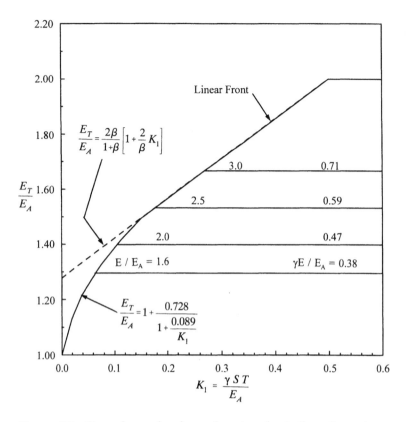

Figure 30 General curved and equations to estimate the voltage at open end of GIS bus.

The voltage at the open end is shown in Fig. 31 for a GIS bus length of 6 m or $T = 0.02\,\mu s$, $E_A = 620\,kV$, and $S = 5000\,kV/\mu s$. Also $Z = 450$ ohms and $Z_C = 60$ ohms. The resultant voltage at the open end, E_T, is 752 kV. After one reflection the voltages at the open end and at the arrester are equal until the arrester operates. They appear to follow the equation developed in Section 2.2.1. That is, if the GIS is treated as a lumped capacitor C, then the equation for the voltages until the arrester operates is

$$e_A = e_T = 2S[t + ZCe^{-t/ZC}] \tag{112}$$

where C is the total capacitance of the GIS bus and Z is the line surge impedance. Therefore $C = 333\,pF$ and $ZC = 0.15\,\mu s$. To determine the time at which the arrester operates, set $t = t_A$ in the above equation and solve for t_A. For the data used in Fig. 31, $t_A = 0.16\,\mu s$. Next, to determine E_T, note that E_T is achieved at a time that is one GIS travel time beyond t_A. Therefore to estimate E_T, solve the above equation for $t_A + T = 0.18\,\mu s$, which gives a value of 752 kV, the same value as obtained for Fig. 31.

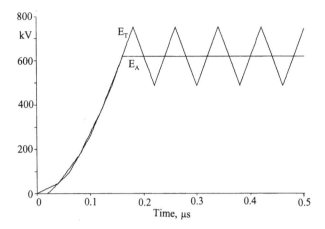

Figure 31 Voltages at arrester and open end of GIS bus.

The previous equations assume that no arrester lead length exists. To modify the equations for this condition, a voltage is added to E_A, that is

$$E_A = E_d + V_{PF} + 2\gamma S T_A \tag{113}$$

If the transformer is connected to the GIS bus by an open-air bus, the surge transmitted to the bus and transformer is increased, since the open-air bus or connection has a larger surge impedance than the GIS bus. However, this increase is normally not significant.

The Stress, Switching. Another very important source and type of stress exists. The operation of disconnecting switches creates an almost vertical front surge that is transmitted and reflected within the station. For normal or slow-speed disconnecting switches, overvoltages are in the range of 1.7 per unit and reach 2.0 per unit in specific cases. For high-speed disconnecting switches, the highest overvoltages may reach 2.5 per unit [23]. This event is not uncommon in an air-insulated station, but the front is rapidly increased or attenuated. However, in a GIS, this attenuation is not present. In the early days of the GIS, failures of transformers were attributed to this source of stress, since the almost limitless frequencies in the surge could excite the transformer at its natural frequency. Today, this stress has been mitigated by improved design of GIS disconnecting switches or the use of a "preinsertion" resistor. This stress has been recognized in the recent IEC standard, IEC 71 [19], where it is called a "very fast front" surge having a time to crest from 3 to 300 ns and containing two frequency components, one from 0.3 to 100 MHz and the other from 30 to 300 kHZ. However, tests for their waveshape and magnitude have not yet been standardized.

Because of the low magnitudes of the very fast front surges and the small times to crest, arresters cannot significantly limit the magnitudes. Thus the design of the disconnecting switch in limiting the magnitude is of paramount importance.

To estimate accurately the very fast front surges, the system and all components must be modeled in detail. Both IEEE and CIGRE have produced guides in the modeling procedure [23–26].

The Strength. To gain insulation strength, the bus is not only insulated with SF_6 but also contained in a circular enclosure to achieve a somewhat uniform field. This field uniformity causes some problems in that almost anything can alter the field and result in a drastic decrease in strength. Of most significance are "free conducting particles," which are very small particles inadvertently introduced during manufacturing or during field assembly. To eliminate the possibility of the introduction of particles during field assembly, some manufacturers provide fully assembled GISs that are transported to the customer's site. Since no domestic manufacturer exists within the USA, and the transportation of a complete GIS is sometimes physically impossible and economically prohibitive, the possibility of problems and failures exists. Also some failures have occurred with free-conducting particles at higher system voltages where the BIL has been reduced below comparable levels at lower system voltages. These problems or failures are not due to lightning or switching overvoltages but to an increased power frequency gradient for the "reduced" BIL units. This factor is apparently responsible in some GISs for the apparent reduced reliability at 500 kV.

Since the field within a GIS is approximately uniform, the strength to alternate waveforms is approximately constant. That is, the BIL is approximately equal to the chopped wave strength. Thus the crest voltage of a surge within the GIS is compared to the BIL. Although standards specify the BSL less than the BIL [27–28], it is suspected that the BSL is approximately equal to the BIL. See Chapter 1 for BILs and BSLs of GIS.

Coordination. Safety margins for GIS are usually a minimum of 20%. That is, the required BIL

$$\text{BIL} = 1.20 E_t \tag{114}$$

where E_t is the crest of the surge voltage within the GIS.

To continue the previous example, assume that the maximum permissible distance between the arrester and the open end of the GIS bus is desired. The BIL of a 230-kV GIS is 950 kV per IEC or either 750 or 950 per IEEE. Assuming a 950-kV BIL, the maximum permissible voltage is $E_t = 950/1.20 = 792$ kV. From the previous example, the maximum value of E_{tmax} is 741 kV. Therefore there is no maximum length of GIS bus; it approaches infinity.

Changing the example to a 750-kV BIL, the maximum permissible voltage is $E_t = 750/1.2 = 626$ kV, $E_T = 764$ kV, and $E_T/E_A = 1.3079$. Then

$$K_1 = \frac{0.089}{\frac{0.728}{E_T/E_A - 1} - 1} = 0.065242 \qquad T = \frac{K_1 E_A}{S} = 0.00762 \tag{115}$$

The maximum distance is only 5.7 meters.

7 COMPARISON WITH IEEE

The calculating or estimating procedure used in the IEEE Guide [17] is first to reduce the station layout to a single-line transformer circuit. See Ref. 17 for the details of this method. The steepness of the incoming surge is reduced dependent on the number of lines following the circuit reduction. The resulting steepness S_R is determined by the equation

$$S_R = \frac{3}{n+2} S \qquad (116)$$

where n is the number of lines. The arrester voltage E_A used is the 10-kA, 0.5-μs discharge voltage plus the arrester lead drop calculated as

$$E_A = E_d + L\frac{di}{dt} = E_d + 2S_R T_A \qquad (117)$$

where L is the inductance of the arrester lead and di/dt is the steepness of the current discharged by the arrester. Per the above equation, this is equal to the second form used in this chapter. To obtain the voltage E_t the following equation is used.

$$\frac{E_t}{E_A} = 1 + \frac{1.92}{1 + 0.385/K_1} \qquad (118)$$

where

$$K_1 = \frac{S_R T_T}{E_A} \qquad (119)$$

This equation includes the effect of power frequency voltage, and the equation is believed to assume a power frequency voltage equal to the line–ground voltage. Previously, for the example in Section 5.3, the IEEE method was used, and for this case it was found to be very conservative, primarily as a result of the reduction method employed. To compare further the IEEE method to that of this chapter, consider a 138-kV system. Assume a transformer BIL of 550 kV, an 84-kV MCOV arrester with a discharge voltage of 267 kV, a safety margin of 20%, an incoming surge having a steepness of 1000 kV/μs, and a V_{PF} of 113 kV, i.e., the line–ground voltage. The maximum permissible voltage at the transformer, E_t is $(550/1.2)(1.10) = 504$ kV. First assuming an arrester lead length of 6 m, the IEEE method results in a maximum separation distance between the arrester terminals and the transformer of $17.8 + 6 = 23.8$ m. The methods of this chapter produce a distance of 26.5 m. If no arrester lead length is assumed, both the IEEE method and the method in this chapter agree with a distance of 26.5 m. Thus, although different in context, the methods agree in this case, and the only debatable issue is the magnitude of the power frequency voltage. Again, the disagreement between these methods when analyzing the circuit of Fig. 27 appears to be the method of circuit reduction.

The IEEE calculation method only considers the transformer. As mentioned in Chapter 12, at present the IEEE method sets the steepness of the incoming surge based on the arrester rating, i.e., 11 kV/μs per kV of MCOV rating to a maximum of 2000 kV/μs.

8 COMPARISON WITH IEC

In the IEC application guide [16], the voltage at any piece of equipment within the station is calculated using the equation

$$E_t = E_d + 2\left(\frac{S}{n}\right)T \qquad (120)$$

where T is the maximum travel time between the arrester (includes the arrester lead length) and the equipment being considered and n is the number of lines. The steepness of the incoming surge, S, is

$$S = \frac{K_c}{d_m + S_L} \qquad (121)$$

where S_L is the span length, and the distance d_m is calculated as in Chapter 12, i.e.,

$$d_m = \frac{1}{(\text{BFR})(\text{MTBF})} \qquad (122)$$

However, d_m is not increased to the next tower location. Instead, the span length S_L is added to d_m. This assures that the incoming surge will arrive at least from the first tower.

As mentioned previously, the safety margins suggested in IEC are 15% for internal insulations and 5% for external insulations. Also, the calculated voltage at the equipment is compared to the BIL and not the chopped wave test level, since IEC does not specify a chopped wave test.

Using the example in the IEC guide, for a 145-kV maximum system voltage, $E_d = 500\,\text{kV}$, $S_L = 300\,\text{m}$, BFR = 1/100 km-years, MTBF = 400 years, $K_c = 675\,\text{kV-km/µs}$, and $n = 2$. The separation distance for internal insulations is 30 m and that for external insulations is 60 m. Therefore, per IEC,

$$d_m = \frac{1}{(1/100)(400)} = 0.25\,\text{km} \qquad S = \frac{675}{0.25 + 0.3} = 1227\,\text{kV/µs} \qquad (123)$$

and therefore

$$\begin{aligned}E_t &= 500 + 2(1227/2)(0.1) = 622\,\text{kV} \quad \text{for internal insulations} \\ E_t &= 500 + 2(1227/2)(0.2) = 745\,\text{kV} \quad \text{for external insulations}\end{aligned} \qquad (124)$$

The required BIL is then

$$\begin{aligned}\text{BIL} &= 1.15(622) = 715\,\text{kV} \quad \text{for internal insulations} \\ \text{BIL} &= 1.05(745) = 782\,\text{kV} \quad \text{for external insulations}\end{aligned} \qquad (125)$$

Per Table 7 of Chapter 1, the selected BILs are 750 kV for internal insulations and 850 kV for external insulations.

Station Lightning Insulation Coordination

To compare to the methods of this chapter, first assume that the arrester lead length is 6 m. Thus $T_A = 0.02\,\mu s$, $T_T = 0.08\,\mu s$, and $T_B = 0.18\,\mu s$. Assuming $V_{PF} = 130\,kV$ and $E = 1200\,kV$, then $E_t = 792\,kV$ and $E_b = 671\,kV$. The required BILs are 864 kV for the transformer and 583 kV for the breaker. The selected BILs are 900 kV for the transformer and 900 kV for the breaker.

The voltage at the transformer using methods of this chapter are larger than for the IEC method, since the method of this chapter (1) includes the effect of the transformer capacitance and (2) includes the effect of power frequency voltage. The voltage at the breaker is less for the methods used in this chapter since the effect of the transformer capacitance are included. As a result, even though in the method of this chapter, the calculated voltage is compared to the chopped wave test voltage, the selected transformer BIL is significantly higher.

To be noted is that the nomenclature in IEC differs from that used here. The calculated voltage at the equipment is called the coordination withstand voltage, the required BIL is called the required withstand voltage, and the selected BIL is called the standard withstand voltage.

The values of the corona constant K_c are given in IEC in an alternate form. That is, the value of A is used where

$$A = \frac{2K_c}{c} \qquad (126)$$

The values of A and the corresponding value of K_c per IEC are shown in Table 17. The values of K_c, the corona constant in IEC, differ slightly from those used here primarily because of round-off error.

The required phase–ground and phase–phase clearances are functions of the BIL and are obtained directly from tables reproduced here as Table 18 [16]. Thus for 850 kV BIL, the clearance is 1.7 m. Using methods of this chapter, the clearance would be $792/605 = 1.3\,m$.

As seen from Table 18, above a 450-kV BIL, the clearances are based on a BIL withstand gradient of 500 kV/m. Below a 450-kV BIL, the gradient reduces steadily to 333 kV/m at a 20-kV BIL. Note that some of these BILs differ from standard values in IEEE. No suggested clearances are present in IEEE standards. Clearances based on switching overvoltages may exceed those for lightning.

Table 17 Corona constant A from IEC and Equivalent K_c

Type of line	A, kV	k_c, kV-km/μs
Distribution lines (phase–phase flashovers)		
with grounded crossarms (flashover at low voltage)	900	135
wood-pole lines (flashover to ground at high voltage)	2700	405
Transmission lines (single-phase flashover to ground)		
single conductor	4500	675
two-conductor bundle	7000	1050
four-conductor bundle	11000	1650
6-to 8-conductor bundle	17000	2550

Source: Ref. 16.

Table 18 Air Clearances per IEC 71

BIL, kV	Clearance, mm, Rod–structure	BIL, kV	Clearance, mm Rod–structure	Conductor-structure
20	60	850	1700	1600
40	60	950	1900	1700
60	90	1050	2100	1900
75	120	1175	2350	2200
95	160	1300	2600	2400
125	220	1425	2850	2600
145	270	1550	3100	2900
170	320	1675	3350	3100
250	480	1800	3600	3300
325	630	1950	3900	3600
450	900	2100	4200	3900
550	1100			
650	1300			
750	1500			

Sources: Ref. 16.

For maximum system voltages at or below 245 kV, the BSL is not normally provided. In this case, in the IEC guide, the switching surges are calculated and then translated to a BIL. For wet insulators, the translation is that BSL/BIL = 0.77. For internal insulation, for GIS, liquid immersed, and solid insulation the assumed BSL/BIL ratios are 0.80, 0.91, and 1.00, respectively. For air clearance and clean insulators, dry, the reverse ratio, BIL/BSL is given by equations

$$\text{for phase-ground } \frac{\text{BIL}}{\text{BSL}} = 1.05 + \frac{\text{required BIL}}{6000}$$
$$\text{for phase-phase } \frac{\text{BIL}}{\text{BSL}} = 1.05 + \frac{\text{required BIL}}{9000} \quad (127)$$

9 SUMMARY

9.1 Voltage with $C_T = 0$, An Open Circuit

At the Transformer or Open Circuit.

$$E_T = K_2\left[E_A + \frac{2S}{n}(T_T + T_A)\right] \quad \text{for } \frac{S(T_T + T_A)}{E_A} = 0 \text{ to } \frac{n}{6}$$

$$E_T = \frac{4K_2}{n+3}[E_A + 2(T_T + T_A)] \quad \text{for } \frac{S(T_T + T_A)}{E_A} = \frac{n}{6} \text{ to } \frac{n+1}{4} \quad (128)$$

$$E_J = E_A + 2ST_A$$

where K_2 is as in Table 5.

Station Lightning Insulation Coordination

At the "Breaker" and at the Junction.

$$E_B = E_A + 2S(T_B + T_A) \qquad E_J = E_A + 2ST_A \tag{129}$$

9.2 Voltage with Transformer Capacitance C_T

At the Transformer and at the Junction.

$$\frac{E_T}{E_A} = 1 + \frac{A}{1 + B/K_1} \qquad \frac{E_J}{E_A} = 1 + \frac{A}{1 + B/K_1} \tag{130}$$

where

$$K_1 = \frac{S(T_T + T_A)}{E_A} \qquad \text{for the transformer}$$

$$K_1 = \frac{ST_A}{E_A} \qquad \text{for the breaker} \tag{131}$$

The value of K_1 is obtained from Table 6. The time to crest of the transformer voltage, t_T can be estimated by the equation

$$t_T = \pi\sqrt{(T_T + T_A)(ZC_T + T_T + T_A)} + \frac{E_A}{S} \tag{132}$$

At the "Breaker."

$$E_B = E_J + 2ST_B \tag{133}$$

$$\text{for } t_A \leq [t_f - 2(T_B + T_A)] \tag{134}$$

$$E_B = \frac{2E}{n} - \frac{2-n}{n} 2S(T_A + T_B) - \frac{2S\tau}{n}\left[1 - e^{-\frac{t_f - 2(T_B + T_A)}{\tau}}\right] \tag{135}$$

$$\text{for } t_A \geq [t_f - 2S(T_B + T_A)] \tag{136}$$

$$\tau = \frac{ZC}{n} \tag{137}$$

9.3 Strength and Coordination

Transformer with Safety Factor of 1.20.

(1) For an incoming surge with a value of t_c less than 60 μs,

$$\text{BIL} = (1.20)E_d \quad \text{if } t_T \leq 3.0\,\mu\text{s and } E_t/E_d \leq 1.10 \tag{138}$$

$$\text{BIL} = (1.20)\frac{E_t}{1.10} \quad \text{if } t_T \leq 3.0\,\mu\text{s and } E_t/E_d > 1.10 \tag{139}$$

$$\text{BIL} = (1.20)E_t \quad \text{if } t_T > 3.0\,\mu\text{s} \tag{140}$$

(2) For an incoming surge having a value of t_c greater than 60 μs or more practically for an incoming surge caused by a shielding failure without a flashover,

$$\text{BIL} = (1.20)E_d$$
$$\text{BIL} = \frac{1.20}{0.83}E_d \tag{141}$$

Breaker—No Safety Margin.

$$\text{BIL} = \frac{E_d}{\delta} \quad \text{if } E_b/E_d \leq 1.15 \tag{142}$$

$$\text{BIL} = \frac{E_d}{\delta} \quad \text{if } E_b/E_d > 1.15 \tag{143}$$

where

$$\delta = e^{-A/8.6} \tag{144}$$

and A is the altitude in km.

Transformer Bushing. The internal insulation BIL is equal to that of the transformer. The external BIL should be equal to or greater than the internal BIL.

10 CONCLUSIONS

1. To determine the BILs and lightning clearances in a high-voltage station, a detailed study using a transient program, e.g., ATP or EMTO, is recommended.
2. For lower voltage stations whose layout is not extensive, the simplified method can be used.
3. The simplified method can also be used to estimate initially the voltages in more complex stations.
4. The simplified method presented here is conservative.

5. Utility standards should be based on generic studies of typical stations. These studies should be performed using a realistic arrester model and, if possible, should include better methods of evaluating the surge voltage waveshape.

6. Multiple lines in a station provide the benefit of decreasing the surge crest voltage and front steepness. However, these lines also collect more surges, and therefore an incoming surge with a larger steepness is required. These two combating features tend to compensate each other. As shown in the examples, the voltages at the transformer tend to increase slightly for multiline stations, and the voltages at other locations tend to decrease.

7. In general, the all-lines-in-service condition requires higher BILs than for contingency conditions. This, however, is dependent on the probabilities of the contingency conditions and therefore may not be true in all cases.

8. The GIS is advantageous in urban areas where land is at a premium and in locations of excessive contamination or at high altitude. In general, because of their compact size and low surge impedance, they are easier to protect than air-insulated stations.

9. In general, the voltage ahead of the arrester, i.e., at the transformer, is greater than the voltage behind the arrester. That is, the arrester provides better protection behind it than ahead of it, except for the maximum voltage attainable.

10. The computer program, SIMP, can be used to quickly and accurately determine the voltages and BILs.

11 REFERENCES

1. W. C. Guyker, A. R. Hileman, W. A. Richter, J. M. DeSalvo, and R. W. Powell, "Insulation Coordination in APS 500-kV Stations," *IEEE Trans.*, Jun. 1967, pp. 655–665.
2. A. R. Hileman, "Insulation Coordination of Air-Insulated Stations," in *Surges in High-Voltage Networks* (Klaus Ragaller, ed.) Brown Boveri Symposium Series, Plennum Press.
3. ANSI/IEEE C57.12.00-1987, "General Requirements for Liquid-Immersed Distribution, Power, and Regulating Transformers."
4. ANSI/IEEE C57.12.14-1989, "Trail-Use Standard for Dielectric Test Requirements for Power Transformers for Operation at System Voltages from 115 kV Through 230 kV."
5. ANSI/IEEE C76.1-1970, "Requirements and Test Code for Outdoor Apparatus Bushings."
6. IEC Publication 137, "Bushings for Alternating Voltages above 1000 Volts."
7. ANSI/IEEE C37.30-1971, "Definitions and Requirements for High-Voltage Air Switches, Insulators, and Bus Supports."
8. NEMA Std SG6, "Outdoor Substations."
9. ANSI/IEEE C37.04-1979, "Rating Structure for AC High-Voltage Circuit Breakers Rated on a Symmetrical Basis."
10. IEC Publication 273, "Dimensions of Indoor and Outdoor Post Insulators and Post Insulator Units for Systems with Nominal Voltages Greater than 1000 Volts."
11. R. L. Witzke and T. J. Bliss, "Coordination of Lightning Arrester Location with Transformer Insulation Level," *AIEE Trans.*, 1950, pp. 964–975.
12. J. M. Clayton and F. S. Young, "Application of Arresters for Lightning Protection of Multi-Line Substations," *AIEE Trans.*, 1979, pp. 566–575.
13. R. W. Flugum, A. R. Hileman, and T. F. Garrity, "Lightning Insulation Coordination for a 600-kV Gas Insulated Cable," IEEE *Trans. on PA&S*, Nov. 1982, pp. 4399–4406.

14. P. M. Balma, R. C. Degeneff, H. R. Moore, and L. B. Wagenaar, "The Effects of Long Term Operation and System Conditions on the Dielectric Capability and Insulation Coordination of Large Power Transformer," *IEEE Trans. on PD*, Paper 96, SM 406-9 PWRD, 1996.
15. F. Mosinski, J. Wodzinski, L. Sikoreski, and J. Ziemcikiewicz, "Electrical Strength of Paper-Oil Insulation Subjected to Composite Voltages," *IEEE Trans. on Dielectrics and Electrical Insulation*, Aug. 1994, pp. 615–623.
16. IEC 71-2, "Insulation Coordination—Part 2: Application Guide," 1996.
17. ANSI/IEEE C62.22, "Guide for the Application of Metal-Oxide Surge Arresters for Alternating-Current Systems," presently under ballot.
18. IEEE 1313-1-1996, "Insulation Coordination—Definitions, Principles, and Rules."
19. IEC 71-1, "Insulation Coordination—Part 1: Definition, Principle, and Rules," 1993.
20. AIEE Working Group Report of the Lightning Protective Devices Committee, "Simplified Method for Determining Permissible Separation Between Arresters and Transformers," *AIEE Trans.* Vol. S82, Special Supplement, 1963, pp. 35–57.
21. M. B. McNulty, "A Generalized Study to Determine the Optimum Location of Lightning Arresters in Power Transmission and Subtransmission Stations," M.S. thesis, Polytechnic Institute of Brooklyn, June. 1964.
22. A. J. Eriksson, K. G. Pettersson, A. Kernicky, R. Baker, J. R. Ochoa, and A. Leibold, "Experience with Gas Insulated Substations in the USA." *IEEE Trans. on PD*, Jan. 1955, pp. 210–218.
23. Fast Front Transient Task Force, IEEE Modeling and Analysis of System Transients Working, IEEE Transmission and Distribution Committee, "Modeling Guidelines for Fast Front Transients," *IEEE Trans. on PWD*.
24. Fast Front Task Force of the IEEE Modeling and Analysis of System Transients Working Group of the T&D Committee, "Modeling Guidelines for Fast Front Transients," *IEEE Trans. on PD*, Jan. 1996, pp. 493–501.
25. Task Force on Very Fast Transients in IEEE Working Group on Modeling and Analysis of System Transients Using Digital Programs, "Modeling and Analysis Guidelines for Very Fast Transients," *IEEE Trans. on PD*, Oct. 1996, pp. 2028–2035.
26. CIGRE Technical Brochure 39, "Guidelines for Representation of Network Elements when Calculating Transients," 1990.
27. IEC 517, "Gas Insulated Stations."
28. ANSI/IEEE C37.122-1983, "IEEE Standard for Gas Insulated Stations."

12 PROBLEMS

1. Shown in Fig. 32 is the layout of the three-line, 138-kV (max. 145-kV) station. The transformer BIL selected is 550 kV, and the internal and external bushing BIL is 650 kV. All remaining station equipment including the bus support insulators are 650-kV BIL. The surge capacitance of the transformer is 2 nF. The dimensions between locations are in meters. The BFR of each line is 1.6 flashover/100 km-year. Span length is 150 meters. Using an 84-kV MCOV station class arrester whose characteristics are given in Table 19, specify the protection of the stations for an MTBF of 100 years. Assume that the incoming surge has a crest voltage of 1200 kV, the total arrester lead length is 6 meters, the bus surge impedance is 450 ohms, and the corona steepness constant K_c is 700. The station is at an altitude of 1000 meters. Also select the clearances. In calculating the arrester current, assume that the power frequency voltage V_{PF} is 112 kV crest. Assume that the incoming surge has a tail time constant such that t_c is less than 50 μs and thus coordinating the

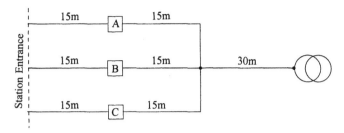

Figure 32 138-kV, three-line station.

arrester lightning discharge voltage with the transformer BSL is not necessary. Also assume that the probability of two lines in service is 0.25 and the probability of 1 line in service is 0.05. In calculating the arrester current, neglect reflections from the struck point, i.e., only use the 1.6 factor.

2. Using the results of problem 1, determine the actual MTBF of the station.

3. Shown in Fig. 33 is a 500-kV station with dimensions in meters. Select the BILs of the station equipment and clearances for the following conditions: Arrester: 318-kV MCOV, with a 10-kA, 0.5-μs discharge voltage of 990 kV. Use Table 20 for 8/20-μs discharge voltages. Lead length = 6 meters. Neglect reflections from the struck point; only use the 1.6 factor. (Calculate the arrester current.) Transformer surge capacitance: 2 nF; MTBF: 300 years; corona constant $K_c = 1000$ kV-km/μs; line: BFR = 0.5 flashovers/100 km-years; span length = 300 meters. Altitude: 1600

Table 19 Voltage–Current Characteristics of an 84-kV MCOV Arrester.

Discharge current, kA	Discharge voltage, kV	Discharge current, kA	Discharge voltage, kV
1.5	209	10	243
3	221	15	260
5	229		

Tabled values for 8/20-μs current impulse, 0.5-μs discharge voltage = 267 kV

Table 20 Voltage–Current Characteristics of a 318-kV arrester.

Discharge current, kA	Discharge voltage, kV	Discharge current, kA	Discharge voltage, kV
1.5	827	15	1001
3	854	20	1043
5	884	40	1115
10	944		

Tabled values for 8/20 μs current impulse, 0.5-μs discharge voltage = 1100 kV.

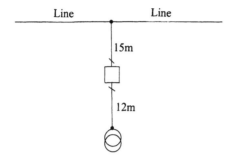

Figure 33 50-kV station tapped onto line.

meters; line and station surge impedance = 350 ohms. Incoming surge: assume a crest of 2300 kV and that this surge is riding atop an opposite polarity power frequency voltage of 300 kV, i.e., $V_{PF} = 300$ kV. Also assume that this surge is caused by a backflash so that $t_c < 50\,\mu s$.

4. For the 138-kV substation of Fig. 34, determine the maximum separation distances L_B and L_T between the arrester–bus connection and the transformer and between the arrester–bus connection and the breaker. Assume that the arrester lead length is 6 meters. The arrester is a constant-voltage arrester having a discharge voltage of 243 kV (84-kV MCOV). The transformer BIL is 450 kV. For the transformer use a safety factor of 1.20 permitting the voltage to reach 1.10 times the BIL. Therefore the maximum voltage at the transformer is (1.10/1.20)BIL. The breaker BIL is 650 kV, and the maximum permissible voltage is 1.15 times the BIL (use no safety factor). The steepness of the incoming surge is 1000 kV/µs, and the crest voltage is infinite. The transformer surge capacitance is 4 nF. Assume that t_A is less than $t_f - 2(T_B + T_A)$. Calculate the maximum permissible distances L_B and L_T for (a) $V_{PF} = 0.0$, (b) $V_{PF} = 80$ kV of opposite polarity to the surge, and (c) $V_{PF} = 80$ kV of the same polarity as the surge. The bus surge impedance is 400 ohms.

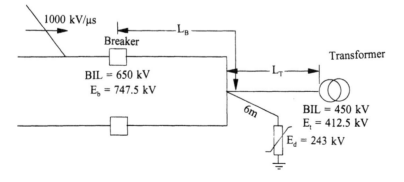

Figure 34 138-kV, two-line station.

5. Prove the following:

$$E_A + 2L\frac{di_A}{dt} = E_A + 2ST_A \tag{145}$$

where L is the inductance of the arrester lead, i_A is the current through the arrester, and T_A is the travel time of the arrester lead. di_A/dt is the current steepness through the arrester lead.

13—Appendix 1
Surge Capacitance

1 TRANSFORMER SURGE CAPACITANCE

The transformer surge capacitance can be estimated from data provided in Refs. 1–3. From the curves in Ref. 1, the total capacitance to ground of the highest voltage winding is primarily a function of MVA, although the capacitance does decrease as BIL increases. As an approximation, the minimum transformer capacitance in nF, C_T, can be represented by the power law equation of the form

$$C_T = A(\mathrm{MVA})^B \qquad (1)$$

where C_T is in nF and MVA is the MVA per phase. The parameters A and B are listed in Table 1.

To illustrate the decrease in capacitance with BIL, the capacitance for 20 MVA per phase is also shown in Table 1. From this table, it appears possible that capacitances as low as 2 nF are possible, especially at higher BILs.

Table 1 Values of A and B of Eq. 1

BIL kV	A	B	C_T for 20 MVA	BIL	A	B	C_T for 20 MVA
110	1.5	0.62	9.6	350	1.1	0.52	5.2
150	1.5	0.58	8.5	450	1.0	0.46	4.0
200	1.4	0.58	8.0	550	0.8	0.51	3.7
250	1.2	0.56	6.4	650	0.6	0.52	2.9

Table 2 Surge Capacitances from Westinghouse

BIL, kV	Surge capacitance to ground, nF	BIL, kV	Surge capacitance to ground, nF
550	2.4 to 3.4	1300	1.7 to 2.3
825	2.2 to 3.0	1550	1.5 to 2.0
1050	2.0 to 2.7	1800	1.4 to 1.8

Additional data as obtained from the Westinghouse Transformer Division are presented in Table 2. The maximum voltage at the transformer occurs for surge capacitances between about 1 and 6 nF, and 2 nF or 4 nF are suggested for use if the actual capacitances are unknown.

2 OTHER EQUIPMENT

The following approximate values are obtained from Refs 1-4. More exact values can be obtained from these references. Because these capacitances are small, and for conservatism, in most cases these capacitances are neglected except in cases where very fast transients are expected, i.e., disconnecting switch operations in gas-insulated stations, fronts of 2 to 20 ns, or breaker operation when energizing a motor, fronts of about 300 ns.

Outdoor Bushings.   200 to 550 pF.

Potential Transformer.   300 to 550 pF.

Current Transformers.   200 to 800 pF.

Circuit Breakers.   (1) Dead tank: 50 pF to ground each side, 6–10 pF across open breaks. (2) Live tank: 5 pF to ground each side, 10 pF across breaks. Note: Some breakers have 650 to 1000 pF across breaks.

3 REFERENCES

1. ANSII/IEEE C37.011-1979, "IEEE Application Guide for Transient Recovery Voltage for AC High-Voltage Circuit Breakers Rated on a Symmetrical Current Basis.'
2. Fast Front Task Force of the IEEE Modeling and Analysis of System Transients Working Group of the T&D Committee, "Modeling Guidelines for Fast Front Transients," *IEEE Trans. on PD*, Jan. 1996, pp. 493–501.
3. CIGRE Technical Brochure 39, "Guidelines for Representation of Network Elements when Calculating Transients," 1990.

13—Appendix 2
Evaluation of Lightning Surge Voltages Having Nonstandard Waveshapes: For Self-Restoring Insulations

1 INTRODUCTION

The BIL of station apparatus and equipment is verified by applying a standard lightning impulse, that is, a lightning impulse having a 1.2/50-µs waveshape. Additional tests on some equipment consist of the application of a 1.2/50-µs wave chopped at either 2 or 3 µs. For example, a circuit breaker must withstand the application of a 1.2/50-µs impulse, chopped at 3 µs, having a crest of 1.15 times the BIL, and the application of a 1.2/50-µs impulse, chopped at 2 µs, having a crest of 1.29 times the BIL. Bushings must withstand the application of a 1.2/50-µs impulse, chopped at 3 µs, having a crest of 1.15 times the BIL. Tests on transformers include a 3-µs chopped wave test at 1.10 times the BIL.

In addition, tests on air and air–porcelain insulation, e.g., air gaps and insulators, almost exclusively employ the standard lightning impulse waveshape to obtain the CFO and the time-lag or volt–time curve.

Thus the lightning impulse strength of all insulations is defined by use of the standard lightning impulse waveshape. However, the waveshape of the lightning surge voltages that appear across the insulation do not resemble either the full wave or the chopped wave 1.2/50-µs impulse. For example, in a station, the typical waveshape of a surge voltage at a location behind an arrester is illustrated in Fig. 1a, while the typical waveshape of a surge voltage at a location ahead of the arrester is

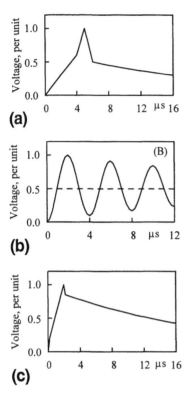

Figure 1 Typical waveshapes of lightning surge voltages (a) behind arrester; (b) ahead of arrester; (c) across line insulation.

depicted in Fig. 1b. The waveshape of the surge voltage across the line insulation caused by a stroke to the tower is shown in Fig. 1c. Some method must be used to permit the estimation of the strength of the insulation, i.e., the CFO or BIL, for the nonstandard waveshapes, knowing only the strength for the standard 1.2/50-μs waveshape. One method often employed is simply to equate the crest of the nonstandard surge voltage with either the BIL or the crest voltage of one of the chopped wave tests. This method, although easy to use, is totally subjective, since the user must decide whether the nonstandard surge has a shape that is best compared to a full wave or to a chopped wave. Some alternate method is desirable that is not subjective and that will provide a consistent means of evaluation. Another fundamental criterion of any such method is that it should be based on the fundamental process of gap breakdown. One method that contains these attributes and that is receiving popular attention is the leader progression model. Another method, derived from it, is the destructive effect method. Both methods result in the calculation of a severity index, which provides a measure of the severity of the surge in relationship to the insulation strength. Thus both methods are generically referred to as severity index methods. The severity index or SI not only provides a measure of severity but also can be used to estimate the required CFO or BIL.

It is the purpose of this appendix to present the theory, concept, and practical application of these severity index methods. These methods only apply to self-restoring insulations, i.e., to air or air–porcelain insulations or to apparatus that is essentially self-restoring. Thus it is applicable to air gaps (clearances), to bus support insulators, to line insulation, to disconnecting switches, and to circuit breakers. It should not be used for any type of non-self-restoring insulation such as that of a transformer or the internal insulation of a bushing. For these insulations or internal insulations, subjective methods of comparison of the stress–strength must still be made. Considering that a typical waveshape of the surge voltage at the transformer is that of Fig. 1b, the usual method is to compare the crest surge voltage to the transformer chopped wave strength of 1.10(BIL) [1]. Thus the surge voltage is permitted to reach a crest value of 1.10 times the BIL. However, normally a safety factor of 15 to 20% is applied so that the criterion is altered to $1.10/1.15 = 0.96$ or $1.10/1.20 = 0.9$ times the BIL. Although this general criterion has been used for many years with apparent success, some caution is required. If the oscillations of the surge voltage at the transformer are not rapidly attenuated, the waveshape is more similar to a full wave, and therefore the crest voltage should be compared to the BIL.

The analysis of nonstandard waveshapes is now receiving increased attention. For those desiring more information, a study of Refs. 2–8 is suggested.

2 THE CONCEPT OF A SEVERITY INDEX

Assume that a surge voltage having a crest of E_C and a nonstandard waveshape impinges across an insulation having an insulation strength described by the CFO. The severity index or SI is defined by the equations

$$\mathrm{SI} = \frac{E_C}{E_{\mathrm{MAX}}} \qquad \mathrm{SI} = \frac{\mathrm{CFO}_{\mathrm{MIN}}}{\mathrm{CFO}} \qquad (1)$$

where E_{MAX} is the maximum crest voltage of the surge having the nonstandard waveshape that may be placed across the insulation given the CFO. E_{MAX} can also be defined as the CFO for the nonstandard waveshape $\mathrm{CFO}_{\mathrm{NS}}$. $\mathrm{CFO}_{\mathrm{MIN}}$ is the minimum CFO for the nonstandard waveshape of surge voltage having a crest of E_C.

To explain by example, assume that the nonstandard surge voltage has a crest voltage of 2000 kV, the CFO is 1800 kV, and the SI is 0.800. Then, for a CFO of 1800 kV, the maximum crest voltage of this nonstandard surge is 2000/0.8 or 2500 kV. Also, the minimum value of the CFO for a crest voltage of the nonstandard surge of 2000 kV is 0.8(1800) or 1400 kV. Thus, as noted, the SI is a measure of the severity of the surge on the insulation. An SI of 1.00 indicates that the surge voltage stress is equal to the insulation strength. An SI of less than 1.00 indicates that the surge voltage stress is less than the insulation strength, and oppositely an SI greater than 1.00 indicates that the surge voltage stress is greater than the insulation strength. But, as noted, the SI is much more definitive than simply providing a "go/no go" decision. That is, if the SI is 0.8, then a 20% margin exists between the stress and the strength—or the BIL could be reduced by 20%. If the SI is 1.20, then the BIL must be increased by 20%.

The SI can also be used to determine margin between the maximum permitted surge voltage and the CFO. that is, from Eq. 1, the maximum permitted surge voltage in per unit of the CFO is

$$\frac{E_{MAX}}{CFO} = \frac{E_C}{SI(CFO)} \qquad (2)$$

or if desired, this can be placed in terms of the BIL, i.e.,

$$\frac{E_{MAX}}{BIL} = \frac{E_C}{SI(BIL)} \qquad (3)$$

The SI is also used to establish the CFO of a nonstandard waveshape for general use. For example, by regression analysis, an equation for the CFO of the nonstandard waveshape of surge voltage across the line insulation, Fig. 1c, was developed for use in the CIGRE method of estimating the lightning performance of transmission lines [9] and is presented in Chapter 10.

3 ALTERNATE METHODS OF ESTIMATING THE SEVERITY INDEX

Several methods of estimating the severity index are in use today. In general, these methods can be divided into those that attempt to model directly the breakdown process and those that are derived from the breakdown process. To gain a more complete understanding of the methods, first consider the breakdown process as illustrated in Fig. 2. Consider a gap with a spacing d across which is applied an impulse voltage. The leader beings its progress across the gap when the voltage

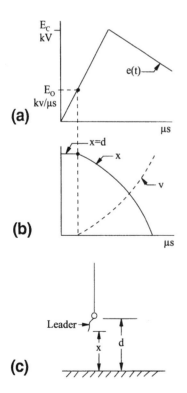

Figure 2 The breakdown process.

gradient exceeds E_0. As the leader proceeds, the voltage across the gap increases, and the distance between the tip of the leader and the ground electrode decreases, thus increasing the voltage gradient across the unbridged gap, x, which in turn increases the velocity v of the leader. As this process continues, the velocity of the leader increases until the leader reaches the ground electrode, at which time gap breakdown occurs.

Models of this breakdown process consist of a single equation for the velocity of the leader. Many equations have been proposed; a summary of these is contained in Refs. 1 and 2. The equation selected by the CIGRE working group for analysis of the voltage across the line insulation [9] is

$$v = k_L e(t) \left[\frac{e(t)}{x} - E_0 \right] \quad (4)$$

where $e(t)$ is the voltage as a function of time, x is the distance of the unbridged gap, E_0 is the gradient at which the breakdown process starts, and k_L is a constant. The calculation procedure consists of determining the velocity at each time instant, finding the extension of the leader for this time instant, determining the total leader length, and subtracting this from the gap spacing to find a new value of x. This process is then continued until the leader bridges the gap.

Per Eq. 1, to determine the SI, two methods can be used. Either the maximum crest voltage of the nonstandard surge for an assigned CFO can be obtained or the minimum CFO for a constant crest voltage of the nonstandard surge can be found.

Consider first that a CFO of 1800 kV is assigned to the insulation and that the maximum crest voltage of the nonstandard surge is to be determined. Further assume that the crest voltage of the surge is E_C. As depicted in Fig. 3a, the process starts by determining, by use of Eq. 4, if breakdown occurs for this surge. If no breakdown occurs, the crest of the nonstandard surge is incrementally increased until the breakdown occurs for E_{MAX} per Fig. 3a. Oppositely, if breakdown occurs for the nonstandard voltage, the crest voltage is incrementally decreased until no breakdown occurs. In either event, the value of E_{MAX} is obtain, and the SI can be found from Eq. 1. Note that the E_{MAX} is the CFO for this nonstandard waveshape or CFO_{NS}.

The SI can also be determined by finding the minimum CFO. The process starts by determining if breakdown occurs for the nonstandard surge of crest E_C. If breakdown occurs, the CFO is incrementally decreased until breakdown does not occur. This CFO is the minimum CFO or CFO_{MIN} as illustrated in Fig. 3b. Oppositely, if breakdown does not occur, the CFO is incrementally increased to find CFO_{Min}. In either case, the SI is found.

This method of determining the SI is called the leader progression model or LPM. Other methods have evolved to estimate the SI based on the LPM. Chief of these is the destructive effective method or DE, first developed by Witzke and Bliss [10–12] in an attempt to estimate the strength of transformer insulation to an oscillatory surge, as illustrated in Fig. 1b. The authors concluded that the transformer insulation could withstand this surge having a crest equal to the chopped wave strength of the transformer. Although this forms the historical basis of the evaluation of transformer insulation, presently the DE method is no longer used for the

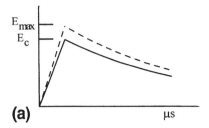

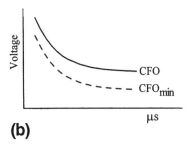

Figure 3 Finding the SI. (a) Varying the crest voltage to find E max. (b) Varying the CFO to find CFO_{MIN}.

evaluation of self-restoring insulation, as explained in the introduction. The DE method was verified by Rusck [2] and further investigated by other authors [3].

The concept of the DE method is based on the idea that there exists a base destructive effect DE_B. If a nonstandard surge contains a DE that exceeds this base DE, flashover occurs, and alternately if the surge contains a DE that is less than this base DE, no flashover occurs. The general equation for the destructive effect is

$$DE = \int_0^\infty [e(t) - V_0]^{k_d} dt \qquad (5)$$

where as before $e(t)$ is the surge voltage. As illustrated in Fig. 4, the voltage V_0 is the voltage below which no flashover can occur, and k_d is a constant.

Several forms of this equation have been used. With $k_d = 1$, the equation is called the equal area criterion. If V_0 is small compared to the surge voltage, and the approximation is made that $V_0 = 0$, the equation decreases in complexity, which leads to simplified equations for typical surge waveshapes (more later).

Using the complete Eq. 5, the method employed to determine the minimum CFO or the maximum crest value of the nonstandard surge is the same as that described for the LPM, except that the DE of the surge is compared to the base DE, DE_B. For example, the crest voltage of the surge is varied until the DE of the surge is equal to the DE_B.

For the approximation $V_0 = 0$, the iterative process is not necessary, as will be demonstrated later.

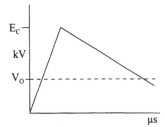

Figure 4 The DE method.

4 DETERMINING THE CONSTANTS OF THE LPM AND DE METHODS

The constants k_L and E_0 for the LPM method and DE_B, k_d V_0 for the DE method are determined by test results using the standard lightning impulse. That is, for the LPM method, since there are two unknowns, any two voltages on the time-lag curve may be selected. This is also true for the DE method with $V_0 = 0$. For the DE method with V_0 not equal to zero, there are three unknowns, and therefore three voltage points on the time-lag must be used. For this development, the full wave strength or the CFO, the 2-μs and the 3-μs chopped wave tests are used. Two types of insulation are considered: (1) air or air–porcelain insulations and (2) apparatus insulations. For air–porcelain insulations, typical time-lag curves indicate that the 2-μ and 3-μs test points are at 1.67 and 1.38 times the CFO. For apparatus insulations, standard tests require a withstand voltage of 1.15 times the BIL for a 3-μs chopped wave. For the circuit breaker, the withstand voltage for a 2-μs chopped wave is 1.29 times the BIL. For these equipments, the 2- and 3-μs voltages are 1.29 and 1.15 times the CFO. In summary, for air–porcelain insulations,

$$2\text{-μs chopped wave} = 1.67(\text{CFO})$$

$$3\text{-μs chopped wave} = 1.38(\text{CFO})$$

For apparatus insulations

$$2\text{-μs chopped wave} = 1.29(\text{CFO})$$

$$3\text{-μs chopped wave} = 1.15(\text{CFO})$$

To develop these constants, the waveshape of the standard lightning impulse is required. This 1.2/50-μs waveshape is approximated as a double exponential, i.e.

$$e(t) = A\left[e^{-\alpha t} - e^{-\beta t}\right] \tag{6}$$

whose crest value is equal to 1.00, and for t in μs,

$$\alpha = 0.0146591 \qquad \beta = 2.46893 \qquad A = 1.03725 \tag{7}$$

4.1 The LPM Method

The constants for the LPM method are developed by first selecting the standard lightning impulse breakdown gradient CFO_g, from which the CFO is obtained for any gap spacing. Knowing the full wave (CFO) and 3-µs chopped wave breakdown voltage, the values of k_L and E_0 are determined, from which the value of the 2-µs chopped wave voltage and the entire time-lag curve can be calculated. The value of these two constants is only dependent on (1) the breakdown gradient CFO_g and (2) the two selected points on the time-lag curve. The constants for both air–porcelain and apparatus insulations for alternate values of CFO_g are shown in Table 1.

The CFO_g varies with gap configuration and polarity. For positive polarity, CFO_g varies from a low of 540 kV/m for a rod–plane gap to a high of about 650 kV/m. For negative polarity, CFO_g varies from a low of 540 kV/m to a high of 750 kV/m, this latter value applying to a rod–plane gap. Since negative polarity predominates, a value used in previous chapters of 605 kV/m could be used. However, to be conservative, a value of 560 kV/m is used, which is the value used previously for positive polarity.

Therefore for a CFO_g of 560 kV/m,

$$k_L = 7.785 \times 10^{-7} \quad E_0 = 535.0 \, \text{kV/m} \quad \text{for air–porcelain insulations}$$

$$k_L = 1.831 \times 10^{-6} \quad E_0 = 551.3 \, \text{kV/m} \quad \text{for apparatus insulations}$$

The resultant time-lag curves for the LPM method are shown in Fig. 5.

4.2 The DE Method with V_0 Not Equal to Zero

To obtain the three constants, the DE for the 2-µs chopped wave voltage, the DE for the 3-µs chopped wave voltage, and the DE for the CFO are equated.

For air–porcelain insulations, the base DE and the constants are

$$DE_B = 1.1506(CFO)^{k_d} \quad k_d = 1.36 \quad V_0/CFO = 0.770 \tag{8}$$

For apparatus insulations,

$$DE_B = 0.3330(CFO)^{k_d} \quad k_d = 1.53 \quad V_0/CFO = 0.845 \tag{9}$$

Table 1 Constants for the LPM Method

	Air–porcelain, 2 µs/3 µs = 1.67/1.38		Apparatus, 2 µs/3 µs = 1.29/1.15	
CFO_g, kV/m	E_0, kV/m	k_L	E_0, kV/m	k_L
500	477.7	9.879×10^{-7}	492.2	2.300×10^{-6}
525	501.7	9.001×10^{-7}	516.9	2.100×10^{-6}
560	535.0	7.785×10^{-7}	551.3	1.831×10^{-6}
605	578.0	6.747×10^{-7}	595.6	1.568×10^{-6}

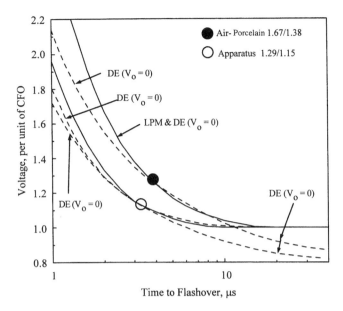

Figure 5 Resultant time-lag curves for the LPM and DE methods for a 1.2/50-μs impulse.

Note that the value of V_0 is given in per unit of the CFO. The resultant time-lag curve is presented in Fig. 5.

4.3 The DE Method with $V_0 = 0$

Since only two constants are required, the DE for the 2-μs chopped wave and the DE for the 3-μs chopped wave are equated. The results are for air–porcelain insulations

$$DE_B = 8.4(CFO)^{k_d} \qquad k_d = 4.4 \tag{10}$$

For apparatus insulations, we have

$$DE_B = 4.6(CFO)^{k_d} \qquad k_d = 5.4 \tag{11}$$

The resultant time-lag curve is shown in Fig. 5. As noted, since $V_0 = 0$, the time-lag curve decreases below the CFO. Thus this method may be adequate for short-duration surge voltages but should produce very conservative values of SI for surge voltages of long time duration.

5 COMPARISON OF METHODS

To provide a comparison of the three methods, two waveshapes as shown in Fig. 6 are used. The equations for these waveshapes are, for Fig. 6a, a linear front and exponential tail:

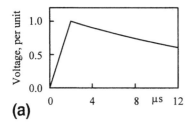

(a)

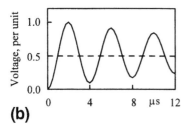

(b)

Figure 6 Waveshapes used for comparison of methods.

$$e(t) = \frac{E_C}{T_f} t \qquad t = 0 \text{ to } T_f$$

$$e(t) = E_C e^{-\frac{t-T_F}{\tau}} \qquad t = T_F \text{ to } \infty \tag{12}$$

For Fig. 6b, an oscillatory surge,

$$e(t) = E\left[1 - e^{-\frac{t}{\tau}} \cos\frac{\pi}{T_f} t\right] \tag{13}$$

where the value of E is such as to produce a voltage of E_C at $t = T_f$.

Tables 2 to 5 compare the results from the three methods for $E_C = 2000$ kV and CFO $= 1800$ kV. Note that the results for the DE method with $V_0 = 0$ are not shown for the oscillatory surge. Because the voltage only decays to about 50% of its original value, the results from this method are invalid. Because the LPM method is directly based on the breakdown phenomena, these results should be most accurate.

Table 2 Comparison of Results: Air–Porcelain Insulation, Linear Front Exponential Tail

Time constant, μs	LPM		DE V_0 not equal to 0		DE $V_0 = 0$	
	SI	E_{MAX}	SI	E_{MAX}	SI	E_{MAX}
20	0.9862	2028	0.9302	2150	0.9838	2033
50	1.0764	1858	1.0565	1893	1.1990	1688
100	1.1161	1792	1.1377	1758	1.3986	1430

Table 3 Comparison of Results: Apparatus Insulation, Linear Front Exponential Tail

Time constant, μs	LPM		DE V_0 not equal to 0		DE $V_0 = 0$	
	SI	E_{MAX}	SI	E_{MAX}	SI	E_{MAX}
20	1.0460	1912	0.9867	2027	1.0835	1846
50	1.1092	1832	1.0689	1871	1.2726	1572
100	1.1093	1803	1.1204	1785	1.4425	1387

Table 4 Comparison of Results: Air–Porcelain Insulation, Oscillatory Surge

Time constant, μs	LPM		DE V_0 not equal to 0	
	SI	E_{MAX}	SI	E_{MAX}
20	0.9276	2156	0.8757	2284
50	0.9950	2010	0.9671	2068
100	1.0417	1920	1.0384	1926

Table 5 Comparison of Results: Apparatus Insulation, Oscillatory Surge

Time constant, μs	LPM		DE V_0 not equal to 0	
	SI	E_{MAX}	SI	E_{MAX}
20	0.9843	2032	0.9259	2160
50	1.0304	1941	0.9896	2021
100	1.0593	1888	1.0384	1926

Comparing the results of the DE methods to the LPM method, the following two points can be made.

1. The DE method with V_0 not equal to zero gives values of E_{MAX} that are from 0.3% lower than the LPM results to about 6% larger. The best comparison occurs for the longer duration surges.

2. The DE method with $V_0 = 0$ gives results that are within 4% of the LPM method for short-duration surges, but for long-duration surges, i.e., a time constant of 100 μs, E_{MAX} is up to 30% lower. These significantly lower values are a result of eliminating V_0. This can be visualized from the time-lag curve, which falls below the CFO.

From these results, it is apparent that the DE method with $V_0 = 0$ should not be used except for short-duration surges or to obtain an initial value of SI.

6 AN EXAMPLE OF THE USE OF THE DE METHOD WITH $V_0 = 0$

To show the possible use of the DE method with V_0 is equal to zero, consider the linear front exponential tail surge voltage waveshape as used in the last section. The DE is

$$\begin{aligned} \text{DE} &= E_C^{k_d} \left[\int_0^{T_f} \left(\frac{t}{T_f}\right)^{k_d} dt + \int_{T_t}^{\infty} e^{-k_d \frac{t-T_f}{\tau}} dt \right] \\ &= E_C^{k_d} \left[\frac{T_F}{k_d + 1} + \frac{\tau}{k_d} \right] \\ &= K_C E_C^{k_d} \end{aligned} \quad (14)$$

Also let the base DE be

$$\text{DE}_B = C_B (\text{CFO})^{k_d} \quad (15)$$

The maximum surge crest voltage occurs when $\text{DE} = \text{DE}_B$, and for this $E_{\text{MAX}} = E_C$. therefore

$$K_C (E_{\text{MAX}})^{k_d} = C_B (\text{CFO})^{k_d}$$
$$E_{\text{MAX}} = \sqrt[k_d]{\frac{C_B}{K_C}} \text{CFO} \quad (16)$$

and the SI is

$$SI = \frac{E_C}{E_{\text{MAX}}} = \sqrt[k_d]{\frac{K_C}{C_B}} \frac{E_C}{\text{CFO}} \quad (17)$$

From this equation the SI can be directly obtained, which shows the advantage of this form of the DE method. Equation 17 is also useful in explaining the two components of the SI. That is, the k_{th} root of K_C/C_B is the term that compares the waveshapes, while the ratio of E_C/CFO compares the crest voltage magnitudes. Now assume a value of SI, e.g., 0.8. To achieve a value of SI of 1.00, i.e., the critical value, then either E_C must be divided by 0.8, which gives the value of E_{MAX}, or the CFO must be multiplied by 0.8, which gives the value of CFO_{MIN}.

7 TIME DURATION REQUIREMENTS

To obtain an accurate value of SI, the nonstandard waveshape should be considered for the entire time duration from the start of the surge to infinity. However, this requires excessive computer time. Sufficiently accurate values of SI can be obtained by limiting the calculations to within a specific time span or to within a specific per unit voltage. For example, to achieve a SI to within 5% of the actual value, the calculation can be limited to voltages that are within about 80% of the crest value for either the linear front exponential surge or the oscillatory surge of Section 5.

Therefore, as a general rule, it is suggested that the SI be calculated for a time interval to which the voltage has decreased to at least 75% of its crest value.

8 GAS-INSULATED STATIONS (GIS)

For GISs, the time-lag curve is essentially flat except for a small turnup at submicrosecond times. Thus the GIS is unaffected by nonstandard waveshapes, and the crest voltage of the surge is directly compared to the BIL of the GIS. In terms of SI, the SI is simply the E_C/CFO or E_C/BIL.

9 MARGINS

As stated previously, the SI, based on the gap breakdown mechanism, provides a consistent method to evaluate nonstandard surge voltages [4]. However, because this method has not been used extensively, and because some measure of safety may be desired, a value less than a critical value of 1.00 is sometimes used. For example, to obtain a 10% margin, an SI of 0.90 can be used.

10 SUMMARY AND RECOMMENDATIONS

1. The severity index, SI, based on gap breakdown phenomena, provides a consistent method to evaluate nonstandard surge voltages. The SI is defined as

$$\text{SI} = \frac{E_C}{E_{\text{MAX}}} = \frac{\text{CFO}_{\text{MIN}}}{\text{CFO}} \qquad (18)$$

where E_{MAX} is the maximum crest voltage of the surge having a nonstandard waveshape that can be placed across the insulation for the given CFO. E_{MAX} can also be defined as the CFO for the nonstandard waveshape, CFO_{NS}. CFO_{MIN} is the minimum CFO for the nonstandard waveshape of surge voltage having a crest of E_C.

2. The severity index can be calculated by one of three methods: (1) the leader progression model, LPM, (2) the destructive effect, DE model with starting voltage $V_0 \neq 0$, and (3) the destructive effect, DE, model with $V_0 = 0$. Of these the LPM model is the most accurate. The DE model with $V_0 \neq 0$ is comparable, but the DE model with $V_0 = 0$ may lead to over 30% inaccuracies for long time duration surges. However, this later DE model may be useful to obtain an SI by a simple equation that is a good approximation for short duration surges.

3. As a general rule, the SI should be calculated for a time interval at which the voltage has decreased to at least 75% of its crest value.

4. As before, the effect of altitude in decreasing the CFO or BIL should be considered.

11 REFERENCES

1. IEEE Std C62.22-1991, "IEEE Guide for the Application of Metal-Oxide Surge Arresters for Alternating Currents Systems."
2. S. Rusck, "Effect of Non-Standard Surge Voltages on Insulation," CIGRE paper 403, Paris, 1958.

3. R. O. Caldwell and M. Darvenzia, "Experimental and Analytical Studies of the Effect of Non-Standard Waveshapes on the Impulse Strength of External Insulation," *IEEE Trans. on PA&S*, vol. 92, pp. 1420–1428, 1973.
4. A. R. Hileman, "Insulation Coordination of Air-Insulated Stations," in *Surges in High-Voltage Networks* (Klaus Ragaller ed.), Brown Boveri symposium Series, Plenum Press, 1980.
5. A. Pigini, G. Rizzi, E. Garbagnati, A. Porrino, G. Baldo, and G. Pesavento, "Performance of Large Air Gaps Under Lightning Overvoltages: Experimental Study and Analysis of Accuracy of Predetermination Method," *IEEE Trans. on PD*, Apr. 1989, pp. 1379–1392.
6. Working Group 33.01 (Lightning), "Guide to Procedures for Estimating the Lightning Performance of Transmission Lines," CIGRE Technical Brochure No. 63, Oct. 1991.
7. P. Chowduri, A. K. Mishra, P. M. Martin, and B. W. McConnell, "The Effects of Nonstandard Lightning Voltage Waveshapes on the Impulse Strength of Short Air Gaps," *IEEE Trans. on PD*, Oct. 1994, pp. 1991–1999.
8. Task Force 15.09 on Nonstandard Lightning Voltage Waves, Lightning and Insulator Subcommittee of the T&D Committee, "Bibliography of Research on Nonstandard Lightning Voltage Waves," *IEEE Trans. on PD*, Oct. 1996, pp. 1982–1990.
9. "Guide to the Procedures for Estimating the Lightning Performance of Transmission Lines," CIGRE Technical Brochure No. 63, 1991.
10. R. L. Witzke and T. J. Bliss, "Surge protection of Cable Connected Equipment," *AIEE Trans.*, vol. 69, pp. 527–542, 1950.
11. R. L. Witzke and T. J. Bliss, "Co-ordination of Lightning Arrester Location with Transformer Insulation Level," *AIEE Trans.*, vol. 69, pp. 964–975, 1950.
12. A. R. Jones, "Evaluation of the Integration Method for Analysis of Non-Standard Surge Voltages," *AIEE Trans.*, vol. 73, pp. 984–990, 1954.

14
Line Arresters

1 INTRODUCTION

The use of line arresters to decrease or eliminate lightning flashovers on transmission and distribution lines was considered in Chapter 10. However, the treatment given was lacking in that it did not discuss the application concerns or the application criteria in sufficient detail to permit the application of these arresters. This chapter is an attempt to correct this deficiency. The method adopted is to present the results of a study of a typical 115-kV, single-circuit line both with and without overhead ground wires, which is then extended to lower voltage lines. This is followed by the development of equations to provide an insight into the phenomena.

The use of arresters on lines to decrease flashovers is not new. In the past, a type of surge arrester called the protector tube was used in the same manner [1]. However, because of its poor reliability, its use by utilities was discontinued. With the advent of the metal oxide arrester with its increased energy capability, and with the development of a nonceramic housing, the use of arresters for protection of lines has received a renewed impetus and popularity.

The first application occurred on a 138-kV tower of the American Electric Power Company [2, 3]. An inordinate number of flashovers occurred at this tower, which was located atop a mountain and in an area where rock formations precluded the use of supplemental grounding. In an attempt to improve the flashover rate, station class arresters were installed at this tower and at adjacent towers. With the success of this application, new arrester designs were developed [4–8] and other applications appeared [9–24]. In these cases, the arresters were used in a retro-fit manner at towers where soil conditions did not permit the use of supplemental grounding and at river-crossing towers. In a similar manner, arresters were installed on one circuit of a double-circuit line to eliminate double-circuit flashovers. In Japan, it is estimated that over 6000 arresters are used in this manner. All these applications have been to lines with overhead ground wires, and thus the shield wires protected the phase conductor from large-magnitude stroke currents.

Most recently, arresters have been applied to lines without overhead ground wires. Thus the arresters are now in a very hostile environment, where large current magnitude strokes can impinge on the arresters. The long-term results of these applications are not yet available. However, the preliminary results are encouraging. No definitive results concerning the arrester failure rates are available, but from informal conversations it appears that rates in the order of less than 1% per year have been achieved.

In all but a few cases, even for transmission lines, the heavy-duty distribution class arresters is used.

A partial list of sampling of the available literature is provided in the Reference section. IEEE and CIGRE committees and working groups are actively studying this application [25, 26], and reports and guides are expected in the future.

2 APPLICATION CONCERNS AND CRITERIA

From an electrical viewpoint, chief among the concerns is the energy discharged by the arrester, which is compared with the capability of the arrester [27–34]. From Chapter 12, the energy capability is probabilistic [35–37], and the probability of arrester failure can be approximated by a Weibull cumulative distribution, i.e.,

$$P_{FA} = F\left(\frac{W_C}{W_R}\right) = 1 - 0.5^{(\frac{Z}{4}+1)^5} \tag{1}$$

where

$$Z = \frac{W_C/W_R - 2.5}{0.375} \tag{2}$$

where W_R is the rated energy capability as supplied by the manufacturer and W_C is the energy capability for a probability of failure of P_{FA}. The rated energy capability is assumed to have a zero probability of failure and be located at 4 standard deviations below the mean.

For the 115-kV example of this chapter, a 76-kV MCOV heavy duty class arrester is assumed with a rated energy capability of 167 kJ. Thus for a 0%, 5%, and 50% probability of failure, the energy capability is 167 kJ, 274 kJ, 316 kJ, and 418 kJ, respectively.

The magnitude of the current discharged by the arrester is also of concern. From Chapter 12, the heavy-duty distribution arrester is capable of discharging a 100-kA, 4/10-μs current impulse. Most certainly, although this capability has not been studied, the current capability must also be probabilistic. However, at present, the current capability can only be stated as 100 kA with a zero probability of failure.

In addition to these electrical concerns, arresters installed on lines must meet general safety criteria. That is, in the event of failure, the ejected parts of the arrester

Line Arresters

must not become a safety hazard to the general public [38, 39]. This general criterion leads to the use of nonceramic housed arresters, which are now in general use.

3 THE SYSTEM STUDIED

The 115 kV single-circuit transmission line studied is shown in Fig. 1. The surge impedance of the ground wires is 339 ohms, that of the phase conductors is 366 ohms, and the mutual surge impedances between the ground wires and phases A, B, and C are 112, 131, and 112 ohms, respectively. Thus the coupling factors for A, B, and C phases are 0.331, 0.386, and 0.331, respectively. The span length is 230 meters, and the measured footing resistance R_0 is assumed at 55 ohms. The soil resistivity is assumed at 1000 ohm-meters giving a ρ/R_o of 18.2 meters. The surge arrester discharge voltage–current characteristics are shown in Table 1.

4 ARRESTER ENERGY—115-kV SHIELDED LINE

For this presentation concerning arrester energy, the time to crest of the stroke current, t_f, is varied dependent on the stroke current. Per Chapter 6, the median of the time to crest conditioned on the stroke current I is

$$t_f = 0.207 I^{0.53} \tag{3}$$

where I is the stroke current in kA and t_f is in µs.

4.1 Shielding Failure, Stroke to Phase Conductor

Using 11 towers, the line of Fig. 1 was set up on the ATP (Alternate Transient Program) and terminates at each end so as to eliminate reflections. An impulse current equal to the maximum shielding failure current of 12.3 kA was applied to phase A of the center tower. Since the current flowing through the footing resistance was less than half of the critical grounding current, $I_g = 21$ kA, the footing resistance was maintained at 55 ohms. Arresters were installed on each of the phases. The energy discharged by the arrester at the struck tower is shown in Fig. 2 as a function of the number of towers with arresters. Two sets of curves are provided, for a time to half value of the stroke current, t_T of 100 µs and 200 µs. In addition, for $t_T = 100$ µs,

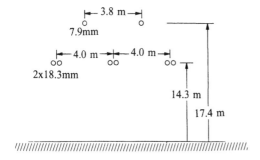

Figure 1 115-kV tower used as an example.

Table 1 Voltage–Current Characteristics, 76 kV Heavy-Duty Distribution Arrester

Discharge current, kA	Discharge voltage, kV	Discharge current. kA	Discharge voltage, kV
0.0001	207.7	1.0	240.0
0.0010	217.7	5.0	263.0
0.0100	221.8	10.0	288.0
0.0500	229.0	20.0	338.0
0.1000	231.1	40.0	392.0

three alternate span lengths are assumed. All curves of each set reach the same energy level when only one tower with arresters is considered, i.e., $n = 1$.

The dotted line curves are drawn assuming that the equal energy is discharged by the installed arresters. For example, for $t_T = 100\,\mu s$, for five towers with arresters, equal energy in all arresters would require that $284\,kJ/5 = 56.8\,kJ$ be discharged by each arrester.

As may be expected, the energy reaches a maximum when arresters are installed on only a single tower. This energy rapidly decreases as additional arresters are installed at other towers. Decreasing span length is beneficial. The resultant arrester energies for 11 towers with arresters and $t_T = 100\,\mu s$ are 36 kJ, 51 kJ, and 62 kJ for spans of 100 m, 230 m, and 400 m, respectively. For $t_T = 200\,\mu s$, the energy for 11 towers with arresters is 81 kJ. For 230-m spans, for the 11 towers with arresters case, the energy discharged for $t_T = 200\,\mu s$ is 14% of that for $n = 1$, while for $t_T = 100\,\mu s$,

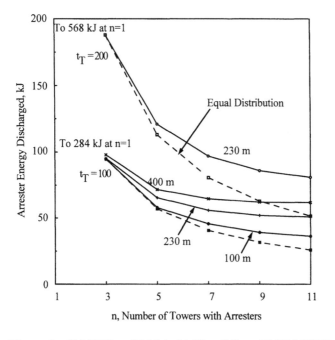

Figure 2 115-kV line, 12.3-kA shielding failure, 76-kV MCOV.

the energy is 18% of that for $n = 1$. Except for $n = 1$ and $t_T = 200\,\mu s$, the discharge energy is significantly below the rated energy of 167 kJ. Except for the 100-meter span, the crest current discharged by the struck arrester is 8.9 kA. Because the time to crest for 12.3 kA stroke current is 0.78 μs, reflections from the adjacent towers arrive at the struck tower before the crest of the stroke current. Thus the arrester current is decreased to about 8.4 kA.

The primary reason for the decrease in energy as additional arresters are added is that the time to half value of the arrester current decreases as shown in Fig. 3. For $n = 1$, the time to half value is approximately equal to that of the stroke current. But as arresters are added, this rapidly decreases and is somewhat influenced by the span length and the time to half value of the stroke current.

The stroke to the phase A conductor results in a voltage to ground that is then coupled to phases B and C. However, the voltage to ground on the ground wire is greater than these coupled voltages. Thus the current through these arresters is opposite in polarity to that of the struck arrester and produces an opposite polarity arrester voltage; interesting but not very important!

For a 230-m span, for three and 11 towers with arresters, Fig. 4 shows the effect of the time to half value of the stroke current. For the case of three towers, with arresters, the energy appears to be linearly dependent, but for 11 towers with arresters, the energy more gently increases. For 11 towers with arresters, the energy for a t_T of 300 μs is 111 kJ, significantly below the rated energy of 167 kJ. For three towers with arresters, the rated energy of 167 kJ occurs for a t_T of 160 μs. Figure 5 shows the increase in the time to half value of the arrester current as a function of the time to half value of the stroke current.

The effect of the footing resistance on the discharge energy is minor. For $t_T = 100\,\mu s$ and a 230-m span, for footing resistances between 20 and 200 ohms, the energy ranged from 96 to 90 kJ.

The shielding failure event or the stroke to the conductor is essentially a single-phase event. That is, the same answers could be obtained by only considering

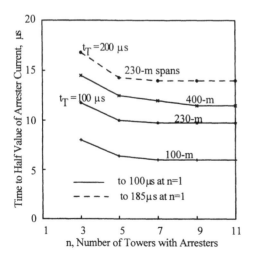

Figure 3 115-kV line, 12.3-kA shielding failure.

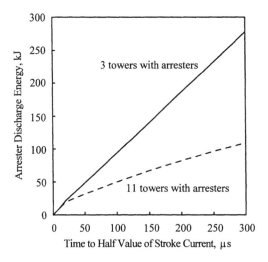

Figure 4 115-kV line, 12.3-kA shielding failure.

arresters on phase A. Arresters on the other phases do not affect the energy discharged by the struck arrester.

Although not shown, a stroke to the phase conductor at the end of the protected line section will result in an increase in arrester discharge energy.

To assess crudely the performance of the arresters for the shielding failure, assume the maximum shielding failure current of 12.3 kA and a t_T of 100 μs, which results in 51 kJ for 11 towers with arresters. Add to this the energy from three subsequent strokes. From Chapter 6, the median current in a subsequent stroke is 12.3 kA with an average time to half value of 37 μs. There is an average of three strokes per flash, i.e., two subsequent strokes. From Fig. 2, the arrester energy

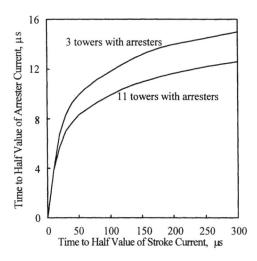

Figure 5 115-kV line, 12.3-kA shielding failure.

Line Arresters

discharged by the first stroke is 51 kJ. From Fig. 4, the energy for each subsequent stroke is about 32 kJ. Thus the total energy is 115 kJ, significantly below the rated energy of 167 kJ. Thus the application is acceptable from a shielding failure standpoint.

4.2 Stroke to Tower

The arrester discharge energy as a function of the number of towers with arresters is presented in Fig. 6 for a 100-kA stroke to the tower, 230-m spans, and $t_T = 100\,\mu s$. The footing resistance at the struck tower and the two adjacent towers were altered from an R_0 of 55 ohms by monitoring the current through the footing resistances and using the equations developed in Chapter 10. A ρ/R_0 of 18.2 meters was maintained. An interesting phenomenon occurs in that the energy through the arresters at the struck tower increases as additional arresters are added. Currents through the arresters at adjacent towers are of opposite polarity to the current through the arrester at the struck tower. These currents from the adjacent arresters flow back to the arrester at the struck and result in the increase in energy. This phenomenon is verified by Problem 10 of Chapter 9 (see the answers).

Since the stroke terminates at the tower, the energy discharged is divided between the three arresters. Arresters on phases A and C discharge equal energies, but because of the difference in coupling factors, the energy discharged by the phase C arrester is about 50 to 65% of the energies of phase A and C arresters. Thus as shown in Fig. 6, from an energy viewpoint, there exist a reason for the use of arresters on all phases.

Strokes can terminate with approximately equal probability at any location along the span. Assuming that the stroke terminates at the midspan, the energies through the adjacent arrester are much less. Per Fig. 7, the energy for a stroke to midspan is only 25% of that for a stroke to the tower.

Unlike the shielding failure event, the arrester energy is a function of the footing resistance, as illustrated in Fig. 8, constructed for 11 towers with arresters, 100-kA

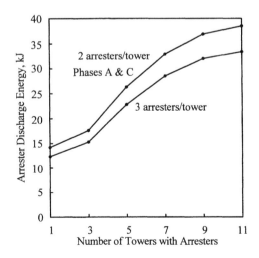

Figure 6 115-kV line, 100-kA stroke to tower.

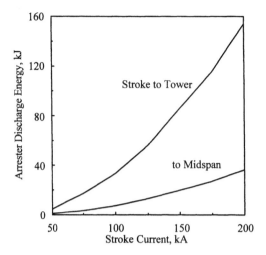

Figure 7 115-kV line, stroke to tower and midspan, $t_T = 100\,\mu s$, three arresters/tower.

stroke, and $t_T = 100\,\mu s$. At 120 ohms, the energy reaches about 52 kJ. The time to half value of the arrester current and the crest arrester current for the same parameters as Fig. 8 are shown in Fig. 9. As noted, arrester currents are moderate, below 10 kA.

The arrester energy is not significantly affected by the time to half value of the stroke current, as shown in Fig. 10.

Assessing the applicability of arresters to the backflash or stroke to tower event, the energies through the arresters seldom exceed 50 kJ, and the arrester currents are small, in the range of 10 kA. Thus the shielding failure event produces more arrester energy than the stroke to tower. However, since the shielding failure is limited to the

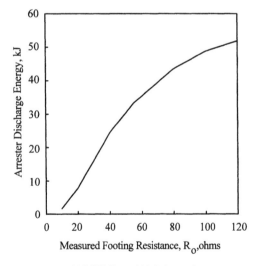

Figure 8 115-kV line, 100-kA stroke to tower, 11 towers with arresters, $t_T = 100\,\mu s$.

Line Arresters

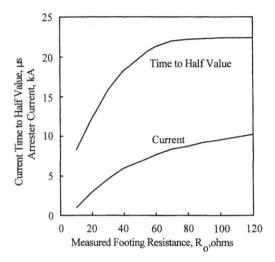

Figure 9 115-kV line, 100-kA stroke to tower, 11 towers with arresters, $t_T = 100\,\mu s$.

maximum shielding failure current, the arrester energies are generally less than the arrester rated energy. Thus the overall conclusion is that arrester energies and currents are readily acceptable, and arresters installed on shielded overhead lines should not be of concern.

5 115 kV UNSHIELDED LINE

To show the comparison to the previous shielding failure, a 12.3-kA stroke is applied to the phase conductor at the tower in the middle of the line. The results are shown in Fig. 11 for the same set of parameters as for Fig. 2. Although the effect of span length is not as prominent as in Fig. 2, the curves are similar. In general, however, the arrester energies are less than for the shielded line.

For the stroke to the phase conductor on a shielded line, the current is limited to the maximum shielding failure current, which is generally in the range of 5 to 10 kA.

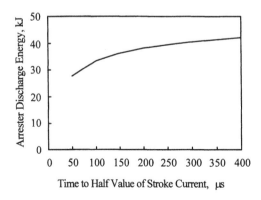

Figure 10 115-kV line, 100-kA stroke to tower, 11 towers with arresters, $t_t = 100\,\mu s$.

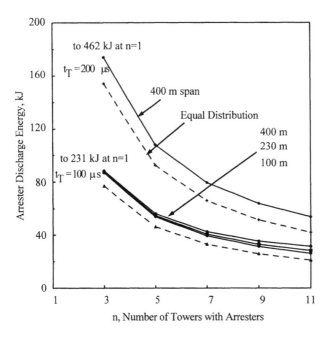

Figure 11 115-kV unshielded line, 12.3-kA stroke at tower.

However, for an unshielded line, the stroke current is unlimited and therefore large energies that exceed the rated energy are possible, although they may not be highly probable. Therefore to assess the arrester failure rate, an alternate method is required. This assessment consists of developing a curve of the time to half value of the stroke current versus the stroke crest current for alternate arrester failure probabilities. These curves for a stroke to the phase conductor at the tower and for a stroke to the midspan are shown in Figs. 12 and 13 for $R_0 = 55$ ohms, $\rho/R_0 = 18.2$ meters, and 230-m spans. The curves are drawn for failure probabilities of 0, 1%, 5%, and 50%, which are for energies of 167, 274, 316, and 418 kJ, respectively. The next step is to calculate the probabilities. One method is illustrated in Fig. 14. The curve is drawn for zero failure probability. The surface to the right of this curve represents the probability that the tail of the stroke current and the magnitude of the stroke current are greater than the points defined by the curve. In equation form,

$$P(W_A \geq 167\,\text{kJ}) = \int_0^\infty \int_f^\infty f(I)f(t_T)dI\,dt_T = \int_0^\infty [1 - F(I)]f(t_T)\,dt_T \quad (4)$$

Performing this task, the probability of exceeding 167 kJ for a stroke to the arrester at the tower is 0.201 and for a stroke to midspan is 0.219. Continuing this process for other values of energy results in the probabilities in Tables 2 and 3.

The third column is the probability of arrester failure given the energy. The fourth column is the probability within the interval from, for example, 167 kJ to 274 kJ, or from 274 kJ to 316 kJ, etc. The fifth column is the average probability of failure in the interval, e.g., for the first interval, the average of zero and 1%

Line Arresters

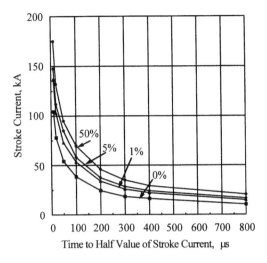

Figure 12 115-kV unshielded line, stroke at tower.

probability. The last column is the multiplication of the fourth and fifth columns. Summing the last column, the probability of failure is 0.07796 for a stroke to the midspan and 0.08442 for a stroke to the arrester at the tower. These probabilities would be lower if more divisions of the probability of failure were made, i.e., add the probabilities of failure of 20%, 70%, etc.

Since half the strokes are assumed to occur at the tower and half at midspan, the average of the two probabilities or 0.08119 should be used.

For the tower of Fig. 1 without ground wires, the number of flashes to the line is 87.7 flashes/100 km-year for an N_g of 6 flashes/km^2-year. Then the number of arrester failures becomes 0.08119 times 87.7 or 7.12 failures/100 km-year.

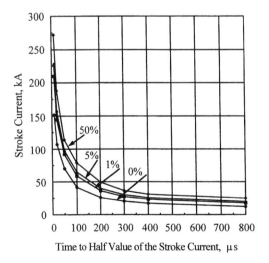

Figure 13 115-kV unshielded line, stroke to midspan.

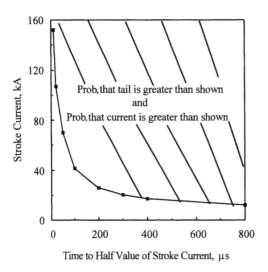

Figure 14 Example of calculation.

Table 2 Probabilities, Stroke at Tower

Arrester Energy, kJ	Probability of exceeding energy	Probability of arrester failure given energy	Probability interval	Average probability given energy	Probability of failure
167	0.201	0.0	0.201–0.151 = 0.050	0.005	0.00025
274	0.151	0.01	0.151–0.137 = 0.014	0.030	0.00042
316	0.137	0.05	0.137–0.097 = 0.040	0.275	0.01100
418	0.097	0.50	0.097	0.750	0.07275
Total probability of failure					0.08442

Table 3 Probabilities, Stroke to Midspan

Arrester Energy, kJ	Probability of exceeding energy	Probability of arrester failure given energy	Probability interval	Average probability of failure given energy	Probability of failure
167	0.219	0.00	0.219–0.148 = 0.071	0.005	0.00036
274	0.148	0.01	0.148–0.128 = 0.020	0.030	0.00060
316	0.128	0.05	0.128–0.088 = 0.040	0.275	0.01100
418	0.088	0.50	0.088	0.750	0.06600
Total probability of failure					0.07796

Considering that in 100 km there are 435 towers with three arresters per tower, the number of installed arresters is 1305. Therefore the arrester failure rate is 7.12/1305 = 0.00546 or 0.55% per 100 km-year. This value appears to compare well with field performance. Note that if only the rated energy is considered, then the probability of failure becomes $(0.201 + 0.219)/2 = 0.210$ and the failure rate becomes $0.210(87.7)/1305 = 0.141$ or 14% per 100 km-years, which illustrates the necessity of considering the probability of failure for alternate energies. In addition, this type of analysis should also consider subsequent strokes in the same manner.

6 ANALYSIS OF ARRESTER ENERGY

Two types of analysis are presented. First, a traveling wave analysis is performed with the objective of estimating the crest arrester current. Next, for longer term effects, the line is considered as an inductance, and finally, the two types of analyses are combined to provide an estimate of the energy.

6.1 Stroke to Ground Wires at Towers

Figure 15 shows the circuit analyzed. Also see Problem 9 of Chapter 9. The ground wire and phase conductor are assumed infinite in length, that is, reflections from adjacent towers are not considered. The arrester discharge voltage is modeled as

$$E_A = E_0 + i_A R_A \qquad (5)$$

The arrester current and the current through the footing resistance are

$$i_A = \frac{2}{D}[R_i(Z_g - Z_m)I - (Z_g + 2R_i)E_0]$$
$$i_R = \frac{1}{D}[Z_m(Z_g - Z_m) + Z_g(Z_c - Z_m + 2R_A)]I + \frac{2}{D}(Z_g - Z_m)E_0 \qquad (6)$$

where

$$D = (Z_g - Z_m)(Z_m + 2R_i) + (Z_g + 2R_i)(Z_c - Z_m + 2R_A) \qquad (7)$$

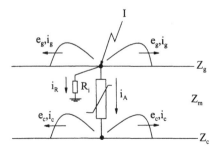

Figure 15 Stroke to ground wire at tower.

These complex equations can be reduced if $Z = Z_g = Z_c$ and the arrester is considered as a constant-voltage arrester, i.e., $E_A = E_0$ and $R_A = 0$. Then

$$i_A = \frac{2}{Z + Z_m + 4R_i}\left[R_i I - \frac{Z + 2R_i}{Z - Z_m}E_A\right] \qquad i_R = \frac{(Z + Z_m)I + 2E_A}{Z + Z_m + 4R_i} \qquad (8)$$

To understand better the equations let $Z_m = 0$. Then the more easily derived equation becomes

$$i_A = \frac{2(R_e I - E_0)}{Z_C + 2(R_A + R_e)} \qquad i_R = (I - i_A)\frac{R_e}{R_i} \qquad (9)$$

where

$$R_e = \frac{R_i Z_g}{Z_g + 2R_i} \qquad (10)$$

The crest surge voltages are

$$e_g = i_R R_i \qquad e_c = e_g - (E_0 + i_A R_A) \qquad (11)$$

Per the 115-kV study results, let $I = 100\,\text{kA}$, $R_0 = 55$ ohms, $\rho = 1000$ ohm-meters, $E_0 = 244\,\text{kV}$, $R_A = 4.4$ ohms, $Z_g = 339$ ohms, and $Z_c = 366$ ohms. Iterating Eqs. 5 and 6 to find R_i of about 24.9 ohms, $i_A = 6.7\,\text{kA}$ and $i_g = 82\,\text{kA}$. Then $E_A = 244 + 6.7(4.4) = 273\,\text{kV}$. Some further modification is necessary, since the median time to crest of a 100-kA stroke is about 2.4 μs, while the travel time of a 230-m span is 0.77 μs. Thus reflections from adjacent spans reduce the current through the arrester. From Chapter 10, K_{SP} is about 0.90, which should be applied to the first portion of the equation for i_A. Then the arrester current is reduced to about 6 kA. For comparison, the arrester current calculated by the ATP ranges from 4.7 kA for 11 towers with arresters to 7.22 kA for one tower with arresters. Therefore the above equations provide a reasonable estimate of the arrester current.

To estimate the arrester energy, a time constant of the arrester current is required. This time constant is essentially that resulting from the footing resistance and the ground wire. From Chapter 10, this time constant is given by the equation

$$\tau = \frac{Z_g}{R_i}T_s \qquad (12)$$

where T_s is the travel time of a span. Therefore the time constant is 10.4 μs and the energy is

$$W_A = E_A i_A \tau = (6 \text{ to } 6.7)(273)(10.4) = 17 \text{ to } 19\,\text{kJ} \qquad (13)$$

From the study performed, the energy varies from 13.9 kJ for one tower with arresters to 33 kJ for 11 towers with arresters. The estimated energy is between these two values and is sufficient for a first estimate.

6.2 Stroke to Phase Conductor at Tower, with Ground Wires or Neutral

Figure 16 shows the circuit analyzed. Also see problem 8 of Chapter 9. The equations are

$$i_A = \frac{1}{D}[Z_c(Z_g - Z_m) + Z_m(Z_c - Z_m + 2R_A)]I - \frac{2}{D}(Z_g - Z_m)E_0$$

$$i_R = \frac{1}{D}[Z_c(Z_g + 2R_i) - Z_m(Z_m + 2R_i)]I - \frac{2}{D}(Z_g + 2R_i)E_0$$

(14)

To simplify, set $Z = Z_g = Z_c$ and $R_A = 0$ with $E_0 = E_A$. Then

$$i_R = \frac{(Z + Z_m)I - 2E_A}{(Z + Z_m + 4R_i)} \qquad i_A = \frac{1}{Z + Z_m + 4R_i}\left[(Z + Z_m + 2R_i)I - \frac{2(Z + 2R_i)}{(Z - Z_m)}E_A\right]$$

(15)

Also for better understanding, if $Z_m = 0$, then

$$i_R = \frac{R_e}{R_i}i_A \qquad i_A = \frac{IZ_c - 2E_0}{Z_c + 2(R_e + R_A)}$$

(16)

Also the crest voltages on the ground wire and phase conductor are

$$e_g = i_R R_i \qquad e_c = e_g + (E_0 + i_A R_A)$$

(17)

To check these equations, use the line data of the previous section. Note that since the shielding failure current is only 12.3 kA, to be expected is that $R_i = R_0$. Then i_R = 9.0 kA and i_A = 10.5 kA. In comparison, the actual value of the arrester current was 8.9 to 9.5 kA and therefore the equations produce a reasonable approximation. The arrester voltage is 290 kV.

To determine an appropriate time constant, the circuit of Fig. 17 is analyzed. The ground wires and phase conductors have been replaced with inductances. To simplify further the inductance and surge impedance of the ground wire and the phase conductor are assumed equal. For an applied wave shape $Ie^{-t/\tau}$, the solution is

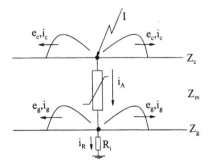

Figure 16 Stroke to phase conductor at tower.

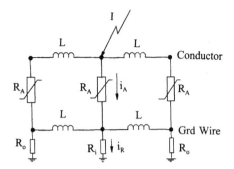

Figure 17 Stroke to phase conductor at tower.

$$i_A = \frac{\alpha^2 - a_1\alpha + a_0}{(\gamma - \alpha)\left(\frac{1}{\tau} - \alpha\right)} e^{t/\tau_1} + \frac{\gamma^2 - a_1\gamma + a_0}{(\alpha - \gamma)\left(\gamma - \frac{1}{\tau}\right)} e^{t/\tau_2} + \frac{\left(\frac{1}{\tau}\right)^2 - \left(\frac{a_1}{\tau}\right) + a_0}{\left(\alpha - \frac{1}{\tau}\right)\left(\gamma - \frac{1}{\tau}\right)} e^{-t/\tau} \quad (18)$$

where

$$a_1 = \frac{R_A + 2(R_0 + R_i)}{ZT_S} \qquad a_0 = \frac{R_A(R_0 + 2R_i)}{(ZT_S)^2}$$

$$\gamma = \frac{2(R_0 + 2R_i)}{ZT_S} \qquad \alpha = \frac{3R_A}{2ZT_S} \quad (19)$$

$$\tau_1 = \frac{1}{\alpha} = \frac{2ZT_S}{3R_A} \qquad \tau_2 = \frac{1}{\gamma} = \frac{ZT_S}{2(R_0 + 2R_i)}$$

and Z is the average of Z_c and Z_g. Also, as before, T_S is the span travel time and τ is the tail time constant of the stroke current. Applying these equations to the shielding failure event for a time to half value of the stroke current of 100 µs.

$$i_A = (0.466 e^{-t/41} + 0.335 e^{-t/0.82} + 0.199 e^{-t/144}) I \quad (20)$$

To develop the equation for the energy, this equation is changed to

$$i_A = (K_1 e^{-t/\tau_1} + K_2 e^{-t/\tau_2} + K_3 e^{t/\tau_3}) I \quad (21)$$

As first approximated by McDermott [40], let the arrester current and discharge voltage be related as

$$i_A = k(e_A)^\alpha \quad (22)$$

where for currents between 5 and 10 kA, α is 7.63. Then the arrester energy is

$$W_A = \int_0^\infty i_A e_A \, dt = \left(\frac{1}{k}\right)^{\frac{1}{\alpha}} \int_0^\infty (i_A)^{1+\frac{1}{\alpha}} dt$$

$$= \left(\frac{1}{k}\right)^{\frac{1}{\alpha}} \Bigg[(K_1 I)^{\left(1+\frac{1}{\alpha}\right)} \int_0^\infty e^{-t\left(1+\frac{1}{\alpha}\right)/\tau_1} + (K_2 I)^{\left(1+\frac{1}{\alpha}\right)} \int_0^\infty e^{-t\left(1+\frac{1}{\alpha}\right)/\tau_2} + (K_3 I)^{\left(1+\frac{1}{\alpha}\right)} \times \int_0^\infty e^{-t\left(1+\frac{1}{\alpha}\right)/\tau_3} \Bigg] \quad (23)$$

$$= \frac{K_1 I E_{A1} \tau_1 + K_2 I E_{A2} \tau_2 + K_3 I E_{A3} \tau_3}{1 + 1/\alpha}$$

where E_{A1}, E_{A2}, and E_{A3} are the discharge voltages for currents of $K_1 I$, $K_2 I$, and $K_3 I$. For $I = 12.3$ kA, for these currents, the discharge voltages are 268, 256, and 240 kV per Eq. 18, and the energy is

$$W_A = 56 + 0.8 + 75 = 132 \, \text{kJ} \quad (24)$$

This should be compared to the actual values of 95 kJ for three towers with arresters. However, for 11 towers with arresters, the energy decreases to 51 kJ. Thus the calculation is very conservative.

6.3 Stroke to Phase Conductor, No Ground Wire

First, to estimate the current, the solution to the circuit of Fig. 18 is

$$i_A = i_R = \frac{I Z_C - 2E_0}{Z_C + 2(R_i + R_A)} \quad (25)$$

Assume a 100-kA stroke to the conductor with $Z_C = 366$ ohms. For currents between 20 and 40 kA, $E_0 = 284$ kV and $R_0 = 2.7$ ohms. Then iterating to find R_i of 24.5 ohms, the arrester current is 85.7 kA. Using the ATP for three towers with arresters, the arrester current is 80 kA, where $R_i = 26$ ohms. The ground current at the two adjacent towers was monitored, and R_i was set at 37 ohms. The current through the footing resistance was 70.5 kA.

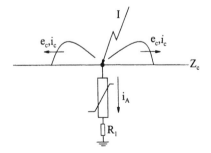

Figure 18 Stroke to phase conductor, no ground wire or neutral.

To estimate the arrester energy, the three towers with arresters circuit of Fig. 19 is analyzed. Again the surge current applied is $Ie^{-t/\tau}$. The resulting arrester current is approximately

$$i_A = \left[\frac{R_2 e^{-t/\tau} + 2R_1 e^{-t/\tau_2}}{2R_1 + R_2}\right] I \tag{26}$$

where both R_1 and R_2 represent current reduced or impulse values of the footing resistance. The time constant τ_2 is

$$\tau_2 = \frac{Z_C}{R_1 + R_2} T_S \tag{27}$$

For the given parameters, $I = 100\,\text{kA}$, $t_T = 100\,\mu\text{s}$,

$$\begin{aligned} i_A &= [0.416 e^{-t/144} + 0.584 e^{-t/4.5}] I \\ i_A &= [K_1 e^{-t/144} + K_2 e^{-t/4.5}] I \end{aligned} \tag{28}$$

and per the equation derived in the previous section,

$$W_A = \frac{K_1 I E_{A1} \tau_1 + K_2 I E_{A2} \tau_2}{1 + 1/\alpha} = 1949 + 92 = 2041\,\text{kJ} \tag{29}$$

where for currents from 20 to 40 kA and beyond, $\alpha = 4.68$.

Using the ATP for three towers with arresters, the energy is 1733 kJ, 18% less than that calculated. However, for 11 towers with arresters, the energy decreases to 750 kJ. Thus while the equation estimates the energy sufficiently accurately for three towers with arresters, it does not estimate the energy for the more practical case for 11 towers with arresters. That energy for the case of 11 towers with arresters is 750/1833 = 43% of the energy for three towers with arresters. This factor can be used to estimate crudely the energy for other cases.

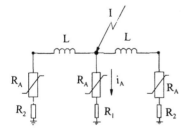

Figure 19 Stroke to phase conductor, no ground wire or neutral.

7 INTERMEDIATE TOWERS WITH NO ARRESTERS

To decrease cost, it is desirable to locate arresters on every second tower or every third tower, etc. However, as noted previously, the shorter distance between arresters decreases the arrester energy. In addition, we must ask whether the arresters on one tower can protect an adjacent tower without arresters. This is the subject of this section, where alternate arrangements are considered. Here the arrester is considered as a constant-voltage arrester, i.e. $E_0 = E_A$, $R_A = 0$, and $Z = Z_g = Z_c$. Only the case of arresters at every second tower is considered.

7.1 Stroke to Conductor, No Ground Wire or Neutral

Consider the circuit of Fig. 20, where a stroke is assumed to terminate on the conductor a travel time of T_1 from a protected tower and T_2 from the unprotected tower. In this specific case $T_2 > T_1$. The high-current or impulse ground resistance is R_1 at TWR1 and R_3 at TWR3. For this example, $T_1 = 0.20\,\mu s$ and $T_2 = 0.567\,\mu s$. The adjacent span has a travel time T_s of $0.767\,\mu s$, i.e., 230 m span. The stroke current is 30 kA and the steepness is 2.5 kA/μs. Also $Z = 366$ ohms, $R_i = R_1 = R_3 = 29$ ohms, and $E_A = 288$ kV. Using the ATP, the voltage at the unprotected tower is as shown in Fig. 21. Note that crest voltage is not attained until 12 μs, the time to crest of the stroke current.

The steepness of the voltage at the stroke terminating point S is

$$S = S_I \frac{Z}{2} \qquad (30)$$

where S_I is the stroke current steepness. This voltage travels both to TWR1, arriving at time T_1, and to TWR2, arriving at time T_2. As illustrated in Fig. 22a, the voltage at TWR1 reaches the arrester discharge voltage at time t_A, which is equal to E_A/S. This produces a negative reflection αS, which arrives at TWR2 at time $2T_1 + T_2 + t_a = T_1 + T_s + t_A$, where α is equal to

$$\alpha = -\frac{Z}{Z + 2R} \qquad (31)$$

This reflection arrives at TWR2 and decreases the steepness. The voltage at this time, E_1, of Fig. 22b is

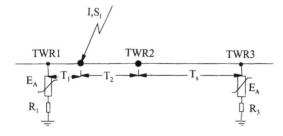

Figure 20 Stroke to phase conductor, no ground wire or neutral.

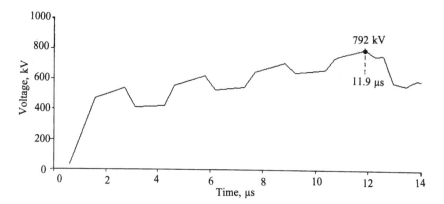

Figure 21 From ATP, stroke to phase conductor, no ground wire or neutral.

$$E_1 = S(2T_1 + t_A) = E_A + 2ST_1 = 471 \text{ kV} \tag{32}$$

Following this voltage, reflections occur from TWR1 and TWR3, and the voltage steadily increases until crest is reached at a time approximately equal to the time to crest of the stroke current, t_f, i.e., 12 μs. This slow increase in voltage can be approximated by the steepness S'', which is the steepness as time approaches infinity, i.e.,

$$S'' = \frac{1+\alpha}{1-\alpha}S = \frac{R_i}{Z+R_i}S \tag{33}$$

Therefore, as an approximation, the crest voltage E_2 a TWR2 is

$$E_2 = E_A + 2T_1 S + \frac{R_i}{Z+R_i} S[t_f - 2T_1 - t_A] \tag{34}$$

Using these equations, the crest voltage is calculated as 839 kV, whereas using the ATP, Fig. 21, the crest voltage is 792 kV, a 6% error.

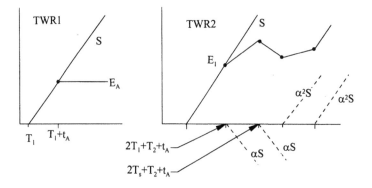

Figure 22 Equation derivation, stroke to conductor, no ground wire or neutral.

Line Arresters

In this calculation, the high current value of the grounding resistance R_i should be used. Therefore the current through the grounding resistance (which is identical to the arrester current) is required, i.e.,

$$i_R = i_A = \frac{2(e_c - E_A)}{Z + 2R_i} = \frac{IZ_c - 2E_A}{Z + 2R_i} \tag{35}$$

where $e_c = IZ/2$.

Eq. 34 can be rearranged to determine the value of T_1:

$$T_1 = \frac{Z + R_i}{Z} \frac{E_2 - E_A}{2S} - \frac{R_i}{2Z}(t_f - t_A) \tag{36}$$

The current steepness of 2.5 kA/μs was used to develop the equations. However, the suggested value of current steepness is $S_{30/90}$, which has a median value of 7.2 kA/μs for the first stroke and a median value of 20.1 kA/μs for subsequent strokes; see chapter 6. The voltage steepness is the current steepness times the surge impedance divided by 2 or for this case $(366/2)(7.2) = 1318$ kV/μs for the first stroke and 3678 kV/μs for subsequent strokes.

For the unshielded 115-kV line assuming the use of eight insulators with a CFO of 707 kV and $R_i = 37.7$, cT_1 (c = speed of light) is zero for both $S = 1318$ kV/μs and $S = 3678$ kV/μs. Thus the tower without arresters cannot be protected except when an unreasonable insulation level is used.

These equations are equally valid for lower voltage lines. For example, consider the line of Fig. 23 without the neutral. For purposes of this example, assume this to be a 13.8-kV tower. For this low CFO line, essentially all strokes will result in flashovers. That is, the voltage across the insulation is $IZ/2$, and therefore the critical current is $2(CFO)/Z$, which for CFOs of 170 and 1017 kV results in currents of 0.93 and 5.5 kA. The probability that these currents are exceeded is 100% and 96%. From Chapter 6, the number of flashes that terminate on the line is 45.8 flashes/100 km-year for an $N_g = 4$ flashes/km²-year, and thus the flashover rate is also 45.8 flashovers/100 km-year. As before, assume $Z = 366$ ohms and a 30-kA stroke with a

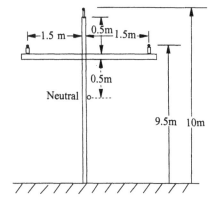

Figure 23 13.8-kV tower.

steepness of 7.2 kA/µs or 20.1 kA/µs. Further, assume the use of a 10.2-kV MCOV arrester having a 25-kA discharge voltage of 46 kV, and assume a span length of 60 meters. As a first case, assume that the pole and crossarm are either steel or concrete. Thus the insulation strength is that of the insulator, or a CFO, negative polarity of about 170 kV. Setting E_c to 170 kV, for $R_i = 0$, 5 and 10 ohms, cT_1 (c = speed of light) is 14, 6 and 0 meters for the first stroke and 5, 2.5, and 0 for the subsequent strokes. Thus the tower with arresters is essentially unprotected. Therefore the flashover rate is 0.5(45.8) = 22.9 flashovers/100 km-years.

Now assume that the pole and crossarm are wood. Flashovers will thus occur phase–phase, and the CFO for 2 meters of wood is about 600 kV. This CFO must be adjusted, since the phase–phase voltage is $(1 - C)$ times the voltage on the struck conductor. For the coupling factor of 0.41, the CFO is changed to 1017 kV. For R_i of 0, 10, and 20 ohms, cT_1 is 111, 97, and 83 meters for the first stroke and 40, 38, and 37 meters for subsequent strokes. Assuming $R_i = 20$ ohms the probability of flashover is $p_1 = 0$ for the first stroke and $p_2 = 0.5(37/60) = 0.308$ for subsequent strokes. The probability of the number of strokes per flash can be found in Chapter 6. Then the total probability of flashover is

$$P_T = \sum_{n=1}^{10} [1 - q_1(q_2)^n] P_n \qquad (37)$$

where $q_1 = 1 - p_1$, $q_2 = 1 - p_2$, n is the number of strokes/flash, and P_n is the probability of n strokes/flash. Performing this calculation, $P_T = 0.315$. The flashover rate is this probability multiplied by 45.8 or 25.4 flashover/100 km-years. Thus with arresters at every second tower, the flashover rate is reduced by 69%. This should be compared to a value of 50% if the intermediate tower is not protected. This result is for a stroke of 30 kA. For larger stroke currents, the value of cT_1 will be reduced. To obtain the total flashover rate, all stroke currents and their probabilities must be considered.

7.2 Stroke to Conductor with Ground Wire or Neutral

Figure 24 shows the circuit considered where $T_1 = 0.20 \mu s$ and $T_s = 0767 \mu s$. For a stroke current I having a steepness S_I the crest voltage on the phase conductor, e_c is $IZ_c/2$, and its steepness S_c on the phase conductor is $S_I Z_c/2$. The crest voltage on the ground wire or neutral, e_g, is the voltage e_c multiplied by the coupling factor Z_m/Z_c, and its steepness S_g is the steepness S_c multiplied by the coupling factor. These voltages travel toward the protected and unprotected towers. At the protected towers, the reflected voltages steepness on the conductor, S'_c, and on the ground wire, S'_g, are

$$S'_c = K_1 S_g + K_2 S_c \qquad S'_g = K_1 S_c + K_2 S_g \qquad (38)$$

where

$$K_1 = \frac{2R_1}{Z + Z_m + 4R_1} \qquad K_2 = \frac{Z + Z_m + 2R_1}{Z + Z_m + 4R_1} \qquad (39)$$

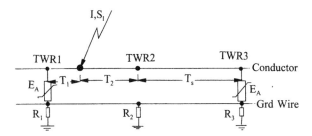

Figure 24 Stroke to phase conductor.

where R_1 is the high-current footing resistance of TWR1. Substituting for e_c and e_g, the reflection coefficients α_c and α_g become

$$\alpha_c = -\frac{Z(Z+Z_m)+2R_1(Z-Z_m)}{Z(Z+Z_m+4R_1)}$$

$$\alpha_g = -\frac{Z_m(Z+Z_m)-2R_1(Z-Z_m)}{Z_m(Z+Z_m+4R_1)} \qquad (40)$$

where for purposes of simplification $Z = Z_c = Z_g$. The original voltages, e_c and e_g, arrive at TWR2 at time T_2 and at TWR1 at time T_1. The negative voltages per Eq. 40 arrive at TWR2 at a time of $T_1 + T_s + t_A$ and decrease the steepness. This is shown in Fig. 25 obtained from the ATP for $R_i = 29$ ohms, $Z_c = 366$ ohms, $Z_g = 339$ ohms, $Z_m = 112$ ohms and for a stroke current of 30 kA and a current steepness of 2.5 kA/μs. Assumed is that there exists no ground at the unprotected tower. The initial voltages on the ground wire, E_{g1}, on the phase conductor, E_{c1}, and across the insulation at the unprotected tower, E_{I1}, are

$$E_{c1} = E_A + 2S_c T_1 \quad E_{g1} = 2S_g T_1 \quad E_{I1} = E_A + 2(S_c - S_g)T_1 = E_A + 2(1-C)S_c T_1 \qquad (41)$$

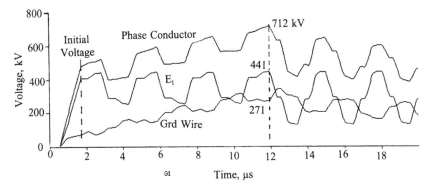

Figure 25 From ATP, stroke to phase conductor with ground wire or neutral.

which are marked in Fig. 25. As noted, the three voltages, ground wire or neutral, phase conductor, and voltage between the phase conductor and ground wire, steadily increases from this time to the time to crest of the stroke current, i.e., 12 μs. An approximate method to include this effect is to estimate the steepness of the voltage at the unprotected tower as time increases toward infinity. Denoting these steepnesses as S_g'' and S_c'', and assuming that the reflection coefficients of Eq. 40 can be used, the equations are

$$S_g'' = \frac{R_1(Z+Z_m)}{Z_m(Z+Z_m) - R_1(Z - 3Z_m)} S_g \approx \frac{R_1}{Z_m} S_g$$
$$S_c'' = \frac{R_1(Z+Z_m)}{Z(Z+Z_m) + R_1(3Z - Z_m)} S_c \approx \frac{R_1}{Z} S_c \qquad (42)$$

Substituting $S_g = (Z_m/Z)S_c$, then approximately $S_g'' = S_c''$. Then the crest voltages, E_c, E_g, and E_I, the latter being the voltage from the phase conductor to the ground wire or neutral, i.e., the voltage across the insulation, are

$$E_c = E_A + 2S_c T_1 + S_c''(t_f - 2T_1 - t_A)$$
$$E_g = 2S_g T_1 + S_c''(t_f - 2T_1 - t_A) \qquad (43)$$
$$E_I = E_c - E_g = E_A + 2(S_c - S_g)T_1 = E_A + 2(1 - C)S_c T_1$$

where t_a is the time required to reach the arrester discharge voltage, i.e., E_A/S_A, where

$$S_A = \frac{(Z - Z_m)(Z + Z_m + 2R_1)}{Z(Z + 2R_1)} S_c \qquad (44)$$

From Eq. 43,

$$T_1 = \frac{E_I - E_A}{2(1 - C)S_c} \qquad (45)$$

These equations are valid when there is no ground at the unprotected tower. If a ground exists, the voltage across the insulation increases. An approximate equation to account for this case is

$$E_I = E_A + 2(1 - C)S_c T_1 + \frac{Z - Z_m}{Z + 2R_2} S_c''(t_f - 2T_1 - t_A) \qquad (46)$$

where R_2 is the footing resistance at the unprotected tower. The added factor is deduced from the observation that at the unprotected tower the reflected voltages are

$$e_c' = -\frac{Z_m}{Z + 2R_2} e_g \qquad e_g' = -\frac{Z}{Z + 2R_2} e_g \qquad e_I' = +\frac{Z - Z_m}{Z + 2R_2} e_g \qquad (47)$$

The revised equation for T_1 becomes

$$T_1 = \frac{E_{\mathrm{I}} - E_{\mathrm{A}} - \dfrac{Z - Z_{\mathrm{m}}}{Z + 2R_2} S_{\mathrm{c}}''(t_{\mathrm{f}} - t_{\mathrm{A}})}{2(1 - C)S_{\mathrm{c}} - \dfrac{2(Z - Z_{\mathrm{m}})}{Z + 2R_2} S_{\mathrm{c}}''} \tag{48}$$

As an example, for the 13.8-kV line with a neutral and with $C = 0.48$, if there exists no footing resistance at the unprotected tower and if the CFO of the unprotected tower is 1017 kV as in the last example, $cT_1 = 212$ meters for the first stroke and 76 meters for subsequent strokes—for all values of R_1. However, if $R_2 = 29$ ohms, then $cT_1 = 120$ meters for the first stroke and 46 meters for the subsequent stroke.

Considering the 115-kV line with 8 insulators having a CFO of 707 kV (negative polarity), cT_1 is 71 meters assuming $R_1 = R_2 = 29$ ohms and a shielding failure stroke current of 12 kA with a steepness of 7.2 kA/μs. Therefore, for the 115-kV line, the unprotected tower is only protected by arresters at the adjacent tower if the stroke is within 71 m of the protected tower. Considering that the span length is 230 m, the tower without arresters is essentially unprotected.

Table 4 compares the calculated voltages and the voltages obtained from the ATP. For the calculation, Z is set to an average value of 353 ohms. Values in parentheses are ATP results.

As concerns the voltage across the insulation, the calculated and ATP results are within about 7%, indicating that the equations can be used to obtain an acceptable estimate.

The arrester and ground crest currents i_{A} and i_{R} are

$$\begin{aligned} i_{\mathrm{A}} &= \frac{2(Z + 2R_1)(e_{\mathrm{c}} - E_{\mathrm{A}}) - 2(Z_{\mathrm{m}} + 2R_1)e_{\mathrm{g}}}{(Z - Z_{\mathrm{m}})(Z + Z_{\mathrm{m}} + 4R_1)} \\ i_{\mathrm{R}} &= \frac{2(e_{\mathrm{c}} - E_{\mathrm{A}} + e_{\mathrm{g}})}{Z + Z_{\mathrm{m}} + 4R_1} \end{aligned} \tag{49}$$

7.3 Stroke to Ground Wire

Similar to the other case studied, Fig. 26 shows a stroke terminating on the ground wire a travel time of T_1 from a tower with arresters TWR1. At the struck point, a surge is created on the ground wire having a steepness S that is

Table 4 Comparison, Calculations Versus ATP

R_1	R_2 at unprotected tower	E_{c1}, initial voltage	E_c, final voltage	E_{g1}, initial voltage	E_g, final voltage	E_{I1}, initial voltage	E_1,* final voltage
0	None	471(459)	471(459)	58 (53)	58 (53)	413 (406)	413 (406)
0	0	(443)	(443)	(0)	(0)	(443)	413 (443)
29	None	471 (477)	878 (712)	58 (70)	465 (344*)	413 (407)	413 (441)
29	29	(457)	(625)	(10)	(175)	(447)	652 (625)

* 271 kV at time final E_1.

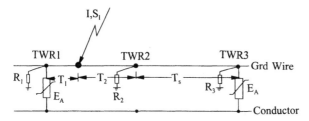

Figure 26 Stroke to ground wire.

$$S = \frac{Z}{2} S_1 \tag{50}$$

This circuit was set up on the ATP, and Fig. 27 shows the voltages at TWR2 on the ground wire, on the phase conductor, and across the insulation. To explain the voltages, the voltage on the ground wire, having a steepness S, accompanied by the coupled voltage on the phase conductor having a steepness CS, travels to TWR1 and TWR2 and arrives at these locations at T_1 and T_2, respectively. At TWR1, Fig. 28a shows that before the arrester reaches its constant value, negative reflections occur from the footing resistance. These reflected steepnesses S'_g and S'_c and transmitted steepnesses S''_g and S''_c are

$$\begin{aligned} S''_g &= \frac{2R_1}{Z + 2R_1} S & S''_c &= \frac{2R_1}{Z + 2R_1} CS \\ S'_g &= -\frac{Z}{Z + 2R_1} S & S'_c &= -\frac{Z}{Z + 2R_1} CS \end{aligned} \tag{51}$$

where C is the coupling factor. Therefore the steepness of the voltage across the arrester, S_A, and the time to achieve the discharge voltage, t_A, are

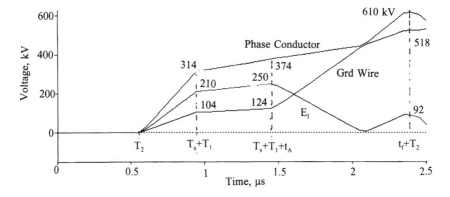

Figure 27 From ATP, stroke to ground wire.

Line Arresters

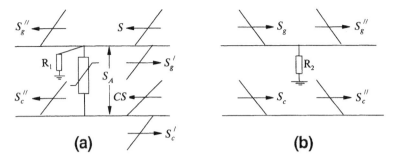

Figure 28 Derivation of equations for Fig. 27.

$$S_A = \frac{2R_1}{Z+2R_1}(1-C)S \qquad t_A = \frac{E_A}{S_A} \qquad (52)$$

When the voltages reach the TWR2, the voltages are again modified by the footing resistance R_2. Assuming that the voltage steepnesses when they arrive at TWR2 are S_g and S_c, per Chapter 9 the transmitted voltages S_g'' and S_c'', as shown in Fig. 28b, are

$$S_g'' = \frac{2R_2}{Z+2R_2}S_g \qquad S_c'' = S_c - \frac{Z_m}{Z+2R_2}S_g \qquad (53)$$

The result of the arrester operation appears at TWR2 at $T_1 + T_s + t_A$, at which time the voltage across the insulation decreases. Placing these voltages and the reflections in equation form,

$$e_g = \frac{2R_2}{Z+2R_2}\left[2ST_1 + \frac{E_A}{1-C}\right]$$
$$e_c = \frac{2R_2}{Z+2R_2}\left[2CST_1 + \frac{C}{1-C}E_A\right] \qquad (54)$$

and the voltage across the insulation, E_1, is

$$e_I = e_g - e_c = \frac{2R_2}{Z+2R_2}[E_A + 2ST_1(1-C)] \qquad (55)$$

and the time T_1 becomes

$$T_1 = \frac{\left(\frac{Z+2R_2}{2R_2}\right)e_I - E_A}{2S(1-C)} \qquad (56)$$

The voltage E_1 checks with the ATP results to within about 1%. Using the above equation for the 115-kV line and assuming $R_1 = R_2 = 29$ ohms, $Z = 353$ ohms, $Z_m = 112$ ohms, $E_A = 288$ kV, and the use of eight standard insulators, i.e., CFO = 654 kV, positive polarity, the values of cT_1 are 226, 164, and 129 meters for stroke currents of 30, 60, and 100 kA with steepness of 23.9, 33, and 42 kA/μs, respectively. The steepness assumed is the median of the maximum steepness for the stroke currents per Chapter 6. As noted for the span length of 230 meters, for a 30-kA stroke, the unprotected tower is protected and further, for the 60- and 100-kA strokes, protection is afforded to 71% and 56% of the span. Thus the combination of the tower footing resistance and arrester produce a remarkable protection of the unprotected tower.

8 BEYOND THE PROTECTED LINE SECTION

Of concern are the voltages beyond the protected line section, since voltages produced within the protected section may be magnified at the unprotected towers. To explain further, assume that the towers on one section of the line have high footing resistances and that arresters are installed at these locations. Adjacent to the last tower in this section are towers whose footing resistance is significantly smaller to the extent that no arresters are applied. In this case there exists a danger that the voltages transmitted by the protected section into the unprotected line may result in a flashover of the towers in the unprotected section. In other words, while the flashover is eliminated in the protected section, it is moved to the unprotected section. The obvious solution is to apply arresters to one or more towers in the low footing resistance section.

8.1 Stroke to Conductor, No Ground Wire or Neutral

Assume that the last tower in the protected section is TWR1 of Fig. 20 and that because the towers to the right have low footing resistance, no arresters are installed. The voltage at the struck tower can be obtained from the previous equations. That is, the crest voltage at the struck tower, at TWR1, E_1 is

$$E_1 = \frac{Z}{Z + 2R_1}(E_A + IR_1) \qquad (57)$$

This voltage continues its travel, unabated, to adjacent towers and appears across the insulation of the unprotected towers. For example, for the 115-kV line with $R_1 = 29$ ohms, assuming a 30-kA stroke current, the voltage at TWR2 and TWR3 is 1000 kV. To decrease this voltage, an arrester should be installed at the next low footing resistance tower. For example, installing an arrester at TWR2 where the footing resistance is 10 ohms produces a voltage at TWR2 and TWR3 of 558 kV, and thus only seven insulators are required.

For the low-voltage line assume a 30-kA stroke current and $R_1 = 29$. The voltage at TWR1 and TWR2 is 790 kV. If an arrester is applied at TWR2 and the footing resistance is 10 ohms, the voltage at TWR3 is reduced to 328 kV. For this line with a CFO of 170 kV, the voltage of 328 kV would result in a flashover. For the other option of a CFO of 1017 kV, even an R_1 of 29 ohms does not produce a sufficient voltage to cause flashover.

8.2 Stroke to Conductor with Ground Wire or Neutral

Assume that a stroke terminates at the tower TWR1 of Fig. 24, where arresters are installed. There are no arresters at TWR2 or TWR3. Assume that the ground wire or neutral is grounded at tower TWR2. This problem can be solved from the development presented in Chapter 9. The voltages on the ground wire, e_g, and conductor, e_c, at TWR1 are

$$e_g = \frac{(Z + Z_m)I - 2E_A}{(Z + Z_m + 4R_1)} R_1$$

$$e_c = e_g + E_A \qquad i_R = \frac{e_g}{R_1} \tag{58}$$

When these voltages reach TWR2 they are modified as follows:

$$e_g'' = \frac{2R_2}{Z + 2R_2} e_g \qquad e_c'' = e_c - \frac{Z_m}{Z + 2R_2} e_g \tag{59}$$

and the voltage across the insulation is

$$e_I'' = e_c'' - e_g'' = \frac{Z - Z_m}{Z + 2R_2} e_g + E_A \tag{60}$$

An interesting observation from Eq. 60 is that as R_2 increases, the voltage across the insulation decreases, and if there exists no grounding resistance at TWR2, the voltage is simply the arrester discharge voltage E_A. This phenomenon is also important when considering induced overvoltages in Chapter 15.

As an example, consider the 115-kV line with $R_1 = 29$ ohms and assume a shielding failure current of 12 kA. Then $e_g = 250$ kV and the voltage across the insulation at TWR2 for $R_2 = 10$ ohms is 450 kV, which is equivalent to about five insulators. Thus arresters are not required on the towers in the unprotected section.

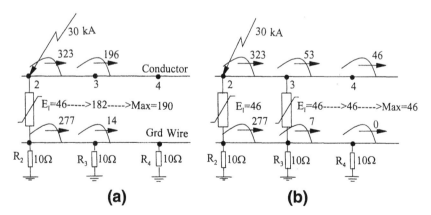

Figure 29 Example of protection beyond protected towers.

Next consider the low voltage line with a neutral and assume the same values of resistance at the struck tower but use a stroke current of 30 kA. Then $e_g = 712$ kV and the voltage across the insulation at TWR2 for $R_2 = 10$ ohms is 396 kV which is acceptable for a CFO of 1017 kV but unacceptable for a CFO of 170 kV. This voltage can be reduced by placing an arrester at TWR2 where $R_2 = 10$ ohms. As shown in Fig. 29a, for a 30-kA stroke to TWR2, the voltage across the insulation at TWR3 is 182 kV. Equations 59 may be used in a chain fashion to obtain voltages at other towers. As this procedure continues, the voltage across the insulation increases. To circumvent this procedure, from Chapter 9, the maximum value of E_I is

$$\text{Max } E_I = e_c - C e_g \tag{61}$$

And therefore the maximum value of E_I is 190 kV. To further reduce the voltage, another set of arresters may be applied at TWR3 as shown in Fig. 3b. The voltages at tower 3, on the ground wire and conductor, e_g'' and e_c'' for incoming surges of e_g and e_c are:

$$e_g'' = \frac{2R_2(e_c + e_g - E_A)}{Z + Z_m + 4R_2} \qquad e_e'' = e_g'' + E_A \tag{62}$$

As illustrated in Figure 29b at tower 4, the voltages across the insulation decreases to 46 kV well within the CFO of 170 kV. The maximum voltage across the insulation remains at 46 kV. Thus for towers or lines with low values of CFO, the initial observation is that arresters should be added to at least 2 towers beyond the protected section. However, note that if a stroke terminates on tower 3, the voltages across the insulation at tower 4 will revert to those at tower 3 of Fig. 29a. Thus for this low CFO, arresters should be placed on all towers.

8.3 Stroke to Ground Wire

Again assume a stroke to TWR1 of Fig. 26. The voltages on the ground wire and on the conductor are

$$e_g = \frac{(Z + Z_m)I + 2E_A}{Z + Z_m + 4R_1} R_1$$

$$e_c = e_g - E_A \qquad i_R = \frac{e_g}{R_1} \tag{63}$$

Also

$$i_A = \frac{2}{Z + Z_m + 4R_1} \left[R_1 I - \frac{Z + 2R_1}{Z - Z_m} E_A \right] \tag{64}$$

To determine the voltages at TWR2, Eqs. 59 are used, resulting in

$$e_I'' = \frac{Z - Z_m}{Z + 2R_2} e_g - E_A \tag{65}$$

Using the 115-kV line as an example, with $R_1 = 29$, $R_2 = 10$ ohms and a 100-kA stroke current, $e_g = 2350$ kV, and the voltage across the insulation at TWR2 is 1230 kV, with a maximum voltage at some distant tower of 1316 kV. To decrease this voltage, an arrester is installed at TWR2. Then for a stroke to TWR2, $e_g = 932$ kV, and the voltage across the insulation at TWR3 for $R_3 = 10$ ohms is 314 kV (maximum = 348 kV), which is equivalent to about four insulators. Thus to guard against flashovers on the unprotected section, arresters may be installed at the next low ground resistance tower.

9 COMMENTS/OBSERVATIONS

9.1 General

1. Economics: The first and most essential part of a proposed application of arresters is the specification of the required degree of line performance. This may differ significantly depending on the service required by a customer. Following the technical feasibility studies to specify the rating, number, and location of the arresters, studies should be made to assess the economics of arresters versus other methods of improving the line performance [41].

2. Other Methods of Improvement: Except for arrester failures, arresters applied to lines decrease the outage rate to zero. However, there are other measures—increasing insulation level, underbuilt ground wires, decreasing footing resistance that can be used to decrease the outage rate that may be more economical. These methods can also be used for low-voltage lines.

3. Calculating Energy: Although some equations are presented to estimate the arrester energy, the analysis is unsuccessful in estimating the energy for practical applications, e.g., for 11 towers with arresters. Therefore, at present, the most practical method is to use a computer program such as ATP or EMTP. Another method is to determine the minimum energy, e.g., for 11 towers, by producing a curve of this minimum energy as a function of the energy for a single tower with arresters.

4. Skipping Towers: Since the arrester energy at the struck tower is a function of the span length, or the distance between arresters, locating arresters on every second or third tower is detrimental. In addition, arresters may not be able to protect the towers without arresters.

5. Beyond the Protected Section: If arresters are only used on a portion of the line, there exists a danger that flashovers can occur at unprotected towers adjacent to the protected section. This is not unexpected. Arresters do have a limited protective range, as developed in Chapters 9 and 13. In general, arresters should be applied to one or more towers in the unprotected section.

6. Arresters Not on All Phases: The suggestion has been made that arresters need only be installed on the phase that has the highest probability of flashover, e.g., on the outside phases of a single-circuit line having a horizontal configuration of phases, or on lower phases of a double-circuit line having a vertical phase configuration. Although this will reduce the flashover rate, it will not reduce it to zero. See Chapter 10 for methods of calculating the flashover rate.

7. Arresters Only on One Circuit: Arresters installed on one circuit, of a double-circuit line will eliminate double-circuit outages. However, they will not eliminate flashovers of the other circuit; see Chapter 10.

8. Distribution: Although arrester application on lines with overhead ground wires is technically feasible, the distribution or low-voltage lines without an overhead ground wire stress the application technology, and only results from field investigations will relieve this concern. While the arresters applied to shielded lines are protected by the overhead ground wire, the arrester applied to the distribution or unshielded lines must withstand the hostile environment of natural lightning. However, on the positive side, distribution arresters have successfully protected equipment and have survived.

9.2 Specific

1. Shielded Lines: In general there exist no technical problems with the application of arresters to shielded lines. The energy discharged by the arresters is within the arrester capability. Since shielding failure currents are limited to about 5 to 10 kA, depending on the line insulation level, arresters may be located at every second tower. Also, depending on the line insulation level, arresters may not be necessary on the unprotected sections. In general, the combination of the ground wire and the arresters provides an excellent protection.

2. Unshielded Lines: The application depends heavily on the line insulation level. To assess the arrester failure rate, a probabilistic method is provided that is based on the observation that the arrester failure energy is a probabilistic function. As an example, for the 115-kV line, a failure rate of 0.55% per year was calculated. For the 115-kV line, arresters located every second tower are possible, and the nonuse of arresters on the unprotected line section is also possible. For low-voltage lines, application from an arrester energy standpoint appears possible. However, the remaining conclusions are dependent on the use of a neutral. For lines without a neutral, arresters located on every other tower do not appear possible, nor does the protection of adjacent unprotected line sections. Thus, depending on the performance desired, arresters may be required on the entire line. The use of a neutral is beneficial in that applying an arrester at every second tower is possible, and arresters can protect the unprotected section.

10 REFERENCES

1. C. H. Shih, T. L. Jones, A. P. Litsky, and L. Panek, "Application of Special Arresters on 138-kV Lines of Appalachian Power Company, *IEEE Trans. on PA&S*, Oct. 1985, pp. 2857–2863.
2. R. E. Koch, J. A. Timoshenko, J. G. Anderson, and C. H. Shih, "Design of Zinc Oxide Transmission Line Arresters for Application on 138-kV Towers," *IEEE Trans. on PA&S*, Oct. 1985, pp. 2675–2680.
3. E. L. Harder and J. M. Clayton, "Line Design Based upon Direct Strokes," Chapter 17 of *Electrical Transmission and Distribution Reference Book*, Westinghouse Electric Corporation, 1950.
4. S. Furukawa, O. Usuda, T. Isozaki, and Y. Trie, "Development and Application of Lightning Arresters for Transmission Lines," *IEE Trans. on PWRD*, Oct. 1989, pp. 2121–2129.
5. K. Ishida, K. Dokai, T. Isozaki, T. Trie, T. Nakayama, Y. Aihara, H., Fujita, and K. Arakawa, "Development of a 500-kV Transmission Line Arrester and Its Characteristics," *IEEE Trans. on PWRD*, Jul. 1992, pp. 1265–1274.

6. T. Yamada, T. Sawade, E. Zaima, T. Erie, T. Ohashi, S. Yoshida, and T. Kawamura, "Development of Suspension-Type Arresters for Transmission Lines," *IEEE Trans. on PWRD*, Jul. 1993, pp. 1052–1059.
7. L. Stenstrom and J. Lundquist, "New Polymer-Housed Zinc Oxide Surge Arrester for High Energy Application," CIGRE SC33 Colloquium, 1993 2.4 IWD.
8. N. Fujiwara, T. Yoneyama, Y. Hamada, S. Ishibe, T. Shimomura, and K. Yamaoka, "Development of a Pin-Post Insulator with Built-In Metal Oxide Varistors for Distribution Lines," *IEEE Trans. on PWRD*, Apr. 1996, pp. 824–833.
9. Y. Nagayama, K. Kohara, M. Takanashi, K., Isumi, S. Shirakawa, and J. Ozawa, "Insulation Coordination Between Zinc Oxide Arrester and Suspension Insulation on the 110-kV Transmission Line," CIGRE SC33 Colloquium, 1989.
10. M. Sato, A. Horide, H. Shibata, Y. Mishima, T. Ichioka, and K. Horrii, "Installation Experience of Lightning Arresters for Transmission Lines," CIGRE SC33 Colloquium, 1989.
11. E. J. Tarasiewvicz, "Lightning Performance of Transmission Surge Arresters on 115-kV Transmission Lines," CIGRE SC-3-95 (WD11)11, IWD.
12. M. G. Comber and R. L. Zinser, "Lighting Protection of Transmission Lines with Polymer-Housed Surge Arresters," CIGRE SC33, Rio de Janeiro, 1996.
13. S. Sadovic, R. Joulie, S. Tartier, and E. Brocard, "Use of Line Sure Arresters for the Improvement of the Lightning Performance of 63 kV and 90 kV Shielded and Unshielded Transmission Lines," *IEEE Trans. on PWRD*, Jul. 1997, pp. 1232–1240.
14. T. Yamada, T. Narita, H. Ota, K. Saito, and E. Zaima, "Field Experience of Line Arresters in TEPCO," CIGRE SC-33 Colloquium, Toronto, Sept. 1997.
15. D. Kundu, "An Approach to Reducing Lightning Outages on 44-kV and 27.6-kV Sub-transmission Circuits using Line Surge Arresters," CIGRE SC-33 Colloquium, Toronto, Sept. 1997.
16. C. Tirado and F. de la Rosa, "Lightning Protection of Transmission Lines with Surge Arresters," CIGRE SC-33 Colloquium, Toronto, Sept. 1997.
17. P. C. V. Esmeraldo, "Considerations About the Use of Line Surge Arresters," CIGRE SC-33 Colloquium, Toronto, Sep. 1997.
18. N. Beldose, "155-kV Line Protectors at Georgia Power Company," *Hi-Tension News*, vol. 57, 1987.
19. R. Reedy, "City of Lakeland Adopts New Design–Protect Lite System with Reduced Pole Heights," *Hi-Tension News*, vol. 63, no. 1, 1992.
20. D. Kundu, "An Approach to Reduce 44-kV Line Outages Due to Lightning," *Hi-Tension News*, vol. 64, no. 1, 1993.
21. P. Giacomo and G. Post, "Wadsworth Utilities Upgrades 69-kV System," *Hi-Tension News*, vol. 64, no. 3, 1993.
22. W. R. Kelly, "Virginia Power Uprate & Retrofit–Lightning Arresters," *Hi-Tension News*, vol. 64, no. 3, 1993.
23. D. Mitchell, "Alabama Power Retrofits 44-kV Transmission Lines to Add Lightning Arresters," *Hi-Tension News*, vol. 65, no. 1, 1994.
24. S. Ito, T. Ichihara, and Y. Ohgi, "Application of Lightning Arresters for Transmission Lines and Its Effect," CIGRE SC33 Colloquium, 1987.
25. A. R. Hileman, L. Stenstrom, C. T. Gaunt, D. Volker, T. Kamamura, and M. G. Comber, "Application of Arresters to Transmission Lines," Task Force Report, CIGRE 33-92(WG11)6IWD.
26. T. Kawamura, L. C. L. Cherchiglia, M. G. Comber, F. de la Rosa, C. T. Gant, M. Kobayashi, J. Michayd, L. Stenstrom, and O. Volcker, "Application of Metal Oxide Surge Arresters to Overhead Lines," CIGRE Task Force 3 of WG 33.11, CIGRE SC33 Colloquium, Toronto, Sept. 1997.

27. L. Stenstrom and M. Mobedjina, "Proposal for a Test Procedure to Determine the Arrester Energy Capability for Zinc-Oxide Surge Arresters, "CIGRE SC33-93(WG11)16, IWD and SC33-93(Coll)3.9, IWD.
28. L. Stenstrom, "Overvoltages and Arrester Currents Along Transmission Line Section Protected by Surge Arresters," CIGRE SC33-95(WG11)10, IWD11.
29. D. S. Birrell, A. Hirany, and B. A. Clairmont, "Requirements for Transmission Line Surge Arresters," CIGRE SC33 Colloquium, Toronto, Sep. 1997.
30. D. S. Birrell, A. Hirany, and B. A. Clairmont, "Energy Testing of Transmission Line Surge Arresters," CIGRE SC33 Colloquium, Toronto, Sep. 1997.
31. L. Stenstrom, "Required energy Capability Based on the Total Flash Charge for a Transmission Line Arrester for Protection of a Compact 420-kV Line for Swedish Conditions," CIGRE SC-33 Colloquium, Toronto, Sep. 1997.
32. W. P. Goch, "Testing of Line Arresters: Class or Application," CIGRE SC33 Colloquium, Toronto, Sep. 1997.
33. M. Kobayashi, H. Sasaki, A. Sawada, and N. Nakamura, "Energy Absorption of Gapless Arresters for Overhead Transmission Lines," CIGRE SC-33 Colloquium, Toronto, 1997.
34. M. Bartkowiak, M. G. Comber, and G. D. Mahan, "Failure Modes and Energy Absorption Capability of ZnO Varistors," *IEEE Trans. on PWRD*, paper PE-135-PWRD-0-12-1997, presented at the 1998 PES Winter Meeting, Tampa, FL.
35. K. G. Ringler, P. Kirby, C. C. Erven, M. V. Lat and T. A. Malkiewicz, "The Energy Absorption Capability and Time to Failure of Varistors Used in Station-Class Metal Oxide Surge Arresters," *IEEE Trans. on PWRD*, Jan. 1997, pp. 203–212.
36. M. L. B. Martinez and L. C. Zanetta, "A Testing method to Evaluate the Energy Withstanding Capacity of Metal Oxide Resistors for Surge Arresters," CIGRE SC-33 Colloquium, Toronto, Sept. 1997.
37. M. L. B. Martinez and L. C. Zanetta, "Comments on the Energy Withstand Capability of Metal Resistors for Surge Arresters," CIGRE SC-33 Colloquium, Toronto, Sept. 1997.
38. R. D. Melchior, J. S. Williams, and N. P. McQuinn, "Fault Testing of Gapless Zinc Oxide Transmission Line Arresters Under Simulated Field Conditions," *IEEE Trans. on PWRD*, Apr. 1995, pp. 786–796.
39. E. Colombo, M. de Nigris, A. Sironi, and M. Cologno, "Failure Mode Tests for Distribution Type Metal-Oxide Surge Arresters with Polymeric Housing," *Trans. on PWRD*, Jan. 1996, pp. 240–252.
40. T. E. McDermott, D. E. Parrish, and D. B. Miller, *Lightning Protection and Design Workstation Seminar Notes*, EPRI TR-000530, Sep. 1992.
41. E. J. Tarasiewicz, "Analysis of Lightning Outage Reduction Versus Cost for Overvoltage Protection of Transmission Lines," CIGRE SC33 Colloquium, Toronto, Sept. 1997.
42. P. P. Baker, and R. T. Manco, "Characteristics of Lightning Surges on Distribution Lines," EPRI Report TR-100218, Dec. 1991.

11 PROBLEMS

1. A 230-kV double circuit line of Problem 1 in Chapter 9 has a footing resistance of 100 ohms with a $\rho = 2000$ ohm-meters. The objective is to reduce the flashover rate to 1.0 FO/100 km-yrs or less.

   1.1 Determine the location of arresters on the line and if arresters can be placed every 2nd tower?

Line Arresters

1.2 If the additional criteria that double circuit flashovers are to be eliminated, determine the location of arresters and if arresters can be placed every 2nd tower?
Assume: $N_g = 4$, Span $= 1000$ ft, CFO $= 1200$ kV, $K_{PF} = 0.4$, $I = 100$ kA, $S_I = 42$ kA/µs and $E_A = 400$ kV   Use SRGKON95 and BFRCIG programs.

2. For an unshielded line the curve of permissible stroke current, I, versus the time to half value of the stroke current for a specific probability of arrester failure can be represented by the equation:

$$It_T = 4000\,kA - \mu s \qquad (66)$$

Assume that all values of t_T-I which exceed this curve result in arrester failure. Assume also that there are 3 arrester per tower, the span length is 200 m and 80 strokes per 100 km-yrs terminate on the line. Find the probability of failure and the failure rate in terms of numbers of arresters failed per 100 km. Calculate the arrester failure rate per 100 km-years for a span length of 200 meters and 3 arresters per tower.

3. Assume the 12-kV line of Figure 33 in Chapter 10 has no overhead ground wire and that the CFO $= 300$ kV. On a line section comprising 3 towers, the high current or impulse footing resistance is $R_i = 25$ ohms. The remaining towers have footing resistances of 10 ohms. Evaluate this application using a stroke current of 30 kA. Let the arrester characteristic $\alpha = 4.7$ and assume the use of 10.2-kV heavy duty polymer distribution arresters, Ohio Brass. Let $Z = 366$ ohms. Also assume that the time to half value of the tail of the stoke current is 77 µs. Span length is 100 ft. Assume that the arrester maximum energy capability is 32 kJ. Evaluate this application and provide comments.

15
Induced Overvoltages

1 INTRODUCTION

Lightning flashes that terminate on the earth or on any adjacent object near the distribution or transmission line induce voltages on the phase conductors, on the ground wires (or neutral), and across the insulation. These voltages or overvoltages were neglected in Chapter 10 when estimating the flashover rate (BFR) of transmission and distribution lines. The purpose of this chapter is to investigate these induced overvoltages for both distribution and higher voltage lines. One may suspect that the effect of induced overvoltage is dominant for distribution lines where the insulation level is low, and where overhead ground wires are not normally employed.

Of some historical interest is that until about 1930, it was believed that all lines should be designed considering only the nearby stroke, i.e., induced overvoltages. Designers believed (1) that the probability of a stroke terminating directly on the line was very remote and (2) that if a stroke did terminate on the line, it was virtually impossible to prevent line flashover. This concept was dispelled in about 1930 when the "direct stroke" theory was presented [1–3]. From that time onward, all line designs, except for low voltage lines, were designed on the basis of a stroke terminating directly on the overhead ground wire.

A considerable volume of work and study has recently taken place in the area of induced voltages. Rusck developed simplifying equations to estimate the maximum induced voltage on a line of infinite extent [4], and Eriksson et al. investigated induced voltages on an 11 kV distribution test line [5]. An IEEE working group presented a paper on estimating performance of distribution lines [6], and most recently McDermott et al. presented new insights on the protection of distribution lines [7].

The calculation of induced overvoltages is composed of two components, (1) the return stroke model with its associated electric field and (2) the coupling model, which employs the fields to obtain the potential on the conductors. Concerning the return stroke model, almost universally the stroke is assumed as a straight vertical channel. The calculation of the vertical and horizontal electric fields is estimated by use of equations for the current and charge. Concerning the coupling model, two basic models are prevalent, one by Agrawal et al. [8] and the other by Chowdhuri and Gross [9–11]. However, several other methods have been proposed [12–14]. In a classic paper by Rachidi et al. [15], the Agrawal et al. model was selected and compared with the more simple Rusck model. The authors reported only a 6% lower voltage using the Rusck equation. The primary purpose of this paper was to investigate the effect of multiple conductors on the induced voltage. For a horizontal configuration of three conductors without ground wires or a neutral, the authors found a 15% lower voltage on the middle conductor. Also in a recent paper, Barker et al. [16] reported on an experiment using rockets to trigger the lightning flash. Thus the termination of the flash is "exactly" known, and with the measurement of the current, the resulting voltages can be compared to any suggested equation. The authors found that the induced voltage is linearly related to the stroke current but exceeds that given by the Rusck equation by 63%. Later unpublished results, for strokes closer to the line, have not confirmed this large difference. In contrast, the results obtained by Eriksson et al [5] on the 11-kV test line verify Rusck's equation.

From oscillographic records, Barker et al. [16] also provided estimates of the waveshape. The median rise time (10–90%) is 1.6 µs, and 5% are less than 1.1 µs. Thus the time to crest is considerably greater than for the first stroke of the flash. The time to half values is very small, in the order of 4 or 5 µs for a stroke current to half value of about 60 µs. Based on these values, the CFO for this nonstandard time to half value is approximately 1.4 times the CFO for the standard lightning impulse waveshape (see Chapter 2).

The primary advantage of Rusck's equation is its simplicity, which can be employed with ease to estimate the induced voltage flashover rate, IVFOR. Therefore (1) the Rusck equation will be used in this chapter and (2) the reduction in voltage caused by multiple conductors will be ignored.

Other useful information can be found in Refs. 17–27. The presentation in this chapter will commence with a theoretical presentation, which will then permit the evaluation of the effect on distribution and transmission lines.

Unless otherwise stated, the examples in this chapter use the following data for the lines.

1. Ground wire or neutral surge impedance $Z_g = 450$ ohms
2. Mutual surge impedance $Z_m = 130.5$ ohms or coupling factor $C = 0.29$
3. Height of phase conductor $h_c = 10$ or 8 meters
4. Height of ground wire or neutral $h_g = 10$ or 8 meters
5. Footing resistance $R = 20$ ohms
6. Ground flash density $N_g = 1$ flash/km^2-year
7. Velocity of first return stroke $v = 0.3$ per unit of the speed of light
8. Striking distance equation = Brown–Whitehead
9. Height of trees or forests, h_t = variable

10. Distance between line and trees or forests, S_{12} = variable

In cases where the ground wire or neutral is higher than the phase conductor, $h_g = 10$ m and $h_c = 8$ m. If the phase conductor is higher than the neutral or ground wire, $h_c = 10$ m and $h_g = 8$ m.

2 CALCULATION OF INDUCED VOLTAGES

To start, assume as shown in Fig. 1 that a single-phase conductor is located at a height of h_c above ground. For any specific stroke current I, the number of strokes that terminate on the conductor can be determined by use of the geometric model as presented in Chapters 6 and 7. That is, with r_{cc} as the striking distance to the phase conductor and r_g as the striking distance to earth or ground, the number of strokes that terminate on the conductor is

$$2N_g L D_g f(I)\, dI \tag{1}$$

where N_g is the ground flash density, L is the length, $f(I)$ is the probability density function of the first stroke current, and D_g is

$$D_g = \sqrt{r_{cc}^2 - (r_g - h_c)^2} \tag{2}$$

Considering all stroke currents, the total number of strokes that terminate on the conductor, N_c, is

$$N_c = 2N_g L \int_3^\infty D_g f(I)\, dI \tag{3}$$

Per Fig. 1, at distances beyond $x = D_g$, all strokes terminate to ground, and these strokes produce induced voltages on the conductor. Rusck [4] shows that for an

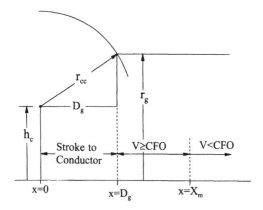

Figure 1 The concept: strokes terminating at distance greater than D_g cause induced voltages.

infinite line, the maximum induced voltage V_c occurs at the line location closest to the stroke terminating point, that is, at $y = 0$ in Fig. 2. It is given by the equation

$$V_c = \frac{30Ih_c}{x}\left[1 + \frac{v}{\sqrt{2-v^2}}\right] = \frac{30Ih_cK_v}{x} \quad (4)$$

where I is the stroke current, v is the velocity of the return stroke in per unit of the velocity of light, and K_v is a convenient notation for

$$K_v = 1 + \frac{v}{\sqrt{2-v^2}} \quad (5)$$

Induced voltages are also developed at other locations along the line, but the time at which this voltage occurs is delayed, since the inducing field must have additional time to reach these locations and the voltage is reduced.

Usually, v is set to an average value of 0.3, although per Wagner's development as presented in Chapter 6, the velocity is a function of the stroke current and can be approximated by the equation

$$v = \frac{0.486}{1 + 27.3/I} \quad (6)$$

However, little difference in voltage exists between using this equation and assuming a constant value of 0.3.

Consider that a line has a specific insulation level or a specific CFO. Equating the voltage to the CFO, the distance X_m beyond which the voltage is lower than the CFO is

$$X_m = \frac{30Ih_cK_v}{\text{CFO}} \quad (7)$$

To explain with the aid of Fig. 1, between $x = D_g$ and $x = X_m$, the induced voltage is greater than the CFO, and flashover results. Beyond $x = X_m$, the induced voltage is less than the CFO, and no flashovers occur.

As noted, both D_g and X_m increase with increasing current, and X_m must be equal to or greater than D_g for any induced voltage to exist. Assuming a CFO of

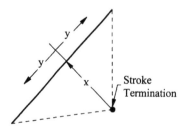

Figure 2 Maximum voltage occurs at $y = 0$.

Induced Overvoltages

250 kV, a conductor height of 10 meters, $v = 0.3$, and using the Brown–Whitehead striking distance equations, Fig. 3 illustrates the concept. The curve of D_g meets the curve of X_m at a current of $I_{SC} = 25$ kA. Only for currents greater than 25 kA does an induced voltage occur. For currents less than 25 kA, no induced voltages occur. Figure 4 shows the induced voltages for currents greater than 25 kA. If the stroke occurs at X_m, the voltage is 250 kV, by definition. But if the stroke occurs at D_g, the voltage increases as shown. For any specific current greater than I_{SC}, the number of strokes that produce a voltage greater than the CFO is equal to the horizontal distance $X_m - D_g$ multiplied by the line length and the ground flash density N_g. Therefore the incremental flashover rate dP is

$$dP = 2N_g L(X_m - D_g)(f(I)\,dI \tag{8}$$

and considering all stroke currents, the number of strokes that produce a voltage greater than the CFO, here denoted as IVFOR or the induced voltage flashover rate, is

$$\text{IVFOR} = 2N_g L \int_{I_{SC}}^{\infty} (X_m - D_g) f(I)\,dI \tag{9}$$

To illustrate this equation, Fig. 5 shows the components of the integrand and the total integrand. Note that the plots are not to the same scale. The area under the curve marked $(X_m - D_g)f(I)$ multiplied by $2N_g L$ is the IVFOR, which is 4.93 flashovers/100 km-years for $N_g = 1$.

Using Eq. 7, the IVFOR can be determined for various values of CFO. That is, the minimum voltage on the conductor is equal to the CFO, and the IVFOR is the number of voltages that equal or exceed the CFO. Thus a curve can be obtained as shown in Fig. 6. Reading the curve, there are 2.82 voltages per 100 km-years that

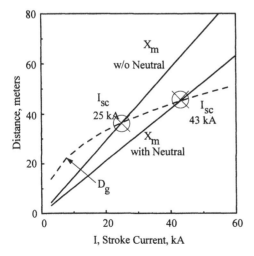

Figure 3 Induced voltage occurs for $I > I_{sc}$.

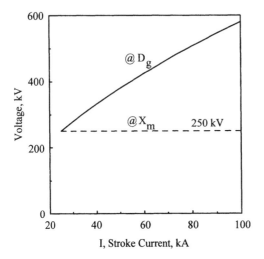

Figure 4 Without neutral, $h_c = 10$.

exceed 300 kV. Or, only considering voltages above 100 kV, 10% are greater than 300 kV and 3.3% of the voltages are greater than 400 kV. This compares well with the previous thoughts that induced voltages are generally 300 kV or less.

3 CONSIDERING THE NEUTRAL OR GROUND WIRE

If a neutral conductor, or a ground wire, or any other conductor is present, and *if this conductor is ungrounded*, the voltage on the other conductor can be calculated

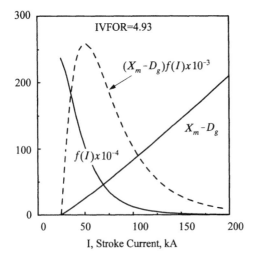

Figure 5 Components of integrand of IVFOR, without neutral, $h_c = 10$.

Induced Overvoltages

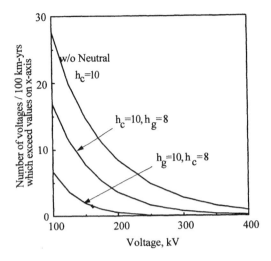

Figure 6 Distribution of voltages.

per Eq. 4. Per Fig. 7, let this other conductor be at height h_g and located directly under the uppermost conductor. Let the voltage on this lower conductor be denoted as V_g. Then the voltage between these conductors, denoted as V_I, is simply $V_c - V_g$ or

$$V_I = \frac{30 I K_v (h_c - h_g)}{x} \tag{10}$$

However, if the other conductor is grounded through a resistance R, the voltage on this conductor decreases, and the voltage between the conductors increases. To obtain an approximation of this voltage, consider the circuit of Fig. 8, where the induced voltages on the conductors, calculated by Eq. 4, are denoted e_c and e_g. From Chapter 9, the resultant voltages are

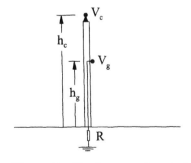

Figure 7 Conductor and neutral.

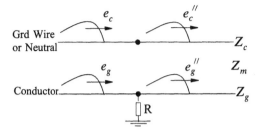

Figure 8 Voltage across insulation increases.

$$e_c'' = e_c - \frac{Z_m}{Z_g + 2R} e_g$$

$$e_g'' = \frac{2R}{Z_g + 2R} e_g \qquad (11)$$

where Z_m is the mutual surge impedance, Z_g is the surge impedance of the ground wire or neutral conductor, and R is the tower footing resistance. Therefore, the voltage V_I becomes

$$V_I = \frac{30 I K_v}{x}\left[h_c - \frac{Z_m + 2R}{Z_g + 2R} h_g\right] \qquad (12)$$

and X_m becomes

$$X_m = \frac{30 I K_v}{\text{CFO}}\left[h_c - \frac{Z_m + 2R}{Z_g + 2R} h_g\right] \qquad (13)$$

Note that the induced voltage across the insulation increases as the footing resistance decreases. Thus, low footing resistance results in higher voltages—just opposite of that for a stroke to the tower!

The IVFOR for this case is

$$\text{IVFOR} = 2 N_g L \int_{I_{SC}}^{\infty} (X_m - D_g) f(I)\, dI \qquad (14)$$

where I_{SC} is the current at which the new value of $X_m = D_g$. Such a case is also illustrated in Fig. 3 for the same parameters as before with the added neutral at a height of 8 meters. As shown, the value of I_{SC} is increased to 43 kA, and thus the IVFOR is reduced to 1.67 flashovers/100 km-years, about 34% of that without the neutral.

The distribution of induced overvoltages across the insulation is also shown in Fig. 6 for the same parameters as before. The overvoltages are significantly reduced from that when the neutral is not considered. There are only 0.76 voltages/100 km-years that exceed 300 kV, a reduction of 73%.

The effect of the CFO on the IVFOR and thus the total flashover rate is shown in Fig. 9 and compared with a similar curve from Ref. 5. For CFOs less than about

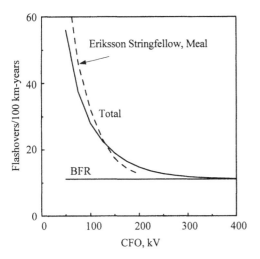

Figure 9 No trees, $h_c = 10$, $h_g = 8$.

200 or 250 kV, the IVFOR is dominant. For this case of the phase conductor above the neutral, the CFO of 250 kV or above is fairly easy to obtain on wood-pole lines. A similar curve provided by McDermott et al. [7] indicates much lower values for the IVFOR. Although all the parameters of the curve are not known, the primary difference is the striking distance equations used. Their equations assumed that $r_c = 10 I^{0.65}$ and $r_g = 0.9 r_c$. If these striking distance equations are used along with the analysis in this chapter, the results are similar to those of Ref. 7. These striking distance equations are similar to the IEEE-92 equations, and as will be shown later, these equations results in significantly lower values of IVFOR.

To this point, all examples or illustrations have considered a neutral located below the phase conductor. The equations are equally valid for the case of an overhead ground wire located above the phase conductor. For example, considering the same parameters as before except interchanging the conductor and the neutral, which is now the ground wire, the IVFOR is 0.159 flashovers/100 km-years, considerably less than for the neutral below the phase conductor. This is primarily caused by the value of X_m per Eq. 13. If the phase conductor is uppermost, the $X_m = 1.05(I)$, whereas if the ground wire or neutral is uppermost, $X_m = 0.66(I)$ and thus, since the curve of D_g is unchanged, the value of I_{sc} increases and is about 92 kA. Figure 6 also contains a curve for this case and illustrates the benefit of the overhead ground wire in reducing the induced voltages. Only 0.045/100 km-years exceed 300 kV.

The example for a conductor without a neutral or ground wire is somewhat impractical, since it did not consider the actual CFO, i.e., the flashover path. From the equations and analysis presented here, flashover between phase conductors does not appear possible, since the voltage difference is small even for phases vertically configured. The other flashover path is to ground or earth and the CFO is large, in the order of 2000 to 3000 kV. Thus flashover, as illustrated by Fig. 6, is remote. The danger in this case comes from strokes terminating on the phase conductor, resulting in very high overvoltages. For example, the voltage on the conductor for a stroke to the conductor is $IZ_g/2$, and therefore the critical current is in the range of 9 to 13 kA. (Approximately 90% of the strokes have currents greater than these values.)

High overvoltages on the line are detrimental from two standpoints: (1) large energies can be discharged through a single adjacent arrester and (2) the incoming surge to the substation can be very large, inhibiting the protection of equipment. Considering the second problem, to reduce the magnitude of the incoming surge, the IEC application guide recommends the use of a protective gap on the towers or poles adjacent to the station. Considering the first problem, protective gaps can be placed on the poles. The gap spacing is set so that the gap flashover occurs at about 300–400 kV. Thus flashover does not occur for most induced overvoltages, and the arrester energy capability is sufficient. However, for those strokes that terminate directly on the line and endanger the arrester, flashovers occur to limit the voltage.

To this point, all calculations have used the Brown–Whitehead equations. Table 1 shows the IVFOR for the alternate striking distance equations for a CFO = 250 kV, $v = 0.3$, $Z_g = 450$ ohms, $Z_m = 130.5$ ohms or C (coupling factor) = 0.29, and $N_g = 1$. The Brown–Whitehead and Love equations result in the same IVFOR. Likewise for the substation and the Young equations. The IEEE-92 equations result in lowest values of IVFOR.

The remaining consideration is the division between the IVFOR and the BFR, which is shown in Table 2 using the Brown–Whitehead equations. For the case of the ground wire above the phase conductor, the BFR is estimated using the CIGRE method with $R_0 = 30$, $\rho = 600$ ohm-meters ($R_i = 20$ ohms), and a span length of 50 meters. For the normal condition, where the neutral is below the phase conductor, the IVFOR is 13% of the total flashover rate. For the neutral or ground wire above the phase conductor, the IVFOR is only 6% of the total. In addition, this configuration provides the lowest total flashover rate. Thus for the normal case, a stroke to the phase conductor is the primary source of flashover.

4 EFFECT OF NEARBY OBJECTS

Objects such as trees or buildings adjacent to the line will shield the line and thus decrease the flashover rate of the line caused by strokes terminating directly on the line. However, these objects now receive more strokes, and the induced voltage flashovers may increase. Thus the total line flashovers, the sum of those caused by direct strokes and those caused by induced voltages, may increase or decrease.

Table 1 IVFOR for Alternate Striking Distance Equations

Equations	$h_c = 10$ m, $h_g = 0$	$h_c = 10$ m, $h_g = 8$ m	$h_c = 8$ m, $h_g = 10$ m
Brown–Whitehead	4.93	1.67	0.159
IEEE-92	2.54	0.57	0.015
Love	4.48	1.54	0.165
Substations (Mousa)	6.10	2.35	0.314
Young	6.22	2.69	0.524

Induced Overvoltages

Table 2 Flashovers/100 km-years for Lines Not Shielded by Trees or Forests

Type	$h_c = 10\,\text{m}, h_g = 0$	$h_c = 10\,\text{m}, h_g = 8\,\text{m}$	$h_c = 8\,\text{m}, h_g = 10\,\text{m}$
IVFOR	4.93	1.67	0.159
BFR	11.15	11.15	2.73
Total	16.08	12.82	2.89

Induced overvoltages on lines by strokes terminating on elevated objects have not been studied, and the direct use of Rusck's equation is questionable. However, since no other method is presently available. Rusck's equation will be used with the caution that future developments may provide improved methods.

To develop the concept, consider the diagram of Fig. 10. The line is depicted at the left having a conductor height of either h_g or h_c, whichever is higher, which for convenience is denoted as h_{gc}. To the right of the line, at a distance S_{12}, are other shielding objects, in the figure described as trees having heights of h_T. If the trees extend to the right and left, they are defined as a "forest." If instead there is only a single line of trees on both sides of the line they are defined as "trees." In the same manner as in Chapter 8, when considering the shielding of stations, an arc of radius r_{cc}, i.e., the striking distance to the conductor, is drawn from the phase conductor (or ground wire), and an arc of radius r_{ct}, the striking distance to the trees, is drawn from the tree tops. Their intersection defines the shielding effect. To illustrate the decrease of D_g and thus the decrease in the number of strokes terminating on the phase conductor, the dashed lines show the value of D_g without the shielding effect of the trees. As noted, the strokes that formerly terminated on the phase conductor now terminate on the trees. These strokes are moved further away and produce less induced voltage, but the number of strokes that terminate at the "first" tree increases. Thus the number of induced voltage flashovers may increase or decrease. If the forest is considered, the arc r_{ct} becomes a horizontal line above the forest. If a

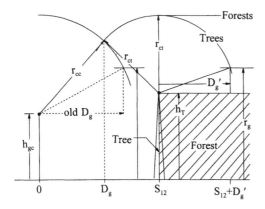

Figure 10 Concept: effect of trees and forests.

single line of trees is considered, the arc r_{ct} continues and meets the striking distance to ground r_g, and the distance D'_g is

$$D'_g = \sqrt{r_{ct}^2 + (r_g - h_T)^2} \tag{15}$$

More formally, the diagram of Fig. 11 can be used to determine the new value of D_g. For the case where $(r_{cc} + r_{ct})$ is greater than S, the resulting equations are

$$\alpha = \tan^{-1} \frac{h_T - h_{gc}}{S_{12}} \tag{16}$$

$$S = \sqrt{S_{12}^2 + (h_T - h_{gc})^2} \tag{17}$$

$$S' = \tfrac{1}{2}(r_{cc} + r_{ct} + S) \tag{18}$$

$$d = \tfrac{2}{3}\sqrt{S'(S' - r_{cc})(S' - r_{cT})(S' - S)} \tag{19}$$

$$S'_{12} = \sqrt{r_{cc}^2 - d^2} \cos\alpha \tag{20}$$

$$D_g = S'_{12} - d\cos\left(\tfrac{\pi}{2} - \alpha\right) \tag{21}$$

These equations are valid for both $h_T > h_{gc}$ and $h_{gc} > h_T$, and when the striking distance to ground is less than the distance represented by the dotted line H_T of Fig. 10. That is,

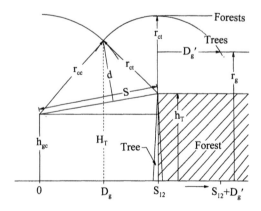

Figure 11 Derivation for trees and forests.

Induced Overvoltages

$$r_g < H_T \qquad H_T = h_{gc} + S'_{12} \tan \alpha + d \sin\left(\frac{\pi}{2} - \alpha\right) \tag{22}$$

The incremental induced voltage flashover rate, Δ (IVFOR) is dependent on trees or forests and on the location of X_m relative to D_g, S_{12} and D'_g. This dependency is better explained by use of Figs. 12a and 12b.

1. For both trees and forests, if a stroke occurs in the region between 0 and D_g or where $X_m < D_g$, all strokes will terminate on the ground wire or conductor. If the ground wire is uppermost, then the BFR should be estimated by the method of Chapter 10. If the phase conductor is uppermost, then the voltage on the conductor is $IZ_C/2$ or the critical current is $2(\text{CFO})/Z_C$. Flashover occurs when this voltage exceeds the CFO. Therefore the number of flashovers or the equivalent of the back-flash rate is

$$\text{BFR} = 2N_g L \int_{I_c}^{\infty} D_g f(I) \, dt \tag{23}$$

where I_c is the current at and above which flashover occurs. For low-voltage lines, I_C is very small, so that essentially all strokes that terminate on the phase conductor result in flashover. Therefore the BFR is

$$\text{BFR} = 2N_g L \int_{3}^{\infty} D_g f(I) \, dI = N_L \approx 2.8 h_c^{0.6} \text{ flashovers}/100 \text{ km-years} \tag{24}$$

where the second equation is from Chapter 6.

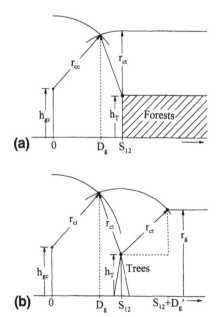

Figure 12 Deriving the Δ IVFOR for (a) forests and (b) trees.

2. For both trees and forests, if a stroke occurs in the region between D_g and S_{12}, all strokes will terminate on the first tree, and thus the IVFOR in this region is zero if $D_g < X_m < S_{12}$.

3. For a forest, if a stroke occurs in the region that is greater than S_{12}, i.e., if $X_m \geq S_{12}$, then the IVFOR becomes

$$\Delta(\text{IVFOR}) = 2N_g L(X_m - D_g) f(I) \tag{25}$$

4. For trees, if a stroke occurs beyond S_{12} and $S_{12} \leq X_m \leq (S_{12} + D'_g)$, then all strokes in this region will terminate on the tree. Therefore

$$\Delta(\text{IVFOR}) = 2N_g L(S_{12} + D'_g - D_g) f(I) \tag{26}$$

5. For trees, if a stroke occurs beyond $S_{12} + D'_g$ and $X_m > (S_{12} + D'_g)$, then

$$\Delta(\text{IVFOR}) = 2N_g L(X_m - D_g) f(I) \tag{27}$$

These equations for IVFOR are valid for the assumptions used in their derivation, i.e., that $S < (r_{cc} + r_{ct})$ and $r_g < H_T$. If these conditions are not achieved, a new set of equations must be employed. These equations are left as one of the problems.

Trees or forests may achieve a height so as to eliminate strokes from terminating on the line. This is illustrated in Fig. 13, where h_T is greater than h_{gc}, r_{cT} is greater than S_{12}, and

$$r_{cc} \leq h_T - h_{gc} + \sqrt{r_{ct}^2 - S_{12}^2} \tag{28}$$

As a final comment, the equations derived for the effect of the trees or forests are sufficiently complex that a computer program becomes necessary. Therefore, the program IVFOR is recommended.

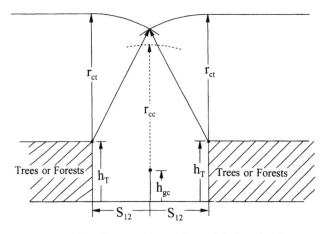

Figure 13 No direct strokes to line, all induced voltages.

5 CORRECTIONS TO THE CALCULATIONS

The IVFOR based on the previous equations is biased. That is, as discussed in Chapter 6, the calculated number of flashes to the line is usually less than that obtained by use of the CIGRE equations. That is, the calculated number of flashes to the line is

$$N_L \text{ (calculated)} = 2N_g L \int_3^\infty D_g f(I)\, dI \tag{29}$$

whereas the number per the CIGRE equation is

$$N_L(\text{CIGRE}) = N_g \frac{(28h^{0.6} + S_g)}{10} \quad \text{flashes/100 km-year} \tag{30}$$

where S_g is the distance between two overhead ground wires. Since the CIGRE equation is accepted and used in the calculation of the BFR, the IVFOR should be corrected by multiplying the calculated IVFOR by the ratio $N_L(\text{CIGRE})/N_L$ (calculated). If the ground wire is higher than the phase conductor, the BFR must be obtained by use of methods of Chapter 10. Since the number of flashes to the ground wire may be less than if trees or forests were not present, the value of N_g used in the calculation of the BFR must also be adjusted. These adjustments are easily performed in a computer program and thus appears another reason for use of the IVFOR program.

6 EXAMPLE—EFFECT OF TREES OR FORESTS

In the following examples, for the case when the ground wire is above the phase conductor, the BFR is calculated using an $R_0 = 30$ ohms, $\rho = 600$ ohm-meters, and a 50 meter span length. The resultant R_i is about 20 ohms.

The effect of the distance between the line and the trees or forest is shown in Fig. 14a for $h_c = 10$, $h_g = 8$, and $h_T = 10$ meters, and in Fig. 14b for $h_c = 8$, $h_g = 10$, and $h_T = 10$ meters. The flashover rate should be compared to the values in Table 2 for no trees or forests, i.e.,

For Fig. 14 a, compare to an IVFOR = 1.67, BFR = 11.5, total = 12.82
For Fig. 14b, compare to an IVFOR = 0.16, BFR = 2.37, total = 2.89

These flashover rates are approached as the distance S_{12} increases. Noticeably and predictably, the IVFOR is greater for trees than for forests, and the maximum flashover rates are larger when the phase conductor is above the neutral. In both cases, after about 20 to 40 m, as S12 increases, the IVFOR decreases and the BFR increases, thus battling each other for control.

The effect of the height of the trees or forests is illustrated in Figs. 14c and 14d, where $h_c = 10$, $h_g = 8$, and $S_{12} = 20$ meters. Here again, as h_T increases, the IVFOR increases and the BFR decreases. For forests, these two factors tend to cancel, giving an approximate constant value of the total flashover rate that is about equal to that for no forests, i.e., 12.82. In the case of trees, the IVFOR is predominant and produces a significantly larger total value than that without trees.

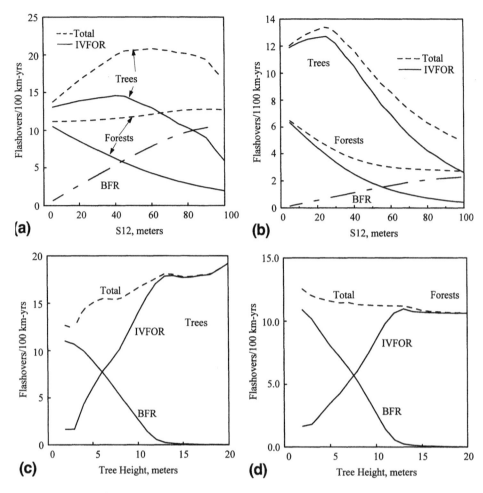

Figure 14 (a) $h_c = 10$, $h_g = 8$, $h_T = 10$; (b) $h_c = 8$, $h_g = 10$; (c) trees, $h_c = 10$, $h_g = 8$, $S_{12} = 20$; (d) forests, $h_c = 10$, $h_g = 8$, $S_{12} = 20$.

A comparison of the IVFOR and BFR is given in Table 3 for tree or forest heights h_T of 10 meters and a distance to the trees or forest S_{12} of 20 meters. As shown in Table 2, for lines without shielding by trees or forests, the BFR dominates. However, from Table 3, the IVFOR dominates, being from 76 to 98% of the total flashover rate. Although there are lines with no trees or forests along the right-of-way, the normal line has some type of shielding and thus to be expected is that the IVFOR will tend to dominate.

As a sanity check, the IVFOR of the 115-kV line of Chapter 14 is 0.0047 without trees or forests, 0.0093 and 0.0048 with forests and trees, respectively, with $h_T = 10$ ms and $S_{12} = 20$ m. Thus induced voltages do not affect the performance of higher voltage lines.

The effect of the CFO on the flashover rate is shown in Fig. 15 for a typical case of the phase conductor above the neutral, $S_{12} = 20$ m, and $h_t = 10$ m. As noted, the BFR is low as a result of shielding of the line. The dotted line is taken from Fig. 9 for

Table 3 Flashovers/100 km-year for Lines Shielded by Trees or Forests

	Forests		Trees	
Flashover rate	$h_c = 10, h_g = 8$	$h_c = 8, h_g = 10$	$h_c = 10, h_g = 8$	$h_c = 8, h_g = 10$
IVFOR	8.60	4.49	13.92	12.6
BFR	2.69	0.24	2.69	0.24
Total	11.29	4.73	16.61	12.88

no trees or forests. In this case the BFR is four times larger. However, the total flashover rates are about equal. Thus, based on the total flashover rate, the effect of trees or forests may increase, decrease, or be equal to that without trees.

7 STRIKING DISTANCE EQUATIONS—A REVIEW

For convenience of the reader, the striking distance equations presented in Chapter 8 on station shielding are listed below.

Young:

$$rg = 27I^{0.32}$$

$$\text{for } h \leq 18\,\text{m}, r_c = r_g \quad \text{otherwise } r_c = \frac{444}{462 - h} r_g \tag{31}$$

Brown–Whitehead:

$$r_g = 6.4I^{0.75}$$

$$\text{for } h \leq 18\,\text{m}, r_c = r_g \quad \text{otherwise } r_c = \left(1 + \frac{h - 18}{108}\right) r_g \tag{32}$$

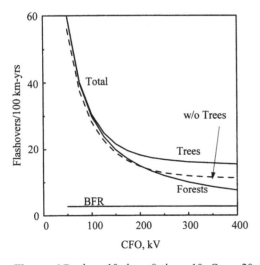

Figure 15 $h_c = 10$, $h_g = 8$, $h_T = 10$, $S_{12} = 20$.

Love:

$$r_g = r_c = 10 I^{0.65} \tag{33}$$

IEEE-1992:

$$r_g = 9.0 I^{0.65}$$
$$\text{for } h \leq 30, r_c = \frac{r_g}{0.36 + 0.17 \ln(43 - h)} \tag{34}$$
$$\text{for } h > 30, \text{ set } h = 30$$

Substations (Mousa):

$$r_g = r_c = 8.0 I^{0.65} \tag{35}$$

while I is the stroke current in kA, h is the height of the conductor, ground wire, or trees in meters, r_g and r_c are the striking distances in meters, and r_c may be either r_{ct} or r_{cc}.

8 PROTECTION AGAINST INDUCED VOLTAGES

Because the steepness of the induced surge is small compared to that for a direct stroke to the line, the protection afforded by surge arresters is improved—or the arrester can provide protection for longer distances. This translates into the thought that arresters need not be applied at every tower or pole. The equations of Chapter 14 can be used to assess this capability. Of most importance is the case of the phase conductor above a neutral. For this case, the equations of Chapter 14 must be reformatted, since they assume a stroke to the conductor and that the surge on the neutral is equal to the coupling factor times this surge. In the present case, as depicted in Fig. 16, a surge is induced on the conductor with a steepness S_c, and a surge is induced on the neutral with a steepness S_g. The surges are assumed to be induced at some point on the line, a travel distance T_1 from a tower with arresters. The objective is to determine the maximum value of T_1 such that the voltage at the unprotected pole is equal to the CFO. This travel time multiplied by the speed of light is the maximum separation distance from the arrester. From Chapter 14, for the case where there is no ground at the TWR2, i.e., R_2 is infinite,

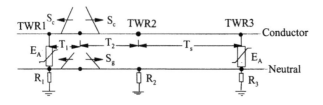

Figure 16 Protection of TWR2 without arresters.

$$T_1 = \frac{\text{CFO} - E_A}{2(S_c - S_g)} = \frac{\text{CFO} - E_A}{2S_c(1-K)} \tag{36}$$

where E_A is the arrester discharge voltage and K is

$$K = \frac{e_g}{e_c} = \frac{S_g}{S_c} \tag{37}$$

If a ground exists at TWR2, the value of T_1 becomes

$$T_1 = \frac{\text{CFO} - E_A - \dfrac{Z - Z_m}{Z + 2R_2} S_g''(t_f - t_A)}{2(S_c - S_g) - \dfrac{2(Z - Z_m)}{Z + 2R_2} S_g''} \tag{38}$$

where Z is the neutral and conductor surge impedance, Z_m is the mutual surge impedance, t_f is the time to crest of e_c, and t_A is the time to reach the arrester discharge. Also,

$$S_g'' = \frac{R_1(1+K)}{K(Z+Z_m) + R_1(3K-1)} S_g \tag{39}$$

and

$$t_A = \frac{E_A}{S_A} \qquad S_A = S_c - \frac{Z_m + 2R_1}{Z + 2R_1} S_g \tag{40}$$

As an example, assume CFO = 250 kV, e_c = 400 kV, e_g = 320 kV, R_1 = 20 ohms, Z = 450 ohms, Z_m = 130 ohms, E_A = 46 kV, and that there is no ground at TWR2, i.e., R_2 is infinity. Assume that t_f = 1.6 µs, S_c = 400/1.6 = 250 kV/µs, and S_g = 320/1.6 = 200 kV-µs. then T_1 = 2.04 µs and cT_1 = 612 meters. Assuming a 50-m span, an arrester should be located at every 5th pole. If a ground exists at TWR2 of 20 ohms, R_2 = 20 ohms, then S_A = 180.6 kV/µs, t_A = 0.255 µs, S_g'' = 14.6 kV/µs, and T_1 = 2.364 µs. Then cT_1 = 709 meters, and for 50 m spans, an arrester is necessary every 14 poles. Therefore, practically, an arrester should be installed at every 5th pole.

9 COMPARISON TO FIELD DATA

Eriksson, Stringfellow, and Meal reported the results of a field study on an 11-kV distribution line in South Africa [27]. A 9.9-km, three-phase, flat configured wood-pole line was constructed having a phase conductor height of 8.5 m. A ground wire (or neutral) was installed 1 m below the phase conductors so as to produce a 500-kV CFO. One end of the line terminated in a counterpoise, while the other end was opened circuited. The ground flash density was 7.5 flashes/km²-year. The terrain was described as undulating grassland with few trees. Over a two-year period, induced voltages were measured on the phase conductor. the maximum induced voltage measured was 300 kV (one in two years) and an average of 13.5 voltages/year exceeded 100 kV.

To compare this to a calculated value, the measured voltage must be corrected, since the authors measured the voltage on the phase conductor, whereas the voltage referred to in this chapter is the voltage between the phase conductor and the neutral. This is illustrated in Fig. 17, where a 100-kV surge on the phase conductor, e_c'', results in a voltage between the phase conductor and the neutral of 89.7 kV. Assuming a 20 ohm footing resistance, a coupling factor of 0.37, and using alternate striking distance equations, the calculated number of voltages exceeding 100 kV varied from a low of 6.5 for the IEEE-92 striking distance equations to a high of 10.5–10.7 for the substation and Young striking distance equations. Values of 9.4 and 8.7 resulted from the use of the Brown–Whitehead and Love equations. In general, considering the assumptions of the unknown parameters, except for the IEEE-92 equations, the comparison appears acceptable, with a 22 to 35% difference.

10 SUMMARY AND CONCLUSIONS

1. For lines without shielding by trees, forests, or other objects.

   (1) For the case of a line without a neutral or ground wire, the CFO to ground is very large, and the primary danger is the direct stroke to the phase conductor, which results in high voltages. The IEC application guide recommends that a protective gap on the line be considered so as to limit the incoming surge to a substation. This gap set for a sparkover of the maximum induced voltage of about 300 kV will also limit the duty on a single arrester employed to protect equipment.

   (2) A neutral below the phase conductor significantly decreases the induced voltage flashover rate. CFOs of 200–250 kV essentially eliminate flashover caused by induced voltages. If the CFO is less than about 200 kV, flashovers are primarily caused by induced voltages.

   (3) A ground wire above the phase conductor provides a further reduction in induced voltages and decreases the total flashover rate.

2. For lines with shielding by trees, forests, or other objects,

   (1) Total flashover rates may increase, decrease, or be equal to those without trees, depending on the tree height and the distance to the trees.

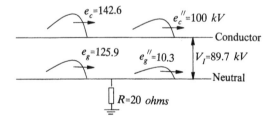

Figure 17 Corrections to apply to field data.

(2) Flashovers are caused primarily by induced voltages; the BFR is reduced as a result of the shielding by the trees.

(3) As compared to a forest, a single row of trees increases the flashover rate.

(4) Trees or forests as remote as 60 meters from the line affects performance.

3. Protection from flashover caused by induced voltages can be provided by line arresters that are widely spaced, e.g., every 5th pole.

11 REFERENCES

1. C. L. Fortescue, A. L. Atherton, and J. H. Cox, "Theoretical and Field Investigations of Lightning," *AIEE Trans.*, vol. 48, Apr. 1929, pp. 449–468 (also *AIEE Lightning Reference Book*, pp. 395–414).
2. C. L. Fortescue, "Direct Strokes–Not Induced Surges–Chief Cause of High-Voltage Line Flashover," *Electric Journal*, vol. 27, Aug. 1930, pp. 439–462, (also *AIEE Lightning Reference Book*, pp. 546–549).
3. A. C. Monteith, E. L. Harder, and J. M. Clayton, "Line Design upon Direct Strokes," Chapter 17, in *Electrical Transmission and Distribution Reference Book*, Westinghouse Electric Corporation, 1950.
4. S. Rusck, "Protection of Distribution Lines," chapter 23 in *Lightning*, vol. 2, *Lightning Protection* (R. H. Golde, ed.), New York, Academic Press, 1977.
5. A. J. Eriksson, D. V. Meal, and M. F. Stringfellow, "Lightning-Induced Overvoltages on Overhead Distribution Lines," *IEEE Trans. on Power Apparatus and Systems*, Apr. 1982, pp. 960–968.
6. IEEE Working Group, "Calculating the Lightning Performance of Distribution Lines," *IEEE Trans. on Power Delivery*, Jul. 1990, pp. 1408–1417.
7. T. E. McDermott, T. A. Short, and J. G. Anderson, "Lightning Protection of Distribution Lines," *IEEE Trans. on PWRD*, Jan. 1994, pp. 138–152.
8. A. K. Agrawal, H. J. Price, and S. Gurbaxani, 'Transient Response of a Multiconductor Transmission Line Excited by a Nonuniform Electromagnetic Field," *IEEE Trans. on Electromagnetic Compatibility*, May 1980, pp. 119–129.
9. P. Chowdhuri and E. T. B. Gross, "Voltage Surges Induced on Overhead Lines by Lightning Strokes," *Proc. IEE*, Dec. 1967, pp. 1899–1907.
10. P. Chowdhuri, "Lightning-Induced Voltages on Multiconductor Overhead Lines," *IEEE Trans. on PWRD*, Apr. 1990, pp. 658–667.
11. P. Chowdhuri, "Analysis of Lightning-Induced Voltages on Overhead Lines," *IEEE Trans. on PWRD*, Jan. 1989, pp. 479–492.
12. M. J. Master, M. A. Uman, Y. T. Lin, and R. B. Standler, "Calculations of Lighting Return Stroke Electric and Magnetic Fields Above Ground," *Journal Geophysical Research*, vol. 23, pp. 227–237.
13. M. J. Master, M. A. Uman, and W. Beadley, "Lightning Induced Voltages on Power Lines: Experiment," *IEEE Trans. on PA&S*, Sep. 1984, pp. 1319–1329.
14. V. Cooray, "Calculating Lightning-Induced Overvoltages in Power Lines: A Comparison of Two Coupling Models," *IEEE Trans. on Electromagnetic Compatibility*, vol. 36, no. 3, 1994, 00179–00182.
15. F. Rachidi, C. A. Nucci, M. Ianoz, and C. Mazzetti, "Response of Multiconductor Lines to Nearby Lightning Return Stroke Electromagnetic Fields," *IEEE Trans. on PWRD*, Jul. 1997, pp. 1404–1411.

16. P. P. Barker, T. A. Short, A. R. Eybert-Berard, and J. P. Berlandis, "Induced Voltage Measurements on an Experimental Distribution Line During Nearby Rocket Triggered Lightning Flashes," *IEEE Trans. on PWRD*, Apr. 1996, pp. 980–995.
17. C. A. Nucci and F. Rachidi, "Experimental Validation of a Modification to the Transmission Line Model for LEMP Calculations," Proc. 8th International Symposium and Technical Exhibition on EMC, Zurich, March 7–9, 1989, pp. 118–126.
18. C. A. Nucci, C. Mazzetti, F. Rachidi, and M. Ianoz, "On Lightning Return Stroke Models for LEMP Calculations," Proc. 19th Conference on Lightning Protection, Graz, Apr. 25–29, 1988, pp. 463–470.
19. C. A. Nucci, F. Rachidi, M. Ianoz, and C. Mazzetti, "Comparison of Two Coupling Models for Lightning-Induced Overvoltage Calculations," Jan. 1995, pp. 330–339.
20. C. A. Nucci, "Lightning-Induced Voltages on Overhead Power Lines, Part 1: Return-Stroke Current Models with Specific Channel-Base Current for the Evaluation of the Return-Stroke electromagnetic Fields," *ELECTRA*, Aug. 1995, pp. 75–102.
21. F. de la Rosa, R. Valdivia, H. Perez, and J. Loza, "Discussion About the Inducing Effects of Lightning in an Experimental Power Distribution Line in Mexico," *IEEE Trans. on PWRD*, Jul. 1988, pp. 1080–1089.
22. S. Yokoyama, K. Miyake, H. Mitani, Takanishi, "Simultaneous Measurement of Lightning-Induced Voltages with Associated Stroke Currents," *IEEE Trans. on PA&S*, Aug. 1983, pp. 2420–2429.
23. S. Yokoyama, M. Miyake, and S. Fukui, "Advanced Observations of Lightning Induced Voltage on Overhead Lines," *IEEE Trans. on PWRD*, Oct. 1989, pp. 2196–2203.
24. E. Cinieri and F. Muzi, "Lightning Induced Overvoltages: Improvement in Quality of Service in MV Distribution Lines by Addition of Shield Wires," *IEEE Trans. on PWRD*, Jan. 1996, pp. 361–372.
25. R. H. Golde, "Lightning Surges on Overhead Distribution Lines Caused by Indirect and Direct Lightning Strokes," *IEEE Trans.*, Jun. 1954, pp. 437–446.
26. M. F. Stringfellow, "The Operating Duty of Distribution Surge Arresters Due to Indirect Lightning Strikes," *Trans. of SAIEE*, vol. 72, part 5, May 1981.
27. A Eriksson and D. Meal, "Lightning Performance and Overvoltage Surge Studies on a Rural Distribution Line," *IEE Proc.*, Mar. 1982.

12 PROBLEMS

1. In Section 4 of this chapter, the incremental IVFOR dependent on the location of X_m relative to D_g, S_{12}, and D'_g was given when $(r_{ct} + r_{cc}) > S_{12}$ and $r_g > h_T$. Perform this same analysis when $(r_{ct} + r_{cc}) < S_{12}$ for both trees and forests.

2. In Section 9 of this chapter, a comparison was made to field data. To perform this comparison, the CFO or voltage across the insulation was set to 89.7 kV, which occurred when the voltage on the phase conductor was 100 kV. Derive the equation for the voltage across the insulation, V_I, given the voltage on the phase conductor, i.e. e''_c.

3. A 3-phase horizontal configured, wood pole line having a phase conductor height of 12 meters has a neutral located 1.5 meters below the phase conductors. The CFO, negative polarity, is 170 kV. Assume $N_g = 4$, $v = 0.3$, $Z_g = 450$ ohms, $Z_m = 162$ ohms and $R = 10$ ohms. Assuming no trees or forests and, using the Brown–Whitehead equations, calculate I_{sc}.

4. A three-phase horizontally configured wood-pole line having a phase conductor height of 10.5 meters has a single overhead shield wire at a height of 12 m. The CFO, negative polarity or positive polarity, is 170 kV. A single line of trees is

located 12 m (S_{12}) from the line. Also $N_g = 4$, $v = 0.3$, $Z_g = 450$ ohms, $Z_m = 162$ ohms and $R = 10$ ohms. Using the programs IVFOR and BFRCIG, find and plot the IVFOR, the BFR, and the total flashover rate versus tree heights of 4 to 20 meters. For the BFR calculations, assume a span length of 60 m, $R_0 = 10$ ohms, $\rho = 200$ ohm-meters, $K_{PF} = 0.7$ and $Z_T = 225$ ohms. Use the Brown–Whitehead equations.

16
Contamination

1 INTRODUCTION

In previous chapters, the insulation requirements for lightning and switching surges have been discussed. The remaining consideration of design is contamination, which is considered in this chapter. For transmission or distribution lines, contamination dictates the type of insulator and the length of the insulator string. In some contamination areas, the insulator length may exceed that required by lightning or switching surges, and therefore the final design of the line is based on contamination. Similarly, in stations, the required length of the insulation, i.e., bus supports, bushings, etc., may be dictated by contamination, so that the BIL of the insulation must be increased when considering contamination.

In some parts of the USA and in some countries, contamination is the dominant design criterion, dictating the design of the line or station insulation. However, fortunately, for most of the world, contamination is not present to the degree that the insulation requirements overshadow the requirements of switching and/or lightning. Also, fortunately, there do exist ameliorating measures, such as alternate insulator types (high-leakage insulators, fog type insulators, and nonceramic insulators) and surface coatings (resistance glazing and room temperature vulcanized silicone) that can be used to decrease the required insulator string length or length of a bus-support insulator.

As for switching surges, either a deterministic or theoretically, a probabilistic method may be used. However, because of many factors, this approach has not been extensively used, and instead a conservative deterministic design rule of minimum strength = maximum stress is normally used.

Although there exists no agreement as to the exact number and type of ceramic insulators to be used in a contaminated atmosphere, the knowledge base on ceramic insulation is extensive. However, about 20 years ago, the nonceramic insulator was introduced and is now extensively used for lines, in stations, and for apparatus bushings. The total knowledge base for these nonceramic insulators is large but not as extensive as that for ceramic insulators. Nevertheless, IEEE and IEC testing standards for nonceramic insulators are now appearing.

The presentation in this chapter begins with a summary of application data concerning ceramic insulators after which the nonceramic insulator is considered. The terms contamination and pollution, like the terms leakage and creepage, are used as synonyms.

2 CERAMIC INSULATIONS

2.1 Mechanism of Flashover

Contamination flashover requires the occurrence of two events: (1) a sufficient degree of the contaminant composed of some ionic soluble salt, delivered to the insulator and deposited on its surface, and (2) a light rain or mist or fog that moistens the surface but does not create a washing effect. Although the contaminant alone creates no problems, this mixture, contaminant and moisture, produces a conducting film such that a current flows through the contamination layer. At locations such as the narrow portion of a post insulator or in the rib area underneath a line insulator, the current is concentrated to the degree that the layer dries, i.e., a dry band is created. The total line-to-ground voltage now appears across these small dry bands, and flashover of the dry bands occurs. These arcs gradually grow outward, and flashover occurs when the arcs extend and meet.

2.2 The Direct Method of Selecting the Insulation

One method of determining the insulation required that obviates, to a degree, the measurement of contamination is the construction of test stations along the proposed right-of-way. Alternate types of insulators and alternate lengths of insulators are located within the test station. The analysis of flashovers on the insulators results in the specification of the required insulation. The obvious disadvantage of this method is the time required to amass sufficient data. (However, the degree of contamination can be acquired and other types of data can be obtained that can then be used with laboratory data.)

A similar direct method is the selection of the insulation based on the performance of lines that traverse the same area. If the performance of these lines is acceptable, the identical insulation may be used on similar voltage lines. If the new line does not have the same system voltage, the number or length of the insulators can be proportionally increased or decreased. This method assumes a linear relationship between contamination performance and string or insulator length, which appears to be justified up to at least 345-kV transmission. The disadvantage of this method is that if the line used for comparison has a lower than required flashover rate, the selection of the insulation for the new line is somewhat undeterminable.

2.3 Determining the Contamination Severity

In general or ideally, contamination can be separated into two classes; industrial and sea. The sea contaminant arrives simultaneously with moisture, whereas the industrial contaminant can increase slowly with time. A contaminant may be NaCl or CaCl thrown up from a road surface, or may be cement dust, fly ash, limestone, or even a gas such as SO_2, as long as it can form a conducting layer in the presence of moisture. In 1972, an IEEE working group published a survey that examined contamination locations and the degree of contamination within the USA [1]. The contamination areas were segregated into general regional areas and small limited localized areas defined as spot contamination locations. The general regional areas were in the Midwest (southern Michigan, northern Indiana, Ohio, Illinois, and western Pennsylvania), the Pacific Coast, Florida, and the Gulf Coast. Spot contamination locations may be close to industrial plants or on lines within about 2 km of the substation. Surprisingly, in general, the most frequent type of contaminant was a mixed one containing both industrial and sea contaminants. The survey showed that at the time, the general countermeasures were increasing the insulator length by 30 to 50%, washing, and applying silicone grease.

In a paper from Japan [2], the authors show that salt contamination decreases rapidly after a distance of 50 km from the sea. Also, beyond about 500 m from an industrial plant, the decrease in contaminants is noticeable.

The degree or severity of the contamination has been specified in three basic ways. For industrial contamination, there is (1) the salt deposit density, SDD, the amount of salt contamination on the insulator surface in units of mg of salt per cm^2 of insulator surface, and there is (2) the moistened layer conductivity in units of microsiemens, µS. For sea contamination, there is the salt salinity, the amount of salt per volume of water, normally in kg per cubic meter of water. To standardize, for industrial contamination, the equivalent salt deposit density or ESDD is used, which is defined as the amount of NaCl that would yield the same conductivity at complete dilution as the non-NaCl salt [3].

The general site severity and its definition per IEEE, CIGRE, and IEC is shown in Table 1 in terms of the ESDD. As noted, several additional classifications are given by CIGRE. As given by the CIGRE working group, the equivalent layer conductivity in µS is approximately 100 times the ESDD in mg/cm^2, and the equivalent salt fog salinity in kg/m^3 is 140 times the ESDD in mg/cm^2. For example, the

Table 1 Contamination Site Severity

Site severity	ESDD, mg/cm^2		
	CIGRE [4]	IEEE [3]	IEC 815 [5]
None	0.0075–0.015		
Very light	0.015–0.03	0–0.03	
Light	0.03–0.06	0.03–0.06	0.03–0.06
Average/moderate/medium	0.06–0.12	0.06–0.10	0.10–0.20
Heavy	0.12–0.24	>0.10	0.30–0.40
Very heavy	0.24–0.48		
Exceptional	>0.48		

equivalent layer conductivity and the equivalent salt spray salinity for an ESDD of 0.05 is $5\,\mu S$ and $7\,kg/m^3$, respectively.

2.4 Strength of Insulation

The strength of contaminated insulation is dependent on so many variables that it is amazing that any general strength characteristics can be assembled. To amplify, the strength is dependent on

1. The type of line or station insulator
2. The type and amount of contaminant and inert binder
3. The configuration of the insulator, I-strings, V-strings, horizontal
4. The length of the insulator string
5. The type of weather conditions, i.e., the type of wetting, droplet size
7. The relative amount of contaminant on the top and bottom of the insulator
8. The testing method

Test Methods

The two primary attributes that are required in the development of a test method are (1) that it should be representative of a service condition and (2) that it should be reproducible. The first of these attributes is never fully achievable, but the second must be achieved. In addition, for efficiency, the test should be accomplished in a short time.

Today there exist as many testing methods as there are nations that perform tests. This has been attributed to the idea that each nation has different contamination problems that require different testing methods. However, the primary though unstated reason is simply that each group of investigators has different approaches to the same problem. In the CIGRE report [4], seven methods are enumerated. In the 1991 edition of IEC 507 [6], only three methods are recognized, the salt fog method and the solid layer method, procedures A and B. Only these three methods are discussed; see the CIGRE report for the others. Per IEC 507, the tests are designed to determine the withstand specific creepage distance defined as the creepage distance in mm per applied voltage in kV rms, line to ground. This is established when three or four tests result in a withstand.

Salt Fog Method. Before about 1970, the only test recognized in international standards was the salt spray test as developed by the English. This test consists of the simultaneous application of the voltage and a salt fog having a specific salinity, defined in terms of kg of NaCl per cubic meter of water. If no flashover occurs within about one hour, a withstand is recorded. The salinity for which three or four tests are withstood is termed the specified withstand salinity. The original objective of the test was to determine the withstand salinity for a specific system voltage, which in many cases today has been changed to determining the withstand voltage for a specific salinity.

This test stimulates sea contamination in which the contaminant and moisture arrive simultaneously.

Solid Layer Method, Procedure A (Wetting Before and During Energization). During this same period the Germans also developed a testing method for industrial pollution, it consists of (1) applying a uniform layer of NaCl and Kieselguhr (diatomaceous earth) to an insulator, (2) drying the insulator until ambient conditions are achieved, (3) moistening the insulator with steam until the layer reaches the maximum conductance, and then (4) applying the voltage and continuing the steam fog. Per IEC 507 standard [6], the mixture consists of 100 g of Kieselguhr, 10 g of silicon dioxide, and 1000 g of water. Also per IEC 507, the mixture may also be 40 g of kaolin (a clay) per 1000 g of water. However, this latter mixture is rarely used. The degree of pollution or contamination is defined as the layer conductivity, which is the layer conductance multiplied by the form factor K_f, which is

$$K_f = \int_0^{L_T} \frac{1}{\pi d} dL \qquad (1)$$

where L_T is the total leakage distance, L is some point along the leakage distance, and d is the diameter of the insulator at this point. To explain, consider Fig. 1. If a voltage is placed across the post insulator of Fig. 1a, the layer conductance is

$$K_T = \frac{I}{V} \qquad (2)$$

This is the total conductance of the layer as distributed on this specific insulator, but the needed conductance is that of the layer. The concept may be better visualized from Fig. 1b, which is a view of a slice of the post insulator at some distance down from the top of the insulator. The contamination layer has a thickness of δ, and the circumference of the insulator is πd. Therefore the resistance of this incremental section is

$$dR = \rho \frac{dL}{A} = \rho \frac{dL}{\pi d \delta} \qquad (3)$$

where ρ is the resistivity. The total resistance R as measured by V/I is

$$R = \frac{V}{I} = \frac{\rho}{\delta} \int_0^{L_T} \frac{L}{\pi d} dL \qquad (4)$$

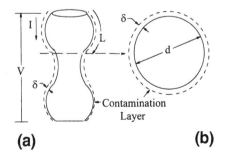

(a)   (b)

Figure 1 Insulator with contamination layer.

Or the layer conductivity K_L is

$$K_L = \frac{\delta}{\rho} = \frac{I}{V} \int_0^{L_T} \frac{1}{\pi d} \, dL = K_T K_f \quad (5)$$

Thus the layer conductivity is only a property of the contamination layer and is not a function of the insulator on which it is applied. It is equal to the total layer conductance multiplied by the form factor K_f. The layer conductivity is usually given in terms of microsiemens, μS, where S is the international symbol for conductance.

The layer conductance is measured by application of a low voltage along with the fog. When the conductance reaches its largest value, the test voltage is applied, with the fog maintained. To determine the form factor, the reciprocal of the circumference is plotted versus the leakage distance. The area under the curve is the form factor.

The test voltage is applied for 15 minutes or to flashover. If no flashover occurs, a withstand is recorded. The layer conductivity for which three of four tests are withstood is termed the specified withstand layer conductivity.

This test, in which the layer is moistened before the voltage is applied, could be viewed as representative of energizing a line with contamination, i.e., a "cold" switch-on. However, it is considered as representative of the general area of industrial pollution.

Solid Layer Method, Procedure B. Developed by IEEE, this test consists of contamination of the insulator by spraying or dipping with a mixture of NaCl and some form of binder such as kaolin or Tonoko. The mixture, per IEC 507, is 40 g of binder per 1000 g of water. The insulator is then dried and cooled to ambient. A constant voltage is then applied simultaneously with or prior to a fog, which may be created by a cold or warm spray or by evaporation of water. The test may consist of determining the CFO or the specific withstand salt deposit density measured in terms of mg of NaCl per cm^2 of insulator surface area. The CFO is determined by use of the up and down method, which consists of a series of about ten tests. As previously stated, for the IEC 507 test, the three out four test withstand method is specified to obtain the specified withstand salt deposit density. However, the CFO method is superior.

This test simulates the slow buildup of contamination of an energized line and is the best representative test of the three, since the test conditions represent the usual contamination event.

This testing method was adopted in the USA since most of the contamination in the USA is of the industrial or mixed type.

General Results from Tests on Cap and Pin Insulators

The general results of the tests that may be employed for a first estimate of the number, length, and type of insulator are provided in Figs. 2 and 3.

From the IEEE Working Group paper [3], for the solid layer method, procedure B, Fig. 2 shows the CFO in kV rms line to ground per meter of connected length as a function of the salt deposit density in mg/cm^2 for standard insulators in a vertical and V configuration. The standard insulator is defined as follows: spacing = 146 mm, diameter = 254 mm, i.e., 146 × 254 mm, creepage or leakage

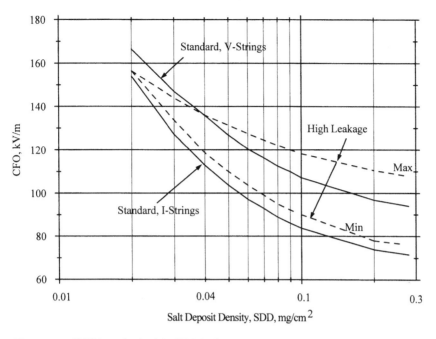

Figure 2 IEEE method with CFO in kV per meter of insulator length. (From Ref. 3.)

distance = 305 mm per insulator. Also show are two dotted curves representing the minimum and maximum values for high-leakage insulators. This variation is dependent on the insulator shape, and therefore a single curve cannot be given. The V-string insulator shows a significant improved performance as compared to the vertical string, primarily because the V-strings are cleaned more easily by artificial fog. Further, in natural conditions, natural cleaning occurs more easily from rain, and the V-string accumulates a lower deposit density for a given contamination level.

The curves for the vertical and V-string may be represented to within about 1% by the following equations.

For vertical or I-Strings,

$$\begin{aligned} \text{CFO (kV/m)} &= 72.3 + \frac{1.64}{C} \quad \text{for } 0.02 < C < 0.04 \, \text{mg/cm}^2 \\ \text{CFO (kV/m)} &= 64.4 + \frac{1.96}{C} \quad \text{for } C > 0.04 \, \text{mg/cm}^2 \end{aligned} \tag{6}$$

For V-strings,

$$\begin{aligned} \text{CFO (kV/m)} &= 106 + \frac{1.22}{C} \quad \text{for } 0.02 < C < 0.04 \, \text{mg/cm}^2 \\ \text{CFO (kV/m)} &= 87.6 + \frac{1.96}{C} \quad \text{for } C > 0.04 \, \text{mg/cm}^2 \end{aligned} \tag{7}$$

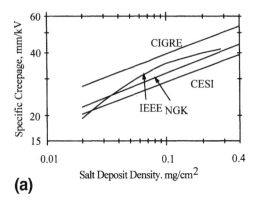

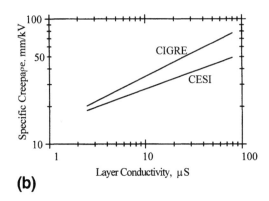

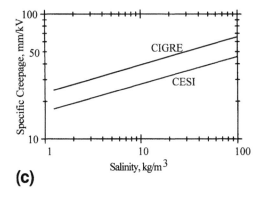

Figure 3. Withstand specific creep for (a) ESDD, (b) layer conductivity, and (c) salt salinity for IEEE, CIGRE, CESI, and NGK. (From Refs. 3, 4, 7, and 2.)

where C is the contamination in mg/cm^2. The standard deviation is stated as 10% of the CFO. Therefore to obtain the equation for V_3, which is equal to the CFO minus 3 standard deviations, the equations should be multiplied by 0.70, where the probability of flashover is 0.135%. In the IEEE Working Group report, V_3 is defined as the withstand voltage. Similarly, other equations may be obtained for other flashover probabilities.

The portrayal of the characteristics in terms of the CFO in kV per meter of insulator length indicates the understanding that the primary specification is the length of the insulator string. The curves or equations are entered with the line-to-ground maximum system voltage. To obtain the actual CFO the CFO/m is multiplied by the insulator string length, which for standard insulators is 0.146 times the number of insulators. For example, for a 230 kV system with a maximum system voltage of 242 kV, the line to ground voltage is $V_{LG} = 139.7$. Assuming vertical strings and a contamination of 0.05 mg/cm^2, the CFO becomes 103.6 kV/m. For a six insulator string, the CFO is 90.8 kV and V_3 is 63.5 kV.

The data from the CIGRE Working Group Report [4] is given in terms of the withstand specific leakage or creepage distance in mm/kV rms line to line voltage as a function of the contamination level. The results are shown in Fig. 3 for the alternate testing procedures where the creepage distance has been converted to the better understood mm/kV rms line to ground voltage, i.e., the actual voltage across the insulator. The curves of Fig. 3a are for the solid layer, procedure B testing method and vertical strings (I-strings). A curve is also shown as generated from Fig. 2. The curves from CIGRE, CESI [7], and NGK [7] can be represented by power law equations. The IEEE curve can be represented by two power law equations. The resulting equations for withstand specific creepage distance L_s in terms of mm of creepage distance per kV, rms, line–ground for the solid layer, procedure B method is of the form

$$L_s(\text{mm/kV}) = A(\text{salt deposit density, mg/cm}^2)^b \tag{8}$$

where the constants A and b are given in Table 2.

The authors of Ref. 2 tested both standard and fog-type insulators and found that Eq. 8 with the parameters of Table 2 adequately represented the test data. The IEEE curve of Fig. 3a applies only for vertical standard insulators, and the equation

Table 2 Constants of Eq. 8 for Cap and Pin Insulators, Solid Layer Method, Procedure B

Source	Insulator configuration	Range mg/cm^2	A	b
IEEE [3]	I-Strings	0.02 to 0.1	86.5	0.374
	I-Strings	0.1 to 0.3	51.4	0.158
	V-Strings	0.02 to 0.1	52.9	0.274
	V-Strings	0.1 to 0.3	37.1	0.122
CIGRE [4]	I-Strings	0.02 to 0.4	66	0.223
CESI [7]	I-Strings	0.02 to 0.4	48	0.220
NGK [2]	I-Strings	0.02 to 0.4	54.4	0.232

Table 3 Parameters of Eq. 9 for All Cap and Pin Insulators, Solid Layer Method, Procedure A

Source	Insulator configuration	Range, μS	A	b
CIGRE [4]	I-Strings	2.5 to 80	14.2	0.387
CESI [7]	I-Strngs	2.5 to 80	14.2	0.28

applies for standard cap and pin insulators in I- or V-configuration. The CESI and CIGRE curves and equations apply to all cap and pin insulators, standard and fog type.

Figure 3b shows the resulting curves from CIGRE and CESI for the solid layer, procedure A test method. The general equation for these curves is

$$L_s(\text{mm/kV}) = A(\text{layer conductivity, μS})^b \tag{9}$$

where the parameters are given in Table 3.

Figure 3c provides similar results for the salt fog method. The general equation is

$$L_s(\text{mm/kV}) = A(\text{salinity, kg/m}^3)^b \tag{10}$$

where the parameters are given in Table 4.

General Results from Tests on Station Insulation

Using the solid layer method, procedure B, the authors of Ref. 2 provided the data to assess the withstand specific creep in mm/kV line to ground as a function of the salt deposit density as shown in Fig. 4. The parameter of the curves is the average diameter D_A as defined in Fig. 4. As noted, the strength of the insulation increases for smaller average diameters. This curve can be used to estimate the specific creep for post insulators and bushings. The general equation for this curve is

$$L_S = A(\text{salt deposit density, mg/cm}^2)^b \tag{11}$$

where A and b are given in Table 5.

Other results from Ref. 7 indicate that the effect of diameter can be approximated by the equations

Table 4 Parameters for Eq. 10 for All Cap and Pin Insulators, Salt Fog Method

Source	Insulator configuration	Range, kg/m³	A	b
CIGRE [4]	I-String	3.5 to 100	23.4	0.224
CESI [7]	I-String	3.5 to 100	16.6	0.28

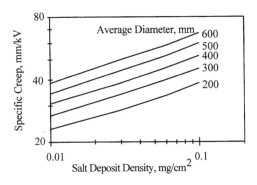

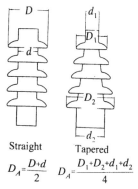

Figure 4 Withstand specific creep for station posts. (From Ref. 2.)

$$\frac{L_c}{L_0} = \left(\frac{D_c}{D_0}\right)^{0.35} \tag{12}$$

where L_c and L_0 are the specific creepage distances corresponding to the average insulator diameters D_c and D_0. In addition, IEC Standard 815 [5] suggests that the creepage distance be increased by 10% for diameters between 300 and 500 mm and by 20% for diameters greater than 500 mm.

Station type insulations tested by the salt fog method indicate a withstand specific creepage distance significantly less than that of the standard insulator, about 30% of that of the standard insulator [7]. In contrast, tests at project UHV

Table 5 Parameters of Eq. 11

Average diameter, mm	A	b
200	63.0	0.220
300	75.8	0.226
400	87.4	0.229
500	103.2	0.240
600	115.6	0.240

Source: Ref. 2.

indicate only a 6% decrease in performance as compared to standard insulators (in terms of kV/m) [8].

Other Results

Linearity, Long Strings. The equations and figures in Section 2.4.2 assume a linearity of the CFO or withstand specific creep distance with insulator string length. Initial testing at Project UHV showed a nonlinearity for string length above about 3 meters (20 standard insulators) for a contamination level of $0.07\,\text{mg/cm}^2$ [9]. That is, the curve of V_3 as a function of insulator length tends to bend over and saturate. A new and larger test building was constructed, and tests in this building indicated that saturation did occur at low contamination levels, $0.02\,\text{mg/cm}^2$ for string lengths of about 3 meters [8]. However, for higher contamination levels, no saturation was evident. To investigate this phenomenon further, outdoor natural contamination tests were made. For these tests, no nonlinearity was found even at low contamination levels. To date, the final answer to the linearity or nonlinearity question remains elusive. The authors of [8] remain convinced that some nonlinearity exists and suggest that the string lengths be increased above that calculated by 2 to 6% for 765/800 kV lines and by 10% for UHV lines.

Creepage or Leakage Distance. Although the creepage distance remains the most important determinant of contamination performance, the design of the insulator remains as an important factor. This is evident from the curves of Fig. 2, where a large band is shown for high-leakage units.

Contamination Uniformity. In natural conditions, because of rain, the contamination severity is usually lower on the top surface of an insulator than on the bottom. However, in the solid layer method, procedure A, the insulators are uniformly coated. Tests reveal that the uniformly coated insulators have lower withstand specific creepage distance, and thus the test method is conservative [7, 10]. For the solid layer method, procedure B, the insulator is sprayed or dipped. Although the uniformity of the coating cannot be confirmed, the accepted result is a somewhat uniform coating.

Compared to Natural Conditions. At Project UHV [8], insulators were tested under natural conditions and showed that the strength as measured by the CFO was equal to or greater than the strength as tested in the fog chamber. Similar results can be found in Ref. 2.

The Contaminant and Binder. In the solid layer test method, a clay is combined with a salt and deposited on the insulator. Universally, the salt is NaCl, and the use of salts such as $CaCl_2$ produced larger flashover voltages. The binder may be kaolin, Tonoko, or diatomaceous earth (fullers earth). The type and amount of clay or binder affects performance. As stated previously, for procedure B, the amount of binder is specified as 40 g/L. The CFO increases by about 25% at 20 g/L and decreases by 25% for 80 g/L [8]. In Ref. 2, the amount of binder is defined in terms of mg of binder per cm^2 of porcelain area. If the flashover voltage is 1.00 p.u. for a deposit of $0.10\,\text{mg/cm}^2$, the flashover voltage increases to 1.12 p.u. for $0.025\,\text{mg/cm}^2$ and decreases to 0.95 for $0.25\,\text{mg/cm}^2$.

2.5 Equivalence of Methods

In theory, there must some type of equivalence of the alternate testing methods. That is, there should be some way to equate an amount of salt deposit density to an amount of layer conductivity to an amount of salinity. In Table 6, the various methods of defining the contamination severity are listed as subjectively defined. The values for CIGRE in parentheses are called reference values, i.e., they are midpoint values. Based on the CIGRE reference values, the correlation becomes

$$1 \text{ mg/cm}^2 \text{ (ESDD)} = 140 \text{ kg/m}^3 \text{(salt salinity)} = 100 \text{ μS (layer conductivity)}$$

For the IEC gradation, the relationship is somewhat different.

2.6 Insulation Coordination

To select the string length and type of insulator, two methods can be used, the deterministic method and the probabilistic method. Because of the many uncertainties as to the contamination level and the strength of the insulation, today the predominant method is the deterministic method, although the probabilistic method is presently being developed.

The Deterministic Method

The deterministic method consists of equating the minimum strength to the maximum stress, and there are two ways to do this. In either case the first step is to select the maximum contamination level. Then one proceeds as in 1 or 2:

1. From the maximum contamination level, the CFO/m is determined and the withstand voltage V_3 is calculated. The insulator string length is found by dividing the maximum system line to ground voltage by V_3. The number of insulators is calculated by dividing the string length by the insulator spacing.

Table 6 Subjective Definitions of Contamination Severity

Pollution levels	Salt fog method, kg/m³	Solid layer method, proc. B, salt deposit density, mg/cm²	Solid layer method, proc. A, layer conductivity, μS
CIGRE			
None	1.25–2.5 (1.75)	0.0075–0.015 (0.0125)	0.75–1.5 (1.25)
Very light	2.5–5 (3.5)	0.015–0.03 (0.025)	1.5–3 (2.5)
Light	5–10 (7)	0.03–0.06 (0.05)	3–6 (5)
Average	10–20 (14)	0.06–0.12 (0.10)	6–12 (10)
Heavy	20–40 (28)	0.12–0.24 (0.20)	12–24 (20)
Very heavy	40–80 (56)	0.24–0.48 (0.40)	24–48 (40)
Exceptional	>80 (112)	>0.48 (0.08)	>48 (80)
IEC 815			
Light	5–14	0.03–0.06	15–20
Medium	14–40	0.10–0.20	24–356
Heavy	40–112	0.30–0.60	36
Very heavy	>160	—	—

2. From the maximum contamination level, the withstand specific creepage distance is determined. The required total creepage distance is found by multiplying this by the maximum system line to ground voltage. The number of insulators is calculated by dividing the total creepage distance by the creepage distance per insulator. The string length can then be obtained by multiplying the number of insulators by the insulator spacing. This is the only method available for cases where only the withstand specific creepage distance is given.

As an example, assume a 230-kV system having a maximum voltage of 242 kV. Therefore, the maximum line to ground voltage is 139.7 kV. Assume a maximum contamination level of $0.10\,\text{mg/cm}^2$. The $146 \times 254\,\text{mm}$ insulator in a V-string is to be used.

1. From Eq. 7, the CFO/m is 107.2 kV/m. Assuming a standard deviation of 10% of the CFO, V_3 is $0.7(107.2\,\text{m}) = 75.0\,\text{kV/m}$. The string length is $139.7/75 = 1.86\,\text{m}$. The number of standard insulators is $1.86/0.146 = 12.8$. Use 13 insulators with a string length of 1.90 m.
2. Using Eq. 8, the withstand specific creepage distance is 28.0 mm/kV. The total creepage distance is $139.7(28) = 3912\,\text{mm}$. The number of insulators becomes $3912/305 = 12.8$ insulators. Use 13 insulators with a string length of 1.90 m.

To determine the required length and BIL of post insulators, Table 1 of Chapter 2 can be used, which is taken from IEC 273 [11] and reproduced here as Table 7.

To demonstrate the use of Table 7 by example, assume the same system voltage and contamination level as before. Further, assume that the average diameter of the post insulator is 200 mm. Using Eq. 11, the required total creepage distance is 5.3 meters. From Table 1, either a 1425-kV BIL, Class I or a 1050 kV BIL, Class II could be used. Since the usual BILs for 230-kV are 900 or 1050 kV, the obvious choice is 1050-kV BIL, Class II.

Table 7 BIL/BSLs of Post Insulators, IEC Publication 273, 1990

			Creep distance, m	
BIL, kV	BSL, kV	Height, m	Class I	Class II
850	NA	1.90	3.10	4.40
950	750	2.10	3.40	4.90
1050	750	2.30	4.00	5.65
1175	850	2.65	4.60	6.50
1300	950	2.90	5.10	7.00
1425	950	3.15	5.60	7.80
1550	1050	3.35	6.20	8.50
1675	1050	3.65	6.35	9.40
1800	1175	4.00	6.90	10.25
1950	1300	4.40	7.65	11.35
2100	1300	4.70	8.25	12.25
2250	1425	5.00	8.70	13.20
2400	1425	5.30	9.20	14.10
2550	1550	5.70	9.80	15.00

The Probabilistic Method

The probabilistic method is formulated in the same manner as for switching overvoltages. As portrayed by Fig. 5a, the strength characteristic is a function of the contamination level. The stress is given by a probability density function that describes the variation of wetted contamination during one year. The probability of flashover per year is given by the convolution of these functions.

To add substance to this description, assume that the variation of wetted contamination can be approximated by a Gaussian distribution function with a maximum contamination level of 0.1 mg/cm^2. Assume that this maximum contamination is located at $\mu_c + 3\sigma$. The standard deviation is usually large, in the order of 150% of the mean [12]. For a σ_c/μ_c of 1.15, $\mu_c = 0.01818$.

Obtaining the strength distribution is more difficult. The concept is illustrated in Fig. 5b. The previous characteristics permit the strength characteristic of voltage vs. flashover probability with salt deposit density as a parameter to be constructed as in Fig. 5b, where a cumulative Gaussian distribution is assumed. However, the desired strength characteristic is in terms of flashover probability vs. salt deposit density. To obtain this, a vertical line is drawn at the line–ground voltage and the flashover probability for alternate salt deposit density is obtained. In more practical terms, the probability of flashover as a function of mg/cm^2 can be directly obtained by

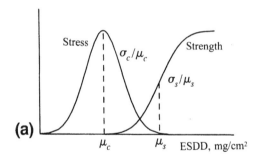

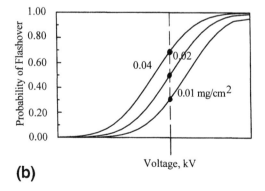

Figure 5 (a) Stress–strength concept, calculating the probability of flashover, (b) obtaining the strength characteristic.

reformulating Eq. 6 or 7 to provide the mg/cm² as a function of voltage, i.e., for Eq. 7,

$$C(\text{mg/cm}^2) = \frac{1.96}{V - 87.6} \tag{13}$$

where V is the applied voltage for the probability desired. For example, for a probability of flashover of 0.50, a voltage of 139.7 kV is used, giving 0.038 mg/cm². For a probability of 0.023, the voltage of 139.7 (0.9) is entered, giving 0.0514 mg/cm². This process is continued until all probabilities are determined. Thus for any stress, the probability of flashover can be determined. Although this represents an "exact method," some approximations or simplifications can be made. Examining several distributions, no continuous distribution function can be found to represent mathematically the entire function. However, in the probability region at and below a probability of about 0.20, the strength characteristic can be approximated by a cumulative Gaussian distribution having a standard deviation of from 18 to 22% of the mean.

Assume that the minimum "strength" is 0.10 mg/cm² and is located at 3 standard deviations below the mean, $\mu_s - 3\sigma_s$. With $\sigma_s/\mu_s = 0.18$, $\mu_s = 0.2174$ mm/cm².

The next step is to determine μ_{sn}, the mean for the strength for $n = 100$ towers, 300 insulators. Using Eqs. 52–543 of Chapter 3, $\mu_{sn} = 0.1065$ mg/cm². Using Eq. 49 of Chapter 3 with the $\frac{1}{2}$ removed,

$$\text{CFOR} = 1 - F\left[\frac{\mu_{sn} - \mu_c}{\sigma_c}\right] = 1 - F\left[\frac{0.1065 - 0.01818}{0.02727}\right] \tag{14}$$
$$= 0.000600 = 6 \text{ flashovers}/10{,}000 \text{ years}$$

where CFOR is the contamination flashover rate. As noted, the flashover rate is very low, which illustrates the conservatism of the deterministic method.

Using the methods of Chapter 3, the required insulator length for an assumed stress distribution can be estimated; see the problems at the end of this chapter.

2.7 Comparison—Number of Insulators

Using the equations developed for the withstand creepage distance, Table 8 compares the number of standard insulators required assuming a vertical string. The contamination severity is the maximum for the classification as taken from IEC 815. In addition, the number of insulators using the withstand specific creep as taken from IEC 815 is also listed. These values are 27.7 mm/kV for light and 34.6 mm/kV for medium contamination.

As expected, there exists a significant variation in the number of insulators. For 550 kV, for light contamination, the number varies from 27 to 37. Discounting CIGRE, the variation reduces to 27 to 31. For the medium classification, the variation is 35 to 49. This clearly demonstrates that experience within the utility must be used to select the proper number of insulators. That is, the test data can only be used as a guide.

Table 9 is developed from the IEEE data using the equations for withstand specific creep and using the IEEE classification and definition of contamination

Table 8 Comparison of the Number of Insulators, Vertical String, Creep = 305 mm/insulator

Max. sys. volt. kV	Light, 0.06 mg/cm²					Medium, 0.20 mg/cm²				
	IEEE	CIGRE	CESI	NGK	IEC	IEEE	CIGRE	CESI	NGK	IEC
145	8	10	7	8	8	13	13	9	10	10
242	14	16	12	13	13	22	21	15	17	16
362	21	24	18	19	19	32	32	23	26	24
550	31	37	27	29	29	49	48	35	39	36
800	46	53	39	43	42	72	70	51	57	52

severity. To check this table, for 345 kV, a minimum of 15 units, I-string, have been used with success by AEP. However, an 18-unit string is more of a standard. At 500 kV, a minimum of 22 units in V-string, and at 765 kV, a minimum of 30 units in V-string have been successfully used. Thus the values per these tables appear correct for low levels of contamination. They are also essentially identical to those presented by the IEEE Working Group [3].

2.8 Effect of Altitude

The previous values are applicable to sea level conditions. The withstand voltage or the CFO at an altitude CFO_A is approximately

$$CFO_A = \delta^m CFO \quad (15)$$

where the CFO is the CFO at sea level. From Chapter 1 the fair weather conditions,

$$\delta = 1.03 e^{-A/8.65} \quad (16)$$

where A is the altitude in km. From IEC 71 [13],

$$m = 0.5 \text{ for standard insulators} \quad m = 0.8 \text{ for fog type insulators} \quad (17)$$

Table 9 Number of Standard Insulators per IEEE

Max. system voltage, kV	Number of standard insulators for contamination severity		
	Very light, 0.03 mg/cm²	Light, 0.06 mg/cm²	Moderate, 0.10 mg/cm²
145	6/6	8/7	10/8
242	11/9	14/11	17/13
362	16/14	21/17	25/19
550	24/21	31/25	38/29
800	35/31	36/37	55/43

First number is for vertical strings; second for V-strings. Creep = 305 mm/insulator.

To apply this to the withstand specific creep, the equation becomes

$$\text{(withstand specific creep)}_A = \frac{\text{withstand specific creep}}{\delta^m} \quad (18)$$

Other investigators [14, 15] had previously suggested a value of m of 0.5, while the author of Ref. 15 showed that m approaches 0.8 for some insulators.

As an example, if the withstand specific creep at sea level is 20 mm/kV and the altitude is 2000 m, then $\delta = 0.8174$, $\delta^m = 0.904$ for standard insulators or $\delta^m = 0.851$ for fog type insulators. Therefore 22.1 mm/kV is applicable for standard insulators and 23.5 mm/kV is applicable for fog type insulators at 2000 m.

2.9 Effect of Contamination on LI and SI Performance

Considering that, as the overvoltage duration increases, the strength of the insulation under contaminated conditions becomes close to that under AC voltage [16], it becomes apparent that the reduction in strength is greater for temporary overvoltages, TOV, than for switching impulses and is greater for switching impulses than for lightning impulses. The possible decrease in strength under the combined stress of AC and these transient overvoltages is presented in CIGRE Technical Brochure 72 [17]. The decrease in strength is remarkable but is dependent on the progress of insulator flashover. That is, if the flashover event under AC voltage has reached the stage where dry bands have formed, the reduction in strength is remarkable. However, if dry bands have not formed, the strength reduction is not as severe. Thus in applying the test data, one must evaluate not only the probability that a lightning stroke occurs during a contamination event but also the probability that the contamination flashover event has progressed to the dry band stage. For the lightning event, rain is assumed, so that beneficial washing of the insulators is occurring. For switching surges, a high-magnitude switching surge occurs as a result of line reclosing caused by the lightning flashover. Also to be considered is that the shunt resistance across each insulator string provided by contamination decreases the switching overvoltages, especially for heavy contamination. Coupled with these observations, there has not been any reported flashover caused by lightning or switching surges on contaminated insulators. Thus the general application of the strength reduction factor is questionable.

With this as a prelude, under the worst possible conditions where dry bands have formed, the strength reductions are severe. From Ref. 17, LI insulation strength is reduced by 20 to 30%. For SI, the strength is reduced by 30 to 60%. The strength for TOV approaches that for AC.

2.10 IEC Recommendations

Suggested withstand specific creep distances for porcelain insulators for alternate levels of contamination are provided in IEC 815 [5] and are shown in Table 10. Again, the specific creep is given in mm/kV (rms, line to ground). The standard further states that

> In very lightly polluted areas, specific creep distances lower than 27.7 mm/kV can be used depending on service experience. 20.8 mm/kV seems to be a lower

Table 10 Suggested Values of Withstand Specific Creep Distances per IEC 815

Pollution level	Examples of typical environments	Withstand specific creep, mm/kV
Light Salt fog salinity: 5 to 14 kg/m^3; solid layer methods: SDD; 0.03 to 0.06 mg/cm^2; layer conductivity: 15 to 20	Areas without industries and with low density of houses equipped with heating plants Areas with low density of industries or houses but subjected to frequent winds and/or rainfall Agricultural areas[a] Mountainous areas All these areas shall be situated at least 10 to 20 km from the sea and shall not be exposed to winds directly from the sea	27.7
Medium Salt fog salinity: 14 to 40 kg/m^3 Solid layer methods: SDD; 0.10 to 0.20 mg/cm^2 layer conductivity: 24 to 25	Areas with industries not producing particularly polluting smoke and/or with average density of houses equipped with heating plants Areas with high density of houses and/or industries but subjected to frequent winds and/or rainfall Areas exposed to wind from sea but not too close to the coast (at least several km distance)[b]	34.6
Heavy Salt fog salinity: 40 to 112 kg/m^3; Solid layer methods: SDD; 0.30 to 0.60 mg/m^3; layer conductivity: 36	Areas with high density of industries and suburbs of large cities with high density of heating plants producing pollution Areas close to the sea or in any case exposed to relatively strong winds from the sea	43.3
Very heavy Salt fog salinity: > 160 kg/m^3 Solid layer methods: SDD; not given; layer conductivity: not given	Areas generally of moderate extent, subject to conductive dusts and to industrial smoke producing particularly thick conductive deposits Areas generally of moderate extent, very close to the coast and exposed to sea-spray of very strong and polluting winds from the sea Desert areas, characterized by no rain for long periods, exposed to strong winds carrying sand and salt and subjected to regular condensation	53.7

[a] Use of fertilizers by spraying or burning of crop residues can lead to a higher pollution level due to dispersal by wind.
[b] Distances from seacoast depend on the topography of the coastal area and on extreme wind conditions.

limit. In case of exceptional pollution, a specific creep distance of 53.7 mm/kV may not be adequate. Depending on service experience and/or laboratory test results, a higher value of specific creep can be used but in some instances the practicability of washing or greasing may have to be considered.

3 NONCERAMIC INSULATORS AND COATINGS

3.1 Background

To this point, the discussion in this chapter has only considered the normal porcelain insulators. However, in the last few decades, chemists have developed new materials, which in service have proved to be superior to porcelain insulation in combating contamination. However, the reported reasons that utilities have used these products are primarily (1) the increased strength to weight ratio, which has permitted compacting and improvement of tower designs, and (2) the resistance to gunshot vandalism. The products are frequently called composite insulators because they are composed of a resin-reinforced fiberglass rod over which weather sheds are formed. The weather sheds can be separately molded and threaded onto the rod or slipped over the rod, or they may be extruded onto the rod. The fiberglass rod provides mechanical strength. Stress or voltage grading rings are usually used for system voltages of 230 kV or higher. The weathershed material may be one of several types, principally,

1. Ethylene propylene diene monomer, EPDM
2. Ethylene propylene monomer, EPM
3. Silicone rubber, SiR

EPDM and EPM are frequently referred to together as EPR, ethylene propylene rubber. Presently, SiR or a combination of SiR and EPDM are dominant. To achieve resistance to tracking and erosion, compounds such as alumina trihydrate, ATH [33], are added to the shed material. The insulators are of the long rod type having no intermediate electrodes and having a small diameter that contributes to their excellent contamination performance. Although the shed material is of primary importance, the shape of the sheds also has an effect. For example, sheds having protected leakage paths provide improved performance [34].

SiR is used in two forms, high-temperature vulcanized, HTV, or room-temperature vulcanized, RTV. The RTV SiR is the type that can be sprayed on porcelain insulators, although it can also be used in constructing SiR insulators.

Although the use of nonceramic insulators is increasing, the ceramic insulator is still dominant. The development of the new nonceramic material has been fraught with problems, and not until the 1980s has it been accepted by most utilities. About 25 years ago when the first nonceramic insulators were produced, many utilities experimented with these units but many failures occurred [35, 36]. At this point, most manufacturers began intensive research that resulted in a second generation of nonceramic insulators whose life expectancy is in the range of 20 to 30 years (porcelain insulators have life expectancies of 40 to 50 years). With the new generation of nonceramic insulators, field results improved.

Because of their weight advantage and superior contamination performance, nonceramic apparatus bushings are being used on circuit breakers, current transformers, and potential transformers. In these latter cases, room temperature vulcanization, RTV, SiR is most often used. Also, nonceramic housings are used on surge arresters. In this case, the advantages are clearly lighter weight and decreased hazard from failed arresters.

According to the results of a survey reported in 1989, approximately 1200 miles of 115- to 765-kV lines are insulated with these nonceramic units [35]. As to operational experience, 96% of the utilities reported good or acceptable performance. The overall failure rate, including both the older and the newer versions of the nonceramic insulators, is 0.43%. The primary reported cause of failure is deterioration (erosion, corona, chalking, and crazing). In contrast to European experience, only one case of brittle fracture failure was reported.

In addition to the SiR coating, a resistive or semiconducting glaze has also been applied to porcelain insulators.

As with ceramic insulators, the primary objective is to obtain sufficient information so as to use these nonceramic insulators properly. Not only is the insulation strength required but perhaps more importantly an assessment of life expectancy or aging is needed.

Finally, in this chapter, the terms nonceramic, composite, and polymer are used as synonyms.

3.2 Field Reports, SiR, and EPDM

Field or service performance has been mixed. For example, during a nine-day period in 1991, 172 contamination outages occurred on lines of the Florida Power & Light Co. [37]. Comparing the performance, the porcelain line posts had an outage rate of 23/100 mile-years, while the silicone posts had none. The polymer called EPDM2 had an outage rate of 36 times that of porcelain and EDPM1 had an outage rate of 118 times that of porcelain. The conclusions were (1) that EPDM1 was not suitable for the weather conditions of FP&L and were removed from inventory, and (2) that both resistive glaze posts and silicone posts had unusually good contamination performance.

Field tests were performed at the Brighton test station along the southwest coast of England from 1983 to 1987 [38]. This station, which has been used to assess the performance of porcelain insulators, is located adjacent to the sea, from which significant salt storms occur. The units installed for tests included insulators for system voltages of 34.5, 230, and 500 kV. Four types of nonceramic insulators were installed. The 34.5-kV insulators had only four flashovers, all of which were on the older design epoxy resin units. The 230-kV units (21 mm/kV leakage), all with rubber sheds, had no flashovers. Flashover performance of the 500-kV units (18 mm/kV leakage) was better than that of the vertical string of porcelain fog type units. The silicone rubber insulators had the best performance, being better than the porcelain fog type units of 60% greater length. All the EPR insulators were generally of equal standard, approximately equal to that of porcelain units of 20% longer length. These tests did demonstrate some flaws in design and indicated that quality control needed improvement. In addition, some degradation of the shed material was found.

Silicone insulators, using HTV silicone sheds and porcelain insulators coated with RTV silicone, were installed in a test installation on the western Swedish coast, which is deemed a severe salt spray location [39]. No flashovers were reported in the 5-to-9-year test period. The authors appear to favor the HTV silicone insulator.

3.3 Field Reports, RTV Coatings

The recommended RTV coating thickness is 20 mils and can be sprayed or brushed on in several coats. It may also be applied over existing coatings. RTV coating may contain a fumed silica or an ATH filler dispersed in a carrier, naphtha, anhydrous cyclohexane, or 1,1,1 trichloroethane. To achieve a proper thickness, three to five spray coats are applied, which requires up to 15 minutes. This can be done live if 1,1,1 is used, but not with naphtha, which is explosive. As reported in Ref. 40, water washing may not be necessary. In addition, the authors state that because RTV coatings suppress leakage currents, wood-pole fires and flashovers are eliminated. The thickness of the RTV coatings have been reported to have some effect on performance [41].

The Milestone Power Station, which is connected to Northeast Utilities' 345-kV lines and located adjacent to Long Island Sound, has experienced flashovers of both line and station insulation caused by sea spray [42]. In an attempt to solve this problem, leakage distances were increased to 1041 cm on the switchgear equipment. However, a 1985 hurricane resulted in another complete 345-kV outage; both line and station equipment flashed over. A laboratory investigation on the use of insulators sprayed with RTV silicone was conducted by first rubbing the RTV with a wet sponge to reduce the hydrophobicity, after which a solid layer of kaolin and NaCl was applied. The solid layer test method was employed modified to obtain a flashover voltage by increasing the applied voltage in steps. These tests were followed by a salt mist test, not salt fog. The results indicated that RTV-silicone-coated Multicone insulators increased the flashover voltage over that of uncoated insulators by 20 to 50%, provided that following flashover the insulators were allowed to rest from two to three days to provide a period for recovery of surface hydrophobicity. If the rest period is shortened below one day, the flashover voltage is still higher than that of the uncoated porcelain but lower than if a longer rest period is used. Tests on uncoated and coated breaker support insulators also showed that RTV increased the flashover voltage. For some unexplained reason, the test results, i.e., the flashover voltages, lacked a significant reproducibility. The sprayed RTV silicone insulation was used at Milestone. In addition, sprayed RTV insulation has also been used at Boston Edison's Pilgram Plant. No subsequent outages have been reported [43].

At PG&E's Moss Landing Test Site, 66 kV was applied to five-unit strings of uncoated and RTV coated insulators [44]. The uncoated string flashed over in 1–3 months; the coated string had no flashovers in 6.5 years. A four-unit coated string flashed over in 3 years.

An IEEE committee report concerning types of insulator coatings including RTV states that RTV coatings offer a superior long-term solution to the contamination problem as proven by their service experience [45].

3.4 Mechanism of Flashover

In an effort to understand the aging and flashover process and thereby to assist in the development of a realistic aging test, a new flashover mechanism has been suggested

[46]. As formulated by the authors, the steps of the flashover mechanism are as follows.

1. When new, the silicon rubber insulator is hydrophobic, i.e. has little or no affinity for water. Thus later droplets arriving on the surface only produce beads and do not form a continuous water layer. Industrial pollution then deposits on the surface and combines with the moisture to form a surface layer.

2. Within a period of 10–12 arc-free hours, the low molecular weight (LMW) chains are driven out of the silicone material and form a hydrophobic layer on top of the pollution layer.

3. Dew/fog/high humidity periods may follow, in which case pollutants are driven through the hydrophobic LMW layer, and salt pollutants dissolve in the water droplets, resulting in a high resistant layer around each droplet.

4. Leakage currents flow through the resistive layer. The resistance decreases as caused by the negative temperature coefficient. However, drying of the layer occurs, which increases the resistance. Finally, equilibrium occurs.

5. Continual wetting occurs, increasing droplet density and reducing the distance between droplets. If distance is short, droplets join to form filaments.

6. Voltage between filaments increases and small or spot discharges occur.

7. The discharges destroy the hydrophobicity, resulting in a wet region, and a conductive path is formed leading to flashover.

3.5 Test Methods

Two types of testing methods are necessary, a method to determine the withstand voltage or the strength of the insulation and a method to assess aging or life expectancy.

Aging Tests. In 1983, the CIGRE Working Group 22.10 [47] published a report that suggested two types of aging tests for nonceramic insulators, one simple and one more sophisticated. The more complex test consists of the application of the following stresses while energized at maximum system line to ground voltage:

1. Solar radiation, i.e., ultraviolet radiation
2. Artificial rain
3. Dry heat
4. Damp heat, near saturation
5. High humidity (saturation) at room temperature
6. Slightly polluted fog.

An example of a cycle of these tests is shown in Fig. 6. The cycle, with changes every 2 hours, lasts for 24 hours. A total test duration of 5000 hours is suggested representing 8 years of life. The alternate simpler but less comprehensive test consists of applying a salt mist (using an atomizing nozzle) having a salinity of $10\,kg/m^3$ to the insulator while under operating voltage for 1000 hours. The salt mist should not directly impinge on the insulator. This fog salinity for 1000 hours creates a degree of pollution that is much greater than that of the salt fog test at a salinity of $10\,kg/m^3$ per IEC 507. A tripping relay is to be set at 1 A.

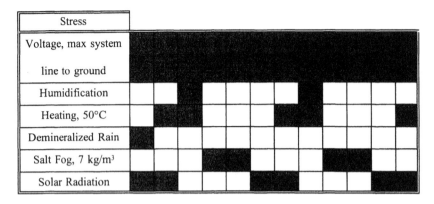

Figure 6 Example of an accelerated weather cycle under operating voltage. Shaded areas, in operation; no shading, out of operation. Each portion of cycle = 2 hours. (From Ref. 47.)

For either test, the insulator passes the test if (1) no more than 3 tripouts occur, (2) there are no punctures of the sheds, (3) the core cannot be seen, (4) there are no major holes, slits, or degradation of the housing.

The simpler test is now specified in the IEEE P1024, Draft Standard [48] and in the IEC Standard 1109 [49] with the additional specification that the voltage applied to the insulator is determined by dividing the total leakage distance of the insulator by 34.6 mm/kV, i.e., the IEC withstand specific creep for medium contamination. However, for medium contamination, IEC lists a salinity of 14–40 kg/m^3, whereas for light contamination, the salinity is 5–14 kg/m^3. To pass this test, there can be no more than three overcurrent trips (1A causes a trip), no tracking, no weather shed punctures, and erosion is not permitted to reach the core.

In testing line post insulators as reported in 1993 [50], the authors adopted a weather cycle test similar to that of Fig. 6 but separated into winter and summer cycles. That is, they employed a salt mist, rain, and ultraviolet radiation along with an applied voltage. The cycle time was 5 hours. One calendar year in service is assumed to be represented by 10 laboratory days of the summer cycle followed by 11 laboratory days of the winter cycle. The insulators were subjected to 6 years of simulated aging.

Insulation Strength Tests. There exists no standard tests to determine the insulation strength for various degrees of contamination. However, a common method of application of a uniform layer is evolving. That is, observations on the insulators, returned from the field showed that a uniform contamination layer was on the insulator surface. Since a new insulator is hydrophobic, any attempt in application of a uniform layer in the laboratory appears impossible without first scrubbing or sandblasting the insulator or applying a chemical agent. However, in an attempt to duplicate nature, the objective of the test method is to develop a method of producing a contaminant layer without destroying the hydrophobic surface. One proposed method is first to apply kaolin with a cotton swab, wet gently, dip the insulator in a slurry of kaolin and NaCl, and finally dry the insulator [51]. Another method [52] consists of spraying small water droplets over the surface, sprinkling Tonoko using a sieve over the surface, drying the insulator, using running tap water to wash the surface, immersing the insulator in a slurry of Tonoko and NaCl, and finally drying the insulator. Another method evolved from the aging tests as described previously [50]. The nonceramic insulators are artificially aged, after which they are dipped in a kaolin and NaCl mixture and then dried. These methods apply to the solid layer method. For the salt fog method, the test is the same as for ceramic insulators.

After the development of a uniform layer, the testing method should ideally follow the testing method for ceramic insulators, i.e. three out of four tests result in a withstand. However, in many cases, to obtain results within a shorter time period, investigators have used a flashover method that can produce approximate results. This method consists of applying a voltage known to be below the flashover voltage. After about 30 minutes, the voltage is increased in approximately 5% steps and held for 5 minutes or until flashover.

Following flashover, the nonceramic insulator temporarily loses most of its hydrophobic property and therefore performs in the same manner as a ceramic insulator. However, after a rest period of several hours, (8–10 hours [46] or 24–72 hours [46]), the insulator recovers its hydrophobicity. As reported with RTV coatings, which should act in the same manner as SiR insulators, even if the rest period is shortened to one day, the strength is higher than that of uncoated porcelain [42]. Considering the application of these units on lines, if a flashover occurs, reclosing within this rest period may result in another flashover.

3.6 Test Results

The authors of Ref. 52 tested EPDM, SiR, and porcelain long rod insulators using the solid layer method to obtain a uniform layer. The withstand voltage, per IEC 507, of these insulators as a function of the salt deposit density is shown in Fig. 7. The withstand voltage is approximately proportional to -0.20 power of the salt deposit density, i.e., about the same proportionality as for porcelain insulators.

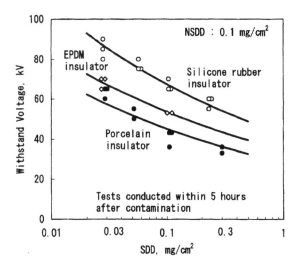

Figure 7 Withstand voltage for composite insulators. (From Ref. 52.)

From this figure, the withstand voltage of the SiR insulator is 50 to 60% greater than that of the porcelain insulator, and the withstand voltage of the EPDM insulator is 20 to 25% greater than that of the porcelain insulator. In this figure, NSDD is the nonsoluble deposit density, i.e., that of Tonoko or kaolin clay. The type and amount of the NSDD also affects the withstand voltage. Large values of NSDD decrease the withstand voltage as also occurs in porcelain insulators. Increasing the thickness of the contaminant layer on the composite insulators reduces the hydrophobicity of the surface, thus leading to a decrease in withstand voltage. The withstand voltage is also affected by the time between the coating and the testing of the insulator, decreasing slightly for shorter times. Sometimes the withstand voltage is reduced following flashover of the insulator, again showing that the time to regain hydrophicity is important.

Following the observations of outages in Florida Power & Light's system [37], tests were performed on line post insulators [50]. The insulators, two EPR, two SiR, and two ceramic, were placed in an accelerated aging chamber and energized at the nominal line to neutral voltage. Winter and summer weather cycles as previously discussed were simulated. The insulators were exposed to 126 days in the chamber, which is estimated to represent six years on the system. The solid layer test method was used but modified to obtain test results efficiently. The nominal line to ground voltage was applied to the insulator and held for 30 minutes. If no flashover occurred, the voltage was increased in 10-kV steps and held for five minutes. Thus the test was directed at obtaining the flashover voltage, and some criticism could be made, since this flashover test does not necessarily result in a withstand voltage. For the ceramic post insulators, flashover occurred at the nominal line to ground voltage. Table 11 shows that the flashover voltages of SiR insulators were greater than those of the EPR insulators. The flashover voltages of the EPR insulators exceeded that of the ceramic by 29 to 48%, while the SiR insulators exceeded the ceramic by 55 to 83%

These authors also reported on a quicker accelerated aging test that consisted of only one week in the aging chamber under the more severe winter cycle. The flashover voltage of these units exceeded those having the longer aging cycle by about 5 to 27% for EPR, 5 to 11% for SiR.

3.7 Resistive Glaze

Concerning the semiconducting glaze insulators, the glaze permits a current of from about 1 to 8 mA to flow to ground, resulting in a somewhat uniform distribution of

Table 11 Flashover Voltages of Composite Post Insulators

Insulator type	ESDD, mg/cm^2		
	0.05	0.10	0.20
EPR1	119	114	105
EPR2	108	106	103
SiR1	146	144	140
SiR2	144	138	124

Source: Ref. 50.

voltage across the insulator. Surface heating tends to maintain the contaminant in a dry state. In some cases, dry bands occur, but they are usually eliminated before flashover. As reported previously, field tests have been successful. The glaze has been shown to be arc resistant. The life expectancy has been questioned. The problem of "cold switch-on" of a line was questioned in a discussion [53–55]. "Cold switch-on" refers to the event where a contaminated line is out of service and then reenergized.

3.8 Summary and Conclusions

Conclusions concerning composite or nonceramic insulators are apparent. The contamination performance in comparison to the ceramic insulator is significantly improved, more for SiR than for the other formulations. For the identical lengths of porcelain and nonceramic insulators, the contamination performance increases in the order of 50% for SiR and about 25% for EPDM or EPR. The other primary advantage of the composite insulator is weight. Spraying RTV silicone onto porcelain insulations also increases the contamination performance by about the same order of magnitude as SiR sheds. Of some consequence is that following flashover on SiR, the insulator requires time for the hydrophobic properties to return before the insulation returns to its normal state. However, even if this rest period does not occur, the contamination performance is improved over that of porcelain. However, there is a possibility that reclosing on a line whose insulation has had a contamination flashover will result in another flashover.

Within the past few years, significant advancement has been made as to the mechanism of flashover and the effect of various parameters. With the mechanisms of flashover established, some improvement in the standard aging tests may be possible, and specification of standard tests to establish the insulation strength become possible.

4 METHODS TO IMPROVE PERFORMANCE

For ceramic insulation, the single most important parameter of contamination performance is creepage or leakage distance. High leakage distance, except for "poorly designed insulators," will improve performance. The goal for both ceramic and nonceramic transmission line insulations is to have acceptable performance while maintaining a short connected length—a length equal to or less than that required for switching or lightning.

For new transmission lines or new substations, consideration should be given to use of nonceramic insulators in both contaminated and noncontaminated areas. In contaminated areas, the pollution performance is paramount. In noncontaminated areas the weight and mechanical strength can be used to decrease capital and maintenance costs.

For an existing line or station, the following measures can be taken:

1. Nonceramic insulators: As for new lines and stations, if contamination is severe, nonceramic insulation is an option, although the cost of retrofitting may be large for transmission lines. SiR insulators appear to be superior to the other formulations.

2. RTV coatings: RTV silicone coated insulators appear to have the same contamination performance as SiR nonceramic units. They are appropriate for both line and station insulators.

3. High leakage or fog type insulators: Insulators such as the fog type units offer increased creepage per unit of insulator length. Creep to length ratios of 2.9 to 4.5 are available.

4. Insulator configuration: For transmission lines, V-string insulators outperform vertical strings.

5. Semiconducting Glaze Insulators: These insulators are appropriate for both station and line insulators. However, RTV coatings are superior.

6 Greasing/washing: As a last resort, the insulators can be greased with silicon or a hydrocarbon (in Greater Britain, about 3 mm thick) and/or washed periodically. Grease must be removed and reapplied periodically—a time-consuming and messy job. RTV coatings are better and generally more cost effective.

Also of interest is Weck's report [56] on the CIGRE 1996 session in which is stated: (1) tests to establish the strength of composite insulators are only necessary for a complete loss of hydrophobicity; (2) in heavy-pollution areas, the creepage distance should be selected as for ceramic insulators; and (3) in light-pollution areas, a substantial decrease in creepage distance is permitted, but the amount of the reduction has not yet been established.

As a final comment, philosophically, the ceramic insulator appears ideal in that the material is arc resistant, is nonporous, has excellent internal strength, is unaffected by ultraviolet radiation, and is nontracking. The newly developed polymer materials are not as arc resistant and are affected by ultraviolet radiation. However, they do have the desired properties of better contamination performance, are lighter in weight, have high flexibility over extreme temperatures, reduce radio noise, result in less maintenance, and have an apparent resistance to gunshot vandalism. These attributes of the nonceramic insulator have gained them ready acceptance throughout the world, and while they are not expected to supplant ceramic insulators completely, they will probably be used for an increased number of lines and stations.

5 OTHER FORMS OF CONTAMINATION

Although not of priority in this chapter, there are other forms of contamination on insulators that may be of importance in specific locations and specific situations. Two of these are the effect of ice and the effect of birds.

5.1 Icing

Ice buildup on insulator strings can decrease the electric strength of lines. In recent papers [28–29], analysis of test results indicated that ice 1.5 cm thick reduces the power frequency withstand voltage below the maximum line to ground power frequency voltage for 230- to 765-kV lines. The number of standard insulators assumed was 12 for 245 kV, 16 for 330 kV, and 33 for 765 kV, maximum system voltages. Tests on ice-covered line and station nonceramic insulators showed a significant effect on shed spacing and found that V-strings performed better than I-strings [8].

Another investigation [30] found that the AC flashover voltage of insulators covered with snow and ice was 25 to 35% lower than that for light contamination. Also SI and LI strength decreased.

5.2 Birds

Birds on transmission lines may produce a stream of defecation that results in insulator flashover. This is a fairly widespread problem. A classic paper was presented in 1971 [31]. With the incentive of discovering the cause of 32 outages on the 500-kV lines, the authors simulated both the composition of the bird defecation and the streaming process. They used this simulated composition and delivery system in a high-voltage laboratory to show that birds that produce streams of defecation can cause line flashovers. In an attempt to prevent bird outages, a bird guard was fashioned and applied to transmission towers. The paper was initially graded as a Conference paper, but after further review it was graded as a Transactions paper.

On the Florida Power and Light system, the author of Ref. 32 estimates that bird streamers caused as many outages as reported for lightning or contamination. He also warns that shorter string length polymer insulators may increase this problem. A bird discourager that many call a "crown of thorns" was constructed and applied to towers.

6 REFERENCES

References Concerning Ceramic Insulators
1. IEEE Working Group, "A Survey of the Problem of Insulator Contamination," *IEEE Trans. on PA&S*, 1972, pp. 1948–1954.
2. I. Kimoto, K. Kito, and T. Takatori, "Anti-Pollution Design Criteria for Line and Station Insulators," *IEEE Trans. on PA&S*, 1972, pp. 317–327.
3. IEEE Working Group on Insulator Contamination, Lightning and Insulator Subcommittee, T&D Committee, "Application Guide for Insulators in a Contaminated Environment," *IEEE Trans. on PA&S*, Sep./Oct. 1979, pp. 1676–1695.
4. Working Group 33.04, Study Committee 33, "A Critical Comparison of Artificial Pollution Test Methods for HV Insulators," *ELECTRA*, no. 64, Jan. 1979, pp. 117–136.
5. IEC Standard 815, "Guide for the Selection of Insulators in Respect of Polluted Conditions," 1986.
6. IEC Standard 507, "Artificial Pollution Test on High Voltage Insulators to be used on AC systems," 1991–1992.
7. M. de Nogris, D. Perin, and A. Pigini, "Design of External Insulation for AC Systems Under Polluted Conditions," CESI Technical Brochure, 1992.
8. K. J. Lloyd and H. M. Schneider, "Insulation for Power Frequency Voltage," Chapter 10 of *Transmission Line Reference Book*, 2d ed., EPRI, 1982, pp. 463–501.
9. M. Kawai, "Insulation for Power Frequency Voltage," chapter 10 of *Transmission Line Reference Book 345 kV and Above*, EPRI, 1975, pp. 296–326.
10. E. Nasser, "Verhalten von Isolatoren bei unterschiedlich verteilter Fremdschicht," *Elecktrotechnischen Zeitschrift*, ETZ-A 84, Jun. 1963, no. 11, pp. 353–335.
11. IEC Standard 273, "Characteristic of Indoor and Outdoor Post Insulators for Systems with Nominal Voltages Greater than 1000 V," 1990, 1992.
12. K. Naito, T. Miano, and W. Nagagawa, "A Study on probabilistic Assessment of Contamination Flashover of High Voltage Insulator," *IEEE Trans. on PWRD*, 1995, pp. 1378–1383.

13. IEC Standard 71-2, "Insulation Co-ordination, Part 2, Application guide," 1996.
14. R. P. Mercure, "Insulator Pollution Performance at High Altitude: Major Trends," *IEEE Trans. on PWRD*, 1989, pp. 1461–1468.
15. V. M. Rudakova and N. N. Tikhodeev, "Influence of Low Air Pressure on Flashover of Polluted Insulators: Test Data, Generalization Attempts and Some Recommendations," *IEEE Trans. on PWRD*, 1989, pp. 607–613.
16. H. Cron and H. Dorsch, "Proportioning Transmission System Insulation to Service Frequency Overvoltages and Switching Surges, with Due Consideration for Loss of Insulation Strength Through Foreign-Body Surface layers," CIGRE paper 402, 1958.
17. CIGRE Working Group 33.07 of Study Committee 33, L. Thione, Convener, "Guidelines for the Evaluation of the Dielectric Strength of External Insulation," Technical Brochure 72.
18. IEEE and CIGRE Working Groups, "Final Report on the Clean Fog Test for HVAC Insulators," *IEEE Trans. on PWRD*, 1987, pp. 1317–1326.
19. C. H. A. Ely and W. J. Roberts, "Switching Impulse Flashover of Air Gaps and Insulators in an Artificially Polluted Atmosphere," *Proc. IEE*, vol. 115, Nov. 1968, pp. 1667–1671.
20. M. de Nigris, d. Perin, A. Pigini, C. S. Lakshminarasimha, and K. N. Ravi, "Guide for the Design of External Insulation from the View of Permanent AC Voltage Under Polluted Conditions," CESI, Jun. 1988.
21. G. Boll, R. Deppe, M. Erich, H. J. Holstien, C. Kneller, R. Meister, J. Richter, H. Roser, H. Weber, and L. Wolf, "Technical Considerations on Three-Phase Transmission at Voltages above 380 kV in Germany," CIGRE Paper 410, 1960.
22. NGK Insulators, "Study of Pollution," T-68069, 1968; "Recent Research on Insulator Pollution," Technical Reports 3A and 3C, 1966.
23. J. S. Forrest, "The Characteristics and Performance in Service of High-Voltage Porcelain Insulators," *IEE*, 1942, vol. 89, pt. II, pp. 600–653.
24. I. Kimoto, T. Fukimura, and K. Naito, "Performance of Heavy Duty UHV Disc Insulators Under Polluted Conditions," *IEEE Trans. on PA&S*, 1972, pp. 311–316.
25. J. S. T. Looms, M. Sforzini, C. Malaguti, Y. Porcheon, and P. Claverie, "International Research on Polluted Insulators," CIGRE paper 33-02, 1970.
26. E. G. lambert, "Contamination Leads to Upgrading of Insulation of PG&E Co. 500-kV Lines, *Electrical World*, Dec. 9, 1968.
27. M. Sforzini, R. Cortina, and G. Marrone, "A Statistical Approach for Insulator Design in Polluted Areas," *IEEE Trans. on PA&S*, 1993, pp. 3157–3166.
28. M. Farzaneh and J. Kiernicki, "Flashover Performance of IEEE Standard Insulators Under Ice Conditions," *IEEE Trans. on PWRD*, 1997, pp. 1602–1610.
29. M. Farzaneh and J. F. Drapeau, "AC Flashover Performance of Insulators Covered with Artificial Ice," *IEEE Trans. on PWRD*, 1995, pp. 1038–1051.
30. H. Matsuda, H. Komuro, and K. Takasu, "Withstand Voltage Characteristics of Insulator Strings Covered with Snow or Ice," *IEEE Trans. on PWRD*, 1991, pp. 1243–1250.
31. H. J. Wester, J. E. Brown, and A. L. Kinyon, "Simulation of EHV Transmission Line Flashovers Initiated by Bird Excretion," *IEEE Trans. on PA&S*, 1971, pp. 1627–1630.
32. J. T. Burham, "Bird Streamer Flashovers on FPL Transmission Lines," *IEEE Trans. on PWRD*, 1995, pp. 970–977.

References Concerning Nonceramic and Resistive Graded Insulators

33. H. Deng, R. Hackam, and E. A. Cherney, "Role of the Size of Particles of Alumina Trihydrate Filler on the Life of RTV Silicone Rubber Coating," *IEEE Trans. on PWRD*, 1995, pp. 1012–1023.
34. R. S. Gorur, E. A. Cherney, and R. Hackam, "Polymer Insulators Profiles Evaluated in a Fog Chamber," *IEEE Trans. on PWRD*, 1990, pp. 1078–1085.

35. H. M. Schneider, J. F. Hall, G. Karady, and J. T. Rendowden, "Nonceramic Insulators for Transmission Lines," *IEEE Trans. on PWRD*, 1989, pp. 2214–2219.
36. IEEE Task Force Report, J. F. Hall, "History and bibliography of Polymeric Insulators for Outdoor Applications," *IEE Trans. on PWRD*, 1993, pp. 376–385.
38. R. G. Houlgate and D. A. Swift, "Composite Rod Insulators for AC Power Lines: Electrical performance of Various Designs at a Coastal Testing Station," *IEEE Trans. on PWRD*, 1990, pp. 1944–1955.
39. A. E. Vlastos, and T. Orbeck, "Outdoor Leakage Current Monitoring of Silicone Composite Insulators in Coastal Conditions," *IEEE Trans. on PWRD*, 1996, pp. 1066–1070.
40. E. A. Cherney, R. Hackam, and S. H. Chim, "Porcelain Insulator Maintenance with RTV Silicone Coatings," *IEEE Trans. on PWRD*, 1991, pp. 1177–1181.
41. H. Deng and R. Hacken, "Electrical performance of RTV Silicone Rubber Coatings of Different Thicknesses on Porcelain," *IEEE Trans. on PWRD*, 1997, pp. 857–866.
42. R. E. Cranberry and H. M. Schneider, "Evaluation of RTV Coating for Station Insulators Subjected to Coastal Contamination," *IEEE Trans. on PWRD*, 1989, pp. 577–585.
43. B. J. Chermiside, "Silicone Transformer Liquids and Insulator protective Coatings," Pennsylvania Electric Association, Jun. 1990.
44. J. Hall, and T. Orbeck, "Evaluation of a New Protective Coating for Porcelain Insulators," *IEEE Trans. on PA&S*, 1982, pp. 4689–4696.
45. IEE Dielectric Electrical Insulation Outdoor Service Environmental Committee, "Protective Coatings for Improving Contamination Performance of Outdoor High Voltage Ceramic Insulators," *IEEE Trans. on PWRD*, 1995, pp. 924–933.
46. G. G. Karady, M. Shah, and R. L. Brown, "Flashover Mechanism of Silicone Rubber Insulators Used for Outdoor Insulation—I and II," *IEEE Trans. on PWRD*, 1995, pp. 1965–1978.
47. CIGRE Working Group 22.10, "Technical Basis for Minimal Requirement for Composite Insulators," *ELECTRA*, May 1983, pp. 89–114.
48. IEEE Draft Standard, P1024, "Draft Standard for Specifying Distribution Composite Insulators (Suspension and Dead-End Type), Mar. 1998.
49. IEC Standard 1109, "Test of Composite Insulators for AC Overhead Lines with a Nominal Voltage Greater than 1000 V," 1991.
50. H. M. Schneider, W. W. Guidi, J. T. Burnham, R. S. Goreur, and J. F. Hail, "Accelerated Aging and Flashover Tests on 138-kV Nonceramic Line Post Insulators," *IEEE Trans. on PWRD*, 1993, pp. 325–336.
51. A. de la O. R. S. Gorur, and J. Chang, "AC Clean Fog Test on Non-Ceramic Insulating Materials and a Comparison with Porcelain," *IEEE Trans. on PWRD*, 1994, pp. 2000–2008.
52. R. Matsuoka, H. Shinokubo, K. Kondo, Y. Mizuno, K. Naito, T. Fujimura, and T. Terada, "Assessment of Basic Contamination Withstand Voltage Characteristics of Polymer Insulators," *IEEE Trans. on PWRD*, 1996, pp. 1895–1900.
53. J. H. Moran and D. G. Powell, "Resistance Graded Insulators—The Ultimate Solution to the Contamination problem?" *IEEE Trans. on PA&S*, 1972, pp. 2452–2458.
54. A. C. Baker, J. W. Maney, Z. Szilagi, "Long Term Experience with Semi-Conductive Glaze High Voltage Post Insulators," *IEEE Trans. on PWRD*, 1990, pp. 502–508.
55. J. H. Moran and D. G. Powell, "Resistance Graded Insulators—The Ultimate Solution to the Contamination Problem? *IEEE Trans. on PA&S*, 1972, pp. 2452–2458.
56. K.-H. Weck, "General Report on Group 33 CIGRE Session," *ELECTRA*, 1994, pp. 71–73, 1996, pp. 83–85.
57. CIGRE Working Group 22.03.01, "Worldwide Service Experience with HV Composite Insulators," *ELECTRA*, May 1990, pp. 71–77.

58. J. T. Burham and R. J. Waidelich, "Gunshot Damage to Ceramic and Nonceramic Insulators," *IEEE Trans. on PWRD*, 1997, pp. 1651–1656.
59. IEEE Dielectric Electrical Insulation Outdoor Service Environmental Committee, "Round Robin Testing of RTV Silicone Coatings for Outdoor Insulation," *IEEE Trans. on PWRD*, 1996, pp. 1881–1887.

7 PROBLEMS

1. A 345/362-kV line is to be built in an area where the maximum contamination is $0.1 \, \text{mg/cm}^2$. V-strings are to be used. Using the IEEE equations, determine the number of 146×254-mm ceramic insulators (305-mm creep/insulator) required using both the deterministic and the probabilistic methods. Assume that the yearly moistened contamination level can be approximated by a Gaussian distribution and that the maximum value is at 3 standard deviations about the mean, with the standard deviation equal to 120% of the mean. Also assume that the insulation strength can be modeled by a cumulative Gaussian distribution having a standard deviation equal to 20% of the mean. Assume 60 towers and design for a contamination flashover rate of 1/100 years.

2. Using the deterministic method, for a 345/362 station, determine the required BIL of ceramic post insulators having a 300 mm diameter for a maximum contamination level of $0.03 \, \text{mg/cm}^2$. Assume the station is at an altitude of 2000 m.

17
National Electric Safety Code

1 INTRODUCTION

In previous chapters the methods and procedures for calculating the strike distances and insulator lengths have been presented vis-à-vis lightning, switching surges, and contamination. However, in most countries, some type of safety code exists that prescribes clearances. These prescribed clearances are minimum clearances and in some cases dictate the design of lines and stations. One such code is applicable in the USA, the National Electric Safety Code (NESC) [1]. This code has also been adopted in other countries and is therefore discussed in this chapter as a typical safety code. The stated purpose of the NESC is the "practical safeguarding of persons during the installation, operation, or maintenance of electric supply and communication lines and associated equipment."

As applied to the electric design of transmission lines, the primary clearances specified in the NESC are

1. Midspan clearance (clearance or strike distance between conductor and ground)
2. Tower strike distance (clearance or strike distance from the conductor to the tower body, arm, truss, i.e., any grounded part)

As applied to the electric design of substations, the primary clearances specified in the NESC are

3. Horizontal clearance (horizontal clearance or strike distance from the conductor or any live part to ground)
4. Vertical clearance (vertical clearance or strike distance from the conductor or any live part to ground)

As a cautionary note, the presentation in this chapter only covers the above subjects so as to obtain an estimate of the clearances. The NESC provides more complete coverage in these areas and in a multitude of other areas. It should be employed for the actual design.

All systems examined in this chapter assume an effectively grounded system.

2 TRANSMISSION LINES—MIDSPAN CLEARANCE

2.1 Reference Heights and Minimum Clearances to 22-kV Phase–Ground

Midspan clearances are derived by first assuming some type of object—a person, truck, etc.—beneath the phase conductor at the point of the lowest clearance, i.e., generally the midspan. The height of this object, called the reference height, is then added to an electrical clearance to obtain the total midspan clearance or strike distance to ground.

Reference heights are given in Table 1 for seven categories as taken from Table 232-3 of the NESC. The minimum midspan clearances for lines having phase to ground voltages between 0 to 750 volts and clearances for lines having phase–ground voltages between 750 volts and 22 kV as obtained from Table 232-1 of the NESC are also shown in Table 1. Categories 3 and 4 are those normally used for design. Note that category 3 is probably derived by assuming that a person carrying some object is standing under the line.

Table 1 Reference Heights and Midspan Clearances

Category	Reference height, meters	Midspan clearances, 0 to 750 volts, meters	Midspan clearance, >750 volts to 22 kV, meters
1. Railroad tracks	6.7	7.6	8.1
2. Roads, streets, alleys, driveways, parking lots	4.3	5.0	5.6
3. Space and ways subject to pedestrians or restricted traffic only	3.0	3.8	4.4
4. Other land, i.e., cultivated, gracing, forest, orchard, traversed by vehicles	4.3	5.0	5.6
5. Water areas not suitable for sailboating or where sailboating is prohibited	3.8	4.6	5.2
6. Water areas suitable for sailboating, areas of			
(a) less than 8 ha	4.9	5.6	6.2
(b) 8 to 80 ha	7.3	8.1	8.7
(c) 80 to 800 ha	9.0	9.9	10.5
(d) over 800 ha	11.0	11.7	12.3
7. Launching or rigging sailboats: Add 1.5 m to heights and clearances of category 7			

Source: Ref. 1.

Additional requirements are (1) for voltages above 50 kV phase to ground, the maximum phase to ground voltage must be used, (2) for voltages exceeding 50 kV, the clearance must be increased by 3% for each 300 m in excess of 1000 m above mean sea level, (3) the clearance is determined for conductor sags using 50°C or the maximum conductor temperature and 0°C conductor temperature with radial ice (with no wind displacement).

2.2 Clearances for Voltages Exceeding 22 kV to 470 kV Phase–Ground

For transmission lines having maximum phase to ground voltages exceeding 22 kV to 470 kV (i.e., maximum system voltages greater than 38 kV to 814 kV), the clearance listed in Table 1 must be increased by 10 mm per kV in excess of 22 kV. The increase in strike distance or clearance, ΔS, in meters, for voltages above 22 kV, is given by the equation

$$\Delta S(\text{meters}) = 0.01(V_{LG} - 22) \tag{1}$$

where V_{LG} is in kV. The 3 requirements as presented in Section 2.1 also apply for this section. The clearances for typical system voltages are presented in Table 2 for Categories 3 and 4.

2.3 The Alternate Method for Voltages Greater than 98 kV Phase–Ground

For voltages greater than 98 kV phase to ground or greater than a maximum system voltage of 169.7 kV, an alternate method of calculating clearances may be used. For voltages greater than 470 kV line to ground or greater than a maximum system voltage of 814 kV, this alternate method must be used.

This alternate method employs the equation

Table 2 Midspan Clearances for System Voltages > 22 kV to 470 kV

Maximum line–ground voltage, kV	Maximum system voltage, kV	ΔS, meters	Midspan clearance, meters	
			Category 3	Category 4
69.9	121	0.48	4.88	6.08
97.6	169	0.76	5.16	6.36
139.7	242	1.18	5.58	6.78
209.0	362	1.87	6.27	7.47
317.5	550	2.96	7.36	8.56
461.9	800	4.40	8.80	10.00

Source: Ref. 1.

$$S(\text{meters}) = bc\left[\frac{aE_{2p}}{500k_g}\right]^{1.667} \quad (2)$$

where S = strike distance to reference object, meters; E_{2p} = statistical SOV per breaker operation, kV, i.e., E_2 for phase peaks; $a = 1.15$, an allowance for $3\sigma_f/\text{CFO}$; $b = 1.03$, an allowance for nonstandard atmospheric conditions; $c = 1.2$, a safety factor; $k_g = 1.15$, the gap factor for conductor to plane gap. To obtain the clearance, the value of S must be added to the reference heights of Table 1.

This equation is derived by first using the deterministic design method of Chapter 2 of minimum strength = maximum stress or

$$V_3 = E_{2p} \quad (3)$$

where E_{2p} is the 2% value using the phase peak method and V_3 is equal to the CFO-3_f. Therefore the required CFO is

$$\text{CFO} = \frac{E_{2p}}{1 - 3(\sigma_f/\text{CFO})} \quad (4)$$

The Paris–Cortina equation [2] from Chapter 2 is

$$\text{CFO} = 500k_g S^{0.6} \quad (5)$$

where k_g is the gap factor, S is the strike distance (meters) and the CFO is the positive polarity switching impulse CFO for sea level conditions. From this equation, the strike distance becomes

$$S = \left[\frac{\text{CFO}}{500k_g}\right]^{1.667} \quad (6)$$

Therefore from Eqs. 4 and 6,

$$S = \left[\frac{E_{2p}}{(1 - 3\sigma_f/\text{CFO})(500k_g)}\right]^{1.667} \quad (7)$$

Letting

$$a = \frac{1}{1 - 3\sigma_f/\text{CFO}} \quad (8)$$

then

$$S = \left[\frac{aE_{2p}}{500k_g}\right]^{1.667} \quad (9)$$

Equation 2 is Eq. 9 multiplied by b and c. With $a = 1.15$, then $\sigma_f/\text{CFO} = 0.044$, a value slightly less than the value of 5% used in the preceding chapters. The NESC equation uses a k_g of 1.15, which agrees with the value given in Chapter 2 for a conductor to plane gap.

The value of S must be increased by 3% for each 300 m of altitude in excess of 450 m. The clearances determined by this method must not be less than the clearance computed for a line to ground voltage of 98 (97.6) kV per Table 2. Per Table 2, for 98 kV line to ground or 169 kV phase–phase, the clearances are 5.16 and 6.36 meters for categories 3 and 4, respectively. These represent the minimum clearances. However, the clearances calculated by this alternate method need not be greater than those listed in Table 2 for the specific system voltage.

2.4 The Statistical Switching Overvoltage

The NESC equation uses the statistical voltage E_{2p} or the 2% value as determined from the distribution of phase peaks, whereas in Chapter 3, E_2 was determined from the distribution of case peaks. The difference between these two values is small, in the order of 5%, and therefore the values of E_2 and E_{2p} may be considered equal.

To review, to obtain a distribution of case peaks, only the maximum value of the overvoltages on the three phases is selected, whereas for phase peaks all three phase overvoltages are considered.

To examine the magnitude of this difference, consider a study of 400 closings of the breaker. Thus there are 400 values of case peaks and 1200 values of phase peaks. Assuming that these values can be approximated by Gaussian distributions, Fig. 1 illustrates the situation. To differentiate the values, E_{2c} is defined as the 2% value for case peaks. Note that there exist eight values of voltage that exceed E_{2c} and 24 values that exceed E_{2p}. If these 24 values are considered in the case peak distribution, they become a 6% value, i.e.,

$$P(\text{SOV} > E_{6c}) = 0.06 \tag{10}$$

Therefore, approximately,

$$E_{2p} = E_{6c} \tag{11}$$

and

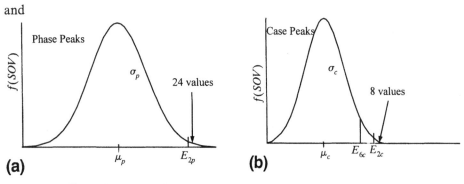

Figure 1 Illustration of (a) phase peak and (b) case peak distributions.

$$E_{2c} = \mu_c + 2.054\sigma_c$$
$$E_{6c} = \mu_c + 1.55\sigma_c \qquad (12)$$

Therefore

$$E_{2p} = E_{2c}\left[1 - 0.50\frac{\sigma_c}{E_{2c}}\right] \qquad (13)$$

from which if σ_c/E_{2c} is 0.09, then

$$E_{2p} = 0.955 E_{2c} \qquad (14)$$

If, instead, the SOV distribution is an extreme value positive skewed distribution,

$$E_{2c} = u_c + 3.902\beta_c$$
$$E_{6c} = u_c + 2.893\beta_c \qquad (15)$$

then

$$E_{2p} = E_{2c}\left(1 - 1.12\frac{\beta_c}{E_{2c}}\right) \qquad (16)$$

And if β_c/E_{2c} is 0.07, then

$$E_{2p} = 0.92 E_{2c} \qquad (17)$$

In both cases, $E_{2p} < E_{2c}$. However, without a large error, E_{2p} and E_{2c} can be considered equal.

2.5 Example of Midspan Clearances

Table 3 illustrates the calculation of midspan clearances for maximum system voltages at or above 362 kV for a range of E_{2p} values. These clearances are valid at altitudes less than 450 m. To obtain the clearances, 3 meters is added to the value of S for category 3, and 4.3 meters for category 4. the minimum clearance required is also listed as obtained from Table 2 for the 169-kV system voltage. The maximum clearance required is listed as obtained from Table 2 for the system voltage.

For example, for the 550-kV system and category 3, for values of E_{2p} or 2.4 and 2.6 per unit, a clearance of 7.36 meters would be required. For values of E_{2p} of 1.5, a clearance of 5.16 is required. For a maximum system voltage of 1200 kV, no minimum and maximum clearances are given, since the NESC requires the use of Eq. 2.

2.6 Comparison with Methods of Chapters 2 and 3

The Gallet et al [3] equation of Chapter 2 is:

National Electric Safety Code

Table 3 Midspan Clearance

Max. system voltage, kV	E_{2p}, p.u.	S, strike distance, meters	Midspan clearance, meters			
			Category 3		Category 4	
			Calculated	Limit	Calculated	Limit
362	3.0	3.21	6.21	Min	7.51	Min
	2.5	2.37	6.37	5.16	6.67	6.36
	2.0	1.63	4.63	Max	5.93	Max
	1.8	1.37	4.37	6.27	5.67	7.47
550	2.6	5.08	8.08	Min	9.38	Min
	2.4	4.45	7.45	5.16	8.75	6.36
	2.2	3.85	6.85		8.15	
	2.0	3.28	6.28		7.58	
	1.8	2.75	5.75		7.05	
	1.6	2.26	5.26	Max	6.56	Max
	1.5	2.03	5.03	7.36	6.33	8.56
800	2.4	8.30	11.30	Min	12.60	Min
	2.2	7.18	10.18	5.16	11.48	6.36
	2.0	6.13	9.13		10.43	
	1.8	5.14	8.14		9.44	
	1.6	4.22	7.22		8.52	
	1.4	3.38	6.38	Max	7.68	Max
	1.2	2.61	5.61	8.80	6.91	10.00
1200	1.8	10.11	13.11		14.41	
	1.6	8.30	11.30		12.60	
	1.4	6.65	9.65		10.95	

Source: Ref. 1.

$$S = \frac{8}{\frac{3400 k_g \delta^m}{CFO_A} - 1} \tag{18}$$

Using the equation,

$$E_2 = V_3 = CFO_A[1 - 3\sigma_f/CFO_A] \tag{19}$$

the strike distance equation becomes:

$$S = \frac{S}{\frac{3400 k_g \delta^m [1 - 3(\sigma_f/CFO_A)]}{E_2} - 1} \tag{20}$$

Assuming $\sigma_f/CFO = 0.05$, $k_g = 1.15$, and sea level conditions, $S = 2.963$ m and if a safety factor of 1.2 is used, $S = 3.555$ m. From Table 3, $S = 3.28$ m, a difference of 8%.

Another interesting item is the midspan clearance which is the strike distance plus the reference height. Assume that a truck having a length of 8 m and a height of

3 m is under the midspan. Also assume that the strike distance from the truck to the conductor at midspan is 3.28 m. From the equation for the gap factor for a conductor-lower structure gap configuration in Chapter 2, the gap factor is about 1.153, which is to be applied to the 3.28 meter strike distance. Without the truck, $S = 6.28$ m and the CFO is 1720 kV. With the truck, $S = 3.28$ m and the CFO is 1140 kV. Thus the strike distance is reduced by 48% but the CFO is only reduced by 34%.

3 TRANSMISSION LINES—TOWER STRIKE DISTANCE DESIGN

3.1 Basic Clearance

The basic clearance or strike distance from the conductor to tower side, arm, or truss specified by the NESC is 75 mm plus 5.0 mm per kV of maximum system voltage (phase–phase) exceeding 8.7 kV. Note that for tower clearances, the reference voltage is the maximum system voltage and not the voltage to ground. In equation form,

$$S(\text{meters}) = 0.075 + 0.005(V_{LL} - 8.7) \qquad (21)$$

where V_{LL} is in kV. For voltages greater than 50 kV, the following equation applies:

$$S(\text{meters}) = 0.280 + 0.005(V_{LL} - 50) \qquad (22)$$

For the preferred values of maximum system voltage, Table 4 gives the minimum strike distances. These clearances must be increased by 3% for each 300 m of altitude above 1000 m.

The above clearances apply to insulators restrained from movement, e.g., V-string insulators, line posts. Where suspension insulators are used and not restrained from movement, the above strike distances apply at the design swing angle. This design swing angle is based on a wind pressure of 6 lbs/ft^2 but may be reduced to 4 lbs/ft^2 for "sheltered" locations, more about this later.

3.2 The Alternate Method

For systems with maximum voltages to ground exceeding 98 kV or exceeding a maximum system voltage of 169.7 kV, an alternate method of determining the strike distance can be used. The equation for the strike distance is

Table 4 Minimum Tower Strike Distances

E_m, kV	S, meters
169	0.88
242	1.24
362	1.84
550	2.78
800	4.03
1200	6.03

$$S(\text{meters}) = b\left[\frac{aE_{2p}}{500k_g}\right]^{1.667} \tag{23}$$

where E_{2p} = statistical SOV per breaker operation, E_2 for the phase peak method; $k_g = 1.2$, the gap factor for the center phase of a tower; $b = 1.03$, an allowance for nonstandard atmospheric conditions; $a = 1.15$, an allowance for $3\sigma_f/\text{CFO}$ for fixed insulators, e.g., V-strings; $a = 1.05$, an allowance for $1\sigma_f/\text{CFO}$ for free swinging insulators. The value of S as calculated by Eq. 23 must be increased by 3% for each 300 m of altitude in excess of 450 m.

The clearance as given by Eq. 23 must not be less than that given in Table 4 for the 169-kV system but need not be greater than that given in Table 4 for the specific system voltage considered.

Equation 23 can be derived in a similar manner as for the midspan clearance. The value of k_g is identical to that of Chapter 2 for lattice towers. Note, however, that for the fixed insulators such as the V-string, the basic equation is $V_3 = E_{2p}$. However, for the free swinging insulator, the basic equation is $\text{CFO} - \sigma_f = E_{2p}$. Thus, for the free swinging insulator, the basic equation appears to be relaxed, but conservatism is added in that the calculated distance applies at the "design swing angle."

The swing angle α is calculated assuming a 290 Pa (6 lb/ft^2, 298.3 kg/m^2) wind pressure, which may be relaxed to 190 Pa (4 lb/ft^2, 19.5 kg/m^2) in areas sheltered by buildings, terrain, or other obstacles. For a wind pressure P and a horizontal span or wind span H, the horizontal force on the conductor caused by the wind pressure, F_{WD}, is

$$F_{WD} = PDH \tag{24}$$

where D is the conductor diameter. The vertical force F_{WT} caused by the weight of the conductor is

$$F_{WT} = WV \tag{25}$$

where W is the weight of the conductor per unit length and V is the vertical or weight span. Therefore the swing angle is

$$\alpha = \tan^{-1}\left[P\frac{D/W}{V/H}\right] \tag{26}$$

For P = wind pressure = 29.3 kg/m^2 or 6 lbs/ft^2; D = conductor diameter in cm or inches; W = conductor weight in kg/m or lbs/ft; H and V in same units of length.

Then

$$\text{Metric:} \quad \alpha = \tan^{-1} 0.293 \frac{D/W}{V/H}$$

$$\text{English:} \quad \alpha = \tan^{-1} 0.50 \frac{D/W}{V/H} \tag{27}$$

3.3 Example of Tower Clearances

In Table 5, examples of calculated tower strike distance are given for maximum system voltages of 362 kV and higher for alternate values of E_{2p}. The values apply for altitudes equal to or less than 450 m. The minimum clearance is that for 169 kV of Table 4; the maximum is that for the system voltage per Table 4.

As discussed, for insulator strings not restrained from movement, the strike distances of Table 5 are for the case when the insulator string is at the design swing angle. In general, the strike distances or clearances for fixed insulators as specified in Tables 4 and 5 are less than the strike distances required by lightning and switching surges. However, for insulators not restrained from movement,

Table 5 Minimum Tower Strike Distances

Max. system voltage, kV	E_{2p}, p.u.	Minimum clearance or strike distance, meters			
		Fixed insulators		Free-swinging	
		Calculated	Limit	Calculated	Limit
362	1.8	1.06	Min	0.91	Min
	2.0	1.27	0.88	1.09	0.88
	2.2	1.49		1.28	
	2.4	1.72		1.48	
	2.6	1.96		1.69	
	2.8	2.22	Max	1.91	Max
	3.0	2.49	1.84	2.14	1.84
550	1.4	1.41	Min	1.21	Min
	1.6	1.76	0.88	1.51	0.88
	1.8	2.14		1.84	
	2.0	2.55	Max	2.19	Max
	2.2	2.98	2.78	2.57	2.78
800	1.4	2.63	Min	2.26	Min
	1.6	3.28	0.88	2.82	0.88
	1.8	3.99	Max	3.43	Max
	2.0	4.76	4.03	4.09	4.03
1200	1.4	5.16	Min	4.43	Min
	1.5	5.79	0.88	4.97	0.88
	1.6	6.45		5.53	
	1.7	7.13	Max	6.13	Max
	1.8	7.84	6.03	6.74	6.03

because of the large swing angle per Eq. 27, these NESC clearances are frequently the determining factors in design.

As an example, consider a 550-kV line having a two-conductor bundle with a subconductor diameter of 46 mm, spacing = 457 mm. The subconductor weight is 3.524 kg/m, which for a V/H of 1.00 gives an NESC swing angle of 21° per Eq. 27. Also assume an altitude of 1000 m. Assuming the use of 24 insulators having a connective length of 3.8 m, a distance of 1.59 m must be added to the strike distance to obtain the arm length. The resulting arm length for alternate values of E_{2p} is shown in Fig. 2a. Note that the maximum required arm length is about 4.5 m; the minimum is about 2.5 m.

Figure 2a also shows a curve calculated by use of the techniques of Chapter 3. Some additional assumptions are

1. SSFOR = 1/100
2. Number towers = 500
3. Conductor height = 18 m
4. Tower width = 1.8 m
5. $\sigma_f/\text{CFO} = 0.05$
6. Wind speed = 56 km/h; design wind speed = 3.6 km/h
7. E_S/E_R per the equation

$$\frac{E_S}{E_R} = 1 - 0.15(E_{2p} - 1) \tag{28}$$

8. The SOV distribution is assumed Gaussian with a standard deviation of

$$\sigma_0 = 0.17(E_{2p} - 1) \tag{29}$$

As noted in Fig. 2a, the NESC clearances dominate for E_{2p} less than about 2.7 per unit. This is the result of the large swing angle required by NESC. The wind pressure should be revised in future additions of NESC.

Using these same assumptions, the curves of Fig. 2b are presented for V-strings. In this case, the NESC clearances are less than those determined by methods of Chapter 3.

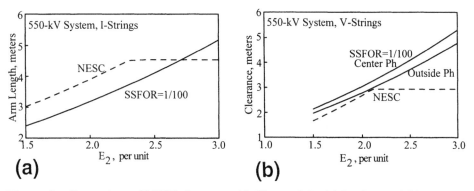

Figure 2 Comparison of NESC clearance with Chapter 3 for (a) I-strings and (b) V-strings.

3.4 Comparison with Methods of Chapter 3

In the development of the probabilistic method in Chapter 3, the statement was made that if a design is based on $V_3 = E_2$, the SSFOR is approximately 1 flashover/100 breaker closings. The question, therefore, arises as to why the NESC clearance is less than the probabilistic method as shown in Fig. 2b? The answer is in the difference in the basic equation for the CFO. That is, the NESC uses the Paris–Cortina [2] equation whereas the formulation in Chapter 3 uses the Gallet et al. Eq. [3], i.e.

$$\text{CFO} = k_g \frac{3400}{1 + 8/S} \tag{30}$$

For $k_g = 1.00$ and CFO $= 1000$ kV, this equation results in $S = 3.333$ m whereas the Paris–Cortina equation produces $S = 3.176$ m, a difference of about 5%.

From Chapter 3, the equation for the strike distance is:

$$S = \frac{8}{\frac{3400(0.96)k_g\delta^m}{\text{CFO}_A} - 1} \tag{31}$$

Also,

$$V_3 = E_2 = \text{CFO}_A[1 - 3(\sigma_f/\text{CFO}_A)] \tag{32}$$

Combining,

$$S = \frac{8}{\frac{3400(.96)k_g\delta^m[1 - 3(\sigma_f/\text{CFO}_A)]}{E_2} - 1} \tag{33}$$

For the center phase of the tower, k_g is about 1.2 and $\sigma_f/\text{CFO} = 0.05$. As an example, assume a 550 kV system with an $E_2 = 1.8$ per unit and sea level conditions. The above equation produces a strike distance of 2.565 m. From Table 5, the midspan clearance is 2.14 m for fixed insulators. Thus, the NESC clearance is about 17% less.

4 TRANSMISSION LINE WORKING CLEARANCE

Of major importance are the rules for operation of electric lines as provided by section 44 of part 4 of the NESC. The clearances or minimum approach distances given within the tables were obtained from IEEE Standard 156-1987 [4] (rewritten as IEEE Standard 156-1995) which form the basis of the NESC. In the IEEE Standard 156 the minimum air insulation distance denoted here as the strike distance, S, is determined first. For phase to ground SOV, the equation used is

$$S = 0.2155(0.01 + a)E_m \tag{34}$$

where E_m is the maximum switching overvoltage in kV, S is the strike distance in cm, and a is a factor to correct for the nonlinear increase in the CFO with S and is a function of the SOV. It is assumed that the strike distance is linear with E_m below 630 kV and thus in this region a is zero. This equation is derived by setting the minimum strength, V_3, equal to the maximum stress, E_m where V_3 is defined as the CFO $-\,3\sigma_f$. The authors gathered all available data for rod-rod gaps from which a minimum curve of V_3 versus S was obtained.

Using regression analysis, two alternate regression equations may be obtained relating S_g to E_m. The first equation is in form of the Paris–Cortina equation, i.e.,

$$S_g = (0.1173 + 0.00172 E_m)^{1.667} \approx (0.00172 E_m)^{1.667} \tag{35}$$

Using the Paris–Cortina equation, the generic equation is

$$S_g = \left[\frac{E_m}{500 k_g (1 - 3\sigma_f/CFO)}\right]^{1.667} \tag{36}$$

Letting $\sigma_f/CFO = 0.05$, the gap factor, k_g, is about 1.37 which is in the range indicated in Chapter 2 of 1.3 to 1.4.

The other regression equation is

$$S_g = \frac{4.43}{\frac{2620}{E_m} - 1} \tag{37}$$

Using the Gallet et al. equation, the generic form becomes:

$$S_g = \frac{8}{\frac{3400 k_g (1 - 3\sigma_f/CFO)}{E_m} - 1} \tag{38}$$

With a maximum error of 8.7% equation 37 may be altered to

$$S_g = \frac{8}{\frac{4000}{E_m} - 1} \tag{39}$$

Using this formulation and assuming $\sigma_f/CFO = 0.05$, then the gap factor, $k_g = 1.38$ which coordinates with that found with the other equation.

A comparison is shown in Fig. 3 of the results from the two regression equations with the data for 550- and 800-kV systems and indicates that equation 35 is preferable.

The phase to ground clearance defined as the minimum approach distance is the strike distance plus an inadvertent movement distance which is approximately 31 cm for system voltages above 72.5 kV and for voltages of 301 to 750 volts. For other system voltages, the inadvertent movement distance is approximately 61 cm.

The phase-phase strike distance is determined by first finding the phase-phase SOV. The equations are

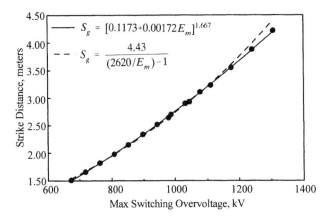

Figure 3 Comparison of equations with data for 550- and 800-kV systems.

$$\text{SOV}_p = 3.375\text{SOV}_g - 3.15 \quad \text{for SOV}_g \text{ of 1.5 to 2.0 pu}$$
$$\text{SOV}_p = \text{SOV}_g + 1.6) \quad \text{for SOV}_g \text{ of 2.0 to 3.0 pu} \quad (40)$$

where SOV_g is the phase-ground SOV in per unit and SOV_p is the phase-phase SOV in per unit. The phase-phase strike distance is the ratio of the SOV multiplied by the phase-ground strike distance. For example if the phase-ground strike distance is 300 cm for a SOV_g of 3.0 pu, the SOV_p is 4.6 pu and the phase-phase strike distance is 460 cm. The phase-phase clearance is the strike distance plus the inadvertent movement distance as described for the phase-ground clearance.

These strike distances are valid for altitudes up to 900 m. For altitudes greater than 900 m the altitude correction factor, C_f is approximately:

$$C_f = 1.02 + 0.03\frac{A - 1200}{300} \quad (41)$$

where A is the altitude in meters. The strike distance is this correction factor times the previously calculated strike distance. That is the altitude correction is only applied to the strike distance.

The maximum required phase-ground strike distances and clearances are presented in Table 6. These clearances are for air, bare hand and clear live line work. For voltages above 72.6 kV with a known maximum SOV, the strike distances and clearances of Tables 7 and 8 may be used. However, if these reduced clearances are used, the SOV must be further controlled by (1) blocking reclosing, (2) use of temporary gaps or arresters at the tower or at adjacent towers, or (3) changing system operation to restrict overvoltages.

In Fig. 4 the clearance required by NESC is compared to that for an SSFOR of 1 flashover per 100 breaker closings as determined by methods of Chapter 3. In calculating this curve, the assumptions are identical to those for Fig. 2. In addition, the maximum switching overvoltage is assumed to be equal to E_2 plus one standard deviation as suggested in Chapter 3. For Fig. 4, for maximum overvoltages greater

Table 6 AC Energized Work, Maximum Strike Distance, S, and Clearance, Cl, in cm [1,4]

Max. system voltage, kV	Phase-ground		Phase-Phase	
	S = min. air insulation distance, cm	Cl = min. approach distance, cm	S = min. air insulation distance, cm	Cl = min. approach distance, cm
0–0.050	Not specified Avoid contact		Not specfied Avoid contact	
0.05–0.300				
0.301–0.750	0.18	31	0.27	31
0.751–15	4.00	65	6.00	67
15.1–36	16	77	25	86
36.1–46.0	23	84	35	96
46.1–72.5	39	100	59	120
72.6–121	64	95	99	129
138–145	78	109	120	150
161–169	91	122	140	171
230–242	128	159	197	227
345–362	228	259	347	380
500–550	311	342	519	550
765–800	422	453	760	791

Source: Ref. 1 and 4.

than about 1.7 pu, the NESC requires greater values, thus limiting the design of lines. For presently existing lines, designed using the deterministic procedure of Chapter 2, the NESC clearances are less than the strike distance. For example for the Allegheny Power System's 500-kV line [5], the maximum switching overvoltage is 2.17 pu and a strike distance of 3.40 m is used. This is greater than the 3.01 m clearance required by the NESC. For the recent Bonneville Power Administration (BPA) 500-kV design

Table 7 AC Energized Work, Phase to Ground Strike Distance S, and Clearance, Cl, in cm [1,4]

Max. system voltage, kV	121		145		169		242		362		550		800	
E_m, pu	S	Cl	S	Cl	S	Cl	S	Cl	S	Cl	S	Cl	S	Cl
1.5											151	182	264	295
1.6											166	197	293	323
1.7											182	213	323	354
1.8											198	229	355	386
1.9											215	247	388	419
2.0	43	74	52	83	61	92	85	116	128	159	234	265	422	453
2.2	47	78	57	88	67	98	94	125	142	174	270	301	—	—
2.4	51	82	62	93	73	104	102	133	153	194	311	342	—	—
3.0	64	95	78	109	91	122	128	159	228	259	—	—	—	—

Source: Ref. 1 and 4.

Table 8 AC Energized Work, Phase-Phase Strike Distance, S, and Clearance, Cl, in cm [1,4]

Max. system voltage, kV	121		145		169		242		362		550		800	
E_m, pu	S	Cl	S	Cl	S	Cl	S	Cl	S	Cl	S	Cl	S	Cl
1.5											193	224	337	367
1.6											234	265	412	442
1.7											277	309	492	523
1.8											322	353	577	607
1.9											370	401	647	697
2.0	78	108	94	124	110	141	153	185	231	261	422	452	760	791
2.2	82	112	99	129	116	147	163	193	246	278	467	301	—	—
2.4	85	116	104	135	122	153	170	201	272	302	519	342	—	—
3.0	99	129	120	150	140	171	197	227	347	380	—	—	—	—

Source: Ref. 1 and 4.

[6], the strike distance is 2.54 m. This line was designed for a SSFOR of 1/10 and $E_2 = 1.7$ pu. Estimating an E_m of 1.8 pu, the required NESC clearance is 2.29 m and thus, the strike distance exceeds the required clearance. Therefore, depending on the parameters, the NESC required clearance may exceed the desired strike distance. Of course, there is no limitation on the strike distance if maintenance is performed when the line is de-energized.

5 CLEARANCES IN STATIONS

Figure 5 illustrates required clearances from "live parts." The guard clearance is the basic clearance. The horizontal clearance is equal to the guard clearance plus 3 feet (0.91 m), which appears to be the length of a person's arm. that is, per Fig. 5, visualize a man standing on the platform with his arm outstretched. The vertical clearance is equal to the guard clearance plus a distance of 8.5 feet (2.6 m), which appears to be the height of a man with a raised arm. Per Fig. 5, visualize a man with his arm upraised standing under the live part.

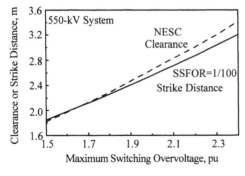

Figure 4 Comparison, NESC and probabilistic method.

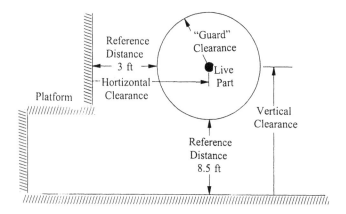

Figure 5 Definition of clearances in stations.

Two tables of clearances are provided, Table 9 for switching surges applies to 362- to 800-kV systems, while Table 10 for lightning applies for all system voltages. The clearance to be employed is the greater of that in Tables 9 and 10.

Unlike the midspan and tower clearances for switching surges, an equation is not provided from which to calculate the clearances. However, from an analysis of the guard clearances, the following equation appears applicable:

$$S(\text{meters}) = 1.09 \left[\frac{1.15 E_m}{500} \right]^{1.667} \quad (27)$$

The clearance per this equation is plotted in Fig. 6 along with the data of Table 9 showing that the equation is a good approximation. Note that in this equation, the gap factor k_g is 1.00, i.e., a rod–plane gap and the multiplier 1.09 represent both nonstandard atmospheric conditions and a safety factor. Also, the maximum SOV is used instead of E_2

The clearances for lightning are given as a function of the BIL. A problem arises if a higher BIL is indicated by contamination, which requires larger clearance even though the clearance strictly applies to lightning. This could be clarified by stating

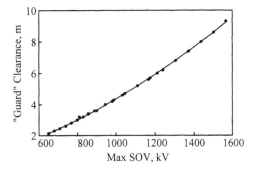

Figure 6 Comparison of regression equation with data.

Table 9 Clearances in Stations Based on Switching Surges for Maximum System Voltages of 362 to 800 kV [1]

Max. system Voltage, kV	Max. SOV per unit	Max. SOV kV	Guard Clearance, m	Vertical clearance to live parts, m	Horizontal clearance to live parts, m
362	2.2	650	2.13	4.7	3.0
	2.3	680	2.30	4.9	3.2
	2.4	709	2.45	5.0	3.4
	2.5	739	2.60	5.2	3.6
	2.6	768	2.80	5.4	3.7
	2.7	798	3.0	5.6	3.9
	2.8	828	3.2	5.8	4.1
	2.9	857	3.4	6.0	4.3
	3.0	887	3.6	6.1	4.5
550	1.8	808	3.2	5.7	4.1
	1.9	853	3.4	5.9	4.3
	2.0	898	3.6	6.2	4.6
	2.1	943	4.0	6.6	4.9
	2.2	988	4.3	6.9	5.2
	2.3	1033	4.6	7.2	5.5
	2.4	1078	4.9	7.5	5.8
	2.5	1123	5.3	7.9	6.2
	2.6	1167	5.6	8.2	6.6
	2.7	1212	6.0	8.6	7.0
800	1.5	980	4.2	6.8	5.1
	1.6	1045	4.7	7.3	5.6
	1.7	1110	5.2	7.8	6.1
	1.8	1176	5.7	8.3	6.6
	1.9	1241	6.2	8.8	7.2
	2.0	1306	6.8	9.4	7.7
	2.1	1372	7.4	10.0	8.3
	2.2	1437	8.0	10.6	8.9
	2.3	1502	8.6	11.2	9.5
	2.4	1567	9.2	11.8	10.0

Source: Ref. 1.

that the BIL is that required by lightning, or better, the clearance could be given as a function of the lightning overvoltage. In this regard, NESC states that "where surge protective devices are applied to protect the live parts, the vertical clearances may be reduced provided the clearance is not less than 2.6 m (8.5 ft) plus the electrical clearance between energized parts and ground as limited by the surge protective devices." The electrical clearance is interpreted to mean the guard clearance. Thus, for example, assume that an arrester rated 209 kV MCOV is applied in a 362-kV station. The discharge voltage is 665 kV and the maximum voltage in the station is 750 kV. Then the vertical clearance is 2.6 m plus the guard or electrical clearance for 750 kV. Assuming that the BIL gradient is 500 kV/m, the vertical clearance becomes $2.6 + 1.5 = 4.1$ m. To compare, for 1050-kV and 1300-kV BILs, Table 10 gives clearances of 5.2 and 5.7 m, respectively. Thus, for the usual case of arresters in the station, the clearance can be reduced.

Table 10 Clearances in Stations Based on Lightning [1]

BIL kV	Guard clearance mm	Vertical clearance to live parts, m	Horizontal clerance to live parts, m
95	101	2.69	1.02
110	152	2.74	1.07
125	177	2.77	1.09
150	228	2.82	1.14
200	304	2.90	1.22
250	406	3.00	1.32
350	584	3.18	1.50
550	939	3.53	1.85
650	1117	3.71	2.03
750	1320	3.91	2.24
900	1600	4.19	2.51
1050	1930[a]	4.52	2.84
1050	2130[b]	4.70	3.00
1300	2600	5.20	3.60
1550	3200	5.70	4.10
1800	3600	6.20	4.60
2050	4200	6.80	5.20

Source: Ref. 1.
[a] For 242-kV max. system voltage.
[b] For a 362-kV max. system voltage.

The guard clearance in terms of the BIL gradient varies from 941 kV/m for 95-kV BIL to 484 kV/m for 1550-kV BIL. For BILs of 750 to 2050 kV, the BIL gradient ranges from 484 to 568 kV/m and averages 518 kV/m. This value is not significantly different from that provided in Chapter 2, where the BIL gradient, taken from IEC Standard 273 [7], is 450 kV/m for BILs of 850 to 2550 kV.

No increase in clearance is specified for high altitude. However, the previous corrections for altitude should be applied.

6 CONCLUSIONS

Working clearances in stations and lines meet the objective of the NESC, i.e., "practical safeguarding of persons during the installation, operation, or maintenance of electric supply and communication lines and associated equipment." In contrast, however, the specification of design clearances or strike distances as in Section 3 appear to be outside the objective of the NESC. Therefore these requirements should be removed.

If the specification of design clearances is maintained in the NESC, the design swing angle needs revision, see Chapter 3.

Working clearances may exceed the strike distance based on a probabilistic design. Thus the NESC clearances may be the limiting design criterion.

Altitude correction factors are in need of coordination throughout the NESC. In addition, as in apparatus standards, the assumption is made that no correction is necessary for altitudes below 1000 m or 450 m. Since all or most data concerning the impulse strength of air or insulators has been corrected to sea level conditions, some correction appears necessary for altitudes above sea level.

7 REFERENCES

1. National Electric Safety Code, American National Standard C2, 1997.
2. L. Paris and R. Cortina, "Switching and Lightning Impulse Discharge Characteristics of Large Air Gaps and Long Insulator Strings," *IEEE Trans. on PA&S*, Apr. 1968, pp. 947–957.
3. G. Gallet, G. LeRoy, R. Lacey, and I. Kromel, "General Expression for Positive Switching Impulse Strength up to Extra Long Air Gaps," *IEEE Trans on PA&S*, Nov/Dec 1975, pp. 1989–1993.
4. IEEE Standard 156-1987 and IEEE Standard 156-1995, "IEEE Guide for Maintenance Methods on Energized Power Lines."
5. A. R. Hileman, W. C. Guyker, H. M. Smith, and G. E. Grosser, Jr, "Line Insulation Design for APS 500-kV System," *IEEE Trans. on PA&S*, Aug. 1967, pp. 987–994.
6. E. J. Yasuda, and F. B. Dewey, "BPA's New Generation of 500-kV Lines," IEEE Trans on PA&S, 1980, pp. 616-624.
7. IEC Standard 273, "Characteristics of Indoor and Outdoor Post Insulators for Systems with Nominal Voltages Greater than 1000 V," 1990.

18
Overview: Line Insulation Design

1 INTRODUCTION

As discussed in the introduction to this book, a study of transmission insulation design results in the following specifications:

1. The phase-to-grounded-tower strike distance, referred to in this chapter as simply the "strike distance"
2. The number and location of overhead shield wires
3. The need for and type of supplemental grounding
4. The number and type of insulators and the insulator string length
5. The need for, rating, and location of line surge arresters
6. The phase–phase strike distance

In specifying these quantities, the stresses imposed by lightning, switching surges, and contamination must be considered. (For lines having system voltages of 230 kV or less, switching surges need not be considered.) Each of these subjects has been discussed in this book and methods have been suggested so that the design values can be obtained. The objective of this final chapter is to compare these specifications for EHV and UHV transmission lines so as to ascertain which stress dominates the design. In this comparison, only single circuit lines that have grounded tower members between the phases are considered. Therefore the specification of the phase–phase strike distance is unnecessary. Also, although the option of line surge arresters is a viable option, it will not be considered in the comparison.

In review, a study of lightning, switching surges, and contamination results in the following specifications:

1. Lightning
    a. The number and location of the overhead shield wires
    b. The need for and the type of supplemental grounding
    c. The insulator string length
    d. The strike distance
2. Switching surges
    a. The strike distance
    b. The insulator string length
3. Contamination
    a. The insulator string length
    b. The number and type of insulators

As noted, the strike distance and insulator string length are each specified in the studies of lightning, switching surges, and contamination. The maximum value of the strike distance and insulator string length as obtained from each of these three study areas is the design specification. Economically desirable is that each of the studies results in identical values of strike distance and insulator string length.

2 COMPARISON OF DESIGN

Except for the more recent Bonneville Power Administration (BPA) 500-kV design [1], all 500- and 765-kV lines have been designed using the deterministic method of Chapter 2. However, there has never been a reported switching surge flashover on any of these lines [2]. The conclusion is that the strike distance can be decreased and that a probabilistic design criterion should be used. In Chapter 3, the probabilistic design for switching overvoltages was developed and compared to the deterministic design. The SSFOR suggested was one flashover per 100 breaker closings, which is becoming a de facto design standard. Using this design value, the strike distance can be significantly decreased from that obtained using the deterministic method.

In 1980, authors from BPA reported that their new 500-kV lines was designed for one flashover per 10 breaker closings, and again no flashovers have been reported [1]. This new design provides further proof that the strike distance can be reduced while maintaining an acceptable switching surge performance.

Economically, reduction of the strike distance can result in considerable savings, estimated in 1980 at $30,000 to $40,000 per km per one meter reduction in strike distance at 550 kV [2]. Assuming that a half-meter reduction in strike distance is possible for a 1000-km system, the savings would be $15 million dollars. The savings for the entire USA 550-kV system approaches half a billion dollars. Thus a strong incentive exists.

However, the question of whether strike distance can be decreased is not fully answered, since the strike distance as used for a transmission line depends on whether lightning, switching, or power frequency voltage dictates design.

This question is better considered by the aid of Fig. 1, where the approximate design requirements of a tower, specified in terms of strike distance, are shown as a function of maximum system voltage for the three criteria, lightning, switching surge, and power frequency voltage. Before reaching any conclusions, each of the design areas will be discussed, after which they will be considered together.

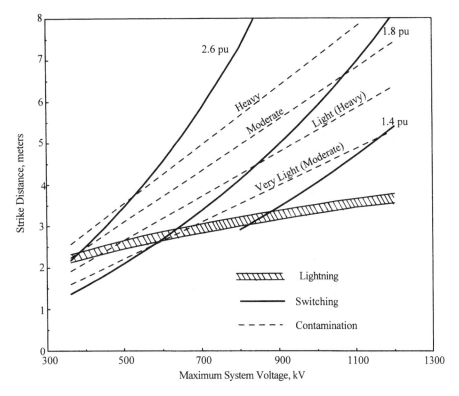

Figure 1 Comparison of requirements from switching surges, lightning, and power frequency voltage.

2.1 Lightning

Using the CIGRE method, the lightning curve or band is constructed for a flashover rate of 0.6 flashover per 100 km-years and a tower footing resistance of 20 ohms with a soil resistivity of 400 ohm-meters. The upper portion of the band assumes a ground flash density of 4.0 flashes/km^2-year and the lower portion, 8.0 flashes/km-year.

The lightning curve is relatively flat, as it should be since if a personality is ascribed to lightning, it does not care whether it hits a 362-kV line, a 550-kV line, or a 1200-kV line. Therefore the lightning requirement should be relatively constant with system voltage. However, tower heights increase and coupling factors decrease with increasing system voltage. These effects, along with the increase in power frequency voltage, combine to produce a gentle increase in the curve.

2.2 Switching Overvoltages

Using the techniques of Chapter 3, the strike distance required for switching overvoltages are shown by curves assuming (1) 500 towers and (2) a Gaussian stress distribution and for statistical overvoltages E_2 of 2.6, 1.8, and 1.4 per unit. An E_2 of 2.6 per unit represents a typical value for high-speed reclosing of breakers without a preinsertion resistor; 1.8 per unit represents a typical value for high-speed reclosing with a single preinsertion resistor; and 1.4 per unit represents a value for a breaker with possibly one or two preinsertion resistors or with controlled closing.

The assumed standard deviation σ_0 of the overvoltage distribution is

$$\sigma_0 = 0.17(E_2 - 1) \tag{1}$$

The switching overvoltage profile is assumed to be

$$\frac{E_s}{E_R} = 1 - 0.15(E_2 - 1) \tag{2}$$

Each of the curves sweeps sharply upward portraying the plot of the strike distance as a function of the CFO.

Interestingly, since 1968, BPA has purchased circuit breakers specified to limit the statistical switching overvoltage to 1.5 per unit. In 1976, a report on field test of breakers from six manufacturers was presented [3]. All breakers tested limited the statistical overvoltage to 1.5 p.u. or less, and four of the breakers limited the statistical overvoltage to less than 1.4 p.u. Five of the breakers had two-step resistors, one had three-step resistor, and four breakers used synchronously controlled closing devices. Using a transient program, studies were performed before these tests but were deemed conservative since they did not simulate all conditions of resistor insertion. If these conditions were simulated, the author estimated that the overvoltages obtained by the simulation would have decreased by 0.05 to 0.15 p.u. Using the 0.5 p.u. decrease, the simulation results indicated that all breakers limited the statistical overvoltage to 1.5 p.u. and three breakers limited the statistical overvoltage to 1.4 p.u. or less. Thus limitation of the switching overvoltage to within 1.5 p.u. has been achieved, and limitation to 1.4 p.u. is possible.

2.3 Power Frequency Voltage

Using the IEEE equations of Chapter 16, the power frequency voltage requirements are shown as a function of the IEEE contamination levels of very light (0.03 mg/cm^2, 20 mm/kV), light (0.06 mg/cm^2, 24 mm/kV), to moderate (0.10 mg/cm^2, 28 mm/kV), and heavy (0.30 mg/cm^2, 32 mm/kV). Use of ceramic 146 × 254 mm insulators in V-strings is assumed. The maximum string length is usually greater than the strike distance. For a 90° V-string, the length of the insulator could be 1.414 times the string length. However, attachment hardware and gusset plates impinge on this distance so that the string length is decreased. Therefore the maximum string length is assumed at 1.25 times the strike distance. As noted, the curve rises in a linear fashion and thus a linear relationship is assumed between string length and specific creep distance.

Two of the curves are labeled with two levels of contamination. The first label refers to ceramic insulators, while the second in parentheses applies to nonceramic insulators. Assumed is that the string length for nonceramic insulators may be 67% of that for ceramic insulators.

2.4 Comparison

Figure 1 provides the overall concept. The strike distance is that for the center phase. However, to examine each standard system voltage level, Table 1 is more useful. From Fig. 1 and Table 1:

362 kV. Lightning requires a strike distance of 2.1 to 2.3 m, which is approximately equal to that for a 2.6 p.u. statistical switching overvoltage, 2.2 m. This strike distance is also appropriate for moderate contamination using ceramic insulators. For heavy contamination, nonceramic insulators appear as an excellent choice. Thus, for this voltage, lightning, switching surge, and contamination requirements require about the same strike distance, and an optimum design is achieved.

550 kV. If the switching overvoltage design is based on a statistical overvoltage of 2.6 p.u., the switching overvoltage dominates the design requiring a strike distance of 4.0 m, whereas lightning requires only 2.5 to 2.7 m. To prevent switching surges from dictating the design, a preinsertion resistor is used in the breaker, decreasing the statistical overvoltage to 1.8 p.u. and decreasing the required strike distance to 2.4 m. Note that the switching surge requirement is now less than that for lightning. At the strike distance of 2.5 to 2.7 m, ceramic insulators could be used in very light to light contamination severities, or nonceramic insulators could be used for moderate to heavy severities. Thus lightning appears to dictate the design.

Table 1 Required Strike Distance, m

Tech. area	Criteria	362 kV	550 kV	800 kV	1200 kV
Switching	1 FO/100 surges	2.6 p.u. = 2.2	2.6 p.u. = 4.0 1.8 p.u. = 2.4	1.8 p.u. = 4.1 1.4 p.u. = 2.9	1.8 p.u. = 8.1 1.4 p.u. = 5.4
Lightning	0.6 FO/100 km-years	$(N_g = 4) = 2.1$ $(N_g = 8) = 2.3$	$(N_g = 4) = 2.5$ $(N_g = 8) = 2.7$	$(N_g = 4) = 3.0$ $(N_g = 8) = 3.2$	$(N_g = 4) = 3.5$ $(N_g = 8) = 3.8$
Power Frequency					
Ceramic					
Very light	20 mm/kV	1.6	2.4	3.5	5.3
Light	24 mm/kV	1.9	2.9	4.2	6.4
Moderate	28 mm/kV	2.2	3.4	5.0	7.4
Heavy	32 mm/kV	2.6	3.9	5.7	8.5
Nonceramic					
Moderate		1.6	2.4	3.5	5.3
Heavy		1.9	2.9	4.2	6.4
SOR	1 unsuccessful reclose/10 yrs	2.6 p.u. = 2.0	2.6 p.u. = 2.8 1.8 p.u. = 2.2	1.8 p.u. = 3.3 1.4 p.u. = 2.8	1.8 p.u. = 4.2 1.4 p.u. = 4.1

Source: From Fig. 1.

800 kV. If the switching overvoltage design is based on a statistical overvoltage of 1.8 p.u., the switching overvoltage dominates the design, requiring a strike distance of 4.1 m, whereas lightning requires only 3.0 to 3.2 m. If the statistical switching overvoltage can be reduced to 1.4 p.u., the required strike distance becomes 2.9 m, which is slightly less than the lightning requirements of 3.0 to 3.2 m. Thus again, lightning becomes important. If the 1.4 p.u. switching surge design is practical, and strike distances of 3.0 to 3.2 are used, nonceramic insulators should be used and are acceptable for light to moderate contamination areas. If the 1.8 p.u. design is used, the strike distance of 4.1 m encompasses heavy contamination conditions for nonceramic and very light to light for ceramic insulators.

1200 kV. For a statistical switching overvoltage of 1.8 p.u. an 8.1 m strike distance is required, far in excess of that required by lightning, 3.5 to 3.8 m. At an 8.1 m strike distance, even ceramic insulation may be used in heavy contamination areas. If a statistical switching overvoltage of 1.4 p.u. is achievable, a 5.4 m strike distance is estimated. This is still greater than that for lightning, and the use of nonceramic insulators is required, except for very light contamination areas.

To summarize

1. For designs at 362 and 550 kV and possibly at 800 kV, switching surges do not dictate design. Rather, lightning is the most important requirement and dictates design.
2. Because of the innovative development of nonceramic insulators, requirements for contamination have been substantially reduced.
3. Unless switching overvoltages are reduced below 1.8 p.u., they become the dominant criterion at 800 kV.
4. At 1200 kV, even with the statistical switching overvoltage held at 1.4 p.u. and the use of nonceramic insulators, switching surges remain as the dominant design criterion.

Thus the conclusion to this point is that at transmission voltages at 550 kV or less or possibly 800 kV or less, lightning remains the dominant design criterion, and only at 1200 kV does switching overvoltage replace lightning as the dominant criterion. From a philosophical viewpoint, this appears reasonable. Switching surges are man-made, so they can be man-controlled, while lightning is a phenomenon of nature that must be accepted and mitigating measures employed.

Returning to the original question of whether strike distances can be reduced, at 550 kV, within the U.S., strike distances of 3.35 to 4.0 m are in common use. Figure 1 and Table 1 indicate that these distances can be reduced to 2.5 to 2.7 m. Further reductions may be possible in areas having lower ground flash densities. For instance, BPA has announced a new advanced design for single- and double-circuit 550-kV lines located in areas having ground flash densities of 0.7 to 1.2 flashes/km^2-yr [1]. Using V-strings, a minimum clearance of 2.5 m is specified. This clearance represents a strike distance of 2.24 m plus a hand clearance around tower members of 0.30 m. This strike distance is based on a statistical switching overvoltage of 1.7 p.u. Based on contamination, eighteen 159 × 280 mm insulators are used. From Fig. 1, the estimated required strike distance for a statistical switching overvoltage of 1.8 p.u. is about 2.4 m, which compares favorably to the BPA design. Therefore the conclusion is that strike distances not only can be reduced by over 0.5 m but are presently being reduced by one utility.

At 800 kV, in the USA, strike distances of about 4.9 m are used for the center phase position. From Fig. 1, based on a statistical switching overvoltage of 1.8 p.u., a strike distance of 4.1 m is shown, a decrease of 0.8 m. Here again, the strike distance can be reduced. If the statistical switching overvoltage is reduced to 1.4 p.u., a reduction to 3.0 to 3.2 m appears possible, a 37% reduction.

Therefore, in answer to the original question, strike distances can be reduced, resulting in considerable savings.

3 COMPARISON BASED ON SOR

Another method upon which to base a comparison, introduced in Chapter 3, is the storm outage rate or the number of unsuccessful reclosures. This method combines the lightning flashover rate and the switching surge flashover rate. That is, the sequence of events is

1. Lightning causes a flashover.
2. The breaker opens to clear the fault.
3. The breaker recloses, producing a switching surge.
4. A flashover occurs caused by the switching surge.
5. The breaker locks open.

Therefore an unsuccessful reclosure occurs, which is called a storm outage. The advantage of this method is that it accounts for areas having a low ground flash density, which would result in a lower number of breaker reclosures. The concept is not new. If there is something new, it is the numerical evaluation. The SOR is calculated by the multiplication of the lighting flashover rate for the entire line by the SSFOR. If the lightning flashover rate or the BFR is in units of flashovers per year and the SSFOR is in units of flashover per 100 breaker operations, then the SOR is in units of unsuccessful reclosures per 100 years. Using the same parameters as before, the curve of Fig. 2 demonstrates the process for a 200-km, 550-kV line having a statistical switching overvoltage of 1.8 p.u. and N_g of 4 flashes/km^2-yr.

Curves similar to those of Fig. 1 are shown in Fig. 3 for an SOR of one unsuccessful reclose in 10 years, assuming a 200-km line. The results are somewhat easier to analyze using Table 1.

362 kV. For a 2.6 p.u. statistical switching overvoltage, strike distance can be reduced from 2.1–2.3 m to 2.0 m, a 10% reduction.

550 kV. For a 1.8 p.u. statistical switching overvoltage, the strike distance can be reduced from 2.5–2.7 to 2.2 m, a 15% reduction.

800 kV. For a 1.8 p.u. statistical switching overvoltage, the strike distance can be reduced from 4.1 m to 3.3 m, a 20% reduction. For a 1.4 p.u. statistical switching overvoltage, the strike distance can be reduced from 3.0–3.2 m to 2.8 m, a 10% reduction.

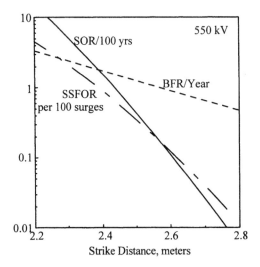

Figure 2 Developing the storm outage rate for a 500-kV, 200-km line, 1.8 p.u. statistical overvoltage, $N_g = 4$. ($E_2 = 1.8$ p.u., sigma/$E_2 = 7.6\%$, $ES/ER = 0.88$, 200 km, 400-m span $n = 500$, $N_g = 4$.

1200 kV. For a 1.8 p.u. statistical switching overvoltage, the strike distance can be reduced from 8.1 m to 4.2 m, a 25% reduction. For a 1.4 p.u. statistical switching overvoltage, the strike distance can be reduced from 5.4 m to 4.1 m, a 25% reduction. However, even the use of nonceramic insulators does not permit the use of strike

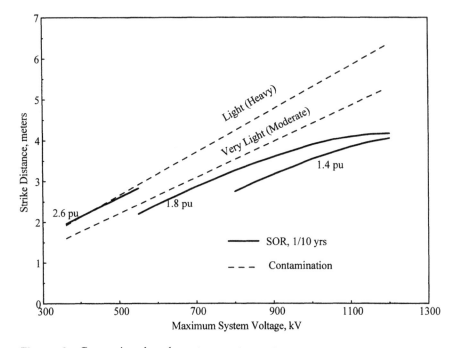

Figure 3 Comparison based on storm outage rate.

distances of 4.1 to 4.2 m, except for light contamination conditions.

For the lightning criteria employed, except for 1200 kV, the use of the storm outage rate of one unsuccessful reclosure in 10 years indicates that strike distance can be further reduced by from 10 to 15%. However, the concept of design based on the storm outage rate may require some modification, since faults caused by lightning may result in voltage dips that decrease power quality. In this case, both the SOR and the lightning flashover rate need consideration.

4 CONCLUSIONS

The general curves of Figs. 1 and 3 are presented to provide an overall view of line insulation requirements. Although admittedly they represent only crude estimates and should not be used for the design of a specific line, they do permit an overall conclusion. Technically, strike distances can be reduced, and economically, large incentives exist to reduce them. The alternate design criterion, the storm outage rate, combines both types of overvoltage, lightning and switching. Comparison of this criterion with the requirements of power frequency voltage illustrates the importance of the latter.

Hopefully this presentation has achieved its unstated primary objective: to show that in general, switching surges should not be and are not the dominant design criterion and that, except at 1200 kV, lightning constitutes the primary concern. This conclusion is partially due to the innovative development of nonceramic insulators but is also due to the control of switching surges. This is not to say that switching surges can be neglected or not considered but is meant to illustrate the progress of the industry within the last 40 years. In this period, more information has been amassed concerning the switching impulse insulation strength and the generation of switching surges than is now known about lightning. In addition, control measures have been evolved to ameliorate the effect of switching surges. In the contamination area, within the last 30 years, research into new materials has concluded with the polymer insulator, which has decreased the length of insulation to the degree that lightning becomes more important. The burden of design innovation now rests with lightning, and in this area, the application of arresters to transmission lines has been initiated and will continue to evolve. Perhaps in the future, the overall design goal of designing the insulation only for the normal power frequency voltage will be achieved.

5 REFERENCES

1. E. J. Yasuda and F. B. Dewey, BPA's new generation of 500-kV lines, *IEEE Trans. on PA&S*, 1980, pp. 616–624.
2. A. R. Hileman, Transmission line insulation coordination, Twenty-eighth Bernard Price Memorial Lecture, South African IEE, Sep. 1979, pp. 3–15.
3. G. E. Stemler, BPA's field test evaluation of 500 kV PCB's rated to limit switching overvoltages to 1.5 per unit, *IEEE Trans. on PA&S*, 1976, pp. 352–361.

Appendix
Computer Programs for This Book

The following DOS programs appear on the disks supplied with this text. These computer programs may be used in some of the problems as specified in the chapters. However, in general, the reader is first requested to solve the problems using the simplified methods as presented in the chapters. These programs are particularly useful when performing practical engineering studies. They may be copied and used as desired.

The initial versions of some of these programs were written for the Electric Power Research Institute (EPRI). The present versions of these programs have been updated. EPRI's permission to include these programs in this book is gratefully acknowledged.

1. BFRCIG99 — Backflash Rate, CIGRE Method — Chapter 10.
   This is the CIGRE method per CIGRE Technical Brochure 63, "Guide to the Procedures for Estimating the Lighting Performance of Transmission Lines", Oct 1991 Help screens are in BFRCIG99.HLP.
2. BFR99 — Backflash Rate, CIGRE Method — Chapter 10.
   An enhanced or investigative program based on the CIGRE method. Several options are available (i.e., simplified method, corona, exact LPM method, alternate time-lag curves, exact consideration of power frequency voltage, counterpoise). The file BFR99.HLP contains the help screens which are called by BFR99.
3. FLASH99 — IEEE method, backflash and shielding failure — Chapter 10.
   A DOS program of the IEEE FLSH17 which was written in BASIC. Help screens in FLASH.HLP
4. SFFOR99 — Shielding Failure Flashover Rate — Chapter 7.
   For transmission lines. Calculates the shielding failure flashover rate.

Computer Programs for This Book

Options on striking distance equations. The file SFFOR99.HLP contains help screens called by SFFOR99.

5. ALPD99 — Shielding Angle Alpha — Chapter 7.
   For transmission lines. Calculates the shielding angle for an inputted or desired SFFOR. The file SFFOR99.HLP contains help screens which are called be ALPD99.

6. SRGKON95 — Calculation of Surge Impendances and Coupling Factors — Chapter 9.
   Calculates self and mutual surge impedances and coupling factors. Also calculates flashover sequence (e.g., 1st, phase A, then phase C, then phase B. For each sequence, provides surge impedances of the "ground" wires and coupling factors which may be used in BFR99). Using this output with BFR99 can give the double circuit flashover rates. No help screens.

7. SRGKON96 — Same as SRGKON95 but does not calculate the flashover sequence.

8. SHIELD96 — Station Shielding — Chapter 8.
   For stations. For masts, 1 to 4, and for shield wires. For an input value of critical current, calculates the distances so that the shielding diagram can be drawn. Help screens are in SHIELD96.HLP, which are called by SHIELD96.

9. SRGBF98 — Incoming surge caused by a backflash — Chapter 11.
   Calculates the steepness and crest voltage of the incoming surge caused by a backflash. Both the simplified method and the more exact method are used. Help screens are contained in SRGBF98.HLP, which are called by SRGBF98.

10. SRGSF98 — Incoming surge caused by a shielding failure — Chapter 11.
    Calculates the steepness and crest voltage of the incoming surge caused by a shielding failure. Both the simplified method and the more exact method are used. Help screens are contained in SRGSF98.HLP, which are called by SRGSF98.

11. SSFOR97 — Switching Surge Flashover Rate — Chapters 3 and 4.
    Calculates the switching surge flashover rate. Help screens in STRIKE97.HLP

12. STRIKE97 — Switching Surge Strike Distance — Chapters 3 and 4.
    Calculates the phase-ground strike distance for a given SSFOR. Help screens in STRIKE97.HLP.

13. PP95 — Phase-Phase Switching Surge Flashover Rate — Chapters 4 and 5.
    Calculates the combined phase-phase and phase-ground switching surge flashover rate and the separate phase-phase and phase-ground flashover rates. Help screens in PP95.HLP.

14. ARR97 — Arrester Selection — Chapter 12.
    Calculates the required minimum arrester MCOV rating. TOVEN.DAT and STDRAT.DAT are data files called by the program. Also MAJHELP.SCR and PRGDISC.SCR are screens used in the program. Help screens are in ARR90.HLP.

15. PPSTR97 — Phase-Phase Switching Surge Strike Distance — Chapter 4 and 5.
    Calculates the phase-phase strike distance and BSL. Uses approximate method. Help screens in PPSTR97.HLP.

16. PPSSFO97 — Phase-Phase Switching Surge Flashover Rate — Chapters 4 and 5.
    Calculates the phase-phase switching surge flashover rate. Assumes phase-ground SOVs have no effect on flashover rate. Help screens in PPSTR97.HLP.
17. SIMP99 — Simplified Equations — Chapter 13.
    Solves the simplified equations developed in Chapter 13. Help screens are in SIMP97.HLP.
18. IVFOR99 — Induced Voltage Flashover Rate — Chapter 15.
    Calculates the voltage induced across the line insulation for a stroke terminating to ground or to trees next to the line. The effect of a single line of trees or a forest can be determined. Help screens in IVFOR.HLP.
19. OPCB99 — Open Circuit Breaker — Chapter 11.
    Calculates the probability (or return period) of a surge caused by subsequent stroke which equals or exceeds the open circuit breakers insulation strength. Help screens in OPCB99.HLP.
20. PROBGAU — Cumulative probability for Gaussian distribution. No help screens.

Index

Arrester
  current through, 341–342, 525, 540–541
  distribution, 539–545
  durability/capability, 503
  energy, 505–508, 510–512, 642–643
  induced overvoltages, 694
  line, 416, 643–670
  models, 514–516
  protective characteristics, 513, 550
  standard ratings, 502
  temporary overvoltages, 506–510, 519–522
  typical ratings, 524
Atmospheric conditions
  corrections, 16–22
  standard, 1

Backflash
  backflash rate, BFR
    CIGRE, 396–398
    IEEE, 408–410
    double circuit, 411–413
    low voltage, 405–408
  grounding, 379–386, 400–402
  nonstandard CFO, 390–393, 627–639
  outages, wood pole, 417–418
  summary, 418–419, 755

[Backflash]
  underbuilt shield wire, 402–404
  voltage across insulation, 373–375
BIL and BSL (*see also* Insulation strength)
  bus support insulators, 169
  bushings, 11, 168, 558
  cables, 13
  circuit breakers, 12, 168, 559
  chopped waves, 15–16
  definition, 2–3
  disconnecting switch, 12, 168–169, 559
  gas-insulated substations, 13
  phase-phase, 14
  standard values, 5, 8, 14
  substation, 12
  summary, 26–28
  tests, 6–8
  transformer, 11, 168, 558, 590
  types, 2

CFO, 9–12, 27, 390–393
Chopped waves, time-lag curves, 15–16, 67–70
CIGRE
  arrester model, 514–516
  backflash rate, 396–398
  contamination, 701

[CIGRE]
   incoming surge, 492–494
   induced overvoltages, 677
   phase–phase, 159
   station insulation coordination, 614–616
Contamination, 701, 718, 728, 756
Corona
   effect of, 354–367, 396, 399–401, 469–470
   steepness constant, k_c, 364, 468
Counterpoises, 382–384, 405
Coupling factor, 345–346, 366–367

Gap factors, 52–59, 139, 167, 180, 187
Gas insulated stations, 13, 607–613
Gaussian distribution table, 91–92
Geometric model, 215–227, 244–251
Grounding, 379–386, 400–402

IEC
   arrester model, 514–517
   arrester standard, 544–545
   clearances, 190–192
   contamination, 718
   estimating SSFOR and strike distance, 129–132
   incoming surge, 490–516
   insulation strength characteristic, 81–83, 127–132
   phase–ground, 187–188
   phase–phase, 159, 189–190
   SOV distribution, 129
   station insulation coordination, 614–616
IEEE
   arrester model, 514–517
   arrester standard, 514–517
   contamination, 701
   incoming surge, 490–492, 494
   national electric safety code, 733
   station shielding, 244
   stroke current distribution, 212
Impulse generator, 22–26
Induced overvoltages, 677
Insulation strength (*see also* BIL and BSL)
   lightning impulse
     air, insulators, towers, 59–70
     wood, 70–76
       fiberglass, 76–77
     phase–phase, 136–145
     power frequency, 79–80
     summary, 82–83

[Insulation strength]
   switching impulse
     gap factors, 52–59
     post insulators, 49–52
     stations, 166–169
     towers, 32–48
     towers, summary, 46–48
Insulation types, 1–2

Lightning flash
   current, 209–214
   geometric model, 221–225
   ground flash density, 215–221
   high altitude, 234
   increase in current with height, 231–234
   mechanism, 197–202
   multiplicity, 214
   negative, positive, 212–214
   number of flashes to ground wire, 228–231
   parameters, 203–209, 214
   striking distance equations, 225–227
   summary, 234–236
   types, 202–203

MTBF, MTBS, 461–464

National electric safety code, 733

Outages, wood pole, 417–418

Separation distance, (*see also* Station insulation coordination), 331–341, 557
Shielding
   substations
     design current, 275–278
     masts, 283–298, 307–309
     shield wires, 280–283, 307
     striking distances equations, 278–280
     summary, 303–307
   transmission lines
     background, 241–244
     Eriksson's model, 251–254
     geometric model, 244–251

Index

[Shielding]
 hillside, hilltop, terrain, effects, 260–261
 new methods, 268–271
 selection of angle, 257–260
 sensitivity, 254–257
 simplified method, SFFOR, 263–265
 subsequent strokes, 265–267
 summary, 269–271
Steepness constant, K_c, 364
Steepness of front, 354–365
Storm outage rate, 757–758
Stroke to tower, 323–351, 373–375, 464–466
Subsequent strokes, 206–207, 265–267, 455–458
Substation insulation coordination
 gas insulated stations, 607–613
 margins, 580–582
 nonsymmetrical layout, 606
 summary, 617
 voltage at breaker, 571–577, 582–583
 voltage at transformer, 563–571, 583–592
Summary
 backflash, 418–419
 contamination, 728
 incoming surge, 494
 induced overvoltages, 694–695
 insulation strength, 82–83
 lightning flash, 234–236
 line arresters, 671–672
 line insulation coordination, 753
 national electric safety code, 733–752
 shielding
  stations, 303–309
  transmission lines, 269–271
 specifying the insulation strength, 26–28
 station insulation coordination, 613
Surge impedance, 313–315, 344, 351–354, 365–366
Switching overvoltage design (*see also* IEC and CIGRE)
 deterministic
  transmission line, 48–49
 probabilistic, phase–ground

[Switching overvoltage design]
 arresters, 172–177
 concepts, 90–100
 estimates, E_2, E_m and σ, 103–104
 estimating the SSFOR, 111–115, 120
 estimating the strike distance, 115–120
 sensitivity, 106–111
 SOV distributions, 100–106
 storm outage rate, 126–127
 substations, estimating SSFOR and strike distance, 163–166
 summary, 754–755
 wind effect of, 121–126
 probabilistic, phase–phase
 arresters, 154–185
 calculating SSFOR, 148–158, 182
 estimating BSL and clearances, 182–184
 estimating SOV, 148
 estimating SSFOR and strike distance, 152–157, 158
 gap factors, 139
 insulation strength, 136–145, 178–180
 sensitivity, 157
 SOV distribution, 145–148
 substations, estimating SSFOR, 178–185

Temporary overvoltages, 506–510, 519–522
Traveling waves
 at discontinuity, 316–323
 lattice diagram, 321–322
 multiple conductors, 343–351
 corona effect of, 354–367

Underbuilt shield wire, 402–404

Velocity of propagation, 313–315

Waveshapes, 3–4, 26, 627–639

CPSIA information can be obtained
at www.ICGtesting.com
Printed in the USA
BVHW012336100619
550635BV00012B/175/P